THE POTATO CROP

The scientific basis for improvement

Series editor

E.H. ROBERTS

Professor of Crop Production, Department of Agriculture and Horticulture, University of Reading

THE POTATO CROP

The scientific basis for improvement

Edited by

P.M. Harris

Department of Agriculture and Horticulture
Reading University

London
CHAPMAN & HALL
A Halsted Press Book, John Wiley & Sons, New York

First published 1978
by Chapman and Hall Ltd
11 New Fetter Lane, London EC4P 4EE

ISBN 0 412 12830 6

Halsted Press, a division of
John Wiley & Sons, Inc., New York

Printed in Great Britain by
Fletcher & Son Ltd, Norwich

Contents

Foreword

The last few decades have witnessed a marked change in attitude to crop research and development. Agronomy is much less empirical in its approaches than it used to be and is now much more influenced by recent developments in crop physiology and soil science. Crop physiologists no longer rely so heavily on classical concepts of growth analysis, but now also use the techniques and discoveries of metabolic physiologists. Whilst plant breeding continues to make increasing use of techniques derived from genetics and cytogenetics, it is now also influenced more by concepts of productivity derived from crop physiology and agronomy. At the same time, the rapid developments we have seen in all aspects of crop protection and agricultural engineering inevitably influence the strategies of agronomists and breeders. Agronomists, breeders and farmers have always had to be concerned with quality as well as productivity, but the concern is now greater because of the more demanding expectations of the consumer, and the new characteristics required by food processors: thus, in addition to such attributes as baking characteristics in wheat, malting quality in barley, shape and flavour in fruit, crop scientists now have to pay attention to such matters as crisping quality in potatoes, protein quality in grain crops, and freezing characteristics of vegetables. In these circumstances the farmers and growers who produce the crops have to be provided with integrated technical advice so that they can make best use of knowledge and resources available; and in order to select appropriate advice in a rapidly changing situation, they need considerable understanding of the technological principles to support their traditional skills in the art of crop husbandry.

It is clear then that whilst the individual disciplines which are applied to crop production are becoming more specialized, the specialists themselves have to be increasingly aware of developments in fields outside their own. Indeed as a symptom of these two trends it is becoming quite common in many of the major international crop research institutes to see research on a crop organized as a team activity; and furthermore much closer links are being developed between research, development, extension and farming in which information flows in both directions.

It was with these considerations in mind that it was thought timely to develop a series of monographs dealing with some of the major world crops. Of course books on individual crops already exist and, within their own objectives, many of these are very valuable. There are, for example, some which deal mainly with husbandry and the current state of the art, and others which consist of a series of specialist chapters reporting aspects of recent scientific research. The main aim of the present series is to produce an integrated and critical approach to the problems involved in improving the production of particular crops, with special attention to the scientific principles involved. They are designed so that the reader is provided with an up-to-date view of all the major factors which influence the production of the crop and the ideas which suggest the means by which further improvements might be obtained. In order to fulfil these objectives in a digestible form, the coverage of the literature needs to be selective rather than encyclopaedic. Nevertheless, the aim has been to be sufficiently comprehensive so that the reader has a useful guide to what the authors consider to be the more important original sources. It is hoped that such an approach will not only find favour with the different kinds of applied scientists who are professionally involved, but will also have a value to workers and students in the pure sciences, many of whom are increasingly seeking a practical relevance to their work. There is no condescension to readers without some scientific training but, assuming some acquaintance with biological concepts, each monograph should be intelligible and useful to management and extension specialists, and indeed to the increasing number of qualified people who are engaged in the essential business of producing the crops in the field.

This volume on the potato provides an excellent illustration of the approach which I have described. It brings together the efforts of leading specialists who not only take us to the forefront of their individual subjects, but who also have a breadth of vision which enables us to see the wider implications; the result is, I believe, not only thought-provoking to professionals in the same disciplines, but also intelligible to those more at home in other scientific fields – or indeed to anyone who takes an intelligent interest in this remarkable crop.

May, 1977 E.H.R.

Preface

One of the major objectives of this book on the potato has been to review the knowledge which is fundamental to successful potato production and upon which further improvement must be based. Underlying this objective is the belief that high yields, efficiently obtained are of prime importance to producers and consumers and to research workers who serve the interests of both. Despite the paradox that the greatest financial returns are often achieved when yields are lower than normal, profitability through scarcity is not in the long-term interest of either producers or consumers.

This attempt to review the knowledge on which the scientific production of potatoes is based has necessitated seeking contributions from a fairly large number of specialists, which is a well-recognized method of dealing with subjects which are too large to be considered in the necessary depth by any one author. However this tendency to specialize within fairly narrow limits carries with it the attendant dangers that few research workers see the complete picture; on the other hand the producer is concerned with the whole system of producing a crop but may not have sufficient knowledge of particular components of the system which may significantly affect the efficiency with which his enterprise is operated. It is the hope that this book, read in its entirety, will provide the basic biological principles on which more productive and economic systems of potato production can be based.

Inevitably the views presented in this book will have been coloured by the particular experiences and enthusiasm of the contributors, and no doubt much has been omitted that might have been included and possibly vice versa; however the overall balance is the responsibility of the editor and any shortcomings in this respect should be attributed to him. It is my hope that anyone reading the book who has constructive suggestions to make will have no hesitation in contacting the editor or contributors.

I am indebted to Professor Roberts for asking me to edit this book and to the contributors who have so willingly cooperated in its production. I would also like to thank Denis Ingram and Kathryn Grant of Chapman and Hall for their valuable help and also Nan Slatter for her secretarial assistance. Thanks are also due to my wife and family for their patient forebearance during the book's preparation. Finally I would like to dedicate this book to all those who have contributed to our knowledge of this fascinating, useful and enjoyable vegetable.

University of Reading P.M.H.

June, 1977

Contributors

E.J. Allen	Department of Agriculture, The Institute of Rural Sciences, Penglais, Aberystwyth SY23 3DD
W.G. Burton	The Food Research Institute, Colney Lane, Norwich, Norfolk
E.G. Cutter	Cryptogamic Botany Laboratories, University of Manchester, Oxford Road, Manchester M13 9PL
G.E. Dalton	Department of Agricultural Economics, North of Scotland College of Agriculture, 581 Kings Street, Aberdeen A89 10D
K. Evans	Rothamsted Experimental Station, Harpenden, Hertfordshire, AL5 2JQ
R.W. Gibson	Plant Pathology Department, Rothamsted Experimental Station, Harpenden, Hertfordshire, AL5 2JQ
D. Gray	National Vegetable Research Station, Wellesbourne, Warwick
P.M. Harris	Department of Agriculture and Horticulture, University of Reading, Earley Gate, Reading RG6 2AT
J.G. Hawkes	Department of Botany, The University of Birmingham, P.O. Box 363, Birmingham B15 2TT
G.A. Hide	Rothamsted Experimental Station, Harpenden, Hertfordshire, AL5 2JQ

J. Holroyd	Weed Research Organization, Begbroke Hill, Yarnton, Oxford OX5 1PF
H.W. Howard	Plant Breeding Institute, Maris Lane, Trumptington, Cambridge
J.C. Hughes	Food Research Institute, Colney Lane, Norwich, Norfolk
R. Jarvis	Ministry of Agriculture, Fisheries and Food, Experimental Husbandry Farm, Boxworth, Cambridge, CB3 8NN
D.H. Lapwood	Rothamsted Experimental Station, Harpenden, Hertfordshire, AL5 2JQ
R.J. Makepeace	Weed Research Organization, Begbroke Hill, Yarnton, Oxford OX5 1PF
J. Moorby	Glasshouse Crops Research Institute, Worthington Road, Rustington, Littlehampton, Sussex
R.K. Scott	Broom's Barn Experimental Station, Higham, Bury St Edmonds, Suffolk
D.L. Trudgill	The Scottish Horticultural Institute, Invergowerie, Dundee DD2 5DA
S.J. Wilcockson	Department of Agriculture and Horticulture, Nottingham, University School of Agriculture, Sutton Bonington Loughborough, LE12 5RD
D.C.E. Wurr	National Vegetable Research Station, Wellesbourne, Warwick

Chapter 1
History of the potato

J.G. HAWKES

1.1 Introduction

The time and place of origin of cultivated plants and their subsequent evolution under domestication have caught the imagination of botanists and agricultural scientists from at least the early days of the last century. The studies of de Candolle, Vavilov and others have shown the need for a synthesis of information from such diverse fields as cytogenetics, history, linguistics, botany and archaeology to help trace the complete pattern of evolution of our ancient crops, many of which were already being cultivated some nine or ten thousand years ago.

The potato is undoubtedly of ancient origin, although our knowledge of its early stages of domestication is not so precise as that of some other crops such as wheat and barley. We know that it was domesticated in South America and that it had been dispersed by man over a considerable area by the time the Spaniards arrived in the sixteenth century.

However, the exact area where it was first grown and the details of its introduction into Europe are still matters of debate.

1.2 The status of the potato in South America

At one time, on the evidence of Gerard (1597), the potato was thought to have come from the colony of Virginia (now North Carolina), in the North American continent. However, all other evidence points away from this hypothesis and places the source of the potato in South America. Data from early Spanish post-conquest chronicles as well as archaeological remains show clearly that the potato was an ancient cultivated crop plant of South America by the time the Spaniards arrived. Furthermore, many wild species occur in South America, and more specifically in the Andes of

Peru and Bolivia, from one or more of which the cultivated potato could have been derived.

Many wild species occur also in Mexico and Central America, as we shall see in more detail in the next chapter. However, these never seem to have been taken into cultivation, and the apparently indigenous forms of *S. tuberosum* in Mexico and Central America all seem to have come from South America, probably in early post-conquest days.

1.3 Evidence for domestication – archaeological data

The most striking archaeological evidence for the antiquity of potato cultivation is afforded by the truly amazing ceramics from the northern coast of Peru belonging to the Mochica, Chimu and Inca cultures (about the fourth century A.D. onwards) (see Fig. 1.1). These ceramics, made in the form of potatoes, have been described and analysed in some detail by Salaman (1937, 1939, 1949). The Mochica peoples, especially,

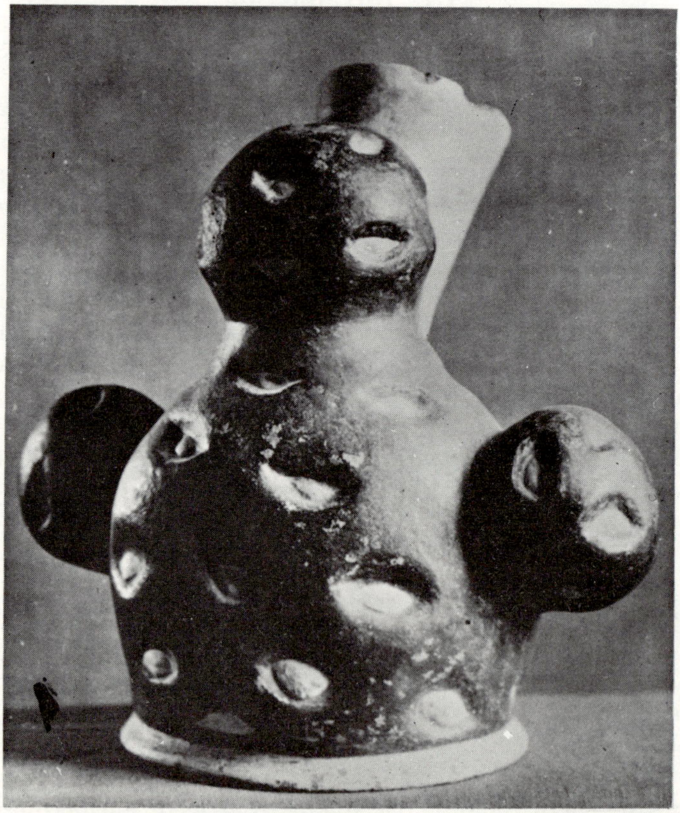

Fig. 1.1 Ancient Peruvian vessel (Mochica culture) representing a potato tuber

were remarkably inventive and portrayed with accuracy and artistic feeling all kinds of birds, fish, mammals and plants, as well as scenes of everyday life, portrait heads and natural objects. Amongst these the vessels depicting potatoes are rather rare, which is not so surprising when one realizes that these were coastal peoples who probably obtained potatoes by trade or barter from the mountains but who did not themselves grow this crop in their warm coastal valley river oases.

Unfortunately, the pottery of the Highland peoples (Tiahuanaco, Chavín, etc.) although very interestingly decorated with abstract patterns and stylized representations of deities, reptiles, birds and felines, did not, in general, depict plants. There is, however, one remarkable exception. This is a series of ceremonial vessels or urns of great size (over 3 ft tall) from the Central Highlands, of Tiahuanaco-Nazca (Huari) style dating from about A.D. 900. These urns, known as 'Pacheco', apparently contained offerings of food and were ceremonially destroyed and then buried. After restoration they were found to depict Andean crop plants such as potatoes, maize, oca (*Oxalis tuberosa*), ulluco (*Ullucus tuberosus*), mashua (*Tropaeolum tuberosum*), and other plants, possibly quinoa (*Chenopodium quinoa*) and peppers (*Capsicum* spp.).

All ceramics representing potatoes are restricted to Peru, and nothing has been recovered from Colombia, Ecuador, Bolivia, Argentina or Chile, even though the potato is certain to have been an ancient crop in those countries also.

Actual remains of the potatoes themselves are found in tombs, storehouses or dwellings, but do not occur with much frequency. A kind of dried potato (chuño or tunta), which is still produced by Indians today by freezing and drying, after trampling and sometimes washing, is known from some archaeological sites (see Hawkes, 1967).

Although certain materials were found from as far back as about 400 B.C., no really ancient remains had been identified with certainty until recently. The difficulty had been to distinguish between the dried samples of potatoes and other tuberous plants earlier than about A.D. 1000 (see Towle, 1961). For instance the 400 B.C. material excavated by Dr Alfred Kidder from the ruins of a house at Chiripa on the shores of Lake Titicaca was described by Dr Towle as closely resembling potatoes which might have been stored as raw tubers or as chuño. Later, in a letter to the present writer she warned that caution should be shown in identifying these tubers as anything more than 'tubers representing Andean root crops, *ullucu*, *oca* and possibly potato'.

The situation at present is now more hopeful. Detailed studies of starch and cell structures using light and scanning electron microscope techniques have enabled us to identify potatoes from the Chilca valley near Lima, radiocarbon dated at an age of 8000 years. These were provided by Frederic Engel (1970) and co-workers, whilst other remains of early food plants excavated by Moseley (1975) have also been examined. Studies on ancient potato, sweet potato, cassava and other plant remains will shortly be published by Martins and Hawkes (see also Martins, 1976).

Such evidence takes the origin of potato cultivation back to an age before the appearance of maize in South America and equates it with the approximate time of agricultural origins in the New World, based on excavations in Mexico and elsewhere.

1.4 The potato at the time of the Spanish conquest

It seems quite clear from what has just been said that the potato was an ancient cultivated crop at the time of the discovery of South America. We do not know the geographical extent of its early cultivation since the archaeological finds are wholly from Peru and at Tiahuanaco in Bolivia, very close to the Peruvian border. Since botanical evidence points to an origin for the cultivated potato in the Peru–Bolivian region, it is only to be expected that, once domesticated, it would have diffused rapidly throughout the high Andes of those two countries. Indeed Weston La Barre (1947) has asserted that the Highland cultures owed their existence to the potato since there are few other plants that can be grown at altitudes of 3500 m or more. Maize does poorly at such height, and although there are certain other tuber crops, as has already been mentioned when describing the 'pacheco' urns, these are grown on a small scale. The small-grained *Chenopodium quinoa* would hardly rank as a main staple food, and the cultivated lupin is similarly a difficult food plant to handle. Unfortunately the archaeological evidence is limited to certain areas only.

1.5 Historical and linguistic evidence

At the time of the conquest in the early sixteenth century we have accounts by chroniclers (Fig. 1.2) of potato cultivation in what are now Colombia, Ecuador, Peru, Bolivia and Chile (for detailed literature sources, see Hawkes, 1967). Even at that early time it seems to have been a variable and intensively cultivated crop, and we can assume with some degree of certainty that many hundreds of varieties were grown over a distribution area not very dissimilar from that in which it is cultivated today.

The native names of the potato also indicate an ancient and widespread cultivation (see Hawkes, 1947), since they differ completely in the main Indian languages that were spoken in the areas where the potato was grown (see Table 1.1). Thus in the Chibcha language of Central Colombia the names *iomza*, *iomuy*, etc., were used; in Quechua the language of the Inca empire, the usual name was *papa*, but this replaced earlier names (*ajsu*, *impari*, etc.) in the languages of the vanquished

Fig. 1.2 Planting and harvesting potatoes in Inca times; from a sixteenth – seventeenth century manuscript by Guamán Poma de Ayala (reproduced by kind permission of the Institut d'Ethnologie, Paris)

tribes. In Bolivia, the Aymará Indians used the words *amka* and *choque*, whilst in Chile, the Araucanians gave it the name *poñi*.

The Spaniards adopted the name *papa* for the potato throughout their western American colonies, and other names as well as the languages themselves have largely died out (although not Aymará or Araucanian). Even so, the old names for the potato in various regions still persist here and there in varietal combinations or in dialect phrases.

In Europe the word *papa* was never adopted. On the contrary, our word is derived from *batata*, the Caribbean Arawak Indian word for *Ipomoea*, the sweet potato. This is because the Spaniards encountered the sweet potato first, and not having a name for a tuber of that sort they used the Indian word *batata*. Subsequently, other tuberous plants which they found in their American colonies were given the same name. *Patata* and potato are obviously cognate forms of *batata*; so the word *papa*, which is still widely used in the whole of Spanish Latin America, never spread outside this area, even though the plant itself is now grown in most other parts of the world.

To sum up this section we can say with some certainty that historical and linguistic evidence clearly corroborate archaeological evidence as to the origin of the cultivated potato in the western part of South America.

Table 1.1 The American Indian names of the cultivated potato

Name	*Language*	*Region*
Iomza *Iomy* *Iomuy*	Chibcha	Colombia, central and northern regions
Pulu *Pulyu*	Colorado, Cayapa	Ecuador, province Esmeraldas
Ajsu	Chinchaysuyo	Peru, department Junín
Impari *Maona* *Mutza* *Zanaro*	Anti	Peru, just north of Cuzco
Papa	Quechua	South America. Originally confined to the Cuzco district of Peru, and later spreading with the spread of the Inca empire
Papa	Yunga (probably derived from Quechua)	Peru, departments Piura, Ancash, Trujillo, Lambayeque
Amka *Choke*	Aymará	Lake Titicaca and La Paz districts (S. Peru, N. Bolivia)
Poñi	Araucanian	Chile, central and southern regions to the Isle of Chiloé

1.6 The introduction of the potato into Europe

The potato appeared in Europe during the last quarter of the sixteenth century. Unfortunately, we know of no account describing its introduction and we must perforce rely on circumstantial evidence only.

Legend has associated Sir Walter Raleigh and Sir Francis Drake with the introduction of the potato into England. There is no factual evidence to substantiate these claims, and much evidence which tends to negate them. Thus Drake saw potatoes in Chile during his world voyage in 1578, but since he did not return to England until 1580 he could not have kept potatoes with him in a living condition for two years (see Drake, 1628). Furthermore, Clusius (1582) who described Drake's plants collected during this voyage from a visit he made to England in 1581 makes no mention of the potato either.

Raleigh is reported to have seen and collected potatoes on one of his voyages, but from a knowledge of the locality we can be certain that these were sweet potatoes and not *S. tuberosum*. Indeed, Raleigh evidently never went anywhere near any area where he might have found potatoes; however, there may be some truth in the statement that Raleigh planted potatoes from England on his estates at Youghal in southern Ireland.

Salaman (1937) quotes evidence that potatoes were bought in 1573 at Seville and concludes that they could have arrived in southern Spain about 1570 and have been cultivated somewhere in the Seville region for some years so as to build up stocks. This date of 1570 is generally quoted as the approximate time of first arrival of the potato in continental Europe. We assume that it came merely as ships' stores and was not at that time regarded as a plant of special significance. From Spain it seems to have spread to Italy, whence it was sent to the famous herbalist Clusius at Vienna by Philippe de Sivry, Prefect of Mons in Belgium, who in turn was sent it from Italy. From this Spanish source the potato spread through Europe as a botanical curiosity and no more.

A separate introduction was recorded in England by the herbalist John Gerard (1596, 1597), who claimed erroneously, as we have already said, to have received it from Virginia in North America. The reason for this is by no means certain but it has been suggested that he confused plants brought by Hariot from Virginia in 1586 (see Hariot, 1588) with the potato which was received on another occasion, between 1588 and 1593 (see Hawkes, 1967, pp. 255–62), probably from a 'prize' Spanish ship captured by English seamen.

1.7 The nature and exact source of the early European potato

Although there is no doubt that the European potato belongs to the species *S. tuberosum* there has been considerable controversy as to what region of South America it came from and thus whether subspecies *tuberosum* from Chile or subspecies *andigena* from the Andes was first brought to Europe.

The Russian botanists, Juzepczuk and Bukasov (1929) regarded the early European potato as having come from Chile and thus having been adapted to the long-day conditions of Europe right from the time of its arrival. Salaman (1946, 1954), however, provided historical and geographical evidence which made an introduction by 1570 from Chile highly unlikely. Since the Andean subspecies *andigena* is adapted to form tubers under the 12 h day of tropical latitudes and does not tuberize in Europe until very late in the season when the natural day length has reduced itself to more or less 12 h in late September and early October, the Russian botanists believed it could not for this reason have been the subspecies first introduced into Europe.

It was nevertheless suggested by Salaman and Hawkes (see Salaman,

1946, 1949, 1954; Salaman and Hawkes, 1949; Hawkes, 1967) that the early European potato came from the Andes, and perhaps from the northern Colombian part. This would have meant that through strong artificial selection in Europe from 1570 to about 1750 when Linneaus described the European *S. tuberosum* the potato would have changed from one subspecies to the other, just as it had presumably done over the previous millenia when it was taken by Indian tribes from the Andes to Chile. This is not so far-fetched as it may seem at first glance, since Simmonds (1966) succeeded in carrying out the same process for a third time during the space of a very few generations of selection for earliness in Britain.

If this hypothesis is correct we might expect to see evidence of subspecies *andigena* in Europe in the late sixteenth and early seventeenth centuries before it became changed into subspecies *tuberosum.* In fact, these changes can be seen very clearly, in herbarium specimens, woodcuts and descriptions (Salaman and Hawkes, 1949).

Andean potatoes when grown in the European long summer days tuberize late and the stems continue to grow, producing tier after tier of flowers and long branched stolons which often shoot up to form new plantlets. The stems are thinner and more frequent than *tuberosum* stems and the leaf is more highly dissected. Other differences (see Hawkcs, 1956) can also be noted. The tuber yield is poor since the night temperatures become too low in September and October for efficient tuberization, and the plants are often killed by frost.

The early descriptions and woodcuts agree very well with the features described above. Thus, Clusius (1601) describes stems up to 7.5 and 10.5 ft long, whilst several authors (Clusius, loc. cit.; Bauhin, 1620) mention that the plants are harvested in November. The many stolons and small tubers of *andigena* when grown in Europe are featured in the early illustrations (see Fig. 1.3; Bauhin, 1620), whilst a water-colour painting of 1598 (Fig. 1.4) sent to Clusius by Philippe de Sivry also shows the dissected leaves and small tubers. Finally, perhaps the most interesting piece of evidence is the herbarium specimen of Caspar Bauhin made in about 1620 which is still very similar to *andigena* (Fig. 1.5).

A final piece of evidence is that the early potato in Europe was regarded as a botanical curiosity and was not generally grown as a field crop until about the mid-eighteenth century, despite the fact that it was a completely accepted cultivated plant in South America. Only in very favoured places in southern France and southern Ireland and elsewhere, with mild winter climates could it be grown as a crop. The reason for this is obvious. The yields were too late and too small, due to the adverse photoperiodic response in respect of tuberization, until selection for earliness had resulted in varieties adapted to the long summer days of western and northern Europe.

LIBER QVINTVS.

LIBER QVINTVS

DE SOLANO.

CAPVT I.

Solanum tuberoſum eſculentum.

I. SOLANVM tuberoſum eſculentum: caulem habet ad duorum triumve cubitorum, rarius in viri altitudinem exurgentem, craſſum, anguloſum, ſtriatum, leuiter hirſutum, in multos ramos infirmos, in terram, niſi adminiculo ſuſtineantur, procumbentes, brachiatum, ex quorum alis pediculi craſſi, anguloſi, flores ſuſtinentes, prodeunt. Folia quæ primo naſcuntur, Barbareæ ſimilia ſunt (quēadmodum figura noſtra & Matthiolo & huic libro addita eleganter, cum omnibus plantæ partibus, exprimit) ex atro purpuraſcentia, ſubhirſuta, cæteris pallide virentibus, abſque pediculo, palmum longis, in ſex, octo, plures & pauciores partes, ex vna coſta dependentes, diuiſis, ex rotunditate oblongis, integris, quibus ſingulis, duo ſextuplo minora interijciuntur, & ſemper pars vna, quæ ſe-

M ptima

Fig. 1.3 The potato in Europe in the early seventeeth century (C. Bauhin, 1620) Note the long branched stolons, small irregular tubers, small daughter plants formed from stolons, many inflorescences and abundant fruits. These features are all typical of short-day adapted *andigena* potatoes when grown under the long summer days of Europe. (The bottom right-hand tuber may be a Jerusalem Artichoke)

Fig. 1.4 Water-colour sketch of the potato sent to Clusius in 1589 by Philippe de Sivry, Prefect of Mons (Plant in Museum, Antwerp). Note the well-dissected leaves and small irregular tubers

We know that plants were raised from true seed at a very early date, thus providing segregants on which selection could act. Clusius (1601) remarks on this and describes the results in some detail; and there are other accounts of selection for earliness in the literature. We thus have in the potato a good early example of selection for physiological characters long before the development of the science of plant breeding.

1.8 The spread of the potato into other parts of the world

We have already mentioned that the potato was introduced into Europe on two occasions, firstly into Spain in about 1570 and secondly into England a little later. This latter introduction took place between 1588 and 1593, with a strong suggestion that 1590 was the actual year of arrival (Hawkes, 1967; p. 261).

From these two introductions it seems likely that the potato spread into nearly every part of the world. From the Spanish introduction it

Fig. 1.5 Caspar Bauhin's herbarium specimen of the potato at Basle, made in approximately 1620. The leaf of this specimen is very close to that of sub-species *andigena* from the Andes of Colombia and elsewhere

diffused through continental Europe and parts of Asia. From the English one, it spread to Ireland, Scotland, Wales and parts of northern Europe, and from one or more of the British sources to most of the British overseas colonies, including what was later to become the U.S.A.

Let us look first at the diffusion of the potato through continental Europe. From Spain it was said to have been taken to Italy by the Carmelite Friars, and Clusius (1601) mentions that it was grown in Italy before 1587 (see also Magazzini de Vallombrosa, 1623). In Portugal potatoes appeared even earlier.

Clusius received potatoes in Vienna, from Italy via Mons in the low countries and sent samples in the succeeding years to botanists in many parts of Germany and Austria. They gradually spread by the end of the seventeenth century to most German-speaking parts of Europe although they only began to be grown on a large scale after about 1780 and did not become really popular until the nineteenth century.

The Swiss herbalist C. Bauhin and his brother J. Bauhin evidently obtained tubers at second hand from Clusius in the late sixteenth century and had sent them to France by about 1600. Potatoes were extensively grown in France by the mid-seventeenth century and probably were more successful as a field crop earlier than in Germany because of the milder autumns in the southern parts of the country. However, there was considerable resistance to their cultivation by all but the very poorest people until Parmentier made them popular in about 1773.

The Slavic nations seem to have obtained their potatoes chiefly from Germany since their words for this plant are derived from German ones, such as *Kartoffel*, *Grundbirne*, etc. There are also words in Czechoslovakia, Bulgaria and southern Russia cognate with Brandenburg, indicating the German source for those regions also. Evidently the general adoption in eastern Europe was late eighteenth to early nineteenth century (see Laufer, 1938).

The potato was first said to have been brought to Russia by Peter the Great at the end of the seventeenth century from Holland, where of course it had become well-known in gardens since Clusius' time. Only small amounts were produced as delicacies for royal banquets up to the mid-eighteenth century but by 1764 it began to be grown around St. Petersburg to a reasonable extent (Lechnovicz, 1970). Its adoption in all parts was greatly speeded up by a royal decree in 1765 and potatoes were then sent to all parts of the country for cultivation.

Now let us turn to potatoes derived from the introduction into England in about 1590. For reasons already stated above the potato began in England, as in most other parts of Europe, by assuming the role of a botanical curiosity and was esteemed as a delicacy in the court of James I. It was not until the mid-eighteenth century that potatoes are recorded as having been grown on a field scale, and this fits in rather well with accounts from other parts of northern Europe. The same time scale is recorded also for Scotland and Wales (Salaman, 1949). On the other hand there is much evidence that the potato was grown on a field scale in Ireland, at least in the south-west, by the early seventeenth century, probably due to the very mild climate which allowed it to survive and set its tubers from October to December. Southwell (1693) for instance, states that his grandfather brought potatoes to Ireland 'who had them from Sir Walter Raleigh [*sic*] after his return from Virginia'. Since Raleigh died in 1618 this would certainly give a last date to the introduction (see also Salaman, 1949).

From Scotland the potato was said to have been taken to Norway in the mid-eighteenth century and possibly thence to Sweden and Denmark, although it seems probable that these last-named countries might have already received samples earlier from German or other sources.

In North America the potato was completely unknown until the early seventeenth century, and there seems to be not a shred of evidence to support Gerard's account of its being native to Virginia or anywhere else on the North American continent. One should perhaps say that wild potatoes are known in the south-western states but this is quite another matter.

In fact the North American continent first received potatoes from England via Bermuda in 1621, where they had been grown since an initial importation from the mother country in 1613. Strangely enough, these first North American potatoes were grown in Virginia, where Gerard some 35 years ago had erroneously assumed them to be native. Many other introductions from England and Ireland were made later in the century, but no records of an introduction from South America were made before Goodrich (1863) obtained some varieties in a Panamá market which were said to have come from Chile.

Potatoes were said to have been taken to India by British missionaries in the late seventeenth century and to China a little earlier. They were grown in Japan by the late seventeenth century and in parts of Africa and the West Indies by about the same period. Nowhere, even in tropical regions, did they become widely grown before the mid-eighteenth century (Laufer, 1938). According to Yen (1961/2) the potato was first introduced into New Zealand in 1769 by the French explorer, de Surville, followed by several small introductions in the nineteenth century. They were adopted widely by the Maoris in the early nineteenth century and were well established by 1840.

Zubeldia *et al.* (1955) have written an interesting account of the potatoes of the Canary Isles which are still very similar to *andigena* varieties from South America; even a triploid clone has been found amongst them. Historical evidence indeed points to the introduction of these potatoes from Peru in the year 1622. This is the only recorded case of potatoes being brought out of South America before the nineteenth century to anywhere except Europe.

References

Bauhin, C. (1620). *Prodromus Theatri Botanici*. Frankfurt-on-Main.
Clusius, C. (1582). *Aliquot Notae in Garciae Aromatum Historiam*. Antwerp.
Clusius, C. (1601). *Rariorum Plantarum Historia*. Antwerp.

Drake, Sir F. (1628). *The World Ecompassed.* (Ed. W.S.W. Vaux.) Hakluyt Society, **16**, 1854. London.

Engel, F.A. (1970). *Curr. Anthrop.* **11**, 55–8.

Gerard, J. (1956). *Catalogus Arborum, Fruticum ac Plantarum,* etc. 1st Edn. London.

Gerard, J. (1597). *The Herball or General Historie of Plantes.* 1st Edn. London.

Goodrich, C.E. (1863). *Trans. N.Y. State agric. Soc.* **23**, 103–39.

Hariot, T. (1588). *A Briefe and True Report of the New Found Land of Virginia.* London.

Hawkes, J.G. (1947). *J. Linn. Soc. (Bot)* **53**, 205–50.

Hawkes, J.G. (1956). *Proc. Linn. Soc.* **166**, 97–144.

Hawkes, J.G. (1967). *J. R. hort. Soc.* **92**, 207–24, 249–62, 288–302, 364–5.

Juzepczuk, S.W. and Bukasov, S.M. (1929). *Proc. U.S.S.R. Congr. Genet. Pl. and Animal Breed.* **3**, 593–611.

La Barre, Weston (1947). *Acta Americana* **5**, 83–103.

Laufer, B. (1938). *Field Mus. Publ. 418, Anthrop. Ser.* **28**, 132.

Lechnovicz, V.S. (1970). *The Potato.* Leningrad, 7–10.

Magazzini De Vallombrosa, Father (1623). *Dell'Agricoltura Toscana.* Vallombrosa, Italy.

Martins, R. (1976). *New Arachaelogical Techniques for the Study of Ancient Root Crops in Peru.* Ph.D. Thesis, University of Birmingham.

Moseley, M.E. (1975). *The Maritime Foundations of Andean Civilization.* California.

Salaman, R.N. (1937). *J R. hort. Soc.* **62**, 61–77, 112–23, 153–62, 253–66.

Salaman, R.N. (1939). *J.R. Anthrop. Inst.* **69**, 109–22.

Salaman, R.N. (1946). *J. Linn. Soc. (Bot)* **53**, 1–27.

Salaman, R.N. (1949). *The History and Social Influence of the Potato.* Cambridge.

Salaman, R.N. (1954). *J. Linn. Soc. (Bot)* **55**, 185–90.

Salaman, R.N. and Hawkes, J.G. (1949). *Proc. Linn. Soc.* **161**, 71–84.

Simmonds, N.W. (1966). *J. Linn. Soc. (Bot)* **59**, 279–88.

Southwell, Sir Robert (1693). *Journal Book of the Royal Society. of London.* 13th Dec., 1693.

Towle, M. (1961). *The Ethnobotany of Pre-Colombian Peru.* Chicago.

Yen, D.E. (1961–2). *Pot. J.* 2–5.

Zubeldia, A., López Campos, G. and Sañudo Palazuelos, A. (1955). *Bol. 33, Inst. Nac. Invest. Agron.* Madrid, 287–325.

Chapter 2
Biosystematics of the potato

J.G. HAWKES

2.1 Introduction

The potato of commerce belongs to a single species, *Solanum tuberosum* L., apart from certain cultivated forms in South America which will be mentioned later and which are generally regarded as distinct. Other well-known cultivated plants in the genus *Solanum* are the egg-plant or aubergine (*S. melongena*), the pepino (*S. muricatum*) and the naranjillo (*S. quitoense*). The chilli peppers (*Capsicum* spp.) and tomato (*Lycopersicon esculentum*) are better known and more widely grown than the last two *Solanum* species mentioned above. Less well-known fruits such as *Cyphomandra* and *Physalis* as well as a number of ornamentals such as *Petunia*, *Schizanthus*, *Salpiglossis* and *Lycium* also belong to Solanaceae. On the other hand, several genera such as *Nicotiana*, *Datura*, *Atropa* and *Mandragora*, have acquired a more sinister reputation because of their poisonous alkaloids. Even so, probably all members of the family possess some alkaloids, even though in small amounts in many cases.

2.2 Brief outline of potato classification

The genus *Solanum*, to which the cultivated potato belongs, is an extremely large one, containing over 2000 species. It extends all over the world except for the far north and south, with a strong concentration of species diversity in South and Central America on the one hand and Australia on the other. In this it mirrors to a large extent the distribution pattern of the family itself.

In addition to *S. tuberosum*, some seven other cultivated species and 154 wild species of potato are generally recognized. The rest of the genus consists of non-tuberiferous species, such as *S. nigrum* (black nightshade) and many spiny herbs and shrubs. The tuber-bearing species are completely confined to the American continent and many are of considerable interest to potato breeders because of their resistance to pests and pathogens and their adaptation to climatic extremes.

There have been many attempts at grouping the potatoes in a satisfactory way within the genus, and some authors have suggested that they should be separated from it altogether. On the whole, such a course of action has not received much support since in floral structure, which is generally considered of prime value in generic distinctions, no species of *Solanum* differs essentially from any other.

The tuber-bearing species, once included within the section *Tuberarium*, are now grouped under the older name *Petota* which seems indeed most appropriate. They are distinguished by actinomorphic flowers, a jointed or articulated pedicel or flower stalk, unbranched glandular or eglandular hairs, no thorns or spines and generally (although not invariably) compound leaves. The typical potato leaf is described botanically as interruptedly imparipinnate, since the major sequence of paired leaflets and terminal unpaired one, is broken or interrupted by smaller folioles or interjected leaflets. The number of pairs of lateral leaflets is often used to characterize a leaf, which can be spoken of as, for instance, 2–3-jugate, indicating 2–3 pairs of main laterals.

Although most species in *Petota* possess underground stolons, which are of course stems botanically, bearing at their tips the characteristic potato tubers, some related species that must be grouped with them on the basis of other morphological characters, lack stolons and tubers altogether (series *Juglandifolia* and *Etuberosa*).

Section *Petota* has been further subdivided for the sake of convenience into groups or series by various authors. In the present work eighteen of these series are recognized.

The series and species are set out below in the form of a list, with notes on natural distribution and breeding value where these are known. Species which are unknown or little studied in the living state are referred to by name only, accompanied by a few brief notes. The cultivated potatoes are listed separately on pp. 45–48.

2.2.1 Classification of potato species

Series I JUGLANDIFOLIA Rydb.

Plants with herbaceous to woody stems, bright yellow flowers and no stolons or tubers. They are included with the true potatoes because of the position of the pedicel articulation at some distance above the base. They bear obvious resemblances to the genus *Lycopersicon*, differing from it in the absence of floral bracts and sterile anther tips. Mountains of Central America and north-western South America as far south as the coastal ranges of Peru and north Chile.

1. *S. juglandifolium* Dun.

A perennial woody climber (liane); leaf pinnate, rugose and hispid above; interjected leaflets few or absent; fruit up to 4.5 cm diam. Costa Rica, Venezuela, Colombia, Ecuador. Forests, scrub woodland and hedges at 1000–3000 m. $2n = 24$.

2. *S. lycopersicoides* Dun.

A bush or shrub, to 2.5 m tall. Leaf bi-pinnatisect, very similar to that of the tomato. Peru. Open ground, from 2800–3150 m. $2n = 24$.

3. *S. ochranthum* Dun.

Habit as for *S. juglandifolium*; leaf pinnate, smooth and velvety; interjected leaflets very numerous. Very large berries to 6 cm diam., resembling a green tomato, although much harder. Colombia, Ecuador, Peru. Forests, scrub woodland and hedges at 1800–3500 m. $2n = 24$.

4. *S. rickii* Correll

A herb up to 50 cm tall, glabrous, with bi-pinnatisect leaves and stellate flowers; pericarp rapidly becoming dry and papery at maturity. North Chile. Rocky dry ravines at 3000 m. $2n = 24$.

Series II ETUBEROSA Juz.

Herbs with very low almost basal pedicel articulation. The plants bear no stolons or tubers, although they are sometimes rhizomatous. Corolla rotate, purplish or blueish. Central Chile as far south as the island of Chiloé; Nahuel Huapi region of South Argentina; Islands of Juan Fernández.

1. *S. brevidens* Phil.

A highly polymorphic species with pubescence varying from dense to almost absent; pedicel articulation 2–4 mm above the base; calyx teeth very short (0–0.5 mm); corolla sky-blue, paling to white at the edges. S. Central Chile, island of Chiloé, Nahuel Huapi region of S. Argentina. Wet forests from sea level to about 1000 m. $2n = 24$.

2. *S. etuberosum* Lindl.

Distinguished by yellowish-green stems and leaves, with short velvety pubescence and often crisped leaflet margins; pedicel articulation 4–5 mm above the base; calyx with well marked teeth (1–1.5 mm); corolla rich purple, very showy. N. Central Chile. Dry mountain forests, from 1250–2500 m. $2n = 24$.

3. S. fernandezianum Phil.
Plant glabrous or glabrescent; pedicel articulation right at the base. Islands of Juan Fernández. Wet forests, from 100–600 m. $2n = 24$.

4. S. palustre Poepp.
Chile. Sea marshes near Valparaiso, rare or possibly extinct.

5. S. subandinum Meigen
Probably closely related to *S. etuberosum*. Chile. Mountain woods and rocks near Santiago, at 1700–2300 m.

Series III MORELLIFORMIA Hawkes

Small tuber-bearing herbs with a strong resemblance to *S. nigrum*, with predominantly epiphytic habit, simple leaves, very small white stellate flowers (less than 1.5 cm diam.), long slender anthers which are slightly coherent laterally, and small few-seeded berries. Mountain forests of Mexico and Guatemala.

1. S. morelliforme Bitt. et Muench
A small epiphyte growing on trees and also on moss-covered walls and rocks in dense shade, never on the ground. Distinguished by the simple leaves, small stellate flowers, slender anthers and small, few-seeded berries. Central to South Mexico; Guatemala. In wet mountain forests from about 2000 to 3000 m. $2n = 24$.

Series IV BULBOCASTANA Rydb.

Small tuber-bearing terrestrial herbs with simple leaves, stellate flowers (1.5–3 cm diam.), short thick anthers and round berries larger than those of *Morelliformia*. Central and South Mexico; Guatemala.

1. S. bulbocastanum Dun.
Leaf with rounded to cuneate base, generally densely pubescent, varying from ovate to linear–lanceolate according to subspecies. Flowers white to deep cream. *Phytophthora* resistant. Both diploid and (apparently) autotriploid forms are known. Central to South Mexico; Guatemala. Woods, grassland, rocks and field borders. Often grows under quite dry conditions. Altitude range 1500–2300 m. $2n = 24, 36$.

1a. Subspecies *bulbocastanum*. Leaf and stem densely pubescent;

leaf ovate in shape, less than 2.5 times as long as broad; corolla white. Central to South Mexico.

1b. Subspecies *dolichophyllum* (Bitt.) Hawkes. Leaf linear–lanceolate, more than 2.5 times as long as broad, attenuate at base and apex; style varying in length from 6 to 10 mm. Corolla white. Central Mexico, States of Morelos and Guerrero.

1c. Subspecies *partitum* (Corr.) Hawkes. Leaf ovate to ovate–lanceolate, less than 2.5 times as long as broad; leaf and stem less densely pubescent than in subsp. *bulbocastanum*; pedicel above articulation and calyx completely glabrous; calyx pale yellow–green; corolla deep cream, the lobes very deeply separated from each other. Guatemala and the State of Chiapas in S. Mexico.

2. S. clarum Corr.
Leaves typically cordate; corolla lobes purple above, white below; base of anther 3–5-lobed. Guatemala, depts. Huehuetenango and Quezaltenango. High mountain forests. $2n = 24$.

Series V PINNATISECTA Rydb.

Herbs with stolons and tubers, imparipinnate to imparipinnatisect leaves and stellate corolla which is white or deep cream coloured, sometimes tinged with purple; berries globular or conical. United States (Colorado) southwards to Central Mexico.

1. S. brachistotrichum (Bitt.) Rydb.
Distinguished from all other species in this series by the characteristic minute 2–3-celled triangular hairs on leaves and stem. Leaflets subsessile to petiolulate, always slightly decurrent on to rachis, 2–3 (–4)-jugate. Corolla slightly mauve-tinted towards acumens, very reflexed. N.W. Mexico. Dry piñon and juniper scrub vegetation at 1750–2500 m. $2n = 24$.

2. S. cardiophyllum Lindl.
Leaflets ovate-cordate to lanceolate, 2–4-jugate, petiolulate, not decurrent (or only occasionally so); calyx acumens 0.5–1.5 mm long, according to subspecies. Blight-resistant. N.W. to S. Mexico, in dry scrub vegetation, field borders, old lava fields, and especially as weeds of cultivation; the tubers of subsp. *ehrenbergii* are edible. Altitude: 1600–2500 m. $2n = 24, 36$.

2a. Subspecies *cardiophyllum.* Leaflets glabrous, shining, dark green above; calyx acumens not more than 0.5 mm long; corolla deep cream to buff-coloured. Anthers short, 5 mm, not attenuate above, often tinged with violet inside. Both diploid and (apparently) autotriploid forms are known, the latter being distinguished generally by very much wider (broadly ovate-cordate) leaflets. Central Mexico.

2b. Subspecies *ehrenbergii* Bitt. Differs from the type subspecies by the paler green, non-shining, oblong–lanceolate to lanceolate leaflets and the frequent presence of hairs of varying lengths on leaves and stems. Calyx acumens 1–1.5 mm long; anthers narrow, generally attenuate. Petals often tinged with mauve towards the tips. Mostly diploid forms are known, although a triploid (presumably autotriploid) has been recorded in San Luis Potosi State. Central to N.W. and W. Mexico.

2c. Subspecies *lanceolatum* (Berth.) Bitt. This does not differ sufficiently from the diploid forms of subsp. *cardiophyllum* to be ranked as a separate species. Distinguished from subsp. *cardiophyllum* by the narrower leaflets (more than 2.5 times as long as broad), the longer peduncle (more than 3.5 cm long) and the 0.5–1.0 mm long calyx acumens. Corolla white or cream-coloured. Central to S. Mexico.

3. S. jamesii Torr.

Plants without typical semilunate pseudostipular leaflets; if present they are pinnatisect and do not clasp the stem. Leaves slightly decurrent, generally pubescent, occasionally glabrous, with (2)–3–4–(5) pairs of leaflets and generally no interjected leaflets. Corolla lobes narrow. N.W. Mexico; S.W. United States; dry scrub vegetation from 1500–2300 m. $2n = 24, 36$.

4. S. hintonii Corr.

Mexico, Mexico State, by stone walls, at 1700 m altitude.

5. S. × michoacanum (Bitt.) Rydb.

Leaf (2)–3-jugate; corolla white, stellate; fruit unknown. A natural hybrid of *S. pinnatisectum* × *S. bulbocastanum*. W. Central Mexico amongst rocks and along field borders, at altitudes of 2000–2100 m. $2n = 24$.

6. S. nayaritense (Bitt.) Rydb.

W. Mexico, State of Nayarit and possibly elsewhere; probably in mountain pine forests; altitude unknown.

7. S. pinnatisectum Dun.

Plant without semilunate pseudostipular leaflets, as in *S. jamesii*.

Distinguished from *S. jamesii* by the glabrous leaves, 6–8 leaflet pairs, generally with a few interjected leaflets and large showy corolla with broad triangular lobes. Berry round to conical. Some blight resistance. Central Mexico. Cultivated fields, waste places and field borders, from 1800–2100 m. $2n = 24$.

8. S. × *sambucinum* Rydb.
A natural hybrid between *S. pinnatisectum* and *S. cardiophyllum* subsp. *ehrenbergii*, which shows segregation to the parental types when grown from seed. Leaf fairly dark green, with 4–5-paired leaflets which are narrowly lanceolate and both petiolulate and decurrent (thus intermediate between the two parents). Corolla with lobes broader than those of *S. cardiophyllum*, narrower than those of *S. pinnatisectum*. Some blight resistance. Central Mexico, States of Querétaro and Guanajuato; a weed of fields and field borders, found in the distribution areas of the two parents, at about 1800–2000 m. $2n = 24$.

9. S. stenophyllidium Bitt.
Leaflets at least six times as long as broad, sub-linear. Little is known of this species, but it may be an extreme southern form of *S. brachistotrichum*. W. Mexico, State of Jalisco, and perhaps elsewhere. In dry grassy places and scrub at about 2000 m. $2n = 24$.

10. S. trifidum Corr.
Leaf 1–3-jugate; stem softly pubescent, provided with hairs and glands; corolla creamy white, with fairly broad lobes. Berries long oval, pointed. Western Mexico (States of Michoacán and Jalisco), in open pine forests, maize fields, roadsides, etc., at altitudes of 2200–2400 m. $2n = 24$.

Series VI COMMERSONIANA Buk.

Tuber-bearing herbs with imparipinnate leaves and stellate corolla with rather broad lobes, generally less than twice as long as broad. Flowers white or mauve-tinted. South America: Bolivia, Argentina, Paraguay, Uruguay, Brazil. All species so far discovered are diploid, although some autotriploid forms occur.

1. S. calvescens Bitt.
Probably related to *S. commersonii*, and especially to subsp. *malmeamum*, from which it may be distinguished by the larger size, upright growth, curved leaves, more acute lateral leaflets, large flowers, and especially by the curious stamens (see also *S. maglia*) in which the anthers and filaments are not clearly delimited from each other, either in form or

colour. Brazil, State of Minas Gerais, and perhaps elsewhere; on shady river banks and in cultivated fields, at altitudes of about 1200 m. $2n = 36$.

2. S. chacoense Bitt.

Leaflets petiolulate, acute to acuminate, the terminal leaflet hardly larger than the laterals; corolla uniformly white; berries globular. An extremely polymorphic species, spreading through South Bolivia, north and central Argentina, Paraguay, Uruguay and South Brazil, generally as a field weed in lowland pastures. The great range of variation has induced many authors to divide it into a large number of micro-species which cannot now be maintained since they are all apparently fertile with each other and intergrade considerably. Some forms with insect and virus Y resistance. $2n = 24$.

2a. Subspecies *chacoense.* Lateral leaflets 2–3 times as long as broad; plains forms are almost glabrous, forming a semi-rosette in the early stages of growth, and also possess very short (not more than 0.75 mm) calyx acumens. Forms of subsp. *chacoense* occurring in the mountain valleys of N.W. Argentina and South Bolivia are more pubescent, with shorter petiolules, longer calyx acumen and more developed stem wings. They rarely form the semi-rosette of basal leaves that is seen in *S. chacoense* in the plains. These forms seem to be due to introgression with the species *S. microdontum* and possibly certain other mountain species, and they might well be considered as a distinct subspecies. Bolivia, Argentina, Paraguay, Uruguay. Waysides, pastures, arable land, scrub and woodland margins, from sea level to 2350 m.

2b. Subspecies *muelleri* (Bitt) Hawkes et Hjerting. Differs from subsp. *chacoense* by the long narrow leaflets, about 3.3–4.3 times as long as broad, the long petiolules, up to 25 mm on the acroscopic side, the very oblique leaflet bases and rather low pedicel articulation. Plant generally glabrous, very occasionally pubescent. Argentina (prov. Misiones), South Brazil (States of Rio Grande, Paraná, Santa Catarina). Grasslands, river banks, waysides, fields, forest margins and clearings, at fairly low altitudes (up to 800 m?).

3. S. commersonii Dun.

Differs from *S. chacoense* in the sessile obtusely rounded leaflets at the apex; terminal leaflet typically much larger than the laterals and the leaf often sublyrate in shape. Peduncle not or once-branched; branches often very short. Corolla white or often tinted purple on the external surface; berry cordate or conical. Occurs in both diploid and (probably) autotriploid forms. $2n = 24$ (36).

3a. Subspecies *commersonii*. Lateral leaflets decreasing rapidly to base of leaf, often markedly decurrent, normally sessile; peduncle once-forked, the branches somewhat contracted; corolla generally purple, the lobes about $1\frac{1}{2}$ times as long as broad or even longer. Most plant breeding and cytological work in the past has utilized this subspecies, often in its triploid forms. Moderately frost-resistant. Coastal belt of Argentina and Uruguay, and coastal regions of South Brazil. Grows in a wide variety of habitats but very frequently in marshy places, fields, river banks, woods and sandy shores (dune slacks), from sea level up to about 400 m.

3b. Subspecies *malmeanum* (Bitt.) Hawkes et Hjerting. Differs from subsp. *commersonii* in the lateral leaflets gradually decreasing to base of leaf, narrowly decurrent and slightly petiolulate; peduncles unbranched or the branches not markedly contracted; corolla always white, the lobes about as long as broad. This subspecies is known in diploid and triploid forms, the latter having been found chiefly in Argentina, prov. Misiones. Like subsp. *commersonii*, it possesses conical berries and is resistant to frost. Has a more inland distribution and is known in N.E. Argentina, Brazil, N. Uruguay and south Paraguay. Inhabits similar places to those in which subsp. *commersonii* is found but seems to prefer shady thickets and woodlands; it has about the same altitude range as subsp. *commersonii*.

4. *S. tarijense* Hawkes
Plant with a dense pubescence of simple and glandular hairs, pleasantly aromatic; calyx acumens well-marked, linear; berry globular, with white raised spots. Some insect resistance. Natural hybridization with blue-flowered mountain species has caused a certain amount of introgression into *S. tarijense* from them. An apparently natural hybrid of this type is *S.* × *berthaultii* (see p. 36) from Cochabamba, Bolivia, which may possibly have resulted from *S. tarijense* × *S. sparsipilum* crosses. C. Bolivia to N.W. Argentina. Scrub and cactus vegetation in dry interandine valleys at altitudes of about 2000–2800 m. $2n = 24$.

5. *S. yungasense* Hawkes
Plant with small (2 cm diam.) pale yellow or deep cream corolla with narrow lobes more than twice as long as broad. This species bears a remarkable vegetative similarity to *S. violaceimarmoratum* (Series Conicibaccata). North Bolivia, sub-tropical forests (Yungas region) at 1100–1900 m. $2n = 24$.

Series VII CIRCAEIFOLIA Hawkes

Tuber-bearing herbs with small leaves, the terminal leaflet enlarged and laterals reduced in size. Corolla white, stellate, less than 1.5 cm diam.; berries narrow-conical. N. Bolivia; hedges and bushy places at high altitudes.

1. *S. capsicibaccatum* Cárd.
Distinguished from *S. circaeifolium* by the softly hairy leaves, 1–2 pairs of lateral leaflets which are more than half the length of the terminal, rather pubescent inflorescence and grooved stigma. C. Bolivia, depts. Cochabamba, Santa Cruz and La Paz, in cloud forest and among bushes and scrub vegetation at 2000–4000 m altitude. $2n = 24$.

2. *S. circaeifolium* Bitt.
Leaf glabrescent, simple, or sometimes 1–2-jugate, the laterals much less than half the length of the terminal leaflet; stigma simple. Bolivia, depts. La Paz and Cochabamba, in cloud forest and scrub vegetation at altitudes of 2500–3900 m. $2n = 24$.

Series VIII CONICIBACCATA Bitt.

Tuber-bearing herbs with generally well-dissected leaves and acuminate leaflets, rotate to ± stellate whitish to generally purple flowers and ovoid–conical to long conical berries. Mexico, southwards to Bolivia. All species grow in humid mountain forests and in other regions of high rainfall.

Conicibaccata is one of the three taxonomic series of wild potatoes to be found both north and south of the Panamá isthmus (together with Juglandifolia and Tuberosa). Many of the species are not very well known and are difficult to cultivate. The polyploid series of species in Conicibaccata, with diploid, tetraploid and hexaploid species, is of great theoretical interest and indicates a possibly rather complex evolutionary history, of which practically nothing is yet known.

1. *S. agrimonifolium* Rydb.
Berries narrow-conical, grooved, 3–4.5 cm long. Corolla large, rotate. Leaf with (6)–7–9 pairs of leaflets and numerous interjected leaflets. S. Mexico to Guatemala. Cloud forests at about 2000–3300 m. $2n = 48$.

2. *S. albornozii* Corr.
A very distinct species, perhaps forming a link between series Conicibaccata and Piurana. South Ecuador, on edges of mountain forests at 2300 m.

3. *S. ayacuchense* Ochoa
Central Peru (dept. Ayacucho), in cloud forests at 3000 m.

4. *S. buesii Vargas*
Very similar to *S. laxissimum*, of which it may be no more than a slight variant. S. Peru in cloud forests at 2400–3600 m.

5. *S. chomatophilum* Bitt.
Leaf 3–4-jugate with 0–4 pairs of small interjected leaflets; leaflets elliptic or broad lanceolate–elliptic; corolla blue, 4 cm diam. with short lobes; filaments pubescent externally; berry ovoid. N. to C. Peru in woods and grassy places, from 2000–4000 m. $2n = 24$.

6. *S. colombianum* Dun.
Corolla rotate-pentagonal; berries broadly ovoid-conical with generally blunt apex, less than 3 cm long; leaves barely acuminate. A very polymorphic species occurring in Venezuela, Colombia and Ecuador. Cloud forests at 2200–3500 m. $2n = 48$.

7. *S. jaenense* Ochoa
A delicate slender plant with 2–3-jugate acuminate leaflets, white rotate corolla and acute fruits to 10 mm long. Endemic to N. Peru (Cajamarca) at 2700 m. $2n = 24$.

8. *S. laxissimum* Bitt. (including *S. santolallae* Vargas)
Leaf 4–6-jugate; corolla pale blue, ± stellate, 3–3.5 cm diam.; filaments pubescent; berries long conical, 2.5–3 cm long. South to Central Peru, growing at the edges of cloud forests, from 1900–3600 m. $2n = 24$.

9. *S. longiconicum* Bitt.
Leaf glabrous, dark green, with acuminate leaflets; interjected leaflets sparse; corolla rotate; berries narrow conical. Costa Rica and Panamá, at 1000–3150 m in mountain forest and páramo vegetation. $2n = 48$.

10. *S. moscopanum* Hawkes
Corolla with flattened or even slightly concave lobes, appearing somewhat 10-pointed. Berries and leaves similar to *S. colombianum* from which this species was probably in part derived. S. Colombia. High altitude forests and clearings at 2900–3400 m. $2n = 72$.

11. *S. multiflorum* Vargas
South Peru, growing amongst scrub and bushes, at about 2100–2700 m.

12. *S. neovargasii* Ochoa
Central Peru, at 2800 m amongst shrubby mountain forests. $2n = 24$.

13. *S. oxycarpum* Schiede
Berries narrow-conical, 3–4 cm long, sharp-pointed; leaflets pubescent,

acuminate; interjected leaflets generally absent; corolla rotate. East C. Mexico. Cloud forests at about 2000–2200 m. $2n = 48$.

14. S. paucijugum Bitt.
Terminal leaflet much longer and broader than the 2–3-jugate laterals which decrease rapidly towards base of leaf; corolla rotate, purple; berries blunt ovoid. Colombia, Ecuador, N. Peru; mountain forests and pastures at altitudes of about 3500–3700 m and perhaps lower also.

15. S. pillahuatense Vargas
South Peru, in forest and scrub at about 2800 m.

16. S. tundalomense Ochoa
Ecuador, in mountain forests at 2900–3200 m.

17. S. urubambae Juz.
S. Peru, in cloud forests and clearings at 2000–2200 m.

18. S. villuspetalum Vargas
Tubers unknown, but may be formed. South Peru, on old cultivation terraces at 2500 m and elsewhere.

19. S. violaceimarmoratum Bitt.
Terminal leaflet much wider, although barely longer than lateral leaflets; leaflets 3–4-jugate; stem often violet-marbled; pedicel swelling into calyx base; corolla stellate. Berries to 2.5 cm long. N. Bolivia, in cloud forests and clearings, pathsides, etc., at 1800–3600 m. $2n = 24$.

20. S. woodsonii Corr.
Costa Rica, Panamá, Venezuela, 3150–4000 m on high páramos (mountain meadows).

Series IX PIURANA Hawkes

Tuber-bearing herbs with shining glabrous or glabrescent leaves which become leathery and with revolute margin when dry; corolla large, rotate, showy, blue to purple or blue and white; berries ovate, with flattened apex. Colombia, Ecuador, N. and C. Peru, in a wide range of habitats.

The confines of this series are not clearly agreed upon by different *Solanum* taxonomists, and no doubt with better knowledge it will be necessary to reduce some species to synonymy or add some from other series. Piurana seems to intergrade into Conicibaccata in the north and into Tuberosa in the south. Diploid and tetraploid species have been reported.

1. S. acroglossum Juz.
Central Peru; habitat unknown; altitude about 3000 m. $2n = 24$.

2. S. paucissectum Ochoa
Terminal leaflet much larger than the 0–2-paired laterals; corolla lilac or lilac and blue. North Peru, dept. Piura, in open places and amongst trees, at 3180 m. $2n = 24$.

3. S. piurae Bitt.
Leaf 2–3(–4)-jugate; 1–3 pairs of interjected leaflets; leaflets lanceolate; pedicels 3–5.5 cm long; corolla blue, to 2.5–3 cm diam.; filaments glabrous. N. Peru, rocky slopes amongst shrubs and grasses at 2500–3300 m. $2n = 24$.

4. S. solisii Hawkes
S. Ecuador, in bushy and grassy places on páramos, at 3500–4000 m.

5. S. tuquerrense Hawkes
Leaf 4–6(–10)-jugate, glabrous or with long spreading hairs; corolla blue, 2–3.5 cm diam.; filaments glabrous. Unique amongst tetraploids in being self-incompatible. Some frost resistance. Colombia, Ecuador, on páramos and amongst shrubs at 3000–3300 m. $2n = 48$.

6. S. yamobambense Ochoa
This species has a strong resemblance to *S. acroglossum*, with which it may perhaps be synonymous. N. Peru, amongst dense undergrowth, at 3160 m.

Series X ACAULIA Juz.

Low rosette-forming herbs (occasionally forming long stems), bearing stolons and tubers; leaves with typically obtuse leaflets auricled at the base on the acroscopic side; peduncle very short or absent; pedicel articulation absent or shown only by a ring of pigment, very rarely well-marked; corolla small, rotate, with very short lobes. Peru, Bolivia, and N.W. Argentina at very high altitudes in alpine meadows, field borders, etc.

1. S. acaule Bitt.
Characters and distribution as for series. Highly resistant to frost. The fairly wide infraspecific variation has caused certain authors to split *S. acaule* into several microspecies. The fertility and range of variability between all forms so far studied makes it advisable, however, to reunite them into the one original species divided into geographical subspecies as follows:

1a. Subspecies *acaule.* Lateral leaflets not much shorter than terminal leaflets, not or only slightly decurrent; pedicel articulation generally invisible or marked by a difference of colour; pubescence of fairly short crisped hairs. Peru to Bolivia and N.W. Argentina, from 2600–4650 m, in alpine meadows, by paths, walls, drainage ditches, cultivated fields, etc. $2n = 48$.

1b. Subspecies *aemulans* (Bitt. et Wittm.) Hawkes et Hjerting. Differs from subsp. *acaule* in the short leaf with much enlarged terminal and 0–4 pairs of decurrent laterals; pedicel articulation well-marked but not abscissing when mature. Pubescence as for subsp. *acaule*. N.W. Argentina, Sierra de Famatina, from 2950–4000 m in similar habitats to subsp. *acaule*. $2n = 48$.

1c. Subspecies *punae* (Juz.) Hawkes et Hjerting. Leaf like that of subsp. *acaule*, but stem very short; pubescence of long weak spreading hairs; articulation of pedicel never visible except by a change of colour, observable on living plant only. C. Peru, perhaps to N. Peru?, at similar altitudes and habitats to subsp. *acaule*. $2n = 48$.

1d. Subspecies *albicans* (Ochoa) Hawkes. Plant apparently similar to subsp. *acaule*, but provided with a dense covering of long white spreading hairs; corolla very pale, almost white. This subspecies is possibly an amphidiploid hybrid of *S. acaule* with a diploid wild species from some other series. N. Peru, dept. Cajamarca, on the páramos or alpine meadows, at 3450 m. $2n = 72$.

Series XI DEMISSA Buk.

A rather heterogeneous group of tuber-bearing species characterized chiefly by the high pedicel articulation and rotate corolla with very short lobes, similar to those of Acaulia. It is probably linked with Tuberosa through the diploid species *S. verrucosum* which I have now placed in Series Tuberosa. Mexico and Guatemala. Only pentaploids and hexaploids are known.

1. S. brachycarpum Corr.
Clearly distinguishable from other species in this Series by the upright stems, rather poorly dissected leaves with petiolulate leaflets and few interjecteds; corolla similar to that of *S. demissum* but paler; berries conical. West, central and south Mexico, in high altitude pine and fir forests at 2300–3350 m. $2n = 72$.

2. *S. demissum* Lindl.
Grows in rosettes or semi-rosettes but sometimes produces a long stem.

Leaflets sessile to sub-sessile with rounded apex. Corolla purple, generally with very short lobes. Highly blight-resistant. Mexico, from Durango State southwards into Guatemala; pine and fir forests from 2650–3800 m. $2n = 72$.

3. *S.* × *edinense* Berth.
Includes all forms that have arisen as hybrids between *S. tuberosum* and *S. demissum.* Lateral leaflets petiolulate, 4–6-paired, with up to 9 pairs of interjecteds; corolla large, with short lobes, much larger than that of *S. demissum.* Blight-resistant.

3a. Subspecies *edinese.* Formed as a natural hybrid in the Edinburgh Botanic Gardens and elsewhere, between *S. demissum* and *S. tuberosum* subsp. *tuberosum.*

3b. Subspecies *salamanii* (Hawkes) Hawkes. Formed as a natural hybrid in Mexico, States of Mexico, Vera Cruz, Tlaxcala, etc., from *S. demissum* and *S. tuberosum* subsp. *andigena*; leaf much more dissected than in the typical subspecies. Weeds of cultivated fields and waste places, at 2600–3500 m. Both subspecies possess $2n = 60$ chromosomes.

4. *S. guerreroense* Corr.
Distinguished from *S. demissum* by the conical fruits and from *S. iopetalum* by the well dissected leaves with 5 pairs of primary laterals and 8 or more pairs of interjected leaflets. Stem short; leaves in a rosette. Some blight-resistance. S.W. Mexico, Guerrero State, in pine-oak forest at 2800–3000 m. $2n = 72$.

5. *S. hougasii* Corr.
Differs from *S. demissum* and *S. guerreroense* in its tall upright habit, white flowers (often tinged purple between petals) longer corolla lobes, and petiolulate leaflets. From *S. iopetalum*, *S. brachycarpum* and *S. guerreroense* it differs in the corolla form and colour, and in the round berries. Bears some similarities to *S. stoloniferum* in series Longipedicellata. Some blight-resistance. West Central Mexico; high pine forests or even above the tree line in alpine meadows, from 2400–4000 m. $2n = 72$.

6. *S. iopetalum* (Bitt.) Hawkes
Habit similar to that of *S. brachycarpum* but flower larger and of a richer deep red–purple colour. East C. Mexico, State of Puebla, at 1900–2100 m. $2n = 72$.

7. *S.* × *semidemissum* Juz.
Differs from *S. demissum* in the upright habit and petiolulate leaflets, and from *S.* × *edinese* in the large terminal and poor leaf dissection

(2–4-leaflet pairs). Is of undoubted hybrid origin, derived possibly from a *S. demissum* × *S. verrucosum* cross (with the participation of an unreduced *S. verrucosum* gamete). Rarely forms berries. Blight-resistant. C. Mexico; field weed or along hedges and waysides, sometimes in pine forests, from 2700–3500 m. $2n = 60$.

Series XII LONGIPEDICELLATA Buk.

Herbs with long creeping stolons; leaves with coarse white hairs, or glabrous. The arched corolla lobes and large acumens give the corolla a circular appearance with acumens standing out sharply from it. Corolla occasionally, however, substellate. Central Mexico to S.W. United States on dry plateaux and mountain slopes, medium altitudes. Most species are tetraploid.

1. *S. fendleri* A. Gray
Plant with rather dense coarse hairs on leaf, stem, peduncles, pedicels and calyx; lateral leaflets sessile to sub-sessile, often decurrent; all leaflets obtuse; corolla deep purple. Some virus resistance. *S. fendleri* grows chiefly under the shade of *Pinus ponderosa* and other trees in damp leaf mould at altitudes of 1700–2850 m. $2n = 48$.

1a. Subspecies *fendleri.* Corolla acumens large, curving gradually into the lobes; corolla semi-stellate in outline.

1b. Subspecies *arizonicum* Hawkes Differs from subsp. *fendleri* by the rotate corolla with rather small acumens and short lobes, and the small, 3 mm long anthers. N.W. Mexico and S.W. United States. Subsp. *arizonicum* is apparently endemic to the Chiricahua mountains in S.E. Arizona.

2. *S. hjertingii* Hawkes
Differs from *S. fendleri* in the glabrous or glabrescent stems, leaves, peduncles, pedicels and calyx, in the generally petiolulate narrow leaflets and the glabrous corolla. Vegetatively extremely similar to *S. cardiophyllum* subsp. *ehrenbergii*, but similar in its floral morphology to *S. fendleri.* N.E. Mexico, States of Coahuila and Nuevo León, in dry piñon scrub (*Pinus cembroides*, *Juniperus monosperma*, etc.), at 1750–2500 m in rather similar habitats to *S. jamesii*. $2n = 48$.

3. *S. papita* Rydb.
Similar vegetatively to *S. fendleri*, but smaller in its growth (generally not more than 10 cm tall) and with small corolla (up to 15 mm diam.) which is whitish with sightly darker acumens. Anthers only 4 mm long. This species is closely related to *S. fendleri* and may be only a southern

subspecies of it; yet it occurs in different habitats, and intergrading forms are not yet known, although this may be due to the few collections made so far in N.W. Mexico. N.W. Mexico, States of Durango and Zacatecas. Growing in dry oak–juniper scrub at 2200–2400 m. $2n = 48$.

4. S. polytrichon Rydb.
Distinguished by the dense spreading pubescence of thick white hairs on the whole plant, and especially on the pedicel, both below and above the articulation; corolla generally white, occasionally mauve; articulation high. Some forms possess sparse leaf pubescence, but the spreading pedicel pubescence is a constant feature. Some blight-resistance. Central and N.W. Mexico, in dry stony and shrubby places, occasionally as a weed of cultivation, at 2000–2400 m. $2n = 48$.

5. S. stoloniferum Schlechtd. et Bché.
A very polymorphic species, differing in degree of leaf dissection and flower colour (white to purple). Constant features are the circular corolla outline with large acumens, the coarse adpressed hairs over all green parts (not spreading as in *S. polytrichon*), the pedicel glabrous or only sparsely hairy above articulation, and the acuminate leaflets. Locally abundant forms (previously described as species) are found in certain regions (e.g. *S. antipoviczii* and *S. ajuscoense* in the higher regions and *S. longipedicellatum* in the lower parts of the Valley of Mexico; *S. tlaxcalense* in the eastern part of the distribution range. Detailed studies are needed here to elucidate the pattern of variability. Shows resistance to blight, insects and viruses. Central Mexico from Toluca to Orizaba. Dry plateaux, valleys and hillsides, chiefly as ruderal plants, from 1800–3000 m. $2n = 48$.

6. S. × vallis-mexici Juz.
This hybrid 'species' is distinguished from *S. stoloniferum* by the dark purple corolla, and broadly obovate to rhomboid terminal leaflet which is larger than the laterals. Formed as a hybrid between *S. stoloniferum* ($2n = 48$) and *S. verrucosum* ($2n = 24$), occurring in the Valley of Mexico and elsewhere, where the altitude ranges of these two species overlap. Woods, fields, waysides, from 2400–3000 m. $2n = 36$.

Series XIII POLYADENIA Buk.

Tuber-bearing herbs with a very dense indumentum of glandular hairs of an objectionable odour; most of these glands possess 2-celled stalks–a feature not known in any other tuber-bearing *Solanum* species. Corolla white, pentagonal to sub-stellate. Berries ovate, cordate or conical, somewhat flattened, with black streaks, especially on the edges. Central to Southern Mexico on dry stony hillsides or in damp rain forests, according to the species.

1. *S. lesteri* Hawkes et Hjerting

Differs from *S. polyadenium* in the more robust habit, thick purple-pigmented stem, dense indumentum of long spreading multicellular hairs on all green parts, in addition to the glandular ones, and long-conical pointed 2-grooved and flattened berries. A distinct species, obviously related to *S. polyadenium*, but occurring under quite different ecological conditions. Endemic to South Mexico (Oaxaca State), in damp mountain forests at 2300 m, growing with *S. brachycarpum*. $2n = 24$.

2. *S. polyadenium* Greenm.

Whole plant covered with very frequent, stalked glands; stalks generally 2-celled. Very sparse multicellular non-glandular hairs also present but not easily visible without a lens. Whole plant more slender and less robust than *S. lesteri*. Corolla white, pentagonal. Insect-resistant and with some blight-resistance. Central Mexico, from Vera Cruz to Jalisco; on dry stony hillsides, by old walls, on old lava and amongst trees and shrubs, from 1900–2000 m. $2n = 24$.

Series XIV CUNEOALATA Hawkes

Small straggling tuberiferous herbs with pinnatifid leaf, the rachis with narrow wedge-shaped decurrent wings between each pair of leaflets; corolla purplish, sub-stellate to rotate with well delimited petal acumens. North Peru to N.W. Argentina and N. Chile, in dry cactus deserts and scrub. Possibly drought resistant.

1. *S. anamatophilum* Ochoa

Peru, dept. Ancash, in sub-xerophytic scrub, with cacti, at 2700 m.

2. *S. infundibuliforme* Phil.

Leaflets roughly lanceolate, linear-lanceolate to linear. Interjected leaflets entirely absent. Corolla pale mauve to purplish, occasionally white, in shape varying from sub-stellate to completely rotate (circular with well-marked acumens, as in Series Longipedicellata). Possibly drought resistant. Central Bolivia to N.W. Argentina and N. Chile; dry cactus and scrub deserts at 2450–4100 m. $2n = 24$.

Series XV MEGISTACROLOBA Cárd. et Hawkes

Rather short stemmed or straggling tuber-bearing herbs whose leaves bear a very enlarged terminal leaflet, with the lateral leaflets much smaller than the terminal or sometimes completely absent. The laterals when present are generally broadly decurrent on to the rachis at the basiscopic side; peduncle very short; pedicel long, with very high articulation; corolla sub-stellate to rotate, purple. N. Peru to N.W. Argentina, growing in waste places, open mountain pastures, etc.

1. S. boliviense Dun.
A rosette or sometimes caulescent species, with generally completely simple leaves and rotate dark purple corolla. Central Bolvia to N.W. Argentina; dry bushy places and cultivated fields, from 2600–3650 m. $2n = 24$.

2. S. × *bruecheri* Corr.
A naturally occurring hybrid, formed probably between *S. acaule* and *S. megistacrolobum* (see Brücher, 1959); some of Correll's specimens would seem to be hybrids of *S. infundibuliforme* × *S. megistacrolobum*, however. N.W. Argentina, amongst rocks and stone walls, etc., from 3500–3700 m. $2n = 36$ (*fide* Brücher).

3. S. chavinense Corr.
A species with 1–2 pairs of lateral leaflets, densely pubescent pedicels and large showy stellate deep purple corolla. N. Peru, amongst rocks, trees and shrubs at 3500–4200 m. $2n = 24$.

4. S. dolichocremastrum Bitt.
Peru, exact locality unknown. Thought by Ochoa to encompass *S. chavinense* and *S. huanucense*.

5. S. hastiforme Corr.
Lateral leaflets reduced to minute decurrent lobes at base of terminal; leaf somewhat hastate; corolla lavender–purple, rotate, with well-marked lobes and distinct acumens. N. Peru, along a bushy rocky stream, at 3200 m. $2n = 24$.

6. S. hawkesii Cárd.
S. Peru. Damp forests at 2000 m.

7. S. huanucense Ochoa
A striking plant with short, 1–2-jugate leaves and very large terminal leaflet. C. Peru, endemic to mountains of prov. Dos de Mayo, in stony rocky places, at 3750 m.

8. S. megistacrolobum Bitt.
Rosette-forming or with straggling stem; terminal leaflet long ovate to

obvate, long-ellipsoid or rhomboid, rounded or obtuse at apex (or with a minute mucron only); laterals lobed, small or absent. *Heterodera* and frost-resistant. S. Peru to N.W. Argentina, high altitude plateaux and mountain slopes, from 2600–4300 m. $2n = 24$.

9. S. raphanifolium Cárd. et Hawkes
Distinguished from *S. megistacrolobum* chiefly by the broadly ovate to orbicular terminal leaflet and the paler sub-stellate to rotate corolla. S. Peru, waste places, hillsides, cultivated fields, etc., from 2800–3800 m. $2n = 24$.

10. S. sanctae-rosae Hawkes
Distinguished from *S. megistacrolobum* and *S. raphanifolium* by the acuminate lateral and terminal leaflets, the small rosette habit, deep blue–purple corolla, with poorly marked acumens and large globose stigma. *Heterodera* resistant. N.W. Argentina, in high mountain pastures, sandy and rocky places, etc., from 2500–3800 m. $2n = 24$.

11. S. sogarandinum Ochoa
Very closely related to *S. megistacrolobum*, of which it may perhaps represent a northern subspecies; differs chiefly in the very robust inflorescence and rotate (not sub-stellate) corolla. Some frost resistance. N. Peru, páramos (high Andean grassland), at 3550 m. $2n = 24$.

12. S. toralapanum Cárd. et Hawkes
Distinguished from *S. megistacrolobum* by the very broadly triangular decurrent rachis wing, or in the simple-leaved forms by the very long spathulate or long obovate leaf blade. C. Bolivia, field margins, rocky and grassy slopes, etc., from 3000–4500 m. $2n = 24$.

Series XVI INGAEFOLIA Ochoa

Tuber-bearing herbs with erect or decumbent habit, thick winged stems, imparipinnate leaves with broadly winged rachis, the wing continuing right to the leaf base, and rotate corolla. Andes of northern Peru.

1. S. ingaefolium Ochoa
A very curious and distinct species with leaves similar to those of the Leguminous genus *Inga*. Leaf rachis with broad wings which swell out from the point of insertion of each leaflet pair and become slightly narrowed towards the next pair below; no interjected leaflets; corolla lilac, rotate. Endemic to the humid mountain forests of N. Peru (Piura) at altitudes of 2800–3000 m.

2. S. rachialatum Ochoa

Differs from *S. ingaefolium* in possessing narrow stem wings, petiolulate lateral leaflets [4–5 pairs instead of 1–3(–4)], interjected leaflets or lobes projecting from the irregularly winged rachis, white corolla, etc. Endemic to the humid lower forest regions of N. Peru (Piura) at altitudes of 1200–1900 m. $2n = 24$.

Series XVII OLMOSIANA Ochoa

Tuber pubescent; leaf 2-jugate with broad irregularly swollen rachis wing hardly differentiated into leaflets or even lobes; corolla cream to pale yellow, stellate, deeply divided into narrow lobes, hooded at apex; berry ovoid to pyriform.

1. S. olmosense Ochoa

Characters as for series. Peru, dept. Lambayeque above Olmos, in rain forest at 1640 m. $2n = 24$.

Series XVIII TUBEROSA Rydb.

This tuber-bearing series contains all the cultivated potato species as well as the wild and weed species most closely related to them. It is characterized by imparipinnate or simple leaves, bifurcate peduncle, rotate to semi-stellate corolla and round berries. Series Tuberosa is found in the Andes of South America and the adjacent coastal belt in temperate and sub-tropical latitudes. One species, *S. verrucosum*, occurs in Mexico.

The wild species will be dealt with first, indicating where appropriate their possible connections with the cultivated ones.

Wild species

1. S. abancayense Ochoa

Peru, department Apurimac, in humid shrub vegetation at 2900–3600 m.

2. S. acroscopicum Ochoa

Distinguished by the completely glabrous leaf, narrow lanceolate petiolulate leaflets and rich purple large showy corolla; berry ellipsoid. Possibly related to series Conicibaccata. S. Peru (Tacna, Moquegua), in mountain forests (?) at 3200–3450 m. $2n = 48$.

3. S. alandiae Cárd.

Lateral leaflets 1–3-paired, decreasing rapidly in size from apex to base of leaf; no interjected leaflets. Related to *S. microdontum*, and possibly derived from it by hybridization. Bolivia (Cochabamba), amongst bushes on the high slopes of the Andes, at 3200 m.

4. S. amabile Vargas
A tall-growing species with dense long spreading hairs, 3–4-jugate leaves, and very large bright lilac to purplish corolla, 3.5–5 cm diam. C. to S. Peru on shrubby mountain slopes, at 2500–3700 m.

5. S. ambosinum Ochoa
Leaf 4–5-jugate with few interjected leaflets and short hairs on all green parts; calyx acumens long, 1.5–3 mm; corolla purple, rotate. C. Peru, on shrubby mountain slopes at 2200–2500 m.

6. S. andreanum Baker
Characterized by 2–3 pairs of acuminate leaflets and purple rotate corolla. S. Colombia, N. Ecuador. Damp high altitude woods and bushy places at 2200–2900 m. $2n = 24$.

7. S. × berthaultii Hawkes
Very glandular insect-resistant species with pale violet–blue pentagonal to sub-stellate corolla. Almost certainly formed as a natural hybrid between *S. tarijense* (Commersoniana) and a blue–flowered mountain species, possibly *S. sparsipilum*. Bolivia, eastern slopes of Andes in rather dry valleys amongst bushes and in waste places from 2100–2700 m. $2n = 24$.

8. S. brevicaule Bitt.
A low-growing bushy species with 2–3(–4)-jugate lateral and few interjected leaflets, decurrent wings from leaf bases and small purple corolla. C. Bolivia, dept. Cochabamba, amongst grass and shrubs, at 2700–4000 m.

9. S. bukasovii Juz.
Low semi-rosette plants with delicate growth; 3–4-jugate leaves and 0–3(–5) pairs of interjected leaflets; sparingly pubescent on all green parts. Flowers purple. Possibly related to *S. multidissectum*. Frost-resistant. C. Peru, in the puna formation (high mountain pastures) at 3000–3900 m. $2n = 24$.

10. S. cajamarcense Ochoa
A very distinctive species, the whole plant possessing a dense coarse pubescence; leaf rough, with undulate-crenulate margins and few lateral leaflet pairs (2–3) which are generally smaller than the terminals;

interjected leaflets numerous; corolla white. N. Peru, dept. Cajamarca, on mountain slopes at about 2600 m.

11. S. canasense Hawkes

Distinguished by the large rotate corolla, highly dissected leaves and the soft adpressed silky pubescence. Leaflets narrow, lanceolate; interjected leaflets numerous. Flowers blue–purple. Some frost resistance. S. Peru. Dry stony places, often amongst bushes, at 3200–4000 m. $2n = 24$.

12. S. candolleanum Berth.

Leaf highly dissected, with 5–6 pairs of narrow lateral leaflets and numerous interjected leaflets; calyx lobes lanceolate; acumens 2–4 mm long; corolla purple. N. Bolivia, dept. La Paz, on rocky mountain slopes, at 2700–3500 m. $2n = 24$.

13. S. cantense Ochoa

Lateral leaflets 2–3-jugate, not decurrent, with no interjected leaflets; corolla white, with hooded acumens. Possibly related to *S. chancayense*. Central Peru, near Canta, amongst stones and bushes. $2n = 24$.

14. S. chancayense Ochoa

Small glabrous plants with 2–3-jugate lateral leaflets and no interjected leaflets; flowers white. Peru; coastal hills (lomas) of Lima and La Libertad departments, from 150–550 m. $2n = 24$.

15. S. chiquidenum Ochoa

Leaves with 1–2 pairs of lateral leaflets, decreasing in size rapidly towards leaf base; no interjected leaflets; coarse shining hairs on both surfaces; calyx sub-glabrous, with linear acumens; corolla white, tinged with lavender, rotate. N. Peru, amongst bushes and in grass, at 2800–3400 m. $2n = 24$.

16. S. coelestipetalum Vargas

Leaf 4-jugate, with frequent interjected leaflets; lower surface densely velvety-pubescent; corolla bright sky-blue. S. Peru, endemic to the Urubamba valley, amongst dry sparse shrubs, at 2400 m.

17. S. colominense Cárd.

A small species with large terminal and rather few small lateral leaflets; corolla dark purple. Bolivia, Cochabamba, damp mountain slopes amongst rocks, at 3400 m.

18. S. contumazaense Ochoa

N. Peru, dept. Cajamarca, at 2840 m. $2n = 24$.

19. S. cyanophyllum Corr.
Ecuador, in forests at 2300 m.

20. S. × doddsii Corr.
Leaf 3–5-jugate, glabrescent, with petiolulate laterals and very few interjecteds. Rest of green parts also ± glabrous; corolla sub-stellate, light lavender. This 'species' is no more than a casual hybrid of *S. chacoense* with the blue-flowered Bolivian species *S. torrecillasense.* Bolivia, Cochabamba, on a rocky wooded slope at 3000 m.

21. S. gandarillasii Cárd.
A very distinct species, entirely glabrous, with enlarged terminal and few, rapidly decreasing laterals; flowers small; calyx with oblong-spathulate, leafy lobes; corolla white. C. Bolivia, eastern Andes, in dry cactus and scrub region from 2000–3000 m. $2n = 24$.

22. S. gourlayi Hawkes
Small rosette or semi-rosette plants with ovate–lanceolate to oblong–lanceolate leaflets which are often slightly decurrent; terminal leaflet broader than laterals; flowers pale purple, semi-stellate. N.W. Argentina. Dry hillsides amongst cactus and scrub vegetation at 2600–4000 m. $2n = 24, 48$.

23. S. gracilifrons Bitt.
Distinguished by the glabrous, very numerous, narrow leaflets at least five times as long as broad and white corolla. C. Peru, dept. Huancavelica. Dry cactus and scrub vegetation at 1900–2000 m. $2n = 24$.

24. S. huancabambense Ochoa
Leaflets oblong–elliptic, obtuse, petiolulate; calyx with coarse hairs; corolla white, sub-stellate, the lobes about 9 mm long × 14 mm wide. N. Peru, dept. Piura, at 1800–3000 m. Bushy places. $2n = 24$.

25. S. humectophilum Ochoa
Peru, dept. Amazonas, in sub-tropical rain forest at 2875 m. $2n = 24$.

26. S. hypacrarthrum Bitt.
Leaf simple or with a few pairs of minute laterals and interjected leaflets; terminal leaflet ± cuspidate at apex; corolla white. C. Peru, dept. Lima, in grassland and amongst bushes at 1800–3400 m. $2n = 24$.

27. S. immite Dun.
Leaflets lanceolate, petiolulate; calyx glabrescent; corolla large, white, rotate, very showy. Central to northern Peru, on stony hillsides, from the coastal lomas (about 150 m) to 2500 m. $2n = 24$.

28. S. jalcae Ochoa
Leaf 2–3-jugate with winged rachis and larger terminal leaflet; leaf pigmented purple below; calyx dark purple; corolla blue, with short lobes and very small acumens. N. Peru, growing on páramos and high alpine pastures from 3100–3400 m. Frost resistant.

29. S. kurtzianum Bitt. et Wittm.
Distinguished by the elliptic to oblong obtuse leaflets with well-marked marginal hairs set on cushions of tissue and visible without a lens; also by the very low pedicel articulation placed in the lower third. Flowers rotate to sub-stellate, white or with central petal streak of bright violet on external surface only. *Heterodera*-resistant and possibly drought-tolerant. W. to N.W. Argentina on dry bushy hillsides, etc., from 1400–3000 m. $2n = 24$.

30. S. leptophyes Bitt.
A low-growing slender species with narrow–lanceolate leaflets 3–4 times as long as broad. Corolla pentagonal to rotate, purple. Leaf dissection and pubescence not as intense generally as with *S. canasense*. S. Peru, N. Bolivia. Dry open places or amongst scrub from 2500–4000 m. $2n = 24$.

31. S. lignicaule Vargas
Stem terete, somewhat woody, with velvety pubescence; leaf softly pubescent, yellow-green. Corolla creamy-white. A very distinct species of unknown affinities. S. Peru, dept. Cuzco, on dry bushy slopes and amongst stones at 3000–3500 m. $2n = 24$.

32. S. lobbianum Bitt.
Stem and peduncle with long spreading hairs similar to those of *S. multidissectum* to which *S. lobbianum* may be related. S. Colombia; exact locality not recorded, probably at about 3000 m.

33. S. maglia Schlechtd.
A large bushy plant with (1)–2–3-paired broadly ovate leaflets, and broader terminal leaflet. Corolla white, rotate. Stamens very characteristic, in a loose barrel-shaped column; anthers and filaments not well demarcated either in colour or form. Central Chile, in the region of Valparaiso, near the sea coast to 120 m; Argentina, prov. Mendoza, in dry valleys at 1500 m. $2n = 24, 36$.

34. S. marinasense Vargas
Distinguished by the leaves which are bright glabrescent and shining above, paler and dull below; 2–3 pairs of broadly lanceolate leaflets. Flowers blue, very showy. A very distinct species, related to series Piurana. S. Peru, bushy places and grassy hillsides from 2000–3500 m. $2n = 24$.

35. S. medians Bitt.

Low-growing straggling plants with enlarged terminal and smaller lateral leaflets in 1–2 pairs with sometimes 1–2 pairs of interjected leaflets. Distinguished from *S. microdontum* by the rhomboid terminal leaflet and from *S. raphanifolium* by the non-decurrent laterals. The forms with 36 chromosomes possess almost orbicular leaves and 0–1 pairs of interjecteds, whilst the diploids possess narrower leaflets and more laterals. Corolla purple. C. Peru, coastal belt and mountains, in scrub vegetation, from sea level to 4000 m. $2n = 24, 36$.

36. S. microdontum Bitt.

Large plants with winged stems and large simple to 2(–3)-jugate leaves, the terminal generally much larger than the laterals; calyx acumens linear, unequal; corolla white, sub-stellate to pentagonal; pubescence coarse, shining, of frequent multicellular hairs. Some *Heterodera* resistance.

36a. Subspecies *microdontum.* More slender than subsp. *gigantophyllum*, with narrow straight stem wings (0–2 mm wide) and terminal leaflets rarely more than 8 cm long. C. to S. Bolivia and north-west Argentina.

36b. Subspecies *gigantophyllum* (Bitt.) Hawkes et Hjerting. Stem more robust, 3–20 mm diam.; wings 2–5 mm wide, undulate and crenulate marginally; terminal leaflet 8–18 cm long. North-west Argentina, southwards to La Rioja province.

Both subspecies grow in high mountain rain forests and amongst shrubs on the eastern slopes of the Andes, from 1000–3000 m. $2n = 24$ (36).

37. S. minutifoliolum Corr.

Leaf with two pairs of large laterals and several small ones, with about twelve pairs of minute sessile interjected leaflets; pubescence of dense yellowish-white hairs; corolla sub-stellate. Ecuador, in dense mountain forests on the eastern slopes of the Andes, at 1500–2900 m.

38. S. mochicense Ochoa

Leaflets 3–4(–5)-paired, sessile to petiolulate; 3–12 pairs of interjected leaflets; hairs quite dense but short and not easily visible; corolla rotate, white. N. Peru, on the lomas (low hills) by the coast, at not more than about 400 m.

39. S. mollepujroense Cárd. et Hawkes

Has affinities with *S. sparsipilum* but differs in the long oblong leaflets and glabrescent leaf surfaces. Bolivia, dept. Cochabamba, in stony bushy places.

40. S. multidissectum Hawkes

Low straggling herbs with generally highly dissected leaf and numerous interjected leaflets. Leaflets broadly ovate, the terminal sub-rotund. Long spreading hairs occur on the stem, rachis and petiolules. Flower fairly large, pale blue. Differs from *S. leptophyes* and *S. canasense* in the wider leaflets and long spreading hairs. Some resistance to frost and *Heterodera*. S. Peru in high Andean pastures, amongst stones, by paths, etc., at altitudes of 3300–4100 m. $2n = 24$.

41. S. multiinterruptum Bitt.

Leaflets sessile to sub-sessile; 4–5 pairs of laterals which are about twice as long as broad, 5–14 pairs of interjected leaflets; corolla very large, up to 5 cm diam. C. Peru, in high mountain grassland, rocky slopes and among bushes, etc., at 3000–4000 m. $2n = 24$.

42. S. neoweberbauri Wittm.

Vegetatively similar to *S. medians* but distinguished by the violet and white corolla. C. Peru, on coastal hills and mountains above. $2n = 24$.

43. S. oplocense Hawkes

Leaf 3–4-jugate with petiolulate but decurrent leaflets and denticulate margin; corolla violet to purple, pentagonal to sub-stellate; anthers strongly tapered from base to apex, often 4-lobed at base; stigma minute (thicker in tetraploid forms), generally not thicker than style apex. S. Bolivia, N.W. Argentina on dry hillsides amongst stones, in cactus and scrub vegetation, at altitudes of 2800–3500 m. $2n = 24$, 48, 72. Little is yet known about the nature of the polyploid races of this species. Resistance to *Heterodera*.

44. S. orophilum Corr.

Leaf with 4–5 pairs of acute to acuminate subsessile laterals and rather few interjecteds; corolla rotate, purple. N. Peru, on rocky bushy slopes at 3500 m.

45. S. pampasense Hawkes

Leaflets lanceolate, covered with hairs of varying lengths, some gland-tipped; calyx acumens linear, 2–4 mm long; corolla pale purple to pale blue, showy; anthers long and narrow. C. to S. Peru (depts. Ayacucho, Apurímac) in dry sub-tropical interandine valleys at 2000–2900 m. $2n = 24$.

46. S. paramoense Bitt.

Plant glabrescent; leaf 3-jugate, with no interjected leaflets; calyx glabrous; corolla violet, up to 4.5 cm diam., rotate. W. Venezuela, on the páramos (high Andean grassland), at 3300 m.

47. S. pascoense Ochoa

Leaflets 2–3-jugate, with a few interjecteds; corolla blue, showy, 3–4.5 cm diam. C. Peru (habitat unknown) at 3500–3600 m. $2n = 24$.

48. S. pumilum Hawkes

Plant with basal leaf rosette and very long peduncle (5–9 cm); rhomboid terminal leaflet much larger than the 2–3-jugate laterals. S. Peru, N. Bolivia; in high Andean grasslands and in stony rocky places, 3500–4000 m.

49. S. × *rechei* Hawkes et Hjerting

Leaf 2–4-jugate, lateral leaflets petiolulate, elliptic–oblong; interjected leaflets generally few; calyx acumens long, linear, 3–5 mm long; corolla white, rotate-pentagonal or sub-stellate, about 3 cm diam.; anthers long narrow, about 7 mm long. Probably hybrid of *S. microdontum* × *S. kurtzianum.* Argentina, endemic to prov. La Rioja, in dry valleys in and around Sierra de Famatina, at about 1100 m. $2n = 24$.

50. S. regularifolium Corr.

Leaves 3–5-jugate; interjected leaflets absent; corolla sub-stellate to pentagonal, pinkish lavender. Probably related to *S. paramoense* and possibly also to *S. colombianum* in series Conicibaccata. Ecuador, on bushy mountain slopes at about 2150 m.

51. S. rhomboideilanceolatum Ochoa

Terminal leaflet larger than the 1–3(–4)-jugate laterals; interjected leaflets few; corolla rich purple. C. Peru, on shrubby and grassy slopes and among rocks and cliffs at 2900–3500 m.

52. S. sandemanii Hawkes

Leaf imparipinnatisect; laterals 2–3-jugate, the upper pair widely decurrent; no interjecteds; calyx lobes often sub-spathulate, very pubescent; seems to be related to *S. infundibuliforme* (series Cuneoalata) on the one hand, and to *S. weberbaueri* (including *S. tacnaense*) on the other. S. Peru, dept. Arequipa, at 2600–3100 m.

53. S. scabrifolium Ochoa

C. Peru, at 2940 m.

54. S. × *setulosistylum* Bitt.

Vegetatively intermediate between *S. chacoense* and *S. spegazzinii*, but with corolla varying from white-stellate to mauve-stellate or sub-stellate. Style setae sometimes present. Leaf often highly dissected, with numerous narrow–ovate interjected leaflets. Young fruits sometimes puberulent. This name has been given to hybrid populations derived apparently from natural crosses between *S. chacoense* and *S. spegazzinii*. N.W. Argentina in scrub and dry stony lands, waysides, etc., at 1600–2100 m. $2n = 24$.

55. S. sparsipilum (Bitt.) Juz. et Buk.
A very polymorphic species found as a weed of cultivated fields and waste places, morphologically very similar to *S. tuberosum* subsp. *andigena* (see below). Distinguished from that species, however, by the smaller more straggling habit, 2–4-jugate leaves with fewer interjected leaflets and rather smaller flower. Its wide distribution may be due to the fact that it was carried by man as a weed with cultivated potatoes. *S. sparsipilum* seems to have played an important role in the evolution of *S. tuberosum*. C. Peru, southwards to C. Bolivia. Cultivated fields, waysides and waste places, from about 2500–4500 m. $2n = 24$.

56. S. spegazzinii Bitt.
A widespread and phenotypically highly variable species. Leaflets long-elliptic, narrowing to each end, often decurrent; interjected leaflets narrow; corolla medium purple to lilac, sub-stellate to pentagonal. *Heterodera* resistant. N.W. Argentina, in dry places in the interandine valleys and basins from 1900–3100 m. $2n = 24$.

57. S. suffrutescens Corr.
Stem stiff, rather woody; lateral leaflets 4-paired with 0–2 pairs of interjecteds; calyx trilabiate; corolla purple, rotate; berries globular. Possibly related to *S. andreanum*. Ecuador, roadside, at 2800 m.

58. S. sucrense Hawkes
This weed and ruderal species is very similar vegetatively to forms of *S. tuberosum* subsp. *andigena*. It differs chiefly from these in the sub-stellate corolla with very long poorly delimited acumens and small calyx, anthers and style. Probably of hybrid origin involving *S. oplocense* in its parentage. *Heterodera* resistant. C. Bolivia (dept. Chuquisaca, Potosi). An arable and garden weed, also found in waste places, from 2500–4000 m. $2n = 48$.

59. S. torrecillasense Cárd.
Lateral leaflets 1–3-paired, with no interjected leaflets; lower leaf surface softly pubescent; corolla rotate, blue-violet to light lavender. Looks very similar to natural hybrids between *S. microdontum* and *S. vernei*. Bolivia, dept. Santa Cruz, on rocky wooded slopes at 2700–3000 m.

60. *S. trinitense* Ochoa
Peru, dept. Cajamarca, at 3450 m.

61. *S. velardei* Ochoa
Peru, dept. Apurímac, in the dry temperate zone at 2450–2900 m.

62. *S. venturii* Hawkes et Hjerting
Leaf with large terminal and smaller 1–3-paired laterals, occasionally simple, with very sparse closely adpressed short 3-celled triangular hairs, visible under a lens only; calyx glabrescent or with a few hairs of same type as leaf hairs; corolla white, rotate to sub-stellate. Related to *S. microdontum* but easily distinguished by the very characteristic pubescence. *Heterodera* resistant. Endemic to N.W. Argentina in high altitude grasslands (pajonales), at 2000–2800 m. $2n = 24$ (36).

63. *S. vernei* Bitt. et Wittm.
A tall robust plant with large leaves and 4–6 pairs of ovate or ovate-lanceolate leaflets covered below with a white cottony or woolly pubescence. Flowers deep purple, often very large. Shows strong resistance to *Heterodera.*

63a. Subspecies *vernei.* Distinguished from subsp. *ballsii* by the apically acuminate petiolulate leaflets, acute ovate interjected leaflets and larger corolla.

63b. Subspecies *ballsii* (Hawkes) Hawkes et Hjerting. Leaflets generally sessile, acute to obtuse apically; interjected leaflets generally shorter, obtuse and sessile; corolla to 3 cm diam.

Both subspecies occur in N.W. Argentina, in high-rainfall cloud forest in semi-open habitats (contrast *S. microdontum*) at 2200–3450 m. $2n = 24$.

64. *S. verrucosum* Schlechtd.
Distinguished by the upright habit, petiolulate leaflets, terminal larger than laterals, well-defined rounded corolla lobes whose margins roll inwards, and white-verrucose berry. Is very probably an ancestral form of all Demissa species, contributing the one genome that they seem to possess in common, and was formerly included in that series. It is the only Tuberosa species which occurs north of the Panamá isthmus. Shows some blight resistance. North-east, central and southern Mexico in pine and fir forests at 2400–3200 m. $2n = 24$.

65. *S. vidaurrei* Cárd.
A delicate species with 3(–4) pairs of lateral leaflets and no (occasionally

1–2 pairs) interjected leaflets; leaflets narrow lanceolate, all pale yellowish-green, and with rather frequent coarse hairs; corolla pentagonal, blue–violet. Seems to have affinities with *S. spegazzinii* and *S. leptophyes*. S. Bolivia, N.W. Argentina, in high altitude, fairly damp grassland, at about 2700–3500 m. $2n = 24$.

66. S. virgultorum (Bitt.) Cárd. et Hawkes
Rather similar to *S. microdontum* vegetatively, but with blue instead of white flowers. N. Bolivia, dept. La Paz, amongst bushes on high mountain slopes, at 2700–4050 m.

67. S. weberbaueri Bitt.
Related to *S. medians* and *S. sandemanii*, but differs from the former in the 2–3-jugate leaf and strongly pubescent pedicel, and from *S. sandemanii* in the rather broad ovate leaflets and absence of strong decurrency. S. Peru, amongst shrubs and rocks, and on stony hillsides, both in the coastal lomas and higher up in the mountains, from 500–3200 m. $2n = 24$.

68. S. wittmackii Bitt.
A wild species with generally a fairly dense glandular pubescence, 6–7 pairs of lanceolate acuminate lateral leaflets and numerous interjected leaflets; calyx acumens linear or spathulate, corolla pale lilac, showy. Material from different collections differs considerably in the amount of glandular pubescence. Central Peru, near Lima. Desert coastal hills (lomas) vegetating only during the season of sea mists (garruas) during the months of July to September, and mountain valleys above. $2n = 24$.

Cultivated species

A. Diploids

1. S. ajanhuiri Juz. et Buk.
This species is similar in many respects to *S. stenotomum*. It differs, however, in the small regular calyx, smaller blue flower, very high pedicel articulation, and stiff leaves. Agrees with *S. stenotomum* in the decurrent bases of the uppermost leaflet pair and in the form of corolla. It was probably formed as a natural hybrid of *S. stenotomum* with *S. megistacrolobum*. Frost resistant. S. Peru and N. Bolivia, at high altitudes. $2n = 24$.

2. S. goniocalyx Juz. et Buk.
Leaves often pale green, shining or not, with large white or pink flowers and regular calyx with a ribbed base. Tubers with bright yellow flesh.

C. to N. Peru, at high altitudes (occasionally grown in S. Peru and N. Bolivia). $2n = 24$.

3. *S. phureja* Juz. et Buk.
Distinguished by the sparsely pubescent leaf, which is shining in the living state, and rather irregular calyx with lanceolate lobes. Tubers yield in 3–4 months under short day conditions and possess no dormancy period. This very widespread and highly variable species was formerly separated into a number of smaller units. These have been found to hybridize without lack of vigour or fertility in the F_2 generation under experimental conditions and hence cannot be considered as distinct. The absence of tuber dormancy indicates that it has become specially adapted to regions that are free from long periods of drought or frost. Venezuela, Colombia, Ecuador, Peru and N. Bolivia. Wet mountain slopes mostly in eastern Andes, at lower altitudes than the other cultivated species, but can also be grown at higher altitudes.

4. *S. stenotomum* Juz. et Buk.
Distinguished from *S. phureja* by the more densely pubescent leaf which is not shining in the living state, tubers produced in 5–6 months or longer and with definite dormancy period. Calyx generally irregular with lanceolate lobes as in *S. phureja*. A very variable species which is possibly ancestral to all the other cultivated potatoes. These have been produced from it either by hybridization (*S.* × *chaucha*, *S.* × *curtilobum*, *S.* × *juzepczukii*, *S. ajanhuiri*, *S. tuberosum*) or by evolutionary divergence at the same level of ploidy (*S. goniocalyx*, *S. phureja*). Central Peru to central Bolivia, at very high altitudes. Some forms are frost-resistant.

B. Triploids

5. *S.* × *chaucha* Juz. et Buk.
All those triploid forms that have been derived from natural crosses between *S. tuberosum* subsp. *andigena* and *S. stenotomum* are grouped here. Some of these, which are most distinct and more widely cultivated, were formerly classed by Juzepczuk and Bukasov (1929) and by myself (1944) as separate species. Many more collections of triploid cultivated potatoes have been made, each of which differs in certain points from the micro-species already described. This is only to be expected when we consider that triploid hybrids could have been formed many times by the crossing of different clones of the two very polymorphic species *S. stenotomum* and *S. tuberosum* subsp. *andigena*. If these hybrids are to be named at all, they should be considered merely as nothomorphic forms of one species. I have therefore retained the name *S.* × *chaucha* since it was the first to be applied to these triploid forms by Juzepczuk and Bukasov. The best way of distinguishing *S.* × *chaucha* from other

cultivated species is by the corolla lobes which are in general about three times as broad as long when spread out flat. Central Peru to Central Bolivia at high altitudes. $2n = 36$.

6. S. × *juzepczukii* Buk.
Distinguished by the semi-rosette habit, long straight leaves, short peduncle (2–4 cm long), pedicels with very high articulation which is indistinct, and small blue corolla (to 2.5 cm diam.) with very short lobes and small acumens.

A natural triploid hybrid between *S. acaule* and *S. stenotomum* (see Hawkes, 1962) which has almost certainly been formed more than once, with different varieties of *S. stenotomum* involved in each case. Central Peru, southwards to South Bolivia at very high altitudes. Frost-resistant. $2n = 36$.

C. Tetraploids

7. S. tuberosum L.
Distinguished from other species of cultivated potato by the pedicel articulation placed in the middle third, short calyx lobes arranged regularly, leaves often slightly arched, leaflets always ovate to ovate–lanceolate, about twice as long as broad, never narrow lanceolate as in some forms of *S. stenotomum* and *S. phureja*. Corolla lobes about half as long as broad. Tubers with well-marked dormancy period.

Two subspecies are now recognized:

7a. Subspecies *tuberosum.* Originally only from the coastal regions of South Central Chile (Island of Chiloé and adjacent mainland). Is distinguished from subsp. *andigena* by the less dissected leaves with wider leaflets, generally arched and set at a wider angle to stem. Pedicel thickened above; corolla often white or pale coloured. Tubers formed under long days or under short days in the tropics at lower altitudes only (500–2000 m). This subspecies was derived from subsp. *andigena* probably on two separate occasions; first in Chile, where subsp. *andigena* was carried by the Indian tribes migrating southwards from the Bolivian Andes; secondly, subsp. *andigena* was brought to Europe after the Spanish conquest where, under similar climatic and daylength conditions to those of Chile, the typical subspecies *tuberosum* was formed, again partly as a result of artificial selection. Cultivated primitively on the southern coast and islands of southern Chile (Chiloé region), now worldwide.

7b. Subspecies *andigena* (Juz. et Buk.) Hawkes. This subspecies may be distinguished by the narrower, more numerous leaflets, which are generally petiolulate, the leaves set at an acute angle to the stem and generally more dissected; pedicel not thickened at apex; tubers formed at high altitudes only (over 2000 m) under short day conditions. This is undoubtedly the ancestral subspecies of *S. tuberosum* formed probably from *S. stenotomum* × *S. sparsipilum* crosses in the Andes of Peru and Bolivia. Andes of Venezuela, Colombia, Ecuador, Peru, Bolivia, N.W. Argentina; also sparingly in Guatemala and Mexico.

Escaped forms of both subspecies have been described under the names of *S. apurimacense*, *S. subandigena*, *S. molinae*, *S. leptostigma*, *S. maglia* subsp. *meridionale*, etc. They differ only in the longer stolons and sometimes unpigmented tubers from the cultivated varieties. $2n = 48$.

D. Pentaploids

8. *S.* × *curtilobum* Juz. et Buk.

Distinguished by semi-rosette habit, straight stiff leaves, very high pedicel articulation and large purple circular corolla 30–35 mm diam. with very short lobes and acumens. Derived from natural crosses between *S.* × *juzepczukii* and *S. tuberosum* subsp. *andigena* (see Hawkes, 1962). Several variations in tuber colour and form are known. Central Peru to South Bolivia at very high altitudes. Frost-resistant. $2n = 60$.

2.3 Distribution and ecology of potatoes

The common cultivated potato, through the efforts of breeders and agronomists, can now be raised in most parts of the world, although in the tropics it needs to be grown in the winter season and even so, cannot be considered at present as well adapted. In fact, cultivated potatoes are generally best adapted to the cool temperate zones of the high altitudes in the Andes (2000–3500 m), at sea levels in temperate regions of North America, Europe, southern Chile and Argentina, and at appropriate altitudes in intermediate latitudes.

The frost-resistant species *S. ajanhuiri*, *S.* × *juzepczukii* and *S.* × *curtilobum* are grown from about 3000 to 4000 m above sea level, whilst *S. tuberosum* subsp. *andigena*, *S. stenotomum* and *S. goniocalyx* yield best at 2500 to 3500 m or a little higher. By human selection in southern Chile and in Europe subsp. *tuberosum* has adapted its photoperiodic requirements to long day lengths, although the temperature conditions cannot be very different from those to which the original subsp. *andigena* was accustomed. The quick maturing and short dormancy species *S. phureja* was evidently selected for a warm temperate frost-free climate in the lower valleys of the eastern Andes but can still be grown at fairly high altitudes in Ecuador and Colombia.

By and large, then, all the cultivated potatoes were basically evolved in cool temperate regions even though some have acquired resistance to frost through hybridization with the frost-tolerant wild species, *S. acaule*

Fig. 2.1 Distribution of cultivated potato species in South America – diploid species

and *S. megistacrolobum*. All historical and archaeological evidence points to the fact that cultivated potatoes were once entirely confined to the high Andes of South America and the coastal strip of central to southern Chile (see maps, Figs. 2.1 and 2.2).

The wild species, on the other hand, are found in a very much wider range of habitats, and over a wider latitude range than the cultivated species, since they are known to occur from the United States (Colorado, Utah, etc.) southwards through Mexico and Central America, as far south as latitude 45° S (see maps, Figs. 2.3 and 2.4). In both sub-continents they are chiefly plants of medium to high altitudes, although in South America they are found in the coastal areas of Peru (growing during the winter fog period) and Chile, as well as the plains of Argentina, Paraguay, Uruguay and southern Brazil.

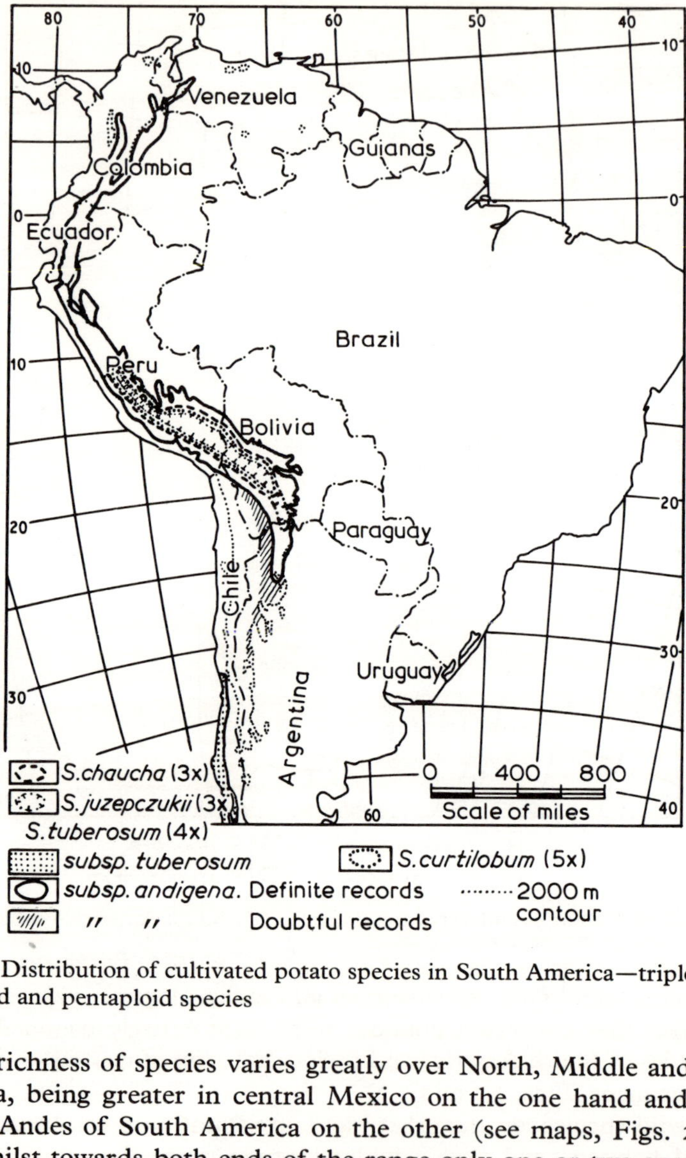

Fig. 2.2 Distribution of cultivated potato species in South America—triploid, tetraploid and pentaploid species

The richness of species varies greatly over North, Middle and South America, being greater in central Mexico on the one hand and in the central Andes of South America on the other (see maps, Figs. 2.3 and 2.4), whilst towards both ends of the range only one or two species are to be found.

Wild species extend through a wide range of plant communities, from the scrub and cactus deserts of Mexico, the southern United States, Peru, Bolivia and Argentina to the high rainfall mountain and cloud forests of the eastern Andean slopes, Mexico and Central America. Others again occur at altitudes of from 3500 to over 4000 m in the high Andes where they exhibit strong frost tolerance. The desert and coastal

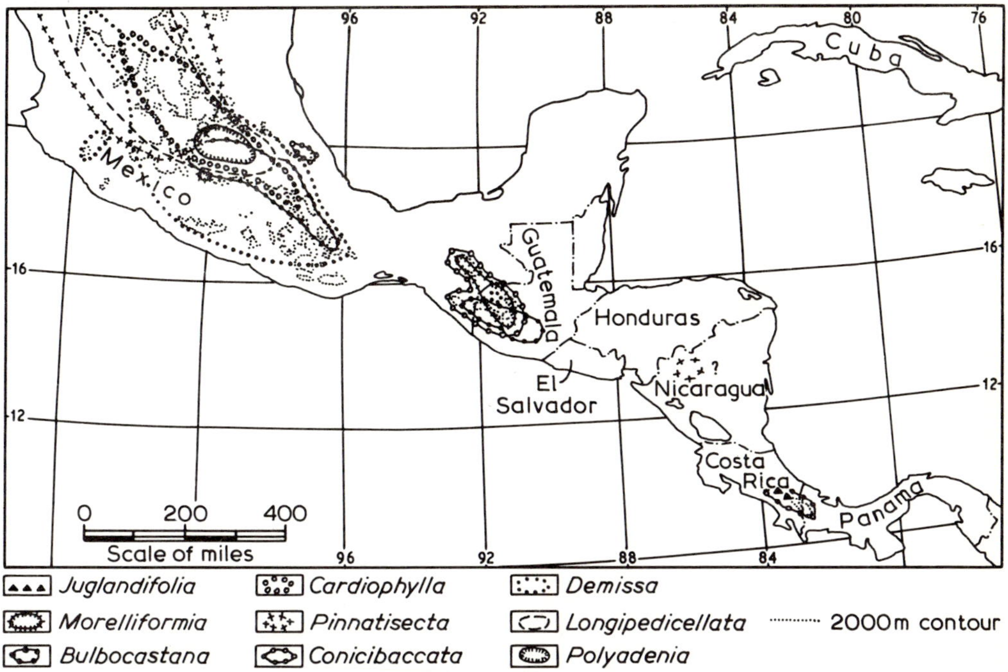

Fig. 2.3 Distribution of series of wild potato species in Mexico and Central America

forms seem to be able to withstand considerable drought and heat, whilst those from the southern pampas of Argentina and adjacent countries are more tolerant of long day length than species from the central regions. This range of tolerance to climate and day length is obviously of interest to the plant breeder in his search for initial material to help extend the climatic range of the cultivated potato.

In addition to their wide range of climatic adaptation, the wild potatoes, in so far as groups of individual species are concerned, show a greater range of adaptation to pests and pathogens than the cultivated potatoes. This, again, makes them of interest and value to breeders, although it would be true to say that at the moment of writing their full potential has not yet by any means been fully realized. However, as will be seen in later chapters of this book, many valuable genetic characters of resistance have been found in wild potatoes and have been incorporated in the newer commercial varieties.

Thus the Mexican *S. demissum* and allied species have conferred a high degree of resistance to late blight, *Phytophthora infestans*. Resistance to potato virus Y and allied strains as well as to various insect pests has been found in the Mexican *S. stoloniferum* and the Argentine *S. chacoense* and allied species in each of the two countries mentioned. Frost resistance

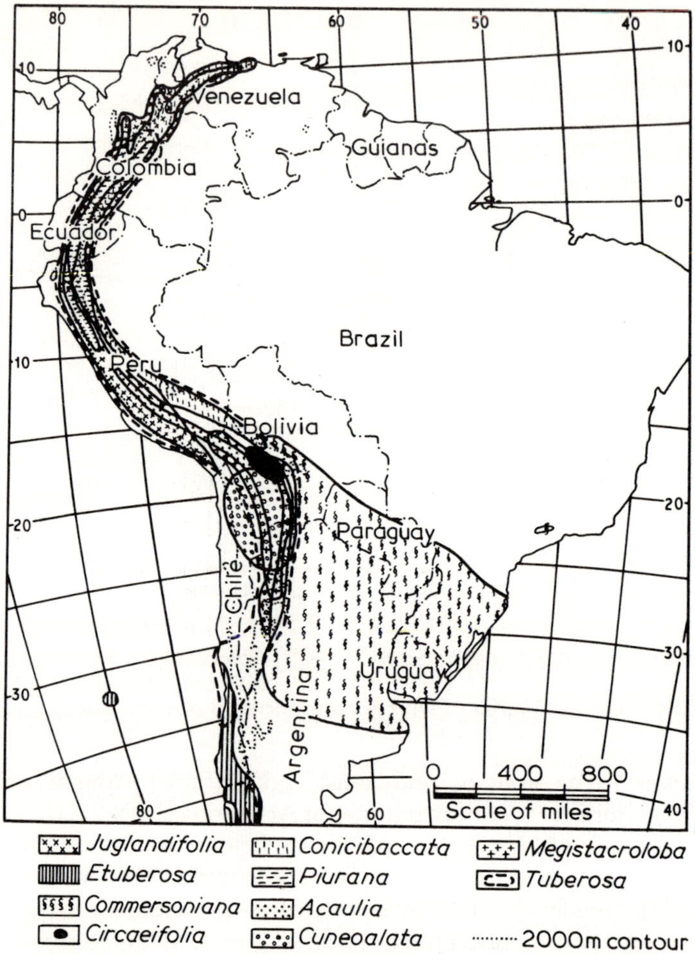

Fig. 2.4 Distribution of series of wild potato species in South America

has been found in a range of wild species, that from the Andean *S. acaule* having already been incorporated in promising breeding lines as well as at least one released variety. This resistance also occurs in some of the primitive cultivated species, as has already been mentioned, as well as resistance derived from another wild species, *S. megistacrolobum*.

What has been accomplished by 'nature' (actually, of course, under cultivation, although not due to conscious direction by man) can also be carried out by the plant breeder. Another probable example of this process is the natural transference of resistance to races of *Heterodera rostochiensis* and *H. pallida* from at least one of the many resistant wild species, *S. oplocense*, into tetraploid weed potatoes and even possibly into cultivated ones in Bolivia.

Many other examples exist of mutations which are potentially useful to the breeder in wild potatoes and indeed also in the primitive cultivated species. A well-known example of this latter is the resistance to *Pseudomonas solancearum* in *S. phureja* and high protein and dry matter levels which have been noted in a number of primitive cultivars from South America (see Hawkes and Hjerting, 1969, for Argentine wild and cultivated species, also Ross and Rowe, 1965, 1969, 1972).

2.4 Species concepts in potatoes

Most cultivated plants are notoriously difficult to classify. The bewildering morphological and physiological variation with few crossability barriers or clearly and easily identifiable discontinuities has more often than not caused the taxonomist to leave cultivated plants severely alone. Where cultivated plants have been the object of study, species concepts have ranged from very narrow ones, in which nearly every variant has been given a Linnaean name at the specific or infraspecific level, and many so-called 'species' have later been found to depend on no more than one or a few gene differences; or, at the other end of the spectrum, an excessive 'lumping' together (in opposition to the work of the 'splitters' mentioned above) by some workers has tended to obscure the very real variation that exists and needs to be analysed. A middle course between these two extremes can perhaps be steered when biosystematic, chemotaxonomic and taximetric studies are used, but unfortunately most taxonomists have on the whole preferred to work with wild species, although there are of course many notable exceptions.

Although genetic variability is high both in wild and cultivated plants the latter seem to possess a complexity of variability which has always been difficult to analyse. In cultivated plants various types of artificial selection have been acting, in addition to natural selection, on a series of species complexes whose history is difficult to unravel. The broad canvas of natural selection has been overpainted constantly by artificial selection, acting often in a contrary or quite unknown direction during prehistoric and historical times from some 10 000 years ago right down to the present. To take an example far removed from potatoes, we have fairly good documentary evidence from the last two centuries on the history of the cultivated rose and we therefore know rather accurately how our garden roses have evolved during this period under the hand of man from sources in Europe, the Middle East and the Far East. Before that time the evidence is fragmentary, but that which we have available to us, helps us to understand the pattern of variability and enables us to set out a valid and useful classification. Had this knowledge not been available our task would have been so much more difficult that it might not have been possible at all. How, then, can one follow over a period of perhaps five to ten millenia the movements of a cultivated plant like the potato or wheat, taken by man from its centre of origin through his wanderings or passed on from tribe to tribe and brought under natural and artificial selection pressures to which, as a wild plant, it had never been subjected? Such a plant might also have hybridized frequently with related wild species in areas other than that to which it was originally adapted and hence acquired further variability, and in fact we have several good examples of this process in many of our major crops. After periods of evolution under domestication the cultigen may have renewed genetic contact with the wild prototype and/or related weed forms; this process also is likely to provide further variation.

So the complex pattern of variability in a cultivated plant derived from its subjection to natural and artificial selection pressures of a complex nature about which we at present know all too little except in general terms, and its acquisition of variability through introgressive hybridization and often also through polyploidization, all add to the difficulty of the taxonomists in classifying it and providing himself with a clear concept of it as a species or a group of species.

A further difficulty with the potato is its high degree of phenotypic plasticity, as well as its genotypic variability. This applies equally to the related wild species as well as the cultivated ones. Thus the potato taxonomist needs to look for those characters that are genetically stable within the species and which are less liable to show a wide range of phenotypic expression.

2.5 Crossability, sterility and breeding behaviour

The pattern of variability in the tuber-bearing *Solanum* species cannot be adequately understood without mention being made of the two alternative methods of reproduction available to them. One of these, the sexual, largely outbreeding reproductive system is particularly common in wild species, less so in cultivated ones. The other way of reproduction is asexually, by means of tubers. Theoretically, potatoes should be able to maintain themselves indefinitely by this means, especially under a favourable stable environment with a minimum of competition from other vegetation. However, if the environment is changing or the competition becomes more intense, the clonally propagated plants will die out and new plants will arise from seed elsewhere. Thus a delicate balance is maintained by means of natural selection between stability of genotype when conditions are stable and a varying series of genotypes for natural selection to act upon when the environment is changing. Such a dual system must be of immense value to potatoes, which are not alone amongst plants in devising a reproductive strategy of this type.

Most potato breeders and geneticists are aware that the majority of wild and cultivated diploid potato species will cross without difficulty and that the resulting F_1's are fertile and vigorous. Even tetraploid by diploid crosses are not very difficult to make, in one direction at least, although the progeny may not be very numerous or indeed fertile. Be this as it may, the 'gene pool' of potatoes is undoubtedly large, which is of considerable advantage to those who wish to transfer characters or blocks of genes from one species to another.

This ease of crossing has led some workers to believe that perhaps all or most potatoes belong to one or a very few species. If crossability is the only criterion for defining a species this comment would be correct. However, in nature, each wild potato species forms a series of interbreeding populations separated from those of other species by geographical and ecological barriers. This seems to be the objective situation, matching also the range of morphological variation and indicating in general that discrete groups of populations exist which fit reasonably well with generally accepted species concepts.

One cannot escape the fact, however, that some wild potato species frequently form hybrids with each other, especially in habitats that have been influenced by man. We shall mention in a future section (p. 64) that the genetic architecture of potato species is so distinct as to reduce fertility and viability in artificial F_2 hybrids between them. This process probably reduces the viability of hybrids in nature but would not prevent gene flow taking place between two or more species. Indeed, we have some evidence for the formation of a limited number of 'fixed' hybrids (see *S.* × *rechei*) or complete sections of introgressed species (mountain forms of *S. chacoense* subsp. *chacoense*) in some instances.

However, field studies strongly indicate that the viability of hybrids in nature is not as high as one would have expected. Studies by Hawkes and Hjerting (1969, pp. 459–60) in Argentina have shown clearly that hybrids are not able to survive poor or adverse conditions since they possess only portions of the adaptive complexes of the parents. In a good year the hybrids will evidently survive as well as the parents, but in a poor year when rainfall is scarce or too abundant the full adaptive mechanism of each parent may enable it to survive, whilst the less well-equipped hybrids will not.

Whatever is the final explanation there seems no doubt that species hybrids, although quite common in some areas and in some years, do not seem to persist for more than a year or two. Thus, although Edgar Anderson's hypothesis of hybrids being able to exist in 'hybrid habitats' is attractive and possibly correct for some groups of plants, it does not seem to fit the facts with potatoes, since parents as well as hybrids can exist easily in those habitats in good years, whilst it is the hybrids, not the parents, which disappear in the bad ones.

We also have some evidence, admittedly needing verification, of species being kept apart temporally by difference of flowering time. An example of this is *S. microdontum* and *S. vernei* in N.W. Argentina (see Hawkes and Hjerting, 1969, p. 460).

Apart from these general examples there is evidence of species evolving under clear cytogenetical barriers amongst the Mexican wild hexaploid species, where F_1 hybrids are completely sterile (Marks, 1955; Hawkes, 1956a). Furthermore, crossability barriers between Mexican and South American species also exist (Gell *et al.*, 1960) as we shall discuss in more detail below.

The majority of potato species are outbreeders, and indeed the diploids are mostly self-incompatible, thus rendering outbreeding obligatory. This outbreeding system helps no doubt to maintain the high degree of infraspecific variability which has already been commented upon.

The odd-numbered polyploids (triploids and pentaploids) which are the results of diploid by tetraploid or tetraploid by hexaploid crosses respectively are of course different 'species' from the normally outbreeding largely sexually reproducing species which we have just been discussing. Many would argue that taxa such as *S.* × *vallis-mexici*, *S.* × *juzepczukii*, *S.* × *chaucha* and *S.* × *edinense* ought not to be called species at all. The pentaploid *S.* × *curtilobum*, formed from a triploid × tetraploid cross (the first parent providing an 'unreduced gamete') is another similar case. Here one is probably justified in giving species names for convenience, even though these taxa are maintained entirely by vegetative reproduction, since they can be easily identified and have a definite distribution area if wild and a place in the agricultural economy if cultivated. They may indeed be continuously replenished by the sort of crossing processes by which they were originally formed.

This same process of unreduced gamete production, which seems quite common in potatoes, has also been the cause for the rather frequent production of triploid forms of diploid species which cannot very easily be distinguished from them and are sterile 'dead ends'. In this case the easiest way of dealing with such triploid forms which merely arise as a result of the union of reduced and unreduced gametes is to give them the same species name as the original diploid. Under favourable conditions, as with *S. maglia*, *S. calvescens*, *S. cardiophyllum* and *S. commersonii*, the triploids form very large colonies. In fact, with the first two species mentioned, they have almost entirely replaced the sexually reproducing diploids.

Returning now to the sexually reproducing species we can see that with almost all the diploids outbreeding is obligatory, through the functioning of genetically controlled self-incompatibility mechanisms. In most tetraploids and all known hexaploids self-pollination is possible, but insect pollinators seem to be needed to transfer pollen. The ratio of cross- to self-pollination is, however, not known, except in *S. acaule*, which seems to be habitually self-pollinated. The hexaploid *S. demissum* also seems to be of this type. An exception to the rule of self-compatibility in tetraploids is provided by the curiously self-incompatible *S. tuquerrense* in series Piurana. Self-compatible diploids such as *S. brevidens*, *S. etuberosum* and *S. morelliforme* also provide an exception to the usual rule of self-incompatibility in this group.

As has already been mentioned, the gene pool of potatoes is extremely large – a fact which is of the greatest value to the breeder. Most species will cross with each other, given a certain amount of patience, although one-way incompatibility is not unknown. However, there are two main groups of potatoes which are very difficult to cross. The first of these includes series Morelliformia, Bulbocastana and Pinnatisecta, all from Mexico and adjacent regions. The second comprises Commersoniana, Circaeifolia, Conicibaccata, Piurana, Acaulia, Demissa, Longipedicellata, Cuneoalata, Megistacroloba and Tuberosa.

We do not possess sufficient knowledge of Ingaefolia or Olmosiana to tell whether or not they would cross with the second group. Polyadenia is slightly fertile with both the large groups mentioned above but there has been little success in using it as a 'bridge' between them.

The two non-tuberiferous series Juglandifolia and Etuberosa lie completely outside any of the potato crossability groups. Indeed Rick (1951) was able to cross *S. lycopersicoides* in Juglandifolia with the tomato (*Lycopersicon esculentum*). The lack of crossability between the first and second groups mentioned above has been broken twice so far, and perhaps will be done again.

The first was by Dionne (1963) who used *S. acaule* as a bridge species in forming hybrids with *S. bulbocastanum* and certain species in Pinnatisecta series. He then formed colchicine-hexaploids from the sterile triploids resulting from this cross and crossed these with *S. tuberosum* (both tetraploids and dihaploids). This work was confirmed also by Hermsen (1966) and Hermsen and Ramanna (1969) who studied the cytology of the various types of hybrid obtained. Hermsen and Ramanna (1975) also obtained *S. bulbocastanum* × *S. verrucosum* hybrids although the value of these hybrids in crossing with *S. tuberosum* has not yet been elucidated.

These results and others confirm the general thesis that most, if not all, of the wild potato species (with the probable exception of series Juglandifolia and Etuberosa, which are not in any case tuber-forming) will sooner or later be brought into the gene pool of the cultivated potato, *S. tuberosum*.

2.6 Cytology of potato species

A polyploid series in potatoes has been known for nearly fifty years. With a base number of twelve, the wild species occur as diploids, triploids,

Table 2.1 Classification and chromosome number of the more important wild and cultivated potato species

Series	Species arranged according to chromosome number ($x = 12$)				
	2x	*3x*	*4x*	*5x*	*6x*
I Juglandifolia	*S. juglandifolium* *S. lycopersicoides*				
II Etuberosa	*S. brevidens* *S. etuberosa*				
III Morelliformia	*S. morelliforme*				
IV Bulbocastana	*S. bulbocastanum* *S. clarum*	*S. bulbocastanum*			
V Pinnatisecta	*S. cardiophyllum* *S. jamesii* *S. pinnatisectum* *S. trifidum*	*S. cardiophyllum* *S. jamesii*			
VI Commersoniana	*S. chacoense* *S. commersonii* *S. tarijense*	*S. calvescens* *S. commersonii*			
VII Circaeifolia	*S. capsicibaccatum*				
VIII Conicibaccata	*S. chomatophilum* *S. violaceimarmoratum*		*S. colombianum* *S. oxycarpum*		*S. moscopanum*
IX Piurana	*S. piurae*		*S. tuquerrense*		

X	Acaulia			*S. acaule*	*(S. acaule)*	*S. acaule*
XI	Demissa				*S.* × *edinense* *S.* × *semidemissum*	*S. brachycarpum* *S. demissum*
XII	Longipedicellata		*S.* × *vallis-mexici*	*S. fendleri* *S. polytrichon* *S. stoloniferum*		
XIII	Polyadenia	*S. polyadenium* *S. lesteri*				
XIV	Cuneoalata	*S. infundibuliforme*				
XV	Megistacroloba	*S. boliviense* *S. megistacrolobum* *S. raphanifolium* *S. sanctae-rosae* *S. toralapanum*	*S.* × *bruecheri*			
XVI	Ingaefolia	*S. rachialatum*				
XVII	Olmosiana	*S. olmosense*				
XVIIIa	Tuberosa (wild)	*S.* × *berthaultii* *S. bukasovii*				

(Continued)

Table 2.1 (Continued)

	Series	Species arranged according to chromosome number ($x = 12$)				
		2x	*3x*	*4x*	*5x*	*6x*
XVIIIa	Tuberosa (wild) (Continued)	*S. canasense*				
		S. gourlayi		*S. gourlayi*		
		S. kurtzianum				
		S. leptophyes				
		S. maglia	*S. maglia*			
		S. microdontum	*S. microdontum*			
		S. oplocense		*S. oplocense*		*S. oplocense*
		S. sparsipilum				
		S. spegazzinii		*S. sucrense*		
		S. vernei				
		S. verrucosum				
XVIIIb	Tuberosa (cultivated)	*S. × ajanhuiri*	*S. × chaucha*		*S. × curtilobum*	
		S. goniocalyx	*S. × juzepczukii*			
		S. phureja				
		S. stenotomum		*S. tuberosum* subsp. *tuberosum*		
				S. tuberosum subsp. *andigena*		

tetraploids, pentaploids and hexaploids, whilst the cultivated series extends to pentaploids only. Some of the best-known species are shown in Table 2.1, set out in columns according to the level of ploidy. An analysis of all those of which the somatic number is known shows that most (74%) are diploid, very few (4.5%) are triploid, rather more (11.5%) are tetraploid, 2.5% are pentaploid, 5% are hexaploid and 2.5% contain several cytotypes. (This survey excludes triploid cytotypes of diploid species.)

It is worth noting that quite a few species occur as more than one cytotype. The triploid forms which are seen in many of the diploid species, for instance, appear to be autotriploids, as has already been mentioned, since there is no evidence for hybridization here, and synthetic autotriploids produced under experimental conditions match the natural ones in every detail (Tarn, 1967). Unfortunately, we know nothing of the origin of the tetraploid cytotypes of *S. gourlayi*, the tetraploid and hexaploid cytotypes of *S. oplocense* or the pentaploid cytotypes of *S. acaule* (Ochoa, personal communication).

All the diploid species so far investigated have shown regular bivalent pairing at meiosis. This continues in F_1 hybrids between them, even when the hybrids themselves are weak and infertile. However, although the chromosomes of different species appear to be homologous from this evidence, there is some reason to believe that homoeologous chromosomes in polyploids also occur.

Turning to the triploid 'species', these all seem quite clearly to be hybridogenic. An apparent exception to the assertion that all triploid species are species hybrids is *S. calvescens*. However, it seems likely that this merely represents the triploid cytotype of a normal diploid whose 24-chromosome forms have not yet been discovered. The origin of most has already been ascertained with a good degree of certainty, e.g. *S.* × *vallis-mexici* (*S. stoloniferum* × *S. verrucosum*; Marks, 1958), *S.* × *bruecheri* (*S. acaule* × *S. megistacrolobum* or some similar species; see Hawkes and Hjerting, 1969, pp. 266, 268), *S.* × *chaucha* (*S. tuberosum* subsp. *andigena* × *S. stenotomum*; Hawkes, 1963), and *S.* × *juzepczukii* (*S. acaule* × *stenotomum*; Hawkes, 1962). These are clearly not species of the general type found in *Solanum*, but they have been retained largely for convenience. Several other naturally-occurring triploid hybrids are known, for which no names have been proposed. In all triploids so far investigated meiosis has been shown to be very irregular and fertility on self-pollination is very low. Some, however, will produce seeds when used as female parents in crosses with diploids and tetraploids.

The tetraploid species, with the exception of *S. tuberosum*, behave as allotetraploids, with 24 bivalents at meiosis. However, *S. tuberosum* subsp. *tuberosum* exhibits a large range of meiotic anomalies, from 1.70 to 5.24 multivalent frequencies per cell, according to various workers (see Swaminathan and Howard, 1953 for details). It is therefore generally assumed to be an autotetraploid (see Section 2.9 below however).

Pentaploids are of a similar nature to the triploids mentioned above and are formed by hybridization, although not always as a result of hexaploid × tetraploid crosses, as might at first sight be supposed. Their origins are generally agreed to be as follows: *S.* × *edinense* (*S. demissum* × *S. tuberosum;* Hawkes, 1963; Ugent, 1967), *S.* × *semidemissum* (*S. demissum* × *S. verrucosum*–providing an unreduced gamete; Hawkes, 1963), *S.* × *curtilobum* (*S.* × *juzepczukii* (unreduced gamete) × *S. tuberosum* subsp. *andigena*; Hawkes, 1962).

The hexaploids are uncommon in South America (*S. moscopanum* and hexaploid cytotypes of *S. acaule* and *S. oplocense*), and their origin and nature is unknown. The Mexican hexaploids, on the other hand, have received some attention, especially *S. demissum* which has been used so much in breeding for *Phytophthora* resistance. The formation of 36 bivalents at meiosis is extremely regular but sterility ensues in the F_1 hybrids between them. This has considerable bearing on genetic relationships between them and will therefore be discussed further in Section 2.8.

2.7 Chemotaxonomic relationships of potatoes

Chemical taxonomy is a discipline that had its origins in the nineteenth century but which has only recently, within the last 20 years, been used extensively to help throw light on taxonomic relationships. Most workers have concentrated their efforts on wild species, but even so a considerable body of literature is building up on the results of chemotaxonomic studies of cultivated plants (see Hawkes, 1968 for literature review).

Useful work is reported on potatoes both in the group of substances generally referred to as secondary metabolites and also with proteins, enzymes, etc. Harborne's papers on potato polyphenols are worth noting here (see Harborne, 1960, 1962). He found (1960) three coumarins and other major polyphenols in the corolla of the two closely related wild species *S. jamesii* and *S. pinnatisectum*, as well as in *S.* × *sambucinum*, which later (Hawkes, 1963) was shown to be a hybrid of *S. pinnatisectum* with *S. cardiophylloum*. No other species were found with these substances in their flowers.

Flavonol glycosides were analysed (1962) in 60 wild species and 39 cultivated clones. Whilst most pigments are distributed widely a particular one, kampferol 3-glucoside, was found only in *S. santolallae* (= *S. laxissimum*). Richness in kampferol glucosides was found in the three series, Conicibaccata, Piurana and Demissa, which are probably taxonomically connected. It is interesting to note also that luteolin 7-glucoside was only found in *S. stoloniferum* and in the hybrid derived from it, *S.* × *vallis-mexici*.

In a paper chromatography study of leaf alkaloids Marks *et al.* (1965) showed the true nature of a presumed hybrid between two wild *Solanum* species, *S. demissum* and *S. stoloniferum.* Using gas–liquid chromatography, in a study of inflorescence hydrocarbons, Mecklenburg (1966) showed interesting taxonomic correlations between her results and those obtained from other methods. Of the three main similarity groups, the first included *S. demissum, S. verrucosum* and *S. stoloniferum.* Group two included *S. trifidum, S. clarum, S. fendleri* and *S. polytrichon*, all Mexican species, whilst group three was much more diverse.

Electrophoretic studies of proteins on paper, starch agar or polyacrylamide gels have also produced results of interest. The results, in a series of characteristic band patterns, have still to be fully interpreted, but on the whole have tended to support taxonomic systems based on morphology alone. Thus, Desborough and Peloquin (1966) compared disc-electrophoretic patterns from 26 wild and cultivated species, finding reasonable comparability between accessions of the same species and clear between-species differences. Interesting results were also obtained in hybrids, where there was segregation for the major bands present in each parental species.

The same workers (1967) studied esterase isozymes by disc electrophoresis. However, since there were fewer of these than in the experiments reported above, it was more difficult to draw taxonomic inferences from the results.

Immunological studies for potato proteins using double diffusion techniques were reported by Gell *et al.* (1956, 1960). The results showed that the tuber-bearing species fell into three groups, according to the type of precipitin spectrum formed in double diffusion plates with *S. tuberosum* antiserum. The maximum number of four bands was given by *S. tuberosum* itself, together with all the rest of the South American tuber-bearing species and the Mexican series Demissa and Longipedicellata. Two bands were shown with Pinnatisecta and Bulbocastana, whilst only one appeared with Morelliformia. These results agree remarkably well with the main crossability groups mentioned in Section 2.5 above.

Later work by Lester (1965) and Hawkes and Lester (1966, 1968) showed that *S. commersonii* could be distinguished serologically from the rest of the South American species and that with immunoelectrophoretic techinques many more subtle differences within the Mexican species can be clarified. For instance, it was possible to demonstrate subspecific differences in *S. bulbocastanum* and to show clear segregation of immunoelectrophoretic characters in seed grown from *S.* × *sambucinum*, thus providing further evidence for its origin as a hybrid between *S. pinnatisectum* and *S. cardiophyllum.*

What little has been published so far demonstrates clearly that chemotaxonomic techniques have thrown considerable light on species relationships in potatoes and could certainly do so even more in the future.

2.8 Evolutionary divergence and genome differentiation in wild potato species

Since most diploid potato species hybridize readily and produce fertile F_1's it is clear that genome differentiation in the conventional sense has not progressed very far. Since F_2 progenies from species crosses often exhibit a range of weak unthrifty plants it seems as though a cryptic structural differentiation has taken place in the chromosomes whereby very small differences (inversions, translocations, etc.) exist which are too small to affect pairing in species hybrids (Stebbins, 1945, 1947). Such F_2 breakdown which indicates differences in the genetic architecture of species is a useful taxonomic tool (Hawkes, 1966).

Although the regular bivalent pairing in the tetraploid species *S. acaule* seems to indicate its strong allopolyploid nature, triploid hybrids between it and diploid species behave cytologically like autopolyploids, thus pointing towards a very small degree of difference between its two constituent genomes (Propach, 1937) and those of the diploid with which it was crossed. Such a hypothesis also accords well with Lester's (1965) conclusions that there are very strong serological similarities between *S. acaule* and species in series Tuberosa and parts of Commersoniana (*S. chacoense*, at least). Stebbins points out that differential pairing may well take place, and to use the terminology of Huskins (1932) when homologues are absent then homoeologues will pair (see Table 2.2, below).

Considering now the other well-known tetraploid, *S. stoloniferum*, the results of Matsubayashi (1955) and Kawakami and Matsubayashi (1956, 1958) indicate that in hybrids with *S. chacoense* something approaching twelve bivalents and twelve univalents are formed. This is thus very different from the *S. acaule* situation and would indicate the presence of one genome in common (designated in Table 2.2. 'A') between

Table 2.2 Suggested genome formulae for certain potato species

Genome	*Taxonomic series or species*
A_1	Series *Tuberosa, Commersoniana, Cuheoalata, Megistacroloba*
A_2A_3	Series *Acaulia*
A_4B	Series *Longipedicellata* (at least in part)
A_1A_4 (B, C, D, etc.)	Series *Demissa*

the two species and a distinct one in *S. stoloniferum* (designated 'B') which is absent in *S. chacoense*. On the basis of these results one might then postulate a genome formula for *S. stoloniferum* of A_4A_4BB, the A_4 being necessary for reasons that will be discussed later. Marks' (1965) results with *S. verrucosum* × *fendleri*, *S. verrucosum* × *S. stoloniferum*, and reciprocals all point in the same direction although in every case there is always a small number of trivalents present, which might thus be taken to indicate that the genome differentiation was not completely clear-cut, and certainly not under the degree of strong genetic control as described by Riley *et al.* (1959) for *Triticum* polyploids.

The cytological analyses made by Swaminathan and Hougas (1954) and by Marks (1955) on *S. demissum* and its hexaploid relatives indicate quite a marked degree of genome differentiation in these species, where there is a tendency to the formation of 24 bivalents and 24 univalents. If we assume, as Marks (1955, 1965) does, that all these hexaploids have two genomes in common, one of these (A_1) being derived from *S. verrucosum*, the second (A_4) being somewhat distinct, and the third genome quite distinct (B, C, D, etc.), then this would accord with the cytology of their hybrids. However, one is bound to conclude, from the results of hexaploid × diploid and other crosses, that the situation may be much more complex than these results indicate, and that it may in the end never be possible to provide a clear-cut picture of genome formulae, even though there is certainly some degree of differentiation amongst the Mexican species.

One may perhaps envisage that in late Cretaceous times the tuber-bearing group of species evolved in an area more or less coincident with present-day Mexico, and that the A genome species migrated southwards before the Central American land bridge was severed in Eocene times. One would then postulate the evolution of the A genome series in South America, independently of the Mexican B, C, D genome series, during the period from Eocene to Pliocene times when North and South America were completely severed from each other.

From Pliocene times onwards the two continents became reunited. This would have given the more vigorous A genome a chance to migrate northwards in perhaps two successive waves; the first of A_4 type by crossing with the native B genome gave rise to the A_4B species (*S. stoloniferum*, for example) and the second more recent one to the hexaploid A_1A_4(B, C, D, etc.) species, incorporating the A_1 genome from the recently arrived *S. verrucosum*. This assumes that the C and D genomes were derived from other Mexican species, perhaps, for example, from *S. oxycarpum* and *S. agrimonifolium*.

Much further work is needed to confirm or destroy these hypotheses and for the moment we can only look on them as not yet proven.

2.9 Evolutionary relationships between cultivated potato species

The origins of the potato as a domesticated plant have already been discussed in Chapter 1. In the present section we shall attempt to look at the evolution of the whole group of cultivated species subsequent to the time of domestication.

The exact nature of the original wild prototype is uncertain, although it is certainly assumed to be a diploid species in series Tuberosa whose natural distribution area coincides with the high Andes from central Peru southwards to central Bolivia. On morphological evidence this prototype may well have been similar to the present-day species *S. canasense*, which occurs in the high basin of Lake Titicaca and the surrounding valleys and plateaux from Cuzco in south Peru to at least central Bolivia at altitudes of 2700–4000 m or more.

1. S. stenotomum The most 'wild-looking', and therefore presumably primitive, species of cultivated potato is the diploid, *S. stenotomum*, which is commonly cultivated at high altitudes but which, except for a few forms, is not frost-resistant. In several forms the leaflets are delicate and narrow, quite similar to those of *S. canasense*, although its tubers, like those of all cultivated potatoes, are much larger, with a wide and interesting range of shape, skin and flesh colour and skin patterning, whilst the wild ones are all smooth-skinned with shallow eyes and a white to dingy brown skin. Occasionally the lenticels may be enlarged and the skin flushed with purple but there are never any colour patternings on the skin of wild potatoes as there so often are in the primitive cultivated varieties of South America.

2. S. phureja An obviously closely-related species to *S. stenotomum* is *S. phureja*, which some workers consider to be no more than a subspecies or set of special variants of the former species. It has very probably evolved in response to agricultural conditions in the warmer eastern valleys of Peru and Bolivia and has a quick growing season of about three months with virtually no dormancy period. Simmonds (1964) believed it to have been formed many times from *S. stenotomum* and to be in fact conspecific with it. It is worthy of note, however, that *S. stenotomum* does not occur in Ecuador, Colombia and Venezuela, where *S. phureja* is very commonly grown, and it could not therefore have been derived from *S. stenotomum* in those countries. Further work is needed, but there is little reason to disagree with the thesis on present evidence that *S. phureja* has certainly been derived from *S. stenotomum* by mutation and selection.

3. S. goniocalyx Whilst the calyx of *S. stenotomum* and *S. phureja* is large, irregular and often bilabiate or trilabiate, that of *S. goniocalyx* is small, regular and with ribbed sepals. It is a species with deep yellow-fleshed tubers and quite broad leaflets, despite its diploid chromosome number. It again seems to have been derived by mutation and selection from *S. stenotomum*.

4. *S.* × *ajanhuiri*. The nature of this frost-resistant high altitude species is only just being elucidated. Recent experimental studies have indicated (Huamán, 1975), that *S.* × *ajanhuiri* was formed from crosses of *S. stenotomum* with the wild frost-resistant species *S. megistacrolobum*. Furthermore, it seems to have originated in northern Bolivia, not far south of Lake Titicaca. Although part of *S.* × *ajanhuiri* is an F_1 hybrid another part may possibly represent a series of F_2 segregates or back-crosses to the *S. stenotomum* parent. The whole process, though occurring under cultivation, must undoubtedly be considered a 'natural' one.

5. *S.* × *chaucha* This name is applied to a series of natural triploid hybrids of *S. tuberosum* subsp. *andigena* and *S. stenotomum*. Sometimes this cross gives rise to a tetraploid progeny through the functioning of unreduced gametes from the diploid parent; at least this is so under experimental conditions, and we assume it to be so in the Andes. Triploids are more common than would be expected from the results of controlled crosses where very few seeds are produced in any case, whether triploid or tetraploid. *S.* × *chaucha* might also be partly formed as an autotriploid, derived entirely from *S. stenotomum* by the union of reduced and unreduced gametes.

The amount of gene flow from diploid to tetraploid, either directly or through the medium of the rather sterile triploids, is also a matter of some interest, although the question is extremely difficult to answer.

This subject has been investigated recently by M.T. Jackson (1975), Jackson *et al.* (1976).

6. *S.* × *juzepczukii* The nature of this high altitude frost-resistant triploid species has been elucidated (Hawkes, 1962). It is a natural hybrid, formed under cultivation, of the wild frost-resistant species *S. acaule* and the cultivated *S. stenotomum*. It is almost completely sterile and has probably been formed more than once (see also Schmiediche, 1977).

7. *S. tuberosum* The nature of subsp. *tuberosum* has already been discussed in Chapter 1. Subsp. *andigena* is the more primitive of the two subspecies and was undoubtedly formed in the central Andes in the region of southern Peru and northern Bolivia. At one time it was thought of as a simple autotetraploid of *S. stenotomum*, since cytologically it behaves as an autotetraploid. However, since there is virtually no conventional genome differentiation in the South American species (with the partial exception of *S. acaule*) one would expect an amphiploid of two diploid species to behave as an autotetraploid in any case. Since the

calyx of *S. tuberosum* differs considerably from that of *S. stenotomum* it was postulated by Hawkes (1956b) and largely confirmed by Cribb (1972) that this tetraploid species was formed as an amphiploid hybrid of *S. stenotomum* × *S. sparsipilum*. The latter is a diploid weed species with a calyx and leaf very much more similar to that of *S. tuberosum*; it occurs a little lower down than *S. stenotomum*, although the two species must presumably have overlapped in altitude range at some time or other. No doubt gene flow of the nature discussed under *S.* × *chaucha* will have enriched the range of variability of subsp. *andigena* once it had been formed.

8. S. × *curtilobum* The nature of this frost-resistant pentaploid species was also elucidated by Hawkes (1962). It has undoubtedly been formed, possibly only once or a very few times, from a cross between *S.* × *juzepczukii* (providing an unreduced gamete) and *S. tuberosum* subsp. *andigena*. (see also Schmiediche, 1977)

Thus the cultivated potatoes form perhaps one of the most complex patterns of species and hybrids of any of the ancient domesticated plants. They seem all to be derived originally from the primitive diploid *S. stenotomum*, either by mutation and selection (*S. phureja*, *S. goniocalyx*), by species hybridization at the diploid level (*S.* × *ajanhuiri*), by diploid × tetraploid crosses (*S.* × *chaucha*, *S.* × *juzepczukii*), by amphidiploidy (*S.* × *tuberosum*) or by triploid × tetraploid crosses (*S.* × *curtilobum*).

References

Brücher, H. (1959). *Züchter* **29**, 149–56.

Cribb, P.J. (1972). Studies on the origin of *Solanum tuberosum* L. subsp. *andigena* (Juz. et Buk.) Hawkes – the cultivated tetraploid potato of South America. Ph.D. Thesis, University of Birmingham.

Desborough, S. and Peloquin, S.J. (1966). *Phytochemistry* **5**, 727–33.

Desborough, S. and Peloquin, S.J. (1967). *Phytochemistry* **6**, 989–94.

Dionne, L.A. (1963). *Euphytica* **12**, 263–9.

Gell, P.G.H., Wright, S.T.C. and Hawkes, J.G. (1956). *Nature, Lond.* **177**, 573.

Gell, P.G.H., Hawkes, J.G. and Wright, S.T.C. (1960). *Proc. R. Soc.* Ser. B, **151**, 364–83.

Harborne, J.P. (1960). *Biochem. J.* **74**, 270–3.

Harborne, J.P. (1962). *Biochem. J.* **84**, 100–6.

Hawkes, J.G. (1944). *Bull. Imp. Bur. Pl. Breed. Genet.*, Cambridge, pp. 142.

Hawkes, J.G. (1956a). *Proc. Linn. Soc., Lond.* **166**, 97–144.

Hawkes, J.G. (1956b). *New Phytol.* **55**, 191–205.

Hawkes, J.G. (1962). *Z. Pflanzenz.* **47**, 1–14.

Hawkes, J.G. (1963). *Rec. Scott. Pl.-Breed. Stn.* 76–181.

Hawkes, J.G. (1966). *Am. Pot. J.* **43**, 81–103.

Hawkes, J.G. (1968) (ed.). *Chemotaxonomy and Serotaxonomy*. Systematics Association, special Vol. 2. Academic Press, London, pp. 299.

Hawkes, J.G. and Lester, R.N. (1966). *Ann. Bot.*, N.S. **30**, 269–90.

Hawkes, J.G. and Lester, R. N. (1968). *Ann. Bot.*, N.S., **32**, 165–86.

Hawkes, J.G. and Hjerting, J.P. (1969). *The Potatoes of Argentina, Brazil, Paraguay and Uruguay—A Biosystematic Study*. Oxford University Press.

Hermsen, J.G.Th. (1966). *Euphytica* **15**, 149–55.

Hermsen, J.G.Th. and Ramanna, M.S. (1969). *Euphytica* **18**, 27–35.

Hermsen, J.G.Th. and Ramanna, M.S. (1975). *Solanaceae Newslett.* **No. 2,** 29.

Huamán, Z. (1975). The origin and nature of *Solanum ajanhuiri* Juz. et Buk. —a South American cultivated diploid potato. Ph.D. Thesis, University of Birmingham.

Huskins, C.L. (1932). *J. Genet.* **25**, 113.

Jackson, M.T. (1975). The evolutionary significance of the triploid cultivated potato, *Solanum* × *chaucha* Juz. et Buk. Ph.D. Thesis, University of Birmingham.

Jackson, M.T., Rowe, P.R. and Hawkes, J.G. (1976). *Am. Pot. J.* **53**, 395.

Juzepczuk, S.W. and Bukasov, S.M. (1929). *Proc. U.S.S.R. Congr. Genet. Pl. Animal Breed.* **3**, 593–611.

Kawakami, K. and Matsubayashi, M. (1956). *Rep. Hyogo Univ. Agric., Ser. Agric.* **2**, 143–8.

Kawakami, K. and Matsubayashi, M. (1958). *Rep. Hyogo Univ. Agric., Ser. Agric.* **3**, 124–30.

Lester, R.N. (1965). *Ann. Bot.* N.S. **29**, 609–24.

Marks, G.E. (1955). *J. Genet.* **53**, 262–9.

Marks, G.E. (1958). *New Phytol.* **57**, 300–10.

Marks, G.E. (1965). *New Phytol.* **64**, 293–306.

Marks, G.E., McKee, R.K. and Harborne, J.B. (1965). *Nature, Lond.* **208**, 359–61.

Matsubayashi, M. (1955). *Rep. Hyogo Univ. Agric., Ser. Agric.* **2**, 25–31.

Mecklenberg, H.C. (1966). *Phytochemistry* **5**, 1201–9.

Propach, H. (1937). *Z. indukt. Abstamm.-u. VererbLehre* **73**, 143–54.

Rick, C.M. (1951). *Proc. Nat. Acad. Sci. U.S.A.* **37**, 741–4.

Riley, R., Chapman, V. and Kimber, G. (1959). *Nature, Lond.* **183**, 1244–6.

Ross, R.W. and Rowe, P. R. (1965). *U.S.D.A. Bulletin 533*, pp. 73.

Ross, R.W. and Rowe, P.R. (1969). *U.S.D.A. Bulletin 533*, pp. 68.

Ross, R.W. and Rowe, P.R. (1972). *U.S.D.A. Publication R1695*, pp. 40.

Schmiediche, P.E. (1977). Biosystematic studies on the cultivated frost-resistant potato species *Solanum juzepczukii* and *Solanum curtilobum*. Ph.D. Thesis, University of Birmingham.

Simmonds, N.W. (1964). *J. Linn. Soc. (Bot)* **59**, 43–56.

Stebbins, G. L. (1945). *Bot. Rev.* **11**, 463–86.

Stebbins, G.L. (1947). *Adv. Genet.* **1**, 403–29.

Swaminathan, M.S. and Howard, H.W. (1953). *Bibliogr. Genet.* **16**, 1–192.

Swaminathan, M.S. and Hougas, R.W. (1954). *Am. J. Bot.* **41**, 645–51.

Tarn, T.R. (1967). The origin of two polyploid species of *Solanum* sect. *Tuberarium*. Ph.D. Thesis, University of Birmingham.

Ugent, D. (1967). *Evolution* **21**, 696–712.

Chapter 3

Structure and development of the potato plant

ELIZABETH G. CUTTER

3.1 Introduction

This chapter attempts to provide a description of the structure of the potato plant, including both external morphology and internal anatomy. Since the existing literature appeared to be particularly deficient on developmental aspects, some re-examination of early stages of development has been made where possible. Notwithstanding the considerable morphological and anatomical literature, it is clear that the structure and development of the potato, admittedly a very complex plant, is by no means fully understood: strangely enough, this applies particularly to the development of the tuber, despite its economic importance. There is considerable scope for further work in this field.

In addition to descriptive studies, experimental work on the control of development, especially of the stolon and tuber, has been reviewed. It is clearly of fundamental importance to achieve some understanding of the factors controlling the development of these organs. Greater association of anatomical studies with these experimental investigations would clearly be beneficial, especially in view of the known effects of hormones on structure in other plants.

In any survey of the morphology of the potato plant, one is struck by the wealth of varietal differences. Since these often seem important, the cultivar used in the work cited in this chapter has been stated wherever possible. In a plant such as the potato, where the development of storage organs and the movement of assimilates into them are of such paramount agricultural importance, it is remarkable that so much remains unknown, or incompletely known, in the field of structure and development.

3.2 Development of the seedling

The potato is relatively rarely grown from seed, except for genetical purposes. Germination is epigeal, i.e. the cotyledons are borne above ground by elongation of the hypocotyl. The radicle emerges from the micropylar end of the seed and develops as a tap root, which soon forms lateral roots (Fig. 3.1). The first foliage leaves are ovate and hairy. When the young plant is only a few cm high, stolons develop in the axils of the cotyledons (Fig. 3.1d), and after penetrating the soil form small tubers (Fig. 3.1c; Hayward, 1938). The tubers which develop from these stolons or from others in the axils of foliage leaves are usually small.

Fig. 3.1 Developing potato seedlings: (a–c) parts of Figs. 272 and 276 of Hayward (1938); (d) seedling of cv. Désirée. x 1.5 Note developing stolons in axils of cotyledons in (d), tubers in (c)

3.3 The root system

A much branched fibrous root system is formed either by the seedling tap root, or by adventitious roots in tuber-grown plants. In early stages of growth the root system is restricted to surface soil, the roots turning downward after extending for some distance horizontally; this leaves the area of subsoil directly below a plant almost free of its own roots (Weaver, 1922).

3.3.1 Structure of the root

Longitudinal sections of the root tip show that the root cap and epidermis have initials in common (Fig. 3.2). Tangential divisions occur in the protoderm to a point about ten cells proximal to the apex, and the epidermis is not clearly differentiated from the root cap until this level. More proximally the epidermal cells are slightly elongated radially. Starch grains are abundant in the more mature cells of the columella of the root cap.

The young root may be diarch (Artschwager, 1918; Hayward, 1938) or triarch (de Vries, 1878) with a wide cortex and an epidermis bearing root hairs. Older roots may have a larger number of protoxylem points. In the adventitious roots examined by the writer (cv. Arran Pilot), the number of protoxylem poles varied from 4 to 6. In the differentiation of the vascular tissues, the protophloem sieve elements become distinguishable first, at a distance of 400–500 μm from the apex. A transverse section of a pentarch root at this stage is illustrated in Fig. 3.3. The future pericycle and endodermis can also be discerned. The inner cortical

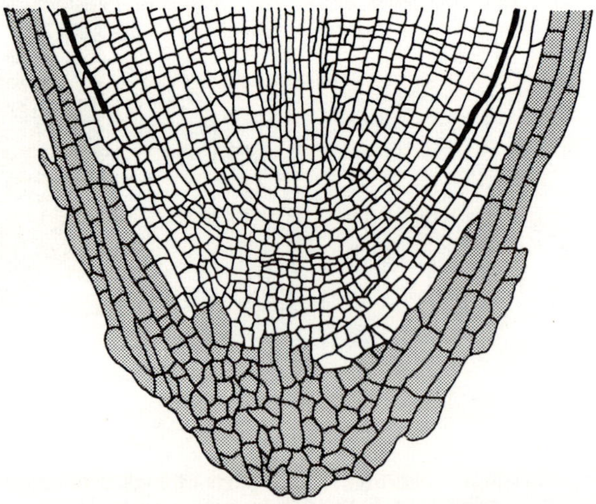

Fig. 3.2 L.S. root apex, showing the common initials of the epidermis and root cap. Cells containing starch grains are shaded; heavy line indicates boundary between epidermis and root cap. x 80

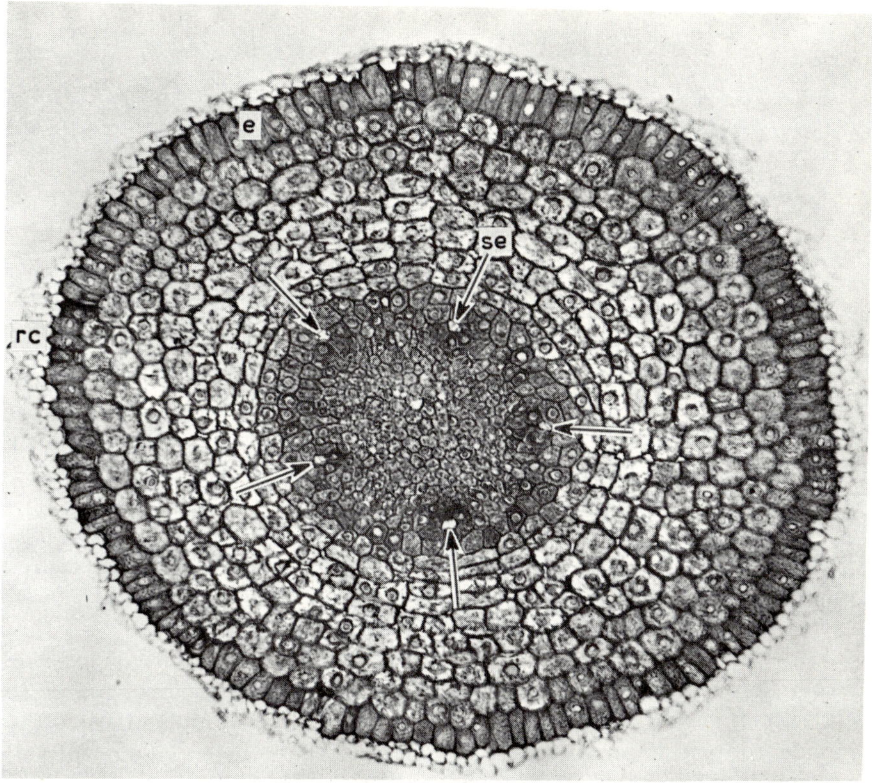

Fig. 3.3 T.S. pentarch adventitious root of cv. Arran Pilot. Maturing sieve elements (se) (arrowed) can be seen at this level. e, epidermis; rc, root cap. x 150

cells are arranged in radial files, the outer are not. Protoxylem elements become mature some distance further from the tip than those of the protophloem.

The endodermis later has well developed Casparian strips. A vascular cambium develops early and the formation of secondary vascular tissues occurs in the same way as in most dicotyledons (see Fig. 3.4). The secondary xylem consists of vessels, which sometimes develop tyloses, and of fibres and parenchyma. The primary phloem becomes crushed; phloem fibres are not present. The cortex persists for some time (Artschwager, 1918; Hayward, 1938).

Lateral root primordia originate in the pericycle (Esmarch-Bromberg, 1919). Adventitious root primordia are formed in a comparable region in underground portions of the stem of tuber-borne sprouts (Hayward, 1938; Ernst *et al.* 1968). Adventitious roots can also be induced to form on the basal surface of tuber slices treated with indoleacetic acid (Miedema, 1973). The root primordia form close to the vascular tissue, but the exact site of origin has not been established.

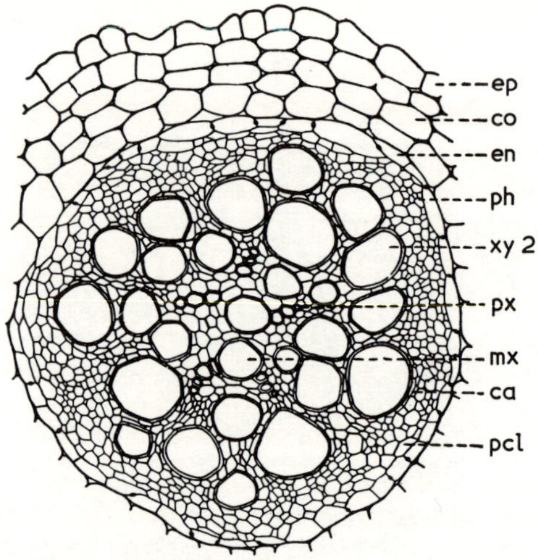

Fig. 3.4 T.S. older pentarch adventitious root. ca, cambium; co, cortex; en, endodermis, showing Casparian strips; ep, epidermis; mx, metaxylem; pcl, pericycle; ph, phloem; px, protoxylem; xy 2, secondary xylem. (Fig. 277 from Hayward, 1938)

Potato roots are attacked by nematodes such as *Heterodora rostochiensis* (see Chapter 12). In susceptible varieties considerable necrosis of the cortex can result, and when the larvae reach the vascular tissues they induce the formation of giant cells, which later extend longitudinally adjacent to the xylem. In resistant varieties, the tissues adjacent to the first giant cells become necrotic and surround the latter, which deteriorate (Huijsman *et al.* 1969). In regions where they abut upon xylem elements, the walls of the multinucleate giant cells develop protuberances which project towards the cell lumen. Apparently in response to increased demand for nutrients by the nematodes the protuberances later extend into the cytoplasm and form an anastomosing network. It is suggested that the giant cells induced by nematode attack constitute a multinucleate form of transfer cell, i.e. cells specialized for absorption (Jones and Northcote, 1972).

3.3.2 Regeneration from roots

Techniques have been devised for inducing the development of adventitious buds on potato roots. One purpose of this is to study periclinal chimeras of buds (see Section 3.4.1). One method involves inducing abundant adventitious root development from etiolated sprouts, and subsequently exposing the roots to light after removing the sprout and all buds. The roots were scraped at a distance of 5 mm from the sprout to a sufficient depth to expose the vascular tissue on the upper side, and small buds later developed just distal to the scraped area (Howard,

1964). In one series of experiments 85% of the scraped roots gave rise to adventitious buds (Claver, 1967). Using roots induced on de-eyed slices of non-dormant tubers (cv. Bintje) treated with auxin, Miedema (1967) subsequently showed that buds developed on both scraped and uninjured roots exposed to light. The exact site of origin of these buds is not known. Further experiments demonstrated wide differences in the ability of different varieties to regenerate buds from roots (Miedema, 1973). Buds formed in this way develop into apparently normal tuber-bearing plants.

3.4 The leafy shoot

The potato is a herbaceous plant with spiral phyllotaxis. Foliage leaves are imparipinnate, with small additional leaflets or folioles intervening between the pinnae. The aerial shoot is initially erect but later becomes partially procumbent (Artschwager, 1918); it dies back at the end of a single season, but the plant perennates by means of underground tubers. The stem is winged, the ribs or wings corresponding in position to the leaves and resulting from unequal decurrence of the edges of the petioles (de Vries, 1878). The compound leaves subtend axillary buds; the shoot shows partial apical dominance. The shoot system is sympodial (Danert, 1957).

Specific morphological features of the leafy shoot differ in the various varieties. British cultivars are described in the *British Atlas of Potato Varieties* (Potato Marketing Board, 1965).

An interesting feature of the morphogenesis of the leafy shoot is that the lateral buds have diverse developmental potentialities. According normally to their position on the stem, these buds may develop either as diageotropic stolons with scale leaves (normally from basal, underground nodes), or as negatively geotropic shoots with foliage leaves (from nodes on the aerial shoot). The factors controlling these alternative pathways of development are considered in Section 3.5.3. A great deal of work has demonstrated both that the habit of the potato plant can be experimentally altered, and that this may have important consequences related to yield. Factors controlling the number of shoots, their rate of growth and subsequent effects on yield are dealt with in Chapters 3 and 7.

Using four varieties, including both earlies and maincrop, Taylor (1953) observed a closely similar mean number of nodes on each main axis (varying between 26.0 and 26.7). Various experimental treatments have been shown to affect stem growth. Thus, thinner stems develop from tubers stored at higher temperatures (Fischnich and Krug, 1963). Treatment of seed pieces with gibberellic acid (GA) increased elongation of the internodes of the aerial stem (Dyson, 1965; Dyson and Humphries, 1966). Stem elongation is also affected by daylength (Digby and Dyson, 1973). Conversely, treatment with growth retardants such as CCC and B_9 decreased growth of the main stem and increased the number and rate of growth of tubers (Dyson, 1965; Humphries and Dyson, 1967a; Digby and Dyson, 1973); results differed in different varieties (Humphries and Dyson, 1967b). Ethylene treatment of pieces of tubers bearing single shoots also resulted in retarded stem growth and thicker stems (Hughes *et al.* 1973). Treatment with thiourea destroyed apical dominance and stimulated the outgrowth of thick shoots (Palladina and Pervova, 1966).

The degree of outgrowth and type of development of lateral shoots in the potato also have important consequences for the final yield. Varieties differ in both the amount of branching and the pattern of branch development (Toosey, 1963); thus there may be varietal differences in the number of stolons per node, the number of tubers per node, and the number of tubers per stolon (Wurr, 1974). To cite an example of such differences, Taylor (1953) found an average of 14.2 nodes with axillary shoots above ground and 11.8 nodes with axillary stolons below ground in cv. Ulster Chieftain, compared with 17.7 nodes above ground and 9.0 nodes with stolons in cv. King Edward. Early varieties tend to produce more nodes subtending stolons than maincrop varieties, but the ratio can be affected by environmental factors. In maincrop potatoes axillary shoots attain their greater foliar area by growing more rapidly than similar shoots in early varieties, rather than by beginning to grow earlier in the season (Taylor, 1953). In all four varieties which he studied, Taylor found that the first axillary shoots to grow out were those at nodes near the base of the stem, other developing later at the upper nodes. In cv. Majestic, the longest lateral shoots were present at the lower nodes, branches becoming smaller acropetally until at several nodes bud outgrowth was inhibited, the lateral shoots at higher nodes being again larger. Treatment with GA altered this pattern of bud outgrowth (Dyson and Humphries, 1966). Lovell and Booth (1969) found a somewhat similar pattern of stolon outgrowth. Fewer nodes subtended stolons under conditions of low nutrient level, and stolon length was reduced. In cv. Arran Pilot, Morris (1966) found that the development of axillary shoots was affected both by the size of the tuber and by the number of sprouts allowed to grow. For example, in small tubers with four sprouts, leafy branches sometimes developed at the fourth node from the base, whereas in large, single-sprout tubers at least 12, and sometimes as many as 20, of the basal nodes subtended stolons, the first leafy shoot occurring above this level. Morris (1966) suggested that competition between sprouts for some growth factor, such as gibberellins, might be important. Further discussion of the factors controlling the development of an axillary bud as a stolon or a leafy shoot will be deferred until Section 3.5.3.

3.4.1 The shoot apex

The shoot apex of the potato is of peculiar interest, because of the various phases in its development. The apex of a lateral stolon gives rise to an elongated, horizontal shoot, forming scale leaf primordia, becomes and remains dormant during the formation, growth, harvest and storage of the tuber, and finally after the breaking of dormancy sprouts and gives rise to an actively growing erect shoot bearing foliage leaves.

Structure and development

Sussex (1955a) surveyed the apical anatomy of six varieties of potato, using dark-grown sprouts. He found that the apical meristem comprised five layers of cells in cv. Kerr's Pink, the whole meristem being highly stratified. This stratification was minimal just after the formation of a new leaf primordium on the apical flank. Of these stratified layers, two constitute the tunica (Steinberg, 1950). In early stages of development, apices of axillary buds may have only two stratified layers. Klopfer (1965b) also found only two layers in apices of embryos and young seedlings. Sussex (1955a) found some differences both in size and structure of apices of different varieties. For example, in cv. Majestic the diameter of a sprout apex at the level of the axil of the youngest leaf primordium was 70–80 μm, whereas it was 140–160 μm in cv. Kerr's Pink and Ulster Chieftain, and 180–200 μm in Arran Banner and two others. The characteristic stratified appearance of the potato shoot apical meristem is shown for cv. King Edward in Fig. 3.5.

In cv. Majestic, Leshem and Clowes (1972) found that the apex of a sprouting eye was flat, whereas the stolon apex was slightly dome-shaped. In cv. King Edward and Arran Pilot, apices of both stolon and leafy shoot are dome-shaped (compare Figs. 3.5 and 3.17) so this probably depends on the variety. Pogorelova (1969) noted that the apex of a dormant bud was flat, but became convex after the onset of active growth.

In the initiation of leaf primordia, periclinal divisions occur in the second layer of cells on the apical flank. The three outermost cell layers participate in the formation of axillary bud primordia (Klopfer, 1965b). These first occur as a tangentially elongated area of cells in the axils of P_3 and P_4 – the third and fourth youngest leaf primordia – (Sussex, 1955a), and in longitudinal section form a 'shell zone', a group or arc of radially elongated cells delimited from the apical meristem by a zone of more differentiated tissue (Fig. 3.6).

Treatment with γ-rays resulted in necrosis of the apex of dormant buds, increasing with the amount of radiation (Mohyuddin and Hamid, 1966; Shimomura and Kuriyama, 1970).

Periclinal chimeras

The occurrence of striking morphological differences between varieties of the potato has already been mentioned. Some of the most divergent

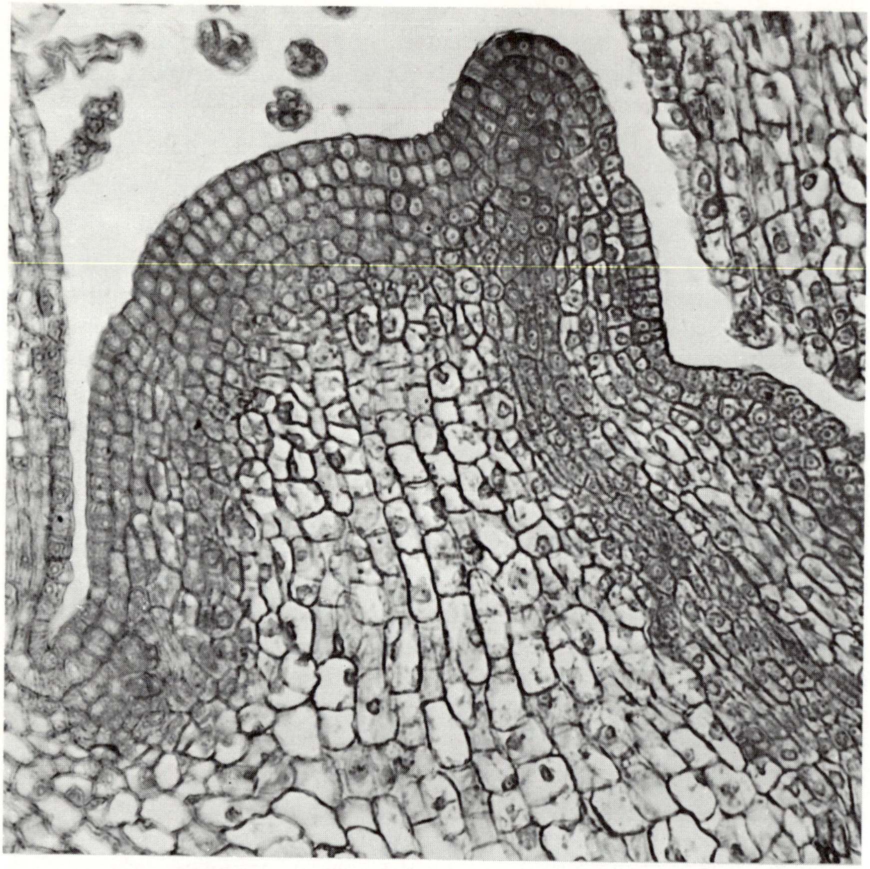

Fig. 3.5 L.S. apex of axillary leafy shoot of cv. King Edward, showing the stratified apical dome. A young leaf primordium with associated procambium is present on the right. x 230

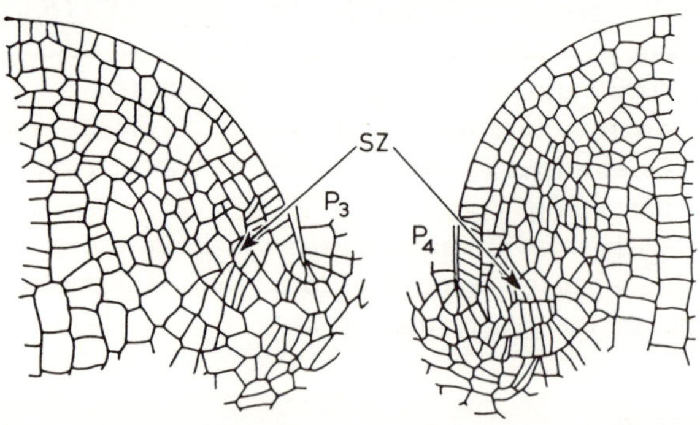

Fig. 3.6 L.S. shoot apex, showing bud primordia in the axils of P_3 and P_4, the third and fourth youngest leaf primordia respectively. The bud meristems are separated from the apical meristem by a shell zone (sz), a zone of periclinally dividing cells (Figs. 13a and b from Sussex, 1955a).

types are, in fact, periclinal chimeras, i.e. bud sports which have originated by a mutation, probably in a single cell, which has divided to form a whole layer of mutant cells in the developing shoot apical meristem (Howard, 1970b). Various methods have been used in attempts to analyse the location of these cells in the meristem. The surface layer of the apical meristem, which gives rise to the epidermis, is designated L_1, the next L_2 and the third layer L_3. Studies of periclinal chimeras thus attempt to discover from which layer of the apical meristem particular structures are formed, and also from how many cells each layer has itself originated.

The first bud sports to be investigated were those involving changes in pigmentation of the tuber. The colour of the tuber depends on anthocyanin in the periderm, formed from L_1 and L_2 (Asseyeva, 1927; Howard, 1970b). Asseyeva (1927, 1933) devised the technique of eye-excision to investigate chimeras. If all eyes are removed from a whole tuber, or from half of a tuber, new adventitious buds may be formed, presumably from deeper tissues (Howard, 1969). There is evidence that these buds originate from one or a few cells derived from L_3 (Howard, 1970a). Using this technique, Asseyeva (1927) obtained normal and leaf mutant forms from halves of the same tuber. She was also able to obtain one variety, for example, one which had pigmentation round the eyes, from another. By the same technique, Crane (1936) demonstrated that cv. Golden Wonder is a periclinal chimera with an inner core which corresponds to cv. Langworthy, since some tubers of Golden Wonder with the eyes excised gave rise to plants which produced Langworthy tubers.

Many leaf-shape bud sports also occur in the potato, for example plants with entire leaves, ivy leaf, holly leaf, and wildings, which have few folioles. It is thought that leaf shape is principally affected by changes in L_2, although mutations in L_1 and L_3 may have a modifying influence (Howard, 1967). The dockenleaf bud sport in cv. Majestic, which has entire leaves, is a periclinal chimera with L_1 unchanged but L_2 and L_3 showing a mutation. Following X-ray treatment, a very interesting shoot was obtained which bore not only normal pinnate, but also entire and sectorial leaves. The arrangement of these leaves on the shoot was such that it could be postulated that L_2 originated from only two initial cells (Howard, 1966).

As already mentioned, the potato apical meristem is considerably stratified. Work on periclinal chimeras of potato, as in other species, has contributed to an understanding of the degree of independence of the outer three layers. Thus Klopfer (1965a) considers that such studies support the view that three independent layers are present, but Howard *et al.* (1963) conclude that L_3 does not have an independent existence. These workers suggest that the apex of axillary buds is produced only by L_1 and L_2. However, anatomical observations of axillary bud formation suggest a more deep-seated origin. Simmonds (1965) concluded from eye-excision experiments that in some varieties L_2 in lateral buds contributes both L_2 and L_3, but that in other varieties the inner layers remain distinct. Other observations suggest that occasional tangential divisions of cells in L_2 may contribute cells to the L_1 or L_3 layers (Howard, 1971, 1972). It seems that here, as in many other facets of work on potatoes, considerable varietal differences may exist.

3.4.2 Meristem culture

Plant pathologists have used meristem culture, often in conjunction with heat treatment, as a means of obtaining virus-free stock of infected varieties. The terminology used in the literature is extremely confused, and often shows complete lack of understanding of apical morphology. As Hollings (1965) has pointed out, the term 'meristem culture' has been applied to pieces of tissue ranging from 0.1 mm to 1 cm or more in size. Strictly, the term shoot apical meristem relates to that part of the apex distal to the leaf primordia, but in fact, although it is now possible, at least in some species, to obtain the development of whole plants from the cultured apical dome only (Smith and Murashige, 1970), the pieces of tissue cultured by plant pathologists usually include the meristem together with at least the two youngest leaf primordia. Hollings (1965) suggests the term 'meristem-tip', but this is too vague to constitute a substantial improvement. That the size of the piece of excised tissue is important in various ways will be shown below.

Methods

Attempts to culture the shoot apex have shown that the potato is apparently one of the more difficult species to grow, some varieties being easier than others. Several different methods have been tried. The potato apex is naturally sterile (Morel *et al.* 1968), and surface sterilization of the shoot tip is all that is required (Morel and Martin, 1955). The general techniques of dissection and excision have been described by Kassanis (1967).

Stace-Smith and Mellor (1968) compared several of the culture media that have been employed. Of the four media they used, that of Murashige and Skoog (1962) was easily the best, both in terms of percentage survival and rate of development of the explants. This medium, which has a relatively high concentration of inorganic salts, was also used (sometimes in a modified form) by Goodwin (1966a), Morel *et al.* (1968) and Mellor and Stace-Smith (1969). Liquid medium was better than agar-solidified medium both with regard to rate of growth and the proportion of rooted buds (Stace-Smith and Mellor, 1968; Mellor and Stace-Smith, 1969). A bridge of filter paper folded into an M, dipping into the culture solution and supporting the explant, is sometimes used (Goodwin, 1966a), and has the advantage that explants can be removed without damage to the root system (Hollings, 1965).

Morel (1964) showed that an exogenous supply of gibberellin is necessary in the culture of potato apices. Without it, a callus-like growth develops. In the presence of GA, a normal shoot is formed but growth stops at 4–5 mm unless the explant is given a high concentration of K and ammonium ions (Morel, 1964; Morel and Muller, 1964). Alternatively, the Murashige and Skoog medium is adequate to support normal growth (Morel *et al.* 1968). It is interesting that the high concentration of K and ammonium ions was not required if the explants were taken from vernalized tubers (Morel and Muller, 1964). A seasonal effect was also noted by Mellor and Stace-Smith (1969), who found that buds excised in spring and early summer formed roots more readily than explants taken later in the year. Explants from sprouts formed in the light grew better than those from dark-grown sprouts (Kassanis, 1967). Both a higher than normal concentration of micronutrients and the presence of a low concentration of 8-hydroxyquinoline stimulated the growth of excised buds (Goodwin, 1966a).

Early difficulties in the transference of cultured plantlets to pots were overcome by grafting them into tomato seedlings (Morel and Martin, 1955), but it has since been found that they can be successfully potted on if covered by a bell-jar (Kassanis, 1967).

Size of explant

Various sizes of explant have been used by different workers. Some examples are: 75–100 μm wide and 150–200 μm long, usually with one leaf primordium (Morel, 1964); 100–250 μm square, sometimes with one leaf primordium (Kassanis, 1957); apical dome with 2–4 leaf primordia (Mellor and Stace-Smith, 1969; Gregorini and Lorenzi, 1974); dome plus 2–3 leaf primordia, of length 300–1000 μm (Stace-Smith and Mellor, 1968); dome plus 1, 2 or 4 leaf primordia, of length 120, 270 and 600 μm respectively (Pennazio and Redolfi, 1974); and the terminal 5 mm of the shoot tip (Thomson, 1956).

The size of the explant is important because it affects both the success of culture and the percentage of virus-free plants obtained. Thus, buds more than 1000 μm long were the first to develop roots in culture; those of less than this length rooted more slowly, and often not within 10 weeks; and explants less than 300 μm long grew slowly and rarely rooted (Stace-Smith and Mellor, 1968). As Kassanis and Varma (1967) have pointed out, some varieties of potato (e.g. Epicure and Arran Comet) have large, domed apical meristems which are relatively easy to excise. The size of the explant is of critical importance in virus eradication (Stace-Smith and Mellor, 1968). For example, pieces of apical meristem of less than 200 μm diameter can be freed from paracrinkle and virus X, whereas plants derived from a large explant still had virus (Kassanis, 1957). Similarly, in another study, the percentage of excised apices still infected with potato virus X was 81.3% for tips of 120 μm length, 96% for 270 μm and 100% for those 600 μm long (Pennazio and Redolfi, 1974). In another investigation, tips more than 1000 μm long failed to give virus-free plants (Šip, 1972). Sometimes cultured apices are grown into plants, and their apices are again excised and cultured; this procedure may be repeated several times (Svobodová, 1964).

Thus, explants cultured in order to eradicate viruses should be as small as is compatible with growth. In this connection, the demonstration that in other species, including other Solanaceae, excised apical meristems devoid of leaf primordia can be successfully cultured to yield whole plants (Smith and Murashige, 1970) is clearly important. The technique of Sussex (1955b), who showed that pieces of the potato apical meristem of very small area would remain viable, left *in situ*, if the rest of the meristem was excised, could be useful in this connection.

Presence of virus in the apex

For some time it has been a debatable point whether virus particles are actually absent from the cells of the apical meristem of infected plants or are merely inactivated by the period of growth in culture. Morel (1964) has suggested that the virus cannot compete for nucleic acid precursors with the actively dividing cells of the meristem.

Electron microscopic observation of cells of the apical meristem and youngest leaf primordia from sprouts of infected tubers revealed bundles of elongated particles arranged in a parallel fashion. These appeared paracrystalline in transverse section, and resembled structures found in infected leaves. It was concluded, therefore, that potato virus X is present in apices of infected plants but becomes eliminated during culture (Pennazio, 1971; Appiano and Pennazio, 1972). In another investigation, homogenates of apical meristems were examined under the electron microscope. Particles of virus X were found in 98% of preparations from meristems less than 100 μm long. No particles were found in 50 meristems of similar size which had been grown for 4 weeks in culture (Krylova *et al.*, 1973). Thus, on present evidence it appears that the meristems become freed from viruses during culture, although further work will be necessary to establish how this occurs.

Heat treatment

Potato tubers which are given heat treatment sometimes give rise to virus-free plants (Kassanis, 1949). This suggested that a combination of heat treatment and culture of the shoot apex might enhance the degree of virus elimination, and this was found to be the case (Thomson, 1956; Morel *et al.*, 1968; Pennazio, 1971; Šip, 1972; Macdonald, 1973). Tubers are usually given fairly high temperatures for periods of a month or more, and apices excised from these grown in culture.

Viruses eliminated by culture

Using the technique of meristem culture, with or without heat treatment, various varieties of potato have been freed of paracrinkle, and potato viruses X, Y, A and S (see references already cited).

3.4.3 Anatomy of the aerial stem

Transverse sections of the apex of the leafy shoot show its spiral phyllotaxis, and the rather uniform meristematic cells of the apical dome (Fig. 3.7a). Owing to the onset of vacuolation in the developing pith and cortex, a ring of still densely staining, meristematic cells, the residual meristem, becomes delimited about 70 μm below the tip of the dome (Fig. 3.7b). Slightly lower down the axis, procambial strands become differentiated in the residual meristem in positions related to young leaf primordia.

Using etiolated sprouts, Sussex (1955a) found that the first sieve tubes could be observed in a leaf primordium four or five plastochrones old and noted that differentiation was continuous and acropetal. The first lignified xylem elements occurred as a discontinuous strand in the stem below a leaf primordium three or four plastochrones old and were observed in the leaf primordia themselves when one or two plastochrones older. Artschwager (1918) considered that the protoxylem elements were the first elements to differentiate from the procambium in developing sprouts, whereas at the tips of older shoots differentiation of the inner phloem occurred before that of the xylem. He had some difficulty in distinguishing differentiating sieve elements, and it seems likely that better differential staining techniques were required. Pogorelova (1969) also reported that the protoxylem differentiated first, groups of external and internal phloem appearing later. The writer found, on the contrary, that in apices of growing lateral shoots the first vascular elements to differentiate were the sieve elements of the abaxial (external) phloem in leaf primordia four or five plastochrones old (Fig. 3.8); these observations thus agree with those of Sussex (1955a), and of the present author, on sprouts. Immature protoxylem elements can be observed only slightly later, and sieve elements of the inner phloem in leaf primordia about six plastochrones old.

Most studies on internal phloem agree that it differentiates later than the external phloem (see references in Esau, 1938). Some workers

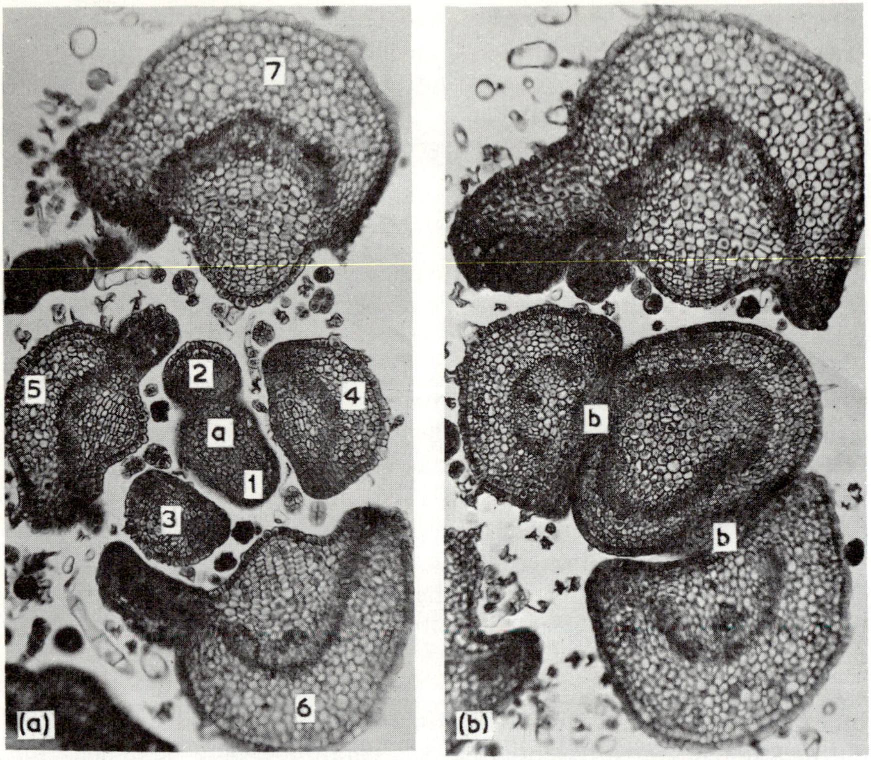

Fig. 3.7 T.S. apex of a leafy shoot of cv. King Edward: (a) at the junction of P_2 with the apex. The apex itself, (a), consists of rather uniform densely staining cells. P_1–P_7 are present in spiral sequence. (b) At the junction of P_5 with the apex. Axillary bud primordia (b) can be discerned in the axils of P_5 and P_6. Procambium is present in the axis, and differentiated vascular elements in P_5. x75

regard the vascular bundles of the Solanaceae as truly bicollateral (Bonnemain, 1970), while others consider that this term, suggesting as it does units of one group each of external phloem, xylem and internal phloem, is inappropriate, since many strands of internal phloem occur in interfascicular regions (Fukuda, 1967). Esau (1938) has described the differentiation of internal phloem in tobacco. Here the procambium on the adaxial side is not sharply delimited from the cells of the ground meristem. Some of the cells of the latter are converted into procambium by longitudinal divisions, and differentiate as strands of internal phloem. Adjacent cells fail to divide, and differentiate as larger parenchyma cells. Although these cells do not remain meristematic, they retain the potentiality to divide and some later give rise to additional strands of internal phloem. Differentiation of internal phloem in the potato appears to be similar; indeed, this description seems to be true for most Solanaceae (Fukuda,

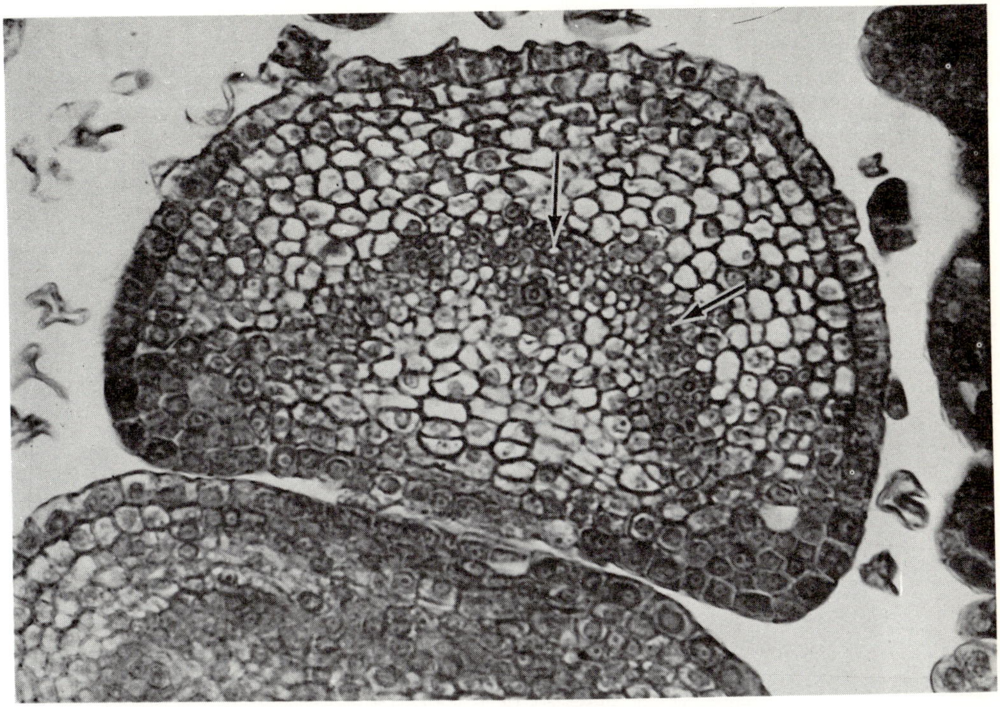

Fig. 3.8 T.S. young foliage leaf primordium, showing maturing sieve elements of the external (abaxial) phloem (arrowed). x 400

1967; Bonnemain, 1970). Extensive anastomosis occurs between the internal and external phloem, especially by means of more or less horizontal sieve elements in the young internodes (Sussex, 1955a; Bonnemain, 1970).

The primary stem has three large and three small bicollateral vascular bundles in internodal regions (Fig. 3.9) (Artschwager, 1918; Esmarch-Bromberg, 1919). Artschwager (1918) stated that a cambium was differentiated in the young stem even before the sieve tubes of the inner phloem became evident. The writer's observations suggest that this is erroneous. However, both fascicular and interfascicular cambium does become discernible in fairly young stems, and functions normally to give rise to secondary xylem centripetally and secondary phloem centrifugally. In interfascicular regions no large vessel members are present (Fig. 3.10) (Artschwager, 1918). Rays are usually uniseriate. Phloem fibres differentiate on the outer periphery of the vascular cylinder (Fig. 3.10) from elements of the primary phloem. The cells between the primary xylem [or, in interfascicular regions, secondary xylem (Fig. 3.10)] and the groups of internal phloem remain parenchymatous and contribute to a region of tissue sometimes called the perimedullary zone (*Markkrone*).

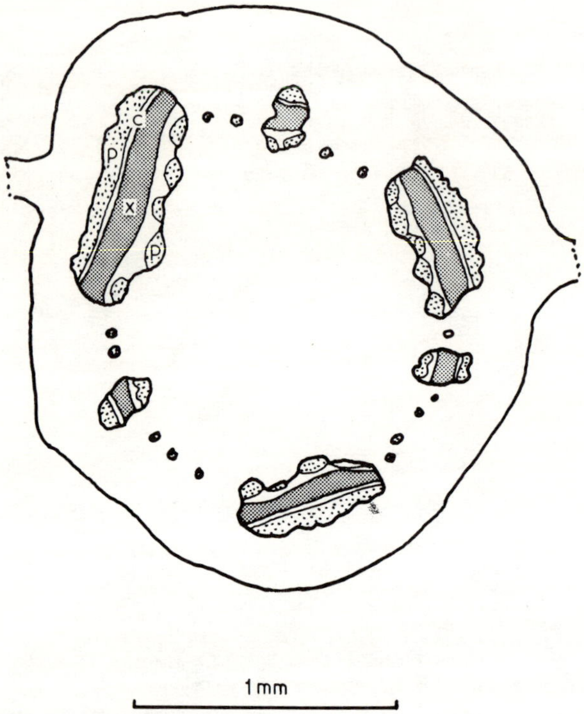

Fig. *3.9* T.S. internodal region of young stem in primary growth. Three large and three smaller vascular bundles are present (c, cambium; p, internal and external phloem; x, xylem). x 18

Artschwager (1918) states that a recognizable endodermis, which may not be distinguishable by its Casparian strips but is identifiable by its regular cell shape and by its content of starch grains, is fairly distinct in the young stem; the writer has not found this evident in the aerial stem. The explanation for this may lie in the observation of Venning (1954) that the starch in the starch sheath was converted to soluble substances during the day and was only visible at night.

The stem is bounded by a single layer of fairly isodiametric epidermal cells, with a waxy cuticle, and a distinctive underlying hypodermal layer (Fig. 3.11). Stomata occur in the epidermis (Artschwager, 1918). Beneath the hypodermis are three to five layers of angular, or sometimes lacunate, collenchyma (Fig. 3.11). A parenchymatous cortex lies internal to this. Starch grains occur in all parenchymatous tissues; those close to sieve elements are the most developed and remain longest (Yoshida, 1970).

It has been reported that stem structure differs somewhat in early and late varieties of potato, late varieties having more superficial stomata, a thicker outer wall of the epidermal layer, and a thicker cylinder of

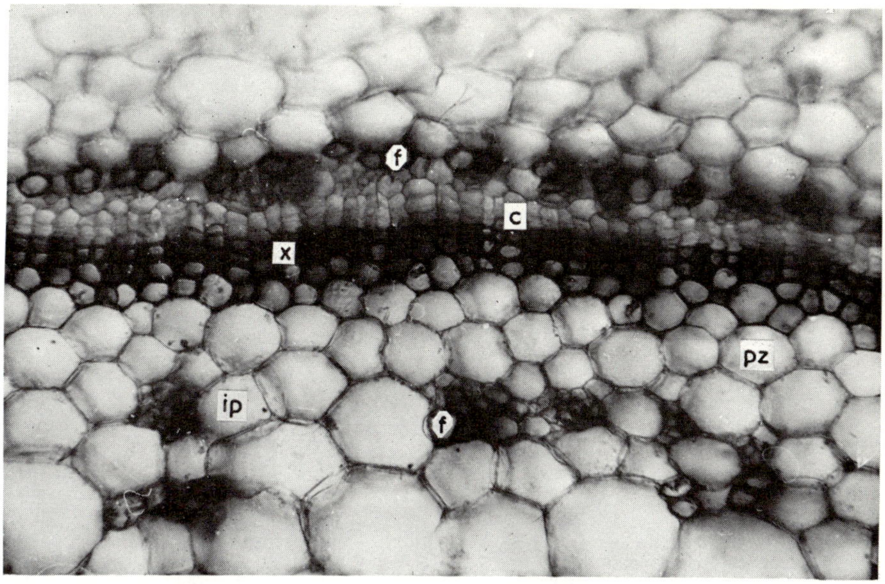

Fig. 3.10 T.S. internode of an older stem with secondary thickening, in an interfascicular region. Small groups of internal phloem elements (ip) are evident. Phloem fibres (f) are present in both external and internal phloem. Parenchymatous cells of the perimedullary zone (pz) are present between the secondary xylem and the internal phloem (c, cambial region; x, xylem). x 160

fibres (Danowitsch and Krinkin, 1929). Stem structure also varied in plants from different environments. Plants grown in the plains of Italy had strongly lignified stems with many fibres; those grown in the mountains had stems with little lignification, a wide cortex and abundant collenchyma (Scaramella Petri, 1956, 1959). It was later shown that the temperature at which the plants were grown could have similar effects. Stems of plants grown at 15° C had a wide cortex and phloem but little xylem, whereas those grown at 20° C had much more secondary xylem with many fibres (Scaramella Petri, 1963b). Presumably these structural characters will be influenced by the general growth of the plant, although this was not really considered.

In experiments on the effects of growth hormones and metallic salts on stem structure, apical parts of tubers (cv. Bintje) were placed in the test solution, and subsequently sprouts were sectioned 20 mm below the apex. More rapid lignification resulted from treatment with IAA, but GA gave a striking increase in the number of lignified xylem elements, which were wide and thick walled, and in the number of phloem fibres. This effect was greatly enhanced by the addition of zinc, although the reason is not understood (Rehman, 1966, 1967). Treatment of sprouts with ethylene resulted in more starch in the cortical cells, and enhanced

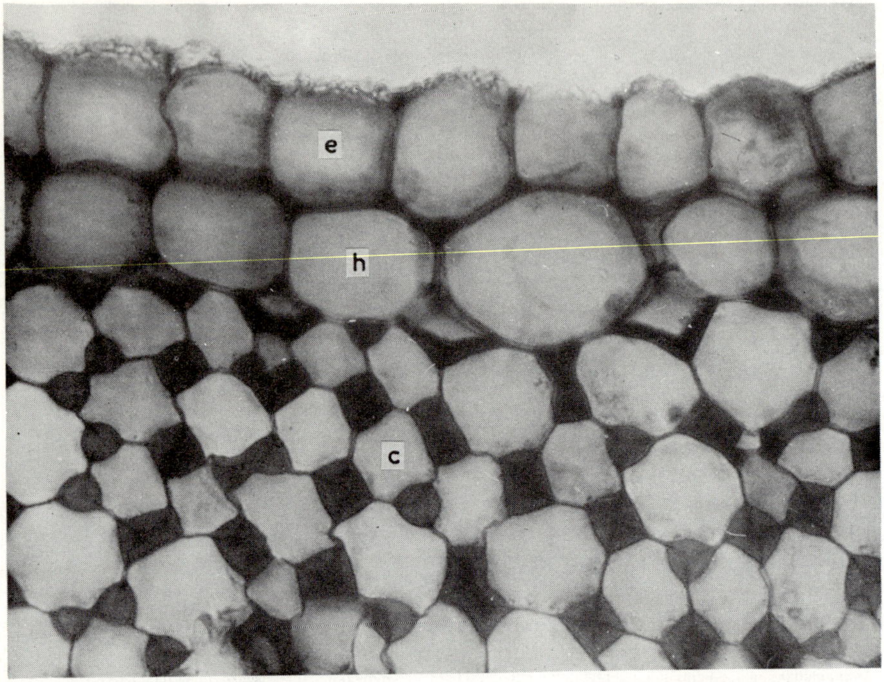

Fig. 3.11 T.S. outer region of a fairly young stem, showing the epidermal layer (e) with cuticle, the hypodermis (h), and the underlying layers of collenchyma (c). x 640

cell division in the vascular cylinder (Hughes *et al.* 1973) although more exact observation seems desirable.

Stems from which buds had been removed formed callus on basal cut surfaces, especially if the vascular tissue was exposed. The callus gave rise to buds (Lauer and Krantz, 1957).

3.4.4 Development and structure of the leaf

The potato foliage leaf is pinnately compound and has small folioles between the pinnae. There are varietal differences in the number of leaflets and of folioles, size, colour, etc. (Howard, 1970b).

As already mentioned, leaf primordia are formed by periclinal divisions in the second layer of cells on the flanks of the apical meristem. Phyllotaxis is spiral. The young leaf primordium grows apically and arches over the shoot apex. At an early stage of development, marginal meristems, which contribute to the growth of the lamina, become evident, and an adaxial meristem, contributing to increase in thickness of the leaf, is also discernible (Fig. 3.12). In the foliage leaf, development of the pinnae takes place basipetally, and each leaflet passes through stages similar to that shown by the young leaf primordium (Artschwager, 1918).

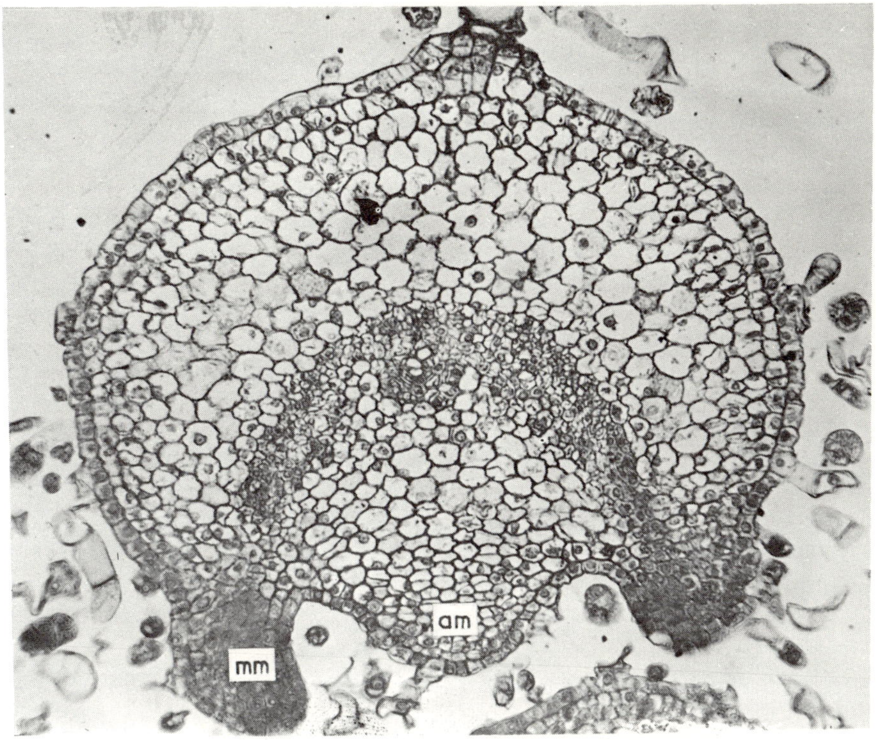

Fig. 3.12 T.S. developing leaf primordium, showing marginal (mm) and adaxial (am) meristems. Differentiating vascular tissues are also present. x 200

In the young leaf primordium only an arc of procambium is present. Differentiation of the first phloem and xylem elements takes place as already described above. Small groups of internal (adaxial) phloem are evident (Fig. 3.13). A study of the ultrastructure of the phloem gave a fairly customary picture. Sieve elements were connected to companion cells by plasmodesmata. Callose and slime were present in the sieve plate pores. Companion cells had many ribosomes, mitochondria and dictyosomes, but did not have dense cytoplasm (Tsutsumi *et al.* 1972).

The vascular tissue of both petiole and midrib forms a semi-circle as seen in transverse section. Beneath the epidermis two or three layers of cells are differentiated as collenchyma. Artschwager (1918) claims that a cambium is formed in the petiole and gives rise to some secondary vascular tissues. Certainly a cambium-like layer is present (Fig. 3.13).

Artschwager (1918) has described and illustrated the early differentiation of palisade and spongy mesophyll in the leaf lamina. Elongation of the future palisade cells begins shortly after vascular tissue is differentiated in the midrib. The mature leaf has a single layer of palisade tissue,

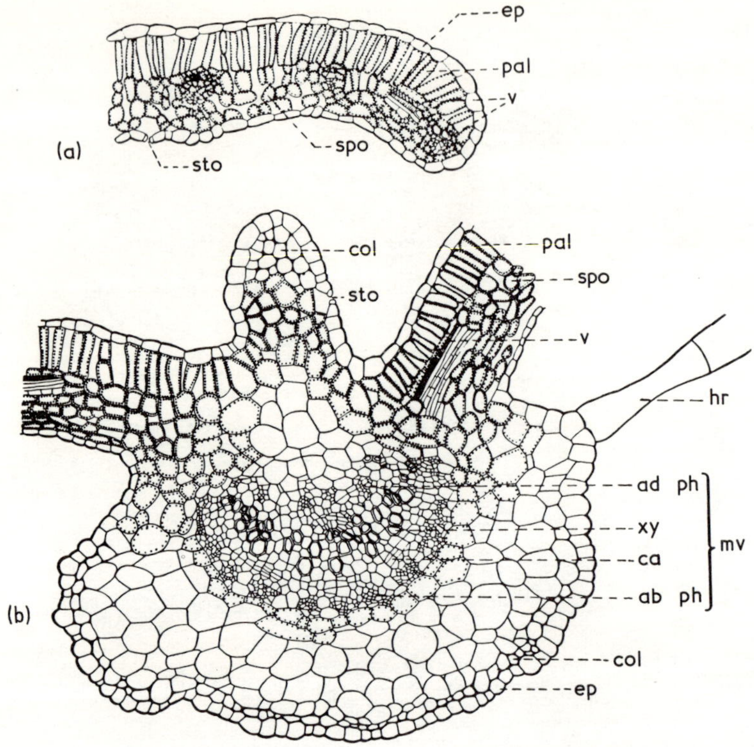

Fig. 3.13 T.S. mature foliage leaf in the region of (a), the lamina, and (b), the midrib. ab ph, abaxial phloem; ad ph, adaxial phloem; ca, cambium or cambium-like region; col, collenchyma; ep, epidermis; hr, hair; mv, midvein; pal, palisade tissue; spo, spongy mesophyll; sto, stoma; v, lateral vein; xy, xylem. (Fig. 288 from Hayward, 1938).

and three to five rows of spongy mesophyll cells, separated by large intercellular spaces (Fig. 3.13). In leaves of plants grown at high light intensity, the palisade tissue consisted of large, more elongated cells rich in plastids. There was also abundant phloem and collenchyma, and many more hairs than in leaves from plants grown at lower intensities (Scaramella Petri, 1963a). In potato leaves injured by air pollution, the palisade cells were the first to become affected and show necrosis, but the spongy mesophyll was also affected later (Hooker *et al.* 1973).

The epidermis consists of a single layer of cells with sinuous walls. Stomata are Ranunculaceous and Cruciferous (anomocytic and anisocytic) and occur on both surfaces, but are much more numerous on the abaxial side of the leaf (Ahmad, 1964). Stomatal number differs in different varieties (Esmarch-Bromberg, 1919). Stomatal length and the number of plastids in the guard cells are positively correlated. Plastid number can be used as an indicator of ploidy level in hybrids (Frandsen,

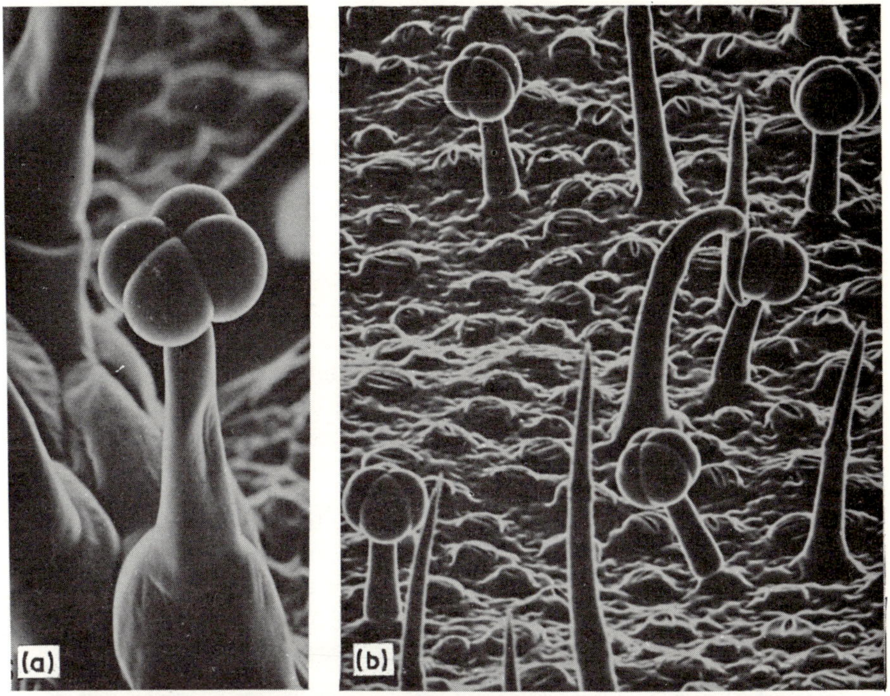

Fig. 3.14 Glandular hairs as seen with the scanning electron microscope: (a) Single glandular hair with 4-celled head on the leaf of *Solanum berthaultii.* x 350. [Fig. 1 from Gibson (1971)]; (b) covering and glandular hairs on *S. tuberosum* cv. Arran Pilot. x 170. (By courtesy of Dr R.W. Gibson and Long Ashton Research Station.)

1968). Both covering and glandular hairs occur on the epidermis; they are initially dense, but become separated by growth of the intervening regions of tissue (Artschwager, 1918). In the genus *Solanum* each individual bears two types of hair. As with other structural characters of leaves, these can be of taxonomic use (Seithe, 1962; Sizova, 1963, 1965a,b). Covering hairs are uniseriate; glandular hairs have either a small, more or less spherical head or a rather ovate multicellular head. Uniseriate covering hairs and two kinds of capitate glandular hairs have been described in cv. Sieglinde (Bancher and Hölzl, 1959). Both types of glandular hair have a single stalk cell. Hairs with a 4-celled head have a neck or collar cell between the stalk and the head; these hairs apparently secrete volatile oil. The second type of glandular hair is many-celled and club-shaped; each head cell contains a protein crystal.

Gibson (1971) has described glandular hairs with a 4-lobed head on the leaves and stems of certain species of wild potato (Fig. 3.14a). These hairs secrete a clear exudate which on contact with air changes to black gummy material. About 30% of aphids placed on these plants became stuck to the exudate within 24 h, and later died, probably of starvation. It was thought that if bred into the cultivated potato, in which they are scarce, these hairs might afford some resistance to aphids (Gibson, 1971; Tarn, 1971). The hairs of *Solanum tuberosum* (Fig. 3.14b) do not seem to have the capacity for capturing and anchoring aphids. However, on hybrids of *S. tuberosum* cv. Pentland Crown and *S. berthaultii* glandular hairs were numerous and aphids were found gummed down by the exudate (Gibson, 1974). Thus these interesting hairs may constitute at least a partial resistance mechanism to aphids which can be bred into the cultivated potato (see also Chapter 12, part 2).

Isolated leaves of the potato will form roots, and these can be stimulated by hormone treatment (Verma, 1966; van Harten, 1972). If the leaf is separated from the plant, together with its axillary bud, this may develop as a stolon or, under certain conditions, as a tuber (see Section 3.6.4).

3.5 The stolon

Potato stolons are lateral shoots, usually those from the most basal nodes below soil level. Typically they are diageotropic shoots with elongated internodes, hooked at the tip and bearing spirally arranged scale leaves (Figs. 3.15 and 3.16) (Kumar and Wareing, 1972). When tubers develop,

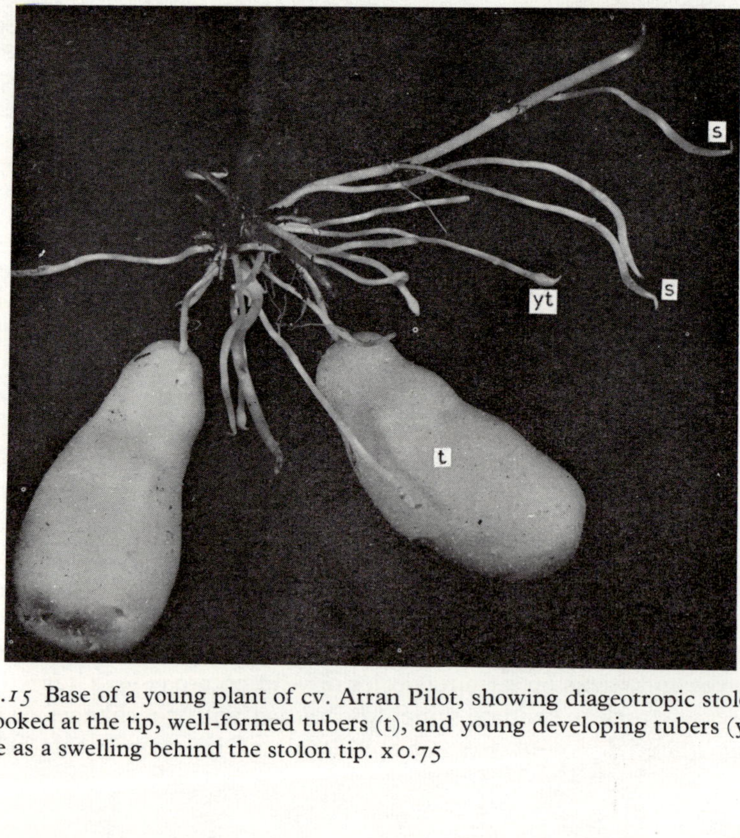

Fig. 3.15 Base of a young plant of cv. Arran Pilot, showing diageotropic stolons (s), hooked at the tip, well-formed tubers (t), and young developing tubers (yt) visible as a swelling behind the stolon tip. x 0.75

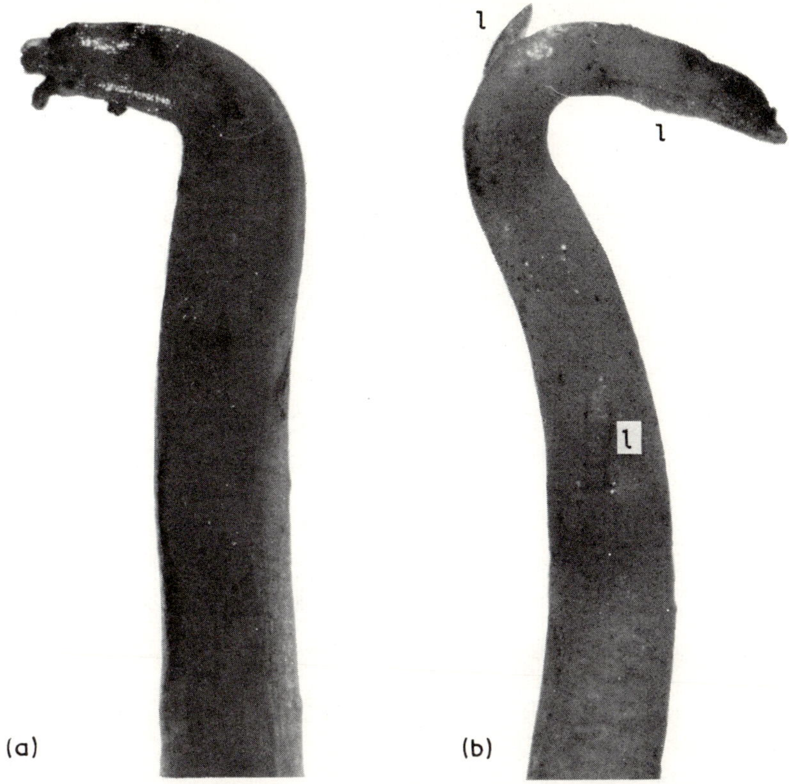

Fig. 3.16 Stolon tips of (a), cv. Arran Banner, and (b), King Edward. Relatively long internodes are present between the scale leaves (l). x 15

they do so from the sub-apical region of stolons (Fig. 3.15). However, tuber formation is regarded as the sum of two separate processes: stolon formation, and tuberization of the stolon tip (Booth, 1959, 1963). These processes are controlled by different factors.

3.5.1 The stolon apex

Booth (1963) described the stolon tip as an attenuated form of a vegetative shoot apex, but except for the type of leaf development there was little evidence of this in the varieties examined by the writer (compare the apices of vegetative shoot and stolon of cv. King Edward in Figs. 3.5 and 3.17). The apical dome of the stolon is perhaps of smaller diameter. Like the apical meristem of the leafy shoot, that of the stolon is conspicuously stratified (Fig. 3.17).

Although no attention seems to have been drawn to this point in the literature, the hooked part of the stolon tip is striking because of the highly meristematic nature of the apical and sub-apical regions (Fig. 3.18, a,b). The epidermal cells in this region are radially elongated

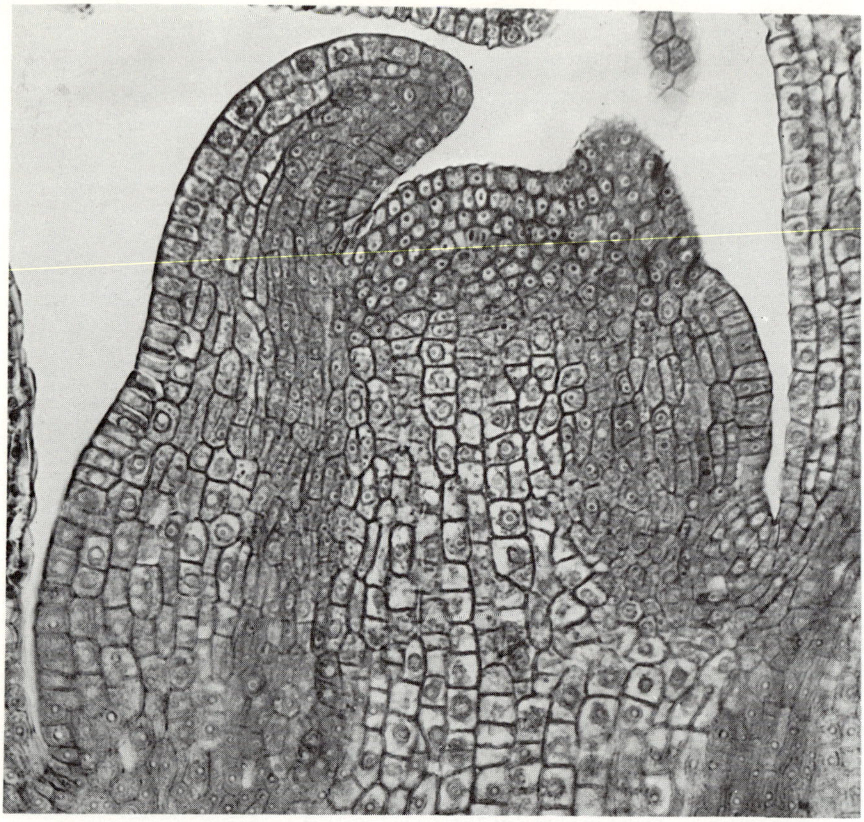

Fig. 3.17 L.S. apex of stolon, cv. King Edward. Compare the stratified apical dome and scale leaf primordia with the section of the apex of a leafy shoot in Fig. 3.5. x 230

(Artschwager, 1924) and vertically compressed (Fig. 3.18c). Booth (1963) noted that the degree of curvature of the stolon tip varied and was related to differences in cell length on the two sides. Cell lengths became equal on opposite sides at about 3 mm from the apex (Fig. 3.19). Booth considered that the orientation of the stolon hook changed periodically, being related to the position of the most recently expanded scale leaf; cells behind this leaf extended more than those in other positions. The writer has observed stolon hooks which were variously related to the position of the last expanded leaf (Fig. 3.20), but it is possible that these may have been in the course of changing their orientation.

Leshem and Clowes (1972) studied the duration of the mitotic cycle in various zones of growing stolon apices by treatment with colchicine, and found the values obtained to be comparable with those recorded for other species. The cells situated on the flanks divided 1.6 times

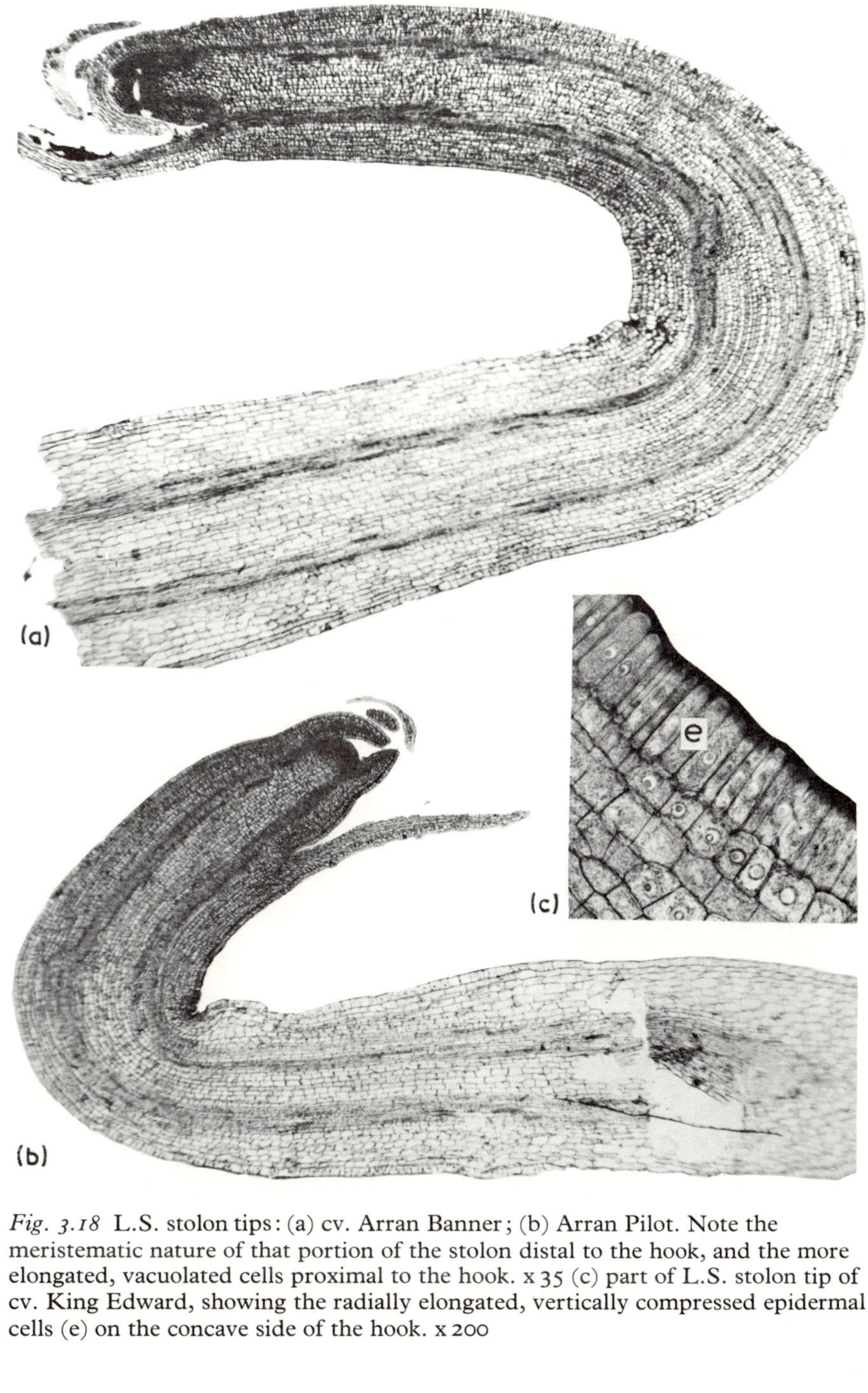

Fig. 3.18 L.S. stolon tips: (a) cv. Arran Banner; (b) Arran Pilot. Note the meristematic nature of that portion of the stolon distal to the hook, and the more elongated, vacuolated cells proximal to the hook. x 35 (c) part of L.S. stolon tip of cv. King Edward, showing the radially elongated, vertically compressed epidermal cells (e) on the concave side of the hook. x 200

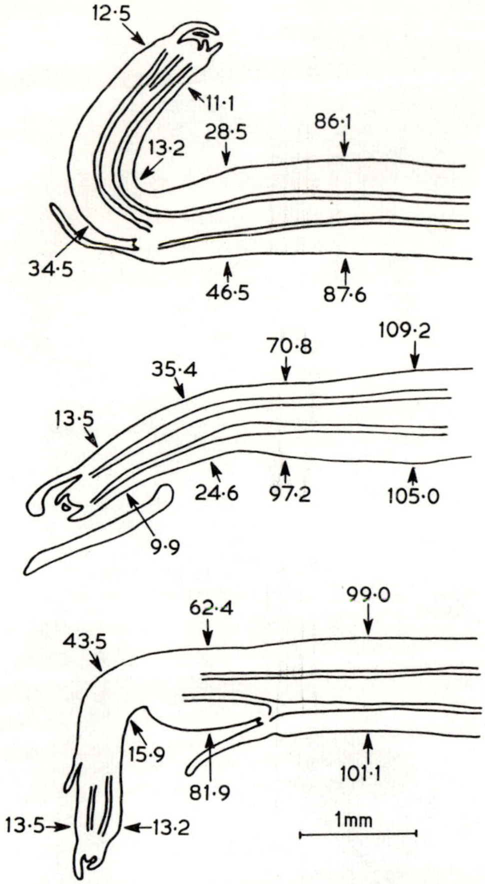

Fig. 3.19 L.S. stolon tips. The numbers indicate the cell length at each location, as an average of the outer three layers of cells. (Fig. 1 from Booth, 1963)

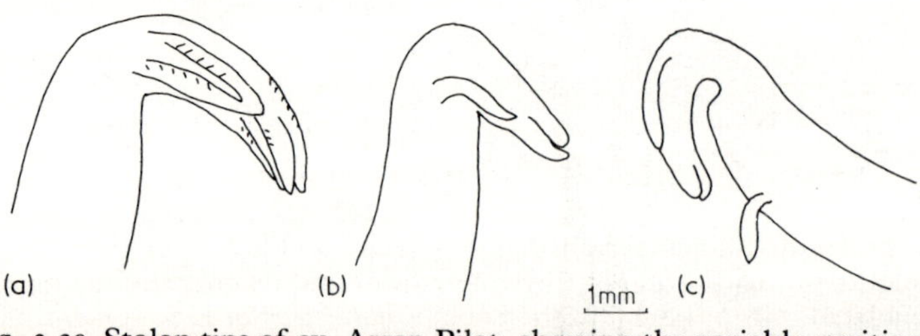

Fig. 3.20 Stolon tips of cv. Arran Pilot, showing the variable position of the scale leaf close to the bend of the terminal hook: (a, b) leaf lateral to the hook; (c) on convex side. x 6

as fast as those at the summit of the apex. As dormancy began, the mitotic cycle lengthened in all regions of the meristem; this was evident before the onset of tuber formation, after which the rate became too low to measure.

3.5.2 Anatomy of the stolon

As in the leafy shoot, a cylinder of densely staining cells, the residual meristem, becomes delimited about 80 μm below the tip of the stolon apex, by vacuolation of the cells of the future pith and cortex. At a relatively early stage of development, intercellular spaces form between the cortical and pith cells. Starch is also found in cortical cells near the procambium, and other cells contain many small crystals of calcium oxalate. Sclereids are reported to occur in the cortex of older stolons (Esmarch-Bromberg, 1919). No collenchyma is present. The pith is of small area compared to the cortex. Stomata and occasional hairs occur in the epidermis (Artschwager, 1924).

The stolon tip is remarkable for the early differentiation of abundant sieve elements. In the specimens observed, about 80 μm below the tip of the apex, the first sieve element of the external phloem was differentiated in relation to the fourth youngest leaf primordium, P_4. The first protoxylem element differentiated only slightly lower down the axis. Both protophloem and protoxylem elements were differentiated below P_5 and P_6, just proximal to their junction with the axis. Despite the rather undifferentiated nature of the scale leaf primordia, therefore, and their relatively early maturation, vascular differentiation in the stolon tip is initially similar to that in the leafy shoot. At about 220 μm below the apex, several sieve elements of the internal phloem had differentiated in the axis below leaf primordia P_3-P_6, and by 500 μm below the tip the internal phloem was the conspicuous tissue (Fig. 3.21). In young stolons subtending developing tubers abundant phloem is present (Fig. 3.22).

In the slightly older developing stolon, separate bicollateral vascular bundles are present. There are relatively few xylem elements, but an extensive development of phloem (Artschwager, 1918). Sieve tube elements are numerous and of considerable diameter (Artschwager, 1924). The end walls are transverse, and indeed in a transverse section of a stolon a number of sieve plates are often seen. Considerable anastomosis takes place between sieve elements, which form a network (Fig. 3.23) (Crafts, 1933). Phloem fibres differentiate both in the external and internal phloem (Fig. 3.24). A vascular cambium eventually develops in fascicular and interfascicular regions, and gives rise to some secondary vascular tissues.

An endodermis consisting of cells smaller than the rest of the cortex, without intercellular spaces, and possessing characteristic Casparian strips, is present (Artschwager, 1924).

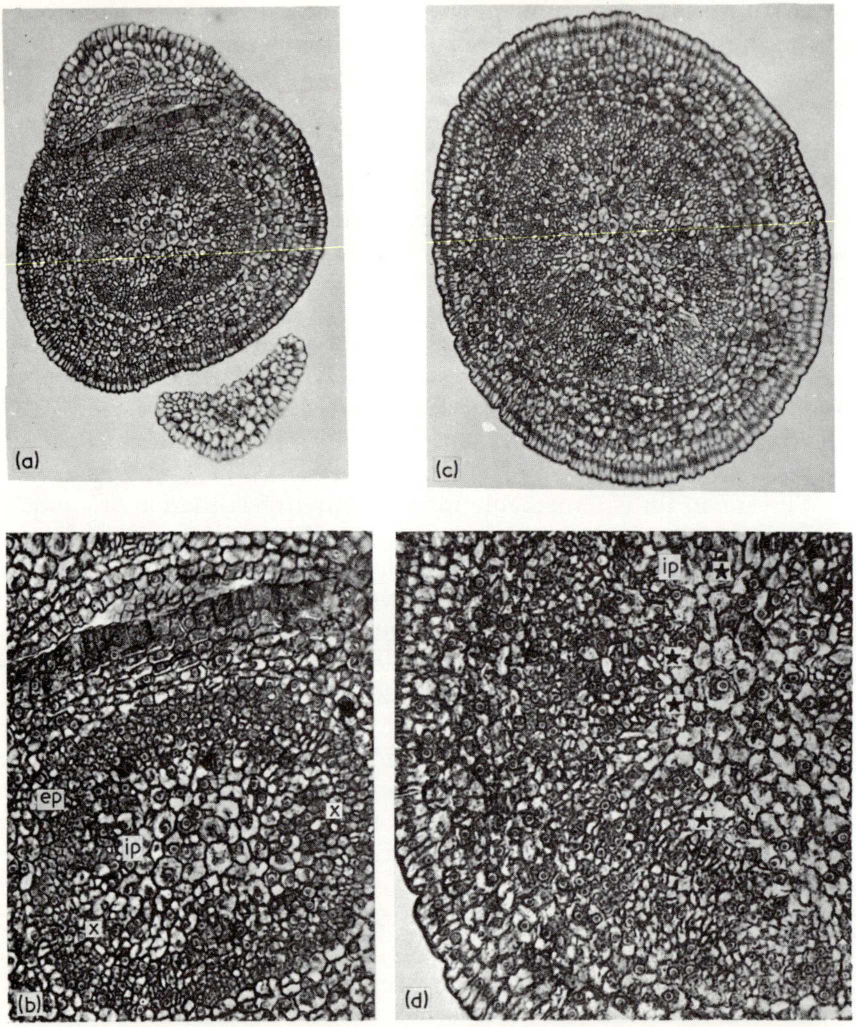

Fig. 3.21 T.S. apical region of stolon, showing differentiation of phloem: (a) approximately 220 μm below the tip of the apex, at the junction of P_6 with the axis. x 75. (b) part of A enlarged, showing differentiated elements of external phloem (ep), xylem (x), and internal phloem (ip). x 150 (c) approximately 500 μm below the tip of the apex, below the junction of P_7. x 75. (d) part of C enlarged, showing that elements of the internal phloem (ip and marked* on the inner side) are now conspicuous. x 150

As in the aerial stem, the elements of the internal phloem, especially in interfascicular regions, are separated from the xylem by parenchymatous cells which comprise the *Markkrone* or perimedullary zone (Fig. 3.24). The *Markkrone* is important in tuber development.

The relative amounts of phloem and xylem in a mature stolon, as seen in transverse section, are shown in Fig. 3.25. Many of the earlier workers were puzzled by the apparent ability of the rather narrow stolon to undertake translocation of all the necessary food materials for a

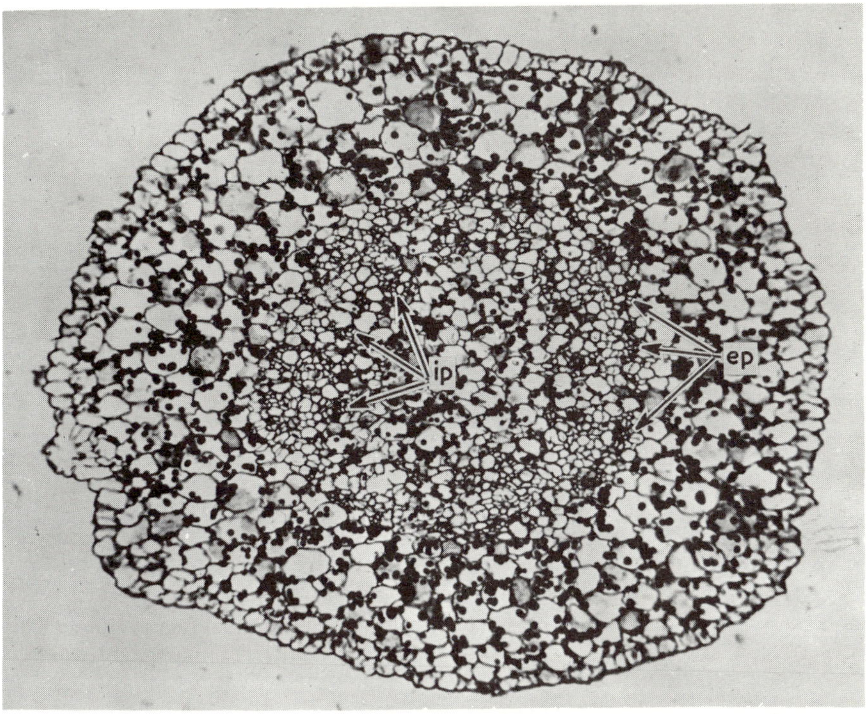

Fig. 3.22 T.S. young stolon below a developing tuber. Abundant internal (ip) and external phloem (ep) is present. Starch grains occur in the pith and cortex. x 100

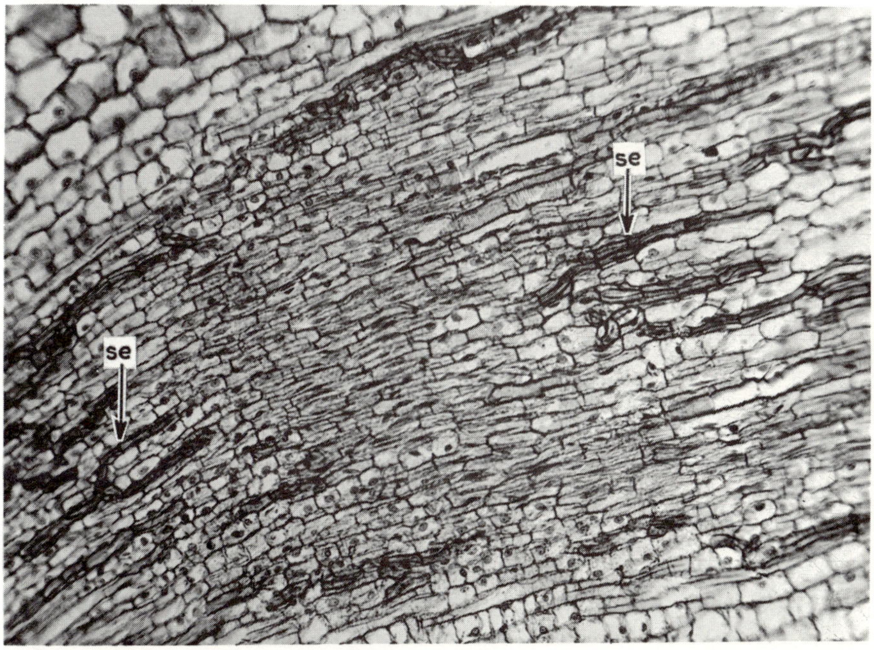

Fig. 3.23 Part of L.S. stolon, showing anastomosis of sieve elements (se). x 150

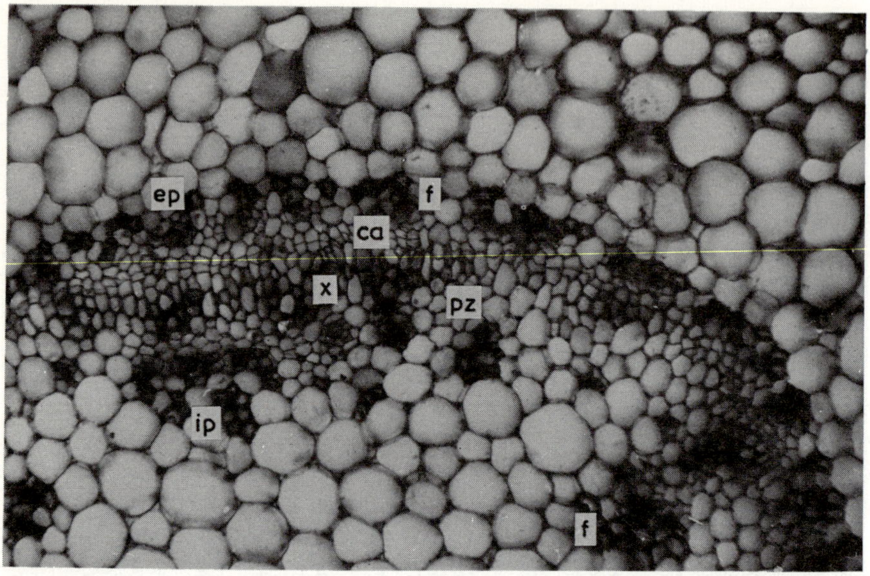

Fig. 3.24 Part of T.S. young stolon of cv. King Edward, showing internal (ip) and external (ep) phloem with fibres (f), the developing cambial region (ca) and the perimeduallry zone (pz) or *Markkrone*. Strands of internal phloem are present in interfascicular as well as fascicular regions. x, xylem. x 160

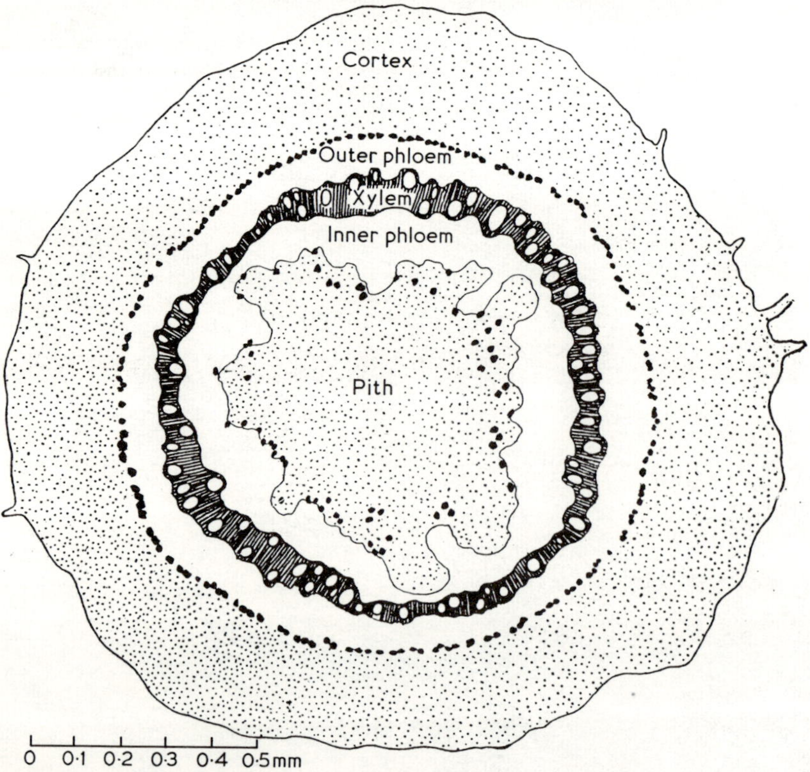

Fig. 3.25 Diagram of T.S. mature stolon, showing the relative areas occupied by the various tissues. (Fig. 6 from Artschwager, 1924)

growing potato tuber. Dixon (1922a,b) estimated the area of phloem in a stolon and the amount of carbohydrate which must be transported along it in a certain time. From this he calculated that a rate of flow of nearly 50 cm h^{-1} would be required. By somewhat similar methods, Crafts (1933) estimated even higher rates of flow. On the basis of existing knowledge at that time, Dixon (1922a, b) did not believe that the phloem could possibly support a rate of flow of this magnitude and therefore argued that transport must have taken place through the xylem. However, these estimates are well within the range of more modern calculations of rate of flow through sieve tube elements (see Crafts and Crisp, 1971). Nevertheless, the potato stolon is clearly an organ capable of and structurally adapted for unusually efficient translocation.

3.5.3 Factors controlling stolon development

Stolons normally develop first at the most basal nodes, and then at progressively higher ones. The number of nodes which subtend stolons, and the length of the stolons, are adversely affected by low levels of nutrients (Lovell and Booth, 1969). Booth (1959) showed that stolons developed at the lower nodes of cuttings taken from the top 10–15 cm of plants of *Solanum andigena*. Thus any axillary bud along the axis can develop as a stolon under appropriate conditions, and stolon development is apparently normally controlled by conditions prevailing in the basal part of the shoot (Kumar and Wareing, 1972). A certain amount of experimental work, largely on *S. andigena*, has been devoted to elucidating the nature of these conditions.

Even in *S. andigena*, which forms tubers only under short days, stolons do not have a comparable requirement and will develop under long day conditions (Booth, 1959). Certain other environmental factors, however, do have some importance. For example, moist dark conditions favour stolon development (Clark, 1921), but only in the presence of a dominant apex (Kumar and Wareing, 1972; Woolley and Wareing, 1972b). Lateral buds will develop as stolons only when they are affected by apical dominance. In decapitated shoots, leafy lateral shoots develop at the top three or more nodes, stolons from lower nodes; but if buds on the decapitated aerial shoot are also removed, the stolons become negatively geotropic leafy shoots (Booth, 1959). The same is true in 4-node cuttings (Kumar and Wareing, 1972). Booth (1959, 1963) noted that stolon development appeared to be promoted by a transmissible substance associated with the proximity of roots; later work showed that roots were not, however, essential (Kumar and Wareing, 1972). Sub-apical 4-node cuttings with all axillary buds except that at the most basal node removed gave rise to stolons if the roots were also removed, but to leafy shoots if the roots were present (Fig. 3.26), suggesting that roots promoted development of leafy shoots. If the bud is subject to apical dominance, however, it may develop as a stolon even in the presence of roots (Kumar and Wareing, 1972).

These interesting experiments all demonstrate that, according to the correlative effects operative at the time, lateral bud meristems can

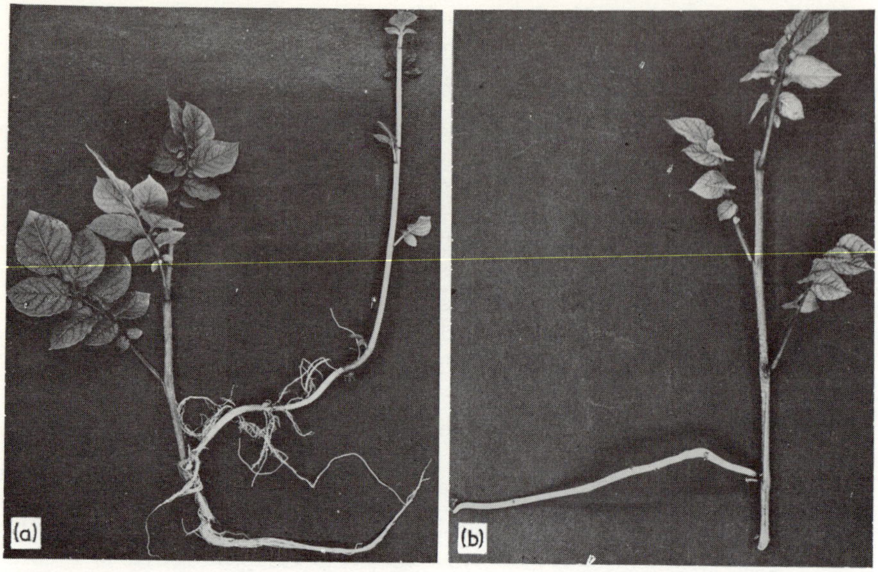

Fig. 3.26 Four-node cuttings of *Solanum andigena* from which the three uppermost axillary buds have been removed: (a) the cutting was allowed to form roots, and the remaining axillary bud developed as a leafy shoot; (b) roots were removed as soon as they appeared, and the bud developed as a stolon. (Plate 1, Fig. 3 from Kumar and Wareing, 1972)

develop either as stolons or as leafy shoots, and, further, that a developing stolon can be converted into a leafy shoot. Further experiments using applied hormones give some indication as to the nature of these correlative effects. Booth (1959, 1963) showed that, whereas in decapitated shoots of *Solanum andigena* the bud at the uppermost node developed as a leafy shoot, if a combination of IAA and GA was applied to the cut surface the bud developed as a diageotropic stolon (Fig. 3.27). After 10–15 days, recovery to a leafy shoot usually occurred, the tip turning up and giving rise to foliage leaves (Fig. 3.28). If a solution of GA was introduced into the base of dark-grown sprouts of the cultivated potato (cv. Majestic) by means of a capillary tube, there was considerable stimulation of stolon formation but the pattern of emergence was changed (Booth, 1963). Application of GA to the stems of intact plants also resulted in the development of aerial stolons in the apical region (Kumar and Wareing, 1972). One-node cuttings cultured in the presence of GA gave rise to thin shoots with axillary stolons (Woolley and Wareing, 1972b). Thus GA, in the presence of natural or applied IAA, seems to stimulate stolon development under a variety of conditions.

Applied cytokinins and gibberellins have antagonistic effects in controlling lateral shoot development. If kinetin or benzyladenine (BA) in carbowax and alcohol was applied to a stolon tip, the apical

Fig. 3.27 Decapitated plants of *Solanum andigena* treated at the cut surface with IAA + GA in lanolin. The uppermost lateral buds have grown out as horizontal stolons, hooked at the tip. (Plate 6, bottom left, from Booth, 1959)

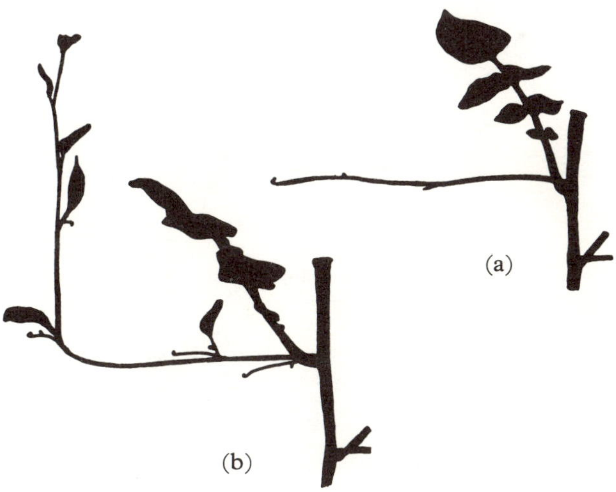

Fig. 3.28 Upper part of decapitated plants of *Solanum andigena* which had IAA + GA applied to the cut surface: (a) development of the uppermost bud as a stolon (compare Fig. 3.26); (b) recovery of a stolon to an erect leafy shoot after 10–15 days. (Fig. 9 from Booth, 1963)

Fig. 3.29 Stolon development in *Solanum andigena*: (a) naturally developing stolons emerging from the soil; (b) after kinetin has been applied to the tip, the stolon turns up and develops as an erect leafy shoot. (Plate 2, Fig. 2 from Kumar and Wareing, 1972)

Fig. 3.30 Decapitated plant of *Solanum andigena* treated with IAA + GA has given rise to a stolon from the uppermost axillary bud. Kinetin was then applied to the tip of the stolon, which 10 days later has continued development as a leafy shoot. (Plate 2, Fig. 3 from Kumar and Wareing, 1972)

hook opened out, and subsequently the shoot became erect and foliage leaves developed (Fig. 3.29). This could be achieved also with stolons induced by decapitation and treatment with IAA + GA (Fig. 3.30). Both light and applied cytokinin were found to be necessary to cause a stolon to develop as a leafy shoot (Kumar and Wareing, 1972). By the use of radioactive BA, it was found that the cytokinin accumulated in the stolon tip after the withdrawal of IAA + GA from a decapitated shoot. It was considered that there was a causal relationship between this accumulation and the subsequent conversion of the stolon to a leafy shoot (Woolley and Wareing, 1972a).

Single-node cuttings cultured in the presence of cytokinins gave rise to leafy shoots, more vigorous than those in control cultures without added hormones. The failure of a lateral bud to grow as a leafy shoot following removal of the roots, mentioned above, could be overcome by the presence of either BA or zeatin (Woolley and Wareing, 1972b).

The results of these various experiments indicate that, in the case of lateral buds which are not subject to apical dominance, a high ratio of cytokinin to gibberellin favours the development of leafy shoots, and a low ratio favours stolon development (Woolley and Wareing, 1972b).

Analysis of endogenous hormones in *Solanum andigena* showed that low light intensity increased levels of acidic gibberellins in leaves of short day (SD) plants. Plants grown under low light intensity formed aerial stolons, and these could be induced in high light intensity plants by application of GA (Woolley and Wareing, 1972c).

The induction of aerial stolons in decapitated shoots of *S. andigena* by application of IAA and GA, described above, occurs to a greater extent when plants are grown under SD conditions. In long day (LD) plants, a high percentage of induced aerial stolons became upright shoots when the IAA/GA mixture was replaced by plain lanolin, but no stolons became converted to leafy shoots in SD plants. Levels of endogenous cytokinins were lower in SD plants (Woolley and Wareing, 1972c). Thus the observations concerning endogenous hormones are compatible with the view that the control of lateral shoot development depends on the balance between cytokinins and gibberellins.

The structural changes accompanying these changes in external form, and their control, have not been examined, and this seems a promising field for future investigation.

3.6 The tuber

As already mentioned, stolon formation usually begins at the lower nodes and progresses acropetally. The first tubers, in turn, usually develop from the lower stolons, and tend to become dominant over those formed later (Plaisted, 1957). The importance of the tubers is indicated by the fact than 75–85% of the total dry matter produced by the plant accumulates there (Ivins and Bremner, 1964). Tubers which ultimately attain the greatest weight are usually produced by the lowest stolons, and indeed most tubers which grew to a diameter of more than half an inch were found to have been formed at the beginning of the period of tuber development (Clark, 1921).

Although tubers are usually formed from the basal stolons, under particular anomalous conditions (for example, very high carbohydrate status of the plant) they may develop from nearly every vegetative bud irrespective of its location on the plant (Werner, 1954). Tubers developed from the sprout apex and from short lateral stolons when the parent tubers were kept at low temperatures (Vöchting, 1902). Various factors can affect tuber yield; these are discussed in Chapter 4 and elsewhere.

3.6.1 Morphology of the tuber

As Artschwager (1924) pointed out, the potato tuber is a modified stem with a shortened (and broadened) axis and rather poorly developed leaves. The 'eye' of the potato tuber, according to Artschwager, is a leaf scar with a subtended lateral bud having undeveloped internodes. He also says that 'each eye contains at least three buds arranged in the form of an obtuse triangle...'. Both of these statements, although broadly correct, are rather misleading. Firstly, the leaf component of the eye is not so much a leaf scar, as a complete scale leaf which has attained little height and which, because of the growth in diameter of the tuber, has a base which is greatly extended tangentially (Fig. 3.31). The 'three buds' in its axil, also mentioned by Esmarch-Bromberg (1919), are not multiple axillary buds, but the original axillary bud of the scale leaf and the second order axillary buds of its first two leaf primordia, which occur (as in the axillary buds of many species) in nearly opposite positions to right and left of the leaf axil; Artschwager's Fig. 5 (1924) in fact shows this quite clearly. Again because of the expansion of the tuber, the first two leaf primordia of the axillary bud have become tangentially separated (Fig. 3.31). Other workers also state that 'many axillary buds' are present in each eye (Reeve *et al.* 1969). The eyes are spirally arranged, as were the scale leaves of the stolon tip from which they have developed. Shallow and deep-eyed varieties are known. The

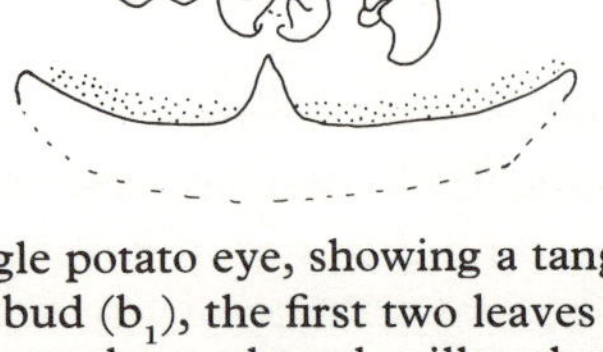

Fig. 3.31 Diagram of a single potato eye, showing a tangentially expanded scale leaf subtending an axillary bud (b_1), the first two leaves of which, to right and left of the scale leaf axil, themselves subtend axillary buds (b_2)

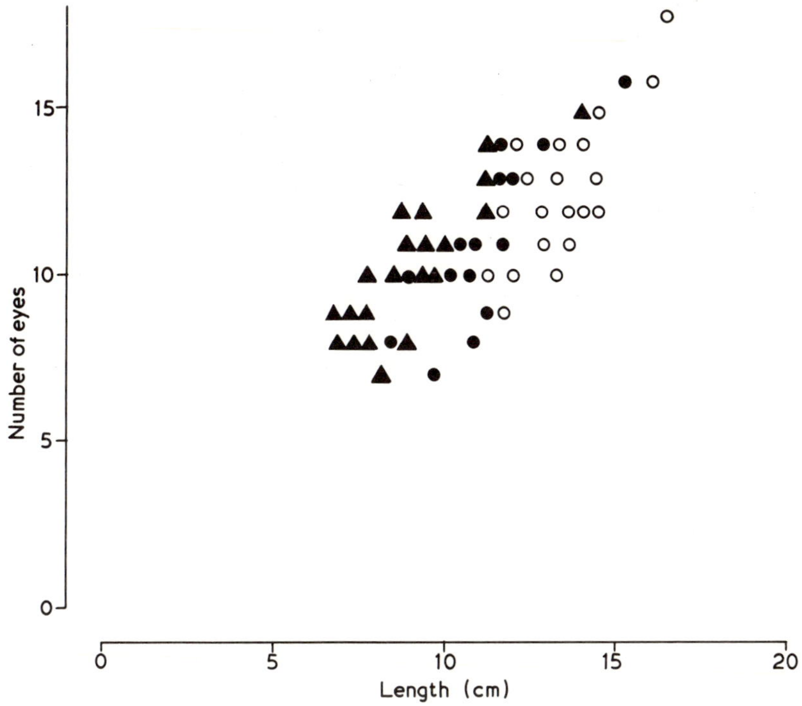

Fig. 3.32 Relationship between tuber length and the number of eyes per tuber. Tubers of 130–150 g are shown by open circles, those of 90–110 g by solid circles, and those of 50–70 g fresh weight by solid triangles. (Fig. 2 from Goodwin, 1967a)

potato tuber is thus morphologically straightforward, a stem with internodes greatly distended in the radial plane, each scale leaf of the phyllotactic spiral subtending an axillary bud with its own spiral sequence of leaf primordia. At the 'heel end' can be found the stolon or the scar of its attachment, and at the 'rose end', slightly offset from the direct axial line, is the apical bud of the stolon (see Fig. 3.34). Buds towards the heel end may be smaller than those in median sites (Goodwin, 1967a).

In tubers of cv. Arran Pilot, an average of one eye per cm length of tuber was found, although small tubers had a larger number of eyes per unit length than larger ones (Fig. 3.32; Goodwin, 1967a). This indicates that tubers continue to elongate, and to incorporate new internodes during a certain period of development. Went (1959) also found a correlation between weight of full grown tubers and the number of internodes. More internodes were present in tubers from plants grown at higher night temperatures.

Upon sprouting of the tuber, the 'apical eye' (i.e. the original stolon apex) resumes growth first (Artschwager, 1924). Temperature can have an important effect on apical dominance in the tuber. This is quickly established by the terminal bud in relatively high temperatures of more than 15° C (Goodwin, 1963; Ivins and Bremner, 1964). On the other hand, at 10° C several buds begin and continue to grow. If the tubers are stored at 1–5° C after lifting for several months, and are then kept at high temperatures, all the buds begin to grow, although eventually some are inhibited (Goodwin, 1963). Thus the temperatures during both storage and sprouting can affect the number of stems per tuber. The factors which lead to sprouting, after the imposition of dormancy on the bud apices, will be considered in Section 3.7.1.

Adventitious shoots occasionally develop from callus on excised pieces of tuber. More callus forms on basal (i.e. heel-directed) surfaces. It was not possible to induce adventitious bud formation at will (Simmonds, 1964).

3.6.2 Formation and early development of the tuber

Surprisingly little accurate information seems to exist in the literature on the earliest stages of tuber formation and development. In view of this, a preliminary re-examination has been made and is reported below.

As has long been known (de Vries, 1878), tuber formation begins in the sub-apical region of the stolon. It is first evident externally as a radial enlargement of this region (Figs. 3.15 and 3.33; Artschwager, 1924). Plaisted (1957) states that the first indication of tuber formation is a thickening of the first internode behind the apical bud of the stolon.

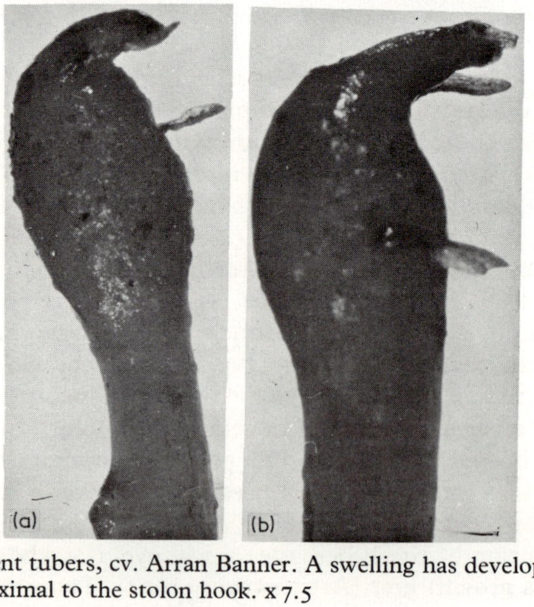

Fig. 3.33 Incipient tubers, cv. Arran Banner. A swelling has developed sub-apically, just proximal to the stolon hook. x 7.5

This statement actually refers, as will be seen below, to the youngest elongating internode of the stolon tip, which is often about the eighth internode proximal to the apical meristem.

It is not by any means certain whether the initial radial expansion of this internode is brought about by cell division or by cell enlargement. Artschwager (1918, 1924) stated that the initial swelling was a result of extensive cell division in the pith region. This was followed by cell division in the cortex and perimedullary zone. Reed (1910) also reported extensive cell division in the pith, although it is not certain whether he was referring to the earliest stages of tuber formation. Reeve *et al.* (1969) also attributed the initial radial enlargement of the stolon tip to cell divisions in the pith in random planes and to cell growth. However, Booth (1963) considered that the initial swelling was attributable to cell enlargement. In his view, there is a loss in polarity first of cell enlargement, and later in the plane of cell division.

At a later stage, the cells of the perimedullary zone (*Markkrone*) divide and apparently give rise to the bulk of the tuber tissue (Artschwager, 1918, 1924).

Goodwin (1967a) states that the continued expansion of the tuber depends on the production of new internodes from the apical bud. Without further elaboration this statement, too, is misleading. Goodwin considers that growth of the buds only stops, and dormancy begins, at lifting or death of the parent plant, and Moorby and Milthorpe (1975) also state that growth of the apical bud and of the tuber itself stops simultaneously, but Leshem and Clowes (1972) have shown that the rate of mitosis in the stolon apex becomes too low to measure soon after the beginning of tuberization. In fact, growth of the tuber depends primarily on expansion of internodes already present in the apical bud, not on the formation of new ones.

Further growth of the tuber gives rise to an elliptical or spherical structure depending on the balance between growth in length and thickness (de Vries, 1878). The shape of the tuber is, of course, a varietal characteristic.

Current observations

These observations largely constitute a clarification, in morphogenetic terms, of obscure statements in the literature; they do not comprise a full study of tuber development.

As already mentioned in Section 3.5.1, the whole of the stolon hook consists of highly meristematic cells, as judged by their staining properties (Fig. 3.18). This includes the elongating internode which is incorporated in the hook itself. Below the hook there is a sudden change to much more expanded, vacuolated cells (Fig. 3.18).

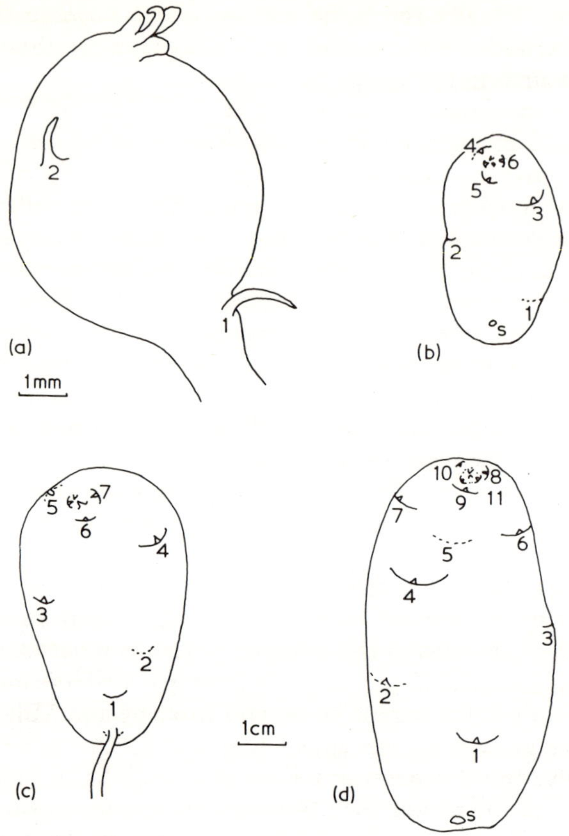

Fig. 3.34 Developing tubers, cv. Arran Pilot: (a) young tuber with the second internode becoming incorporated. The stolon hook has almost straightened out. x 5. (b–d) Tubers comprising 6, 8 and 11 internodes respectively. x 1/2. (s. stolon or scar of its attachment. Eyes numbered in order of formation; dotted eyes on other side of tuber)

The first sign of tuber formation, as mentioned by earlier authors, is a swelling of the sub-apical region of the stolon tip. This occurs at about the hooked region. Observation of developing tubers of increasing size shows that initially only one internode is involved in this swelling. In slightly older, more swollen tubers a second internode, distal to the first, is also incorporated in the developing tuber, so that a scale leaf may be present at or near the base of the developing tuber and another about half-way along its length (Fig. 3.34a). At about this stage, because of the considerable radial expansion of these two internodes, the hook becomes straightened and the apical bud of the stolon is situated in a more or less terminal position on the young tuber (Fig. 3.34a). In cv. Arran Pilot, tubers at this stage measured about 1.5 × 1.0 cm. Some tubers about this size incorporated four internodes. External examination of tubers of increasing size revealed that an increasing number of internodes became incorporated into the tuberous portion of the axis. For example, tubers respectively 4.5 × 2.5 cm., 6.0 × 3.5 and 8.0 × 4.0 cm comprised 6, 7–8 and 11 internodes (see Fig. 3.34b–d). Internodes are shorter towards the rose end. Thus tuberization proceeds acropetally, involving some longitudinal extension and very considerable transverse expansion of successive internodes.

Transverse sections of stolon tips show about nine scale leaf primordia above the hook. Similar sections of very young developing tubers show a swollen region, with starch-containing parenchymatous cells, distal to the junction of the eighth or ninth youngest leaf primordium with the axis (Fig. 3.35a). In tubers of this size, about 1.0–1.5 mm in diameter, a cylinder of differentiating vascular tissue can be seen. Internal and external phloem is abundant (Fig. 3.35b), xylem elements rather sparse and widely spaced (Fig. 3.35a). Longitudinal sections of stolon tips show about nine files of cortical and about twelve (less regular) files of pith cells above the bend of the hook, and about the same number below. Comparable sections of developing tubers in early stages seem to show no increase in the number of files of cortical cells; about half the tubers sectioned seemed to show some increase in the number of files of pith cells (Fig. 3.36), but these are rather irregularly arranged, and also the number of files seen in a section depends to some extent on the presence of internal phloem in that particular plane. The pith cells did seem to be wider, however, and contained starch (Fig. 3.36a). Thus the writer tends to agree with Booth (1963) that early tuber development is primarily attributable to enlargement of the pith cells, although cell division probably rapidly follows. Deposition of starch grains varied considerably in the young tubers observed, cv. Arran Pilot showing conspicuous starch grains (Fig. 3.36a) which were not present in Arran Banner.

Periclinal divisions giving rise to the phellogen occurred in both the epidermis and hypodermis at an early stage of tuber formation (Fig. 3.37). There was some evidence that division of the hypodermis usually occurred first, but certainly both layers of cells underwent division within a short time.

3.6.3 Structure of more mature tubers

In the literature on the more mature tuber there is still disagreement concerning the relative contributions of cell division and cell expansion. Reed (1910) found that nearly all the pith cells divided, and that the parenchyma between the protoxylem and internal phloem (the *Markkrone*) also divided actively; as a result the phloem became broken up into many strands. Lehmann (1926) noted that growth of the tuber mainly depended on both division and enlargement of the storage parenchyma.

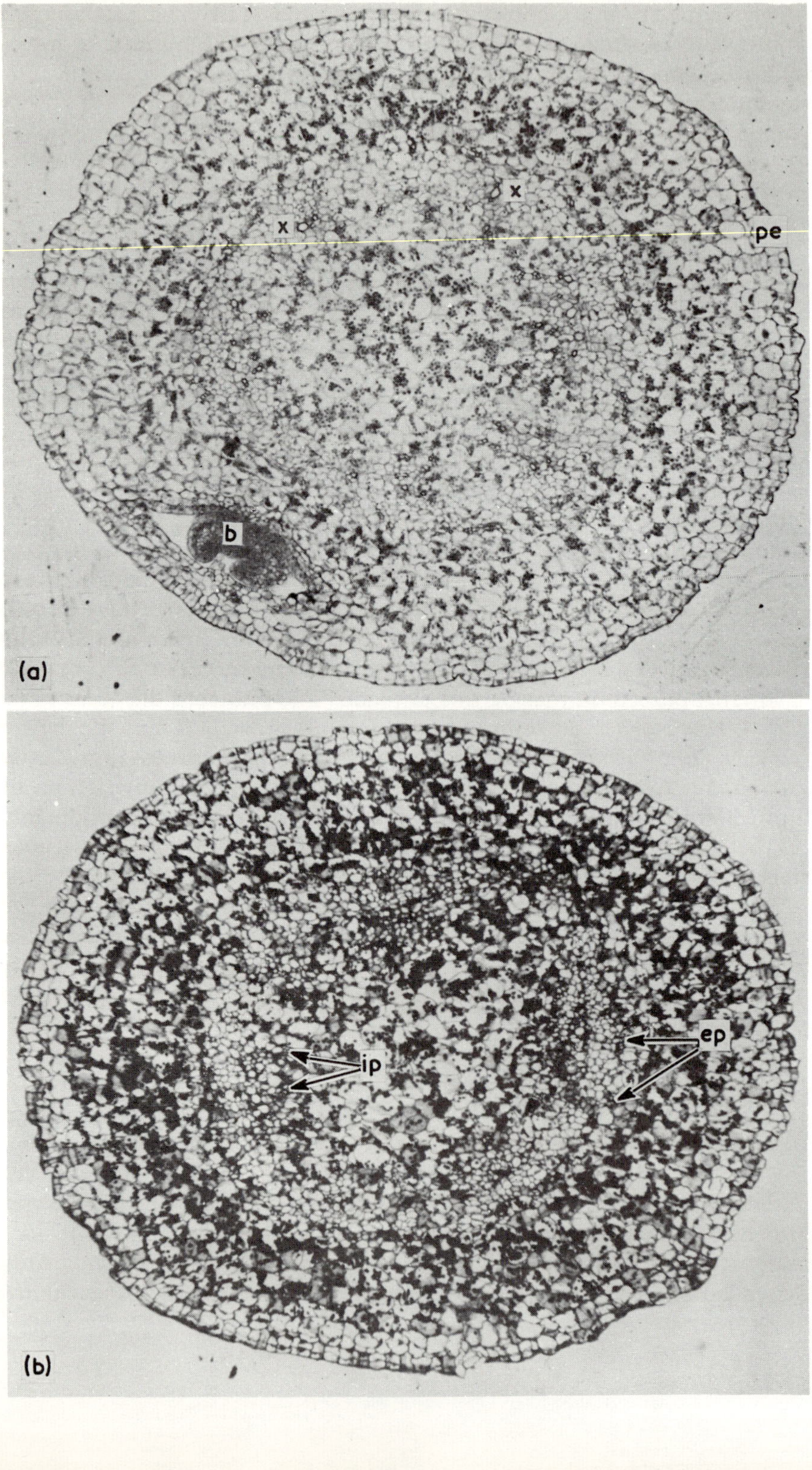
x
x
pe
b
(a)
ep
ip
(b)

Others consider that cell division in the perimedullary region is more important than that in the pith (Reeve *et al.* 1973a, b). Certain quantitative observations have been made. For example, tubers of cv. Kennebec and Russet Burbank increased in size 225- to 260-fold during their development. Cell size increased from 15- to 20-fold in the former and 7- to 18-fold in the latter, indicating that there must have been considerable cell division early in tuber development (Reeve *et al.* 1973a). These workers considered that cell enlargement becomes increasingly important in tubers above 30–45 g, but Plaisted (1957) found that cell division continued in tubers up to 200 g (cv. Cobbler). He observed a 500-fold increase in cell number in tubers between 37 mg and 200 g, and a 10-fold increase in average cell size. Plaisted calculated that there was a daily increase in cell number per tuber of 11.7%, and considered that tuber growth was more directly related to increase in cell number than to increase in cell volume.

Cell size increases with tuber size (Lehmann, 1926) and tubers of 1 cm diameter have already undergone a considerable amount of cell enlargement (Reeve *et al.* 1973a). Cell enlargement is thought to become the dominant mode of growth in tubers over 30–40 g (Reeve *et al.* 1973b). There is a decreasing gradient in cell size of storage parenchyma from heel to rose end of mature tubers (Reeve *et al.* 1971).

Reeve *et al.* (1973b) state that the major source of new tissue in growing tubers is procambial activity giving rise to the external and internal phloem; they further imply that phloem strands bring about an increase in number of storage parenchyma cells. These observations seem to reveal a fundamental misunderstanding of the nature and function of procambium and the phloem to which it gives rise.

The tuber, like the stolon from which it develops, contains an abundance of internal and external phloem strands (Fig. 3.38). Reeve (1967) considers that the region of internal phloem comprises at least three-quarters of the total tuber tissues. Certainly this is an exaggeration if one considers only the sieve elements and not all the surrounding parenchyma. Some workers, however, include the perimedullary zone and other tissues in the internal phloem region (Reeve *et al.* 1969). The phloem strands in a mature tuber comprise only about 5% of its volume (Reeve *et al.* 1973b). As Reed (1910) pointed out, the phloem becomes 'broken up' into many strands separated by parenchyma. These strands undergo anastomosis.

Fig. 3.35 T.S. young developing tuber of cv. Arran Pilot: (a) at junction of ninth scale leaf primordium and its axillary bud, b. Starch grains are conspicuous in pith and cortex. The scattered xylem elements (x) are stained dark. Periderm (pe) is beginning to develop by periclinal divisions in the epidermis and hypodermis. x 75. (b) Slightly lower down. In this section the staining demonstrates the phloem elements (ep and ip). x 100

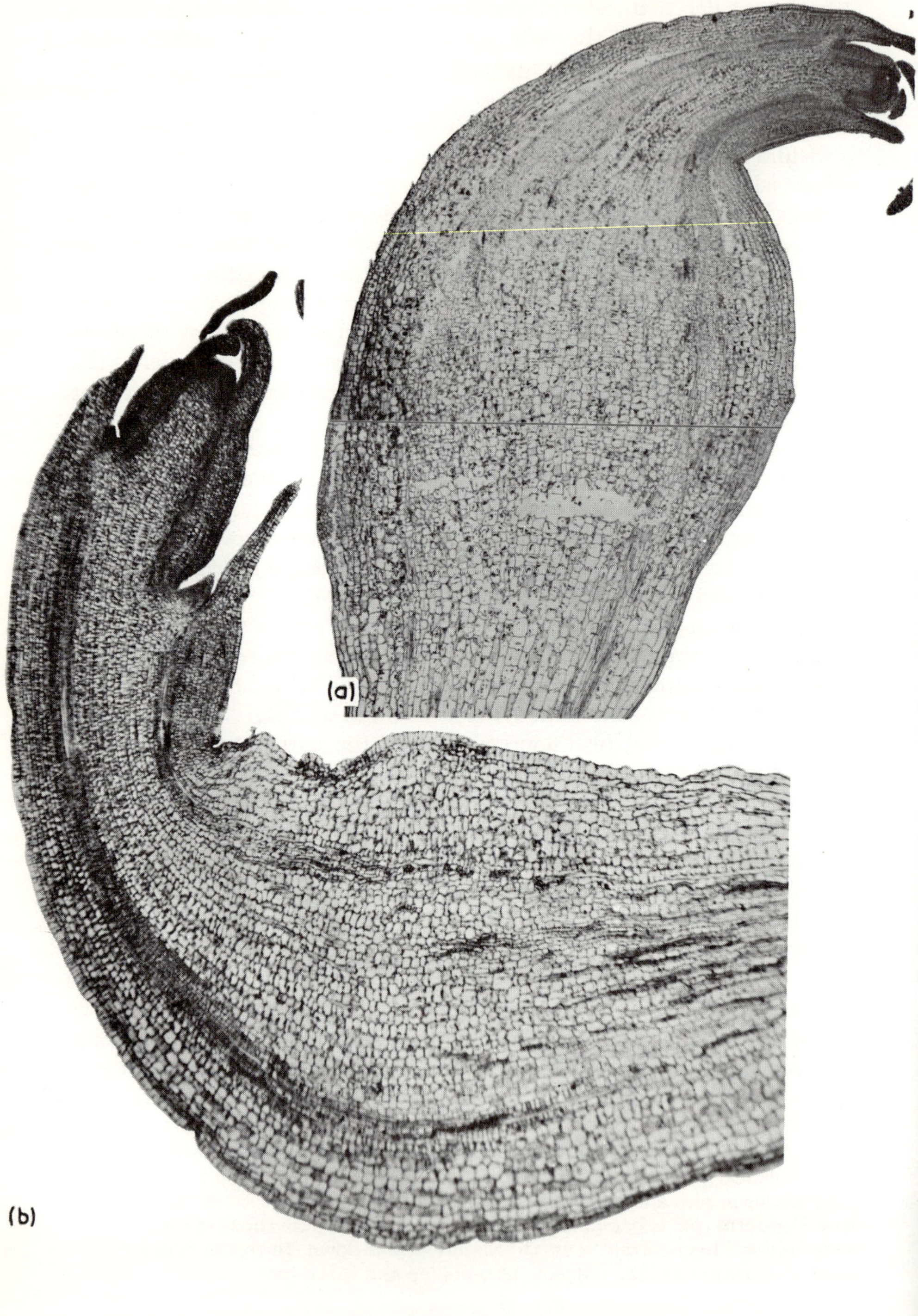
(a)
(b)

In the mature tuber the proportion of xylem is small. The protoxylem elements gradually become separated from the metaxylem (Artschwager, 1924). There is no continuous cambium, although some workers consider that there is a discontinuous one which produces a few conducting elements (Lehmann, 1926).

Former views that much of the tuber was of cambial origin seem to be erroneous. There is rather general agreement that division of the perimedullary zone makes a major contribution to the tuber tissues. This region is said to be of procambial origin (Reeve *et al.* 1969, 1970), but see the discussion of the differentiation of internal phloem in Section 3.4.3.

An endodermis with Casparian strips is reported in early stages of tuber development (Reeve *et al.* 1969), but, as a result of the great expansion of the tissues within it, gaps gradually appear where the endodermis has failed to keep pace with this active cell division, and in later stages no trace of an endodermis remains (Artschwager, 1924).

The cells of the storage parenchyma are more or less isodiametric (Lehmann, 1926). In aged tubers callose often develops in the storage parenchyma cells. It occurs in the form of spherical or elliptical structures, and various staining reactions demonstrate its similarity to sieve tube callose (Wodicka and Wenzl, 1971).

The thickness of the walls of storage parenchyma cells increases considerably during growth. Frequent primary pit fields are present (Reeve *et al.* 1973c). An analysis of the cell walls shows that they are composed of pectic substances (55%), cellulose (28%), protein (10%) and hemicellulose (7%). There is a high proportion of galactose in the pectin polysaccharides. This wall composition resembles that found generally in young growing tissues (Hoff and Castro, 1969).

Protein

Cubical protein crystals occur in parenchyma cells of the potato, especially those which are not rich in starch. They occur in cells at the periphery of the tuber in all varieties, and also in the head cells of glandular hairs (Hölzl, 1965). Pure protein occurs in greater amount in peripheral regions than in the interior of the tuber (Fischnich and Heilinger, 1959). Protein crystals do, however, occur in the pith cells, and these are the first to disappear during storage. They seem to comprise normal reserve protein in crystalline form (Hölzl and Bancher, 1958).

Fig. 3.36 L.S. young developing tubers of cv. Arran Pilot. The axis proximal to the stolon hook is distended radially. In (a) starch grains are present in these cells. At this stage in tuber development the hook has not yet straightened out. (a) x25; (b) x40.

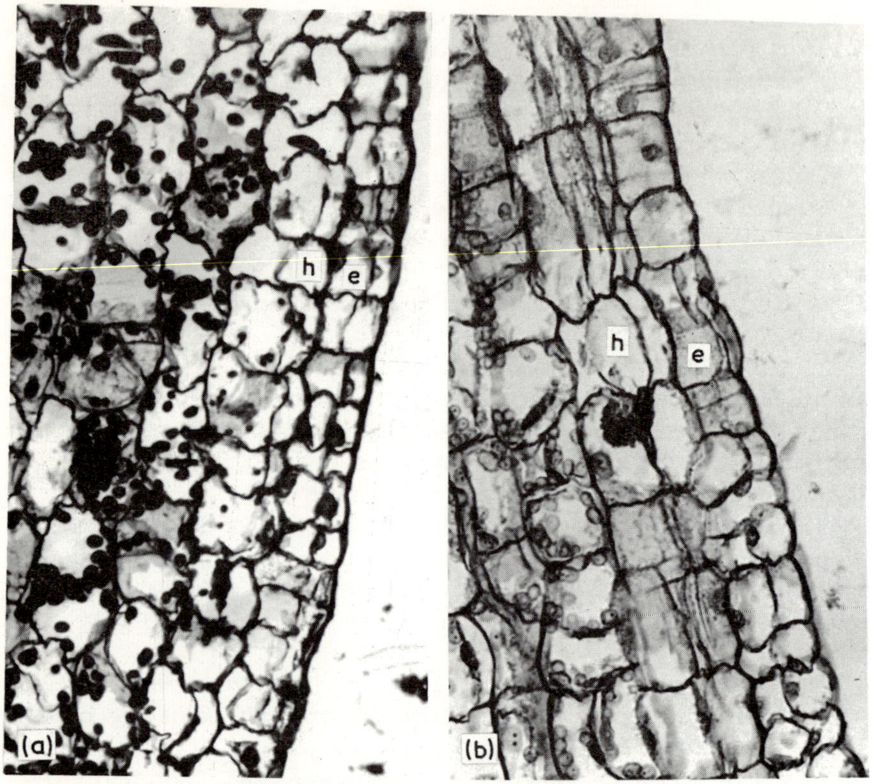

Fig.3.37 Section of young developing tubers showing the inception of periderm by periclinal divisions in the epidermis (e) and hypodermis (h): (a) T.S.; (b) L.S. x 300

The protein crystals of the potato are unique because of their large size – usually about 10 μm, but they may occasionally reach 25 μm. These crystals can now be isolated from potato tissues. They consist of protein with some amino acid component, chiefly lysine (14%) (Hoff, 1971). A relatively high-molecular RNA is present (Hölzl, 1965; Hoff, 1971). It has been suggested that this could be stabilized messenger RNA, required during the onset of active growth following the breaking of dormancy (Hoff, 1971). The protein crystals are known to disappear during sprouting.

Starch

Storage parenchyma cells contain starch grains with an eccentric hilum and a series of lamellae, thought to represent layers of starch. These are more distinct in tubers from early plantings (Salunkhe and Pollard, 1954). Starch grains may attain a length of up to 100 μm. A proportion

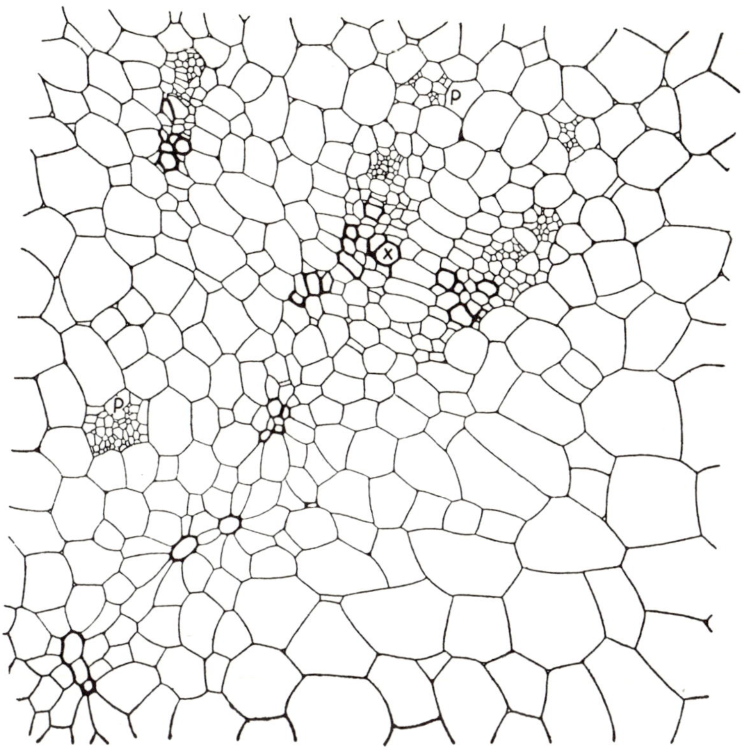

Fig. 3.38 Part of T.S. of a mature tuber showing the vascular tissues (p, phloem; x, xylem). (Fig. 2 from Artschwager, 1924)

of compound grains occurs (Fischnich and Heilinger, 1959; Banks *et al.*, 1973).

Starch content of tubers is 10–25% of fresh weight, with a maximum of 35%, depending on the variety. Cells with the highest starch content are those adjacent to vascular bundles in the tuber (Fischnich and Heilinger, 1959). Starch content is reduced by storing tubers at low temperatures (Ohad *et al.*, 1971). Formation of starch grains proceeds much more rapidly in early varieties than in late ones (Jager and Jakovlev, 1939). Varieties having tubers with a high starch content were also composed of larger cells, and there was a relationship between cell size and mealiness (Barrios *et al.*, 1961). Methods for extracting starch grains from tubers have been devised (Reeve, 1967). In cv. Russet Burbank and White Rose, considerable differences in the size of starch grains in different regions of the tuber were noted (Reeve *et al.*, 1970).

Starch grains were formed in tubers when sugars such as sucrose, glucose or maltose were supplied to the cut ends of stems (Hori, 1954). When plants were supplied with $^{14}CO_2$, autoradiographs showed the greatest amount of label along the margin and at the hilum end. This is regarded as supporting the view that growth occurs by apposition (Badenhuizen and Dutton, 1956).

Starch grains are formed by amyloplasts (Hori, 1954). They are surrounded by a membrane derived from the plastid envelope (Hölzl, 1965). Storage of tubers in the cold leads to disintegration of this membrane (Ohad *et al.*, 1971). The amyloplasts themselves have a thylakoid and grana structure similar to that of chloroplasts, but less well developed. They also contain both DNA and RNA. After tubers have been exposed to light for several days, typical chloroplast ultrastructure develops in the amyloplasts (Badenhuizen and Salema, 1968).

The fine structure of potato starch has been investigated by various techniques. When starch grains were ruptured by ultrasound, they appeared to have a smooth surface, with internal concentric lamellations. The grains may be made up of a system of radial submicroscopic units, probably fibrillar (Gallant *et al.*, 1972). A structure comprising radially organized fibrillar units (microfibrils) was previously proposed for Lintnerized potato starch (Sterling and Pangborn, 1960), although it is not certain whether the microfibrils are real or artifacts (Leonard and Sterling, 1972). (Lintnerized starch has been treated with cold, dilute mineral acid for a short time.) Freeze-etching indicated a granular structure of native starch (Leonard and Sterling, 1972). Other workers propose that potato starch grains consist of a core consisting of crystalline amylose, surrounded by a jacket of layers of micelles of amylopectin (Gruber *et al.*, 1972). Various models of starch grain structure have been proposed (Sterling and Pangborn, 1960; Frey-Wyssling, 1974).

Periderm

At an early stage of tuber development, periclinal divisions occur in the epidermal and hypodermal layers (Fig. 3.37) (Esmarch-Bromberg, 1919; Reed 1910; Reeve *et al.*, 1969). Artschwager (1918) considered that the resulting phellogen originated from the inner row of daughter cells produced by division of the hypodermal layer. He believed that the hypodermis and epidermis give rise to separate, distinctive periderms.

Usually 6–10 layers of cork are present, but this can be modified by environment. For example, tubers of cv. Russet Burbank from plants grown at different temperatures had varying numbers of cell layers in the periderm. At 45–50° F the periderm consisted of 7–12 layers of cells, and was 100–140 μm thick; at 65–70° F, these figures were respectively 9–25 and 120–250, and at 75–80° F there were 10–28 layers of periderm, 125–270 μm thick. Tubers from plants grown at the higher temperatures were well-russeted, i.e. the periderm was cracked and divided into patches of thick tissue (Yamaguchi *et al.*, 1964). Perhaps these structural effects were a response to altered hormone levels in the plant; however, factors controlling periderm formation are little understood.

The outer phellem cells of the potato tuber, when observed under the electron microscope, contain numerous ribosomes and mitochondria, but no plasmalemma or tonoplast. The cell walls seem to consist of layers of different electron density, and it is considered that autolysis of the cytoplasm may yield wall-thickening material (Rainbow and White, 1972).

Lenticels are formed in the periderm at sites below the original stomata (Artschwager, 1918). It is possible to estimate the number of lenticels in a tuber by placing it in a solution of methylene blue under vacuum. The resulting pattern of blue spots when analysed showed a range between 74 and 141 lenticels per tuber (Wigginton, 1973).

The number of lenticels per unit of tuber surface can be influenced by size of tuber, soil type and weather. Moisture conditions in the soil at the time of tuber formation were particularly important (Meinl, 1966). The relationship of the lenticels with underlying tissues is also affected by soil moisture. Under dry conditions a suberized layer forms below the complementary cells of the lenticel. Increased soil moisture, and water content of the tuber, leads to swelling of the cortical cells of the tuber and eventual rupture of the suberized layer, resulting in proliferating open lenticels (Pérombelon and Lowe, 1975). Lenticel development was also affected by immersing the tubers in GA solutions (Chirilei *et al.* 1963).

As seen under the scanning electron microscope, lenticels of potato tubers are usually circular, forming a crater in the surface of the tuber. In the centre there may be a raised plateau or a dome. Pores can be seen between the cells in the centre, where gaseous exchange is thought to occur. Waxy outgrowths resembling fine threads occurred on cells from the sides of these pores. It is thought that these outgrowths, which were observed in eleven varieties, may function in the regulation of water loss (Hayward, 1974).

Wound periderm

The potato tuber responds to wounding by the formation of periderm, and this 'wound periderm' confers an important degree of resistance to some bacterial and fungal diseases. Wounding tissue usually stimulates cell division (Lipetz, 1970). This is generally preceded by deposition of suberin at the cut surface. The type of wounding may have an effect, extensive abrasion resulting in a deep-seated periderm, and smooth peeling giving rise to a superficial periderm (Fig. 3.39) (Artschwager, 1927). The distance of the phellogen from the wounded surface is also

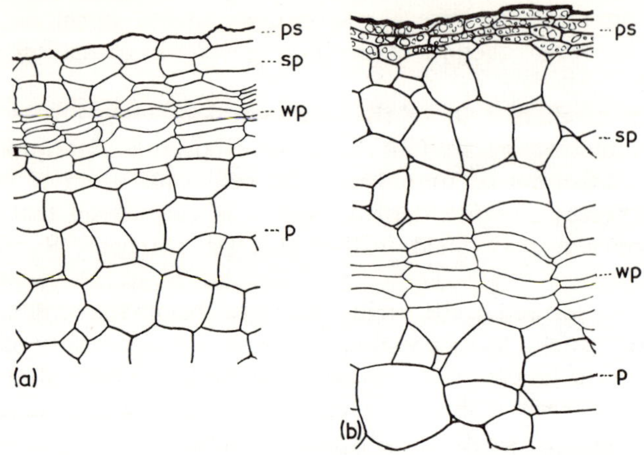

Fig. 3.39 Wound periderm in cv. Early Rose: (a) the tuber was injured by smoothly peeling off the skin; (b) in peeling the surface was considerably abraded. Wound periderm (wp) is formed further away from the surface. p, storage parenchyma; ps, peeled surface; sp, suberized parenchyma at the surface of the cut; wp, wound parenchyma. (Part of Fig. 3 of Artschwager, 1927)

affected by the water content of the tissue (Rosenstock, 1963). The formation of wound periderm is affected by temperature and humidity, being more rapid at higher temperatures and high humidity (Artschwager, 1927; Wigginton, 1974). Oxygen partial pressures lower than those in the atmosphere are inhibitory (Wigginton, 1974). Suberization can occur after only one day at 21–35° C, being followed by periderm formation one or more days later. The layers of cork cells become impregnated with suberin. At the same time starch disappears from the layers of cells adjacent to the cut surface (Artschwager, 1927; Priestley and Woffenden, 1923).

Priestley and Woffenden (1922) have claimed that both natural and wound periderm formation follow the same causal sequence of events:

(1) Blocking of a parenchymatous surface by deposition of suberin or cutin in the presence of air;

(2) Accumulation of sap at this blocked surface; and

(3) Development of a phellogen. Priestley and Woffenden (1922) tested the significance of this sequence in various ways. If they prevented access of air to the wounded tissue, e.g. by removing a cylinder of tuber tissue with a cork borer, replacing it *in situ* and sealing the surface, cellular proliferation occurred but not the deposition of suberin or formation of an organized meristem. If they covered the cut surface with paraffin wax a phellogen meristem formed, but no suberin. For the development of a phellogen, therefore, blocking of the cut surface seems to be essential.

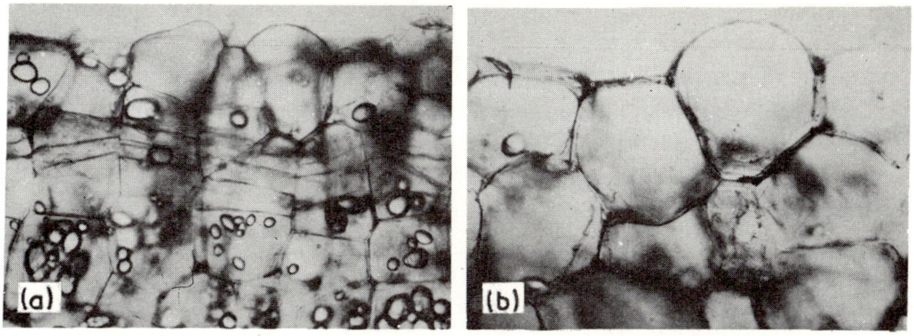

Fig. 3.40 Effect of tris buffer on wound periderm: (a) cell formation at the wounded surface after 5 days in water. x 125. (b) inhibition of cell formation in similar tissue after 5 days in tris buffer. x 190. (Figs. 1 and 2 from Kahl *et al.* 1969)

More recently, other experimental studies have succeeded in separating phellogen formation and suberin deposition. Mitotic activity resulting from wounding could be totally suppressed by placing the tissue in tris buffer (tris – (hydroxymethyl-) aminomethan), but synthesis of suberin proceeded normally (Fig. 3.40). If the tris ions were washed out of the tissue slices mitosis could occur (Kahl *et al.* 1969). Using thin slices of tubers of cv. Saskia and others, it was possible to obtain proliferation in the surface layers of cells with inhibition of suberin synthesis. The necessary conditions included separating the tissue without bruising it, rinsing the slices to remove adhering injured tissues, and maintaining 100% humidity. The inhibition of suberin synthesis was reversible (Lange *et al.* 1970).

The nature and amount of wound periderm varies in different varieties (Priestley and Woffenden, 1923; Artschwager, 1927; Nielsen, 1968). Tubers grown in humus had thin skins, while those from sand had the thickest. In wounded tubers maintained under the same conditions, cv. Alpha and Bintje had 10–12 layers of suberized cells, Up-to-Date and Ackersegen 7–8 and Kennebec 5–6. Tubers of Ackersegen formed wound periderm more slowly than the other varieties (Nielsen, 1968).

Although these and other studies have been carried out in an attempt to discover the factors leading to development of wound periderm, little or no attempt seems to have been made to study the factors controlling the formation of normal periderm in the potato tuber. Consideration of the small amount of evidence relating to other species has led to the suggestion (Cutter, 1969) that formation of phellogen may be inhibited by hormones produced by an actively growing shoot apex or young shoot. Since various growth inhibitors can promote tuber formation (see Section 3.6.4) and thus periderm formation, with accompanying cessation of apical growth, this view may be worth testing. It is also possible that the 'sap' which Priestley and Woffenden (1922) considered accumulated at the suberized surface could be a hormone of some kind. The statement of Lipetz (1970) that cutting or wounding of many fruit tissues is accompanied by an increase in ethylene production may be of particular interest, since ethylene can induce tuber formation in the potato (see Section 3.6.4). Further work is needed to elucidate factors affecting periderm formation.

3.6.4 Factors controlling tuber formation

Although tubers develop from stolon tips, not all stolons form tubers (Svensson, 1962). Separate factors regulating tuber formation must therefore be involved. Investigation of the effects of various environmental factors led to the view that a tuber-inducing stimulus was formed in the potato plant. The nature of this stimulus has been investigated both by application of exogenous substances such as various hormones or growth inhibitors, and by extraction and analysis of endogenous substances from induced and non-induced plants.

Environmental effects

Tuber formation occurs earlier at low temperatures (Mes and Menge, 1954; Slater, 1963, 1968; Ivins and Bremner, 1964) and is delayed at high temperatures (Slater, 1963). Gregory (1956) also found, using cv. Kennebec, that tuber yield was good if plants were grown in short days (SD) with low night temperature, but no tubers were formed in SD with high night temperature. Went (1959) showed that temperature affected not only the yield of the treated plants, but also that of their progeny. When tubers produced on plants grown at low temperature were used as seed, the resulting plants had more and heavier tubers than controls. This effect, which resembles that observed in the field (e.g. Scaramella Petri, 1959) could even be traced two generations back. Went considered it unlikely that it could be due to storage of a tuber-forming hormone in the tuber piece, since development of tubers on the new plants occurred long after this had been exhausted. This interesting effect, which Went compared to that of vernalization, remains unidentified.

Really low temperature treatment may also be effective in inducing tuber formation. For example, even in young rapidly growing plants, tubers can be induced to develop by maintaining the plants for 7 days at 7 °C or less (Burt, 1964). Treatment of sprouts with alcohol at − 2 to − 8 °C prior to culturing segments also hastened tuber formation, the optimum treatment being − 4 °C for 5 min. (Blanc, 1973). However, this was clearly a very abnormal situation and it is dubious how relevant these results are to normal agricultural practice.

Fig. 3.41 Cuttings of *Solanum andigena* taken from plants grown in (a) long days, and (b) short days. The lowermost axillary bud has developed as a stolon in long days, and a tuber in short days. (Plate 1, Figs. 1 and 2 from Kumar and Wareing, 1973)

There is also a quantitative effect of photoperiod in the potato. Plants grown in SD form tubers earlier than those kept in long days (Chapman, 1958; Madec and Perennec, 1959; Slater, 1963; Ivins and Bremner, 1964). In *Solanum andigena* short photoperiods are an absolute requirement for tuber formation (Fig. 3.41) (Booth, 1959). In the cultivated potato, however, plants maintained under long days (LD) make active vegetative growth but do ultimately form tubers, some 3–5 weeks later than plants kept under SD (Chapman, 1958). Plants given photoperiods of 9 h formed tubers in 26–30 days. At least 14 short photoperiods were required for induction to occur, and the stimulus could also disappear after 14 days of non-inductive conditions (Gregory, 1956; Chapman, 1958). The region of perception of the stimulus was the terminal bud, including leaves less than 5 cm long. The stimulus was apparently only slowly transmitted through other parts of the plant, since in decapitated plants allowed to develop two shoots, of which one was kept in SD and one in LD, tubers formed only on that half of the plant kept under SD (Chapman, 1958).

The nature of this photoperiodically controlled stimulus to tuber formation has been investigated in various ways. Both Gregory (1956) and Chapman (1958) showed that it could be transmitted across a graft union. If scions from SD plants were grafted on to LD stocks, the latter began to form tubers in 14 days; if LD scions were used no tubers were formed. In single-node segments of stem excised from induced plants and grown in sterile culture, a tuber developed from the axillary bud; only vegetative shoots developed in the case of segments from non-induced plants (Gregory, 1956; Chapman, 1958). This work has recently been confirmed (Forsline and Langille, 1976). Gregory (1956) concluded that a graft-transmissible tuber-inducing stimulus was formed under specific conditions of photoperiod and temperature, and occurred throughout the plant. Madec (1963) considers that the stimulus is produced in tuber tissues, the relative role of tuber and foliage in the inductive process being variable. In seedlings, however, tubers must be induced by the leaves only (Madec and Perennec, 1959). It has been shown by further grafting experiments that the foliage of a tomato scion cannot induce tuber formation in a potato stock, but can support the development of tubers which have been previously induced (Madec and Perennec, 1959). Tizio (1964a) claims that tuber formation is controlled not only by the postulated stimulatory factor in the leaves, but also by an inhibitory factor synthesized by the roots. Application of auxins to stem fragments, which retarded root formation, also resulted in precocious tuberization (Tizio, 1964b).

Kumar and Wareing (1973) have carried out a series of ingenious grafting experiments on *Solanum andigena*, in which tuber formation can be induced by 10 SD, although maximal tuber induction only occurs after 20 SD cycles. This species has an absolute SD requirement for tuber formation. They showed that, as in the cultivated potato, SD scions grafted on to LD stocks induced tuber formation in the stocks in two weeks, and that none was formed in LD/LD grafts. Cuttings made from plants maintained in SD also formed stolons with small tubers at the basal node, whereas no tubers formed on cuttings of LD plants (Fig. 3.41). Kumar and Wareing concluded that a stimulus to tuber formation was formed under SD conditions, and extended their experiments to investigate the movement of this substance within the

Fig. 3.42 Cuttings from plants of *Solanum andigena* grown in short days: (a) normally oriented; (b) inverted. In each case a tuber has developed at the lowest node with respect to gravity. (Plate 1, Figs. 5 and 6 from Kumar and Wareing, 1973)

plant. In a group of plants, the lower half was maintained under SD, while the upper half was layered and given LD. Control plants were given SD throughout. Cuttings made from the upper half of the treated plants, which had been kept in LD, formed tubers, showing that the stimulus could move upwards. Further experiments demonstrated that it could move upwards through a graft union. But one of the most interesting findings was that tubers would develop only at what was gravitationally the lowest node, even if this was morphologically the most apical. Sub-apical cuttings were planted the right way up or inverted; in either case tubers developed at whatever was now the lowest node with respect to gravity (Fig. 3.42) (Kumar and Wareing, 1973).

Okazawa and Chapman (1962), using the cultivated potato, also concluded that the movement of the stimulus was not polar; nor was that of gibberellins, believed to inhibit tuber formation. Using two-shoot plants of cv. Red McClure, one shoot maintained in SD and one in LD, these workers showed that not only did partial severance or girdling of the SD shoot result in delay of tuber formation, but similar treatment of the LD shoot slightly hastened tuberization. It is thus thought that tuber formation may be regulated by the relative activity of several substances rather than the absolute concentration of a single substance (Slater, 1968; Hammes and Nel, 1975), and that the substances involved may be the as yet unidentified tuber-forming stimulus, and natural gibberellins (Okazawa and Chapman, 1962). The point has been made, however, that some positive stimulus to tuber formation must exist in induced plants, since it would be difficult to explain the results of grafting experiments solely in terms of an inhibitor, such as gibberellin, present in non-induced plants (Gregory, 1956; Kumar and Wareing, 1974).

Experiments in culture

Barker (1953) showed that isolated nodes of etiolated potato sprouts could be grown in aseptic culture. He obtained outgrowth of roots and stolons, some of which later gave rise to tubers. Other workers have utilized this technique for growing isolated nodal pieces (e.g. Gregory, 1956; Chapman, 1958), and cultured stolons or isolated stem segments have been used in studies of the effects of hormones on tuber formation (Palmer and Smith, 1969b, 1970; Smith and Palmer, 1970; Garcia-Torres and Gómez-Campo, 1973; Palmer and Barker, 1973).

Effects of hormone treatment

Auxins such as IAA, NAA or 2,4-D seem to increase the size and sometimes earliness of tubers (Madec and Perennec, 1959; Harmey *et al.* 1966), but not to have any inductive effect. Indeed, IAA was inhibitory at certain concentrations (Kumar and Wareing, 1974).

Treatment of potato plants with gibberellic acid delayed tuber formation even under SD (Slater, 1963). Fewer tubers developed per plant, and there was a tendency for tubers to grow out as stolons (Lovell and Booth, 1967). Cuttings of induced plants of *Solanum andigena* also failed to form tubers when treated basally with a solution of GA (Kumar and Wareing, 1974). Tuberization was retarded or prevented when GA was supplied to nodal segments in culture (Harmey *et al.* 1966; Tizio, 1972).

This inhibitory effect of GA on tuber formation has led to the experimental application of growth retardants such as (2-chloroethyl)-trimethylammonium chloride (CCC), which affect the synthesis of endogenous gibberellins, to test whether they have a stimulatory effect. When tubers were treated with CCC, or the plants were watered with this substance, tubers were formed about 5–10 days earlier than in controls, depending on concentration (Gifford and Moorby, 1967; Tizio, 1969). Tuberization also occurred earlier in cultured segments of sprouts given CCC in the dark (Tizio, 1972). In other experiments, CCC was either applied as a soil drench to LD plants or applied in solution to plugs of tuber. In both instances, tuber formation was induced (Kumar and Wareing, 1974; Hammes and Nel, 1975).

In plants of *Solanum andigena* sprayed with abscisic acid (ABA), tuber formation was promoted, and in the cultivated potato, cv. Ulster Premier and Ulster Prince, a greater weight of tubers was obtained (El-Antably *et al.* 1967). Using *S. tuberosum* cv. White Rose in LD, Smith and Rappaport (1969) were unable to induce tuber formation by spraying the plants with ABA or by treating the stolon tips in culture. In other experiments with cultured stolons, ABA prevented tuber formation (Palmer and Smith, 1969b). More recently, however, Wareing (personal communication) has succeeded unequivocally in inducing tuber formation with ABA in isolated stem segments. In single node cuttings from induced plants of *S. andigena* with the leaf attached, the axillary bud rapidly forms a tuber. If the leaf is removed, however, and the segment is kept in the dark, the bud grows out as an elongated, etiolated shoot (Fig. 3.43a). If ABA is supplied in solution through the base of a leafless segment, the axillary bud develops as a tuber (Fig. 3.43b). When GA is similarly supplied, the bud develops as an elongated shoot with an apical hook (Fig. 3.43c). Kinetin causes some swelling of the basal internode (Fig. 3.43d) but does not lead to tuberization. Under these conditions, therefore, ABA can substitute for the effect of the SD-induced leaf and cause tuberization of the axillary bud.

Excised stolons of *S. tuberosum* grown in aseptic culture formed tubers if supplied with a cytokinin. Of those tested, kinetin was the most effective, inducing 80–100% tuberization. Controls did not form tubers (Fig. 3.44) (Palmer and Smith, 1969a, b). A certain period of continuous treatment was required. If stolons were maintained on medium with kinetin for up to 3 days, and then transferred to medium without it, they did not form tubers. Stolons transferred after 4 days or longer, however, showed increasing tuberization, although this did not become visible until after 10 to 12 days in culture. Kinetin-induced tuber formation could be completely inhibited by a temperature of 35 °C (Palmer and Smith, 1969b, 1970). By supplying kinetin-8-^{14}C to cultured stolons it was shown that the labelled material accumulated at the site of tuberization, although even more label was found at the base of the stolon, perhaps because of the proximity of the label in the medium. The extracted labelled material did not have the same R_f value as kinetin (Smith and Palmer, 1970).

Recent experiments with excised nodal segments of induced and non-induced potato plants, cv. Katahdrin, showed that the presence of kinetin in the culture medium could substitute for the stimulatory effect of induction on tuber formation in the axillary buds (Forsline and Langille, 1976).

The results of these experiments suggest that a natural cytokinin could be the tuber-forming stimulus. When applied to the basal end of cuttings, however, kinetin inhibited tuber development at high

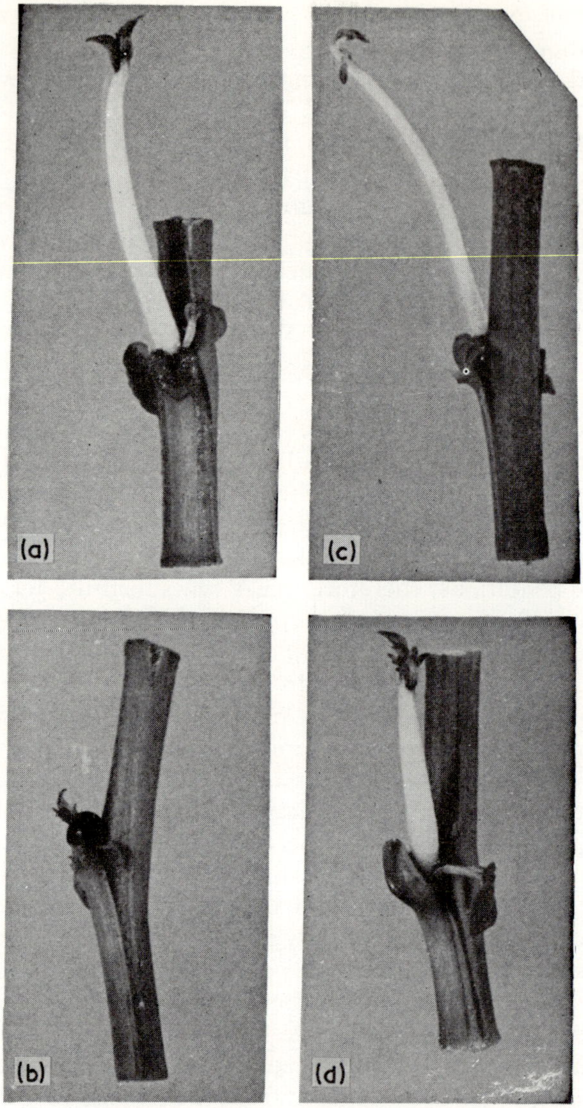

Fig. 3.43 Single node cuttings from induced plans of *Solanum andigena*, i.e. grown in short days, maintained in the dark with the leaf removed: (a) control; the bud grows out as an etiolated shoot; (b) treated with ABA; the bud develops as a tuber; (c) treated with GA; the bud develops as an elongated shoot with an apical hook; (d) treated with kinetin; the bud develops as a shoot with slight swelling of the basal node. (By courtesy of Professor P.F. Wareing)

concentrations although it slightly stimulated it at the lowest concentration used (Kumar and Wareing, 1974). Since the cuttings were from induced plants, however, these results need not be at variance with the

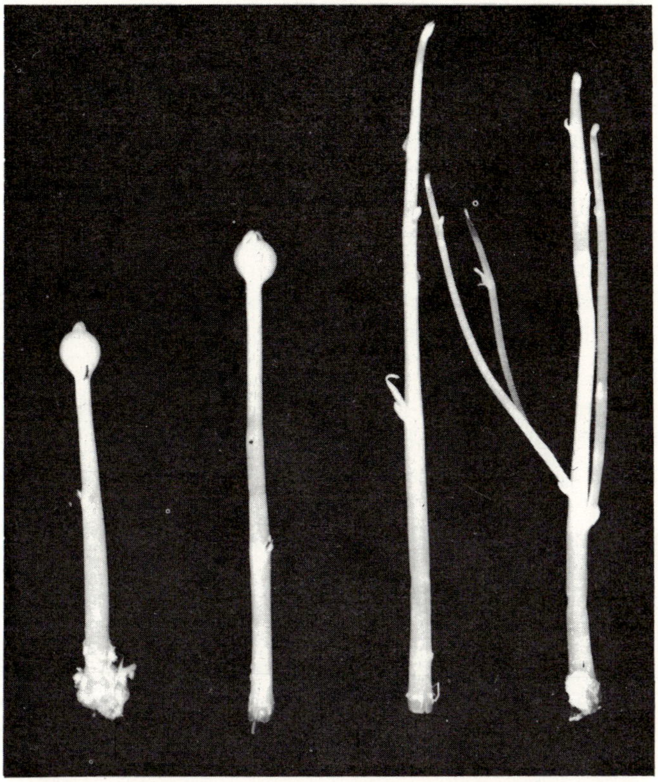

Fig. 3.44 Cultured stolons of cv. Norgold Russet. Right, grown in control medium; left, grown for 30 days in the presence of 2.5 μg ml $^{-1}$ kinetin. Tubers have developed behind the stolon tips. (Fig. 3 from Palmer and Smith, 1970)

postulated occurrence of an endogenous cytokinin capable of inducing tuber formation.

It is gradually becoming apparent that the site of hormone treatment may be an important factor affecting the result obtained. Together with the use of *S. andigena*, which requires SD for tuber formation, and *S. tuberosum*, which has no such absolute requirement, the widely differing treatments may account for the conflicting results that are sometimes obtained.

There is some evidence that ethylene, also, may play a role in tuber formation. When sprouts of cv. Arran Pilot bearing several stolons were treated with gaseous ethylene, extension growth was inhibited and swelling of all rapidly growing regions occurred. Sub-apical swellings on stolons were similar morphologically to normal tubers, but contained no starch (Catchpole and Hillman, 1969). Ethrel, a substance which releases ethylene, led to an increase in tuber formation when applied to the soil (Garcia-Torres and Gómez-Campo, 1972). In cultured stem segments, also, ethrel enhanced tuber formation and counteracted the effects of applied GA (Garcia-Torres and Gómez-Campo, 1973). On the other hand, cultured stem segments supplied with 2-chloroethylphosphonic acid, another substance which releases ethylene, formed stolons which were generally swollen but without localized tuberous swellings (Palmer and Barker, 1973). Thus ethylene may play some part in the process of stolon and tuber formation, but is probably not the sole stimulus involved. These observations also emphasize the need for accompanying anatomical work, to establish the degree to which the induced tuber is structurally normal.

Carbohydrate metabolism

A number of workers have concluded that sucrose is required for, but does not induce, tuber formation (Gregory, 1956; Madec and Perennec, 1959; Madec, 1963). In cultured stem segments, tubers were formed in the presence of 8%, but not 2%, sucrose (Harmey *et al.* 1966). In other experiments with cultured stolons, however, tubers did not develop in the presence of sucrose unless kinetin was also supplied. For the induction of tuber formation by kinetin, 6, 8 or 10% sucrose was required, 2 or 4% being inadequate (Palmer and Smith, 1970).

Tuber formation seems to be associated with a high concentration of soluble sugars in the stolon tip, and with conditions which lead to this (Slater, 1963, 1968; Burt, 1964). In cultured stolons treated with kinetin the level of reducing sugars decreased, as compared with controls or ethylene-treated stolons. This coincided with the rapid synthesis of starch in cultures with kinetin (Palmer and Barker, 1973).

Lovell and Booth (1967) considered that the first sign of starch deposition was a reliable indicator of tuber initiation. Starch accumulated in a localized sub-apical region of the stolon tip prior to any external sign of tuberization. Palmer and Smith (1969a) noted that starch accumulated in the apical region of kinetin-treated stolons, but not in control stolons. They considered that this provided some evidence that cytokinins may act by mobilizing metabolites to the site of tuber formation. It has already been noted that starch was absent from tuber-like structures induced by ethylene (Catchpole and Hillman, 1969). Smith and Palmer (1970) have suggested that kinetin may exert its stimulatory effects on tuber formation by promoting the activity of starch synthetase and suppressing that of starch hydrolase. Gibberellic acid, on the contrary, may have the converse effect on these enzymes, thus inhibiting tuber formation.

Analysis of endogenous hormones

Endogenous hormones present in the potato plant have been extracted and identified, and more recently the effects of various environmental

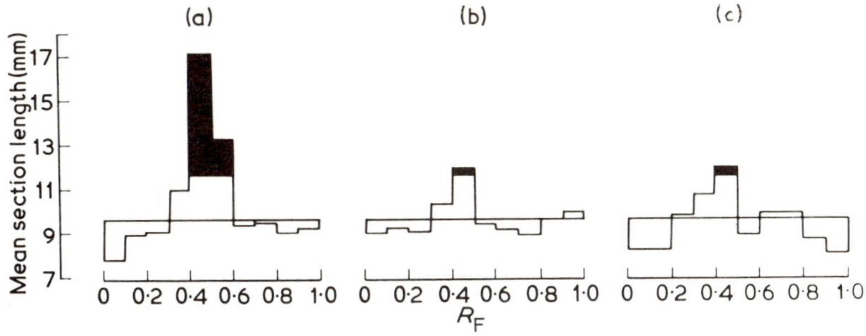

Fig. 3.45 Effects of photoperiod on gibberellin levels in plants of *Solanum andigena*: (a) long days; (b) short days; (c) 3 weeks under short days, followed by 12 long days. The assay was the pea epicotyl test. Shaded areas indicate significant differences from controls at the 1% level. (Fig. 1 from Kumar and Wareing, 1974)

conditions on the levels of endogenous hormones have been studied.

Indoleacetic acid was extracted from potato tubers and identified by chromatography (Booth and Wareing, 1958). As will be shown in Section 3.7.4, gibberellins have been extracted from potato tubers. Under storage conditions, potato tubers also produce ethylene (Creech *et al.* 1973). The inhibitor abscisic acid (ABA) has also been extracted from potato tubers (Cornforth *et al.* 1966).

Although it is considered that the tuber-forming stimulus is present in the tuber as well as in aerial parts of the plant, neither seedlings nor cuttings have attached tubers, yet both can form tubers. Thus there is no evidence that any of these substances isolated from tubers is the elusive tuber-inducing stimulus in question.

The hormones extracted from the leaves have chiefly been gibberellins. By means of thin layer chromatography, at least four zones or peaks of gibberellin-like activity were found in leaves of *Solanum andigena* (Railton and Wareing, 1973a). Levels of gibberellin were significantly lower in SD than in LD plants. Plants transferred from SD to LD showed levels of gibberellin similar to those in plants kept under SD throughout (Fig. 3.45) (Kumar and Wareing, 1974). When plants which had had 5 SD cycles were irradiated with red light for 30 min during the long dark period of the sixth SD cycle, there was an increase in the content of gibberellins so that levels approached those of LD plants, apparently because of an increased rate of synthesis. There were also changes in the relative amounts of the four peaks (Railton and Wareing, 1973c).

In further experiments with *S. andigena*, tritiated gibberellin A_1 was applied to leaves of LD plants; half of the plants were then kept under LD, half under SD, and gibberellins were extracted periodically. Conversion of the labelled gibberellin to a more polar acidic compound occurred under both photoperiodic régimes (Railton and Wareing, 1973b). The gibberellin-like activity of the leaves seemed to be always associated with extracts of isolated chloroplasts (Railton and Wareing, 1973a).

Although the occurrence of higher levels of gibberellins in LD plants is compatible with the failure of such plants to form tubers, in view of the known inhibitory effects of GA on tuber formation, it must be recognized once more that a difficulty exists in attributing tuber formation to a negative condition such as the lower concentration of an inhibitor. As Kumar and Wareing (1974) have pointed out, low levels of gibberellin in SD scions are unlikely to lead to reduction of gibberellins in LD stocks, yet in SD/LD grafts these form tubers. This positive stimulus to tuber formation as yet remains as elusive as the flowering hormone; further work should be devoted to its elucidation. Recent work has shown, however, that the levels of two out of four cytokinins extracted from potato plants were significantly higher in induced than in non-induced plants. Above-ground parts of induced plants contained approximately 10 times the amount of one such substance than was present in non-induced plants after 2 days. Both substances reached a maximum in induced plants after 4 days in above-ground parts, and after 6 days in below-ground parts, tuber initiation occurring after 8–10 days (Forsline and Langille, 1975). Cytokinins may therefore be one necessary part of the positive stimulus to tuber formation, as indeed experiments in culture suggest.

3.7 Dormancy and its control

Growth of the buds on potato tubers is subject to a number of rather different controls. During the development of the tuber there is no true bud dormancy, but correlative inhibition is operative. After harvest other mechanisms supervene (Madec and Perennec, 1969). One of these is classed as true dormancy, that is, a state in which the tuber will not sprout if stored at a temperature below the optimum. According to the variety, the period of dormancy may be from 18 to 33 weeks, usually 20–23 weeks. However, immediately after harvest potato tubers cannot be induced to sprout even under optimum environmental conditions. This is usually termed the rest period, a somewhat unsatisfactory term, and is of 5 to 19 weeks duration in different varieties (Emilsson, 1949). It is not always clear whether authors are referring to rest or true dormancy, as thus defined. In the so-called rest period buds fail to grow even under optimum temperature and light conditions, and there appears to be an internal 'clock' which triggers cellular events leading to sprouting (Rappaport and Wolf, 1968a, 1969). Tuan and Bonner (1964) consider that while the genome of dormant buds is largely repressed, it is not yet established that this is the primary cause of dormancy. Others believe that the tuber may be temporarily unable to supply the buds with metabolites essential for growth (Madec and Perennec, 1969).

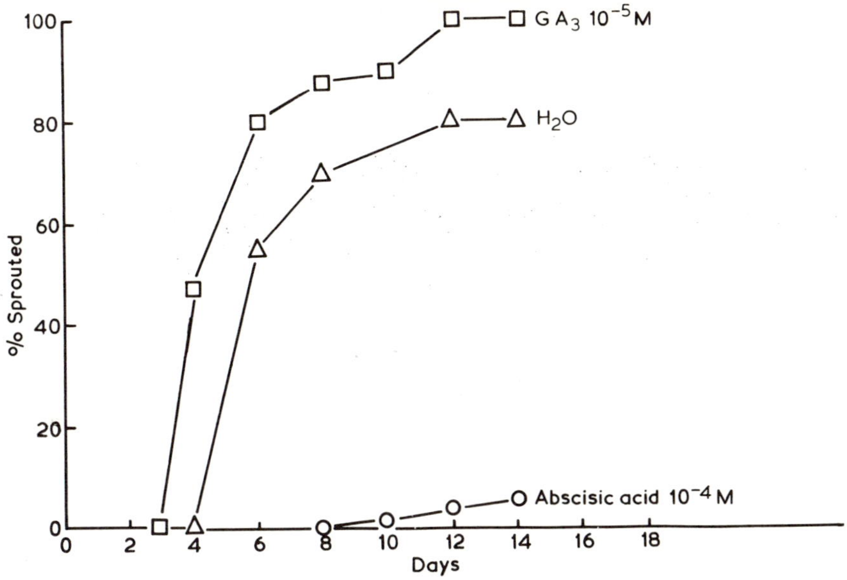

Fig. 3.46 Effect of abscisic acid and GA on sprouting of excised potato buds. Solutions were applied daily to the buds. (After Madison and Rappaport, 1968)

Treatment of excised plugs of tuber tissue with ABA inhibits sprouting, whereas GA is slightly stimulatory (Fig. 3.46). Indeed, it is considered that dormancy and sprouting of the potato tuber are probably controlled by a balance of inhibiting and promoting substances (Bruinsma and Swart, 1970); it seems likely that much more work will have to be devoted to the elucidation of the mechanisms involved.

3.7.1 Changes occurring at sprouting

When the rest period is terminated and the buds of the tuber resume growth, various changes occur both in the bud meristem itself and in the tuber.

Effects on the tuber

When the buds start to grow, starch is broken down in the tuber. This ceases if the buds are removed. Thus if a hormone is involved the tuber must require a continuous supply of it to maintain starch breakdown (Edelman *et al.* 1969). These workers suggest that such a hormone might be required for synthesis of an enzyme, and that the rate of its synthesis might control further sprout growth. In view of the involvement of gibberellin in the control of sprouting, it seems worth considering whether interaction between gibberellin and α-amylase synthesis may be implicated, as in the case of barley endosperm. Studies along similar lines to those on barley could be undertaken with the potato.

During dormancy crystals are formed in peripheral cells of tubers.

It is thought that they may be formed within vesicles of the endoplasmic reticulum. The crystal-containing bodies may be sites of protein storage; they disappear after sprouting (Marinos, 1965).

Changes in the bud meristems

Changes in activity of the bud meristems when the rest period is terminated have been studied principally by labelling with radioactive isotopes, mainly the precursors of RNA and DNA. Usually ^{3}H-uridine or ^{3}H-thymidine was applied to apices of buds on plugs of tuber tissue, and in some experiments GA was also applied. Buds were subsequently sectioned and histoautoradiographs made. Using this method, RNA synthesis was detected in the smaller plugs (0.2 × 0.2 cm) 6 h after excision, and a low level of DNA synthesis 12 h after excision. On larger plugs, RNA synthesis was first detected at 12 h and DNA synthesis at 24 h after excision. Buds treated with GA showed more labelling in the first 12 h (Fig. 3.47); DNA synthesis was accelerated in the period from 6 to 12 h after excision (Rappaport and Wolf, 1968b, 1969). In all these experiments synthesis of RNA preceded that of DNA (Rappaport and Wolf, 1968a, 1969), and both preceded cell expansion and cell division in the buds. Rappaport and Wolf (1969) consider that since RNA synthesis increased after excision of the plugs whether GA treatment was given or not, it is doubtful whether gibberellin plays a primary role in the termination of rest. However, gibberellin may be synthesized in the cells along the cut surface of tuber tissue (Rappaport and Wolf, 1968b).

Treatment of buds with ABA considerably reduced the incorporation of labelled uridine and thymidine. If ABA and GA were applied together, ABA overcame the ability of GA to promote synthesis of RNA and DNA. This is thought to indicate that ABA has a direct effect on the synthesis of nucleic acids, rather than an indirect one such as blocking synthesis of gibberellins (Shih and Rappaport, 1970).

Mitosis apparently began in the procambium of labelled buds and progressed into the apex (Shih and Rappaport, 1970).

Treatment of tubers with ethylene chlorhydrin led to incorporation of label in the buds (Tuan and Bonner, 1964; Rappaport and Wolf, 1968a). The rate of RNA synthesis doubled within 2 days after the end of treatment, and by 10 days had risen to 130 times the level in dormant buds. Rate of DNA synthesis was similar but slower (Tuan and Bonner, 1964).

Tuan and Bonner extracted chromatin from tubers and from dormant buds or those which had been given ethylene chlorhydrin treatment. Chromatin from tubers was quite ineffective in supporting DNA-dependent RNA synthesis, and that from dormant buds was inactive,

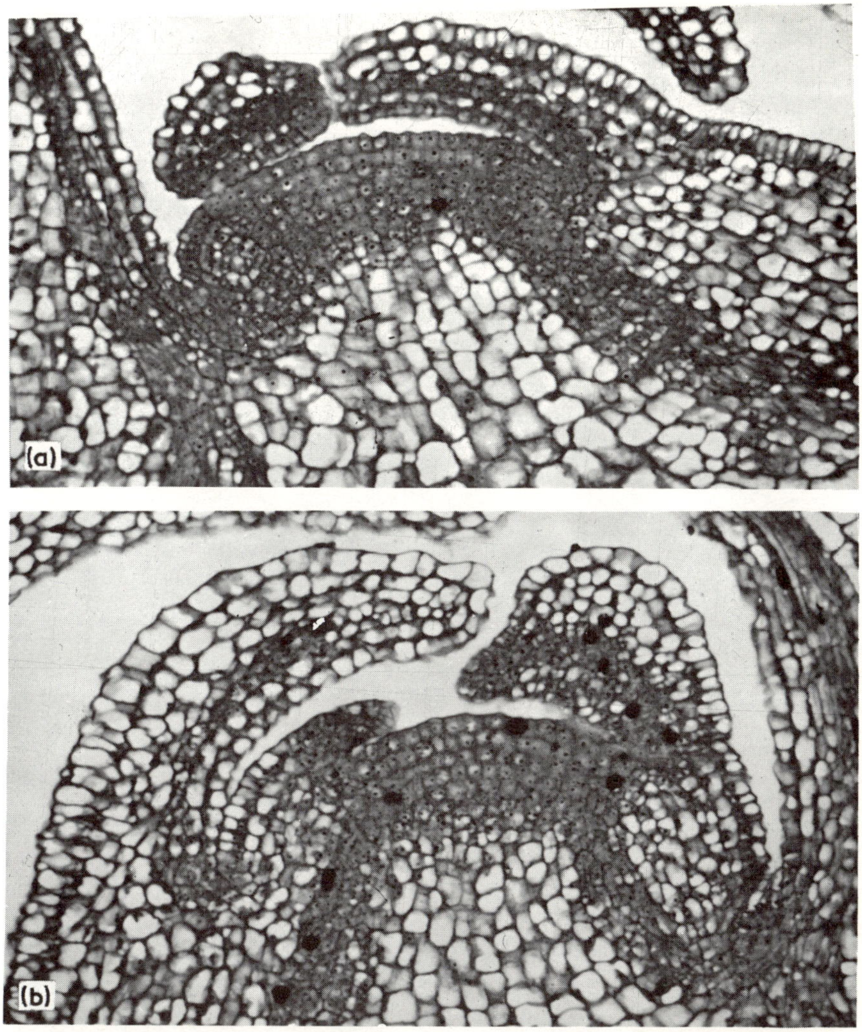

Fig. 3.47 Autoradiographs of L.S. shoot apex after being supplied with 3-thymidine for 12 h: (a) water control; (b) treated with GA. (Plate 5 from Rappaport and Wolf, 1969)

whereas chromatin from buds after treatment was 10-fold more effective. The RNA which is synthesized in buds after treatment with ethylene chlorhydrin is DNA-dependent, since actinomycin-D inhibits its synthesis *in vivo*. Tuan and Bonner concluded that the genome of dormant potato buds is largely repressed.

It is thus rather surprising that RNA synthesis in activated buds always seems to precede DNA synthesis.

Marinos (1967) has studied the fine structure of dormant potato

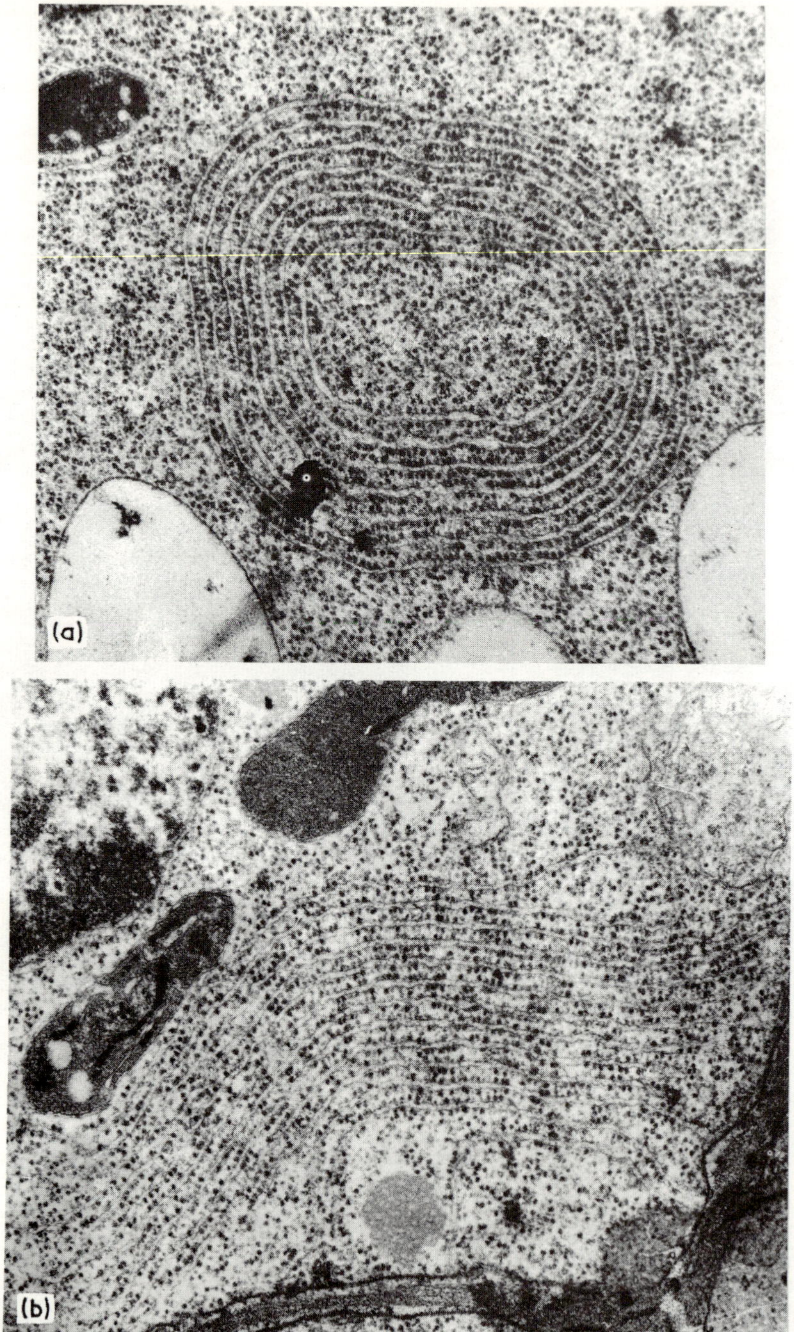
(a)
(b)

apices, cv. Sebago. Plastids contain osmiophilic droplets and starch grains. Intraplastid bodies, believed to be sites of accumulation of nucleic acids and protein, are prominent in the apical dome and young leaf primordia of dormant buds, but tend to disappear when sprouting begins, along with phytoferritin granules. In resting buds of cv. White Rose potatoes, striking concentric configurations of endoplasmic reticulum (ER) were observed (Shih and Rappaport, 1971). These consisted of 7 to 13 ER lamellae with associated ribosomes (Fig. 3.48a). Such configurations were never observed in cells of buds on plugs which had been excised for more than 6 h. They tended to open out and by 12 h after excision only typical ER was observed. Thus this opening out seems to coincide with the acceleration of RNA synthesis. Treatment of buds with ABA or GA stimulated the opening and extension of these concentric structures (Fig. 3.48b).

Treatment with ABA led to increased vacuolation (Fig. 3.49c), first noticeable about 1 h after treatment, whereas GA-treated buds were not strikingly different in ultrastructure from controls (Fig. 3.49a, b) (Shih and Rappaport, 1971). Rappaport (1972) suggested that ABA might therefore have some active role, rather than the passive one of merely keeping cells quiescent. It is also thought that this effect of ABA may explain the failure of GA completely to reverse its inhibitory action (Shih and Rappaport, 1971).

3.7.2 Endogenous substances in tubers

Gibberellins and gibberellin-like substances have been extracted from potato peelings. Using the dwarf maize bioassay, Smith and Rappaport (1961) showed that the level of endogenous gibberellins remained low during the rest period and increased towards the end of this time. Boo (1962), on the other hand, found an increase in endogenous gibberellin-like substances during the last part of the rest period. His results have been questioned on the grounds that with the aim of removing inhibitors he extracted with chloroform, and may have discarded some of the gibberellins present along with his chloroform extract (Rappaport *et al.* 1965).

In other analyses, gibberellin-like substances were detected in the neutral fraction obtained from potato buds and peelings; these substances stimulated elongation of dwarf peas but were inactive on dwarf maize (Hayashi *et al.* 1962; Hayashi and Rappaport, 1966).

Fig. 3.48 (a) Concentric configuration of endoplasmic reticulum with ribosomes in a cell of a dormant potato bud. x 34 000; (b) open configuration of endoplasmic reticulum from a cell of a bud after 1 h of GA treatment. x 35 000. (Figs. 2b and 3a from Shih and Rappaport, 1971)

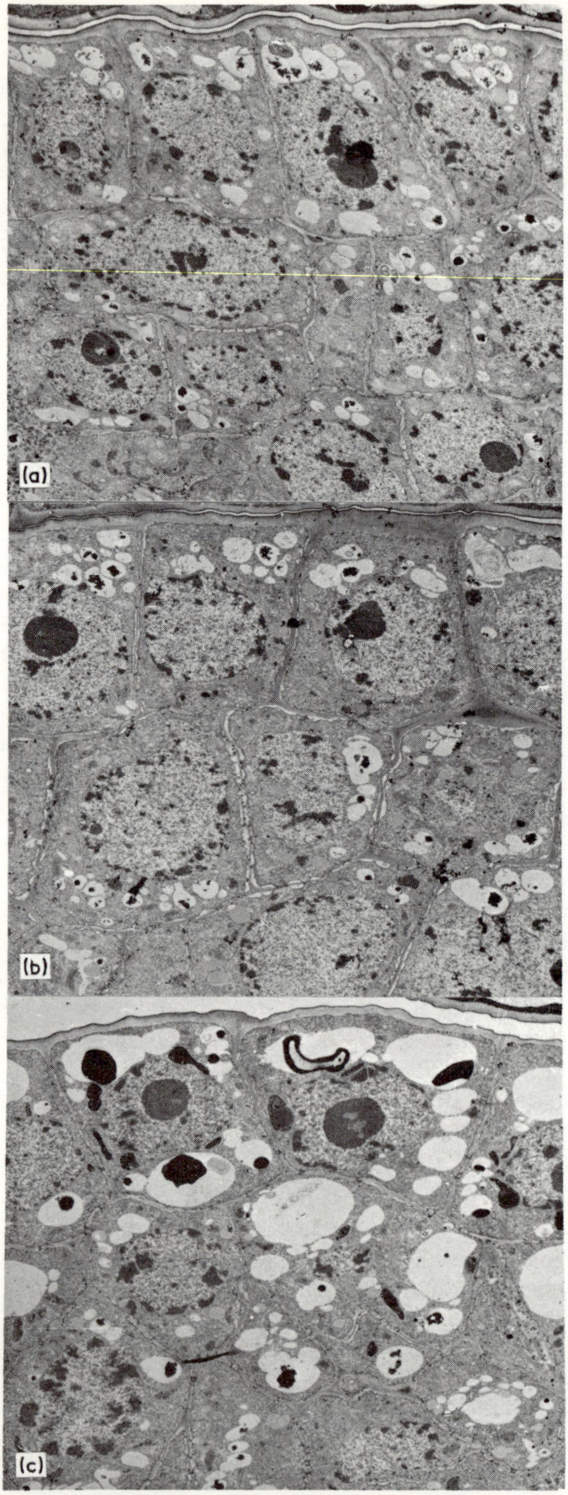
(a)
(b)
(c)

One of the most interesting findings was that gibberellin-like substances increased markedly following wounding of the tuber tissues. It was considered that the amount of such substances produced by an excised plug of tuber tissue would be sufficient to trigger changes in the buds associated with termination of dormancy (Rappaport and Sachs, 1967). Thus this may account for the effects of wounding on bud growth.

3.8 Development and structure of the flower

Danert (1957) has described the cymose inflorescence of the potato. The vegetative shoot is a sympodium; each portion terminates in an inflorescence, vegetative growth being continued by the bud in the axil of the last true foliage leaf. This shoot appears to be laterally displaced. Danert interprets the last leaf of the main axis as a bract subtending the reproductive shoot, which is concaulescent with the parent axis. A second part of the inflorescence is formed below the terminal flower without a subtending bract. In this way, helicoid monochasia are formed below the terminal flower. The flowers are actinomorphic and hypogynous (Artschwager, 1918; Jones, 1939). According to Young (1923) the calyx originates as a marginal ring with five lobes, but Sattler (1973) reports for the flower of *Solanum dulcamara* that the sepals are initiated separately in spiral sequence and the calyx tube is later formed by growth of the tissue between these primordia; the author found the latter to be the case also in *S. tuberosum* cv. King Edward. At maturity the calyx is green and hairy and consists of five joined sepals. The five petal primordia form in close sequence and the corolla tube develops later than the calyx tube. The five stamens alternate with the petals and are borne on the corolla tube. The anthers fuse postgenitally, enclosing the pistil (Sattler, 1973). At maturity the stamens have short, stout filaments and long anthers. Pollen is shed through pores at the tips of the anthers (Jones, 1939).

There are usually two carpels, the primordia of which are formed simultaneously. These fuse to form a syncarpous, bilocular, superior ovary with a long style and a bilobed stigma. The mature fruit is a green berry with axile placentation, which often fails to develop in cultivated potatoes. The carpels are orientated obliquely to the median plane of the flower (Fig. 3.50) (Jones, 1939).

The sepals and petals each have from three to five vascular strands, the median strand being the largest (Fig. 3.51) (Artschwager, 1918; Hayward, 1938). Rather large cells containing crystal sand (numerous small crystals of calcium oxalate) are common in the calyx and carpels (Young, 1923). Marginal meristems are evident in the developing petal primordia (Fig. 3.51).

Fig. 3.49 Electron micrographs of part of the apical dome of potato buds 12 h after they were excised and treated with (a) water; (b) GA; and (c) ABA. Large vacuoles are present in cells treated with ABA. x 3 800. (Fig. 1 from Shih and Rappaport, 1971)

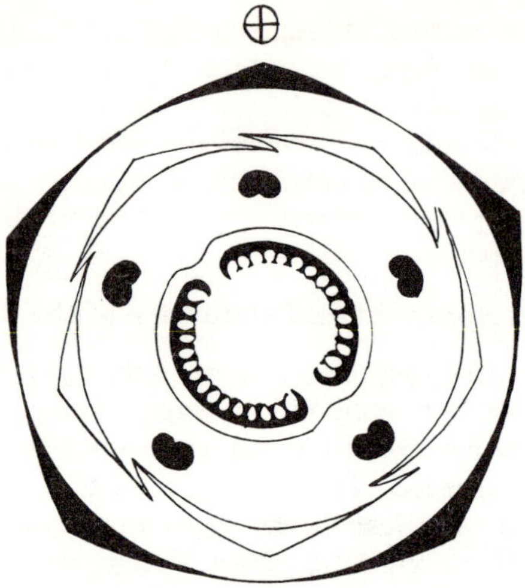

Fig. 3.50 Floral diagram of potato. (Fig. 56F from Jones, 1939)

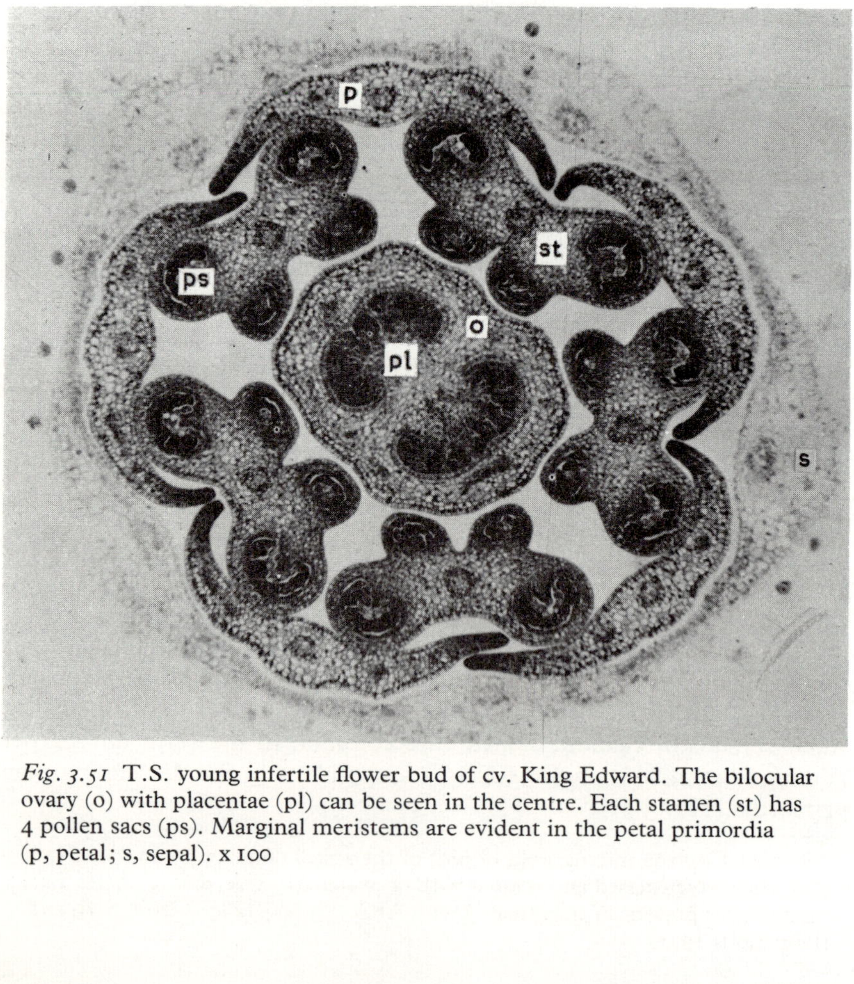

Fig. 3.51 T.S. young infertile flower bud of cv. King Edward. The bilocular ovary (o) with placentae (pl) can be seen in the centre. Each stamen (st) has 4 pollen sacs (ps). Marginal meristems are evident in the petal primordia (p, petal; s, sepal). x 100

The stamens have a single vascular strand. In the developing anthers the sporogenous tissue forms a mass of densely staining cells, horseshoe shaped in cross-section (Fig. 3.51). The parietal layers are not modified to aid in dehiscence, which takes place through terminal pores. The tapetal cells are large (Young, 1923; Hayward, 1938). A glandular tapetum has been reported (Davis, 1966). Meiosis in the anther takes place before that in the ovule (Clarke, 1940).

Pollen sterility occurs quite frequently, and depends partially on environmental factors. One cause is apparently abnormal division of the pollen mother cells resulting from higher temperatures. Under such conditions abortive pollen grains were formed (Stow, 1927). The proportion of abnormalities in meiosis may also be increased by virus and fungus diseases (Khan, 1951).

Numerous ovules are formed, covering the placenta. A single archesporial cell develops in the hypodermis, distinguishable by its dense contents and deep staining (Young, 1922, 1923; Rees-Leonard, 1935). The archesporial cell develops directly as a megaspore mother cell. Of the four megaspores formed by meiosis, three degenerate, the megaspore at the chalazal end remaining functional (Rees-Leonard, 1935; Hayward, 1938). The megaspore increases in length and undergoes nuclear division, the nuclei moving to opposite ends of the embryo sac, which grows rapidly and becomes curved. These two nuclei divide simultaneously, then the four resulting nuclei divide again, producing four nuclei at each end of the embryo sac. Walls are formed and a seven-celled megagametophyte results. The antipodal cells are usually ephemeral (Lamm, 1937; Davis, 1966), and degenerate early, before fusion of the polar nuclei. The synergids elongate and develop the filiform apparatus. The polar nuclei fuse before fertilization. These events seem to be similar in at least three varieties, Irish Cobbler, Earlaine and Khatadrin (Rees-Leonard, 1935; Clarke, 1940; Williams, 1955). Ovules with more than one embryo sac have occasionally been reported (Young, 1922; Lamm, 1937).

Fertilization has been observed 36 h (Clarke, 1940) or 40–45 h (Williams, 1955) after pollination, doubtless depending on environmental conditions. The usual type of double fertilization takes place. The endosperm nucleus divides 60–70 h after pollination; an early free-nucleate stage has not been observed (Clarke, 1940; Williams, 1955).

Various degenerative changes may occur at different stages during ovule development; it seems likely that this may account in part for the poor capacity for seed-setting of many commercial varieties, but more work is required (Rees-Leonard, 1935; Clarke, 1940). Evident degenerative changes in the developing flower buds, sometimes described as 'blasting', include cessation of growth, development of a yellowish colour, and wilting. Both the embryo sac and surrounding nucellus may shrivel, and degeneration of the mature megagametophyte often occurs (Young, 1923; Rees-Leonard, 1935). Degenerative changes in the embryo sac and ovule seem to result from unfavourable environmental conditions, and are much more uniform than those in the anther (Young, 1923).

Abscission of the flower often occurs before fruits and seeds become mature. An abscission layer is formed at a region where there is a joint in the pedicel. Weinheimer and Woodbury (1967) have described the abscission layer in cv. Russet Burbank and Menominee. They found that the abscission zone was not well defined. The protective layer developed when the buds were still small, and consisted of a large mass of small meristematic cells. A separation layer could not be discerned, but at the time of abscission whole areas of cells disintegrated and collapsed. The abscission zone can be located externally by a hairy swelling, due to cell enlargement, just below an indentation in the pedicel. This was evident at an earlier stage of development of the flowers of cv. Russet Burbank, and Weinheimer and Woodbury (1967) considered that it might be associated with their premature abscission. These workers thought that abscission in Russet Burbank might be by breakdown of the cells in the separation layer of the middle lamella only, whereas in Menominee there was a complete dissolution of the cells in this region. When bud blasting occurs the upper part of the abscission layer splits, and callus is formed over the broken surface (Hayward, 1938).

3.9 Embryogenesis

Embryo development is of the Solanad type, in which the terminal cell of the two-celled proembryo divides by a transverse wall (Fig. 3.52) (Maheshwari, 1950; Wardlaw, 1955: Davis, 1966).

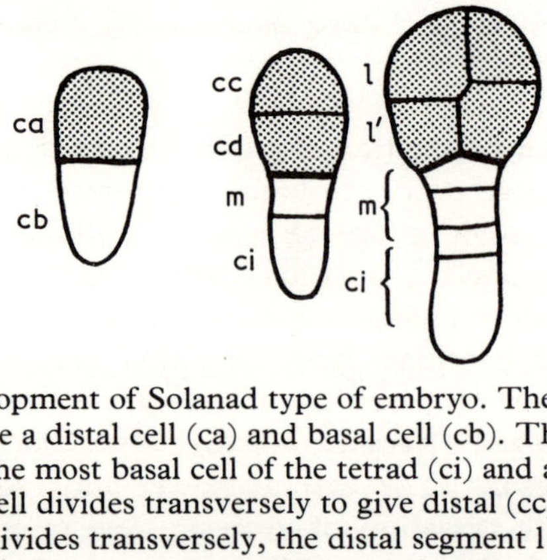

Fig. 3.52 Early development of Solanad type of embryo. The zygote divides by a transverse wall to give a distal cell (ca) and basal cell (cb). The latter divides transversely to give the most basal cell of the tetrad (ci) and an intermediary cell (m). The distal cell divides transversely to give distal (cc) and subdistal (cd) segments. When cc divides transversely, the distal segment l and adjacent segment l′ are formed. (Figs. 55A, F and G from Wardlaw, 1955)

The first division of the zygote is transverse and occurs 4 or 5 days after pollination (Clarke, 1940; Williams, 1955). By further division a 4-celled linear proembryo is formed about 7 days after pollination (Hayward, 1938; Clarke, 1940; Williams, 1955). The basal cell plays a relatively minor role in subsequent development, but divides to form a short suspensor of two or more cells (Hayward, 1938; Maheshwari, 1950; Wardlaw, 1955; Williams, 1955). There is no hypophysis (Souèges, 1922).

The location and planes of subsequent cell division are highly variable in Solanaceous embryos (Johansen, 1950). Souèges (1922) considers that the terminal cell of the proembryo gives rise to the cotyledonary part of the embryo, the next cell to the hypocotyl and initials of the central cylinder, and so on, but, in agreement with more recent views on embryology in general, Bhadurim (1936) found that definite parts of the embryo were not always formed from a particular cell of the 4-celled proembryo.

Whatever the precise sequence of cell divisions, a club-shaped embryo is formed by 10 days, and a spherical embryo by 12 days, after pollination (Williams, 1955).

In cv. Chippewa, seed failure is characterized by enlargement and proliferation of the tissue surrounding the embryo sac; these cells spread through the area normally occupied by the developing embryo. Placental tissue also proliferates, increasing the diameter of the ovary, which develops into a parthenocarpic fruit (Williams, 1955).

3.9.1 Induction of embryoids

Considerable work has been devoted in recent years to the experimental induction of embryo-like structures, often called embryoids, from developing pollen grains in cultured anthers. These embryoids can develop into plantlets and eventually mature plants. The first successful reports related to pollen of *Datura* (Guha and Maheshwari, 1964, 1966) and *Nicotiana* (Nitsch and Nitsch, 1969), both members of the Solanaceae. It might therefore be expected that pollen of the cultivated potato would be equally amenable to this treatment, but unfortunately this has not proved to be the case. Since haploid embryoids capable of developing into mature plants can be obtained by this technique, it is potentially an important one for the plant breeder.

Work with *Solanum* to date relates to various species, but has not always resulted in the formation of haploid plantlets. Harn (1972) obtained callus from cultured anthers of *S. nigrum*, from which plantlets developed. Of 98 plants which developed, about half were haploid, and the remainder diploid, triploid or aneuploid. There was some evidence of divisions in the microspores (pollen grains), so the callus may have been of pollen origin. In experiments of this kind, callus often develops from the

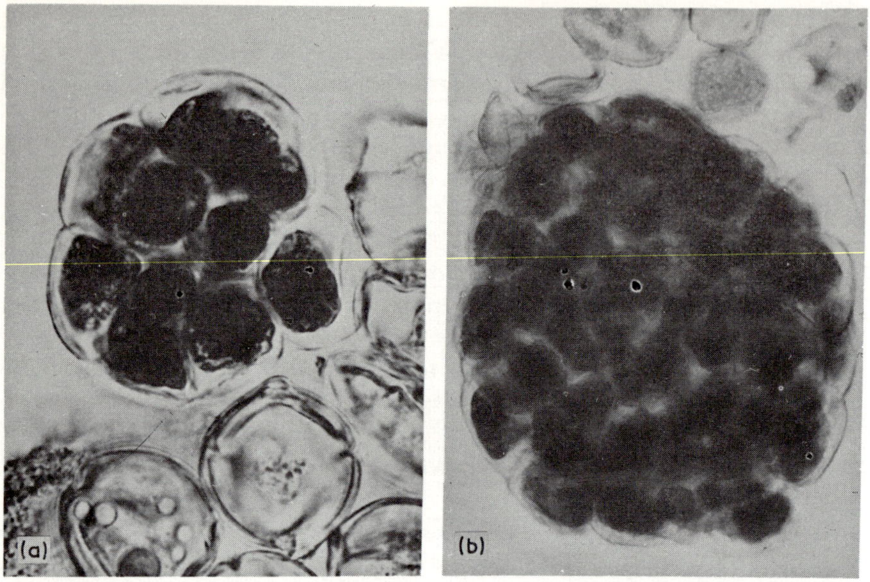

Fig. 3.53 Pollen embryoids of (a) cv. Maris Piper; (b) Pentland Crown. Considerable cell division has taken place. Degenerating pollen grains can also be seen. (Fig. 1 from Dunwell and Sunderland, 1973)

connective of the anther, which is diploid somatic tissue. Irikura and Sakaguchi (1972) also obtained callus which proliferated from inside the locules of anthers of the wild tuberous species *S. verrucosum* when cultured on media with added hormones. In 4 out of 290 anthers of *S. verrucosum* small plantlets developed, and 2 haploid plants with 12 chromosomes were obtained. Anthers of *S. tuberosum* gave rise only to callus. In other experiments with *S. tuberosum*, cv. Hansa, several embryoids were obtained from callus originating from one anther (Kohlenbach and Geier, 1972). However, the callus had proliferated from connective tissue.

Dunwell and Sunderland (1973) have successfully induced embryoid formation in pollen of *S. tuberosum*, cv. Pentland Crown, Record and Maris Piper, by culturing anthers at or just prior to the stage of the first mitosis in the pollen grains. The success rate was not high, and in most of these experiments embryoids did not progress beyond the multicellular stage (Fig. 3.53), but in one anther of cv. Pentland Crown a plantlet, and in three others roots, emerged after about 3 weeks of culture. These were haploid. Callus formation took place in all three varieties, mostly from somatic tissues. In potato, an average of less than five pollen embryoids were produced per anther, whereas in tobacco they may be numbered in thousands (Dunwell and Sunderland, 1973).

In cv. Pentland Crown, tetraploid plantlets were derived from anther

Fig. 3.54 (a) Dihaploid plantlet derived from anther culture of cv. Pentland Crown. The plantlet was grown to maturity, when it was found to be tetraploid. (b) Tetraploid plantlet of cv. Pentland Crown. Note small developing tubers. (By courtesy of Dr N. Sunderland.)

culture (Fig. 3.54). The precise origin of these is not known; they could either have originated from pollen grains or from the anther wall. The potato has a high proportion of diploid pollen with 48 chromosomes instead of the normal 24 (Sunderland, personal communication).

Nitsch (1969) showed that in *Nicotiana* the stage of development of the flower from which the anthers were excised was important. With potato, Dunwell and Sunderland (1973) found that the stage of development was far more critical than the composition of the medium. With greater attention to this aspect, therefore, further work on anther culture should provide a source of haploid and dihaploid plants for potato breeders.

3.10 Conclusions

It is clear that the control of growth and development in the potato plant, as in others, depends on correlative effects between one part of the plant and another. As yet, too little is known about the structural relationships of these various parts, and the possible channels of transport between them. For example, little is known about the vascular connections between particular leaves and individual tubers. More work in this field is likely to prove helpful.

In the foregoing account of the structure and development of the potato plant, particular attention has been devoted to the stolon and tuber, having regard to the commercial importance of the latter. Much work has been devoted to the factors which control stolon and tuber formation and development. Although some of this work seems far removed from actual agricultural application – for example, the use of magnetic fields or of extremely low temperatures, or the aseptic culture of stolon tips – it is clearly valid to seek basic knowledge by any means at hand. Ultimately, however, techniques must be devised for enhancing tuber formation, growth and development which can actually be applied in the field. The most promising current line of investigation seems to be the application of various hormonal substances, together with a study of extracted endogenous hormones. It seems clear that gibberellins are closely involved in the control of tuber formation. Paradoxically, gibberellins stimulate the development of stolons – the usual site of subsequent tuber formation – but inhibit tuber formation itself. Tuber formation seems to be associated with a low level of endogenous gibberellins, but the results of certain grafting experiments make it difficult to accept that this condition alone is sufficient to lead to tuberization. More probably the balance between gibberellins and some other substance(s) is important. For example, Hammes and Nel (1975) have suggested that a balance between endogenous gibberellins and inhibitors may determine the active gibberellin level in the plant, and hence affect tuberization. As already pointed out, both cytokinins and inhibitors of extension growth such as ethylene and abscisic acid are also somehow implicated.

In the normal development the formation of a tuber just behind the stolon tip ensures that it is subtended by a stem that possesses a considerable complement of phloem, which plays a fundamental role in the ensuing translocation of food materials into the tuber. Although tubers can be induced to develop in other locations, such as from axillary buds on the leafy shoot, by certain environmental conditions or by application of hormones, it may be questioned whether the development of such tubers would ultimately be limited by lack of phloem. Thus it may be quite important to assess the anatomical aspects of responses to hormone treatments; there is certainly much scope for more work in this field.

Acknowledgements

I am greatly indebted to Dr Subha Basu for making some of the preparations on which some of the original illustrations are based, to Mr G.

Grange for taking many of the photographs, and to a number of people for supplying copies of previously published photographs. The source of these illustrations is acknowledged in the captions. I am also grateful to Miss Mary Gregory, of the Jodrell Laboratory, Kew, for a number of references to recent anatomical literature, and to Dr S. Carnegie, Department of Agriculture and Fisheries for Scotland, for supplying some material with developing tubers.

References

Ahmad, K.J. (1964). *Lloydia* **27**, 243–50.
Appiano, A. and Pennazio, S. (1972). *J. gen. Vir.* **14**, 273–6.
Artschwager, E.F. (1918). *J. agric. Res.* **14**, 221–52.
Artschwager, E.F. (1924). *J. agric. Res.* **27**, 809–35.
Artschwager, E.F. (1927). *J. agric. Res.* **35**, 995–1000.
Asseyeva, T. (1927). *J. Genet.* **19**, 1–26.
Asseyeva, T. (1933). *Trudy prikl. Bot. Genet. selek.* **27**, 135–218.
Badenhuizen, N.P. and Dutton, R.W. (1956). *Protoplasma* **47**, 156–63.
Badenhuizen, N.P. and Salema, R. (1968). *Revta Biologia, Rio de Janeiro* **6**, 139–55.
Bancher, E. and Hölzl, J. (1959). *Protoplasma* **50**, 356–69.
Banks, W., Greenwood, C.T. and Muir, D.D. (1973). *Stärke*, **25**, 331–5.
Barker, W.G. (1953). *Science* **118**, 384–5.
Barrios, E.P., Newsom, D.W. and Miller, J.C. (1961). *Proc. Am. Soc. hort. Sci.* **78**, 413–20.
Bhadurim, P. N. (1936). *Bot. Gaz.* **98**, 283–95.
Blanc, A. (1973). *C.r. hebd. Séanc. Acad. Sci.* **276**, sér. D, 2665–7.
Bonnemain, J-L. (1970). *Revue gén. Bot.* **77**, 5–51.
Boo, L. (1962). *Svensk bot. Tidskr.* **56**, 193–6.
Booth, A. (1959). *J. Linn. Soc. (Bot)* **56**, 166–9.
Booth, A. (1963). In *The Growth of the Potato*, ed. Ivins, J.D. and Milthorpe, F.L. pp. 99–113. London: Butterworth.
Booth, A. and Wareing, P.F. (1958). *Nature* **182**, 406.
Bruinsma, J. and Swart, J. (1970). *Potato Res.* **13**, 29–40.
Burt, R.L. (1964). *Eur. Potato J.* **7**, 197–208.
Catchpole, A.H. and Hillman, J. (1969). *Nature* **223**, 1387.
Chapman, H.W. (1958). *Physiologia Pl.* **11**, 215–24.
Chirilei, H., Chirilă, C. and Curticăpeanu, G. (1963). *Lucr. Grăd. bot. Buc.* 1961–2 (2), 651–65.
Clark, C.F. (1921). *U.S. Dept. Agric. Bull.* **958** (27 pp.)
Clarke, A.E. (1940). *Am. Potato J.* **17**, 20–5.
Claver, F.K. (1967). *Revta Fac. Agron. Univ. nac. La Plocta* **43**, 45–54.
Cornforth, J.W., Milborrow, B.V. and Ryback, G. (1966). *Nature* **210**, 627–8.
Crafts, A.S. (1933). *Pl. Physiol.* 8, 81–104.
Crafts, A.S. and Crisp, C.E. (1971). *Phloem Transport in Plants.* San Francisco: Freeman.
Crane, M.B. (1936). *J. Genet.* **32**, 73–7.
Creech, D.L., Workman, M. and Harrison, M.D. (1973). *Am. Potato J.* **50**, 145–50.
Cutter, E.G. (1969). *Plant Anatomy: Experiment and Interpretation. Part 1. Cells and Tissues.* London: Edward Arnold.

Danert, S. (1957). *Züchter* **27**, 22–3.
Danowitsch, N.D. and Krinkin, F. Ch. (1929). Abstract in *Bot. Zbl.* **19**, 452 (1931).
Davis, G.L. (1966). *Systematic Embryology of the Angiosperms*. New York: Wiley.
Digby, J. and Dyson, P.W. (1973). *Potato Res.* **16**, 159–67.
Dixon, H.H. (1922a). *Nature* **110**, 547–51.
Dixon, H.H. (1922b). *Report Br. Assoc. Adv. Sci.* (Hull) 1922, pp. 192–203.
Dunwell, J.M. and Sunderland, N. (1973). *Euphytica* **22**, 317–23.
Dyson, P.W. (1965). *J. Sci. Fd Agric.* **16**, 542–9.
Dyson, P.W. and Humphries, E.C. (1966). *Ann. appl. Biol.* **58**, 171–82.
Edelman, J., Jefford, T.G. and Singh, S.P. (1969). *Planta* **84**, 48–56.
El-Antably, H.M.M., Wareing, P.F. and Hillman, J. (1967). *Planta* **73**, 74–90.
Emilsson, B. (1949). *Acta agric. suec.* **3**, 189–282.
Ernst, J.M., Perry, B.A. and Blackhurst, H.T. (1968). *J. Rio Grande Vall. hort. Soc.* **22**, 53–7.
Esau, K. (1938). *Hilgardia* **11**, 343–424.
Esmarch-Bromberg, F. (1919). *Landw. Jbr* **54**, 161–266.
Fischnich, O. and Heilinger, F. (1959). *Angew. Bot.* **33**, 49–70.
Fischnich, O. and Krug, H. (1963). In *The Growth of the Potato*, ed. Ivins, J.D. and Milthorpe, F.L. pp. 72–8. London: Butterworth.
Forsline, P.L. and Langille, A.R. (1975). *Physiologia Pl.* **34**, 75–7.
Forsline, P.L. and Langille, A.R. (1976). *Can. J. Bot.* **54**, 2513–16.
Frandsen, N.O. (1968). *Theor. Appl. Genet.* **38**, 153–67.
Frey-Wyssling, A. (1974). *Schweiz. landw. Forsch.*, **13**, 385–93.
Fukuda, Y. (1967). *J. Fac. Sci. Tokyo Univ.* Sect. III, Botany, **9**, 313–75.
Gallant, D., Degrois, M., Sterling, C. and Guilbot, A. (1972). *Stärke* **24**, 116–23.
Garcia-Torres, L. and Gómez-Campo, C. (1972). *Potato Res.* **15**, 76–80.
Garcia-Torres, L. and Gómez-Campo, C. (1973). *Potato Res.* **16**, 73–9.
Gibson, R.W. (1971). *Ann. appl. Biol.* **68**, 113–19.
Gibson, R.W. (1974). *Potato Res.* **17**, 152–4.
Gifford, R.M. and Moorby, J. (1967). *Eur. Potato J.* **10**, 235–8.
Goodwin, P. (1963). In *The Growth of the Potato*, ed. Ivins, J.D. and Milthorpe, F.L. pp. 63–71. London: Butterworth.
Goodwin, P.B. (1966a). *J. exp. Bot.* **17**, 590–5.
Goodwin, P.B. (1967a). *J. exp. Bot.* **18**, 78–86.
Gregory, L.E. (1956). *Am. J. Bot.* **43**, 281–8.
Gregorini, G. and Lorenzi, R. (1974). *Potato Res.* **17**, 24–33.
Gruber, E., John, K., Alloush, S. and Schurz, J. (1972). *Microscopica Acta* **72**, 189–200.
Guha, S. and Maheshwari, S.C. (1964). *Nature* **204**, 497.
Guha, S. and Maheshwari, S.C. (1966). *Nature* **212**, 97–8.
Hammes, P.S. and Nel, P.C. (1975). *Potato Res.* **18**, 262–72.
Harmey, M.A., Crowley, M.P. and Clinch, P.E.M. (1966). *Eur. Potato J.* **9**, 146–51.
Harn, C. (1972). *Caryologia* **25**, 429–437.
Harten, A.M. van (1972). *Potato Res.* **15**, 73–5.
Hayashi, F., Blumenthal-Goldschmidt, S. and Rappaport, L. (1962). *Pl. Physiol.* **37**, 774–80.
Hayashi, F. and Rappaport, L. (1966). *Pl. Physiol.* **41**, 53–8.
Hayward, H.E. (1938). *The Structure of Economic Plants*. New York: Macmillan.
Hayward, P. (1974). *Planta* **120**, 273–7.
Hoff, J.E. (1971). *HortScience* **6**, 190.
Hoff, J.E. and Castro, M.D. (1969). *J. agric. Fd Chem.* **17**, 1328–31.

Hollings, M. (1965). *A. Rev. Phytopathol.*, **3**, 367–96.
Hölzl, J. (1965). *Protoplasma* **60**, 446–79.
Hölzl, J. and Bancher, E. (1958). *Öst. bot. Z.* **105**, 385–407.
Hooker, W.J., Yang, T.C. and Potter, H.S. (1973). *Am. Potato J.* **50**, 151–61.
Hori, S. (1954). *Bot. Mag., Tokyo* **67**, 57–62.
Howard, H.W. (1964). *Nature* **203**, 1303–4.
Howard, H.W. (1966). *New Phytol.* **65**, 284–7.
Howard, H.W. (1967). *Radiat. Bot.* **7**, 389–99.
Howard, H.W. (1969). *Genetica* **40**, 233–41.
Howard, H.W. (1970a). *Potato Res.* **13**, 220–2.
Howard, H.W. (1970b). *Genetics of the Potato Solanum tuberosum.* London: Logos Press.
Howard, H.W. (1971). *Potato Res.* **14**, 91–3.
Howard, H.W. (1972). *Potato Res.* **15**, 374–7.
Howard, H.W., Wainwright, J. and Fuller, J.M. (1963). *Genetica* **34**, 113–20.
Hughes, D.L., Takahashi, B., Timm, H. and Yamaguchi, M. (1973). *Am. Potato J.* **50**, 437–44.
Huijsman, C.A., Klinkenberg, C.H. and den Ouden, H. (1969) *Eur. Potato J.* **12**, 134–47.
Humphries, E.C. and Dyson, P.W. (1967a). *Eur. Potato J.* **10**, 116–26.
Humphries, E.C. and Dyson, P.W. (1967b). *Ann. appl. Biol.* **60**, 333–41.
Irikura, Y. and Sakaguchi, S. (1972). *Potato Res.* **15**, 170–3.
Ivins, J.D. and Bremner, P.M. (1964). *Outlook on Agriculture* **4**, 211–17.
Jager, F.Kh. and Jakovlev, M.S. (1939). *Dokl. (Proc.) Acad. Sci. U.S.S.R.* **23**, 491–4.
Johansen, D.A. (1950). *Plant Embryology.* Waltham, Mass: Chronica Botanica.
Jones, M.G.K. and Northcote, D.H. (1972). *J. Cell Sci.* **10**, 789–809.
Jones, S.G. (1939). *Introduction to Floral Mechanism.* London and Glasgow: Blackie and Son Ltd.
Kahl, G., Rosenstock, G. and Lange, H. (1969). *Planta* **87**, 365–71.
Kassanis, B. (1949). *Nature* **164**, 881.
Kassanis, B. (1957). *Ann. appl. Biol.* **45**, 422–7.
Kassanis, B. (1967). In *Methods in Virology*, Vol. **1**, ed. Maramorosch, K. and Koprowski, H. pp. 537–66. New York and London: Academic Press.
Kassanis, B. and Varma, A. (1967). *Ann. appl. Biol.* **59**, 447–50.
Khan, Shams-ul-Islam (1951). *Cytologia* **16**, 124–30.
Klopfer, K. (1965a). *Z. PflZücht.* **53**, 67–87.
Klopfer, K. (1965b). *Flora B*, **156**, 50–77.
Kohlenbach, H.W. and Geier, T. (1972). *Z. PflPhysiol* **67**, 161–5.
Krylova, N.V., Stepanenko, V.I. and Reifman, V.G. (1973). *Acta Virol., Prague* **17**, 172.
Kumar, D. and Wareing, P.F. (1972). *New Phytol.* **71**, 639–48.
Kumar, D. and Wareing, P.F. (1973). *New Phytol.* **72**, 283–7.
Kumar, D. and Wareing, P.F. (1974). *New Phytol.* **73**, 833–40.
Lamm, R. (1937). *Svensk bot. Tidskr.* **31**, 217–20.
Lange, H., Rosenstock, G. and Kahl, G. (1970). *Planta* **90**, 109–18.
Lauer, F.I. and Krantz, F.A. (1957). *Am. Potato J.* **34**, 158–64.
Lehmann, R. (1926). *Planta* **2**, 87–131.
Leonard, R. and Sterling, C. (1972). *J. Ultrastruct. Res.* **39**, 85–95.
Leshem, B. and Clowes, F.A.L. (1972). *Ann. Bot.* **36**, 687–91.
Lipetz, J. (1970). *Int. Rev. Cytol.* **27**, 1–28.
Lovell, P.H. and Booth, A. (1967). *New Phytol.* **66**, 525–37.

Lovell, P.H. and Booth, A. (1969). *New Phytol.* **68**, 1175–85.
Macdonald, D.M. (1973). *Potato Res.* **16**, 263–9.
Madec, P. (1963). In *The Growth of the Potato*, ed. Ivins, J.D. and Milthorpe, F.L. pp. 121–30. London: Butterworth.
Madec, P. and Perennec, P. (1959). *Eur. Potato J* **2**, 22–49.
Madec, P. and Perennec, P. (1969). *Eur. Potato J.* **12**, 96–115.
Madison, M. and Rappaport, L. (1968). *Pl. Cell Physiol., Tokyo* **9**, 147–53.
Maheshwari, P. (1950). *An Introduction to the Embryology of Angiosperms*. New York: McGraw-Hill.
Marinos, N.G. (1965). *Protoplasma* **60**, 31–3.
Marinos, N.G. (1967). *J. Ultrastruct. Res.* **17**, 91–113.
Meinl, G. (1966). *Flora B*, **156**, 419–26.
Mellor, F.C. and Stace-Smith, R. (1969). *Can. J. Bot.* **47**, 1617–21.
Mes, M.G. and Menge, I. (1954). *Physiologia Pl.* **7**, 637–649.
Miedema, P. (1967). *Euphytica* **16**, 163–6.
Miedema, P. (1973). A physiological study of adventitious bud formation in potato. *Thesis, Groningen. Centre for Agricultural Publishing and Documentation, Wageningen.* Also *Agric. Res. Reports* **787**.
Mohyuddin, M. and Hamid, M.A. (1966). *Pakist. J. scient. ind. Res.* **9**, 392–4.
Moorby, J. and Milthorpe, F.L. (1975). The potato (*Solanum tuberosum* L.). In *Crop Physiology. Some Case Histories*, ed. Evans, L.T. London: Cambridge University Press.
Morel, G. (1964). *Revue hort.* (Paris) **261**, 733–40.
Morel, G. and Martin, C. (1955). *C. r. hebd. Séanc. Acad. Agric. Fr.* **41**, 472–5.
Morel, G., Martin, C. and Muller, J-F. (1968). *Annls Physiol. vég., Brux.* **10**, 113–39.
Morel, G. and Muller, J-F. (1964). *C. r. hebd. Séanc. Acad. Sci.* **258**, 5250–2.
Morris, D.A. (1966). *Eur. Potato J.* **9**, 69–85.
Murashige, T. and Skoog, F. (1962). *Physiologia Pl.* **15**, 473–97.
Nielsen, N.K. (1968). *Acta Agric. scand.* **18**, 113–20.
Nitsch, J.P. (1969). *Phytomorphology* **19**, 389–404.
Nitsch, J.P. and Nitsch, C. (1969). *Science* **163**, 85–7.
Ohad, I., Friedberg, I., Ne'eman, Z. and Schramm, M. (1971). *Pl. Physiol.* **47**, 465–77.
Okazawa, Y. and Chapman, H.W. (1962). *Physiologia Pl.* **15**, 413–19.
Palladina, T.A. and Pervova, Y.A. (1966). *Fiziologiya rastenii* **13**, 283–9.
Palmer, C.E. and Barker, W.G. (1973). *Ann. Bot.* **37**, 85–93.
Palmer, C.E. and Smith, O.E. (1969a). *Nature* **221**, 279–80.
Palmer, C.E. and Smith, O.E. (1969b). *Pl. Cell Physiol.* **10**, 657–64.
Palmer, C.E. and Smith, O.E. (1970). *Pl. Cell Physiol.* **11**, 303–14.
Pennazio, S. (1971). *Riv. Ortoflorofruttic. ital.* **5**, 446–52.
Pennazio, S. and Redolfi, P. (1974). *Potato Res.* **17**, 333–5.
Pérombelon, M.C.M. and Lowe, R. (1975). *Potato Res.* **18**, 64–82.
Plaisted, P.H. (1957). *Pl. Physiol.* **32**, 445–53.
Pogorelova, O.V. (1969). *Farmacija*, SSSR **18**, 52–6.
Potato Marketing Board (1965). *British Atlas of Potato Varieties.*
Priestley, J.H. and Woffenden, L.M. (1922). *New Phytol.* **21**, 252–68.
Priestley, J.H. and Woffenden, L.M. (1923). *Ann. appl. Biol.* **10**, 96–115.
Railton, I.D. and Wareing, P.F. (1973a). *Physiologia Pl.* **28**, 88–94.
Railton, I.D. and Wareing, P.F. (1973b) *Physiologia Pl.* **28**, 127–31.
Railton, I.D. and Wareing, P.F. (1973c). *Physiologia Pl.* **29**, 430–3.
Rainbow, A. and White, D.J.B. (1972). *New Phytol.* **71**, 899–902.
Rappaport, L. (1972). *Proc. 18th Int. hort. Cong.* **5**, 143–55.

Rappaport, L., Blumenthal-Goldschmidt, S. and Hayashi, F. (1965). *Pl. Cell Physiol., Tokyo* **6**, 609–14.

Rappaport, L. and Sachs, M. (1967). *Nature* **214**, 1149–50.

Rappaport, L. and Wolf, N. (1968a). *Proc. int. Symp. Plant Growth Substances*, Calcutta, ed. Sircar, S.M. pp. 79–88.

Rappaport, L. and Wolf, N. (1968b). In *Biochemical Regulation in Diseased Plants or Injury*, pp. 203–11. *Phytopathological Society of Japan*, Tokyo, 1968.

Rappaport, L. and Wolf, N. (1969). *Symp. Soc. exp. Biol.* **23**, 219–40.

Reed, T. (1910). *Ann. Bot.* **24**, 537–48.

Rees-Leonard, O.L. (1935). *Bot. Gaz.* **96**, 734–50.

Reeve, R.M. (1967). *Am. Potato J.* **44**, 41–50.

Reeve, R.M., Hautala, E. and Weaver, M.L. (1969). *Am. Potato J.* **46**, 361–73.

Reeve, R.M., Hautala, E. and Weaver, M.L. (1970). *Am. Potato J.* **47**, 148–62.

Reeve, R.M., Timm, H. and Weaver, M.L. (1971). *Am. Potato J.* **48**, 450–6.

Reeve, R.M., Timm, H. and Weaver, M.L. (1973a). *Am. Potato J.* **50**, 49–57.

Reeve, R.M., Timm, H. and Weaver, M.L. (1973b). *Am. Potato J.* **50**, 71–8.

Reeve, R.M., Timm, H. and Weaver, M.L. (1973c). *Am. Potato J.* **50**, 204–11.

Rehman, S. (1966). *Bull. Soc. Bot. N. Fr.* **19**, 44–53.

Rehman, S. (1967). *Mém. Soc. bot. Fr.* (1967) 155–64.

Rosenstock, G. (1963). *Beitr. Biol. Pfl.* **38**, 275–319.

Salunkhe, D.K. and Pollard, L.H. (1954). *Proc. Am. Soc. hort. Sci.* **64**, 331–4.

Sattler, R. (1973). *Organogenesis of flowers*. University of Toronto Press.

Scaramella Petri, P. (1956). *Nuovo G. bot. ital.* **63**, 20–31.

Scaramella Petri, P. (1959). *Eur. Potato J.* **2**, 153–64.

Scaramella Petri, P. (1963a). *Agricoltura ital., Roma*, **12**, 481–490.

Scaramella Petri, P. (1963b). *Eur. Potato J.* **6**, 242–57.

Seithe, A. (1962). *Bot. Jb.* **81**, 261–335.

Shih, C.Y. and Rappaport, L. (1970). *Pl. Physiol.* **45**, 33–6.

Shih, C.Y. and Rappaport, L. (1971). *Pl. Physiol.* **48**, 31–5.

Shimomura, H. and Kuriyama, E. (1970). *Rep. Tokyo Coll. Pharm.* **20**, 708–12.

Simmonds, N.W. (1964). *Am. Potato J.* **41**, 129–36.

Simmonds, N.W. (1965). *J. Hered.* **56**, 139–42.

Šip, V. (1972). *Potato Res.* **15**, 270–3.

Sizova, M.A. (1963) *Trudy prikl. Bot. Genet. Selek.* **35**, 118–33.

Sizova, M.A. (1965a). *Trudy prikl. Bot. Genet. Selek.* **37**, 51–108.

Sizova, M.A. (1965b). *Trudy prikl. Bot. Genet. Selek.* **37**, 109–28.

Slater, J.W. (1963). In *The Growth of the Potato*, ed. Ivins, J.D. and Milthorpe, F.L. pp. 114–20. London: Butterworth.

Slater, J.W. (1968). *Eur. Potato J.* **11**, 14–22.

Smith, O.E. and Palmer, C.E. (1970). *Physiologia Pl.* **23**, 599–606.

Smith, O.E. and Rappaport, L. (1961). *Adv. Chem. Series*, **28**, 42–8.

Smith, O.E. and Rappaport, L. (1969). *Am. Potato J.* **46**, 185–91.

Smith, R.H. and Murashige, T. (1970). *Am. J. Bot.* **57**, 562–8.

Souèges, R. (1922). *Bull. Soc. bot. Fr.* **69**, 163–78; 236–41; 352–65; 555–85.

Stace-Smith, R. and Mellor, F.C. (1968). *Phytopathology* **58**, 199–203.

Steinberg, C. (1950). *Nuovo G. bot. ital.* **57**, 319–34.

Sterling, C. and Pangborn, J. (1960). *Am. J. Bot.* **47**, 577–82.

Steward, F.C. and Caplin, S.M. (1951). *Science* **113**, 518–20.

Stow, I. (1927). *Jap. J. Bot.* **3**, 217–37.

Sussex, I.M. (1955a). *Phytomorphology* **5**, 253–73.

Sussex, I.M. (1955b). *Phytomorphology* **5**, 286–300.

Svensson, B. (1962). *Eur. Potato J.* **5**, 28–39.

Svobodová, J. (1964). *Abstracts 10th int. bot. Congress*, 485–6.
Tarn, T.R. (1971). *Nature* **234**, 425–6.
Taylor, C.E. (1953). *Ann. appl. Biol.* **40**, 778–88.
Thomson, A.D. (1956). *Nature* **177**, 709.
Tizio, R. (1964a). *C. r. hebd. Séanc. Acad. Sci.*, Paris **258**, 6503–6.
Tizio, R. (1964b). *C. r. hebd. Séanc. Acad. Sci.* Paris **259**, 428–31.
Tizio, R. (1969). *Eur. Potato J.* **12**, 3–7.
Tizio, R. (1972). *Potato Res.* **15**, 257–62.
Toosey, R.D. (1963). In *The Growth of the Potato*, ed. Ivins, J.D. and Milthorpe, F.L. pp. 79–95. London: Butterworth.
Tsutsumi, F.V., Olivas, E.E. and Alvarez-Fuertes, G. (1972). *Revta lat.-am. Microbiol.* **14**, 153–64.
Tuan, D.Y.H. and Bonner, J. (1964). *Pl. Physiol.* **39**, 768–72.
Venning, F.D. (1954). *Phytomorphology* **4**, 132–9.
Verma, S.C. (1966). *Eur. Potato J.* **9**, 259–60.
Vöchting, H. (1902). *Bot. Ztg* **60**, 87–114.
Vries, H. de (1878). *Landw. Jbr*, **7**. 591–682.
Wardlaw, C.W. (1955). *Embryogenesis in Plants*. London: Methuen and Co. Ltd.
Weaver, J.E. (1922). *Development and Activities of Roots of Crop Plants*. Carnegie Institute, Washington.
Weinheimer, W.H. and Woodbury, G.W. (1967). *Am. Potato J.* **44**, 402–8.
Went, F.W. (1959). *Am. J. Bot.* **46**, 277–82.
Werner, H.O. (1954). *Am. Potato J.* **31**, 375.
Wigginton, M.J. (1973). *Potato Res.* **16**, 85–7.
Wigginton, M.J. (1974). *Potato Res.* **17**, 200–14.
Williams, E.J. (1955). *Bot. Gaz.* **117**, 10–15.
Wodicka, B. and Wenzl, H. (1971). *Potato Res.* **14**, 150–3.
Woolley, D.J. and Wareing, P.F. (1972a). *Planta* **105**, 33–42.
Woolley, D.J. and Wareing, P.F. (1972b). *New Phytol.* **71**, 781–93.
Woolley, D.J. and Wareing, P.F. (1972c). *New Phytol.* **71**, 1015–25.
Wurr, D.C.E. (1974). *National Vegetable Research Station 24th Annual Report* 1973, pp. 65–6.
Yamaguchi, M., Timm, H. and Spurr, A.R. (1964). *Proc. Am. Soc. hort. Sci.* **84**, 412–23.
Yoshida, M. (1970). *Mem. Fac. Agric. Hokkaido Univ.* **7**, 209–16.
Young, W.J. (1922). *Am. J. Bot.* **9**, 213–14.
Young, W.J. (1923). *Am. J. Bot.* **10**, 325–35.

Chapter 4

The physiology of growth and tuber yield

J. MOORBY

4.1 Introduction

In the past decade there have been several books and reviews which have considered the physiology of potatoes (Ivins and Milthorpe, 1963; Burton, 1966; Milthorpe and Moorby, 1967; Moorby and Milthorpe, 1975). The first two of these were concerned, in part, with trying to put the study of the potato crop on a firm fundamental basis. The second two reviews reported how this work was proceeding. It seems appropriate for the purposes of this book to describe the present state of our knowledge of potato physiology and, probably more important, to discuss how our attitudes and research priorities have changed and suggest areas where we appear to need further information.

Discussions such as this have usually started with the dormant tuber and subsequent development of sprouts during storage. I believe that basic physiology of sprout growth is now well understood and I intend to devote more attention to non-storage growth. The first, and major part of the chapter will be concerned, therefore, with growth in the field and the development of the crop of tubers. Physiological aspects of storage will then be considered and sprout growth discussed in terms of its management for different purposes.

4.2 Growth in the field

4.2.1 Pre-emergence growth

The potato is different from most other major annual crops in that it is usually propagated vegetatively. The seed tuber bears a number of sprouts, some or all of which may be growing. It is these growing sprouts which produce the daughter plant and the subsequent crop and, unless the growing sprouts are damaged the others do not resume growth (Morris, 1966b).

As with any other stem, a sprout grows by the production and expansion of a succession of internodes. Root and stolon primordia start to develop at the nodes, but their growth is usually limited until after planting. The major external factor controlling the expansion of the internodes appears to be light. Sprouts grown in the dark differ from those grown in the light by lacking chlorophyll, having longer internodes and a 'hooked' rather than erect apex. In addition, they have smaller leaves and a lower dry weight to fresh weight ratio, i.e. a greater water content per unit dry weight. Very little light is required to prevent these effects, less than 10 W m^{-2} of visible radiation usually being sufficient. The presumption is that this light effect is mediated through phytochrome, but gibberellins may also be involved. The application of gibberellic acid to sprouts grown in the light results in growth which is similar to that which occurrs in darkness. Morris (1967b) suggested, however, that light and gibberellic acid produced similar results by different means with the latter causing an increase in the rate of breakdown of tuber reserves; presumably in a similar fashion to the well understood system which is in the aleurone layers of cereal seeds (Lang, 1970; Jones, 1973).

When a tuber is planted the rate of sprout growth increases rapidly. Several factors are involved in this increase; for example, the greater supplies of water and mineral nutrients in the soil than in the mother tuber and the absence of light. If, however, tubers are not planted but stored in the dark and supplied with a mineral nutrient solution they do not grow as fast as those which have been planted in moist compost at the same temperature (Morris, 1966b). It must be concluded that a whole complex of internal and external factors are involved. One internal factor which might be involved, but has not been studied in potato, is a possible supply of growth substances from the root primordia which start to grow actively after planting (Jones, loc, cit.; Hall, 1973).

Rather better understood are the mineral nutrient relations after planting. Headford (1962) found that although root growth was rapid when emerging plants were supplied with a nutrient solution all the N in the emerging sprouts after 15 days at 20° C could be accounted for by transfer from the mother tuber. It seems unlikely, however, that there was no uptake of N in this period especially when there is often a net loss of mineral nutrients from the plant. This can amount to 20% of the K and 50% of the P in the two weeks after planting (Moorby, 1968). This loss can have a marked effect on the subsequent growth and development of the plant. Morris (1967a) showed that nutrient solution supplied to tubers planted in Perlite increased the rate of sprout emergence and also the shoot: root dry weight ratio (suggesting the relief of a deficiency situation) when compared with tubers supplied with distilled water. In addition, similar treatments can increase the number of above-ground stems produced by each tuber (Moorby, 1967). This could have been caused by an effect of the ions on the apical dominance relationships within and between the emerging sprouts (Phillips, 1969). Effects such as these suggest that the emerging sprouts compete for the available water and nutrients in the mother tubers and soil. Where there is little variation in the size of the growing sprouts planted, which is often the case, the number of shoots which emerge, S, and the mean weight, W, are a function of the population density, D, with

$$S = aD^b \,; \; W = a'D^{-b'} \tag{4.1}$$

b varies from 0 at full to 1 at no competition and both a and b vary with variety (Morris, 1967a). As will be seen below, the number of stems per planting position can have repercussions on the development of the crop canopy and tubers but there might also be more immediate effects on the performance of the first leaves to emerge above the soil surface.

In barley, for example, withholding nitrate for only 4 days after planting reduced the size and rate of photosynthesis of the first leaf (Dale, 1972). The performance of the first leaf was dependent on adequate supplies of N and the reserves in the endosperm were inadequate to maintain these (Dale *et al.* 1974). The consequently reduced supplies of assimilates from the first leaf can, in turn, affect the development of the second leaf and hence the effects on the development of the plant can be considerable. The relatively greater reserves of N in the mother tuber might be expected to mitigate against so drastic an effect in potatoes, but the possibility should be considered when interpreting the effects of fertilizer usage, especially during the early growth of the crop.

In view of the preceding discussion it might be expected that it would be possible to produce significant effects by controlling the concentration of nutrients in seed tubers. As will be discussed below, it is not easy to produce large differences in the nutrient concentrations. In one set of experiments tubers grown in soil with no added N had a N concentration of 1.1% whereas the application of 201–301 kg ha^{-1} only increased the concentration to 1.6%. In addition, however, the added fertilizer affected the rate of production of the seed tubers and hence their maturity. It is not surprising therefore, that although the crop produced from the seed tubers with the higher N concentration emerged faster and had a more rapid canopy development, the effects on tuber yield were inconsistent (Gray, 1973, 1974).

An effective way of supplying a greater amount of nutrients to the emerging sprouts is to control the sprout development in such a way that a greater proportion of mother tuber reserves are available to each sprout. Under these conditions, it is possible to show that functions analagous to those in Equation 4.1 can be used to describe the positive relationships between the degree of competition and the number of sprouts on a tuber. An increase in temperature intensified this competition (Morris, 1967a) but, in spite of this, there was a more rapid emergence at higher temperatures because of the increased rate of growth. In one cultivar the rate of emergence increased linearly over the range 0–25 °C by 0.0067 day^{-1} °C^{-1} (Sadler, 1961).

As the length of growing sprouts increases throughout storage changes appear to take place in the apical meristems which allow them to grow more rapidly after planting. Sadler (loc. cit.) showed that at 25 °C the rate of emergence G, day^{-1}, was related to the length of the longest sprout L, mm, by

$$G = 0.167[1 - 0.82 \exp(-0.065 L)] \tag{4.2}$$

and a similar relationship was also found by Burrage (1965). The internal changes in the sprout apex which bring about this effect are not understood, but it is possible to produce sprouts which do not behave in this manner and which produce abnormalities when planted.

One possibility, especially when the old tubers are planted in cold soil, is to cause tuberization before emergence ('little potato'). It is, indeed, possible for tuberization to occur without planting and for this process to be repeated through four tuber generations (Claver, 1971).

A second abnormality is the loss of geotropic response by the emerging sprouts and the production of 'coiled-sprouts'. In this situation sprouts grow horizontally or coil and some internodes swell and split. There can be a loss of apical dominance in the sprout system resulting in the eventual emergence of a large number of stems at each planting position, delayed tuberization and a late crop of mainly small tubers (Moorby and McGee, 1966; Catchpole and Hillman, 1975). Under some conditions the effects are greater with deep planting and compacted soil (Lapwood *et al.*, 1967) or in the presence of *Verticillium nubilum* (Pitt *et al.*, 1965). The most important effect is, however, that associated with sprout development (Moorby and McGee, loc. cit.; Cox, 1970; Ali *et al.*, 1970) and which may involve the action of ethylene (Catchpole, personal communication).

Branch and stolon growth

During pre-emergence growth the apical dominance relationships within the sprouts seem to be less pronounced than in storage growth and the stolon primordia and some axillary buds start to grow. The number of shoots which emerge above the ground is, therefore, almost always greater than the number of growing sprouts planted.

The lateral shoots which develop beneath the soil surface grow either

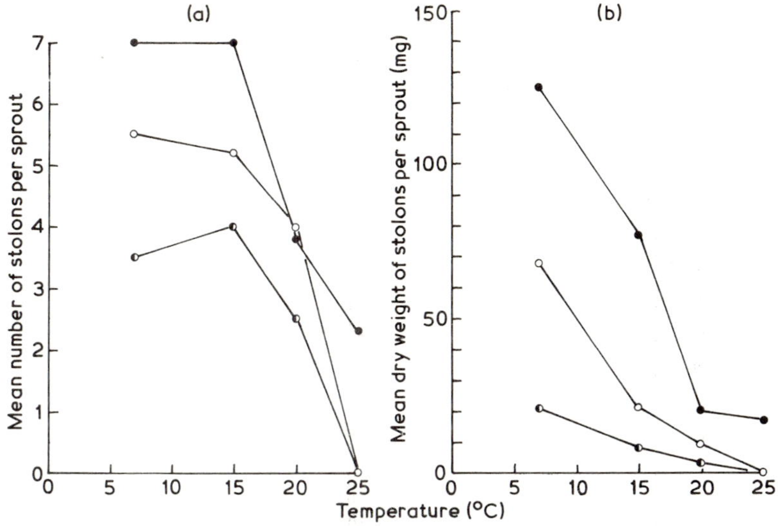

Fig. 4.1 The effect of temperature and number of sprouts per tuber on (a) the mean number of stolons per sprout and (b) the mean dry weight of stolons produced by each sprout in cv. Arran Pilot. Closed, open and pied circles are 1, 3 and 6 sprouts per tuber respectively (after Morris, 1967a)

vertically and emerge to form leafy shoots, or diageotropically to form stolons with scale leaves. These stolons may, or may not, produce tubers. The stolons can change into shoots if they emerge above the soil surface. It is very unusual to find normal stolons exposed to light. When the vertical shoots emerge from the soil the number and dry weight of the stolons they bear are inversely related to the temperature and number of sprouts per tuber (Fig. 4.1). It appears therefore, that stolon growth is reduced when inter-sprout, and apparently intra-sprout competition is increased (see discussion above). This conclusion was confirmed in the experiments illustrated in Fig. 4.1 by the observation that the stolons formed the greatest proportion of the total dry weight of the sprout system at 7° C and when there was only one sprout per tuber (Morris, 1967a).

Conditions after stem emergence also influence the growth of stolons but the competition with the other meristems still continues. Hence, conditions which favour the growth of shoots, in contrast to tuber growth, enhance stolon growth also; for example increased supplies of mineral nutrients (Svensson, 1962; Lovell and Booth, 1969). The first stolons are usually initiated at the basal nodes of a sprout and further initiation occurs in acropetal progression. There is a lag period before extension growth starts and the later a stolon is initiated the longer the delay and the slower the rate of growth (Fig. 4.2) (Lovell and Booth, loc. cit.).

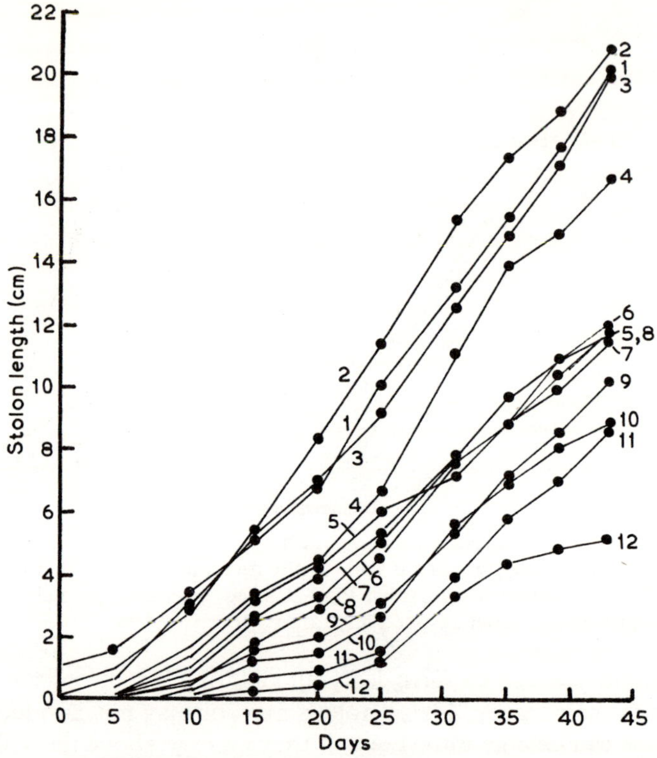

Fig. 4.2 The relationship between stolon elongation and the time of stolon initiation in cv. Majestic. Stolons are designated 1 to 12 in order of formation and values are the means of 14 plants for each stolon group (after Lovell and Booth, 1969)

There are obvious problems in understanding the control of stolon growth; why are stolons usually formed only at the base of the stem and how is the diageotropic growth maintained? There are technical problems which complicate the study of these questions in *S. tuberosum* where the stolons are produced underground and most work, therefore, has used *S. andigena*. In this species the stolons are often borne above the soil where they can be manipulated easily, especially in long days when there is no tuber formation.

Booth (1959, 1963) showed that when shoots of *S. andigena* are decapitated the lateral buds grow out as normal vertical leafy shoots. If however a mixture of indole acetic acid and gibberellic acid was applied to the decapitated stump the laterals developed into stolons. Booth also showed that it was possible to induce stolon formation at aerial nodes of undecapitated plants if moist *Sphagnum* was wrapped around the node to encourage root formation. He speculated that the reason for the usual restriction of stolons to the base of the plant might be the proximity of the roots as a source of growth substances. Further work by Kumar and Wareing (1972) suggested that the roots were acting as a source of cytokinins but that these only favoured the production of stolons if the lateral buds were subject to apical dominance. If the apical dominance was removed the kinins were involved in the conversion of the stolons to leafy shoots. This work was confirmed and extended by Woolley and Wareing (1972) who showed that a ^{14}C-labelled kinin, 6-benzylaminopurine could replace the effect of the roots and that it accumulated in the stolon tips prior to their conversion to leafy shoots. They suggested that the presence of a stem apex, or applied indoleacetic acid and gibberellic acid, diverted the kinins moving from the roots to the stem apex or treated region. In the absence of either of these factors the kinins could move into the stolon apices and lead to the formation of leafy shoots.

Wareing and his co-workers suggest that all lateral shoots are potential stolons. The below-ground laterals grow as stolons in the absence of light because of the increase in gibberellin concentration and the diversion of kinins to the apically dominant regions. Above the soil surface they suggest that the concentration of gibberellins is usually too low to permit stolon growth. When apical dominance is released the concentration of gibberellins in the stolons is low, and the diversion into them of kinins from the roots causes them to change into leafy shoots and grow vertically.

4.2.2 Post-emergence growth

Throughout the period of emergence growth the plant is dependent on the carbohydrate reserves of the mother tuber. After emergence leaf expansion is rapid because most of the primordia which expand into leaves are already present in the shoot apex. The plant rapidly becomes autotrophic, usually by the time 200–400 cm^2 of leaf have been produced (Headford, 1961; White, 1961), but the transfer of reserves from the mother tuber continues until the latter is almost completely depleted or until microbial infection brings about decay. Throughout this period there is a net loss of dry weight from the mother tuber. There is, however, a continual turnover of materials and it is possible to demonstrate the incorporation of photosynthetically incorporated ^{14}C into ethanol–insoluble reserves in the mother tuber at a time when the latter has lost 80% of its original dry weight (Moorby, 1968).

As the plant becomes autotrophic its sensitivity to environmental factors such as temperature and radiation and the supplies of water and mineral nutrients increases. Because of the wide range of species and varieties of cultivated potatoes the range of responses to these factors is enormous. If we are to take advantage of this adaptive ability we must obtain some understanding of its extent, of the nature of the responses and of the interactions between them (cf. Simmonds, 1971). The amount of work in this field is small and at the present time there is little value in trying to provide more than a progress report. Necessarily, there will be exceptions to this description and these will become more frequent, and may possibly lead to completely different interpretations, as the work progresses. Many of the responses change after the plants form tubers and in this section I will restrict my comments to events before tuber initiation. I will consider the responses of the plants plus tubers in a later section.

Effects of temperature, radiation and photoperiod

The basic responses of leaf and stem growth to radiation, photoperiod and temperature have been described for a few cultivars, but these are insufficient to permit more than simple generalizations. Experiments in controlled environment rooms using standard European cultivars suggested different temperature optima for leaf and stem growth, 20 and 25° C respectively (Borah and Milthorpe, 1962; Bodlaender, 1963; Marinus and Bodlaender, 1975). Within the range 0.81 to 4.68 MJ m^{-2} d^{-1} stem elongation was on inverse function of the radiation received by the plants but branch production was stimulated. The rate of leaf production was slightly reduced by the high radiation levels, possibly because of the reduced stem elongation. However, this was a minor factor compared with the increased branching and hence the total leaf area was a positive function of insolation.

A similar result was obtained by Sale (1973a) using the American cultivar Sebago in field experiments in the Murrumbigee Irrigation Area in Australia. The mean daily total insolation was about 25 to 30 MJ m^{-2} and when this was reduced by 34% by shading the stem length and maximum leaf area were increased and specific leaf weight (g cm^{-2}) was reduced by about 30%.

It can be seen that the radiation received by the plants can alter the canopy structure and hence light interception and photosynthesis but it would be dangerous to generalize any further from these two sets of results. The potato is very sensitive to the quality of light. Experience has shown that when grown in artificial light the potato requires rather more supplementary i.r. radiation to produce normal stem elongation and leaf expansion than do most other crops. In addition, in the experiments in controlled environment rooms discussed above, the photoperiod was 16 h compared with the 10.5 to 14 h in the field experiments. A reduction in the photoperiod can reduce the stem length (Bodlaender, 1963; Krug, 1963; 1965; Caesar and Krug, 1965) and the degree of reduction is dependent on variety. European maincrop varieties tend to be more responsive to changes in photoperiod than do early varieties, but this may not be true of all *Solanum* species and cultivars (cf. Alvey, 1965). Although most tuberous *Solanum* species originated in regions of low latitude with relatively short days breeding has tended to reduce the photoperiodic response of European and North American varieties because most of the growth in these regions takes place in relatively long photoperiods (Simmonds, 1971; Glendinning, 1975).

It is convenient here to consider changes in the growth substances in potato plants since one of the major determinants appears to be photoperiod. Okazawa (1959, 1960) demonstrated a lower concentration of gibberellins in potato plants grown in short than in long days. It has already been shown that exposure to short days results in shorter stems and a reverse effect, an increase in stem length, can be produced if the plants are treated with gibberellic acid (Lovell and Booth, 1967).

More recent work by Railton and Wareing (1973 a–c) using *S. andigena* has confirmed the early findings and shown that differences can be detected after only two short days. It was possible to separate the gibberellins into four fractions which reacted differently to changes in day length. The gibberellins appeared to be synthesized in the leaves and this synthesis was sensitive to daylength. Exposure of leaves to red light during the long dark periods increased the synthesis of gibberellins by short day plants and suggested an involvement of phytochrome. This result also suggests a possible reason for the effects of light quality on stem extension mentioned above (see also Hammes and Beyers, 1973).

An increase in the concentration of cytokinins in short days (and also lower night temperatures) has been reported by Forsline and Langille (1975). They found an increase after 2 short days in 2 of the 4 cytokinins detected in the above-ground parts of the plants and a maximum concentration in the below-ground parts after 6 short days.

The response of potatoes to low temperatures has been studied in some detail in connection with two phenomena; the promotion of tuberization and frost resistance. Low temperatures retard vegetative growth and this seems to be important in controlling the onset of tuberization. It will be discussed further in this context. In addition, low temperatures are of major importance in delimiting the distribution of the potato crop at high latitudes and altitudes (Burke and Gusta, 1974). The *Solanum* species vary in resistance to frost from species which can survive exposure to temperatures of -5 to $-10\,^{\circ}$C, e.g. *S. acaule*, to *S. tuberosum* subsp. *tuberosum* which cannot usually survive below 0 to $-1\,^{\circ}$C (Richardson and Weiser, 1972). The way in which damage is caused by frost is not fully understood, but it seems to arise when extra-cellular water freezes. This usually occurs slightly below $0\,^{\circ}$C because of supercooling and the presence of solutes. Damage can be caused by the loss of water from the cytoplasm, i.e. a form of desiccation, but there is also a decrease in the electrical impedance of the tissue which suggests a disruption of the cell membranes and leakage of ions (Hayden *et al.* 1972). A major difficulty in the production of commercially successful varieties capable of withstanding frost is the development of a rapid, preferably non-destructive, test for frost-hardiness which can be used in screening procedures. Once this is available sufficient gentic variation seems to exist for a successful breeding programme to be mounted (Estrada Ramos, 1974).

Effects of water

The tissue water relations of potato leaves appear to be similar to those of other crops (Fig. 4.3) (Shepherd, 1973). There appears to be a slight curvilinear relationship between leaf water potentials as measured with a thermocouple psychrometer and relative leaf water content, and a linear relationship between the latter and the osmotic potential of the cell sap expressed by a pressure of 7000 J kg $^{-1}$. The differences between the osmotic and water potentials shown in Fig. 4.3 are the sum of the turgor pressures (positive or negative) and the negative water potentials. If the tissues have not wilted, the latter are usually small compared with

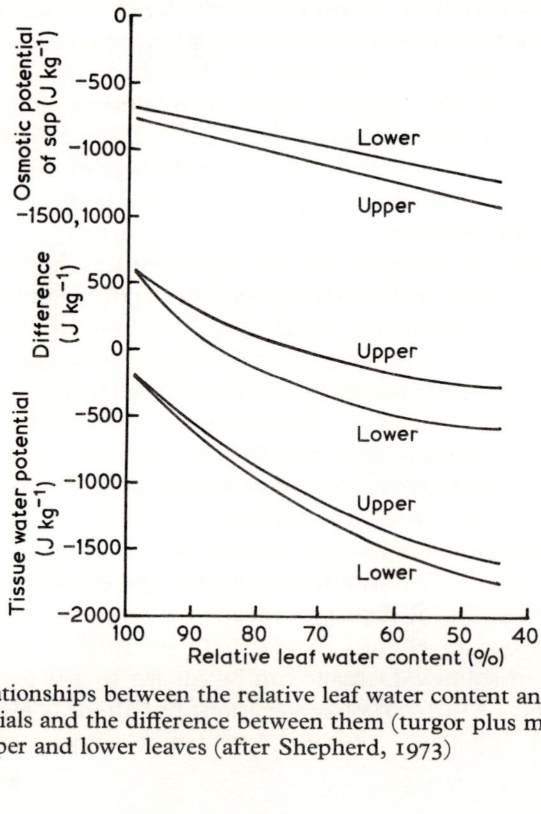

Fig. 4.3 The relationships between the relative leaf water content and osmotic and water potentials and the difference between them (turgor plus matric potentials) of upper and lower leaves (after Shepherd, 1973)

the former. The upper, i.e. younger, leaves in the two crops had more negative osmotic potentials than the lower, older leaves. The turgor pressures of the upper leaves were greater and this was thought to be the reason for the greater water potentials. Results consistent with these, but over smaller ranges of relative leaf water content, have been obtained using a pressure bomb to measure leaf water potentials (Moorby *et al.*, 1975) and a gravimetric method to estimate tuber water potential (Epstein and Grant, 1973).

There can be large diurnal fluctuations in the relative water content of leaves caused by the inability of plants to absorb water at a rate sufficient to replace that lost by transpiration (Werner, 1954). These fluctuations are accompanied by higher leaf temperatures relative to air temperatures in periods of high evaporative demand and limited supplies of water (Lomas *et al.*, 1972). When there are ample supplies of water the evaporative cooling caused by transpiration can maintain leaf temperatures close to ambient. These authors working in dry conditions in Israel with mid-day air temperatures of 30–32 °C found leaf temperatures of up to 40 °C. These could be reduced very rapidly to equal or less than the air temperatures by overhead irrigation. Similar results were obtained by Sanders and Nylund (1972a, b) who also found that mist irrigation maintained leaf turgor, reduced stomatal resistance and allowed haulm growth to be sustained for a longer period.

The water potential gradients in the plant can be changed within minutes of transpiration starting or ceasing. These effects can reverse the direction of water and ion flow through the xylem cells in stolons (see Section 4.2.3) and can lead to the extraction of water from tubers and changes in stolon turgidity and tuber volume and fresh weight (Macmillan, 1891; Engel *et al.*, 1964; Meinl, 1965; Baker and Moorby, 1969; Krauss and Marscher, 1974). The opposite effect can be seen under conditions of low soil moisture and high atmospheric humidity e.g. after spraying. There can then be flow, e.g. of a herbicide, in the xylem from the leaves to the tubers (Headford and Douglas, 1967). Although the changes in tuber water potential are similar to those in leaves, the fluctuations do not appear to be as violent. They have large effects on tuber growth and biochemistry but these are discussed below.

Moorby *et al.* (1975) and Munns and Pearson (1974) showed that drought increased the leaf resistance to the diffusion of water vapour as measured with a diffusion porometer. There were also associated decreases in the rate of photosynthesis and the amounts of photosynthates exported from the leaves. The rate of photosynthesis seemed to be particularly sensitive to a reduction in leaf water potential, but there did not appear to be any direct effect of the water deficits on the translocation processes. In both these sets of experiments it was possible to show a decrease in the rate of photosynthesis at leaf water potentials of about

-500 J kg^{-1} whereas many other crops show no decrease at -1000 J kg^{-1}.

The importance of the control of water loss by stomata and the onset of this control at very small leaf water deficits is illustrated by two other, very different, sets of data. Endrödi and Rijtema (1969) showed that the ratio of the actual transpiration of a potato crop in The Netherlands to the potential transpiration was reduced at leaf water potentials below only -100 J kg^{-1}. In contrast, Waggoner and Simmonds (1966) investigated the behaviour of *S. tuberosum* plants bearing the recessive mutant 'droopy'. The stomato on plants showing this character do not close, and the plants continue to transpire, even after they wilt. Waggoner and Simmonds (loc. cit.) suggested that water deficits could also influence photosynthesis by some means other than stomata. Moorby *et al.* (1975) came to a similar conclusion. Because of an inability to detect effects of low leaf water potentials on the activities of either ribulose diphosphate carboxylase or carbonic anhydrase, they suggested that the diffusion of CO_2 through the liquid phase of leaves might be sensitive to water deficits.

It appears that the major effects of water deficits are exerted through the reduction in the rate of photosynthesis and the smaller supply of carbohydrates. The effect of these factors on tuber growth is best discussed in a later section (Section 4.2.4).

The effect of mineral nutrients

Any growth response or biochemical change in the tubers must be dependent on the concentration of the ions in the tubers (Kunkel *et al.*, 1973). It can be seen, therefore, that some understanding of the patterns of ion uptake and transport are basic to any understanding of how the yield and quality of potatoes are influenced by fertilizer usage.

A considerable proportion of the ions which enter the haulm in the early stages of growth probably come from the mother tuber but as growth proceeds an increasing proportion come from the soil. Roots appear to grow rapidly after planting. In experiments with plants grown at a rectangular spacing of 80 cm without ridging ^{32}P was easily detected after 80 days from sites 30 cm deep and 20 cm from the plant; and after 100 days from sites 80 cm deep and 60 cm from the plant (McCorquodale and Moorby, 1969). Other experiments with ^{32}P suggested that in a sandy soil approximately equal amounts of phosphate were absorbed from the ridge in which the potatoes were planted, the soil beneath the ridge and the soil beneath the furrow. In contrast, in a heavier clay, a rather larger proportion of the ^{32}P was absorbed from the ridge (Newbould *et al.*, 1968).

Root distribution is affected markedly by plant spacing and Emmond (1971) found little evidence of the absorption of phosphate from adjacent hilled rows which were more than 30 cm from the test plants. In similar experiments, however, there was an increased yield from plants when nitrate was applied at distances of up to 90 cm from the test plants.

Direct comparisons between the results of Newbould *et al.* (loc. cit.) and McCorquodale and Moorby (loc. cit.) and those of Emmonds (loc. cit.) are not possible. The former authors used carrier-free ^{32}P-labelled K_2HPO_4 where the amount of chemical added was insufficient to affect root growth. Emmond, however, applied 27 g of ^{32}P-labelled $NH_4H_2PO_4$ to an area of approximately 0.4 m^2 and this could have increased root growth and ^{32}P uptake. Similarly, Emmond applied large amounts of NH_4NO_3 and could follow its dispersal through the soil and it is impossible to decide whether the response measured should be attributed to root growth or the movement of nitrate.

The most comprehensive data on the absorption of mineral nutrients by potatoes are those of Carpenter (1963) which cover several years and cultivars and many rates of fertilizer application. Some of Carpenter's data are shown in Figs. 4.4 and 4.5. These are typical of the sort of changes in the mineral content of haulm and tubers which occur throughout the season although the absolute amounts vary with the fertilizers applied.

The absorption of all the elements measured continued throughout the season (Fig. 4.4). The amounts of N, P and K in the haulm increased to a maximum and then fell slightly. The amounts in the tubers increased throughout the season and at the end of the experiment were approximately the same as in the haulm. If the experiment had been extended the loss would probably have continued (cf. Moorby, 1968). This decrease in the haulm is especially noticeable in plants with long growing periods e.g. *Solanum andigena* cv. Renacimiento (Ezeta and McCollum, 1972). The concentration of N and P in the haulm (Fig. 4.5) decreased throughout the season whereas that of K rose to a maximum before decreasing. This difference from N and P probably resulted from the large amounts of K applied to these plants. It was not apparent when smaller amounts of K were applied. The concentration of all three ions in the tubers decreased slightly in the early harvests but changed little over most of the season (Fig. 4.5).

The behaviour of Mg was very different from that of N, P and K. There was a less obvious decrease in the amount in the haulm and that in the tubers accounted for only about 26% of the total amount in the plant. The concentration of Mg in the haulm increased slightly throughout the season whereas that in the tubers decreased and was less than 20% of that in the haulm.

These changes in the concentrations and amounts of N, P, K and Mg can be explained in terms of the relative mobilities of the ions in the plant. As the foliage ages and leaves senesce and abscind there is some loss of material from the plant but there is also a transfer of some

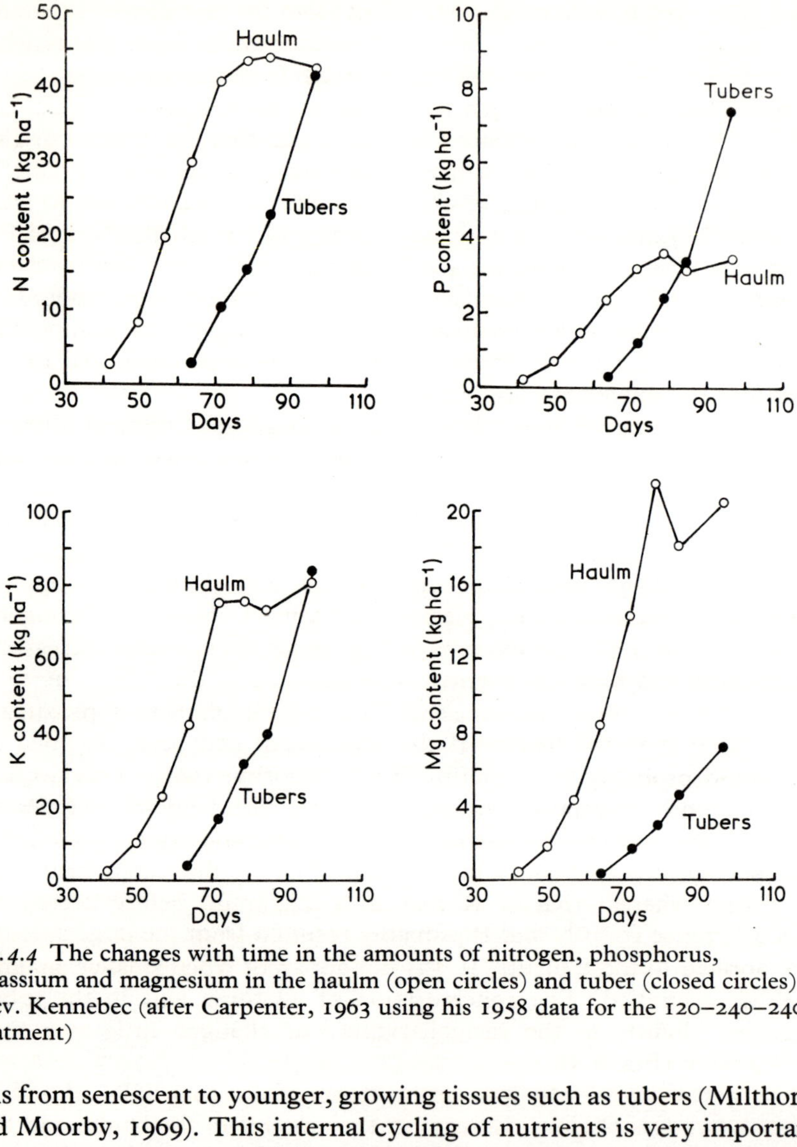

Fig. 4.4 The changes with time in the amounts of nitrogen, phosphorus, potassium and magnesium in the haulm (open circles) and tuber (closed circles) of cv. Kennebec (after Carpenter, 1963 using his 1958 data for the 120–240–240 treatment)

ions from senescent to younger, growing tissues such as tubers (Milthorpe and Moorby, 1969). This internal cycling of nutrients is very important. Some substances such as N, K and especially P seem to be readily mobilized and can be transferred to the tubers, so leading to the decrease in the concentration of these ions in the haulm (Moorby, 1968; Dyson and Watson, 1971). Some of the materials transferred are part of the normal turn-over in leaves but a considerable proportion might be recently absorbed materials which are transported to the foliage in the transpiration stream and immediately recycled to the tubers via the phloem (Baker and Moorby, 1969). Ions such as Mg^{24} enter the phloem

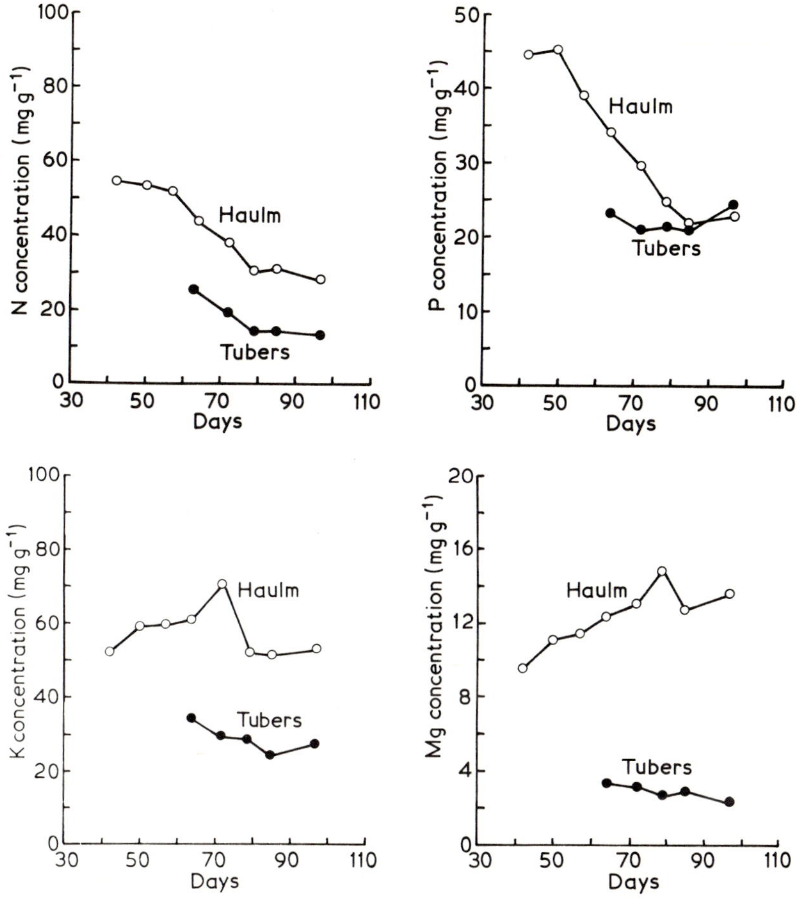

Fig. 4.5 The changes with time in the concentrations of nitrogen, phosphorus, potassium and magnesium in the haulm (open circles) and tubers (closed circles) of cv. Kennebec (after Carpenter, 1963 using his 1958 data for the 120–240–240 treatment)

less readily, are not exported from the haulm to the same extent, and hence the concentration of Mg in the haulm tends to increase throughout the season.

There appears to be little or no redistribution of Ca^{2+} and other alkaline earth cations. These appear to be supplied to the tubers exclusively in the xylem. Since in transpiring plants any movement through the xylem of the stolons is from the tubers to the rest of the plant, it is not surprising that the movement of ^{89}Sr (used as a tracer for Ca) into tubers could only be detected in periods of darkness (Baker and Moorby, loc. cit.). More direct evidence which substantiates the transport of Ca in the xylem in response to gradients in water potential has been obtained by Krauss and Marschner (1973, 1974). They investigated the effects of light and darkness on the movement of ^{45}Ca through stolons, and independently controlled the relative humidity around the tubers and remainder of the plant. This lack of calcium transport in the phloem limits the period during which movement can occur, and the usually high relative humidity around the tubers also restricts movement in the xylem. The result is a low concentration of Ca in the tubers even when calcium is being absorbed freely by the plant. For example, in one compilation, the range of concentrations of Ca in the haulm is 5.8 to 41.2 mg g^{-1} and in the tubers only 0.2 to 2.9 mg g^{-1}. (McIlrath (1972) quotes K and P concentrations in the ranges 10.5 to 39.6 and 1.1 to 4.9 mg^{-1} respectively and Carpenter (loc. cit.) in four years experimentation using four varieties and a wide range of fertilizer application obtained a range of tuber N concentrations at harvest of only 11.1 to 18.6 mg g^{-1}. These 2- to 4-fold differences between tubers and haulm contrast markedly with the 14-fold range in Ca concentration. Although the variation in the ionic content of tubers is relatively small, it is very important in the control of tuber growth and quality.

The distribution of phloem-mobile ions within tubers should be reasonably uniform because there are many phloem anastomoses in the stolons and a complete network of sieve tubes throughout the tubers. Ca again appears to be anomalous, with abnormally high concentrations in the cortex, i.e. exterior to the vascular ring (Bretzloff, 1971; Bretzloff and McMenamin, 1971). This is not too surprising if the calcium ions enter via the xylem along a gradient in water potential; the outer tissues of tubers are more likely to be at a lower water potential than the pith. There may, however, be other, more subtle, effects since there is some evidence that the 'peel' and 'flesh' of tubers obtain some ions from different depths in the soil (Ellis, 1966).

A technique which is often used to monitor the nutrient status of plants is foliar analysis. This can be applied to potatoes (MacKay *et al.*, 1966; Fong and Ulrich, 1969a, b; 1970, 1974; Ulrich and Fong, 1970, 1973). Its success as a technique depends on the maintenance of a regular balance between haulm and tuber growth and on the careful choice of tissues to be analysed. Within these limitations, foliar analysis can work well, and it is possible to define recommended concentrations of various ions. Tests such as these tend to break down, however, if applied indiscriminately to crops grown by different methods in a range of environments. For example, the concentration of some ions in leaves is affected by the use and type of irrigation (Kalra, 1971; Sanders, Nylund & Quisumbing, 1972; Sanders *et al.*, 1972) and plant spacing (van Burg, 1967; Bodlaender and Reestman, 1968). In addition, there can be competitive effects between ions, e.g. Cl^{-1} can reduce the amount of NO_3^- and total amount of N in the haulm but has no effect on protein N (Murarka *et al.*, 1973).

A wide variety of quality factors are associated with the concentrations of ions in tubers (Hughes, 1974). The most obvious relationship, but possibly not the most straightforward, is between the nitrogen and protein concentrations and a considerable effort is being expended on breeding varieties with higher concentrations of protein. Such breeding is difficult because of negative correlations between protein content and other quality factors (see discussion in Smith, 1974). There is little necessity for higher protein concentrations in potatoes used for human food in developed countries since modest amounts of existing varieties can supply sufficient protein (albeit deficient in methionine) to satisfy human requirements (Burton, 1974a). The position may be different in underdeveloped countries, but in such situations there are often associated energy deficiences which cannot be satisfied easily by potatoes unless there are very large intakes. An alternative use for potatoes with high concentrations of protein would be as animal feed. This latter use has not received much attention in recent years but could now be worth further attention in countries like the U.K. which are self-sufficient for potatoes but have to import animal feedstuffs.

Mineral nutrients interact with many other internal and external factors to affect tuber quality (Hughes, 1974). If the effects of any one factor are to be understood it will be necessary to undertake much more rigorous and detailed experimentation than has been usual in the past. For example, a very large set of field trials concerned only with the effect of fertilizers on yield (of wheat) could assign only 6% of the variation to the imposed treatments; the remaining variation was environmental in origin (Milthorpe, 1969). A similar situation appears to exist with the effects of fertilizer usage on potato quality (Birch *et al.*, 1967). It seems pointless, therefore, to investigate the effects of, say, KCl and K_2SO_4 on the dry matter content of tubers in field experiments at several sites and in different years when nothing is known about the potassium status of the soils or the other environmental conditions during the experiments.

Future progress will probably depend on obtaining some quantitative insight into the nature of these very complex interactions. The most promising way of obtaining this insight appears, at the present time, to be the use of simulation methods. Mathematical models of ion uptake and fertilizer response already exist and are being developed continuously. Some of these are sensitive to some environmental variables and can be used to predict the overall concentrations of nutrients in plants (Greenwood *et al.*, 1971; Greenwood *et al.*, 1974) and others try to predict the concentrations in individual organs (Milthorpe and Moorby, 1974; Milthorpe *et al.*, 1974; Nye *et al.*, 1975). It should be possible eventually to use models similar to these for potatoes and to include, in addition, functions which give some indication of various quality factors.

Post-emergence competition between plants

Section 4.2.1 included some discussion of the competition between the emerging sprouts and between the stems and stolons. This competition continues after emergence and results in the formation of smaller individual shoots.

Discussion of the growth of individual plants and the interference between them is made difficult by the growth habit of the potato. Each mother tuber bears one or more sprouts which often branch. It is rare, therefore, to find plants consisting of a single above-ground stem. Instead, there is usually a group of stems all arising from a single mother tuber (Wurr, 1975). This collection of stems is often termed an 'hill'. The above-ground stems might arise directly from the mother tuber, or several might be produced by the branching of a sprout. Much confusion has been caused by different definitions of what constitutes an individual. I believe that the simplest and most logical definition is to consider each stem which arises below ground, whether from the mother tuber or another stem, to be an individual (Moorby, 1967). When defined in this way each stem probably has a root system and stolons and it has leaves. If isolated, it could easily survive and bear tubers and fruit. Any branches which arise above ground are considered to be parts of a single individual because they have common systems of roots and stolons. A final, practical, advantage to this definition is that the number of stems present in any situation can be determined easily without disturbing the plants. This is impossible if a definition is used which relies on whether shoots originated directly from separate sprouts on a mother tuber i.e. so-called main stems. In some instances there is little difference between the number and behaviour of main and above-ground stems (Allen and Wurr, 1973). In others, where the mother tubers had large well-developed sprouts, there are large differences (Moorby, 1967).

The crop as a whole is composed, therefore, of many groups (or hills) each made up of several individual stems. Since each of these is potentially capable of a separate existence, and probably relies on its own leaves and roots for supplies of carbohydrates, mineral nutrients and water, it is not surprising that these individuals compete with each other. Hence, as the number of stems per unit area is increased, the total amount of haulm produced also increases. There is, however, a decrease in the size of each individual, i.e. the mean haulm weight per stem is reduced (Toosey, 1963; Moorby, 1967; Svensson and Naglicka, 1975). These effects have large repercussions on the growth of tubers and will be discussed below.

4.2.3 Tuber initiation

Until this point, in the discussion, we have been concerned with the growth of leaves, stems, stolons and roots. Once the tubers are initiated the growth of all the other organs is retarded and the tubers become the dominant meristems and sinks for organic and inorganic nutrients. It is important, therefore, that the factors controlling tuber growth, i.e. the economically important part of the plant, be understood.

As discussed by many authors, there have been two sets of opinions concerning the control of tuber initiation. Both have had their advocates for at least 50 years, but the mechanism of tuber initiation is still unknown. The situation in the early 1960s was discussed by Slater (1963) who favoured the suggestion that tuberization resulted from an accumulation of substrates at the stolon apices and Madec (1963) who favoured the involvement of a specific tuber-forming substance. In the past decade evidence in support of both theories has been produced.

In many ways the argument is specious since it is usually difficult, if not impossible, to be sure of the actual time of tuber initiation. In simple experimental systems, for example, in cuttings or explants, which are harvested after a few days or can be observed readily, timing can be reasonably precise. In larger plants this is impossible, and in the field tuber initiation is usually spread over two weeks or more. The tuber growth curve is sigmoid with, usually, a long linear phase. The initial curvilinear portion can however be of variable duration and the apparent time of tuber initiation estimated by extrapolation to zero weight of tubers can be in error (cf. Fig. 4.6). What is more important

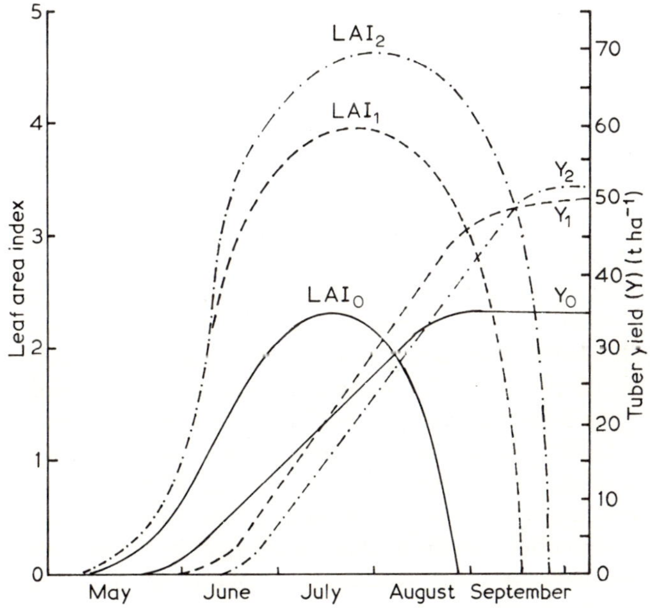

Fig. 4.6 The change with time in the leaf area index and tuber yield of cv. Majestic grown at three rates of nitrogen supply (N_0 N_1 and N_2) in England (after Ivins and Bremner, 1965)

in most situations is the ability of the plant to support tuber growth after initiation. This is dependent on the presence of sufficient foliage to produce the necessary assimilates and adequate supplies of water and mineral nutrients. If tuber initiation occurs before the plant can supply these necessities, as in the 'little potato' disorder, it is impossible to produce an economic crop. Many of the arguments appear, therefore, to be very academic since even if there is a specific tuber-forming substance its expression is dependent on an adequate supply of substrates. This conclusion is emphasized by the large increases in average yields which have been achieved (about 70% in Great Britain in the last 50 years) whilst these arguments have been in progress.

The most likely explanation is probably that tuber growth is initiated by a complex of factors, any one of which can be limiting in a particular set of circumstances. It is, however, generally accepted that there is a balance between the growth of tubers and the rest of the plant and anything which favours the growth of one will retard the growth of the other. However, many of the treatments which can change this balance also have quite specific effects on particular growth substances and it is difficult to disentangle cause, effect and side-effect. The situation is well illustrated by the relationships between gibberellins and tuber initiation.

Gibberellic acid stimulates the growth of the haulm of *S. tuberosum* and *S. andigena* and retards tuber initiation in both species (Humphries, 1958; Okazawa, 1959, 1960; Humphries and French, 1960, 1961; Booth, 1963; Lovell and Booth, 1967; Smith and Rappaport, 1969; McCorquodale, 1971; Kumar and Wareing, 1974; Hammes and Nel, 1975). Treatment with gibberellic acid can inhibit tuber growth after initiation and this is accompanied by a diversion of assimilate movement from the tubers to the meristems of the haulm (Lovell and Booth, 1967). A change in the reverse direction is seen when plants initiate tubers (Lovell and Booth, loc. cit.; Moorby, 1968, 1970; Booth and Lovell, 1972) and this is accompanied by the deposition of starch, often before any visible signs of tuber formation. Lovell and Booth (loc. cit.) think that the gibberellic acid acts primarily by inhibiting starch deposition and that this affects the ability of the tuber initials to act as sinks.

Growth retardants can hasten tuber initiation (Dyson, 1965; Bodlaender and Algra, 1966; Gifford and Moorby, 1967; Humphries and Dyson, 1967; El-Antably *et al.*, 1967; Gunasena and Harris, 1969, 1971) and some of these are considered to prevent gibberellin synthesis or action. The only anomalous result is that of El-Antaby *et al.* (loc. cit.) who found an increase in tuber initiation of plants of *S. andigena* treated with abscisic acid. However, Railton and Wareing (1973d) found that the application of abscisic acid to the leaves of the same species resulted in an increase in the gibberellin content of up to 100-fold. It is not too surprising, therefore, that Smith and Rappaport (1969) and Palmer and Smith (1969b) could find no effect of abscisic acid on tuber initiation in either whole plants or stolon explants of *S. tuberosum*. McCorquodale (1971) found that aseptically cultured stolons of *S. tuberosum* did not form tubers in the presence of gibberellic acid and that there was no effect of the growth retardants CCC nor B9 nor of abscisic acid on tuber initiation. It appears, therefore, that the site of action of these growth regulators is not the stolon tip but the site of formation of the gibberellins. Gibberellins appear to be formed in the leaves and less can be extracted in short than in long days. It is uncertain whether this is caused by a reduction in synthesis or a greater rate of turnover of the gibberellins but it is unlikely to be a result of reduced growth since the effect can be seen after only two short days (Railton and Wareing, 1973c).

All these results suggest that the gibberellins are formed in the leaves and then move into the stolons. When the gibberellin concentration in the stolon apices is reduced below some critical value, tubers are initiated. This reduction can be hastened by short days or the application of some growth regulators. There is, however, a paradox in this interpretation. There is usually a positive correlation between growth rate and gibberellin concentration but a reduction in concentration in the stolons results in an increase in the growth rate of the meristematic regions i.e. the new tubers. This increase is maintained by a diversion of assimilates and other materials to the tubers, and associated increases in the photosynthetic and net assimilation rates.

This apparent paradox may result from the interaction of two processes. The reduction in gibberellin concentration in the stolons is accompanied by similar decreases in the gibberellin concentration in the haulm and in the growth rate of the haulm. There will, therefore, be a reduction in the demand for assimilates in the haulm and more will be available for the stolons. We revert, therefore, to the other theory of tuber initiation which emphasizes the importance of the supply of substrates to the stolons. It is appropriate, therefore, at this point, to discuss the evidence in favour of this second theory.

It is generally accepted that there is an inverse relationship between haulm and tuber growth. For example, if haulm growth is encouraged by the application of early irrigation or fertilizers, especially N, tuber initiation is delayed. The eventual rate of tuber growth and its duration are, however, greater because of the increased size and persistence of the haulm (Ivins and Bremner, 1964; Gunasena and Harris, 1968; Krauss and Marschner, 1971; Dyson and Watson, 1971). In contrast, tuberization is encouraged if haulm growth is inhibited, for example by low temperatures (Borah and Milthorpe, 1962; Bodlaender, 1963; Burt, 1964a, 1965; Gregory, 1965; Slater, 1968).

The available evidence suggests that before tuber initiation the haulm is the dominant sink for assimilates but that at the time of tuber initiation there is a major diversion of assimilates to the stolons and tuber initials. Burt (1964a) obtained some evidence suggesting an accumulation of assimilates before tuberization but Lovell and Booth (1967) could find no accumulation of ethanol-soluble sugars but a big increase in starch concentration which could be reversed by treatment with gibberellic acid. We have, therefore, a chicken and an egg situation with a close association between phenomena but insufficient evidence to resolve cause and effect.

There now seems to be abundant evidence for the existence of a tuber-forming substance. The formation of this substance is favoured by short days and it can move across graft unions to promote the formation of tubers on 'non-induced' long day stocks (Gregory, 1956; Okazawa and Chapman, 1963; Kumar and Wareing, 1973). The evidence that high concentrations of gibberellins prevent tuberization has been discussed above. In addition there have been suggestions that many other substances are necessary for tuberization. For example, Palmer and Smith (1969a, 1970) and Smith and Palmer (1970) have suggested that kinins promote tuberization and that this promotion can be inhibited by abscisic acid (Palmer and Smith, 1969b). They believe that the kinins promote tuber initiation by inhibiting the action of 'starch hydrolase' so permitting the deposition of starch. They were able to show an accumulation of starch after kinin treatment but before any visible signs of tuber formation. However, there was no evidence for an increase in protein synthesis. In support of this work Forsline and Langille (1975) have shown an increase in the concentration of two compounds with cytokinin activity in *S. tuberosum* grown in short days with low night temperatures. In contrast, however, McCorquodale (1971) was unable to find any effect of kinins on tuber initiation by stolons grown in *vitro* and Tizio and Biain (1973) have also questioned the importance of kinins.

Other substances which have been implicated in tuber initiation are coumarin (Stallknecht, 1972) and ethylene (Catchpole and Hillman, 1969; Garcia-Torres and Gomez-Campo, 1972, 1973). Also, changes occur during the storage of tubers which favour the onset of tuberization. The longer tubers are stored, the shorter the time between planting and tuberization. If storage is continued for a sufficient period, tuberization occurs before planting and this effect can be repeated through four tuber generations (Claver, 1971). This effect of duration of storage can also be seen when stolons are excised from sprouting tubers and placed on aseptic media (McCorquodale, 1971). In spite of these effects, the mother tuber is not essential for tuberization as shown by Bodlaender and Marinus (1969) and the tuberization of plants grown from seed.

This mass of apparently contradictory evidence illustrates the futility at the present time of trying to give any definitive explanation of the mechanism of tuber initiation. Except in a few instances, there appears to be no absolute requirement for any particular set of conditions or chemicals. For example, *S. andigena* is normally considered to require exposure to short days before it will initiate tubers but Simmonds (1971) showed that it was possible by selection for a few generations to produce plants which were virtually insensitive to day length. A full understanding of tuber formation is probably dependent on our obtaining a much deeper understanding of the detailed processes involved; the decrease in the rate of elongation of the stolons and swelling of the sub-apical internodes, and starch accumulation.

4.2.4 Tuber growth and crop yield

Growth of individual tubers

The first detectable sign of tuber initiation seems to be the onset of starch deposition (see previous section) and this is followed by the radial expansion of the stolon about 12 internodes behind the apex (in cv. Arran Pilot). As the apical bud continues to grow the younger internodes start to expand, i.e. the apical bud usually contains about 12 leaves. More than half of these leaves subtend lateral buds which also grow and produce leaves and lateral buds. These are the future eyes of the tuber which become dormant in acropetal succession whilst the tuber is still growing (Goodwin, 1967a).

The growth of the tuber tissue is by both cell division and expansion (Artschwager, 1924; Plaisted, 1957; Booth, 1963; Reeve *et al.*, 1969; Reeve *et al.*, 1973a, b). Plaisted (loc. cit.) showed that as tubers increased in weight from 37 mg to 200 g the number of cells increased 500-fold whereas mean cell volume increased only 10-fold. He thought that both processes continued throughout tuber growth and that cell division was the more important of the two. Reeve *et al.* (loc. cit.) found that cell division and expansion were not uniform throughout the tuber and that there were more cell divisions in the perimedullary and cortical regions of the tuber, i.e. those derived from the procambium, than in the pith. The greatest cell enlargement also occurred in these procambial derivatives. They believe that increase in cell number tended to precede cell enlargement, and that most of the growth of tubers after they had reached 30–40 g (cv. Kennebec and Russet Burbank) was by cell enlargement. There was, however, continued cell division in the phloem, to keep pace with the increase in tuber size, and some new cells were produced around these phloem strands and particularly near the developing tuber buds. These sites of cell division are nearest the ends of the phloem system through which the assimilates are transported to the tuber. All their work indicated a decrease in the mean cell volume from the stem end of the tubers, i.e. the oldest internodes, to the younger bud end. Similar gradients were found in the specific gravity and total solids and nitrogen content of tubers with the highest concentrations in the older tissues nearest the stolon. There also appear to be higher concentrations in the cortex and perimedullary tissue than in the pith and periderm with the highest concentrations near the vascular ring (Reeve *et al.*, 1970). Quarmby and Ratkowsky (1972) confirmed these results and Quarmby (personal communication) has been able to construct a mathematical model which can predict the dry matter distribution within the tuber from discrete tissue samples and also produce mean values for whole tubers.

Individual tubers on a plant appear to grow at varying rates and the largest at any time need not necessarily be the fastest growing at the time of measurement (Sadler, 1961; Moorby, 1967). When plants are exposed to $^{14}CO_2$ there is little relationship between the amount of ^{14}C transported to the tubers in 24 h and their fresh weight (Lovell and Booth, 1967; Moorby, 1968, 1970). The tubers which obtained the greatest proportion of the mobile ^{14}C accumulated much of it in ethanol insoluble forms, presumably starch, and it is tempting to assume that these are the tubers with the greatest rate of growth at that particular time. The transport system *per se* does not appear to impose any restriction on the rates of tuber growth (Gray and Smith, 1973; Moorby and Milthorpe, 1975; Munns and Pearson, 1974; Moorby *et al.*, 1975). Variations in the growth rate appear to be associated with the duration of cell division in the tubers, the fastest growing tubers being those which had rather longer periods of cell division so producing more cells capable of expanding and more phloem through which the assimilates could be moved (Reeve *et al.*, 1973b).

The assimilated carbon is delivered to the tubers as sucrose and is then used to synthesize starch. The specific gravity of the various tuber tissues would suggest that the sucrose is not transferred far from the phloem before being converted to starch. The various enzymes concerned with starch synthesis have been investigated by Pressey, 1969, 1970; Ohad *et al.*, 1971; Sowokinos, 1971, 1973; Hawker *et al.*, 1972; Isherwood, 1973 and Moorby *et al.*, 1975. There is little evidence of a close control of starch synthesis by changes in the activities of these enzymes except possibly a decrease in the concentration of sucrose synthetase as the plants age and the supply of assimilates is reduced. This conclusion is supported by the observation that starch synthesis can continue throughout the storage of tubers (Isherwood, loc. cit.) and even after planting during a period when there is a net loss of material from the tuber (Moorby, 1967). In contrast to the continued ability to synthesize starch, there is a decrease in the *in vitro* activity of ribosomal preparations indicating a decrease in protein synthesis (Cocucci *et al.*, 1972).

An unusual type of tuber growth which is closely correlated with starch synthesis is second growth. Second growth appears to be initiated by circumstances which temporarily restrict the supply of assimilates available for tuber growth such as periods of water deficit and high temperatures or treatment with gibberellic acid (Lugt, 1960; Choudhuri and Ghose, 1963; Ruf, 1964; Bodlaender *et al.*, 1964; Lugt *et al.*, 1964; Yamaguchi *et al.*, 1964; Sabalvoro, 1965; McCorquodale, 1966; Lovell and Booth, 1967; Booth and Lovell, 1972). When tubers which have been prevented from growing by treatments such as these, once again receive adequate supplies of assimilates, growth and the starch synthesizing system seem to be unable to resume their activity except in restricted parts of the tuber, usually those adjacent to the buds. Alternatively, some of the buds resume growth as stolons and initiate tubers at a later date. There seems to be an unusually rapid onset of maturity in progressively older internodes followed by a resumption of growth in parts of the tuber – resulting in elongated or mis-shapen tubers, or as stolons – leading to chain tuberization. Second growth is very similar to 'sugar-end potatoes' where cells in the basal region of the tubers have high sugar and low starch contents which cause problems for processers (Weaver *et al.*, 1972; Iritani *et al.*, 1973; Iritani and Weller, 1973).

The tuber regions which produce second growth are, according to Reeve *et al.* (loc. cit.), those in which cell division continues for longer periods than in the rest of the tuber. This is an interesting correlation. We know little about the factors controlling the growth of individual tubers, and especially the relative importance and control of cell division and expansion. Experiments with callus derived from tubers have shown that growth is dependent on supplies of many growth substances, macro- and micronutrients and vitamins (Anstis and Northcote, 1973). Results obtained from callus cultures should be treated with caution, however, since they may not be applicable to the more structured situation in a tuber. For example, experience suggests that the optimum auxin concentration for callus growth may be much too high for tuber growth. Also interesting is the apparent dependence of active starch synthesis on a continued supply of assimilates. Both phenomena seem to be involved in the problems of tuber initiation and growth and both merit further study.

The growth of tubers as a crop

The tubers on a plant compete with each other for the substrates they need for growth. They are partially buffered from reduced supplies of mineral nutrients by the transfer of considerable amounts of materials from the haulm (see Section 4.2.2). This leads to the relative constancy of the concentrations in tubers of mobile nutrients such as nitrogen and phosphorus (Kunkel *et al.*, 1973). However, some competition for

N was suggested by the work of Bodlaender and Reestman (1968) in which there were bigger interactions between nitrogen supply and plant density at the later rather than the early harvests. Competition between tubers for water is indicated by effects on the dry matter content of tubers, i.e. their specific gravity. For example, Bleasdale and Thompson (1969) believed that high dry matter content at high plant densities was caused by the greater depletion of the soil water by the large tuber populations. There is also a complicating factor of tuber size, and Ifenkwe *et al.* (1974) found that the maximum dry matter content was in tubers about 4 to 5 cm in diameter.

Bremner *et al.* (1967) showed that potato plants can compete with each other for light. In addition, about 90% of the dry matter which moves into tubers is dependent on current assimilation (Moorby, 1970). It seems reasonable, therefore, to suppose that the tubers compete for the assimilates which are produced. This provides a means for the agronomic manipulation of tuber size and yield. As was described in Section 4.2.2 the mean size of each stem decreases as the number of stems per hill, or per unit area, is increased. At the same time, there is little effect of the plant population on the number of tubers produced by each stem. There is, therefore, a reduction in the amount of foliage capable of supplying assimilates to each tuber and the mean tuber size is decreased (Toosey, 1963; Bleasdale, 1965; Bremner and Taha, 1966; Moorby, 1967; Sharpe and Dent, 1968; Allen, 1972; Thompson and Taylor, 1974). The overall result of these effects is an asymptotic approach to a maximum yield per unit area. Although the overall mean tuber size is decreased at higher plant densities, this is achieved by increases in the number of smaller tubers. The number of tubers in the larger grades remains relatively constant and this is unusual in a competitive situation (Bleasdale, 1965; Moorby, 1967). The tubers which eventually form the harvested crop are initiated over a period of about two weeks. (Others are initiated later, but rarely grow to an appreciable size and any starch they contain is often reabsorbed by the plant). Some tubers start to grow immediately after initiation. In most competitive situations the dominance established in this way would be maintained, but here these first formed tubers eventually stop growing and allow others to start. Why this occurs is not known, but the work of Reeve *et al.* discussed above provides some suggestions. If their observations that most the late growth of tubers is by cell expansion proves generally applicable, there must be a maximum size that the tubers can attain. This will be determined by the supply of substrates and the number of cells capable of expansion. Once all the available cells have expanded and been filled with starch the tuber must stop growing. In addition to these effects, the observed variations in the growth rates of individual tubers could be caused by variations in the numbers of dividing cells but local variations in soil temperature, water supply or the concentrations of growth substances might also be involved. The growth of the crop of tubers is finally terminated by the senescence of the haulm or its destruction chemically or by pathogens.

Since about 90% of the dry matter accumulated by the tubers is from photosynthesis which occurs after tuber initiation it might, at first, be expected that there would be a close relationship between tuber yield and total insolation. However, there are so many possible side effects that it is rather surprising that any correlations have been found. Scholte Ubing (1958) found such an interrelationship but there was also an interaction with the water use of the crop which could have caused secondary effects (see Section 4.2.2 and Penman, 1963). A closer relationship might be expected to exist over shorter periods and Soltanpur and Moore (1969) claim to have found a linear relationship between mean relative growth rates (R) measured over weekly periods and the total short wave radiation. Unfortunately, the equation they used to estimate the relative growth rate was incorrect since they used $R = \mathrm{d}W_{\mathrm{total}}/\mathrm{d}t \cdot 1/W_{\mathrm{haulm}}$ instead of $R = \mathrm{d}W_{\mathrm{total}}/\mathrm{d}t \cdot 1/W_{\mathrm{total}}$. The first equation would estimate a constant proportion of the relative growth rate if the ratio $W_{\mathrm{total}} : W_{\mathrm{haulm}}$ was constant throughout the experiment. This would not be so in a growing crop. The equation used would, however, estimate the net assimilation rate ($\mathrm{d}W_{\mathrm{total}}/\mathrm{d}t.\ 1/A$) if W_{haulm} was proportional to the leaf area (A). Insufficient data is given to confirm this point, but it would seem unlikely in a period when there is usually a net loss of dry matter from the haulm to the tubers.

The best evidence for a correlation between yield and radiation comes from the work of Sibma (1970) and Sale (1973a, b). Sibma found a good correlation between the total radiation received between three weeks after emergence and harvest. However, it can be seen from the partial correlation coefficients he quotes that the length of growing season was rather more important than the average daily radiation. This might be expected from the larger haulm which was probably produced in the high radiation conditions. Yet another complicating factor could have been an effect of radiation on the number of tubers which were formed. Sale (loc. cit.), for example, thought that the major effect of differences in solar input was exerted about the time of tuber initiation when shaded plants produced fewer tubers more than 1 cm in diameter than did plants exposed to full sunlight. Gray and Holmes (1970) found a similar effect but also approximately equivalent decreases in the net assimilation rates and dry weight increments of tubers on plants which were shaded for a period after the tubers had started to grow. In view of all these complicating factors, it is surprising that it is possible to obtain

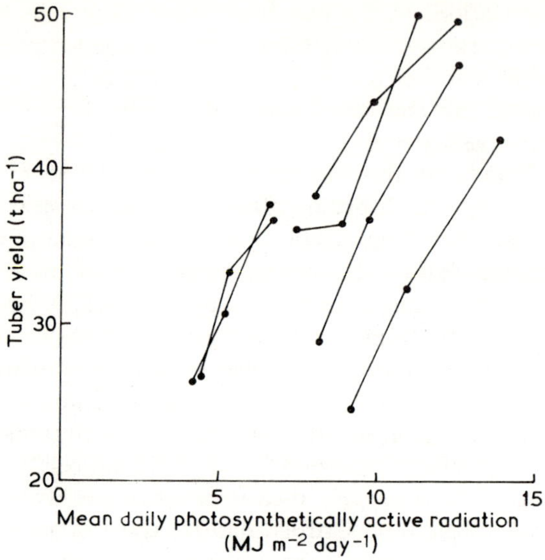

Fig. 4.7 The relationship between tuber yield and the mean daily total of photosynthetically active radiation in some experiments using cv. Sebago in the Murrumbigee Irrigation Area, Australia (after Sale, 1973b)

any relation between tuber yield and the mean daily radiation (Fig. 4.7). It should be noted, however, that these data (from Sale, 1973b) show a different relationship for each of his six crops. That is, factors peculiar to each crop determine the nature of the relationship. Sale considered these to be the number of tubers which grow to a reasonable size and the time of onset of rapid tuber growth, both of which are sensitive to the conditions at the time of tuber initiation. Effects of radiation on the rate of tuber growth were less important but the persistence of the foliage had an effect.

Although these general correlations exist over the whole season, during the period of tuber growth there are more complex changes in the distribution of growth in the plant and in the pattern of carbon assimilation. There appear to be increases in the rates of photosynthesis and net assimilation at about the time of tuber initiation. At the same time the rate of haulm growth starts to decrease (Fig. 4.8) (Milthorpe, 1963; Gifford and Moorby, 1967; Meinl, 1967; Moorby, 1970; Sale, 1973a). The net assimilation rate then declines to a minimum value at about the time of maximum leaf area and finally increases again as the leaves senesce. It was argued by Moorby (1970) that this final increase probably resulted from an increase in the rate of photosynthesis in the remaining leaves because of the continued demand for assimilates from the tubers. This conclusion has been supported by the work of Sale (1973a, b, 1974) who found a similar increase in net assimilation

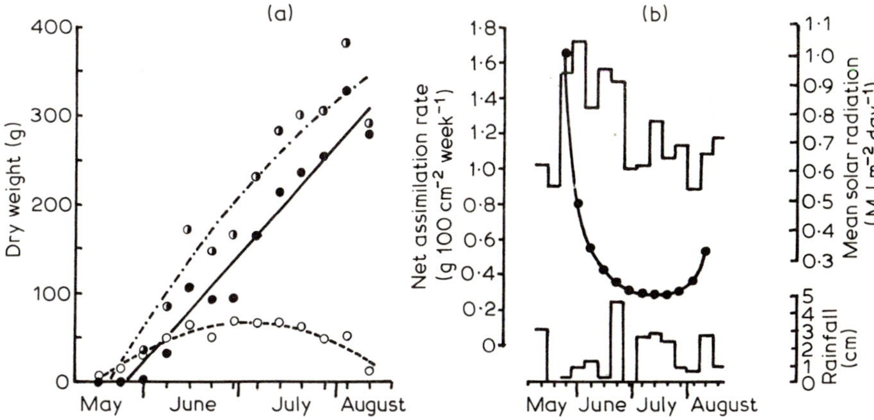

Fig. 4.8 (a) The changes with time in the weight of the haulm (o), tubers (●) and entire plant (◑) of a crop of Arran Pilot grown in England. (b) The changes with time in the net assimilation rate of the crop shown in Fig. 8a and the mean solar radiation and rainfall data for the same period (after Moorby, 1970)

rate together with decreases in leaf area and rate of photosynthesis per unit ground area.

Another point of interest in Fig. 4.8 is the apparent insensitivity of the rates of tuber growth and net assimilation to short term fluctuations in the incoming radiation. Again, it was argued that this was caused by the demand from the tubers exerting some control over the rate of photosynthesis. More direct evidence of such a feedback control comes from observations that if tubers were removed or their growth retarded by low soil temperatures, the net assimilation rate and rate of photosynthesis was reduced (Burt, 1964b, 1966; Nösberger and Humphries, 1965). The same dominance of the tubers over events in the haulm can be seen when tomato scions are grafted on to potato root stocks (Khan and Sagar, 1969) and when low-yielding potato scions are grafted on to high-yielding potato root stocks (Börger *et al.*, 1956). In contrast, partial defoliation has no effect on the rate of tuber growth (Humphries, 1969). The system seems, therefore, to have a considerable homeostasis unless the tubers are removed. If this occurs the other meristems on the plants are unable to take advantage of the greater availability of assimilates and photosynthesis declines. There has been no critical examination of the nature of this control in *Solanum*. In *Capsicum* it appears to be related to increases in both the stomatal and mesophyll resistances to CO_2 transfer (Hall, 1975) and in *Medicago* to effects only on the mesophyll resistance (Hodgkinson *et al.*, 1972).

One obvious way to increase yield would seem to be to prolong the period of photosynthesis. The general antagonism between haulm and tuber growth has been mentioned in several parts of this chapter and is well illustrated in Fig. 4.6. Here, the application of N increased the maximum leaf area index and the persistence of the haulm. It also delayed the onset of tuber growth, but once established both the rate and duration of tuber growth were increased. The effect on duration of growth was probably a result of the greater leaf area duration ($_{t_0}\int^{t_1}A$, where t_0 and t_1 are the tissues of emergence and final senescence of the haulm) although this was correlated with the maximum leaf area index.

Bremner and Radley (1966) and Bremner and Taha (1966) produced evidence for a linear relationship between tuber yield and the leaf area duration and found the best statistical fit to the relationship when they assumed that all leaf area indices above three were three. This is a tacit assumption that there was no increase in the efficiency of carbon fixation and utilization when plants produced leaf area indices greater than three. This would depend very much on the structure of the crop canopy and amount of solar radiation. It is not surprising, therefore, that Gunasena and Harris (1968, 1969, 1971) found no improvement to the relationship when the same restriction was applied. They thought the difference may have arisen because they used a more dense plant population but other factors might also have been involved.

Any further understanding of the mechanisms controlling tuber growth must await further information on the control processes which operate in, and on, tubers and which have been discussed above. In addition, we need to know much more about the photosynthetic system of potatoes. This will require knowledge of the nature of the crop canopy and how it is modified by environmental variables such as radiation, temperature and the supplies of mineral nutrients and water. The photosynthetic behaviour of potato leaves appears to be similar to that of other species with little varietal variation (Meinl, 1969). The rate of photosynthesis per unit area of young leaves is low, rises to a maximum at about the time of flowering or tuber initiation and thereafter declines (Meinl, 1967; Winkler, 1971). The behaviour of crop canopies appears to follow a similar time course but there was no further increase in the rate of canopy photosynthesis with levels of incoming photosynthetically active radiation of more than about 200–225 W m $^{-2}$ (Sale, 1974). This is rather low, even for C_3-crops. Attempts by Winkler and Sale (loc. cit.) to construct carbon balances for the crop have proved useful and should now be extended to examine the effects of a range of environmental variables and different genotypes. These will necessitate investigations of canopy structure and especially carbon distribution and use in growth and respiration. There is a close correlation between age and respiration rate (Burton, 1964; Winkler, loc. cit.) but this might be related to the usual correlation between respiration and the amount of carbon fixed in the previous photoperiod (Sale, 1974; Ludwig *et al.*,

1975). Respiration rates are probably also affected by temperature and tuber age and the onset of dormancy. Relationships between carbon fixation and tuber yield are, therefore, considerably more complex than are indicated by Burton's (1964) preliminary calculations using limited data from experiments in controlled environments. Further work in this complicated field will probably necessitate the use of sophisticated simulation techniques. Several mathematical models are available which can be used to simulate canopy photosynthesis (cf. Setlík, 1969; Evans, 1975). Unfortunately, there are, as yet, no published models which can accommodate the feedback controls of photosynthesis which appear to originate in the tubers and are such an important factor in the potato crop.

4.3 Sprout growth in storage

If tubers are allowed to grow undisturbed the buds they bear become dormant in acropetal succession. Finally, the apical bud becomes dormant and the tuber stops growing at about the same time. Tubers can, however, be harvested at any time; they might be still growing, but have reached a commercially useful size (e.g. for 'earlies'), or they might have stopped growing and be fully dormant (e.g. 'maincrop'). In many instances growth is stopped by haulm destruction to prevent the production of large tubers or infection by pathogens. Tuber growth is then stopped abruptly and the buds became dormant. However, the tubers are not 'mature' in the sense that they are easily damaged and have special biochemical characteristics. A precise definition of an immature tuber is impossible at present, but in general terms it is one without a fully developed periderm and in which the starch synthesizing enzymes are fully active. After dormancy has been established the behaviour of all these types of tuber is similar but the timing of events might differ. This can lead to effects on the following crop if the tubers are used for seed (Gray, 1973) but the effects can be avoided by suitable management of the tubers during storage (Goodwin *et al.*, 1969a, b; Holmes and Gray, 1972).

Dormancy and sprout growth have recently been discussed at length (Moorby and Milthorpe, 1975) and I will be content here to outline the events which occur and the factors which control them; incorporating where necessary any new information. In particular, I will try and show which aspects of sprout growth have to be controlled or emphasized in order to produce tubers most suited to their eventual use.

The rates of respiration and metabolism in dormant tubers are low (Burton, 1964, 1974b) and the observed increase in oxygen uptake at the end of dormancy is mainly caused by an increase in bud respiration (Workman and Twomey, 1967). Respiration can also be increased by treatment with ethylene (Reid and Pratt, 1972; Rylski *et al.*, 1974). The latter authors showed that ethylene could shorten the dormant period but that it also inhibited sprout elongation and Burton and Meigh (1971) thought that ethylene and other volatile substances produced by stored potatoes were not important in the control of dormancy. They may, however, have significant effects on sprout growth (see Section 4.2.1).

One of the most important factors in the control of dormancy appears to be the inhibition of protein synthesis. The cessation of dormancy is accompanied by an increase in the rates of nucleic acid and protein synthesis and there are associated increases in the concentrations of gibberellins and a decrease in the concentration of abscisic acid (Madison and Rappaport, 1968; Rappaport and Wolfe, 1969; Shih and Rappaport, 1970). The gibberellins and abscisic acid appear to act independently on the synthesis of nucleic acids rather than, say, abscisic acid inhibiting gibberellin synthesis. Increasing concentrations of gibberellins also promote the breakdown of starch to reducing sugars, but these do not appear to have any direct effect on the release of buds from dormancy (Clegg and Rappaport, 1970).

The first bud to be released from dormancy and start to grow is usually the apical bud. This is usually the largest bud on a tuber and the last to stop growing in the previous season. The lateral buds are then released from dormancy in basipetal succession. Their subsequent growth depends very much on the environmental conditions, and especially the temperature (Goodwin, 1967b; Morris, 1966a, 1967b). Two processes appear to operate: the sprouts compete for substrates from the mother tuber, and an apical dominance situation develops with the shorter sprouts gradually being inhibited by the larger (Fig. 4.9). At temperatures of the order of 20° C apical dominance is established so rapidly that often only one sprout, usually the apical, grows. The shorter laterals remain inhibited unless the dominant sprout is damaged. How the dominance of the largest sprouts is exerted is unknown, but some form of growth inhibitor appears to be involved (Goodwin and Cansfield, 1967). Because of the intersprout competition the mean dry weight per growing sprout, D(g) and mean sprout length were positive functions of the tuber weight, W(g) and negatively related to the number of sprouts per tuber, N (Fig. 4.10). The linear relationship $D = 563.4 - 336.4\ N + 5.761\ W$ accounted for 94.5% of the variance in Arran Pilot (Morris, 1966a). Competition was absent at 7° C but was apparent at higher temperatures where the faster growth rate was dependent on larger supplies of substrates. The phloem system which extends throughout the storage tissue of the mother tuber ensures that depletion of the reserves is uniform throughout the tuber and is not restricted to the regions adjacent to the sprouts (Morris, 1966a; Edelman and Singh, 1966; Edelman *et al.*,

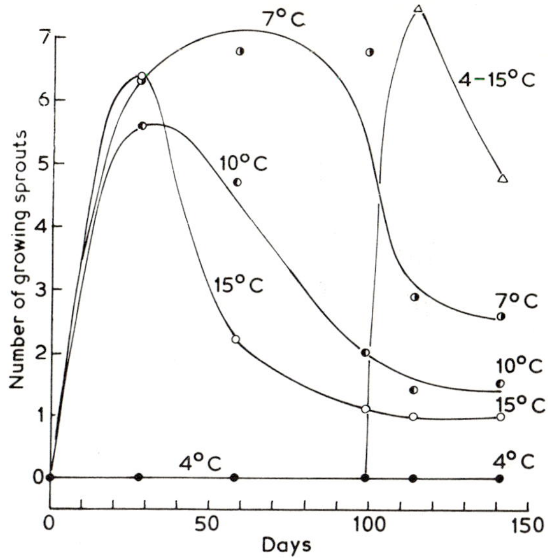

Fig. 4.9 The changes with time in the number of growing sprouts on tubers of cv. King Edward transferred from 4° C to 7, 10, and 15° C on day zero and 15° C on day 99 (unpublished data of D.A. Morris)

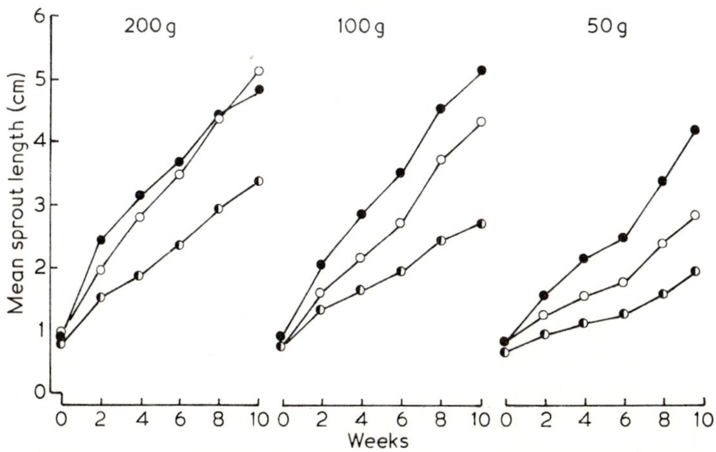

Fig. 4.10 Effect of tuber size (200, 100 and 50 g fresh weight) and number of sprouts per tuber on sprout elongation at 15° C of cv. Arran Pilot. Closed, open and pied circles are 1, 2 and 4 sprouts per tuber respectively (after Morris, 1966a)

1969). Although there is ample evidence of intersprout competition we do not know for what the sprouts compete. It is unlikely to be carbohydrate; there are ample supplies in the mother tuber and in many instances the sprouts accumulated starch. Headford (1961) suggested that there might be competition for nitrogen and Morris (1966b) found that competition could be decreased by supplying a solution of mineral nutrients. Tubers contain very low concentrations of calcium (see Section 4.2.2) and insufficient mobilization of this can lead to necrosis of the elongating regions of sprouts and eventual death of the sprout apices (Dyson and Digby, 1975). Apical death can, in turn, release the other sprouts from apical dominance, a sequence of events which has been ascribed to physiological age.

The chemical changes which occur during storage have been studied extensively, especially those concerned with carbohydrate metabolism. The physiological basis of the changes is the breakdown of reserve materials to provide substrates which can be transported and used to support sprout growth. The concentration of sugars produced in storage varies markedly with variety and because of this some cultivars are more suitable for processing than others. The breakdown starts at the end of dormancy apparently in response to a signal, probably hormonal, which seems to originate in the sprouts. Some evidence suggests that gibberellins can hasten the breakdown of tuber reserves and it is easy to demonstrate the disappearance of starch from tuber tissue treated with gibberellic acid (Morris, 1967b). Low temperatures also favour the breakdown of starch and since sprout growth is slow at low temperatures sugars, especially sucrose, accumulate (Isherwood, 1973; Pollock and ap Rees, 1975a, b). This 'sweetening' of tubers at low temperatures appears to occur through the action of sucrose phosphate synthetase and the sucrose produced is hydrolysed by invertase (Pressey and Shaw, 1966; Moll, 1968; Pressey, 1970; Pollock and ap Rees, loc. cit.). Isherwood (1973) has shown that when tubers are transferred from 10 to 2 °C and back to 10 °C the sum of starch and sucrose and respiratory CO_2 did not change but there were large changes in the relative amounts of each. The changes were brought about by the relative effects of temperature on the starch–sugar interconversions and the rate of respiration. In addition to these changes in carbohydrates there are analogous changes in the nitrogenous constituents of tubers. During sprout growth proteins are hydrolysed to amino acids and proline appears to be of particular importance (cf. Emilsson and Lindblom, 1963).

It is quite easy to see how this information on sprout growth and biochemical changes can be put to use. For example, in the production of early potatoes the aim is to produce an economically useful crop as quickly as possible. The information given in earlier sections indicated that tuber initiation is hastened by the use of large sprouts and that relatively low stem densities provided a high proportion of the yield in the larger grades of tubers. These conditions can be most easily satisfied by sprouting tubers at temperatures of 15 °C or more in the light to produce tubers with one well developed, apically dominant, sprout. These can then be planted at a rate which will give the optimum plant population. The large sprouts will emerge and form tubers rapidly but are unlikely to produce a large maximum yield. Early tuber initiation is not important in the production of maincrop potatoes. It may, indeed, be a disadvantage because it usually restricts haulm growth and reduces the final yield. The aim with maincrop potatoes in therefore a large population of stems which will only initiate tubers after the production of a large leaf area index. This can be achieved by storing tubers at cool temperatures to produce many small sprouts and planting them so as to produce the required plant population. In sprouting seed tubers we are not too concerned with the biochemical changes which accompany sprout growth. In the storage of ware tubers, however, the biochemical changes are of prime importance and sprout growth is usually only an unfortunate necessity. Most users require tubers with low sugar concentrations and with minimal loss of weight in storage. This is usually achieved by storage at about 10° C. Sprout growth, which can be appreciable at this temperature, is often prevented by the use of growth inhibitors such as isopropyl-*m*-chlorocarbanilate (CIPC). There is necessarily an accumulation of sugar as storage proceeds and this is unacceptable for many processes. It can be corrected by storage for a period at a higher temperature to 'recondition' the tubers, i.e. some of the sugars are used in increased respiration and others are converted to starch.

4.4 Conclusions

The work of the past decade has confirmed many of the theories which were under discussion in the mid-1960s but it has also shown the limitations of others. The basic principles of sprout growth are now well understood and it should be possible to produce good instructions on how to manage seed tubers so that they are in the optimum condition for their subsequent use. There are, however, two complicating factors. The first is the problem of tuber maturity or physiological age. These terms seem to be used in the way we are told a drunk uses a lamp post – more for support than illumination. This state of affairs will continue until we know much more about what constitutes tuber maturity. However this is discussed in detail in Chapter 8. The second problem is the present impossibility of predicting the weather, other than in very general terms. This should not be used as a reason for delaying progress, however, since any physiological and agronomic help that can be given to growers is better than nothing. The problems of weather arise also in the later growth of the crop. We can now provide descriptions of how the crop reacts to some of the environmental variables. We are still ignorant of many of the interactions between variables although it is these interactions which provide the basic control over the growth of the crop. I have suggested that the usefulness of simple field experimentation has now to be questioned. Simple experiments on the interactions between environmental variables are almost impossible and we will probably have to use simulation techniques to make progress on this problem. We already have some information which could be used in this way and there are some commendable moves to obtain more data by means of compatible experiments in different environments.

There is one final major problem: the nature of the crop itself. The potato is a very adaptable plant and because of this it can grow in a wide range of environments and produce crops with many different attributes. This versatility is obviously very useful, but it also raises obstacles to our aim of a fuller understanding of the reactions of the crop. Much of our present knowledge has been obtained simply to explain known reactions of the crop to various situations. The aim should now be to a more coherent quantitative understanding of the behaviour of the crop which can be used in a predictive fashion and which will allow us to make better assessments of the potentialities of unusual genotypes or the possibilities for the crop in new environments.

References

Ali, M.A., Lennard, J.H. and Boyd, A.E.W. (1970). *Ann. appl. Biol.* **66**, 407–15.

Allen, E.J. (1972). *J. agric. Sci., Camb.* **79**, 315–21.

Allen, E.J. and Wurr, D.C.E. (1973). *Potato Res.* **16**, 10–20.

Alvey, N.G. (1965). *Eur. Potato J.* **8**, 1–13.

Anstis, P.J.P. and Northcote, D.H. (1973). *J. exp. Bot.* **24**, 425–41.

Artschwager, E. (1924). *J. agric. Res.* **27**, 809–35.

Baker, D.A. and Moorby, J. (1969). *Ann. Bot.* **33**, 729–41.

Birch, J.A., Devine, J.R., Holmes, M.R.J. and Whitear, J.D. (1967). *J. agric. Sci., Camb.* **69**, 13–24.

Bleasdale, J.K.A. (1965). *J. agric. Sci., Camb.* **64**, 361–6.

Bleasdale, J.K.A. and Thompson, R. (1969). *Eur. Potato J.* **12**, 173–87.

Bodlaender, K.B.A. (1963). In *The Growth of the Potato ed.* Ivins, J.D. and Milthorpe, F.L. pp. 199–210. *Proc. 10th Easter Sch. agric. Sci.*, Univ. Nottingham, London; Butterworths.

Bodlaender, K.B.A. and Algra, S. (1966). *Eur. Potato J.* **9**, 242–58.

Bodlaender, K.B.A. Lugt, C. and Marinus, J. (1964). *Eur. Potato J.* **7**, 57–71.

Bodlaender, K.B.A. and Marinus, J. (1969). *Neth. J. agric. Sci.* **17**, 300–8.

Bodlaender, K.B.A. and Reestman, A.J. (1968). *Neth. J. agric. Sci* **16**, 165–76.

Booth, A. (1959). *J. Linn. Soc., Bot.* **56**, 166–9.

Booth, A. (1963). In *The Growth of the Potato*, ed. Ivins J.D. and Milthorpe, F.L. pp. 99–113. *Proc 10th Easter Sch. agric. Sci.*, Univ. Nottingham, London: Butterworths.

Booth, A, and Lovell, P.H. (1972). *New Phytol.* **71**, 795–804.

Borah, M.N. and Milthorpe, F.L. (1962). *Indian J. Pl. Physiol.* **5**, 53–72.

Börger, H., Huhnke, W., Köhler, D., Schwanitz, F. and von Sengbusch, R. (1956). *Züchter* **26**, 363–70.

Bremner, P.M., El-Saeed, E.A.K. and Scott, R.K. (1967). *J. agric. Sci., Camb.* **69**, 283–90.

Bremner, P.M. and Radley, R.W. (1966). *J. agric. Sci., Camb.* **66**, 253–62.
Bremner, P.M. and Taha, M.A. (1966). *J. agric. Sci., Camb.* **66**, 241–52.
Bretzloff, C.W. (1971). *Am. Potato J.* **48**, 97–104.
Bretzloff, C.W. and McMenamin, J. (1971). *Am. Potato J.* 48, 246–54.
Burg, P.F.J. van (1967). *Neth. Nitrogen Tech. Bull.* **4**, 1–30
Burke, M.J. and Gusta, L.V. (1974). I. pp. 22–60 in *Report of the International Potato Centre's Planning Conference on Cold Hardiness* Lima: C.I.P.
Burrage, S.W. (1965). *The growth of early potatoes*. Ph.D. Thesis, University of, Nottingham.
Burt, R.L. (1964a). *Eur. Potato J.* **7**, 197–208.
Burt, R.L. (1964b). *Aust. J. biol. Sci.* **17**, 867–77.
Burt, R.L. (1965). *Eur. Potato J.* **8**, 104–14.
Burt, R.L. (1966). *Aust. J. biol. Sci.* **19**, 711–14.
Burton, W.G. (1964). *Eur. Potato J.* **2**, 90–101.
Burton, W.G. (1966). *The Potato*. H. Veenman and Zonen, Wageningen.
Burton, W.G. (1974a). *Potato Res.* **17**, 374–409.
Burton, W.G. (1974b). *Potato Res.* **17**, 113–37.
Burton, W.G. and Meigh, D.F. (1971). *Potato Res.* **14**, 96–101.
Caesar, K and Krug, H. (1965). *Eur. Potato J.* **8**, 28–32.
Carpenter, P.N. (1963). *Bull. Maine agric. Exp. Stn.* **610**, 1–38.
Catchpole, A.H. and Hillman, J. (1969). *Nature* **223**, 1387.
Catchpole, A.H. Hillman, J. (1975). *Potato Res.* **18**, 282–9.
Choudhuri, H.C. and Ghose, S. (1963). *Eur. Potato J.* **6**, 160–7.
Claver, R.K. (1971). *Rev. Fac. Agron. La Plata*, **47**, 65–73.
Clegg, M.D. and Rappaport, L. (1970). *Pl. Physiol.* **45**, 8–13.
Cocucci, S.M., Garlaschi, F.M. Bianchi, E. and Marre, E. (1972). *Physiologia Pl.* **27**, 220–5.
Cox, A.E. (1970). *Potato Res.* **13**, 332–41.
Dale, J.E. (1972). *Ann. Bot.* **36**, 967–79.
Dale, J.E., Felippe, G.M. and Marriott, C. (1974). *Ann. Bot.* **38**, 575–88.
Dyson, P.W. (1965). *J. Sci. Fd Agric.* **16**, 542–9.
Dyson, P.W. and Digby, J. (1975). *Potato Res.* **18**, 290–305.
Dyson, P.W. and Watson, D.J. (1971). *Ann. appl. Biol.* **69**, 47–63.
Edelman, J., Jefford, T.G. and Singh, S.P. (1969). *Planta* **84**, 48–56.
Edelman, J. and Singh, S.P. (1966). *J. exp. Bot.* **17**, 696–702.
El-Antably, H.M.M., Wareing. P.F. and Hillman, J. (1967). *Planta* **73**, 74–90.
Ellis, F.B. (1966). *Agric. Res. Council Letcombe Lab. Ann. Rep.* **16**, 47–8.
Emilsson, B. and Lindblom, H. (1963). In *The Growth of the Potato*, eds. Ivins, J.D. and Milthorpe, F.L. pp. 45–62. *Proc. 10th Easter Sch. agric. Sci.*, University of Nottingham, London: Butterworths.
Emmond, G.S. (1971). *J. Am. Soc. hort. Sci.* **96**, 445–7.
Endrödi, G. and Rijtema, P.E. (1969). *Neth. J. agric. Sci.* **17**, 283–99.
Engel, K.H., Meinl, G. and Raeuber, A (1964). *Z. angew. Met.* **4**, 364–9.
Epstein, E. and Grant, W.J. (1973). *Agron. J.* **65**, 400–4.
Estrada Ramos, N. (1974). App. II pp. 61–5. In *Report of the International Potato Centre's Planning Conference on Cold Hardiness*, Lima: C.I.P.
Evans, L.T. (1975). *Crop Physiology – Some Case Histories*. Cambridge: University Press, Cambridge.
Ezeta, F.N. and McCollum, R.E. (1972). *Am. Potato J.* **49**, 151–63.
Fong, K.H. and Ulrich, A. (1969a) *J. Am. Soc. hort. Sci.* **94**, 341–4.
Fong, K.H. and Ulrich, A. (1969b). *J. Am. Soc. hort. Sci.* **94**, 356–9.
Fong, K.H. and Ulrich, A. (1970). *Commun. Soil Sci and Pl. Anal* **1**, 43–55.

Fong, K.H. and Ulrich, A. (1974). *J. Am. Soc. hort. Sci.* **99**, 334–7.
Forsline, P.C. and Langille, A.R. (1975). *Physiologia Pl.* **34**, 75–7.
Garcia-Torres, L. and Gomez-Campo, C. (1972). *Potato Res.* **15**, 76–8.
Garcia-Torres, L. and Gomez-Campo, C. (1973). *Potato Res.* **16**, 73–9.
Glendinning, D.R. (1975). *Potato Res.* **18**, 256–61.
Gifford, R.M. and Moorby, J. (1967). *Eur. Potato J.* **10**, 235–8.
Goodwin, P.B. (1967a). *J. exp. Bot.* **18**, 78–86.
Goodwin, P.B. (1967b). *J. exp. Bot.* **18**, 87–99.
Goodwin, P.B., Brown, A., Lennard, J.H. and Milthorpe, F.L. (1969a). *J. agric. Sci. Camb.* **73**, 161–6.
Goodwin, P.B., Brown, A., Lennard, J.H. and Milthorpe, F.L. (1969b). *J. agric. Sci., Camb.* **73**, 167–76.
Goodwin, P.B. and Cansfield, P.E. (1967). *J. exp. Bot.* **18**, 297–307.
Gray, D. (1973). *Ann. appl. Biol.* **75**, 83–91.
Gray, D. (1974). *J. agric. Sci., Camb.* **82**, 363–9.
Gray, D. and Holmes, J.C. (1970). *Potato Res.* **13**, 215–19.
Gray, D. and Smith, D.J. (1973). *Potato Res.* **16**, 293–5.
Greenwood, D.J., Cleaver, T.J. and Niendorf, K.B. (1974). *J. agric. Sci., Camb.* **82**, 217–32.
Greenwood, D.J., Wood, J.T., Cleaver, T.J. and Hunt, J. (1971). *J. agric. Sci., Camb.* **77**, 511–23.
Gregory, L.E. (1956). *Am. J. Bot.* **43**, 281–8.
Gregory, L.E. (1965). In *Encyclopaedia of Plant Physiology*, **15i**, 1328–54.
Gunesena, H.P.M. and Harris, P.M. (1968). *J. agric. Sci., Camb.* **71**, 283–96.
Gunasena, H.P.M. and Harris, P.M. (1969). *J. agric. Sci., Camb.* **73**, 245–59.
Gunasena, H.P.M. and Harris, P.M. (1971). *J. agric. Sci., Camb.* **76**, 33–52.
Hall, A.J. (1975). *Fruit Excision, Defloration and Photosynthesis in Capsicum.* Macquarie University, Ph.D. Thesis.
Hall, R.H. (1973). *Ann. Rev. Pl. Physiol.* **24**, 415–44.
Hammes, P.S. and Beyers, E.A. (1973). *Potato Res.* **16**, 68–72.
Hammes, P.S. and Nel, P.C. (1975). *Potato Res.* **18**, 262–72.
Hawker, J.S., Ozbun, J.L. and Preiss, J. (1972). *Phytochem.* **11**, 1287–93.
Hayden, R.E. Dionne, L. and Fensom, D.S., (1972). *Can J. Bot.* **50**, 1547–54.
Headford, D.W.R. (1961). *Sprout growth of the potato.* Ph.D. Thesis, University of Nottingham.
Headford, D.W.R. (1962). *Eur. Potato J.* **5**, 14–22.
Headford, D.W.R. and Douglas, G. (1967). *Weed Res.* **7**, 131–44.
Hodgkinson, K.C., Smith, N.G. and Miles, G.E. (1972). *Aust. J. agric. Res.* **23**, 225–38.
Holmes, J.C. and Gray, D. (1972). *Potato Res.* **15**, 220–35.
Hughes, J.C. (1974). *Potato Res.* **17**, 512–47.
Humphries, E.C. (1958). *Ann. appl. Biol.* **46**, 346–51.
Humphries, E.C. (1969). *Proc. Int. Symp. Tropical Root Crops.*, University of West Indies, **1**, **II**, 34–45.
Humphries, E.C. and Dyson, P.W. (1967). *Eur. Potato J.* **10**, 116–26.
Humphries, E.C. and French, S.A.W. (1960). *Ann. appl. Biol.* **48**, 177–88.
Humphries, E.C. and French, S.A.W. (1961). *Ann. appl. Biol.* **49**, 331–9.
Ifenkwe, D.P., Allen, E.J. and Wurr, D.C.E. (1974). *Am. Potato J.* **51**, 233–42.
Iritani, W.M. and Weller, L. (1973). *Am. Potato J.* **50**, 223–33.
Iritani, W.M., Weller, L. and Russell, T.S. (1973). *Am. Potato J.* **50**, 24–31.
Isherwood, F.A. (1973). *Phytochem.* **12**, 2579–91.
Ivins, J.D. and Bremner, P.M. (1964). *Outl. Agric.* **4**, 211–17.

Ivins, J.D. and Milthorpe, F.L. (1963). *The Growth of the Potato. Proc 10th Easter Sch. agric. Sci.* University of Nottingham, London: Butterworths.

Jones, R.L. (1973). *Ann. Rev. Pl. Physiol.* **24**, 571–98.

Kalra, G.S. (1971). *Influence of three levels of soil moisture and three levels of soil fertility on the uptake of nutrients and histological changes in potato plants.* Ph.D. Thesis, University of Maine.

Khan, A. and Sagar, G.R. (1969). *Ann. Bot.* **33**, 753–62.

Krauss, A. and Marscher, H. (1971). *Z. Pflanzenernährung u. Bodenkunde*, **128**, 153–68.

Krauss, A. and Marscher, H. (1973). *Z. Pflanzenernährung u. Bodenkunde*, **136**, 229–40.

Krauss, A. and Marscher, H. (1974). *Z. Pflanzenphysiol.* **137**, 116–23.

Krug, H. (1963). *Gartenbauwissenschaft*, **28**, 515–64.

Krug, H. (1965). *Eur. Potato J.* **8**, 14–27.

Kumar, D. and Wareing, P.F. (1972). *New Phytol.* **71**, 639–48.

Kumar, D. and Wareing, P.F. (1973). *New Phytol.* **72**, 283–7.

Kumar, D. and Wareing, P.F. (1974). *New Phytol.* **73**, 833–40.

Kunkel, R. Holstad, N. and Russell, T.S. (1973). *Am. Potato J.* **50**, 275–82.

Lang, A. (1970). *Ann. Rev. Pl. Physiol.* **21**, 537–70.

Lapwood, D.H., Hide, G.A. and Hirst, J.M. (1967). *Pl. Path.* **16**, 61–3.

Lomas, J., Schlesinger, E., Zilka, M. and Israeli, A. (1972). *J. appl. Ecol.* **9**, 107–19.

Lovell, P.H. and Booth, A. (1967). *New Phytol.* **66**, 525–37.

Lovell, P.H. and Booth, A. (1969). *New Phytol.* **68**, 1175–85.

Ludwig, L.J. Charles-Edwards, D.A. and Withers, A.C. (1975). In *Environmental and Biological Control of Photosynthesis*, ed. Marcelli, R. pp. 29–36. The Hague: Junk.

Lugt, C. (1960). *Eur. Potato J.* **3**, 307–24.

Lugt. C., Bodlaender, K.B.A. and Goodijk, G. (1964). *Eur. Potato J.* **7**, 219–77.

McCorquodale, A.J. (1966). *Second-growth of potato tubers in relation to water deficit and the movement of photosynthate.* M.Sc. Thesis, University of Nottingham.

McCorquodale, A.J. and Moorby J. (1969). In *Root growth*, ed. Whittingham, W.J. pp. 395–6. *Proc. 15th Easter Sch. agric. Sci.* University of Nottingham, London: Butterworths.

McCorquodale, I.A. (1971). *Tuberization of potato stolons in aseptic culture.* Ph.D. Thesis, University of Nottingham.

McIlrath, W.J. (1972). In *Biology Data Book*, ed. Altman, P.L. and Dittmer, D.S. Vol. I, *Fed. Am. Soc. Exp. Biol.* pp. 339–406. Bethesda.

Mackay, D.C., Maceachern, C.R. and Bishop, R.F. (1966). *Soil Sci. Soc. Am. Proc.* **30**, 73–6.

Macmillan, C. (1891). *Ann. Nat.* **25**, 462–5.

Madec, P. (1963). In *The Growth of the Potato*, ed. Ivins, J.D. and Milthorpe F.L. pp. 121–30. *Proc. 10th Easter Sch. agric. Sci.* University of Nottingham, London: Butterworths.

Madison, M. and Rappaport, L. (1968). *Pl. Cell Physiol.* **9**, 147–53.

Marinus, J. and Bodlaender, K.B.A. (1975). *Potato Res.* **18**, 189–204.

Meinl, G. (1965). The Growth and Water Supply of Individual Organs of the Potato. In *Water Stress in Plants*, ed. Slavik, B. Prague: Czech. Acad. Sci.

Meinl, G. (1967). *Photosynthetica* **1**, 51–6.

Meinl, G. (1969). *Photosynthetica* **3**, 9–19.

Milthorpe, F.L. (1963). In *The Growth of the Potato*, eds. Ivins, J.D. and Milthorpe,

F.L. pp. 3–16. *Proc. 10th Easter Sch. agric. Sci.* University of Nottingham, London: Butterworths.
Milthorpe, F.L. (1969). *Aust. J. Sci.* **32**, 345–9.
Milthorpe, F.L. and Moorby, J. (1967). *Proc. 3rd Trienn. Conf. Eur. Ass. Potato Res.* Zurich.
Milthorpe, F.L. and Moorby, J. (1969). *Ann. Rev. Pl. Physiol.* **20**, 117–38.
Milthorpe, F.L. and Moorby, J. (1974). *An Introduction to Crop Physiology.* pp. 182–7. Cambridge: Cambridge University Press.
Milthorpe, F.L., Moorby, J., and Morgan, J.M. (1974). In *Mechanisms of Regulation of Plant Growth*, ed. Bieleski, R.L., Ferguson, A.R. and Cresswell, M.M. pp. 895–900. R. Soc. New Zealand, Wellington.
Moll. A. (1968). *Flora, Jena* **159**, 277–92.
Moorby, J. (1967). *Eur. Potato J.* **10**, 189–205.
Moorby, J. (1968). *Ann. Bot.* **32**, 57–68.
Moorby, J. (1970). *Ann. Bot.* **34**, 297–308.
Moorby, J. and McGee, S. (1966). *Ann. appl. Biol.* **54**, 159–70.
Moorby, J. and Milthorpe, F.L. (1975). (*Solanum tuberosum* L.) (1975) *Crop Physiology—Some Case Histories*, ed. Evans, L.T. pp. 225–257. Cambridge: Cambridge University Press.
Moorby, J. Munns, R. and Walcott, J. (1975). *Aust. J. Pl. Physiol.* **2**, 323–33.
Morris, D.A. (1966a). *Eur. Potato J.* **9**, 69–85.
Morris, D.A. (1966b). *Inter-relationships between sprout and mother tuber in the potato.* Ph.D. Thesis, University of Nottingham.
Morris, D.A. (1967a). *Eur. Potato J.* **10**, 296–311.
Morris, D.A. (1967b). *Planta* **77**, 224–32.
Munns, R. and Pearson, C.P. (1974). *Aust. J. Pl. Physiol.* **1**, 529–37.
Murarka, I.P. Jackson, T.L. and Moore, D.P. (1973). *Agron. J.* **65**, 868–70.
Newbould, P., Howse, K.R. and Elliot, J.G. (1968). *Agric. Res. Council. Letcombe Lab. Ann. Rep.* **19**, 51–2.
Nösberger, J. and Humphries, E.C. (1965). *Ann. Bot.* **29**, 579–88.
Nye, P.H., Brewster, J.L. and Bhat, K.K.S. (1975). *Pl. Soil* **42**, 161–70.
Ohad, I., Friedburg, I., Ne'eman, Z. and Schramm, M. (1971). *Pl. Physiol.* **47**, 465–77.
Okazawa, Y. (1959). *Proc. Crop. Sci. Soc. Japan* **28**, 129–33.
Okazawa, Y. (1960). *Proc. Crop Sci. Soc. Japan* **29**, 121–4.
Okazawa, Y. and Chapman, H.W. (1963). *Physiologia Pl.* **16**, 623–9.
Palmer, C.E. and Smith, O.E. (1969a) *Nature* **221**, 279–80.
Palmer, C.E. and Smith, O.E. (1969b). *Pl. Cell Physiol.* **10**, 657–67.
Palmer, C.E. and Smith, O.E. (1970). *Pl. Cell Physiol.* **11**, 303–14.
Penman, H.C. (1963). In *The Growth of the Potato*, ed. Ivins, J.D. and Milthorpe, F.L. pp. 191–8. *Proc. 10th Easter Sch. agric. Sci.* Univ. Nottingham, London: Butterworths.
Phillips, I.P.J. (1969). In *Physiology of Plant Growth and Development*, ed. M.B. Wilkins, pp. 165–202. London: McGraw-Hill.
Pitt, D., Hardie, J.L., Hall, T.D. and Graham, D.C. (1965). *Pl. Path.* **14**, 14–22.
Plaisted, P.H. (1957). *Pl. Physiol.*, **32**, 445–453.
Pollock, C.J. and ap Rees, T. (1975a). *Phytochem.* **14**, 613–17.
Pollock, C.J. and ap Rees, T. (1975b). *Planta* **122**, 105–7.
Pressey, R. (1969). *Pl. Physiol.* **44**, 759–64.
Pressey, R. (1970). *Am. Potato J.* **47**, 245–51.
Pressey, R. and Shaw, R. (1966). *Pl. Physiol.* **41**, 1657–61.
Quarmby, A.R. and Ratkowsky, D.A. (1972). *CSIRO Fd Res. Rep.* **75**, 1–42.
Railton, I.D. and Wareing, P.F. (1973a). *Pl. Physiol.* **28**, 88–94.

Railton, I.D. and Wareing. P.F. (1973b). *Pl. Physiol.* **28**, 127–31.
Railton, I.D. and Wareing, P.F. (1973c). *Pl. Physiol.* **29**, 430–3.
Railton, I.D. and Wareing, P.F. (1973d). *Planta* **112**, 65–9.
Reeve, R.M., Hautala, E. and Weaver, M.C. (1969). *Am. Potato J.* **46**, 361–73.
Reeve, R.M., Hautala, E. and Weaver, M.C. (1970) *Am. Potato J.* **47**, 148–62.
Reeve, R.M., Timm, H. and Weaver, M.L. (1973a) *Am. Potato J.* **50**, 49–57.
Reeve, R.M., Timm, H. and Weaver, M.L. (1973b). *Am. Potato J.* **50**, 71–8.
Rappaport, L. and Wolfe, N. (1969). *Symp. Soc. exp. Biol.*, **23**, 219–40.
Reid, M.S. and Pratt, H.K. (1972). *Pl. Physiol.* **49**, 252–5.
Richardson, D.G. and Weiser, C.J. (1972). *Hort Sci.* **7**, 19–22.
Ruf, R.H. Jr. (1964). *Am. Potato J.*, **41**, 377–81.
Rylski, I., Rappaport, L. and Pratt, H.K. (1974). *Pl. Physiol.* **53**, 658–62.
Sabalvoro, E.S. (1965). *The effect of moisture stress on second-growth of potato tubers.* M.Sc. Thesis, University of Nottingham.
Sadler, E. (1961) *Factors influencing the development of sprouts of the potato.* Ph.D. Thesis, University of Nottingham.
Sale, P.J.M. (1973a). *Aust. J. agric. Res.* **24**, 733–49.
Sale, P.J.M. (1973b). *Aust. J. agric. Res.* **24**, 751–62.
Sale, P.J.M. (1974). *Aust. J. Pl. Physiol.* **1**, 283–96.
Sanders, D.C. and Nylund, R.E. (1972a). *Am. Potato J.* **49**, 123–37.
Sanders, D.C. and Nylund, R.E. (1972b). *Am. Potato J.* **49**, 187–95.
Sanders, D.C., Nylund, R.E. and Quisumbing, E.C. (1972). *Am. Potato J.* **49**, 218–26.
Sanders, D.C., Nylund, R.E., Quisumbing, E.C. and Shetty, K.V.P. (1972). *Am. Potato J.* **49**, 243–54.
Scholte Ubing, D.W. (1958). *Landbouwk. Tydschr. 's-Grar.* **70**, 455–64.
Setlik, I. (ed.) (1969). In *Prediction measurement of photosynthetic productivity. Proc. IBB/PP Tech. Meeting*, Trebon, PUDOC Wageningen.
Sharpe, P.R. and Dent, J.B. (1968). *J. agric. Sci., Comb.* **70**, 123–9.
Shepherd, W. (1973). *J. exp. Bot.* **24**, 1003–13.
Shih, C.Y. and Rappaport, L. (1970). *Pl. Physiol.* **45**, 33–6.
Sibma, L. (1970). *Neth. J. agric. Sci.* **18**, 125–31.
Simmonds, N.W. (1971). *Trop. Agric. Trin.* **48**, 291–9.
Slater, J.W. (1963). In *The Growth of the Potato*, ed. Ivins, J.D. and Milthorpe, F.L. pp. 114–20. *Proc. 10th Easter School agric. Sci.*, University of Nottingham, London: Butterworths.
Slater, J.W. (1968). *Eur. Potato J.* **11**, 14–22.
Smith, O. (1974). Nutritive values of Potatoes, A Review. App. 5 in *Planning Conference on Potato Quality*. Lima: CIP.
Smith, O.E. and Rappaport, L. (1969). *Am. Potato J.* **46**, 185–91.
Smith, O.E. and Palmer, C.E. (1970). *Physiol. Pl.* **23**, 599–606.
Soltanpur, P.N. and Moore, F.D. (1969). Agron. J. **61**, 967–8.
Sowokinos, J.R. (1971). *Am. Potato J.* **48**, 37–46.
Sowokinos, J.R. (1973). *Am. Potato J.* **50**, 234–47.
Stallknecht, G.F. (1972). *Pl. Physiol.* **50**, 412–13.
Svensson, B. (1962)., *Eur. Potato J.* **5**, 28–39.
Svensson, B. and Naglicka, I. (1975). *Potato Res.* **18**, 105–8.
Thompson, R. and Taylor, H. (1974). *Potato Res.* **17**, 51–63.
Tizio, R. and Biaiń, M.M. (1973). *Phyton* **31**, 3–13.
Toosey, R.D. (1963). In *The Growth of the Potato*, eds. Ivins, J.D. and Milthorpe, F.L. pp. 79–95. *Proc. 10th Easter Sch. agric. Sci.*, University of Nottingham, London: Butterworths.
Ulrich and Fong, K.H. (1970). *Commn. Soil Sci. and Pl. Anal.* **1**, 141–54.

Ulrich, A. and Fong, K.H. (1973). *Commun. Soil Sci. and Pl. Anal.* **4**, 413–26.
Waggoner, P.E. and Simmonds, N.W. (1966). *Pl. Physiol.* **41**, 1268–71.
Weaver, M.L., Hautala, E., Nonaka, M. and Iritani, W. (1972). *Am. Potato J.* **49**, 376–82.
Werner, H.O. (1954). *Nebraska Univ. Agric. Exp. Sta., Res. Bull.*, **176**, 1–39.
White, A.C. (1961). *The role of the mother tuber in the growth of potato shoots.* B.Sc. Thesis, University of Nottingham.
Winkler, E. (1971). *Potato Res.* **14**, 1–18.
Woolley, D.J. and Wareing, P.F. (1972). *Planta* **105**, 33–42.
Workman, M. and Twomey, J. (1967). *Proc. Am. Soc. hort. Sci.* **90**, 268–74.
Wurr, D.C.E. (1975). *Potato Res.* **18**, 83–91.
Yamaguchi, M., Timm, H. and Spurr, A.R. (1964). *Proc. Am. Soc. hort. Sci.* **84**, 412–3.

Chapter 5

Mineral nutrition

P.M. HARRIS

In this chapter an attempt will be made to illustrate the changes with time in the accumulation of mineral elements and their concentration in the dry matter in various parts of potato plants grown with what was presumed to be an adequate nutrient supply. This will be followed by a discussion of the way in which nutrients may affect growth and yield. Particular attention has been paid to the effects of nutrients on leaf area and to the way in which nutrients affect early tuber growth, the rate of tuber growth over the main period of tuber bulking and the cessation of tuber growth associated with leaf senescence. The nature of the relationship between tuber yield and plant nutrients will be discussed together with factors which may affect this relationship. The estimation of nutrient requirement will be briefly considered.

The literature has not been exhaustively reviewed and for a more comprehensive coverage reference should be made to Smith (1968), while Holliday (1963) has also reviewed the fertilizer needs of the crop. Most of the discussion will be concerned with nitrogen (N), phosphorus (P) and potassium (K), which as Watson (1963) has pointed out, have large general effects on growth that vary with the amount supplied over a wide range, while the other essential mineral elements have specific biochemical roles and relatively 'small effects on growth within a narrow range of supply'. Of these three major nutrients, it is perhaps inevitable that N has received most attention, for the widespread use of substantial amounts of P and K has ensured that many soils in areas where potato production is important, have accumulated these nutrients to the point where crops have become relatively insensitive to their addition as fertilizers. In many areas excess winter rainfall and the tendency to grow potatoes in intensive arable farming systems ensures that N will frequently limit yield.

5.1 Nutrient accumulation, distribution and concentration

5.1.1 N, P and K

Data presented in Figs. 5.1–5.6 have been calculated from data collected by Gunasena (1969) on the variety, Craig's Royal, grown on a sandy loam soil in Southern England. The crop received a normal commercial rate of fertilizer application rotavated into the seed-bed prior to planting.

The dry matter yields of the three main components of yield: leaves, stems and tubers (Fig. 5.1) plotted against time, emphasize the dominant role of tubers as sinks for assimilate over the greater part of the growing season. The weight of stolons and easily recoverable roots contributed relatively little to total yield, falling from 13%, 44 days after planting to 0.3%, 98 days later.

The concentration of N, P and K in dry matter varied with time and yield component (Fig. 5.2). The concentration of N was always highest

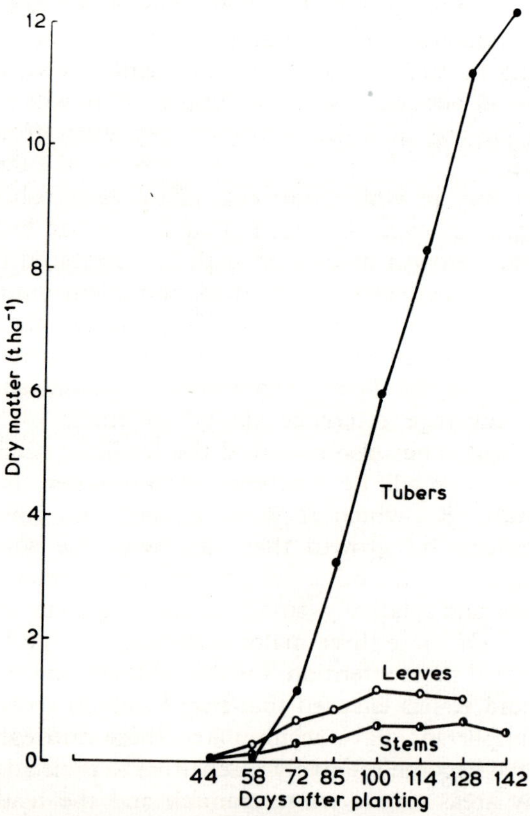

Fig. 5.1 The accumulation of dry matter in leaves, stems and tubers. Drawn from data of Gunasena, 1969. Craig's Royal 1966

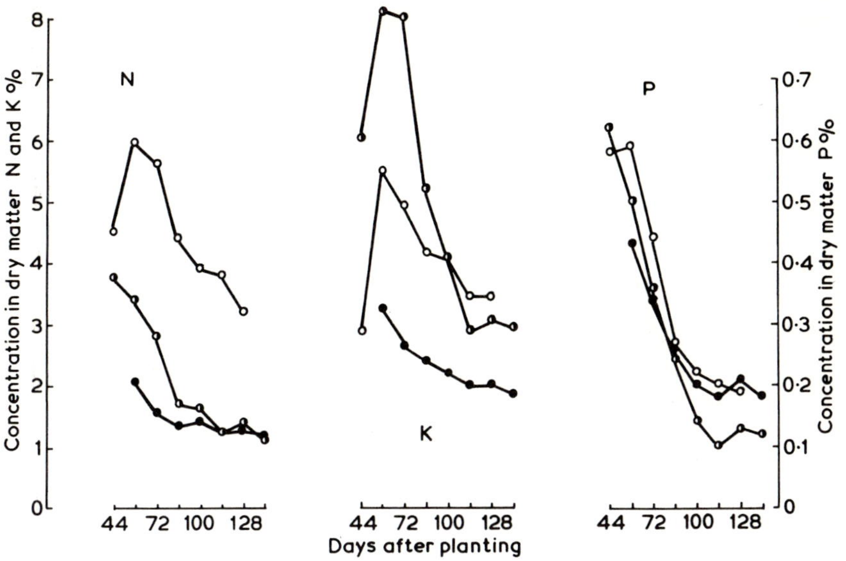

Fig. 5.2 The N, K and P content of the dry matter of leaves, stems and tubers. Craig's Royal 1966. Drawn from data of Gunasena, 1969

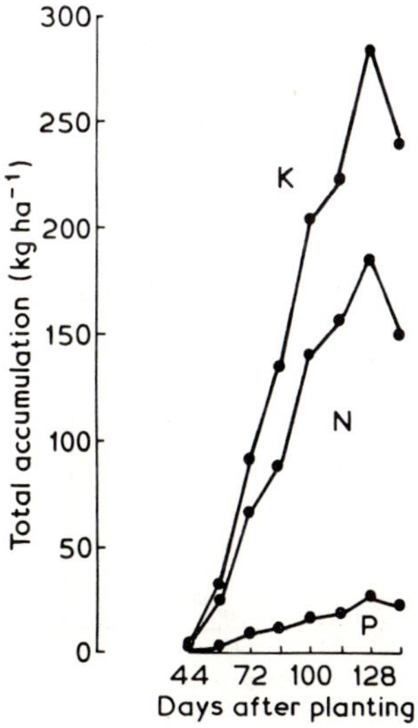

Fig. 5.3 The total accumulation of N, P and K. Craig's Royal 1966. Drawn from data of Gunasena, 1969.

in the leaves and reached a peak value of 6% early in the life of the crop, the highest K concentration occurred in the stems with peak values of over 8% early in the season. Contents of P were about one-tenth of those of the other two nutrients and varied much less between the various yield components.

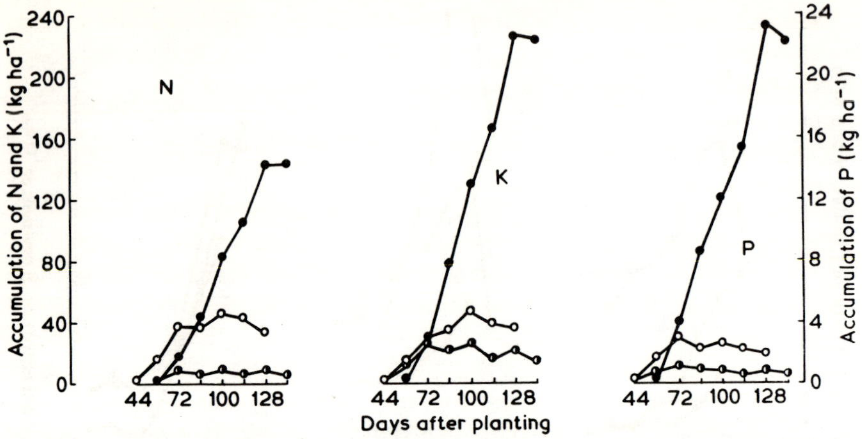

Fig. 5.4 The accumulation of N, K and P in leaves, stems and tubers. Craig's Royal 1966. Closed circles = tubers, half closed = stems, open circles = leaves. Drawn from data Gunasena, 1969.

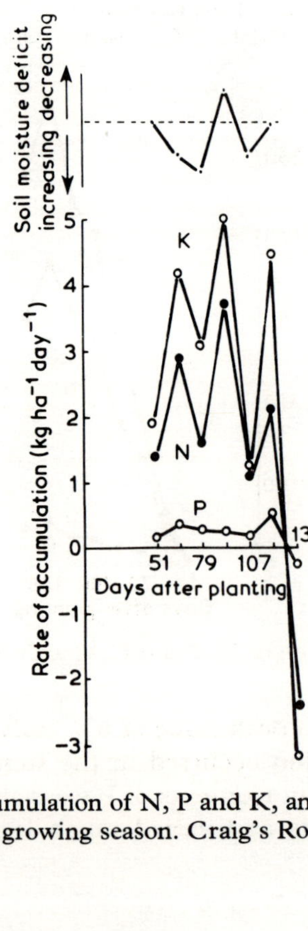

Fig. 5.5 The rate of accumulation of N, P and K, and the changes in soil moisture deficit over the growing season. Craig's Royal 1966. Drawn from data of Gunasena, 1969

The accumulation of the three major elements by the whole crop is shown in Fig. 5.3 and accumulation in the three main yield components in Fig. 5.4. They illustrate the large quantities of N and K and the much smaller quantity of P accumulated. Within 14 days of their initiation, tubers become the dominant sink for all three nutrients. Maximum accumulation of nutrients occurred 128 days after planting, when tubers accounted for 77%, 89% and 79% respectively of the total N, P and K.

The rate of uptake varied with time and nutrient being lowest with P and somewhat higher for K than N, peak values were reached in mid-season for N and K, but somewhat later for P. Considerable fluctuations in the rate of nutrient uptake were superimposed upon this general pattern, decreased rates coinciding with periods when the soil moisture deficit was increasing (Fig. 5.5). The crop was irrigated when necessary and the relatively low soil moisture deficits at which nutrient uptake was considerably reduced has a bearing on the sensitivity of the crop to soil water stress which is referred to later in this chapter and in Chapter 6. In the first 'trough' the N and P accumulation in tubers (Fig. 5.6) was unaffected, whilst that in the stem and leaves was reduced to values approaching or falling below zero, a similar trend was evident with P. In the second short period of increasing soil water stress, whilst the rate of nutrient accumulation in tubers was reduced, it remained

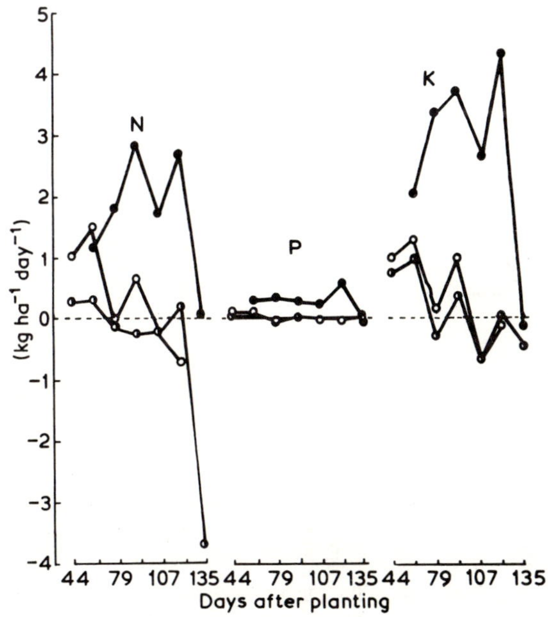

Fig. 5.6 The rate of accumulation of N, P and K in leaves, stems and tubers over the growing season. Craig's Royal 1966. Closed circles = tubers, half closed = stems, open circles = leaves. Drawn from data of Gunasena, 1969

Table 5.1 Selected characteristics of various potato cultivars grown in various parts of the world (adapted from Ezeta and McCollum, 1972)

Location *Variety*		*Peru Renacim-iento*	*Colombia 3 varieties*	*Utah Russet Burbank*	*California White Rose*	*Maine Green Mountain*	*Maine 4 varieties 4 years*	*North Carolina Pungo*	*Netherlands Mentor*	*England Craig's Royal*
Plant density plants ha $^{-1}$ × 10^3		33.3	45.6	?	43.0	35.9	35.9	37.7	40.0	35.9
Planting to tuber initiation		125	44 – 51	60	60 – 65	60	48 – 60	50	48	58
Planting to harvest		180	109	138	119 –124	112	89 – 97	100	120	142
Total dry matter production t ha $^{-1}$		12.6	11.9 – 18.6	9.9	10.6 – 12.7	7.9	4.5 – 8.8	7.5	10.2	12.7
Tuber dry matter/ Total dry matter		0.70	0.77– 0.82	0.77	0.75– 0.85	0.75	0.60– 0.80	0.83	0.95	0.96
Rate of tuber growth kg $^{-1}$ ha $^{-1}$ day $^{-1}$		224	280 –360	170	180 –290	141	132 –204	350	184	144
Rate of uptake	N	2.5	2.5 – 4.0	2.2	3.2 – 3.3	4.2	1.7 – 5.1	3.3	1.7	2.2
kg ha $^{-1}$ day $^{-1}$	P	0.22	0.38– 0.45	0.25	0.30– 0.48	0.39	0.14– 0.50	0.22		0.30
	K	6.6	4.7 – 5.6	2.6	3.8 – 6.3	5.8	2.6 – 6.9	4.5		3.34
% tuber dry matter	N	0.9	1.5	1.4	1.1 – 1.3	1.8	1.1 – 1.8	1.6	1.1	1.2
at harvest	P	0.18	0.2	0.14	0.15	0.18	0.17– 0.26	0.25	—	0.18
	K	2.6	1.3	1.4	1.5 – 1.7	1.8	1.5 – 2.5	2.1	—	1.8
Nutrients removed	N	80		107		107		100	107	147
in tubers kg ha $^{-1}$	P	16		11		11		16	—	22
	K	230		107		107		131	—	220

fairly high, while values for leaves and stems were negative, suggesting that nutrients were translocated to tubers from other plant parts when water restricted nutrient uptake. It would appear that tubers effectively competed with other plant parts when environmental factors limited nutrient supply.

These short term fluctuations in the rate of nutrient uptake did not affect the rate of dry matter accumulation in the tubers which remained linear from 72 to 128 days after planting, but there is some suggestion that the growth of leaves and stems may have been affected. Nutrients did not appear to limit the rate of tuber growth in this experiment.

A comparison of the nutrient characteristics of various cultivars grown in N. and S. America and in Europe are given in Table 5.1, drawn largely from data presented by Ezeta and McCollum (1972). The rates of tuber growth and nutrient uptake refer to maximum rates, not sustained over the whole period.

5.1.2 Ca, Mg and S

The total and tuber accumulation of Ca, Mg and S with time are shown in Fig. 5.7 for the variety 'Green Mountain' grown in Maine (Hawkins,

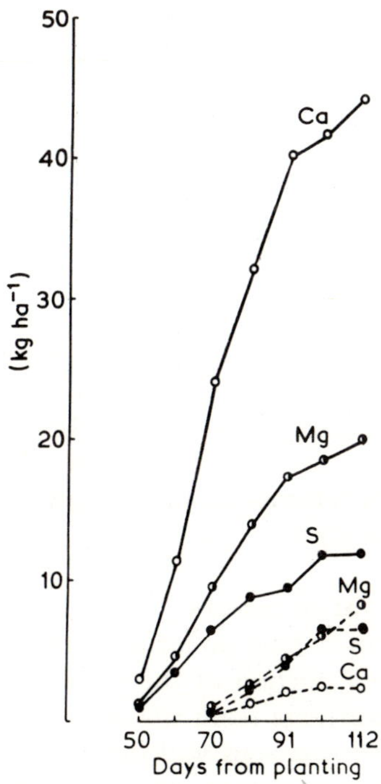

Fig. 5.7 The accumulation of Ca, Mg and S in the potato crop, variety Green Mountain. Full lines = total crop, broken lines = tubers. Drawn from data of Hawkins, 1946.

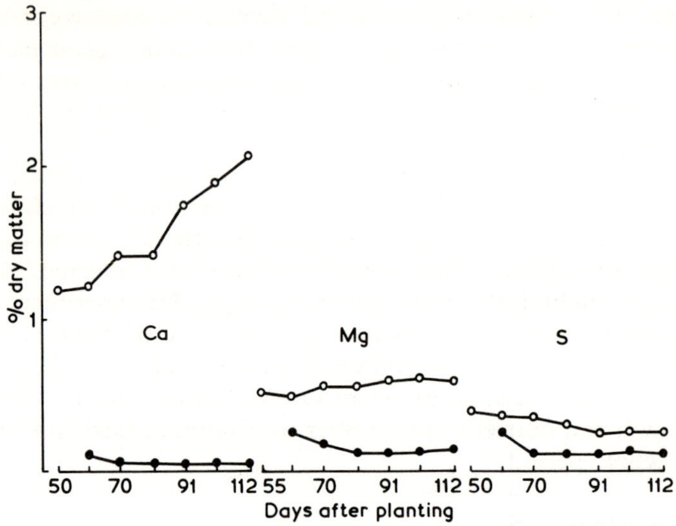

Fig. 5.8 The Ca, Mg and S concentration in the dry matter of leaves, stems, stolons and roots (open circles) and tubers (closed circles). Drawn from data of Hawkins, 1946.

1946). The quantity of Ca accumulated was twice that of Mg and four times the quantity of S, however only 6% of the Ca was in tubers compared with 41% and 55% of magnesium and sulphur respectively. For all three nutrients, tubers appear to be a less dominant sink than for N, P and K (see Chapter 3). Variation with time in the concentration of these nutrients in tubers and haulm (leaves + stems) are shown in Fig. 5.8.

5.1.3 Micronutrients

The range in the concentration of a number of micronutrients in young mature leaves taken from crops where the tubers are 'half grown' have been found by Geraldson *et al.* (1973) to be as follows, expressed as parts 10^{-6} in the dry matter: Iron 70–150, Boron 30–40, Zinc 20–40, Manganese 30–50. Williams *et al.* (1960) found that at Rothamsted a 20.1 t ha $^{-1}$ crop of tubers removed 44 g copper, 42 g manganese, 0.74 g molybdenum and 99 g of zinc.

5.1.4 Nutrient removal in tubers

It is perhaps of some value to have an estimate of the quantity of nutrients removed in the harvested portion of the crop, for while this is not the only source of loss of nutrients from the soil, it is an important source. The quantity of nutrient removed will depend upon the yield and the concentration of nutrient, and both can vary considerably from site to site and year to year. The estimates in Table 5.2 are made for the amount of nutrients removed per tonne of fresh tubers.

Table 5.2 Mineral nutrients removed t $^{-1}$ of fresh tubers

Element	*(1) (kg)*	*(2)*	*(3)*
Nitrogen	2.68		2.36
Phosphorus	0.62		0.53
Potassium	3.93	4.67	4.55
Calcium	0.07	0.2	
Magnesium	0.22	0.13	
Sulphur	0.21	0.48	
Zinc	0.001 8	0.005	
Copper	0.001 43	0.002 19	
Manganese	0.001 34	0.002 1	
Iron	0.004 2		
Boron	0.000 62		
Molybdenum		0.000 037	
Sodium		0.23	

Data calculated from the following sources: (1) Kunkel *et al.*, (1973); (2) Cooke (1967); (3) Gunasena (1969).

Table 5.3 The effect of N on yield and the removal of N, P and K t $^{-1}$ of tuber fresh weight

Treatment N (kg h $^{-1}$)	*Yield tubers (t ha $^{-1}$)*	*N*	*P (kg t $^{-1}$)*	*K*
0	11.95	2.81	0.50	5.41
94	19.57	3.26	0.46	5.24
188	22.74	3.57	0.45	4.73

Data calculated from Gunasena (1969).

The data from the different sources agree quite well in terms of the orders of magnitude involved, but a considerable degree of variability must be accepted. For example, nutrient concentration is not independent of yield as is shown in Table 5.3 where yield increases induced by N fertilizer resulted in a decrease in K and P and an increase in N removed per tonne (t) of tuber fresh weight. In data calculated from Gunasena (1969) the effect of season, variety and the rate and timing of N and K fertilizer resulted in the following ranges in the removal of mineral nutrients in fresh tubers: N 2.28–3.57, P 0.40–0.62, K 3.70–5.41 kg t $^{-1}$.

5.2 Relationships between mineral nutrients, growth and yield

Milthorpe (1963) showed that except where water is limiting, tuber growth rate is linear for most of the growing season, departing from linearity only for a short period after tuber initiation, when tuber growth rate increased with time and for a short period at the end of the growing season, when growth rate declined during the period when leaves senesced relatively rapidly. Tuber yield is therefore largely determined by the average tuber growth rate during the linear phase of bulking and the duration of this period.

Ivins and Bremner (1965) were able to show that tuber bulking rates increased with increasing values of leaf area index (*L*, the area of leaf per unit area of ground) until most of the incident radiation was intercepted which, under the conditions of the experiments, occurred when *L* was approximately 3. The duration of the linear bulking phase is determined by factors which bring about tuber initiation and which affect the time lag between initiation and the onset of the linear phase of tuber growth, and factors which bring about the death of leaves. Fertilizers may therefore be expected to exert their primary effect on yield by affecting the development and the maintenance of leaf area and by promoting or retarding the onset of the linear phase of tuber bulking. Nutrients may also have some effect upon the photosynthetic efficiency of leaves.

5.2.1 Nutrient effects on leaf area and efficiency

In a pot experiment, Watson and Wilson (1956) showed that N, P and K increased dry matter yield mainly by increasing leaf area per plant, while effects on leaf efficiency (*E*, the net increase in dry matter per unit area of leaf per unit of time) were relatively small and erratic. It was noted that the stage of growth at which *L* was most affected differed with nutrient: thus N greatly increased *L* throughout the life of the crop, P increased leaf area in the early stage of growth, but hastened the senescence of leaves, thus depressing *L* in the later stages of growth, while K, conversely, had little effect in the early stages of growth, but increased *L* considerably later in the season and delayed the senescence of leaves. These effects are shown in Fig. 5.9.

In a pot experiment reported by Humphries and French (1963), N had the greatest effect on *L*, an increase associated with an increase of leaves on lateral rather than on main stems, and although N at first increased the survival of leaves, later leaf senescence was accelerated. K did not in this experiment affect leaf survival, but it did increase leaf efficiency, an effect not found with either P or N.

In the field, the level of soil nutrient supply can be less readily controlled than in pots and in British, and presumably in other intensively farmed temperate agricultural conditions, N is usually the most important nutrient limiting yield, owing to the build-up of P and to a lesser extent K reserves in the soil following intensive fertilizer use. Thus in southern England, Gunasena and Harris (1968, 1971) showed that N fertilizer applied in the seed-bed substantially increased *L* over most of the growing season, while the effect of K was much smaller and was confined to the

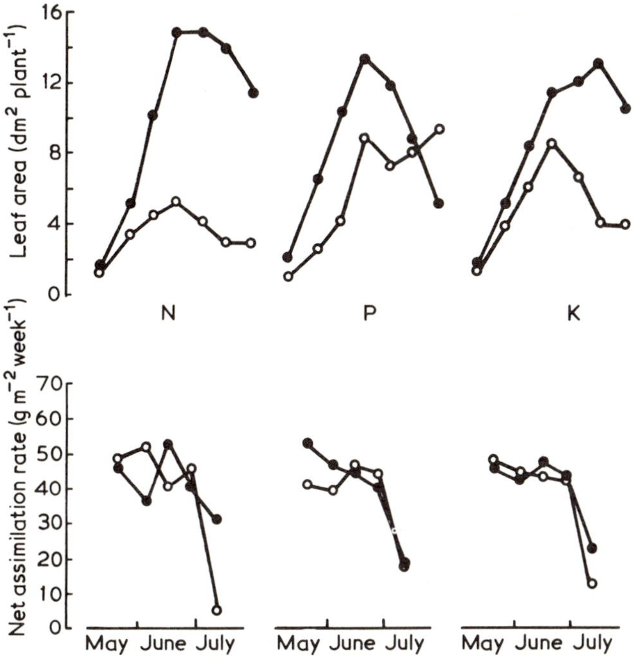

Fig. 5.9 The effects of N, P and K on leaf area and net assimilation rate of potatoes grown in pots. Closed circles with nutrient, open circles without. Redrawn from Watson and Wilson, 1956

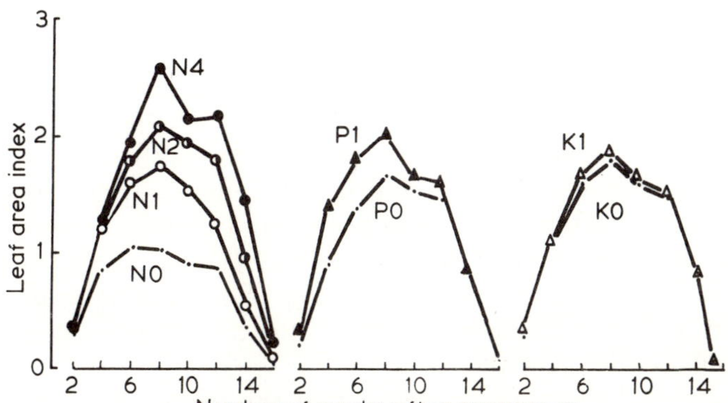

Fig. 5.10 The effects of N, P and K on changes with time in leaf area index of potatoes, variety King Edward. (No = 0, N1 = 94, N2 = 188 and N4 = 376 kg N ha^{-1}, Po = 0, P1 = 156 kg P ha^{-1}; Ko = 0, K1 = 82 kg K ha^{-1}). Redrawn from Dyson and Watson, 1971

middle of the season. Dyson and Watson (1971) working on a heavier soil at Rothamsted also showed that increases in L due to P and K were small compared with the response to N (Fig. 5.10). N increased L from

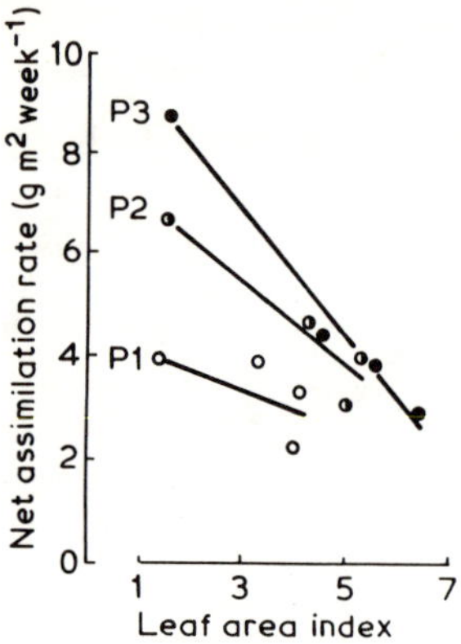

Fig. 5.11 Relationships between net assimilation rate and leaf area index at three different plant densities; values are means for the period 10–24 June. Variety Maris Page. (P1 = 34 602, P2 = 46 132 and P3 = 64 599 tubers ha^{-1}). From Ngugi, 1972

an early stage of growth, although only the largest amount prolonged the life of leaves; the effect of K was only marked at the end of the growing season when the growth of leaves was prolonged, while P only increased *L* up to four weeks from emergence. The timing of the nutrient effects on *L* in the field thus agreed with the earlier date from pot experiments.

Watson (1958) showed that owing to the mutual shading of leaves, *L* tends to be negatively correlated with *E*. Dyson and Watson (loc. cit.) were able to show negative linear correlations between *L* and *E* which could vary over the growing season. Ngugi (1972) also found a negative effect of N on *E* associated with an increase in *L*, but the slope of the linear relationship was affected by plant population (Fig. 5.11).

The fact that *E* and *L* are not independent limits their practical value. However it is commonly found that leaf area duration (*D*), the integral of *L* over time is often closely and positively correlated with both total and tuber dry matter yield (Gunasena and Harris, 1968, 1969, 1971; Dyson and Watson, 1971). while in other cases a modification in the calculation of *D* to exclude values of *L* in excess of 3 has improved the relationship (Radley, 1963).

There can be little doubt however that the major effect of supplying nutrients which are most likely to be limiting yield is on the development and maintenance of a leaf canopy capable of intercepting most of the incident radiation.

5.2.2 Mineral nutrition and early tuber growth

It has been pointed out (Ivins and Bremner, 1965) that high levels of nutrients and especially of N may delay the apparent initiation of tubers and thus the onset of the linear phase of tuber growth. It is not clear

Table 5.4 The effect of N on various yield components of potatoes grown in pots approximately nine weeks after planting

	Level of N kg ha $^{-1}$			
	0	*67*	*135*	*269*
		dry matter g pot		
Tops	2.6	2.8	3.1	2.7
Roots	0.76	0.70	0.82	0.65
Tubers	6.54	2.01	2.49	2.16

Data from Sommerfeldt and Knutson, 1968.

to what extent such apparent delays are due to the direct effect of nutrient ions on tuberization or whether it is due to the indirect effects associated with the stimulation of other plant parts such as leaves and stems, which reduces the supply of assimilate to the stolon tips. It was pointed out in Chapter 3 that sampling in the field has usually been too infrequent to establish accurately the date of tuber initiation and that in any case tubers are not initiated simultaneously but over a period of time.

The effect of N on the early growth of potatoes grown in pots (Sommerfeldt and Knutson, 1968) is shown in Table 5.4; the data suggest that the increase in the ratio of tops and roots to tubers is due to a decrease in tuber yield rather than to any increase in the yield of other plant parts. However Krauss and Marschner (1971) showed that the effect of N on early tuber growth was not independent of the size

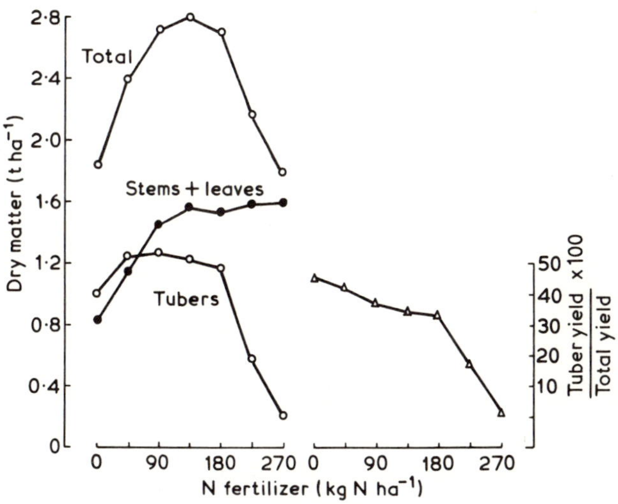

Fig. 5.12 The effect of N fertilizer on the dry matter yield of various components and on the proportion of total yield formed by tubers at an early stage of growth (51 days after planting), in the variety Maris Page. Harris, unpublished data.

of the container in which they were grown. In large containers, where there was no obstacle to root growth, a continuous supply of N before tuber initiation resulted in extremely long stolons and prevented tuber initiation; in smaller containers with a comparable N supply, tuberization took place in 45 days. Tripathi (1973) found that nitrate ions stimulated growth and the formation of new roots, but had no effect on tuberization, however when phosphate ions were added in addition to nitrate, stolon growth was stimulated and tuberization delayed.

In a field experiment on sandy loam soil with the variety 'Maris Page', Harris (unpublished data) was able to show (Fig. 5.12) that at an early stage of growth with rates of N in excess of 180 kg ha $^{-1}$, tuber dry matter yield fell dramatically while the yield of leaves and stems remained practically constant. Thus it would appear from both pot and field experiments that the negative effects of N on early tuber growth are not necessarily related to the stimulation of other yield components. However the absence of data on all plant parts (e.g. roots) make it impossible to draw absolutely firm conclusions.

5.2.3 Rate of tuber growth and nutrient supply

It can be readily shown that the rate of tuber growth is affected where nutrients are deficient (Fig. 5.13). According to Ivins and Bremner (loc. cit.) L values of 3 are sufficient to ensure maximum tuber growth rates. It is difficult to draw any firm conclusions about this; for example in an experiment where L varied between 3 and 6 for much of the growing season, Gunasena and Harris (1971) found a linear relationship between yield and D, which accounted for 85% and 75% of the variation in yield, of a second early and a maincrop respectively, whereas in a later experiment, Ngugi (1972) found that over a period of 4 weeks tuber growth rates remained unaffected by variation in L between 3.7–5.7, with values

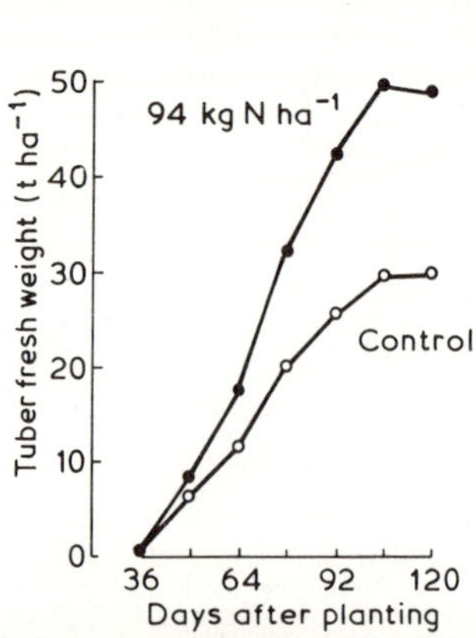

Fig. 5.13 The effect of N fertilizer on the rate of tuber growth, variety Craig's Royal. Data from Gunasena and Harris, 1969

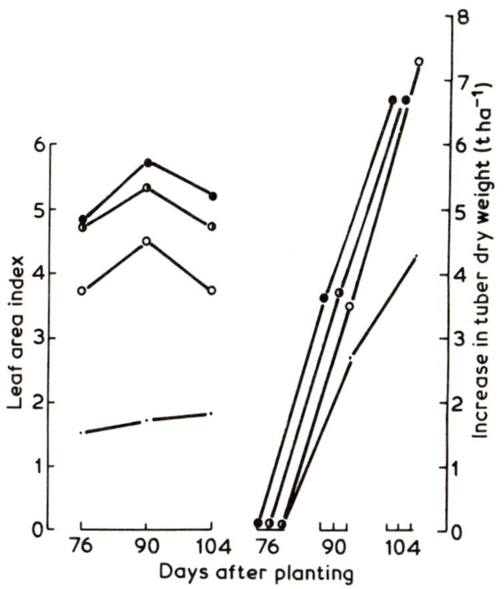

Fig. 5.14 The effect of N fertilizer on leaf area index and tuber growth over a period of 4 weeks (point = nil, open circles = 99.5, half closed circles = 199, closed circles = 299 kg N ha^{-1})

Table 5.5 The effect of rate and time of application of N on tuber numbers and yield in the variety Craig's Royal

N treatment	*Tuber number per plant*	*Total tuber fresh weight (kg ha^{-1})*
Nil	15.0	30.0
In seed-bed*	20.9	51.2
After tuber initiation*	16.4	54.8

*Mean 94 and 188 kg N ha^{-1}
From Gunasena and Harris, 1969

of *L* below 2, there was a considerable reduction in the rate of tuber growth (Fig. 5.14).

It is possible that where nutrients increase tuber growth rates, which are correlated with increases in values of *L* considered to be above those necessary to intercept most of the incident radiation, that some other factor such as sink strength (Watson, 1971) may also be increased. For example, it is plausible to suggest that an increase in tuber numbers would increase the sink for carbohydrates and yield if this is sink limited. In an experiment in which the rate and time of application of nitrogen were varied (Gunasena and Harris, 1969), increasing the rate of N increased tuber numbers, but for comparable rates of N application, delaying application until after tuber initiation increased yield and decreased tuber numbers (Table 5.5). Thus tuber numbers would not appear to be an important factor limiting total tuber yield, a conclusion in agreement with Edward (1971).

5.2.4 Nutrients and the cessation of tuber growth

It was shown earlier that N and K may delay, and P accelerate leaf senescence. It is commonly found (see Section 5.4.1) that high rates of N fertilizer only begin to assert positive effects on tuber yield at a late stage of growth; as suggested earlier such high rates may have delayed early tuber growth and may not have affected tuber growth rate during the main bulking period because such high rates of fertilizer are associated with supra-optimal values of L. Prolongation of L at the end of the season may not be the consequence of a decline in the rate at which leaves senesce but of increases in the maximum values of L attained; this is shown in Table 5.6 and Fig. 5.10. These data agree with the suggestion of Ivins and Bremner (1965) that the maintenance of L at the end of the growing season is due to the attainment of high peak values. However, early irrigation treatments which enhance maximum values of L may give rise to earlier death of the crop canopy and bring forward the cessation of growth (see Chapter 6), while high maximum values of L achieved by the increasing the plant population may not prolong leaf growth (Section 5.4.6).

In studying the relationship between N and senescence in varieties of contrasting maturity classifications, Kerketta (1976) found that the varieties senesced in the order – Maris Page, Arran Comet and Pentland Crown. The following factors would appear to be involved:

(a) The varieties appeared to differ in the degree of N depletion that could be tolerated in leaves before death occurred. The N content of the terminal leaflet on each leaf on three main stems was measured at

Table 5.6 The effect of N on maximum and final recorded values of leaf area index and on the rate of decline of leaf area index over a period of 28 days, variety King Edward

	Leaf area index		
N ($kg\ ha^{-1}$)	*Maximum*	*Final*	*Rate of decline* ($L\ week^{-1}$)
0	1.34	0.61	− 0.12
79	2.16	1.00	− 0.29
158	2.47	1.19	− 0.32

From Harris, 1960.

Table 5.7 Initial mean values of terminal leaflet N and final values for varieties of contrasting longevity

Variety	*Longevity*	*Mean terminal leaflet N % dry matter*	
		Initial	*Final*
Maris Page	Short	6.28	3.51
Pentland Crown	Long	5.88	2.56

From Kerketta, 1976.

14 day intervals throughout growth. Table 5.7 gives the mean values at the beginning of the season and at the last sampling before senescence occurred, for the earliest and the latest maturing varieties. The earliest maturing variety appeared less able to tolerate a depletion in N content before senescence occurred.

(b) Senescence also appeared to be related to the distribution of N within the plant (Fig. 5.15). Compared with the earliest maturing variety, the longest lived variety accumulated a greater proportion of its N in leaves and roots and a lower proportion in tubers. It is possible that leaf survival may be associated with factors which reduce the effectiveness with which tubers compete for N with other plant parts.

A factor of considerable importance in bringing about the rapid decline in L after peak values have been reached could be the increasing demands of tubers for mineral nutrients. It was shown earlier that the concentration of N in tubers tended to remain fairly constant with time, whereas the concentration in other plant parts falls fairly rapidly; thus internal competition for nutrients could be one factor involved in senescence. The removal of tubers has been shown to prolong the life of leaves (Nosberger and Humphries, 1965).

(c) Kerketta's experiments also suggested a possible connection between root growth and senescence. Although the root data were fairly crude, and consisted only in measuring the dry weights of the easily recoverable roots, it can be seen in Fig. 5.16 that a decline in root growth appeared to precede the decline in L. It is possible that the effects of tuber growth on other plant parts are first manifest in the growth of roots; this in turn could affect the uptake of plant nutrients and reduce the supply to the crop, thus accentuating the effects of competition for nutrients between leaves and tubers. A reduction in leaf growth could lead to a reduction in the assimilate available for roots accumulated in face of the competition for assimilate with the tubers. A reduction in root growth might also influence the supply of cytokinin to leaves (Skene, 1975) which in turn could enhance leaf senescence. This would suggest that it might be possible to arrest senescence by applying nutrients

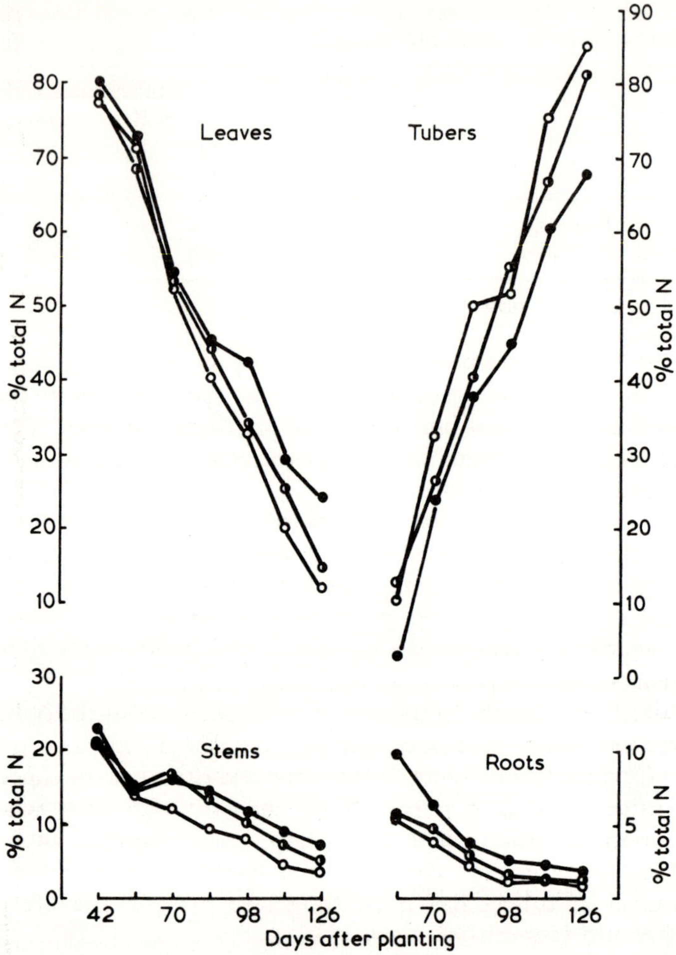

Fig. 5.15 The effect of variety on the distribution of N between various plant parts. (open circles = Maris Page, half closed circles = Arran Comet, open circles = Pentland Crown). From Kerketta, 1976

and cytokinin to leaves in the latter stages of growth. However the foliar application of N and cytokinin failed to achieve this objective in Kerketta's experiments. It was shown that N could successfully be incorporated into the plant by spraying the leaves with urea, but there was no evidence concerning the fate of cytokinin in the plant, although the cytokinin used (6-benzylaminopurine) did appear to have transient effects on L.

5.3 Fertilizer requirements

In order to apply optimal rates of plant nutrients, usually N, P and K, the basic biological information required is the relationship

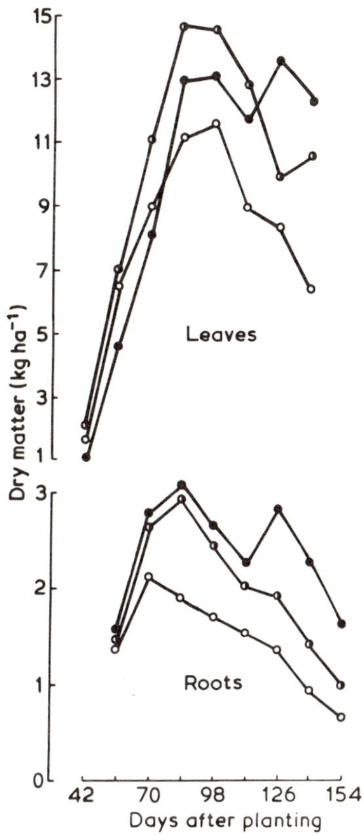

Fig. 5.16 The effect of varieties of contrasting longevity on the time trend in the accumulation of dry matter in roots and leaves. Symbols as in fig 5.15. From Kerketta, 1976

between nutrient and yield and some indication of the nutrient status of the particular field in which the crop is to be grown. The effect of the nutrients on the quality of the crop must also be taken into account. It should be possible to define the general form of the relationship between plant nutrients and yield from well conducted field trials, but the nutrient status of the soil must be determined or estimated for each site. Unpredictable events occurring during the growing season, such as the incidence of pests and diseases, or the incidence and quantity of rainfall may alter the yield nutrient relationships, although some opportunities may exist for assessing and correcting the nutrient status of the crop during the course of growth.

5.3.1 Relationships between fertilizers and yield

In a classic paper, Crowther and Yates (1941) examined all the experimental results available to them with nutrients at three or more levels;

they found that the response curve was an exponential, following the pattern proposed by Mitscherlich (1913) whose formula, relating level of fertilizer to yield they adopted with slight modifications, namely:

$$y_1 = y_0 + d(1 - 10^{-kx})$$

where y_0 = yield without the nutrient in question, d = the limiting response, y = the yield with a dressing of the nutrient in x amount, k = a value which determines the rate at which response falls off with successive increments of nutrients and which was assumed to be constant for each of the three principal classes of fertilizer.

In essence, yield bears an asymptotic relationship to nutrient supply. A weakness of the experimental data on which the response curve was based was that the levels of nutrients tested were too low, and that interactions between nutrients had to be ignored because they could not be estimated satisfactorily. In a later review, Boyd (1961) presented evidence which indicated that the interactions between nutrients were larger and more consistent than had been supposed and that they affected both the magnitude of the response to a given level of nutrient and the degree to which the response falls off with increasing dressings. Table 5.8 taken from Boyd's paper (loc. cit.) showed how the response to a nutrient could be affected by variations in the basal dressing of the other two nutrients tested. In the absence of any basal dressing, the response decreased sharply at higher levels of application. With increasing basal dressings, the response was much greater and fell off less sharply.

Boyd concluded that the concept of an exponential response curve did not fit the facts and that for each nutrient the response curve rises to a maximum and then falls off again. The level of a nutrient at which the response is maximized would vary, Boyd argued, with many factors but particularly with the amount of other nutrients present in the soil and applied as a basal dressing.

Table 5.8 The effect of the level of P and K on the response to N. Potatoes t ha^{-1}. Mean of 30 experiments.

	P and K		
	P0K0	*P1K1*	*P2K2*
Yield without N	16.8	22.6	24.3
Response to N			
N1–N0	1.2	2.5	3.3
N2–N1	−0.5	0.5	1.2

N1 = 30.0 kg N ha^{-1}; P1 = 27.3 kg P ha^{-1}; K1 = 77.8 kg K ha^{-1}.
From Boyd, 1961.

Holliday (1963) agreed with this – commenting that 'anyone who has handled comprehensive fertilizer data will realize that the actual curves show a point of inversion'. Holliday suggested that it is usual to use a quadratic curve of the form $y = a + bx - cx^2$ or a square root function $y = a + bx - cx^{0.5}$ where y = yield, x = the level of plant nutrient, and a, b and c are constants. The latter equation has the merit that it is not symmetrical about the point of inversion. Summarizing 51 experiments carried out between 1956 and 1962 in Britain, Birch *et al.* (1967) found that square root functions fitted their data better than quadratic functions. Sharpe (1969) compared these two functions using data from 95 experiments reported by Boyd and Dermott (1964), and found that on statistical grounds it was difficult to select which function best fitted the experimental data, but concluded that on biological grounds the square root function was more appropriate; presumably because there is little biological evidence to suggest that after an optimal level of fertilizer has been applied, yield would decrease at the same rate as it increased before the optimum was reached.

More recently, Boyd (1970) and Fitts (1974) have suggested that compared with a curve, as good or better fit to data relating yield to fertilizer level might be obtained by two straight lines. Boyd has reviewed the evidence for this type of response curve for a number of crops. The investigation was prompted by experiments with sugar beet, where the general form of the relationship between fertilizer N and sugar yield had a steeply-rising portion with a fairly abrupt transition to an almost linear and almost horizontal portion when larger amounts of N were applied. This relationship became evident when experiments were grouped according to the amount of N required to reach the transition point.

One of the reasons for the failure of voluminous experiments with fertilizer to reveal the exact shape of the response curve has been the paucity of levels tested and the relatively narrow range of the response curve investigated: defects which are only now beginning to be remedied. Boyd commented that there was little information for more than four amounts of N on potatoes, but that many tests with four amounts indicate that this crop would be expected to behave similarly to the sugar beet crop. He suggested that on the rising part of the response curve increases in dry matter yield of about 30–35 kg dry matter kg N can be expected; sites rich in N may show linear responses wholly on the upper part of the curve.

Holliday *et al.* (1965) published data on the response of potatoes to sulphate of ammonia, combined with various rates of hoof and horn meal; the latter behaved in a similar way to the inorganic fertilizer and by using all the combinations of the two sources of N a comprehensive response curve to N was obtained (Fig. 5.17), the data are presented

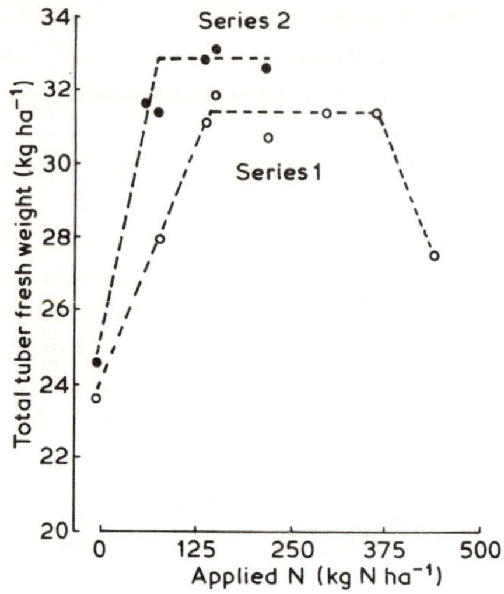

Fig. 5.17 The relationship between applied N and the fresh weight yield of tubers. Drawn from data of Holliday *et al.*, 1965.

for 2 series of experiments each the mean of 3 experiments. The data is consistent with a linear discontinuous form of response with a sharp transition from a linear rising portion to a flat portion. In the first series, where a particularly wide range of N levels were applied, there is evidence that at very high levels, yield may drop sharply.

Boyd (1970) reported the results of several groups of potato experiments in which four or more levels of P were tested in amounts up to 220 kg P ha^{-1}. The results for four soil series are shown in Fig. 5.18, each are the mean of from 6–20 experiments with 6- or 9-fold replication per site. As with the N response curve, the transition from the rising portion to the horizontal portion appears to be fairly well defined. It was suggested that the rate of increase per unit of P is probably related to the method of placement; in all those with small rates, the fertilizer was broadcast on the seed-bed before planting 'on the flat', on many of the more responsive sites, the fertilizer was applied more effectively by broadcasting over the ridges before planting. Unpublished data (Boyd, private communication) from an experiment in which a level of 218 kg P ha^{-1} was applied on a soil with a high phosphate status indicated that this application resulted in a sharp decrease in yield. However on soils very deficient in P in Alaska (Lauglin *et al.*, 1974) in experiments in which 7 levels of P were tested in increments up to 340 kg P ha^{-1}, there was no suggestion that the relationship between P and K was anything other than curvilinear, however the growing season was short and it is possible that the length of growing season could affect the shape of the yield response curve to P.

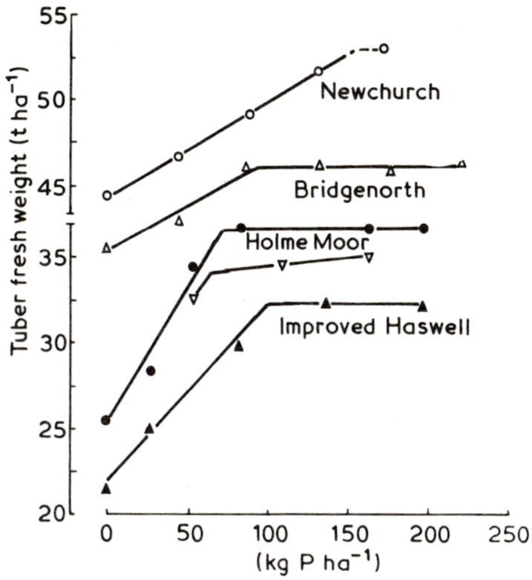

Fig. 5.18 The effect of P fertilizer on potatoes. From Boyd, 1970

According to Boyd (1970) data for K is scanty, but experiments with four amounts suggests that a response curve of similar form to P could be expected.

Fitting functions to fertilizer responses provides a useful means of summarizing data, but the calculated constants are usually unique to the data from which they were calculated and cannot be interpreted biologically. Such curves are also frequently of limited economic value, for while different equations may account for a similar proportion of the variation in the yield data, the optimal rates of fertilizer calculated from them may differ considerably. The simplicity of the discontinuous linear functions referred to has obvious attractions, not least being the ease with which the data can be handled and, given simple price information, decisions made concerning whether or not to apply fertilizer, and if any is justified, the optimal level can be chosen with precision. However, the biological basis for this form of response curve has not been established.

5.3.2 Fertilizer and tuber quality

Most of the data on fertilizer response refers to tuber fresh weight yield, and usually includes total yield and the yield of a particular size fraction over some minimum riddle size. Fertilizers usually increase the size of tubers and often the yield of small tubers remains fairly constant over a a fairly wide range of fertilizer treatments (Fig. 5.19(a)).

Dry matter yield is usually less responsive than fresh weight yield, for N and K especially tend to decrease dry matter percentage. Inkson and Reith (1967) showed that P had little effect on the dry matter percentage of tubers and that the optimal rate of fertilizer P was similar to that

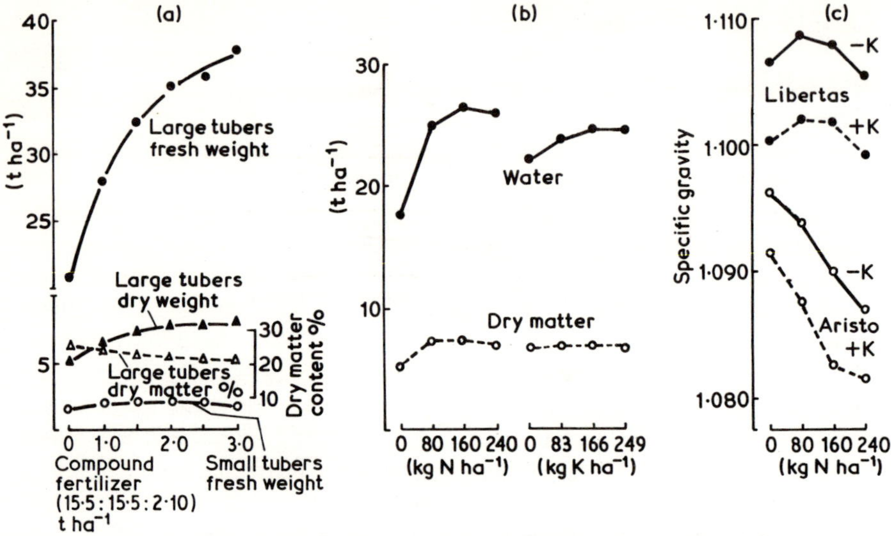

Fig. 5.19 (a) The effect of a complete N:P:K ferilizer on the fresh weight yield of large (ware) and small tubers, and on the dry matter yield and content of large tubers. From Rodger and Robertson, 1970. (b) The effect of N and K on the dry matter yield and yield of water, variety Aristo. (c) The effect of N and K (nil and 166 kg K ha $^{-1}$) on the specific gravity of two varieties of potatoes. (b) and (c) from Schippers, 1968

for fresh weight, N and K however decreased dry matter content and the optimal rates were lower for dry matter production. In Fig. 5.19(a) the effect of a complete fertilizer on tuber fresh and dry weight yields are shown, and in Fig. 5.19(b) the effect of N and K on the fresh and dry matter yield of the variety, Aristo. Schippers (1968) concluded that with more than 80 kg N ha $^{-1}$, the increase in yields in his experiments were due almost entirely to water, while the only effect of K was to increase the yield of water.

In Fig. 5.19(c) the effect of N and K on tuber specific gravity is shown; there is some evidence of an interaction between these nutrients in that the effect of N was curvilinear at low levels of potash and approached linearity at higher levels. Schippers (loc. cit.) also showed that the effect of N and K on specific gravity may be influenced by variety. A full discussion of factors affecting tuber quality is given in Chapter 13.

5.4 Factors influencing the yield response to fertilizer

It is unfortunately clear that in spite of the publication of data from innumerable fertilizer trials, there is still considerable uncertainty concerning the shape of the yield response curve, a situation which is further complicated by the variability in response induced by the environment or attributable to the way in which fertilizers are used; some of these factors are considered here.

5.4.1 Length of growing season

Yields of 4 levels of N relative to the highest are plotted against time in Fig. 5.20; all levels increased yield compared with the control, but excepting the control, yield was initially negatively related to the level of N applied. As time progressed, the yield of the lower levels of N fell relative to that of the highest level, but if allowance is made for some fluctuation in the data, only at the end of the growing season were yields beginning to become ordered in accordance with the rate of N application. Thus responses to high rates of N would only be expected if the growing season were long, which is in agreement with the effects of N on the growth of leaves and tubers outlined earlier. Circumstantial evidence (Pizer *et al.*, 1961) suggests that when the growing season was curtailed by blight (*Phytophora infestans* L.) the response of tubers to N and K was halved, while the response to P was unaffected; again this is in agreement with the probable timing of the effects of nutrients on leaf growth. However, Birch *et al.* (1967) did not find that the incidence of blight affected the response to N.

The length of the growing season may be arbitrarily shortened by harvesting the crop at an immature stage as in the production of potatoes, for an early market; in which case fertilizer rates tend to be modified in comparison with those recommended for crops harvested at maturity. Commonly, levels of P are similar but the rates of N and K reduced (Ministry of Agriculture, Fisheries and Food, 1973) which again would seem to follow logically from the timing of the effects of nutrients on leaf area. Ekeberg (1972) showed that in Norway, when planted before May 20th, the optimum rate of N was 120 kg ha^{-1}, this fell to 80 kg

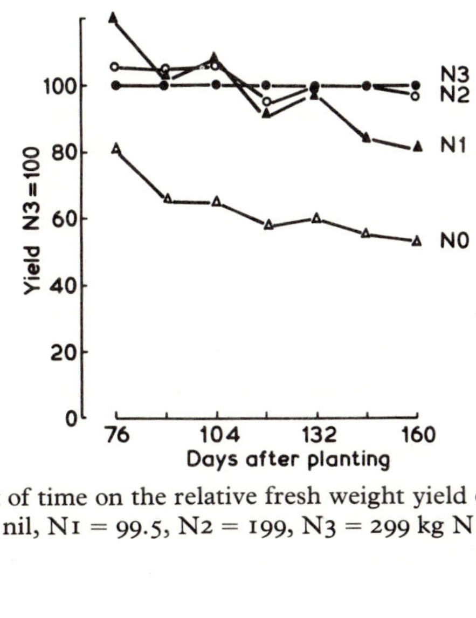

Fig. 5.20 The effect of time on the relative fresh weight yield of tubers at 4 levels of N (N0 = nil, N1 = 99.5, N2 = 199, N3 = 299 kg N ha^{-1}). From data of Ngugi, 1972

ha^{-1} where planting was delayed until late May and to 40 kg ha^{-1} when planted in early June. Gately (1971) also found that responses to N in a series of trials carried out in Ireland, were greater at the earlier sown sites.

5.4.2 Fertilizer placement

It has been suggested that the most efficient placement of fertilizer is that which provides for an adequate supply of soluble nutrients in a well-aerated zone of moist soil occupied by actively absorbing roots at periods of growth when the demands for plant nutrients are most acute.

Batey and Boyd (1967) have reviewed experiments in which various methods of placement have been compared. They pointed out that considerable attention has been given to the placement of fertilizer for the potato crop because the efficiency of fertilizer use is partly determined by the ultimate position of the fertilizer in the ridge, which in turn depends upon how and when the fertilizers are applied and the planting method. The placement methods they described, which are applicable to crops grown in single row ridges include the following systems.

The fertilizer may be broadcast 'on the flat', i.e. before the land has been ridged and the crop planted. The earlier the application relative to planting and the 'rougher' the soil conditions the greater the depth of incorporation of the fertilizer into the final seed bed, and subsequently through the ridge. Fertilizer applied in this way is more likely to be localized around and above the seed tuber. A traditional method of application is to broadcast the fertilizer over ridged land before planting, on 'splitting' the ridges, the fertilizer is concentrated mainly just below the tuber. Various attachments to planting machines are available which place the fertilizer in various zones in the ridge; with contact placement much of the fertilizer is located in contact with the tuber or developing sprout, whereas with band placement it is separated from the fertilizer by a zone of soil, the width of this buffer between fertilizer and tuber and its orientation may be varied according to the design of the placement mechanism. It is also possible to inject fertilizer solutions or anhydrous ammonia into the ridge after planting.

It has been shown that the form of the yield response to fertilizer can be altered by the method of placement. The response curve may differ for fertilizer broadcast on the flat and broadcast over the ridges, the degree of curvature appears to be more pronounced for placed than for broadcast fertilizer. Thus small and moderate amounts of fertilizer tend to be utilized more efficiently when placed than when broadcast, and less placed fertilizer is needed to reach maximum yield. Where conditions are unfavourable to placement, yield may decrease sharply at high levels of fertilizer application, for example on sandy soils deficient in N and K in low rainfall areas; here the high level of fertilizer required can in conjunction with low soil moisture content considerably damage the germinating plants and delay emergence.

In experiments with compound fertilizers containing N, P and K, Widdowson *et al.* (1967) again found that the form of the response curve was affected by method of fertilizer placement, working the fertilizer well into the seed-bed would appear to be the best method at higher application rates; and unlike the experiments reported by Boyd *et al.* (1968), where maximum yields were similar regardless of method of placement, the data suggest that the yield potential may be increased where high rates of fertilizer are broadcast.

These experiments pose important problems in connection with fertilizer experiments, for presumably the method of application should change with the level of fertilizer used, but in most experiments this important effect is not taken into account.

Placement of fertilizer at some depth below the seed tuber has been shown to be beneficial under circumstances where soil moisture is limiting the uptake of nutrient located in the superficial layers of the soil near the seed tuber (Holliday and Draycott, 1968). This is more fully discussed in Section 5.4.4.

5.4.3 Timing fertilizer application

Usual practice is to apply most or all of the fertilizer at or just before the time of planting. However Radley (1963) suggested that under good environmental conditions, e.g. a plentiful supply of plant nutrients, assimilates tend to be used in the production of foliage rather than tubers and that this accounts for the apparent delay in tuber initiation associated with the use of fertilizers applied in the seed-bed. Ivins (1963) suggested that it might be possible to increase the duration of tuber bulking without sacrificing bulking rate by withholding N, or part of it, until tubers had been initiated, and in this way increase final tuber yield.

Experiments testing this hypothesis have been reported on a sandy loam soil in the south of England (Gunasena and Harris, 1968, 1969, 1971). Large benefits in delaying the application of all or part of the N fertilizer were associated with seasons when rainfall after the application of N in the seed-bed was sufficiently high to bring about leaching. The beneficial effects of delaying the application of both N and K fertilizers were associated with increased recoveries of these nutrients in the crop attributed to the avoidance of leaching losses. These effects are shown in Table 5.9. In these experiments it was estimated that 5–7 cm of drainage occurred after the seed-bed application of fertilizers.

In subsequent experiments, where leaching after fertilizer application did not occur, then delaying the application of all or part of the fertilizer until after tuber initiation did not result in any marked effects on the

Table 5.9 The effect of time of application of N and K on total tuber yield and the apparent recovery of these nutrients from fertilizer

		N					*K*			
Year	*Level of N (kg ha⁻¹)*	*Time of application* Nil	Early	Late	Split	*Level of K (kg ha⁻¹)*	*Time of application* Nil	Early	Late	Split
				Craig's Royal (t ha⁻¹)						
1966	125	36	52	48	–	156	50	52	54	–
1967	Mean of 94 and 188	30	49	56	–	–	–	–	–	–
1968	220	38	67	82	82	183	62	67	71	77
				Pentland Dell (t ha⁻¹)						
1968	220	43	74	83	82	235	69	74	78	75
				Apparent recovery from fertilizer %						
				Craig's Royal						
1966	125		59	87	–	156		26	47	–
1967	94		72	83	–	–		–	–	–
	188		55	84	–	–		–	–	–
1968	220		74	121	109	183		38	20	51
				Pentland Crown						
1968	220		82	109	106	235		48	46	62

Early = application at planting, late = after tuber initiation, split = half on each occasion. From Gunasena and Harris, 1968, 1969, 1971.

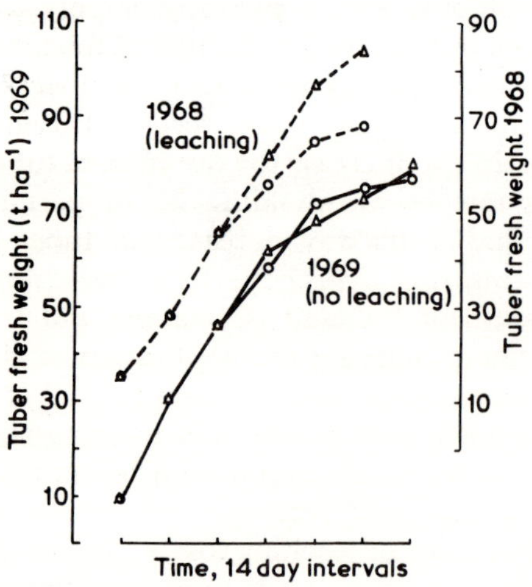

Fig. 5.21 The effect of time of application of N on tuber yield in a year when leaching occurred after planting (broken lines) and when no leaching occurred (full lines) triangles = N applied part at planting and part after tuber initiation, circles = N applied at the time of planting only). Data from Gunasena, 1969 and Ngugi, 1972

response to N (Fig. 5.21). There was little evidence to suggest that delaying the application of N had any marked effect on the period between the apparent initiation of tubers and the establishment of the linear phase of tuber bulking. Although the varieties differed in the 2 years, in the absence of leaching delaying the application of N reduced the uptake of N and the development of leaf area over most of the growing season.

Gunasena and Harris (1968, 1971) were also able to show some yield benefit from delaying or splitting the application of K, which like N, was attributed to reduced leaching losses. It is perhaps not surprising that no yield advantages were shown from delaying the application of all or part of a complete N, P and K fertilizer on Bunter sandstone (Culpin and Greaves, 1956) owing to the importance of a good supply of P early in the life of the crop when root development is insufficient to obtain adequate supplies of this relatively immobile nutrient (Wiersum, 1961).

It was suggested earlier that there could be a link between N and leaf senescence, and that competition between tubers and leaves for N might be a factor involved with the rapid decline in leaf area and the eventual termination of growth; it was further suggested that this train of events was initiated by a decline in root growth. It would seem logical to suggest that appropriately timed applications of N to the foliage might facilitate the late uptake of N and alleviate the internal competition for N, thus promoting tuber growth by delaying the senescence of leaves. However, in many instances no effect has been found from the application of various forms of N as foliar sprays. Experiments on the University of Reading farm in S.E. England, using both ammonium nitrate (unpublished) and urea (Kerketta, 1976) as foliar sprays, failed to show any effect on tuber yield. In a dry year foliar applications of urea to potatoes (Holliday, 1970) failed to affect yield; Colombies (1975) found that foliar sprays of amino acids or urea had no significant effect on yield. This is perhaps not surprising as an increasingly ineffective root system would affect the supply of other essential nutrients and growth regulators such as cytokinin which are thought to be manufactured in roots. However Kerketta (1976) showed that the combined application of cytokinin (6-benzylamino purine) and urea had only transient effects on leaf area index and yield, while unpublished experiments by the author in which higher concentrations of cytokinin (200 parts 10^{-6} in 10% ethanol) applied frequently with or without urea failed to elicit any response in the yield of tubers. While Kerketta (loc. cit.) showed that foliar applied urea was effectively taken up and distributed in the crop, it is not known how effectively foliar applications of cytokinin are absorbed and transported in the plant.

It is, of course, possible that growth regulators other than cytokinin might be involved in addition to a wider range of mineral nutrients; Thimmegowda *et al.* (1975) for example reported that foliar sprays of N, P, K, trace elements, naphthalene acetic acid and triodobenzoic acid at the flower stage increased tuber yield substantially compared with a water spray.

Another approach to solving the problem of deficiencies of N in the latter part of the growing season has been the use of slow-release forms of N such as nitroform, and sulphur-coated urea. Generally such forms have compared unfavourably with quicker acting forms of N. Lorenz *et al.* (1972) showed that slow-release N sources did not increase yield or N absorption during the later stages of growth. If the decline in nutrient uptake with time is at least partly associated with a cessation of root growth due to competition from tubers, then slow release forms of N would be at a disadvantage, for a decline in the nutrient supply might be expected to increase the growth of tubers relative to foliage thus reducing the maximum value of L, a factor shown earlier at least in conncection with N nutrition to be important in the maintenance of L at the end of the growing season.

5.4.4 Organic manures

A wide range of organic manures may be applied to crops ranging from materials such as hoof and horn meal which have a high N content and behave in a similar way to inorganic fertilizers, to bulky organic manures resulting from a mixture of dung, urine and bedding or slurries of dung, urine and water, where animals are housed without the possibility of bedding material becoming mixed with excreta. Here attention is focussed on 'farmyard manure' the traditional bulky organic manure resulting from cattle over-wintered in yards bedded with straw. Such manures have been held in high esteem by farmers, and in Britain, when the material is available, the potato crop receives priority. However, while the importance of such materials has declined in Britain with the increased use of inorganic fertilizers, they may still be of considerable significance elsewhere; according to Sokolov and Orlovskii (1971), organic fertilizers, of which farmyard manure is the principal, play an 'extraordinary role' in the cultivation of the tundra and taiga virgin soils, not only as a source of nutrients, but also in improving the physico-chemical properties of soil. In temperate ragions too, the benefits of bulky organic manures are widely believed to derive from their physical rather than from their nutritional properties. Boyd (1961) showed from survey data collected in Britain, that farmers did not adjust their rates of fertilizer when farmyard manure was used, indicating that the manure was valued for properties other than its content of mineral elements.

Crowther and Yates (1941) showed that the responses of crops to fertilizer N, P and K were considerably reduced when grown in the presence of 25 t ha^{-1} of farmyard manure however they pointed out that the most remarkable feature concerning the effect of farmyard manure was the small difference between the responses to N in the absence of farmyard manure and its presence. The responses to P and K on the other hand were substantially reduced by farmyard manure. Similar observations were made by Edwards *et al.* (1956).

Boyd (1959) re-examined the experimental evidence and concluded that the response to N in the absence of farmyard manure was very dependent on the basal supply of P and K; in the presence of farmyard manure it is less dependent. Thus the effect of farmyard manure was to increase the response to N applied alone, while given a basal dressing of P and K, farmyard manure reduced the response to N. The nature of the responses to P were similar to those for N, but the overall level of response was lower. Where farmyard manure was applied there was a marked reduction in the response to K when compared with large responses without farmyard manure. Boyd concluded that the effect of farmyard manure could be adequately accounted for in terms of the nutrients N, P and K without having recourse to any other effects. The value of 10 t of farmyard manure was estimated to be equivalent to about 15 kg N, 9 kg of K and 31 kg of P.

Some experiments have shown that as the level of N, P and K supplied as inorganic fertilizer was increased, then the response to farmyard manure was progressively reduced until at a sufficiently high level of fertilizer the response to farmyard manure was eliminated (Table 5.10). In a series of 10 experiments carried out on the Leeds University farm in one of the drier parts of England (long term mean rainfall per annum 61 cm) on medium textured glacial soils overlying magnesium limestone, it was shown (Holliday *et al.*, 1965) that the effects of farmyard manure could not always be eliminated by high levels of inorganic fertilizer. Over a period of 10 years while some treatments changed, there was always a standard combination of treatments consisting of 0 and 25 t ha^{-1} of farmyard manure factorially combined with 0, 79 and 158 kg N ha^{-1}. A uniform high rate of P and K was applied to all treatments. The responses to N in the presence and absence of farmyard manure are shown in Fig. 5.22(a) for the means of the 5 wettest and 5 driest years. In the 5 driest years the 2 N response curves converged, indicating that the effect of farmyard manure could be eliminated in the presence of a sufficiently high level of N, P and K applied as a fertilizer as shown in previous experiments. In the 5 driest years the curves were approximately parallel indicating that farmyard manure was having an effect which could not be reproduced by inorganic fertilizers. Internal evidence suggested that there was no response to levels of N in excess of 158 kg N ha^{-1}. Given the high level of P and K basal to all treatments, any response to farmyard manure at this high level of N could be considered

Table 5.10 Total tuber yield (t ha $^{-1}$) response to farmyard manure at various levels of N, P and K supplied by fertilizers.
(a) Response to 37.5 t ha $^{-1}$ of farmyard manure

N	*P*	*K*	*tubers*
	(kg ha^{-1})		*(t ha^{-1})*
0	0	0	7.0
34	20	47	1.2
68	39	93	0.5

(b) Response to 25 t ha $^{-1}$ of farmyard manure

N	*P*	*K*	*P*	*K*
(kg ha^{-1})	*(kg ha^{-1})*			
	55	156	82	233
	tubers (t ha^{-1})			
0	7.5		8.0	
79	2.3		0.5	
158	2.3		− 2.3	
236	2.0		1.0	

(a) Data from Reith and Inkson (1958), (b) Data from Harris, unpublished.

to be due to effects which are not readily explained in terms of the quantity of N, P and K supplied by the organic manure. The term 'extra effect' was coined by Holliday *et al.* (1965) to describe this effect and values plotted against the maximum calculated potential soil moisture deficit occurring in each of the 10 years [Fig. 5.22 (b)] show a positive relationship between the extra effect of farmyard manure and increasingly dry soil moisture conditions.

Pot experiments (Harris, 1960) suggested that the beneficial effect of farmyard manure in dry years could not be attributed to any significant increase in available soil moisture for the equivalent of 25 t of farmyard manure ha^{-1} in pots increased available soil moisture by about 0.2 cm, sufficient for about 1 day of transpiration.

In these experiments farmyard manure was applied in the bottom of the ridges and the inorganic fertilizers then broadcast before the tubers were planted and the ridges split back to cover the seed. The explanation advanced for the beneficial effect of farmyard manure in dry seasons, was that nodules of farmyard manure in the soil were capable

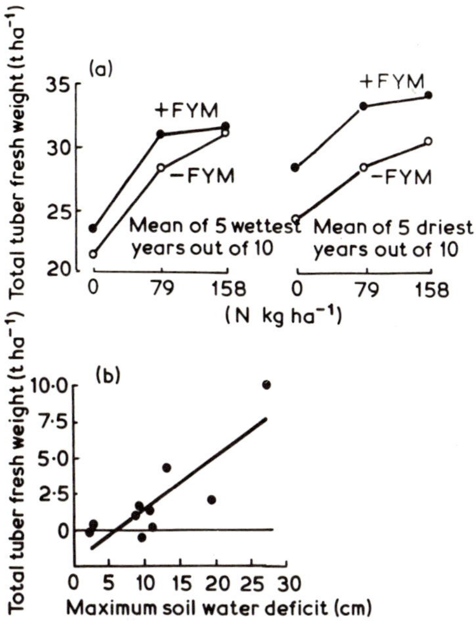

Fig. 5.22 (a) The effect of N with and without farmyard manure on tuber yields in wet and dry seasons. From Holliday, 1970. (b) The effect of soil water conditions on the response of potatoes to farmyard manure at levels of NPK believed to be sufficient for maximum yield. From Holliday *et al.* 1965

of retaining sufficient moisture to permit roots to absorb N and other nutrients from within their structure at a time when the moisture in the surrounding soil was insufficient to permit the uptake of nutrients from the inorganic fertilizer. It would be expected that the dry season or extra effect of farmyard manure could be reproduced by inorganic nutrients equivalent to those available in the latter, provided some means of making them more available under dry conditions could be devised.

Holliday (1970) has summarized the experiments which tested this hypothesis. Two methods were attempted, the first a foliar application of N applied at predetermined soil moisture deficits proved to be unsuccessful in the one exceptionally dry year in which it was tested and because of this it was abandoned. The second method adopted was to place nutrients at a greater depth into the soil profile where the loss of moisture is less rapid. In the 3 years 1961–3 potato experiments similar to those in the 10 year series were completed, with the addition of a treatment in which the estimated available plant nutrients in 25 t ha^{-1} of farmyard manure was placed 15–18 cm below the seed tuber in 2 bands, each 20 cm from the centre of the row at planting time. This placement was achieved by applying liquid fertilizer through hollow tines. Although there were some differences between the N response curve in the presence of farmyard manure compared with that in the presence of the deep placed liquid fertilizer, both produced yields appreciably greater than that achieved with conventionally placed fertilizer. The deep placed fertilizer largely simulated the effects of farmyard manure; a result in agreement with an experiment reported by Warren and Johnston (1961) in which fertilizers which were dug in substantially reduced hitherto unexplained yield advantages found when farmyard manure was applied to sugar beet. Further experiments established different soil moisture regimes and the growth of tubers grown without fertilizer or with shallow (5 cm) or deep (18 cm) placed liquid fertilizer also showed that advantages accrued from deep placed fertilizer under a dry soil moisture regime but not under a wet regime. Growth analysis studies showed that the seasonal leaf area index curves of the deep–dry regime treatment was similar to the shallow and deep placement on the wet regime. In contrast, the shallow–dry season leaf area index curve attained a lower ceiling by the end of July and decreased more rapidly during August and September.

On the basis of these experiments, Holliday suggested that the nature of crop responses to irrigation needed further examination remarking that: 'It is difficult to avoid the conclusion that at these low deficits the irrigation response is almost entirely due to the water making the plant nutrients more available'.

5.4.5 Rainfall

During the growing season

Rainfall may interfere with the uptake of plant nutrients as already shown, and may also limit the response of crops to nutrients which have already been taken up. Rainfall is also an important factor affecting the incidence of disease, particularly such important diseases as *Phytophthora infestans* L. The relationship between rainfall during the growing season and crop response to fertilizer may therefore be expected to be complex. Consequently it is difficult to arrive at any general conclusions, which would in any case be as difficult to implement in practice owing to the well-known hazards involved in predicting the season's weather sufficiently far in advance to be able to take it into account in planning fertilizer levels.

The response of potatoes (variety King Edward) over a period of 10 years on glacial drift soils overlying magnesium limestone in central England (Holliday *et al.*, 1965) varied from 11.7 ton ha^{-1} when the maximum soil moisture deficit was less than 2.5 cm to 2.7 ton ha^{-1} when the maximum calculated deficit was 25 cm. Cooke (1967) pointed out that in most summers in England there will be little leaching of nitrate from the top soil under crops and the effect of moister weather is likely

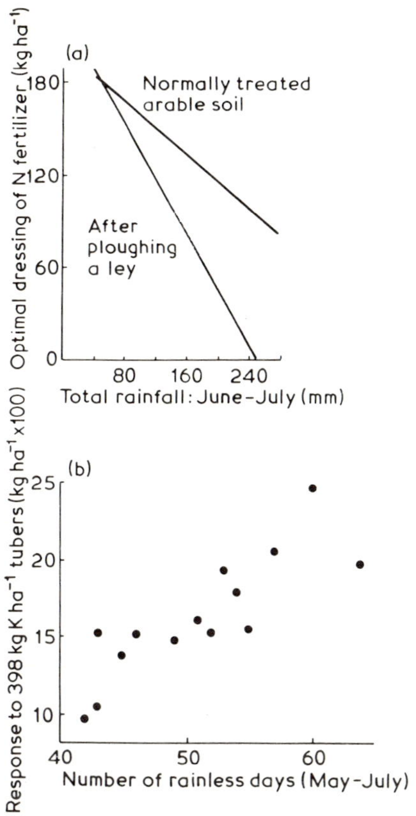

Fig. 5.23 (a) The relationship between June–July rainfall and the optimal application rate of N for potatoes. From Kuipers, 1902. (b) The relationship between the number of rainless days and the response of potato tubers to 398 kg K ha^{-1}. From Van der Paauw, 1958

to be an increase in N supply. However, it was shown earlier that the crop is vulnerable to the effects of leaching when the soil is at or near field capacity and rain falls after planting and the application of fertilizer but before the crop canopy has developed appreciably. While the response to N would appear to be greater in wet than in dry years, the optimal rate of application has been shown (Kuipers, 1962) to increase with a decrease in the June–July rainfall [Fig. 5.23(a)].

With the early variety Home Guard, harvested before reaching maturity, Evans (1974) found that where lack of moisture did not affect growth the optimum level of N was less than at non-irrigated sites in dry seasons. This finding is in agreement with the suggestion in the previous section that one of the first effects of increasing soil moisture deficit is to decrease the N supply to the crop.

The possibility of coupling fertilizers in which the N is in the ammonium form with nitrification inhibitors offers some prospects of reducing leaching losses of N in early Spring (Gasser, 1970). Van Burg *et al.* (1967) showed that even without the addition of a nitrification inhibitor, anhydrous ammonia injected 1–3 weeks before planting was more effective than conventional fertilizing with ammonium nitrate applied at the same time. When injected after growth had commenced, however, (Harris, 1970) anhydrous ammonia was no more effective than ammonium nitrate applied at the same date.

Van der Paauw (1958) showed that potash fertilizers had the biggest effect on yield in dry years [Fig. 5.23(b)], which was attributed to a reduction in the availability of soil K. Hackett (1968) has shown that when K is deficient, root growth and especially the development of laterals, may be markedly retarded, which could affect the ability of the crop to remove water from the soil. K may also influence the water economy of crops through its influence on stomatal movements (Peasles and Moss, 1966).

Simpson (1956) found that little of the fertilizer P applied to potatoes was recovered in dry years, and high yields were obtained without applied P even in soils low in soluble P. This may be related to an increase in the P absorbed by plants in the warmer soil conditions associated with drier weather; when irrigated, the need for phosphatic fertilizer appeared to decrease. That soil moisture conditions can have a marked effect on the accumulation of P is shown in unpublished data on the effect of irrigation on the uptake of P by sugar beet grown on alluvial soil on the Reading University farm in the Thames valley in England; irrigating with 3 cm of water rapidly increased the accumulation of P by 57% while total dry matter yield was increased by only 12% over the same period. There was a water table within root range and presumably the extra P accumulated was associated with an increase in its availability in the upper nutrient rich part of the soil profile.

Winter rainfall

Large fertilizer dressings of N leave residues of nitrate in the soil and more is released as crop residues decay. Reserves of nitrate are easily leached from the topsoil and the residue of nitrate left to benefit a following crop depends on the amount of rain in winter. Birch *et al.* (1967) pointed out that in Britain there are two types of variation in winter rainfall; the inherent regional variations in climate and the year-to-year variations in weather conditions. Allowance for the first can take the form of regional recommendations, although there is likely to be a degree of association between rainfall and cropping system which affects N requirements. Allowance for seasonal variation requires assessment and decision in late winter. Where winter rainfall (October–March) exceeded 30 cm, 30–40 kg more N ha^{-1} was required than when winter rainfall was less than 30 cm. In Holland it has been shown (Van der Paauw, 1963) that 40 kg ha^{-1} more soil N is available in early spring after a dry winter than after a wet one.

It would be useful if, instead of relying on empirical relationships between factors such as rainfall and soil N status and also between previous cropping and soil N status, if a soil N index could be determined to give an objective characterization of the status of the soil at the beginning of the growing season on which sound advice could be based with regard to the level of N fertilizer to apply. In a recent review, Robinson (1975) concluded that except under exceedingly uniform soil and climatic conditions the definition of a predictive soil N index is not possible for field crop conditions. One of the principal reasons for the lack of success of predictive soil N index being the climatic variations between sites and on the same site in different years which dictates the frequency, duration and intensity of wetting/drying cycles in the topsoil which control the production of available natural N to the crop growing on a particular soil.

5.4.6 Plant density

It has been pointed out (Van Burg, 1967) that most experiments on the effect of plant density have been conducted at one rate of fertilizer application, but that where the interaction between fertilizer and plant density had been investigated, some had shown that an increase in the level of N or N, P and K had a greater effect at high than at low crop densities, while in other instances no interaction had been found.

In the rather special circumstances of seed-potato production in Holland where haulms were destroyed in mid-July, and using seed 40–50 mm diameter (Van Burg, loc. cit.), the response to N increased with an increase in plant density from 40 000 to 50 000 plants ha^{-1}, and the response to N was brought forward in time by close planting. In a second experiment however, where plant density was varied by seed size as well as spacing, the interaction was marked only with small seed (35–40 mm) and was absent with large tubers (50–55 mm). There was an overall positive relationship between tuber yield and mainstems ha^{-1} but only with small seed was there a positive interaction between number of stems and N supply. It is to be expected that any factor increasing plant numbers would be more likely to have a positive effect on yield in the early stages of growth.

In a further experiment with a maincrop variety, where the foliage was allowed to die naturally, it can be seen (Fig. 5.24) that with respect to final tuber yield, the response to N was dependent on plant density. It would appear from these experiments that one of the effects of increasing plant density was to increase the proportion of the dry matter accumulated

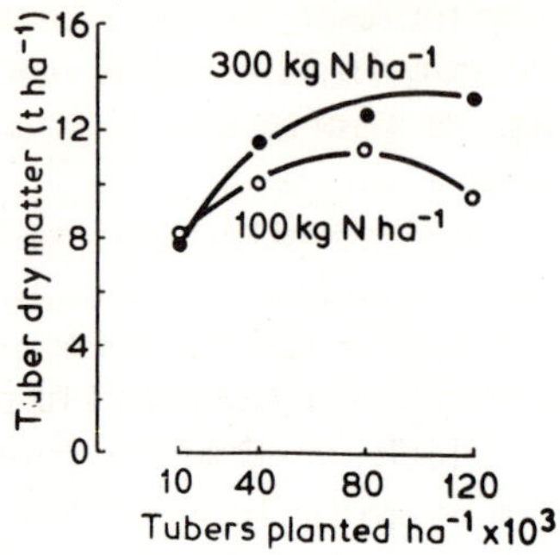

Fig. 5.24 The effect of plant density and N on tuber dry matter yield. From data of Van Burg, 1967

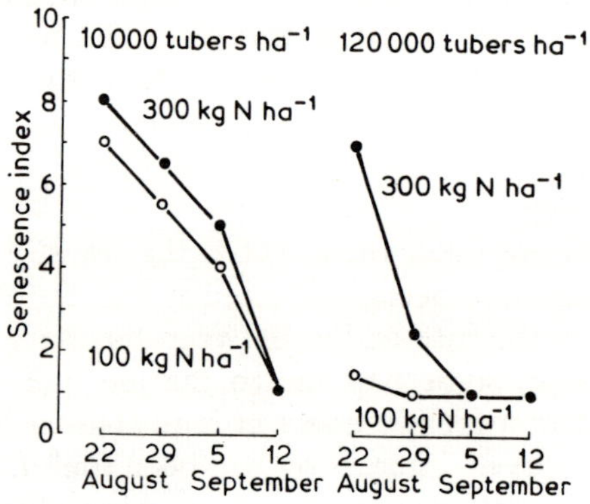

Fig. 5.25 The effect of N and plant density on haulm senescence. 10 = no die back, 1 = haulms compeletly dead. From Van Burg, 1967

in tubers, but this tendency was offset when extra N was supplied. At high plant densities, extra N seemed to permit the investment of more dry matter in leaf production which, given a long enough growing season, eventually increased tuber yields. While haulm died back rapidly, Van Burg was able to show that N delayed the senescence of foliage, an effect particularly marked at the highest plant density, which treatment otherwise tended to advance the senescence of the crop. These effects are shown in Fig. 5.25.

Bodlaender and Reestman (1968) found that with the late variety Alpha, the optimal plant density was increased by 100 kg N ha^{-1} compared with no fertilizer N, but that a further increase in N supply did not change the optimal plant density. Increasing the N supply was most effective at high plant densities, especially after a long growing period. Favourable interactions between N and plant density were not usually

Table 5.11 The effect of plant density and N on the total tuber fresh weight yield of potatoes, variety Maris Page

Level of N (kg ha $^{-1}$)	Plants (ha $^{-1}$) 34 602	46 132 (tubers kg ha $^{-1}$)	64 599
0	43.9	45.5	46.5
99	59.0	76.6	72.1
199	79.1	86.6	81.6
299	75.0	87.4	92.7

Data from Ngugi, 1972.

obtained at early liftings, and this would apply to later varieties where the growing season was prematurely curtailed by disease.

While Ngugi (1972) failed to show any significant interaction between nitrogen fertilizer and plant density, the data in Table 5.11 show that up to a N application of about 200 kg ha $^{-1}$, the optimal tuber density was 46 000 ha $^{-1}$, with 300 kg ha $^{-1}$ the highest yield was achieved with a density of 65 000 tubers ha $^{-1}$.

In Ngugi's experiments there was some suggestion that while increasing the level of N at any one plant density increased L and reduced E, the higher the plant density the higher the value of E for any particular value of L; this was particularly marked at low values of L (Fig. 5.11). Ngugi was not able to find any correlation between tuber numbers and E which might have explained the effect and it is therefore probable that plant density effects on the spatial arrangement of the crop canopy are implicated, although it is difficult to see why these effects should be more pronounced at low than at high values of L.

The effect of plant density on the development of L at four rates of N is shown in Fig. 5.26. While in this experiment the plant density treatments were confounded with blocks, the data suggest that in the absence of fertilizer N, plant density had little effect on L, and effects on L became more pronounced as the level of N increased; however only at the highest level of N was there a positive increase in L with the highest plant density. It is also interesting to note that the effects of plant density were only marked in the first part of the growing season, and that once L began to decline, positive effects of plant density on L were either absent or small.

From this limited evidence, only tentative suggestions can be put forward to indicate the nature of the yield relationships between plant density and N.

(a) Positive interactions between N and plant density might be expected where N increases the number of mainstems per unit area

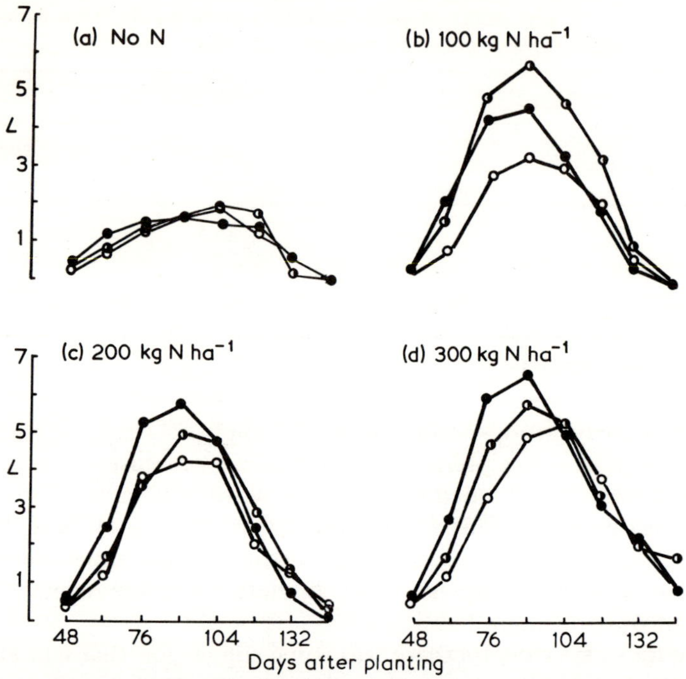

Fig. 5.26 The effect of plant density on the development of L in a crop of second early potatoes (variety Maris Page) at four levels of N. (Open circles = 34 602, half closed circles = 46 132, closed circles = 64 599 plants ha^{-1}). Redrawn from Ngugi, 1972

at high plant densities especially when crops are harvested before reaching maturity.

(b) The ability of plant density to increase L and therefore yield, would appear to be dependent upon the level of the N supply.

(c) The tendency for the effects of plant density on L to be confined to the early part of the growing season when L is increasing could be complimented by the tendency of high rates of N to prolong the active life of the haulm,

(d) There is some indication (Ngugi, 1972 and Fig. 5.11) that the photosynthetic efficiency of the crop canopy (E) might be affected by plant density, an effect which does not seem to be attributable to effects on tuber numbers, but which might be associated with an improved spatial arrangement of leaves.

For these reasons it would seem that for maximum yields, both high plant densities and high levels of N might be needed. However, much more evidence is required before firm conclusions can be reached. For example, the effect of plant density on the rate of loss of water from the soil, and the effect this may have upon the response to plant nutrients could be important (see Chapter 6), while varieties because of their effect on stem numbers, tuber numbers, the structure of the crop canopy and the longevity of the foliage, could also have important effects on the interaction between plant nutrients and plant density. Most of the above discussion has been limited to N, evidence concerning the interaction between plant density and other nutrients would seem to be scanty.

5.4.7 Variety

In making recommendations for fertilizer use on potatoes in Britain (Ministry of Agriculture, Fisheries and Food, 1973), separate recommendations are given for maincrop and early canning potatoes; this classification does embody varietal differences, but a major factor influencing fertilizer need is the differing lengths of growing season. From evidence presented in earlier sections, it would be expected that tuber yields would respond increasingly to N as the season advanced; however varietal and climatic effects could be important. Ivins and Bremner (1965) for example found that early varieties proved to be more responsive to N than maincrops, for their leaf growth is more often limiting. The apparent delay in tuber initiation associated with nitrogenous manuring, they maintained, is never appreciable in a true early crop, i.e. for an early variety planted early, because N uptake is more limited by low temperatures and because environmental conditions generally are less suitable for leaf growth early in the season.

Apart from extreme differences combining varietal effects and artificially curtailed growing seasons, there appears to be limited information on the interaction between variety and fertilizer. This is perhaps not surprising, for as pointed out earlier, most fertilizer trials have tested insufficient levels of fertilizer, and comprehensive fertilizer trials involving a number of varieties in factorial trials would involve impracticable numbers of treatments. However, it might be expected that varieties could affect the response to fertilizers due to varietal differences in the rate at which leaves senesce, to differences in the partitioning of nutrients between different plant parts and perhaps as shown earlier, differences in the N content needed to support leaf growth. Ekeberg and Rønsen (1973) have, for example, found that the varieties Kerrs Pink, Pimpernel and King George V responded to 140 kg K ha^{-1} while Parnassia and Ora only responded to 70 kg K ha^{-1}. The responsive variety Kerrs Pink had a higher tuber K content than the less responsive variety Parnassia, which suggests that a possible reason for the greater response to K might be a reduced ability of foliage to compete with tubers for a limited nutrient supply. Johansson *et al.* (1973) found that the optimal economic rate of N was 75 kg N ha^{-1} for the varieties Bianella and Prevalent and twice that for the variety Kaptah.

It was shown earlier that varieties may differ in their root development and this in turn might affect the ability of a variety to exploit soil phosphate reserves. However Schooneveldt *et al.* (1968) found no varietal difference in response to phosphate manuring on crops grown on a responsive reclaimed sandy heath soil.

5.4.8 Soil pH

Apart from the affect of soil reaction on soil borne diseases such as potato scab (*Steptomyces scabies*), the major effect on yield is probably associated with the availability and uptake of plant nutrients. Bolton (1971) found in liming experiments carried out on the contrasting soil types at Rothamsted and Woburn in England, that yields of potatoes were similar at all pH values above 5 when phosphate and K were sufficient, but when K was not given the more acid soils grew the larger crops. In both experiments the response to K was least on the most acid plots because liming decreased yields when K was not given. This lime-induced K deficiency was explained by liming increasing the proportion of K absorbed on the exchange complex, and thus decreasing the concentration of K in the soil solution; K ions can displace adsorbed Ca from cation exchange sites on the soil more easily than Al which is the predominant exchangeable ion in acid soils. At Woburn, but not at Rothamsted, the response to phosphate was larger on the unlimed plots.

Laughlin *et al.* (1974) found that in experiments carried out in Alaska, the addition of lime raised soil pH from 4.4 to 5.8 and yield increased. Ca content of foliage and tubers was increased and Mn and Zn decreased; the concentrations of Al, Ba, B, Cu, Fe and Sr were depressed in the foliage and Ba in the tubers. The uptake of N, K, P, Ca, Mg and Na were increased with lime application.

That the potato crop can tolerate very low pH values was demonstrated in experiments on a sphagnum peat soil (Maclean *et al.*, 1967), where 6.7 t limestone ha^{-1} increased yield in the year of application on a soil with a pH of 3.7, but where in three subsequent years, yields were as satisfactory on unlimed plots.

For a detailed review of the literature, the reader is referred to Smith (1968).

5.4.9 Soil productive potential

Holliday (1955) suggested that the response to fertilizers is not only dependent upon the nutrient status of the soil but on its inherent fertility – the latter being made up of those soil characteristics which affected the ability of the soil to supply water to the crop, e.g. organic matter, soil texture, and depth of the soil profile. In a later paper (Holliday, 1963) it was shown that where soils used in potato fertilizer experiments were classified according to their inherent fertility or production potential, that soils of high potential justified the application of higher rates of fertilizer than soils of lower yield potential.

5.5 The estimation of nutrient requirement

The estimation of nutrient requirement must be based on a knowledge of the yield response to nutrients and an assessment of the nutrient status of the soil on which the crop is to be grown. It has been shown that there in some considerable doubt as to the former and while reasonable estimates can be made of the P and K status of soils, no suitable method of analysis has been found on which to assess soil N status. Recommended rates of P and K are adjusted to take the soil status of these nutrients into account, although it is customary to give some P and K even on soils of high status, in order to ensure against the occasional chance of obtaining substantial responses on such soils, and in order to replace the nutrients removed in the crop. It would seem reasonable to avoid running down the soil's reserves of these nutrients where their removal by leaching is not significant and where yield reductions due to a supra-optimal supply are perhaps unlikely. Cooke (1967) has also drawn attention to the important fact that fertilizer residues often give higher yields of crops however much fresh fertilizer is given.

In order to assess the N status of the soil as a basis for modifying rates of N application, it has been found necessary to resort to historical records of cropping, stocking, the use of organic manures and winter rainfall and relationships established by experiment between them and response to N fertilizer. It is unfortunate that this is the case as this is the nutrient to which the largest yield responses are often obtained and the one where excess could lead to a reduction in yield. Thus both under- and over-estimates of N requirement may lead to substantial reductions in yield and also to a waste of unutilized fertilizer which is unlikely to have a residual effect in those areas where, over part of the year, leaching is substantial.

Of the three important major nutrients, it is for N in particular that it is tempting to look to foliar analysis for an objective method of estimating the N status of the crop. Providing a diagnosis is made early enough, the crop would appear to be responsive to applications of N given at least up to the time of tuber initiation. Tyler *et al.* (1961) estimated the nitrate–nitrogen content of petioles taken from the fourth leaf below the growing tip of the plant (usually the youngest fully expanded leaf) at two-weekly intervals throughout the growing season. They found that early samplings were more accurate in predicting likely yield response to N fertilization, and that there was often time for supplementing the crop with N. The wide range of petiole nitrate–nitrogen contents that were found in plants during growth is shown in Fig. 5.27(a).

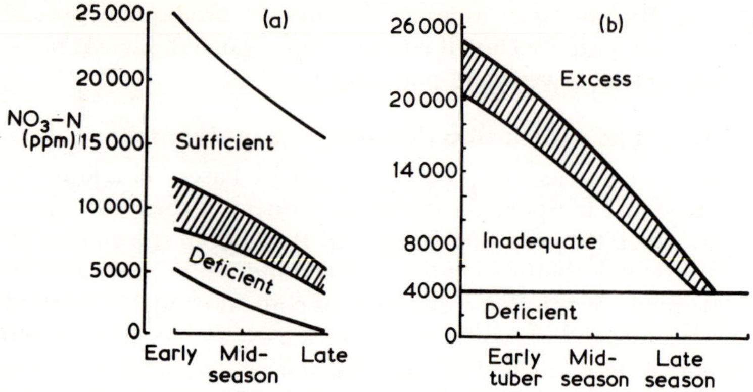

Fig. 5.27 Ranges of values for petiole NO_3 – N content (dry matter basis) and suggested critical values for (a) California, mainly variety White Rose. From Tyler *et al.*, 1961. (b) Idaho, variety Russet Burbank. From Preston Jones and Painter, 1974. The shaded portion indicates as adequate supply of nitrogen

Table 5.12 Plant analysis guide for potatoes grown under Californian conditions

Stage of growth	*Nutrient (Parts 10^{-6})*	*Content in dry matter of petiole of fourth leaf from tip*		
		Deficient	*Intermediate*	*Sufficient*
Early season	NO_3–N	8 000	10 000	12 000
	PO_4–P	1 200	1 600	2 000
	K%	9	10	12
Mid-season	NO_3–N	6 000	7 500	9 000
	PO_4–P	800	1 200	1 600
	K%	7	8	9
Late season	NO_3–N	3 000	4 000	5 000
	PO_4–P	500	800	1 000
	K%	4	5	6

From Geraldson *et al.*, 1973.

Values for the nutrient content of petioles which may be used as guides to the nutrient status of the crop based on experimental work carried out in California are given in Table 5.12. To what extent these values are applicable to other environments and other varieties is not certain. That they may not be for N at least is shown in Fig. 5.15(b) for the variety Russet Burbank grown in Idaho (Preston Jones and Painter, 1974). Van Burg (1966) found that for seed potato production in the Netherlands, with the variety Bintje, the nitrate–nitrogen content of the 'tops' needed to be 4.0% of the dry matter at the beginning of June, when plants were 10–15 cm tall in order to ensure a concentration of 0.6%

at haulm destruction, which was found to be the critical level. It was however difficult to make a sharp distinction between an adequate and an inadequate N supply.

Gately (1971) found in 19 trials with the variety Kerr's Pink, that where the nitrate–nitrogen content of the plant in the early stages of growth exceeded 0.45% of the dry matter of the tops, little response to extra fertilizer N could be expected.

Doll *et al.* (1971) showed that while critical levels for the nitrate–nitrogen content of the dry matter of petioles could be established in their experiments, values decreased rapidly with time and therefore the age of plants needed to be known precisely before the data could be interpreted. On the other hand, soil nitrate levels, while also decreasing with time, tended to be more stable and it was suggested that soil nitrate level might be a more stable criterion of N need than the nitrate content of petioles. Their experiments indicated that supplemental N fertilizers should be applied if soil nitrate levels fell below 20 parts 10^{-6} 10–12 weeks after emergence.

It is unlikely that any one criterion will adequately represent both the N status of the crop and also the likely response to additions of N fertilizer. Low nitrate levels in the crop may be associated with a depletion of soil moisture in the nutrient rich layers of soil, and it has been shown (e.g. Fig. 5.5) how sensitive the rate of nutrient accumulation may be to soil water conditions. The ability of the crop to respond to additions of N fertilizer may also be dependent upon the age of the crop. Williams (1969) has stressed the importance of measuring the nitrate content of the crop and the soil and also the water content of the soil in order to be able to draw sensible conclusions concerning the N status of the crop and the need for additions of N fertilizer.

Given that the N status of the crop can be characterized and due attention given to the soil environment, the practicability of estimating the criteria and implementing the findings have been enhanced by the development of a rapid diagnostic technique for the estimation of nitrate–nitrogen in the field (Williams, loc. cit.). However the quantity of N required to correct a deficiency would need to be determined experimentally although the level required would probably be dependent upon the length of the growing season, which is difficult to predict. Whether or not the crop could be irrigated and the chance incidence of rainfall would also be expected to have a considerable influence on the resolution of the problem. The evidence also suggests that different criteria may have to be developed for particular areas and varieties.

5.6 Concluding remarks

It has been shown that the potato crop accumulates large quantities of N and potash and somewhat smaller quantities of P during the course of growth; these are the elements which are most likely to be deficient in normal agricultural soils and which must often be supplied in fairly large amounts if maximum yields are to be approached. A prerequisite for the accurate assessment of fertilizer requirement is a knowledge of the way in which the crop responds to additions of limiting nutrients and of the nutrient status of the soil on which the crop is grown. It is evident that, in spite of the large number of fertilizer trials that have been carried out, the form of the response curve is not known with any degree of certainty and that for N, the nutrient most likely to limit yield severely in many cases, the nutrient status of the soil cannot be determined with any great degree of accuracy.

In order to determine the form of the response curve more fertilizer trials are needed in which six or more levels of fertilizers are tested and, to complement this work, the biological basis of the response curve (or curves) needs to be established. In particular, this should attempt to assess the role which nutrients play in the efficiency with which incident radiation is trapped by the crop, together with effects on the distribution of assimilate within the crop and any influence nutrients may have on the photosynthetic efficiency of the crop canopy, either directly or indirectly via effects on 'sink capacity'. However, it is hardly surprising to find that the biological basis of the response to fertilizers is not known, when the physiological determinants of yield are so incompletely understood.

It is also evident that the form of the yield response curve to mineral nutrients may be affected by many factors, such as plant density, water supply, the timing and placement of nutrients, the length of the growing season and variety. There appears to have been little or no systematic attempt to quantify the effect of such variables on the response to fertilizers, no doubt because to the complex experimentation that would be involved.

Substantial increases in yield can only be achieved by linking early tuber initiation (or the onset of the linear bulking phase of growth), with high rates of tuber growth and a prolongation of the activity of the crop canopy until growth is curtailed by external environmental factors rather than by internal factors bringing about the death of the plant. At present, early tuberization would seem to be inimicable to prolonged tuber growth; hence the crop would seem to be unable to exploit to the full the growing period permitted by the environment. Mineral nutrients have a role to play in determining the onset of tuber initiation (or the linear phase of bulking), the rate of tuber growth, and senescence; however, the distribution of dry matter between the various plant parts, the disposition and utilization of N (and other nutrients) within the crop, and the regulation of growth by plant hormones are involved in the duration of tuber growth, and mineral nutrition must be considered in conjunction with such factors. A more complete understanding of these factors would enable more rational experimentation to proceed and could give rise to new treatment combinations, e.g. of well timed applications of mineral nutrients and growth regulators, which may raise the yield potential of the crop.

It is perhaps worth-while making the observation that in many published accounts of fertilizer experiments, the inclusion of information on environmental factors such as rainfall and potential evaporation and plant factors, such as plant density and the physiological state of the seed tuber, would considerably increase the usefulness of the data.

With regard to the assessment of the nutrient status of the soil, attention has been drawn to the well-known difficulties presented by N; for this nutrient, the use of foliar analysis for diagnosing the status of the crop has attractions. However it is likely that recommendations could vary considerably with region, variety, plant population and other variables, while the quantity of nutrients needed to remedy any deficiencies revealed and the efficacy of post-planting fertilizer dressings needs more attention.

References

Batey, T. and Boyd, D.A. (1967). *NAAS Quart. Rev.* **78**, 47–56.

Birch, J.A., Devine, J.R., Holmes, M.R.J. and Whitear, J.D. (1967). *J. agric. Sci., Camb.* **69**, 13–24.

Bodlaender, K.B.A. and Reestman, A.J. (1968). *Neth. J. agric. Sci.* **16**, 165–76.

Bolton, J. (1971). *Rothamsted Report for 1970*, **2**, 98–110.

Boyd, D.A. (1959). *J. agric. Sci. Camb.* **52**, 384–91.

Boyd, D.A. (1961). *Proc. Fertil. Soc.* **65**, London.

Boyd, D.A. (1970). *Pot. Symp.* **9**, 461–73.

Boyd, D.A. and Dermott, W. (1964). *J. agric. Sci. Camb.* **63**, 249–63.

Boyd, D.A. Hill, J.R. and Batey, T. (1968). *Expl. Husb.* **16**, 13–20.

Colombies, H. (1975). *Pomme de Terre Francaise*, **368**, 5–6. (In *Field Crop Abstracts* (1976) **29**, 3058).

Cooke, G.W. (1967) *The Control of Soil Fertility*. Crosby Lockwood and Son Ltd.

Crowther, E.M. and Yates, F. (1941). *Emp. J. exp. Agric.* **9**, 77–97.

Culpin, S. and Greaves, D.W.B. (1956). *Expl. Husb.* **1**, 44–7.

Doll, E.C., Christenson, D.R. and Wolcott, A.R. (1971). *Am. Potato J.* **48**, 105–12.

Dyson, P.W. and Watson, D.J. (1971). *Ann. appl. Biol.* **69**, 47–63.

Edward, P.D. (1971). *Rothamsted Report for 1970*, **1**, 98–9.

Edwards, H.G.A., Watkin, J.E. and Webber, J. (1956). *Expl. Husb.* **1**, 25–43.

Ekeberg, E. (1972). *Forskning og Forsøk i Landbruket* **23**, 181–201.

Ekeberg, E. and Rønsen, K. (1973). *Forskning og Forsøk i Landbruket* **24**, 209–19.

Evans, C. (1974). *Expl Husb.* **27**, 99–103.

Ezeta, F.N. and McCollum, R.E. (1972). *Am. Potato J.* **49**, 151–63.

Fitts, J.W. (1974). Proper soil fertility evaluation as an important key to increased crop yields. In *Fertilisers, crop quality and economy*, ed. Hernando Fernandez V. Amsterdam: Elsevier.

Gasser, J.K.R. (1970). Some biological effects of ammonia injected into soils. In *Anhydrous Ammonia*, Proc. symposium, National College of Agricultural

Engineering eds. Gasser, J.K.R., Angus, G.F., Kilgour J. and Marriott, T.A. London: IPC Business Press Ltd.

Gately, T.F. (1971). *Potato Res.* **14**, 84–90.

Geraldson, C.M., Klacan, G.R. and Lorenz, O.A. (1973). Plant analysis as an aid in fertilising vegetable crops. In *Soil Testing and Plant Analysis*, eds. Walsh, L.M. and Beaton J.D. Revised edition. Soil Science Society of America.

Gunasena, H.P.M. (1969). Studies on the growth of the potato with particular reference to the efficient use of nitrogen and potassium. Ph.D. Thesis, Reading University.

Gunasena, H.P.M. and Harris, P.M. (1968). *J. agric. Sci. Camb.* **71**, 283–96.

Gunasena, H.P.M. and Harris, P.M. (1969). *J. agric. Sci. Camb.* **73**, 245–59.

Gunasena, H.P.M. and Harris, P.M. (1971). *J. agric. Sci. Camb.* **76**, 33–52.

Hackett, C. (1968). *New Phytol.* **67**, 287–300.

Harris, P.M. (1960). An investigation into the action of farmyard manure on yield with particular reference to the potato crop. M.Sc. Thesis, Leeds University.

Harris, P.M. (1970). Anhydrous ammonia for potatoes. In *Anhydrous Ammonia*, Proc. Symposium, National College of Agricultural Engineering, eds. Gasser, J.K.R., Angus, G.F., Kilgour, J. and Marriott, T.A. IPC Business Press Ltd.

Hawkins, A. (1946). *J. Am. Soc. Agron.* **38**, 667–81.

Holliday, R. (1955). *Agric. Progr.* **30**, 42–51.

Holliday, R. (1963). In *The Growth of the Potato* eds. Ivins, J.D. and Milthorpe, F.L. 248–64. London: Butterworths.

Holliday, R. (1970). Soil profile moisture and nitrogen availability. In *Nitrogen nutrition of the plant*, ed. Kirkby, E.A. Leeds University.

Holliday, R. and Draycott, A.P. (1968). *J. agric. Sci. Camb.* **71**, 413–18.

Holliday, R., Harris, P.M. and Baba, M.R. (1965). *J. agric. Sci. Camb.* **64**, 161–6.

Humphries, E.C. and French, S.A.W. (1963). *Ann. appl. Biol.* **52**, 149–62.

Inkson, R.H.E., and Reith, J.W.S. (1967). *Trans. Meeting of Commissions II and IV, Int. Soc. Soil Sci.* (1966) 377–84.

Ivins, J.D. (1963). Agronomic management of the potato. In *The Growth of the Potato*, eds. Ivins, J.D. and Milthorpe F.L., pp. 303–10.

Ivins, J.D. and Bremner, P.M. (1965). *Outlook on Agric.* **4**, 211–17.

Johansson, O., Jonsson, L. and Svensson, E. (1973). *Lantbr Högsk. Meddelanden* A207.

Kerketta, R. (1976). Growth, nitrogen nutrition and leaf senescence in the potato. Ph.D. Thesis, Reading University.

Krauss, A., and Marschner, H. (1971). *Z. Pflanzenernährung und Bodenkunde*, **128**, 153–68.

Kuipers, S.F. (1962). *Stikstof* **6**, 3–19.

Kunkel, R., Holstad, N. and Russell, T.S. (1973). *Am. Potato J.* **50**, 275–83.

Laughlin, W.M., Martin, P.F. and Smith, G.R. (1974). *Am. Potato J.* **51**, 393–402.

Lorenz, Q.A., Weir, B.L. and Bishop, J.C. (1972). *J. Am. Soc. hort. Sci* **97**, 334–7.

Maclean, A.J., Jasmin, J.J. and Halstead, R.L. (1967). *Can. J. Soil Sci.* **47**, 89–94.

Milthorpe, F.L. (1963). In *The Growth of the Potato*, eds. Ivins, J.D. and Milthorpe, F.L. pp. 3–16. London: Butterworths.

Ministry of Agriculture, Fisheries and Food (1973). Fertilizer recommendations. *Bull.* **209**, London: H.M.S.O.

Mitscherlich, E.A. (1913). Bodenkunde fur Land-und Forstwirte, Berlin 1913. [Quoted by E.W. Russell (1961) *Soil Conditions and Plant Growth.* 9th edn. Longmans.]

Ngugi, D. (1972). A study of the physiological basis of yield in the potato crop, with particular reference to the efficient use of nitrogen. Ph.D. Thesis, Reading University.

Nosberger, J. and Humphries, E.C. (1965). *Ann. Bot. N.S.* **29**, 579–88.
Peaslee, D.E. and Moss, D.N. (1966). *Soil Sci. Soc. Am. Proc.* **30**, 220–3.
Pizer, N.H., Wright, A.H., Caldwell, R.H., Hargrave, J., Burgess, G.R. and Cory, V. (1961). *J. agric. Sci. Camb.* **56**, 197–211.
Preston Jones, J. and Painter, C.G. (1974). *Curr. Information Series* **240**, University of Idaho College of Agriculture.
Radley, R.W. (1963). In *The Growth of the Potato*, eds. Ivins, J.D. and Milthorpe, F.L. pp. 211–20. London: Butterworths.
Reith, J.W.S. and Inkson, R.H.E. (1958). *J. Agric. Sci., Camb.* **51**, 218–24.
Robinson, J.B.D. (1975). *Special Publ.* **1** Commonwealth Agricultural Bureau.
Schippers, P.A. (1968). *Eur. Potato J.* **11**, 23–33.
Schoonneveldt, J.C. van, Ippel, H., Neutel, H. (1968). *Landbouwk. Tijdschr. Wagen.* **80**, 63–9 (quoted in *Field Crop Abstract* S. (1969). 22, 1244).
Sharpe, P.R. (1969). The determination and economic analysis by regression techniques of functional relationships in the potato crop. Ph.D. Thesis, Reading University.
Simpson, K. (1956). *J. Sci. Fd Agric.* **7**, 745–53.
Skene, K.G.M. (1975). Cytokinin production by roots as a factor in the control of plant growth. In *Development and Function of Roots*, eds. Torrey, J.G. and Clarkson, D.T. London: Academic Press.
Smith, O. (1968). In *Potatoes: Production, Storing, Processing*, ed. Smith, O., pp. 183–241. Connecticut: The Avi Publishing Company.
Sokolov, A.V. and Orlovskii, N.V. (1971). *Acad. Sci. USSR*. Translated 1974, Israel Program for Scientific Translations.
Sommerfeldt, T.G. and Knutson, K.W. (1968). *Am. Potato J.* **45**, 231–7.
Thimmegowda, S., Muniyappa, T.V., Prasad, T.W.R. and Gowda, B.N.S. (1975). *Current Res.* **4**, 114–15. Ubiversity of Agricultural Science, Bangalore, Karnatka, India. In *Field Crop Abstract* (1976) **29**, 3063.
Tripathi, B.K. (1973). *C. R. hebd. Séanc. Acad. Sci. D*, **277** (15) 1467–9.
Tyler, K.B., Lorenz, O.A. and Fullmer, F.S. (1961). *California Agric. Exp. Station* Bulletin 781.
Van Burg, R.F.J. (1966). Diagnosing the nitrogen status of seed potatoes. *Proc. Third Triennial Conf. Eur. Ass. Potato Res.* 167.
Van Burg, P.F.J. (1967). Relation of rate of nitrogen fertilisation, seed spacing and seed size to yield of potatoes. Netherlands Nitrogen Technical Bulletin (4) The Hague.
Van Burg, P.F.J., Van Brakel, G.D. and Schepers, J.H. (1967). The agricultural value of anhydrous ammonia on arable land. Experiments 1963–66. Netherlands Nitrogen Technical Bulletin (3) The Hague.
Van der Paauw. F. (1958). *Pl. Soil* **9**, 254–68.
Van der Paauw, F. (1963). *Pl. Soil* **19**, 324–31.
Warren, R.G. and Johnston, A.E. (1961). *Rothamsted Report for 1961*, 45–8.
Watson, D.J. (1958). *Ann. Bot. N.S. London* **22**, 36–54.
Watson, D.J. (1963). In *The Growth of the Potato*, eds. Ivins, J.D. and Milthorpe, F.L., pp. 233–47. London: Butterworths.
Watson, D.J. (1971). In *Potential Crop Production*, eds. Wareing, P.F. and Cooper, J.P. pp. 76–88. London: Heinmann.
Watson, D.J. and Wilson, J.H. (1956). *Ann. appl. Biol.* **44**, 390–409.
Widdowson, F.V., Penny, A. and Williams, R.J.B. (1967). *J. agric. Sci., Camb.* **69**, 247–53.
Wiersum, L.K. (1961). *Pl. Soil* **15**, 189–92.
Williams, R.J.B. (1969). *Chem. Ind.* 1735–6.

Chapter 6
Water

P.M. HARRIS

The dependence of high tuber yields upon an adequate supply of soil water is well known and can be illustrated by the good correlation often shown between rainfall and yield. Yield and rainfall data collected over an eight-year period in central England from healthy stocks of Majestic seed on crops receiving similar cultural treatments each year (McDermott and Ivins, 1955) suggest that there is a linear relationship between total yield and the May–September rainfall (Fig. 6.1). The increase in yield of 1.4 t ha^{-1} for each cm of rainfall is similar to yield responses found in irrigation experiments in Britain. The data indicate that water supply must be a major cause of year-to-year variation in yield.

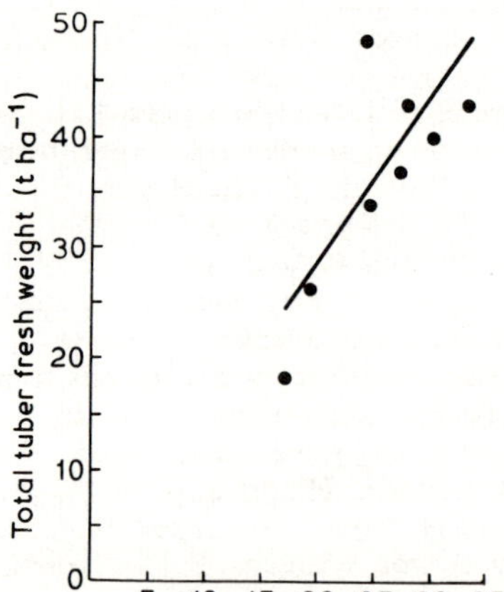

Fig. 6.1 The relationship between yield and rainfall of Majestic potatoes grown in Central England (1947–54). From McDermott and Ivins, 1955

This chapter reviews the relationship between water supply and yield and considers reasons for the sensitivity of the crop to water supply, in the hope that ways might be suggested in which irrigation practice might be improved or the effects of water stress minimized.

6.1 Yield and water supply

In a review of potato irrigation experiments, Thompson (1970) found that the response of total tubers per cm of water applied fell most frequently between 0.5 and 1.0 ha^{-1} cm^{-1}, with individual values ranging from negative to well over 2t ha^{-1}. Consideration of some of the higher responses suggested that the potential response could be in the region of 2–2.5 t ha^{-1} cm^{-1}.

Penman (1971) has estimated in a more formal way the actual and potential response of tuber yield to water supply from experiments on a sandy loam soil in the south of England. Starting with the soil at field capacity and in the absence of rain or irrigation, the soil will be dried by transpiration and in the early stages of this drying process plant growth will not be affected. Penman put forward the hypothesis that this period of unrestricted growth will last up to a threshold of soil dryness and beyond that threshold there will be a complete check to growth. This threshold was called the limiting deficit (D_l), defined as the rainfall equivalent of water that must then be added to restore the soil to field capacity. It represents the amount of water stored in the soil profile that the soil itself can contribute to plant growth. For rain (R), and irrigation (I), the deficit (D) at any time is:

$$D = E_T - (R + I)$$

where E_T = the potential evaporation rate. The hypothesis is that while D is less than D_l growth is unchecked and while D is more than D_l and still increasing then growth is zero. Penman extended the argument to include the concept of active evaporation (E_A) contributing to plant growth, thus $E_A = E_T - D_m + D_l$, where D_m is the maximum deficit reached during the period considered. It is important to note that as the profile is rewetted, whatever the value of D all the evaporation is active.

The further hypothesis was advanced that the growth rate, measured in terms of total dry matter yield is proportional to the potential transpiration rate when water supply is non-limiting, and that total growth (Y) is proportional to the total active evaporation, hence:

$$Y = kE_T \text{ while } D_m \text{ is less than } D_l,$$

otherwise

$$Y = k(E_T - D_m + D_l).$$

Penman maintained that to obtain a measure of the limiting deficit is

Table 6.1 Response of potatoes to irrigation, tubers as harvested, t ha $^{-1}$

Year	*I* (*cm*)	*Yo*		*Y1–Yo*		*(Y1–Yo)/1*		*Approximately 4k*	
		N1	*N2*	*N1*	*N2*	*N1*	*N2*	*N1*	*N2*
		Early potatoes (a) Ulster Chieftain							
1951	5.6	9.3	9.5	9.5	11.4	1.7	2.0		
1952	6.9	14.7	15.2	9.1	11.5	1.3	1.7	1.8	2.0
1953	5.1	24.5	29.5	4.9	6.6	1.0	1.3		
		(b) Arran Pilot							
1960	3.8	22.7	26.0	4.5	6.5	1.2	1.7		
1961	10.2	14.6	17.2	18.7	20.2	1.8	2.0	1.8	2.1
1962	7.6	7.2	7.0	6.0	8.5	0.8	1.1		
		Maincrop potatoes–Majestic							
1954	5.6	34.5	44.0	−2.0	−2.2	−0.4	−0.4		
1955	16.0	26.5	28.0	20.1	15.3	1.3	1.6	1.8	2.0
1956	1.3	34.8	38.3	1.5	1.2	1.2	0.9		

I = irrigation applied in cm, Y*o* = yield in the absence of irrigation, Y*1* = yield in the presence of irrigation. N1 = 62.5, N2 = 125 kg N ha $^{-1}$ 1951–1956, N1 = 75, N2 = 150 kg N ha $^{-1}$ 1960–2
Data from Penman (1971).

the major technical objective of irrigation experiments. It can be estimated in various ways one of which is to adjust values of D_l in the calculation of E_A until a plot of Y against E_A gives a straight line.

Data for experiments carried out over nine years on three varieties of potatoes are given in Table 6.1.

The value of the limiting deficit was estimated to be 2.5 cm while for the lower level of N $k_1 = 0.45$ and for the higher level $k_2 = 0.50$ t dry matter ha^{-1} cm^{-1}. On the assumption that $4k$ represented the value for tuber fresh weight yield, the actual responses to irrigation were often close to this value, suggesting that the maximum theoretical response to irrigation of well fertilized potatoes is in the region of 2 t ha $^{-1}$ cm $^{-1}$.

Arkley (1963) showed that there was a linear relationship between the amount of dry matter produced and the amount of water transpired (Tr) by a particular species of plant under given conditions of climate and soil fertility: i.e. $Y = kTr$. The regression constant k was dependent on the plant species and is the amount of dry matter produced per unit of water transpired, the reciprocal of the transpiration ratio. Arkley pointed out that the growth rate of a plant, given an adequate supply of nutrients and water, increases within limits with increasing light and heat; these are closely correlated and as they increase so does evaporation, and thus the rate of transpiration with a coincidental increase in growth. However, radiation which exceeds the plant's ability to utilize, or the advective energy of dry air, or both, may cause evaporative stresses that are entirely independent of the stress caused by the growth promoting light and heat. Thus if the ratios of dry matter production to transpiration are to be compared for different climatic conditions, then the equation for each experiment must be corrected to account for advective energy and excess radiation.

A correction for climate based on mean monthly relative atmospheric humidity made it possible for Arkley to combine the data on yield and transpiration from different years and different localities, the relationship between yield and transpiration being as follows:

$$Y = k_f\, Tr/100 - H$$

where k_f is a constant for a given kind of plant when the soil is kept at a uniform high level of fertility during the entire growth period and H is the mean daily relative atmospheric humidity in %.

While Arkley did not present any data for potatoes, Rijtema and Endrödi (1970) found that in the Netherlands there was no close relationship between transpiration and yield in comparisons involving different years and different varieties; but that there was a very close relationship between total dry matter yield and the ratio transpiration over vapour pressure deficit (Fig. 6.2). However, Arkley showed that this relationship was dependent on temperature and that the use of relative vapour pressure deficit took this into account and was fundamentally more sound. It is possible that variation in temperature between seasons was not a significant factor in the data from the Netherlands.

Arkley was also able to show that the transpiration ratio (transpiration/yield) was affected by soil fertility and approached a minimum as the level of fertility increased. He was able to incorporate a correction factor for fertility into the equation relating yield to transpiration. This agrees with Penman's observation of the effect of N on the value of k and with

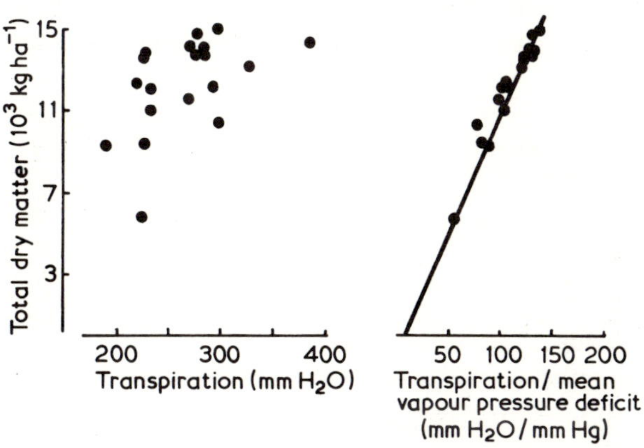

Fig. 6.2 The relationship between the total dry matter yield ha^{-1} of potatoes and (a) transpiration (b) transpiration/vapour pressure deficit. From Rijtema and Endrödi, 1970.

his comment that the better the standard of farming the greater is the return from added water.

Penman's approach to the relationship between transpiration and yield presents some difficulties, for it deals with potential rates of transpiration rather than with actual rates. It is known that growth rate declines as soil moisture is depleted (Hagan *et al.*, 1959), but the constancy of the ratio between yield and actual transpiration and therefore its independence of the availability of soil water is probably due to the fact that transpiration and growth decline simultaneously. Penman proposes that there is a sudden cessation of growth when a limiting deficit of 2.5 cm has been reached and that any transpiration that may occur at higher soil water deficits is not related to growth. It is difficult to reconcile this with the fact that many soils are able to supply such a quantity of available water from the top 15 cm of soil and, while the crop is believed to be shallow rooted (Section 6.2.1) no data have been found in the literature suggesting that roots are confined to such shallow depths. If transpiration does not cease at this depth then it becomes difficult to reconcile Rijtema and Endrödis' (1970) data (Fig. 6.2) with the concept of a limiting deficit. However, it is possible to offer an explanation for a relatively rapid cessation of growth at a fairly small moisture deficit if it is accepted that growth is very dependent on a current supply of N and that this supply is greatly reduced when the relatively shallow ploughed layer, where most of the organic matter and fertilizer are to be found, is dried out (Section 6.2.3).

A further estimate of the response of tuber yield to soil moisture deficit is given by Gregerson and Jørgensen (1973), their data relating tuber fresh weight yield and tuber dry matter yield to soil moisture deficit are given in Fig. 6.3. Yield was linearly related to soil moisture deficit for every cm increase in soil moisture deficit 3.3 t ha^{-1} fresh tubers and 0.8 t ha^{-1} of tuber dry matter were lost. Fig. 6.4 also illustrates a linear relationship between mean soil water potential and the yield of tubers.

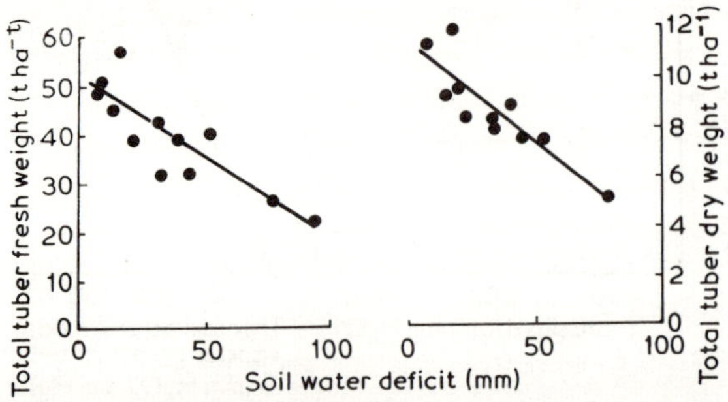

Fig. 6.3 The relationship between soil moisture deficit and (a) total tuber fresh weight yield (b) total tuber dry matter yield. From Gregerson and Jørgensen, 1973

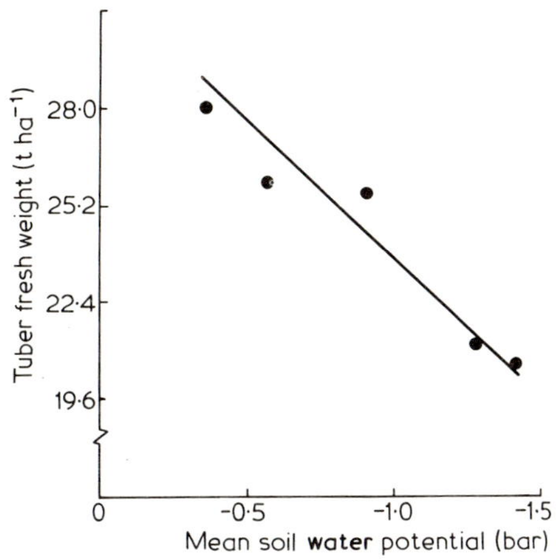

Fig. 6.4 The relationship between soil water potential and tuber fresh weight yield. From Hagan *et al.*, 1959

6.2 Factors contributing to the potato crop's sensitivity to water stress

The internal water balance of the crop is a function of the rate of uptake and loss of water, while the sensitivity of the crop to internal water stress might differ from that of other crop species. This section deals with some of the more obvious factors involved.

6.2.1 Root factors

Data on root characteristics of crops are difficult to obtain: there is consequently a lack of information concerning the distribution of roots in space and time in the field and how such distribution patterns may be affected by genotype and the environment.

Comparative data are relatively rare, but Corey and Blake (1953), showed that on Sassafras loam the effective rooting depth of potatoes was 56 cm compared with about 77 mm for maize and more than 90 cm for tomatoes. On this soil the amount of water available to the potato crop was estimated to be about 10 cm. On the same soil type but at a different location it was rather surprisingly found that the effective rooting depth of potatoes and wheat were approximately the same (53 and 51 cm, respectively). While there was no evidence for potatoes, it was shown with the other crop species that soil type had a large effect on available water, partly due to effects on rooting depth and partly to differences in soil water characteristics.

Durrant *et al.* (1973) compared the growth of roots of sugar beet, potato and barley by observing growth in the field through glass panels and related root growth to changes in soil moisture. Within 10 days of planting, potato tubers developed 12–18 similar sized roots that soon fanned out into the soil beneath neighbouring plants; short laterals were developed, first near the tuber and then further away. Most of the roots were in the top metre of soil with only one or two roots penetrating deeper. The depth of observed root growth was generally related to, but 10–15 cm deeper than, the maximum depth of soil moisture extraction. The mean (1969–71) maximum depth of soil moisture extraction (cm) by roots was: potatoes 92, sugar beet 78 and barley 93. Thus water was extracted from greater depths than the more drought tolerant sugar beet crop and was on a par with barley, again not particularly drought sensitive. However, while sugar beet and barley eventually extracted similar amounts of water from each horizon, potatoes extracted less, especially from below 60 cm (Table 6.2).

Fulton (1970) found that while the yield of potatoes was restricted by soil moisture stress in excess of 0.5 bar 15 cm below the soil surface during late May and June, watering for maize could be delayed until the stress was 5 bars at 40 cm, while for tomatoes irrigation could be timed when the soil moisture tension was 1 bar at 25 cm. In spite of the greater sensitivity of the potato crop to soil moisture stress, the maximum root extension was almost identical for all three crops, and reached depths of over 120 cm. Late in the season 60% of the potato and tomato roots were located in the top 68 cm of soil, whereas 60% of the maize roots were located in the top 56 cm of soil. Thus, differences in the soil moisture regime necessary for maximum yield of these crops could not be explained by the position of the major part of the root systems. Fulton speculated that the potato roots might have a relatively low capacity for water absorption so that almost all the root system must have access to water at low tension in order to produce maximum yield.

Under semi-arid conditions in Lebanon, Fuehring *et al.* (1966) reported

Table 6.2 Maximum amount of water removed from each soil horizon mean values for 3 years, 1969–71

Depth	*Crop*		
	Potatoes	*Sugar beet*	*Barley*
	mm/10 cm horizon		
30–40	15	16	16
40–50	13	14	16
50–60	11	13	13
60–70	9	13	13
70–80	7	12	12
80–90	5	12	11
90–100	5	10	8

Data from Durrant *et al.*, (1973).

that 7 days after irrigating potatoes, and with some removal of water taking place at a depth of 100 cm, the crop was on the point of yield reduction due to water stress (yields were reduced if the irrigation interval exceeded 1 week). At this point 75% of the 'available' water still remained in the root zone and if irrigation was not applied the amount of water transpired by the crop was considerably reduced; there was a linear reduction with time in stomatal opening in each irrigation cycle.

On a sandy loam soil in southern England, Mohindra (1975) found that at normal plant densities, the maximum rooting depth in July of sugar beet, spinach beet, kale and potatoes were 91, 91, 96 and 76 cm respectively.

Lesczynski and Tanner (1976) found that the roots of Russet Burbank potatoes grown on Plainfield loamy sand were essentially restricted to the plough layer (0–30 cm) and decreased rapidly below this depth. The roots below the plough layer contributed a maximum of 15% of the total root length per plant for all sampling dates. The root density in the plough layer generally ranged from 2–6 cm of roots cm^{-3}, roots less than 0.2 mm diameter accounted for about two-thirds of the total root length. Field observations by Mohindra (loc. cit.) also suggested that compared with spinach beet and kale, potatoes had much finer roots. This may be of some significance, for Gardener (1965) indicated that finer roots may offer greater resistance to water uptake.

Lesczynski and Tanner (loc. cit.) suggested that the lack of any significant rooting below 30 cm could be a result of frequent irrigation in their experiment. Jean and Weaver (1924) showed that different irrigation treatments could alter the rooting habit of the potato crop; the maximum penetration being some 18 cm deeper on 'lightly' watered compared with a 'fully' watered crop. However, the irrigation need was determined empirically and as the fully watered crop yielded only 53% of the lightly watered crop, it would seem that the former was overwatered.

Danielson (1967) has discussed the general relationships between irrigation practice and root development. He concluded that since root growth is related to the general vigour of the plant, maximum development in any region of the soil should take place when optimum conditions are maintained in all regions. Under certain circumstances restriction of surface soil water may allow a crop to survive later drought periods because of decreased evapotranspiration resulting from restricted foliage development (see Section 6.3.4).

Unpublished data (Steckel, 1975) suggest that water may have some influence on the rate of root penetration down the soil profile; in root observation chambers the mean time taken for the roots of four varieties of potatoes to reach a depth of 60 cm was 51 and 54 days for the dry and wet soil water regimes, respectively.

Undoubtedly many other factors may influence rooting behaviour,

De Roo and Waggoner (1961), for example, showed that the pans developed by normal tillage associated with potato production could restrict root development, although on a coarser textured soil than a Merrimac sandy loam, some compaction seemed to benefit root growth.

6.2.2 Some physiological responses to water stress

These factors have been dealt with in detail in Chapter 3 and are only briefly and partially discussed here. While it was shown in lysimeter experiments in Australia (Shepherd, 1972) that leaf water potential fell less rapidly in potatoes than in clover and ryegrass on days of high potential evaporation (*Ep*), the rate of evaporation was reduced to a much larger extent for the potato crop than for a mixed crop of grass and clover. The reduced evaporation from the potato crop was attributed to a much greater sensitivity of potato leaf diffusion resistance to a decrease in leaf water potential (Table 6.3). Endrödi and Rijtema (1969) also drew attention to the high value of leaf water potential (about 1 bar) at which actual transpiration began to fall below the potential rate. This would suggest that stomata begin to close at low values of soil moisture potential, consquently frequent irrigation to maintain the soil close to field capacity is needed. Indeed Shepherd (loc. cit.) showed that wilting often occurred on days of persistently high *Ep* in potatoes, clover and ryegrass; this could be prevented by continuous trickle irrigation to ensure that the soil was always at or above field capacity, even though when wilting was observed, the areas had been heavily watered in the previous two days.

Burrows (1969) showed that in comparison with sugar beet the ratio of actual transpiration to potential transpiration fell more rapidly with increasing soil water deficit (Fig. 6.5). It was suggested that the ability of potatoes to reduce transpiration rates by lowering the conductivity of leaves through stomatal closure might prove advantageous to other growth processes; however this would seem unlikely for Moorby *et al.* (1975) found that the rate of photosynthesis appeared to be very sensitive

Table 6.3 Average leaf diffusion resistance during 5 days of moderate and 5 days of high potential evaporation (E_p)

Species	*Moderate* E_p		*High* E_p	
	0900–1000	*1200–1400* (*h*)	*0900–1000*	*1200–1400* (*h*)
	(*s cm* $^{-1}$)			
Potato	2.9	4.9	3.3	6.8
Clover	1.6	2.4	1.2	3.7
Ryegrass	2.0	2.8	1.9	2.6

Data from Shepherd (1972).

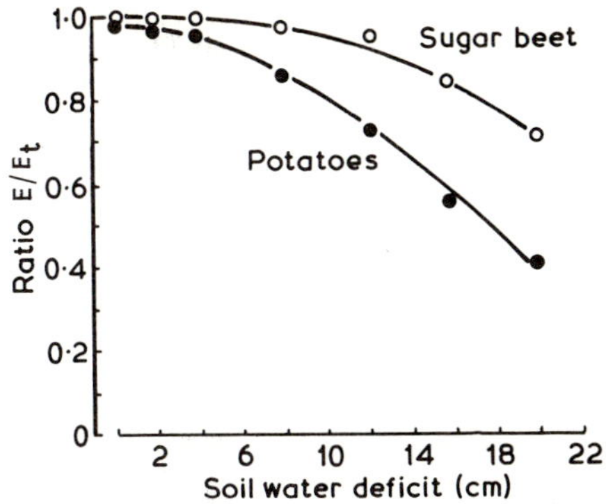

Fig. 6.5 The relationship between soil water deficit and relative transpiration rate in sugar beet and potatoes. From Burrows, 1969 . (E = actual transpiration rate, Et = Potential transpiration rate.)

to small increases in leaf resistance. Epstein and Grant (1973) however, failed to find any clear relationship between leaf diffusive resistance and leaf relative water content; they suggested that under the stress conditions of their study, the stomata were probably open and the response to soil moisture deficits was not reflected by the porometer measurements. Under irrigated but not unirrigated conditions they found differences in the leaf relative water content (RWC) between varieties which they ascribed to differences in growth habit; on the other hand they quoted work by Werner (1954) which showed little variation in RWC between six varieties of potatoes when soil water was readily available and large differences when soil water became deficient; a common finding appeared to be that stomatal recovery was slow after a drought, and as stress became more pronounced, the plants ceased to respond rapidly to changes in evaporative demand.

6.2.3 Nutrition

The importance of a steady supply of nutrients, particularly N, has been dealt with at some length in Chapter 5, where it was shown that the uptake of nutrients appeared to be particularly sensitive to soil moisture conditions. It is perhaps instructive here to draw attention to some work on the grass crop, which is equally sensitive to soil moisture stress and responds to irrigation at low soil moisture deficits (Tayler, 1965). This crop, traditionally believed to be shallow rooted, was shown by Tayler to be able to extract water from depths of at least 90 cm; while Penman (1956) has shown that while water may be non-limiting for transpiration at soil moisture deficits up to 10 cm, growth could be markedly reduced when soil moisture deficits were only half this value. A similar result was reported by Garwood and Williams (1967a) who also tested the hypothesis (Garwood and Williams, 1967b) that when the water available to plants from the surface soil has been exhausted, the depression in growth can be attributed more to a deficiency of plant nutrients than

Table 6.4 The effect of surface and deep application of fertilizer on the yield of ryegrass (kg ha^{-1})

Soil water deficit	*cm*	*July 30* *3.3*	*September 8* *9.1*	*Percentage of N applied recovered on 8 September*
Surface application	0	1540	480	0
(30 July + 13 August)	NPK	1620	580	4.2
Injected at 46 cm	0	1550	470	0
(30 July + 13 August)	NPK	1440	1320	30.7

Data from Garwood and Williams (1967b).

to a lack of water. They were able to substantiate the hypothesis by showing that when the surface soil was dry and a soil water deficit of 5 cm existed, there was no response to surface applied N, but that injection of N into moist soil at a depth of 46 cm produced a marked increase in growth (Table 6.4). The major nutrient found to be deficient in the subsoil was N.

The review by Holliday (1970) referred to in Chapter 5 (Section 5.4.4) has shown similarly that there is good reason to believe that at least part of the sensitivity of the potato crop to small reductions in soil water status could be due to the indirect effect of soil water stress on nutrient availability. It is possible, although there is little evidence, that this effect could be exaggerated by the practice of growing the crop in ridges where much of the fertilizer may be concentrated (particularly if broadcast over the ridges or on the flat before ridging up) and which because of their open texture and the exposure of a greater surface area to the atmosphere, are prone to dry out more rapidly than a similar volume of soil in an unridged state. Penman (1971) has shown the limiting deficit to be 2.5 cm for the potato crop, grown on sandy loam soil at Woburn. According to Salter (1969) the available water capacity per 10 cm on a soil of this textural class is 1.75 cm. A soil mositure deficit of 2.5 cm would suggest that the top 14 cm of the profile has been depleted of available water. It is not difficult to envisage that most of the organic matter and fertilizer will be concentrated in this zone and that once depleted of water, growth could cease from a shortage of nutrients. The close relationship found between yield and active evaporation by Penman might at least partly be explained in nutritional terms. To what extent this might contribute to the sensitivity of the crop to soil moisture conditions under more extreme climates with higher rates of evaporation is uncertain.

6.3 The effective use of water

6.3.1 Frequency of irrigation

In the review by Singh (1969) it was concluded that large soil water deficits should not be allowed to build up at any stage of growth if maximum yields are to be obtained.

Table 6.5 Yield and efficiency of water use as affected by different irrigation treatments, mean 1955–6

*Soil water tension at which water applied** *(atm)*	*Water applied*† *(cm)*	*Number of irrigations*	*Yield*‡ *Total* *(t ha⁻¹)*	*Yield*‡ *Marketable*	*Ware* *(%)*	*Yield response per cm water applied*§ *Total* *(t ha⁻¹ cm⁻¹)*	*Yield response per cm water applied*§ *Marketable*
(1) 0.3	15.24	14.6	21.29	14.64	69	1.09	0.93
(2) 0.6	10.16	7.0	16.09	9.64	60	1.14	0.90
(3) 1.2	10.16	4.0	10.89	4.88	45	0.51	0.33
(4) 2.4	8.69	2.5	8.75	3.97	45	0.30	0.27
(5) 4.8	1.90	0.5	6.68	2.16	32	–	–

*Measured by tensiometers or gypsum blocks placed 15 and 30 cm deep.
†Enough water applied at each irrigation to bring the top 50 cm of soil to field capacity.
‡Converted from bushels per acre on the assumption that 1 bushel = 27.2 kg.
§Treatment 5 taken as control, responses and water applied calculated by subtracting the yield and water applied to this treatment from treatments 1–4.
Data from Jones and Johnson, 1958.

Jones and Johnson (1958) carried out irrigation experiments in which the crop was protected from rainfall and in which treatments were included where irrigation began at different soil water tensions; Table 6.5 gives data for five of the treatments which illustrate a number of important points connected with irrigation frequency. It is clear that total yield declined as the irrigation interval increased and that one consequence of decreasing the frequency of irrigation is usually a decrease in the quantity of water applied.

Marketable yield is of paramount importance and the data emphasize the importance of the ample supply of water needed to maintain the quality of the crop. The severity and timing of soil moisture deficits can affect the number and size of tubers, hence the proportion of the crop reaching a marketable size. The shape of tubers (Robins and Domingo, 1956) and the appearance of tubers affected by superficial diseases such as potato scab (*Streptomyces scabies*) (Wellings and Rosser, 1968) may also be affected. Dry matter content or the specific gravity of tubers are important criteria of quality and here the effects of irrigation (and rainfall) are complex. Irrigation will affect tuber size and maturity and both will affect tuber dry matter content. Where high soil temperatures have been reduced by irrigation then positive effects of irrigation on the specific gravity of tubers have been reported (Motes and Greig, 1969). Increasing the uptake of nutrients such as N and K by irrigation may be expected on the other hand to depress dry matter content. Irrigation or rainfall late in the growing season would appear to be particularly significant in decreasing dry matter content (Murphy and Goven, 1959; Sanders *et al.*, 1972).

However, the yield of marketable produce per unit area, while being

Table 6.6 The effect of frequency of Irrigation on yield and Irrigation efficiency

		Assumed effective rooting depth (cm)		*Irrigation regime (% depletion of available water). Water applied (in brackets) cm. Most frequently watered → least*	*Yield (t ha⁻¹) corresponding to treatment in previous column. Most frequently watered → least*				*Response to irrigation (t ha⁻¹) cm⁻¹). Most frequently watered → least*			*Grade*	*Source*
New Jersey U.S.A.	1953 Katahdin	23		33%(15.2)50%(15.2)67%(20.3)nil	27.13	*29.10*	27.61	19.53	0.50	0.63	0.40	US	Blake *et al.*, 1955
				50%(12.7)67%(10.2)		*27.24*	26.87	19.53		0.60	0.72	No. 1	
	1954 Katahdin	23		33%(20.32)50%(19.0)67%(20.3)nil	*34.58*	32.52	33.53	19.63	*0.73*	0.68	0.68	US	
				50%(17.8)67%(15.2)nil		*34.45*	32.32	19.63		*0.83*	*0.83*	No. 1	
Canada sandy loam	1953 Irish Cobbler	15		25%(10.1)50%(8.4)75%(7.5)nil	*17.44*	16.82	14.43	8.78	0.86	*0.96*	0.75	Total	Fulton and Murwin 1955
	1954 Irish Cobbler	15		25%(11.7)50%(6.7)75%(5.0)nil	*15.06*	13.65	10.16	5.90	0.78	*1.16*	0.85	Total	
U.K. sandy loam over Bunter pebble beds	1960 Arran Pilot	38		25%(11.4)50%(8.4)75%(10.7)nil	*26.61*	25.35	23.09	17.32	0.81	*0.95*	0.54	Total	Peeler *et al.*, 1966
	1961 Arran Pilot			25%(9.6)50%(8.9)75%(10.2)nil	*20.58*	17.57	19.08	13.55	*0.73*	0.45	0.54	Total	
	1962 Arran Pilot			25%(10.4)50%(7.6)75%(6.6)nil	23.84	*26.61*	22.09	16.06	0.75	*1.39*	0.91	Total	
	1963 Arran Pilot			25%(6.3)50%(6.3)75%(3.0)nil	25.10	*26.61*	22.34	21.84	0.52	*0.76*	0.17	Total	
	1960 Ulster Prince	38		25%(11.4)50%(8.4)75%(10.7)nil	22.09	23.34	*24.35*	17.82	0.37	*0.66*	0.61	Total	
	1961 Ulster Prince			25%(9.6)50%(8.9)75%(10.2)nil	16.06	16.57	*17.32*	10.54	0.57	*0.68*	0.66	Total	
	1962 Ulster Prince			25%(10.4)50%(7.6)75%(6.6)nil	16.82	*18.57*	18.32	12.80	0.39	0.76	*0.84*	Total	
	1963 Ulster Prince			25%(6.3)50%(6.3)75%(3.0)nil	*26.10*	23.34	24.60	18.57	1.19	0.75	*2.01*	Total	
U.K. sandy loam over Bunter pebble beds	1960 King Edward	38	P1	25%(14.6)50%(15.7)75%(13.5)nil	45.18	*46.43*	43.17	37.15	0.55	*0.59*	0.44	Total	Peeler *et al.*,
			P2		*52.46*	49.4	50.20	39.15	*0.91*	0.65	0.81		
			P1		35.14	*39.16*	35.14	27.36	0.53	*0.75*	0.58	Ware	1966
			P2		37.15	38.90	*39.66*	28.11	0.62	0.69	0.85		
	1961 King Edward	38	P1	25%(17.5)50%(17.5)75%(15.0)nil	47.94	49.45	*51.20*	48.69	−0.04	0.04	*0.17*	Total	
			P2		*50.95*	50.20	49.70	47.19	*0.21*	0.17	0.17		
			P1		30.27	33.13	*35.39*	29.87	0.03	0.19	*0.37*	Ware	
			P2		32.63	*32.88*	30.62	25.60	0.40	*0.42*	0.33		
	1962 King Edward	38	P1	25%(14.0)50%(14.0)75%(14.0)nil	*46.69*	44.68	43.42	40.41	*0.45*	0.30	0.21	Total	
			P2		49.95	*51.45*	47.69	39.91	0.72	*0.82*	0.55		
			P1		*35.14*	34.89	33.88	29.62	*0.39*	0.38	0.30	Ware	
			P2		35.64	*37.65*	35.14	28.36	0.52	*0.66*	0.48		

	1960 Majestic	38	P1	25%(14.6)50%(15.7)75%(13.5)nil		*41.41*	38.65	40.91	34.14	*0.50*	0.29	*0.50*	Total	
			P2			46.69	*48.44*	39.16	38.40	0.56	*0.64*	0.20		
			P1			36.90	33.38	*37.15*	29.62	0.50	0.24	*0.56*	Ware	
			P2			36.65	*42.17*	34.64	32.38	0.29	*0.62*	0.17		
	1961 Majestic	38	P1	25%(17.5)50%(17.5)75%(15.0)nil		39.15	39.91	*41.16*	40.66	−0.09	−0.04	*0.03*	Total	
			P2			*48.19*	42.42	41.67	38.65	*0.54*	0.21	0.20		
			P1			28.61	30.37	*31.37*	29.87	−0.07	*0.03*	0.10	Ware	
			P2			*37.15*	29.37	30.62	27.11	*0.57*	0.13	0.23		
	1962 Majestic	38	P1	25%(14.0)50%(14.0)75%(14.0)nil		*36.14*	33.63	33.88	33.13	*0.21*	0.03	0.05	Total	
			P2			38.90	*40.91*	36.14	33.38	0.39	*0.54*	0.20		
			P1			28.11	25.35	26.86	25.85	*0.20*	−0.03	0.07	Ware	
			P2			30.12	30.12	26.86	26.10	*0.29*	*0.29*	0.05		
			P1 = 27 679 P2 = 36 905 plants ha^{-1} The quantities of water applied for the 50% and 75% treatments are estimates based on information given in the paper											
U.K. sandy loam	1969 Maris Page		Daily (22.0) Weekly (20.7)		Nil	72.20	*80.60*	46.4		1.17	*1.65*		Total	Harris (unpublished)
Holland			pF 2.5 (33.5) pF 3.4 (13.0)		Nil	*43.65*	27.15	10.19		1.00	*1.30*		Total	Rijtema
						43.65	23.95	6.80		1.10	*1.32*		Primary	1966

a most important criterion on which to judge the success of a particular agronomic practice, is not the sole criterion of efficiency, for there may under certain circumstances be factors of production which are as important as area in limiting output. Water is often available only in limiting quantities, and as the demands for water escalate this is increasingly likely to be the case. Efficiency expressed as yield per unit of water applied must therefore be taken into account. In Table 6.5 it can be seen that while the two wettest regimes gave similar returns of both total and marketable yields per unit of water applied, the returns then fell dramatically as the irrigation interval was increased.

Where labour is expensive the frequency with which a crop is irrigated will be an important criterion to be considered as a factor determining irrigation frequency, particularly if irrigation involves manually moving portable spray lines. Such operations are not infrequently the cause of the under utilization of irrigation equipment observed in environments where irrigation is not an absolute essential and is not, therefore, carefully integrated into the farming system. Table 6.5 shows how the number of irrigation operations declined as the irrigation interval was stretched.

In Table 6.6 data are presented which enlarge the basis on which the above deductions were made. The experiments selected were those where a non-irrigated control was included which made it possible to calculate the response per unit of water applied. The data are taken from North America and Europe, and are mainly confined to light soils. The frequency with which yield or the yield response per unit of water applied is maximized by a particular irrigation treatment is shown in Table 6.7 for all experiments included in Table 6.6 which have three irrigation treatments in addition to a control. Some of the differences between treatments are quite small. It would seem that irrigating when 50% of the water estimated to be available to the crop in the soil has

Table 6.7 The frequency with which yield or water use efficiency maximized by irrigation regime

	Number of comparisons	*Frequency of irrigation*			
		Most *a*	*Intermediate* *b*	*Least* *c*	*Shared*
		Number of occasions parameter maximized			
Yield	22	11	7	4	0
Total	22	11	7	4	0
Ware	16	4	7	4	1 (*a/b*)
Water use efficiency					
Total yield	22	6	11	4	1 (*a/c*)
Ware yield	16	3	7	5	1 (*a/b*)

Data derived from Table 6.6.

been depleted would offer the best chance of maximizing the return per unit of water applied.

Some experiments included data on the number of irrigations carried out which showed that when irrigated at approximately 25%, 50% and 75% of the estimated depletion of available soil water, the average number of irrigation applications needed were 6.7, 4.4 and 3.5 respectively. Compared with more frequent irrigation, irrigating when the available soil water content has been depleted by 50% would seem to offer considerable advantages in reducing substantially the number of irrigations required and of maximizing the return per unit of water applied without giving any substantial reduction in yield per unit area.

Most of the experiments presented in Table 6.6 were not concerned with interactions between irrigation and other agronomic factors; it is, therefore, interesting to note the important effect of plant density on the response to irrigation in the experiments of Peeler *et al.* (1966). Based on the data derived from this experiment (Table 6.6), the effect of variety and plant density on the yield response per unit of water applied is shown in Table 6.8; the values given are maximum values irrespective of irrigation regime. The mean effect of increasing plant density by 33% was to increase the total yield response from 0.32 to 0.61 t ha $^{-1}$ cm $^{-1}$, and that of ware from 0.38 to 0.54 t ha $^{-1}$ cm $^{-1}$. When averaged over three years, response to irrigation in the variety Majestic would seem to be particularly dependent upon plant population.

The data illustrate the danger of drawing firm conclusions concerning the benefits to be derived from introducing a new technique, such as irrigation, and evaluating it against agronomic factors fixed by traditional practice. Similarly, one of the dangers inherent in frequent irrigation, where the soil is maintained close to field capacity, is the risk of drainage, particularly on sandy soils where N may be leached from the root zone, yet few experiments testing the frequency of irrigation have also tested the effects of supplying additional fertilizer N (see Fig. 6.7).

Table 6.8 The effect of variety and plant density on the yield response per unit of irrigation water applied (t ha $^{-1}$ cm $^{-1}$).

	Total yield of tubers				*Marketable yield*			
Variety	*King Edward*		*Majestic*		*King Edward*		*Majestic*	
'000 Plants ha $^{-1}$	*28*	*37*	*28*	*37*	*28*	*37*	*28*	*37*
1960	0.59	0.91	0.50	0.64	0.75	0.85	0.56	0.62
1961	0.17	0.21	0.03	0.54	0.37	0.42	0.03	0.57
1962	0.45	0.82	0.21	0.54	0.39	0.66	0.20	0.29
Mean	0.40	0.65	0.25	0.57	0.50	0.60	0.26	0.49

Data derived from Peeler *et al.*, 1966.
Note, the values are maximum values irrespective of irrigation treatment–see Table 6.6.

Daily replacement of water loss

Experimental evidence discussed in Section 6.2.2 would suggest that some benefit might be obtained through reducing irrigation frequency to the ultimate by replacing the amount of water lost by transpiration on a daily basis. Data are given in Table 6.9 from an unpublished experiment (Harris, 1969) on a sandy loam soil at Reading, in the south of England, in which daily irrigation was compared with irrigation at weekly intervals. The crop was protected from rainfall by the use of mobile transparent covers. The irrigation treatments were applied around midday when leaf water stress should be approaching maximum values. In this experiment daily irrigation proved to be less effective in terms of response per cm water applied than weekly irrigation. More work however is needed to interpret this result. Much of the water applied daily seemed to be intercepted by the foliage and this might affect the efficiency of the treatment. Work on sugar beet at Reading (Sparrow, 1972; Harris, unpublished) has shown that the application of very small quantities of water to the leaves of sugar beet at midday can give rise to improved efficiency of water use, but that this did not occur with potatoes.

It is possible that daily irrigation might be more effective under environmental conditions giving rise to higher rates of potential transpiration. Middleton *et al.* (1975) working in central Washington in a climate with an annual rainfall of 18 cm. in hot, windy and sunny weather in the cropping season, found that reducing the frequency of furrow irrigation from 4 days to alternate furrows every other day (i.e. half of each row was wetted every day) decreased the yield of U.S. no. 1 grade tubers; however this was attributed to excessive leaching of nutrients. While there were no direct comparisons of irrigation frequency with sprinklers, the authors suggested that comparison between experiments in 1969, when water applications were made every 2–3 days, and 1971 with daily irrigation to replace the evapotranspiration from the previous day, suggested that the latter improved water distribution and gave higher and more uniform production of ware tubers.

Table 6.9 The effect of frequency and quantity of water applied on the total tuber yield of the variety Maris Page, on a sandy loam soil (1969)

Treatment	*Total tuber yield ($t\ ha^{-1}$)*	*Water applied (cm)*	*Response to irrigation ($t\ ha^{-1}$)*	*Response per cm water ($t\ ha^{-1}\ cm^{-1}$)*
No irrigation	46.4	—	—	—
Half requirement daily	59.8	11.0	13.4	1.22
Full requirement daily	72.2	22.0	25.8	1.17
Half requirement weekly	62.8	10.4	16.4	1.58
Full requirement weekly	80.6	20.7	34.2	1.65

'Requirement' estimated from a nearby evaporation tank.
Harris, unpublished data.

It is interesting to note that at St. Paul in Minnesota (Sanders and Nylund, 1972b) in 1967, a year classified as one of 'high' (water) stress, daily mist irrigation increased yield over a non-irrigated control treatment by 35%, while weekly applications of water by furrow irrigation increased yield by 27%. Mist irrigation was shown to improve the turgor of leaves and to reduce stomatal resistance. However, the rates of water application were not comparable and it was suggested that the soil moisture levels were increased by misting to a point where aeration may have limited growth (Sanders and Nylund, 1972a). Estimates of the water lost by evaporation and transpiration over the whole of the growing season were made for the four irrigation treatments tested (nil, furrow, mist and furrow + mist) for two seasons (1968 and 1969); in Fig. 6.6 yield per plot is plotted against these estimates. In spite of the small number of comparisons it seems clear that per unit of water evaporated, mist irrigation is a very inefficient means of using water to increase yield. This is confirmed by experiments in Iowa (Peterson and Weigle, 1970) where conventional irrigation (applying 2.5 cm of water at a rate of 0.43 cm h^{-1} whenever the water in the soil was reduced to 50% of field capacity) was compared with 'air conditioning irrigation' (0.13 cm water applied between 1100 and 1600 h central daylight time whenever the temperature rose above 26.7 °C; this was supplemented when necessary by normal sprinkler irrigation to maintain soil moisture content as with conventional irrigation). Tuber yields and water applied are presented in Table 6.10 for these treatments over a three year period (meaned over four varieties). While a non-irrigated control treatment was not included which would have made it possible to calculate the yield response per unit of water applied, the tuber fresh weight yield per unit of water

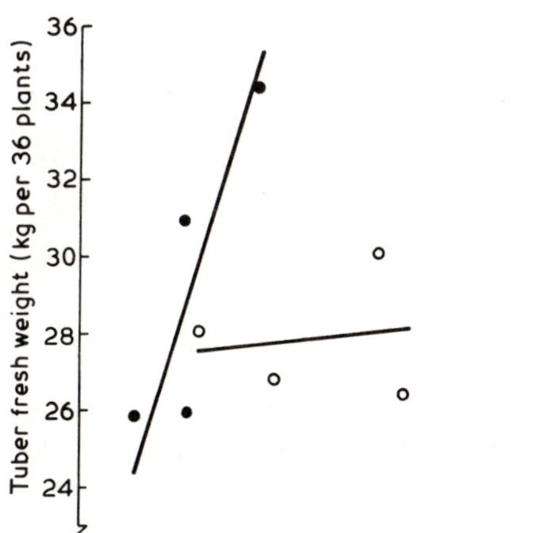

Fig. 6.6 Relationships between estimated evapotranspiration and yield of tubers. Solid circles without mist irrigation, open circles with mist irrigation. Drawn from data of Sanders and Nylund, 1972a, b

Table 6.10 The effect of conventional and 'air conditioning' irrigation on the mean yield of four varieties of potatoes in Iowa.

Irrigation treatment	*Year*			*Mean*
	1966	*1967*	*1968*	*1966–8*
		Tuber fresh weight (t ha⁻¹)		
Conventional	26.6	38.6	34.2	33.1
'air conditioning'	40.0	38.1	43.1	40.4
		Water applied (cm)		
Conventional	27.9	25.4	22.9	25.4
'air conditioning'	154.9	93.9	127.0	125.2
		Tuber yield (ha⁻¹ cm⁻¹)		
Conventional	0.95	1.52	1.49	1.30
'air conditioning'	0.26	0.40	0.34	0.32

Data calculated from Peterson and Weigle (1970).

applied (meaned over three years) fell from 1.3 t ha^{-1} cm^{-1} with conventional irrigation to 0.3 t ha^{-1} cm^{-1} with the air conditioning treatment. The additional yield response for the extra water applied in the latter treatment was a minute 0.073 t ha^{-1} cm^{-1}.

If an important function of irrigation is to maintain a supply of nutrients and particularly N to the crop, then complete wetting of the soil profile might not be essential, on the assumption that the nutrients are concentrated in the superficial layers of the profile. A reduction in the quantity of water applied on each occasion might therefore be expected to increase the yield response per unit of water applied.

Data presented in Table 6.9 show that halving the estimated water requirement for daily or weekly irrigation treatments halved the response (total tubers) and gave similar responses per unit of water applied. However, Carter (1962) in experiments in the south of England (Luddington) found that irrigating with 2.5 cm of water applied at the marble stage and thereafter with 2.5 cm whenever the soil moisture deficit reached 5.0 cm reduced the number of irrigations required and increased yield of tubers, compared with applying 2.5 cm of water whenever the soil moisture deficit reached 2.5 cm.

6.3.2 Stage of growth and water supply

The literature on the response of potatoes to water at different stages of growth has been reviewed by Salter and Goode (1967). They defined the term 'moisture sensitive period' as 'a particular period during the development of a plant in which the plant is, or appears by its observed responses to be, more sensitive to moisture conditions than at other stages of development'.

They concluded that an adequate supply of water is required by the crop from the time of tuber initiation almost until the tubers are mature if good yields of high quality potatoes are to be obtained. The experiments reviewed suggested that a plentiful supply of water before tuber initiation increased the number of tubers per plant, whereas a plentiful supply after this stage of growth resulted in an increase in tuber size.

In varieties which have a tendency to initiate many tubers, such as King Edward, the consequences of irrigating from the time of emergence can, in extreme cases, because of the resulting increase in tuber numbers, decrease the size of individual tubers to such an extent that the yield of ware may be reduced by irrigation (North, 1960).

Llewelyn (1967) carried out a detailed investigation on the effect of different water regimes on the growth and development of the potato crop on a medium loam soil in central England. A non-irrigated control treatment, receiving natural rainfall, was compared with irrigation regimes commencing at times determined by the stage of growth. The data indicate that irrigation regime could have a large influence on yield but that this was dependent upon variety and year. In general, irrigation before tuber initiation led to smaller increase in yield than irrigation commencing at later stages of growth. This was especially the case with King Edward, where the yield of marketable tubers was particularly unresponsive to early irrigation; although yields were not reduced as in farm trials with this variety (North, loc. cit.). King Edward was more responsive to irrigation than Majestic (see Section 6.3.5). In general, ware and total yield were affected to a similar extent by irrigation regime with Majestic but not with King Edward.

Growth studies showed that irrigation began before tuber formation increased the investment of dry matter is leaf production and that this in turn lead to an increase in the rate of tuber bulking in the early stages of growth. However the treatment also induced early leaf senescence so that this superiority was not maintained; supplying extra N in 1964 did not affect this situation. Leaf persistence appeared to be less affected by early irrigation in Majestic and this variety gave a good response to this treatment in 1964. A study of the interrelationships between leaf persistency and root and tuber development and with N nutrition, as affected by variety and water supply, could repay further study for reasons outlined earlier (Chapter 5).

6.3.3 Method of irrigation

The three basic methods of applying water to the soil for use by crops are surface, sub-surface and overhead irrigation, and as pointed out by Withers and Vipond (1974), each method has inherent advantages and hazards which affect its value for any particular situation. The problems is complex and has been dealt with in detail elsewhere (Israelsen and Hansen, 1962; Hagen *et al.*, 1967; Withers and Vipond, loc. cit.,). In some circumstances where local conditions permit, natural sub-irrigation may be practiced. However, overhead irrigation or furrow irrigation (because of the ubiquitous practice of growing the crops in ridges) are the major methods used.

Equally good yields may be obtained from either method (Haddock, 1961) although rather more water may need to be applied with furrow irrigation. In one case, on a sandy soil in central Washington (Middleton *et al.*, 1975) a comparison of the two methods showed that furrow irrigation required approximately four times as much water as sprinkler irrigation and yielded considerably less. In this experiment in addition to a basic irrigation treatment calculated to supply enough water to meet the estimated daily evapotranspiration (ETI) additional treatments (ETI.25, ETI.5, ETI.8) supplying 1.25, 1.5 or 1.8 more than the basic rate were used to approach minimum applications needed to offset distorted sprinkler distribution of water resulting from wind and to determine the effect of extra water on fertilizer N requirements. It was shown that when the crop was irrigated at 2–3 day intervals (1969), treatment ETI gave lower yields than treatments ETI.25 and ETI.5, but when water was applied daily to replace transpiration from the previous day (1971), there was no benefit from increasing the quantity of water applied. However, in this situation there was an interaction between the quantity of water supplied at each irrigation and the rate of N applied (Fig. 6.7); increasing the quantity of water gave reduced yields at the lowest rate of N supplied, an effect which could be eliminated if sufficient fertilizer N was applied which suggested that at least on this sandy soil, the adverse

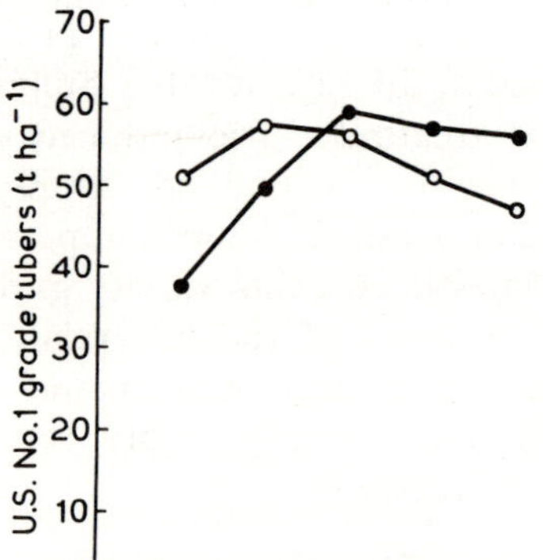

Fig. 6.7 The effect of nitrogen fertilizer on tuber fresh weight yield with sufficient water applied by sprinklers to meet evaoptanspiration needs (open circles) and with 1.8 times as much water (closed circles). Redrawn from Middleton *et al.*, 1975

effects of excess water could be attributed to the leaching of N; a conclusion supported by Penman's (1971) observations on a sandy loam soil under temperate conditions.

The use of mist irrigation has been referred to previously (Section 6.3.1). It offers interesting possibilities of increasing yields beyond those possible with other methods of applying water; however more experimental work is needed to evaluate the magnitude of any such advantages and to assess whether the extra investment in equipment would be justified. Assessment against trickle irrigation applied daily would be particularly interesting.

6.3.4 Crop cover

Factors affecting the rate of development of crop cover (% of the soil surface covered by leaves when viewed directly from above) will affect the rate of water loss from the soil. Rijtema (1966) concluded that row crops covering more than 50% of the soil act as a full crop cover in relation to transpiration. Tanner and Lemon (1962) suggested that up to a crop cover of about 55%, increase in crop cover is proportional to an increase in evapotranspiration. Unpublished data (Stanhill, 1958) suggests that leaf cover above 50% still had a slight effect on the ratio of actual to potential evapotranspiration when soil deficits exceeded 2 cm but not at lower deficits. The relationship between crop cover and the ratio of actual to potential evapotranspiration was curvilinear, the effect decreasing as plant cover increased. However Mohindra (1975) obtained relationships which differed considerably from those of Stanhill.

Endrödi and Rijtema (1969) showed that in deriving the effect of soil cover on transpiration under 'normal' weather conditions, that systematic deviations occurred when the weather was either extremely wet or extremely dry. Stanhill (loc. cit.) took both leaf cover and soil moisture conditions into account in estimating soil moisture deficits. However, there are a number of factors which can affect the rate of loss of water from bare soil, for example, bulk density and the intensity of evaporating conditions (Stanhill, 1955). As suggested earlier, ridging the soil in potato production could enhance the loss of water by evaporation from the soil. Cultivations which expose moist soil to the atmosphere in the period before much leaf cover has developed will also increase the speed with which water is lost from the soil, as will any weed growth which increases the ground surface covered by vegetation. The substitution of herbicides for cultural weed control may be of some value in conserving soil water during the early stages of growth; Shotton (1966) reported investigations into the use of herbicides in potato crops on experimental husbandry farms in Britain and found that on two farms herbicides gave generally higher yields than cultivation in the absence of irrigation; on sandy soil irrigated yields were normally higher when herbicides were not employed.

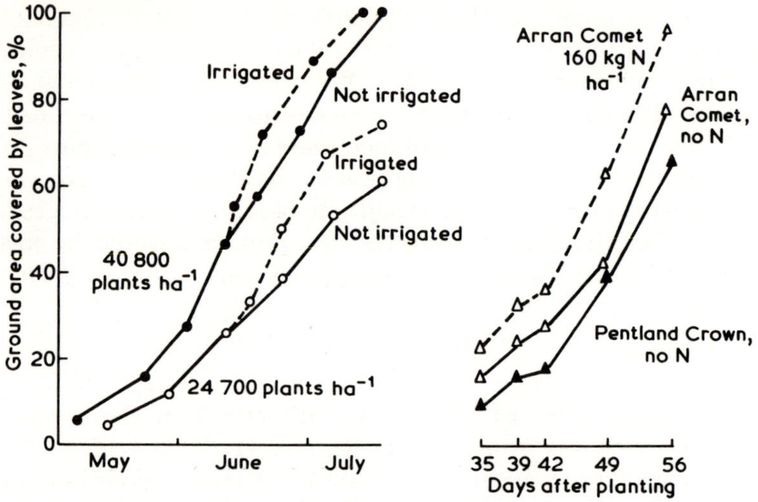

Fig. 6.8 The effect on the ground area covered by leaves (%) of (a) plant density and irrigation (from Mohindra, 1975) and (b) variety and nitrogen (Harris, unpublished data)

Many agronomic factors influence the rate at which the ground is covered by leaves, for example, variety, the physiological state of the seed tuber, plant density, fertilizer and irrigation can all have marked effects; some of these are illustrated in Fig. 6.8.

The effect of such treatments on plant–soil–water relationships need not necessarily be merely confined to their effects on leaf cover development and the potential rate of water loss. In the sugar beet crop, Farazdaghi (1968) and Harris (1972) have shown that while increases in plant density increased the rate at which the soil was covered by leaves, the potential rooting depth was also increased, which, under certain circumstances, led to a delay in the onset of wilting as plant density increased and to a reduction in the response of the crop to irrigation. More recently Mohindra (1975) investigated the influence of plant density on the water economy of a number of crop species including the potato; he found (Table 6.11) that of the four species tested, only the tap-rooted crops were able to increase their depth of rooting in response to an increase in plant density; where soil moisture conditions were not limiting, the maximum rooting depth of the potato crop remained unaffected by plant density. However, where the crop was not irrigated, the lowest plant density (24 700 plants ha^{-1}) appeared less able to deplete the available moisture from the profile down to a depth of 30 cm than the two higher plant densities (40 800 and 80 000 plants ha^{-1}).

It would appear that the major effect of increasing plant density on the water economy of the potato crop would be to increase the rate of

Table 6.11 The effect of plant density and species on maximum rooting depth (cm) in July

Crop	*Number of plants (ha $^{-1}$)*			
	24 700	*40 800*	*80 000*	*222 200**
Sugar beet		46	91	107
Spinach beet		76	91	107
Kale		96	96	96
Potatoes	76	76	76	

*This plant density was difficult to achieve, actual plant densities were: sugar beet 207 000, spinach beet 204 100 and kale 164 900 plants ha $^{-1}$.
Data from Mohindra (1975).

loss of water from the soil profile in the period before a complete crop cover is obtained, with little compensatory increase in the volume of water from which water could be extracted. The more rapid depletion of water in the upper nutrient rich part of the soil profile might also be expected to accentuate the need for and response to irrigation at high plant densities, more especially as the demand for nutrients is increased by increasing plant density (Chapter 5). The tendency for higher plant densities to give larger responses to irrigation is in agreement with this conclusion (e.g. Howe and Rhoades, 1948; Peeler *et al.*, 1966; Wellings, 1972).

6.3.5 Variety

Different varieties have been included in a number of irrigation experiments; some have shown little or no difference in response to irrigation, particularly when un-irrigated yields are expressed as a percentage of irrigated yields (e.g. Kopetz, 1955 quoted by Singh, 1969; Gregersen and Jørgensen, 1973) in other cases varietal differences in responsiveness to irrigation have been commented upon (e.g. Pratt *et al.*, 1952; Peeler, 1966; Llewelyn, 1967; Peterson and Weigle, 1970).

Pratt *et al.* (loc. cit.) noted that varieties which set large numbers of tubers per hill such as Chenango, failed to produce as well proportionately without irrigation as did varieties such as Katahdin and Kennebec that set few tubers per hill; however the interaction between irrigation and variety was significant in only a few of the many comparisons made over different sites and years. Peeler (loc. cit.) suggested that the yield of Arran Pilot was rather more sensitive to small soil water deficits than Ulster Prince and that this might be attributed to the tendency of Arran Pilot to set a larger number of tubers.

There appears however to be little firm evidence connecting the number of tubers set with sensitivity to drought, and while Llewelyn (1967) showed that King Edward, which sets more tubers than Majestic was also more sensitive to drought than the latter, alternative explanations have been put forward. Ageeb (1968) found that the difference in drought resistance between these varieties could be partly explained by differences in root growth. King Edward had a less branched root system than Majestic and the roots were more concentrated in the surface layers of soil, which resulted in growth of the latter being checked at a higher soil water potential. However, although there is no evidence, it is possible that greater tuber numbers may increase competition for assimilates between roots and tubers and reduce the supply available for root production. Ageeb also found that King Edward plants depleted soil water more rapidly than Majestics, which was ascribed to the taller and more open habit of growth of the former leading to a more rapid rate of evaporation from the leaves.

In unpublished experiments, Stecker (1976) compared the yield of four varieties under irrigated and non-irrigated conditions, root growth was observed in a glass sided root chamber. Data from the experiment are given in Table 6.12 which show that while there were large apparent differences in the responsiveness of varieties to irrigation, this was not correlated with the speed with which roots penetrated the soil profile. However it has been shown (Section 6.2.1) that this need not necessarily be the most important criterion of root growth for the ability of roots to deplete any one layer of soil of 'available' water is also limited and a more thorough exploitation of any one horizon might occur at the expense of a rapid penetration in depth.

The importance of the influence of variety on tuber numbers and the need to take this into account when considering wehther or not to irrigate in the early stages of tuber growth, had been referred to in Section 6.3.2.

Howard (1976) has drawn attention to the large differences in performance of breeding stock raised in good seed districts of high rainfall from their performance when tested in areas of relatively low rainfall.

Table 6.12 Varietal sensitivity to water stress and rate of root growth

Variety	*Unirrigated as % irrigated total tuber yield*		*Time taken for the roots of the unirrigated crop to reach a depth of 60 cm (days)*	
Most sensitive varieties				
Maris Piper	41.4	42.2	50	48.0
Pentland Crown	43.0		46	
Least sensitive varieties				
Majestic	49.2	50.0	60	56.5
King Edward	50.8		53	

Adapted from unpublished data kindly provided by J. Steckle.

Apparently little attention has been given to breeding for drought resistance, although Saleem and Shafi (1966) working in West Pakistan claimed that certain *Tuberosum* × *Andigena* and *Tuberosum* × *S. commersonii* hybrids showed better drought resistance than *Tuberosum* varieties. Atsmon (1973) in a general review on breeding for drought resistance in field crops has pointed out that not withstanding the great economic importance of breeding for drought conditions, it has been little investigated, a negligence which may be attributed to the subject's great complexity.

A possible reason for neglecting this aspect of breeding in the potato, apart from the complexity of the problem it undoubtedly shares with other crops, must be attributed to the fact that centres of potato production and research are largely located in temperate areas with reasonable quantities of rainfall. However, recent experience in Europe where lack of rain has severely affected potato yields, may perhaps stimulate more interest in breeding for drought tolerance; even in the absence of such extreme conditions, the fact that it has been estimated that in England there is a need for irrigation for early crops nine years and for maincrops eight years out of ten (Smith and Meads, 1961), suggests that drought tolerance is an aspect of potato breeding which deserves more attention.

6.3.6 Control of stomatal opening

The possibility of manipulating stomatal aperture by chemical means in order to increase the yield of crop per unit of water transpired has been considered by Waggoner (1969) who concluded that while reducing stomatal aperture was unlikely to increase yield per unit area, there was a possibility that yield per unit volume of water transpired could be increased.

Fulton (1967) reported experiments carried out in Canada in 1965 to test the effectiveness of the monomethyl ester of decenyl succinic acid for restricting stomatal opening in potatoes and to determine whether the size of the stomatal opening affected the amount of water used by the crop. The effect of the chemical and a control on stomatal aperture and evaporation for a non-irrigated and irrigated crop are shown in Fig. 6.9. It can be seen that the chemical brought about marked reductions in stomatal aperture on both irrigated and non-irrigated crops, but that the size of the stomatal aperture had no effect on the amount of water used. The yield of potatoes was unaffected by the application of alkenyl succinic acid but irrigation increased yield from 9.4 to 20.2 t ha^{-1}.

The variation in stomatal width (0.7–4.2 μm) were similar to those found for potatoes grown in the field in central England (0.1–5 μm) by Burrows (1969), however, as shown earlier (Section 6.2.2) in these experiments the potato was able to reduce rates of transpiration by

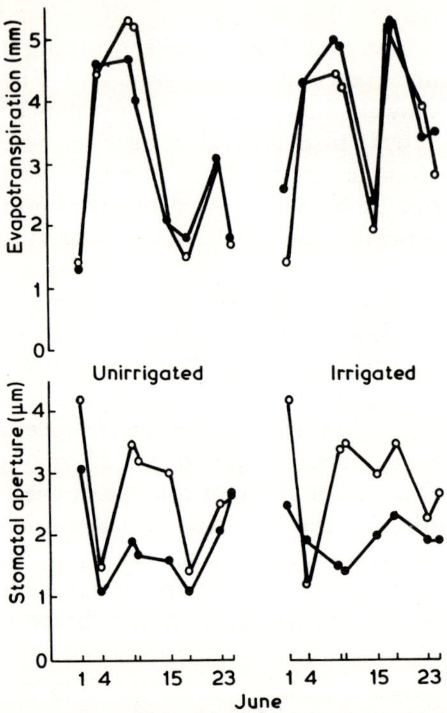

Fig. 6.9 The effect of the monomethyl ester of decenyl succinic acid on evapotranspiration and stomatal aperture of field-grown potatoes with and without irrigation (open circles control, closed circles treated) Drawn from data of Fulton, 1967

stomatal closure. Sanders and Nylund (1972a) were able to show that mist irrigation decreased stomatal resistance during the period of misting, and compared with conventional irrigation this could result in increased yields (Sanders and Nylund, 1972b).

There is much interest in the role of plant growth regulators in plant water stress (Livne and Vaadia, 1972; Vaadia 1976). Not only are growth regulators such as cytokinin and abscissic acid involved in stomatal movements, but they are also involved in the longevity of leaves and with the retention or loss of leaves under conditions of water stress. Fischer and Hagan (1965) have drawn attention to the sensitivity of leaf area to water stress and the possibility of using growth regulators not only to affect stomatal behaviour but also to reduce leaf shedding, which offer interesting possibilities for testing on the potato crop. In spite of the numerous experiments carried out with the application of a wide range of growth regulators to the crop, little work seems to have been directed towards testing the use of appropriate growth regulators in controlling the water status of the crop.

It has been shown, for example, that the use of the growth retardant cycocel in the wheat crop (Humphries *et al.*, 1967), increased yield under dry conditions apparently due to an improvement in the root system. However in an unpublished experiment (Harris, 1969) cycocel applied as a foliar spray to potatoes failed to have any effect on the yield of either irrigated or unirrigated plots of Maris Page.

6.3.7 Disease

Water can have a considerable bearing on the incidence and effect of diseases on any crop and the potato is no exception. Relationships between water stress and disease have been reviewed by Cook (1973), Duniway (1973) and Boyer (1973). The spread of late blight (*Phytophthora infestans*) is affected by the humidity of the air in the crop canopy which clearly may be affected by irrigating while it has been shown (Farrell *et al.*, 1969) that *Phytophthora infestans* can cause abnormal stomatal behaviour in invaded leaf tissue. Common scab (*Streptomyces scabies*), a disease of somewhat less importance, is sensitive to soil water conditions at certain stages of growth. However, these considerations are beyond the scope of this chapter and reference should be made to Chapter 11.

6.3.8 The estimation of irrigation need

The various methods of calculating the water status of the soil on which to base decisions concerning when to irrigate and how much water to apply, are dealt with in detail in the general reference cited in Section 6.3.3 and are not therefore discussed here. However, a few points are briefly made. The first is that any method of calculation is probably preferable to guess-work, however experienced the 'guesser' may be. It is perhaps pertinent here to quote Jean and Weaver (1924):

'One of the most important problems of crop production under irrigation where an ever increasing amount of tillable land is placed under the ditch and the supply of water is limited, is the proper use of the latter. At present the time of application of irrigation water and the amount used are determined entirely empirically, much water often being wasted, yields sometimes actually reduced, and other injurious results ensuing from the use of an over supply. In these experiments water was supplied at times designated by the owner of the land, a man of many years' experience in irrigation, whose judgement in such matters has been attested by highly successful crop production'. It is salutary to note that the yield of 'fully' watered plots of potatoes in one of their experiments was 160 bushels per acre, while the 'lightly' watered treatment yielded 303 bushels per acre. Penman (1952) has shown in experiments on sugar beet in which weather based estimates of irrigation need were compared with the estimates made by two experienced farmers, that the maximum yield was obtained from the former and that experienced practice differed widely, either applying too much or too little water.

Rijtema (1966) argued that while methods used to calculate potential evaporation underestimated the true value, this was not important in irrigation practice, partly because under conditions of low density and low soil fertility, a large variation in evapotranspiration can exist without reducing total dry matter production substantially, partly because generous efficiency factors are usually allowed for the efficiency with which irrigation water is applied, and also because, especially on clay soils, to irrigate the crop to maintain transpiration at the potential rate would require an impractically short time interval between irrigations while soil aeration may then become a limiting factor. The first factor is unlikely to be relevant to profitable potato production and more recently Enrödi and Rijtema (1969) and Burrows (1969) have pointed out that, for irrigation practice as well as in hydrological investigations, it is important to determine the actual as well as the potential water use of a crop from meteorological data. Owing to the fall of actual transpiration relative to potential transpiration under conditions where the rate of evaporation is relatively high, even when the soil moisture potential is high (Section 6.2.2) then it is possible to argue that it might be useful to have a more accurate estimate. However there is little doubt that in many practical situations the use of a rain gauge coupled with existing methods of estimating potential evaporation would represent an advance in the rational use of water for irrigation.

The possibility of taking future rainfall into account in determining whether to delay irrigation or to reduce the quantity of water applied to maize in central Missouri has been investigated by Hashemi and Decker (1969). They showed that this could result in significant savings in irrigation water and reduce the frequency of irrigation. The application of this idea might be usefully explored in the potato crop both to save water and effort and to prevent the yield reductions which have been demonstrated when rainfall closely follows irrigation (Penman, 1972). However, in order to maximize the use of equipment it is usual to irrigate over a period of time, in which case irrigation must start earlier than the planned deficit, in order to prevent this being considerably exceeded over part of the area before the cycle is completed. As the major cause of yield loss might be attributed to leaching of N, at least on sandy soils, then application of a top dressing of N after such leaching has occurred might be worth considering. However such refinements would necessitate the assessment of actual rather than potential transpiration rates if irrigation requirement is estimated from meteorological data.

6.4 Irrigation to control temperature

While the major use of irrigation is to reduce water stress in the crop, it may also be used to control extremes of temperature. The use of irrigation for frost protection has been reviewed by Landers and Witte

(1967), who point out that protection is normally only profitable where frosts in the growing season are infrequent, such as winter frosts in Mediterranean climates and spring and autumn frosts in temperate regions. Protection depends on the principle that 334.9 J g^{-1} of latent heat are released as water freezes to offset the heat lost by the crop to its cooler surroundings. The ice formed does not insulate the crop from injury and a film of continuously freezing water must be maintained by the sprinklers on the surface of the ice. This local release of heat holds the plant near optimum conditions.

Wetting a comparatively dry soil prior to a frost increases the heat capacity of the soil and improves its ability to absorb and release heat. Given high daytime solar radiation and little chilling of the ground by evaporation or cold irrigation water, the amount of solar heat stored in the soil will be increased (Landers and Witte, loc. cit.). Thus at night, more heat is released from the soil to counteract radiational cooling. Irrigation applied at the rate of 8.4 mm every day was successfully used on a sandy soil in central England to raise the air temperature and increase the yield of early potatoes by 7.5 t ha^{-1} (Harvey, 1963).

The reduction of soil temperature due to irrigation has been demonstrated in the potato crop by Lorenz (1950) and Box *et al.* (1963). Peterson and Weigle (1970) compared low volume irrigation around midday when the temperature was high with the usual practice in Iowa of applying 2.5 cm of water at 3–4 day intervals when the rainfall was inadequate. The following temperature reductions were obtained on an arid day when the sprinklers were operating: soil temperature (at 5.1 cm) 3.3 °C, leaf temperature 6.1 °C, and air temperature (within the canopy 30.5 cm from the soil surface) 4.5 °C. It was not possible however to determine to what extent yield increases resulting from this treatment could be ascribed to the observed reductions in temperature or to effects of stomatal aperture due to increased humidity. Sanders and Nylund (1972a) also found that mist irrigation decreased air temperature within the leaf canopy, leaf temperature (especially in the central part of the canopy) and soil temperature.

While it is not possible to estimate the beneficial effect of reducing the temperature of the environment and of the crop by various forms of irrigation, the additional use of mist irrigation to reduce temperatures below those of conventionally irrigated crops would appear to involve the use of very large quantities of water to produce relatively small increases in yield (Section 6.3.1).

6.5 Conclusion

The crop has been shown to be very sensitive to small soil water deficits and for near maximum yields this deficit should not be allowed to exceed approximately 50% of the available water within the root range of the crop. However where potential evaporation rates are very high, wilting has been observed even when the soil is at field capacity.

There would appear to be a good relationship between transpiration and yield although the relationship can be improved by taking relative humidity into account. In some cases it has been shown that while transpiration may occur when a limiting deficit has been reached, such transpiration does not contribute to growth. This may be attributed to the sensitivity of the crop to interruption in the nutrient supply consequent upon the drying out of the upper nutrient rich portion of the soil profile.

Good responses to irrigation may be expected to be within the order of $1 - 2$ t ha^{-1} cm^{-1}, with maximum possible responses in the order of $2 - 2.5$ t ha^{-1} cm^{-1}. Generally the crop responds to irrigation throughout the whole growing period, although irrigation before tuber initiation may adversely affect yield, or reduce the marketable yield response per unit of water applied in varieties which tend to initiate large numbers of tubers.

Reasons for the sensitivity of the crop to small soil water deficits probably include the tendency of stomata to close at relatively low leaf water potentials and for the rate of photosynthesis to be sensitive to increases in stomatal resistance. However data are limited and conflicting. Relative to other crops, rooting depth is not particularly shallow, but the crop does not appear to be able to deplete each layer of soil down the profile of water as effectively as other crop species with which it has been compared. This limitation may partly be related to root diameter which appears to be smaller than other crop species in the few comparisons that have been made, and partly to the numbers of roots per unit volume of soil, however this latter is dependent on plant population. Varietal differences in the rate of root extension and the ability to deplete the water available in the upper layers of soil have been observed; the rate of water loss may also be affected by characteristics of the crop canopy. The relatively high nutrient requirements of the crop and the sensitivity of nutrient supply to the water status of the nutrient rich superficial layers of the soil, would also appear to be an important factor in the relationship between water supply and crop growth.

A number of factors which could affect the efficiency with which water is used by the crop have been considered. The frequency with which water is applied appears to be of considerable importance, but evidence for the ultimate reduction of the interval between watering to a day is sparse and has given conflicting results. The application of mist irrigation during the hottest part of the day has given beneficial effects on yield, but largely at the expense of efficient water use measured in terms of the response of tuber yield per unit of water applied. Careful monitoring of soil water use, either by direct measurement or by calculation from meteorological data, is required to make the most effective use of irrigation – to avoid yield reductions through the over- or under-application of water. Yield reduction following excessive water use would seem largely, at least on sandy soils, to be due to the leaching of N beyond the root range of the crop. The maximum benefit from irrigation, is only achieved where factors such as nutrient supply and plant population are not limiting yield.

The possibility of increasing the yield per unit of water used by the crop through chemicals to reduce stomatal aperture has received little attention but what evidence there is does not appear to be very encouraging. The use of growth regulators to control the effects of water stress on leaf area and stomatal behaviour also appears to have received little attention. Similarly, breeding for drought resistance has been neglected; this may be partly due to the complexity of the problem and partly because the crop tends to be most important in regions with sufficient rainfall to give reasonably reliable yields. However, even in such regions, yield in most years is limited by soil water supply. Varietal differences in response to soil moisture conditions would seem to exist and investigations of varietal differences in stomatal behaviour, rooting characteristics and the relationship between water, nutrient supply and growth, would seem to offer interesting avenues to investigate with the objective of reducing the very close dependence of yield upon an adequate and well distributed supply of rain.

References

Ageeb, D.A. (1968). The basis for the different responses to drought of Majestic and King Edward potatoes. Ph.D. Thesis, University of Nottingham.

Arkley, R.J. (1963). *Hilgardia* **34** (13) 559–84.

Atsmon, D. (1973). In *Agricultural Genetics*, ed. Moav, R. pp. 157–76. Chichester: John Wiley.

Blake, G.R., Brill, G.D. and Campbell, J.C. (1955). *Am. Potato J.* **32**, 327–31.

Box, J.E., Sletten, W.H., Kyle, J.H. and Pope, A. (1963). *Agron. J.* **55**, 492–4.

Boyer, J.S. (1973). *Phytopathology* **63**, 466–72.

Burrows, F.J. (1969). *Agric. Meteorology* **6**, 211–26.

Carter, A.R. (1962). Irrigation experiments. *Luddington Experimental Station Report for 1961.*

Cook, R.J. (1973). *Phytopathology* **63**, 451–8.

Corey, A.T. and Blake, G.R., (1953). *Soil Sci. Soc. Am. Proc.* **17**, 314–17.

Danielson, R.E., 1967. In *Irrigation of Agricultural Lands*, eds. Hagan, R.M., Haise, H.R. and Edminster, T.W. *Agronomy* (11). pp. 390–424. American Society of Agronomy.

De Roo, H.C. and Waggoner, P.E. (1961). *Agron. J.* **53**, 15–17.

Duniway, J.M. (1973). *Phytopathology* **63**, 458–66.

Durrant, K.M.J., Love, B.J.G., Messem, A.B. and Draycott, A.P. (1973). *Ann. appl. Biol.* **74**, 387–94.

Endrödi, G. and Rijtema, P.E. (1969). *Neth. J. agric Sci.* **17**, 283–99.

Epstein, E. and Grant, W.J. (1973). *Agron. J.* **65**, 400–4.

Farazdaghi, H. (1968). Some aspects of the interaction between irrigation and plant density in sugar beets. Ph.D. Thesis, University of Reading.

Farrell, G.M., Preece, T.F. and Wren, M.J. (1969). *Ann. appl. Biol.* **63**, 265–75.
Fischer, R.A. and Hagan, R.M. (1965). *Expl. Agric.* **1**, 161–77.
Fuehring, R.H.D., Mazaheri, A., Bybordi, M. and Khan A.K.S. (1966). *Agron. J.* **58**, 195–8.
Fulton, J.M. (1967). *Can. J. pl. Sci.* **47**, 109–11.
Fulton, J.M. (1970). *Can. J. soil Sci.* **50**, 92–4.
Fulton, J.M. and Murwin, H.F. (1955). *Can. J. agric. Sci.* **35**, 552–6.
Gardener, W.R. (1965). *A. Rev. Pl. Physiol.* **16**, 323–42.
Garwood, E.A. and Williams, T.E. (1967a). *J. agric. Sci., Camb.* **68**, 281–92.
Garwood, E.A. and Williams, T.E. (1967b). *J. agric. Sci., Camb.* **69**, 125–30.
Gregersen, A. Jørgensen, V. (1973). *Saertryk Af Tidsskrift For Planteavl*, **77**, 611–20.
Haddock, J.L. (1961). *Am. Potato J.* **38**, 423–34.
Hagan, R.M., Haise, H.R. and Edminster, T.W. (1967). Irrigation of agricultural lands. *Agronomy* (11). American Society of Agronomy.
Hagan, R.M., Vaadia, Y. and Russell, M.B. (1959). *Adv. Agron.* **11**, 77–98.
Harris, P.M. (1972). *J. agric. Sci., Camb.* **78**, 289–302.
Harvey, P.N. (1963). *J. Minist. of Agric.* **70**, 214–15.
Hashemi, R. and Decker, W. (1969). *Agric. Meteorology*, **6**, 245–57.
Holliday, R. (1970). In *Nitrogen nutrition of the plant*, ed. Kirby, E.A. pp. 189–200 University of Leeds.
Howard, H.H. (1976). Personal communication.
Howe, O.W. and Rhoades, H.F. (1948). *Proc. Soil Sci. (Soc. Am.* **19**, 94–8).
Humphries, E.C., Welbank, P.J. and Williams, E.D. (1967). *Nature* **215**, 782.
Israelsen, O.W. and Hansen, V.E. (1962). *Irrigation principles and practices.* 3rd Edn. Chichester: John Wiley and Sons.
Jean, F.C. and Weaver, J.E. (1924). Root behaviour and crop yield under irrigation. Carnegie Institution of Washington D.C. Publication 357.
Jones, S.T. and Johnson, W.A. (1958). *Proc. Soil Sci. Soc. Am.* **71**, 440–5.
Kopetz, L.M. (1955). Quoted by Singh, G. (1969) (loc. cit.).
Landers, J.N. and Witte, K. (1967). *Irrigation for frost protection.* eds. Hagan, R.M., Haise, H.R. and Edminster, T.W. Irrigation of Agricultural Lands, 1037–57.
Lescznynski, D.B. and Tanner, C.B. (1976). *Am. Potato J.* **53**, 69–78.
Livne, A. and Vaadia, Y. (1972). In *Water Deficits and Plant Growth*, ed. Koslowski, T.T. Vol. 3, pp. 255–75. London: Academic Press.
Llewelyn, J.C. (1967). The influence of water regimes on the growth and development of the potato crop. Ph.D. Thesis, University of Nottingham.
Lorenz, G.A. (1950). *Am. Potato J.* **27**, 396–407.
McDermott, N. and Ivins, J.D. (1955). *N.A.A.S. Q. Rev.* **27**, 106–8.
Middleton. J.E., Cline, T.A., Roberts, S., McNeal, B.P., James, D.W. and Carlisle, B.P. (1975). Irrigation and fertiliser management for efficient crop production on a sandy soil. College of Agriculture Research Center, Washington State University, Bulletin 811.
Mohindra, M.K. (1975). The influence of plant population on crop water economy. Ph.D. Thesis, Reading University.
Moorby, J., Munns, R.M. and Walcott, J. (1975). *Aust. J. pl. Physiol.* **2**, 523–33.
Motes, J.E. and Greig, J.K. (1969). *J. Am. Soc. hort. Sci.* **94**, 510–11.
Murphy, H.J. and Goven, M.J. (1959). *Bull Maine agric. Exp. Stn.* **583**.
North, J.J. (1960). *J. R. agric. Soc.* **121**, 7–20.
Peeler, C.H.; Harvey, P.N. and Rosser, W.R. (1966). *Expl Husb.* **14**, 30–42.
Penman, H.L. (1952). *J. agric. Sci.* **42**, 286–292.
Penman, H.L. (1956). In *The growth of leaves*, Proceedings of the third Easter

School in Agricultural Science, University of Nottingham, pp. 170–177, ed. Milthorpe, F.L., London: Butterworths.

Penman, H.L. (1971). *Rothamsted Experimental Station Report for 1970*, Part 2, 147–70.

Peterson, E.L. and Weigle, J.L. (1970). *Am. Potato J.* **47**, 94–8.

Pratt, A.J., Lamb, J.Jnr., Wright, J.D. and Bradley, G. (1952). Yield, tuber set and quality of potatoes. *Cornell University agric. Exp. Station*, Bulletin 876.

Rijtema, P.E. (1966). Transpiration and production of crops in relation to climate and irrigation. *Inst. Land and Water Management Res.*, Technical Bulletin **44**, Wageningen, The Netherlands.

Rijtema, P.E. and Endrödi, G. (1970). *Neth. J. agric. Sci.* **18**, 26–36.

Robins, J.S. and Domingo, C.E. (1956). *Agron. J.* **48**, 488–92.

Saleem, R.M. and Shafi, M. (1966). *West Pakistan J. agric. Res.* **4**, 99–120.

Salter, P.J. (1969). *Report National Vegetable Research Station 1968*, 16–28.

Salter, P.J. and Goode, J.E. (1967). *Commonwealth Agricultural Bureaux, Research Review* **2**.

Sanders, D.C. and Nylund, R.E. (1972a). *Am. Potato J.* **49**, 123–37.

Sanders, D.C. and Nylund, R.E. (1972b). *Am. Potato J.* **49**, 187–95.

Sanders, D.C., Nylund, R.E., Quisumbing, E.C. and Shetty, K.V.P. (1972). *Am. Potato J.* **49**, 243–54.

Shepherd, W. (1972). *Water Resources Res.* **8**, 1092–5.

Shotton, F.E. (1966). *Proc. 8th Br. Weed Control Conf.* **1**, 21–7.

Singh, G. (1969). *Am. Potato J.* **46**, 398–403.

Smith, L.P. and Meads, D. (1961). *N.A.S.S. Q. Rev.* **54**.

Sparrow, O.H.G. (1972). Developments from the study of the effects of the application of small quantities of water to the leaves of sugar beet at mid-day, Ph.D. Thesis, Reading University.

Stanhill, G. (1955). *Nature* **176**, 82–3.

Stanhill, G. (1958). The estimation of evapotranspiration from meteorological data. Unpublished communication, National Vegetable Research Station, Wellesbourne.

Steckel, J. (1975). Private communication.

Tanner, C.B. and Lemon, E.R. (1962). *Agron. J.* **54**, 207–12.

Taylor, R.S. (1965). *Outlook on Agric.* **4**, 234–42.

Thompson, R.H. (1970). Irrigation of the potato crop. A dissertation submitted as part of the final examination for an Honours degree in Agriculture. University of Reading.

Vaadia, Y. (1976). *Phil. Trans. Roy. Soc. Ser. B*, **273**, 513–22.

Waggoner, P.E. (1969). Environmental manipulation for higher yields. In: *Physiological aspects of Crop Yield*, eds. Eastin, J.D., Haskins, F.A., Sullivan, C.Y. and Van Bavel, C.H.M.

Wellings, L.W. (1972). *Exp. Husb.* **22**, 39–50.

Wellings, L.W. and Rosser, W.R. (1968). *Pl. Pathol.* **18**, 1–5.

Werner, H.O. (1954). Influence of atmospheric and soil moisture conditions on diurnal variations in relative turgidity of potato leaves, University of Nebrasta College of Agriculture, Agricultural Experimental Station Research Bulletin 176.

Withers, B. and Vipond, S. (1974). *Irrigation: design and practice.* London: B.T. Batsford Ltd.

Chapter 7

Plant density

E.J. ALLEN

Potato seed tubers show a wide variation in size and possess a variable number of growing points (buds) arranged in groups (eyes) over their surface. At the end of dormancy some buds develop into sprouts and after planting a variable proportion of these sprouts develop into main stems. Thus, a potato plant consists of a variable number of stems, the exact number depending upon the size and treatment of the plant's parent tuber. The individual stems of a plant are wholly dependent upon their parent tuber for growth until emergence and a considerable time may elapse after emergence before the stems become fully independent. Thus, the competitive interference between stems of a potato plant changes considerably with time and this together with the variation in number of stems per plant necessarily results in the effects of plant density being complex. As variation in the number of stems per plant may be associated with variation in their size and performance it is clearly unsatisfactory to assess density in this crop on a simple plants per unit area basis.

The establishment and use of a sound unit of plant density is particularly important in potatoes for seed costs amount to between 30% and 50% of total growing costs (Potato Marketing Board, personal communication) and density affects the number of tubers set which markedly affects the size distribution of the tubers. For all outlets of potatoes, the size of the tubers is an important quality criterion and emphasizes the importance of obtaining the correct plant density. In practice, plant density in the potato crop is manipulated through the number and size of the seed tubers planted and there is a need to establish the most appropriate unit of density which should be capable of explaining the effects of all component parts of density, namely seed size, within-row spacing (and their combination in seed rate) number of stems and spatial arrangement, on total and graded yields of potatoes. There is also a need to consider whether the response to plant density or indeed the most appropriate unit of plant density is affected by physiological age of the seed tubers, for, as shown in Chapter 8, tuber age markedly affects growth and development. It is, therefore, logical to begin a consideration of plant density effects in the potato with the growth and development of the seed tuber. The possible units of density may then be assessed and the most appropriate one used to illustrate the effects of plant density.

7.1 Development of the seed tuber

It is necessary to consider the morphology of the seed tuber and of the sprouts and stems which result from its growth and development. At all stages in this developmental programme it is essential to identify factors which may cause variation in the numbers of sprouts and stems per seed tuber and in their growth.

7.1.1 Eyes

Each seed tuber possesses a number of eyes arranged in the phyllotactic spiral over its surface (Fig. 7.1). Milthorpe (1963) explained that the apical bud is part of the primary vegetative axis of the stolon and the lateral buds arising from it are secondary axes, some of which become main buds in the eyes of the tuber as the wave of expansion described

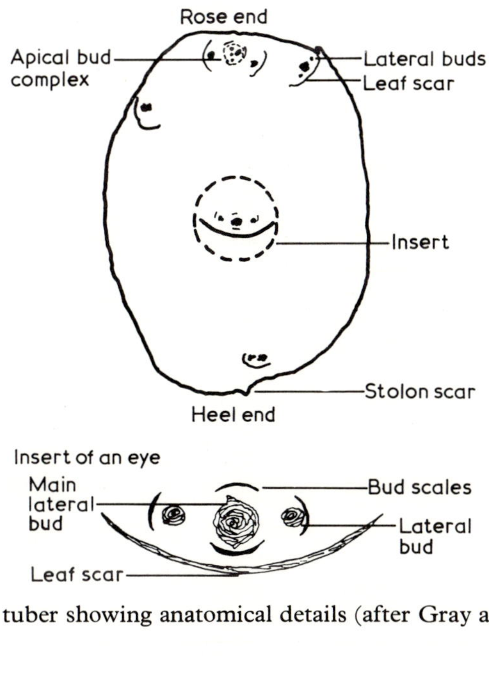

Fig. 7.1 A potato tuber showing anatomical details (after Gray and Bleasdale, 1972)

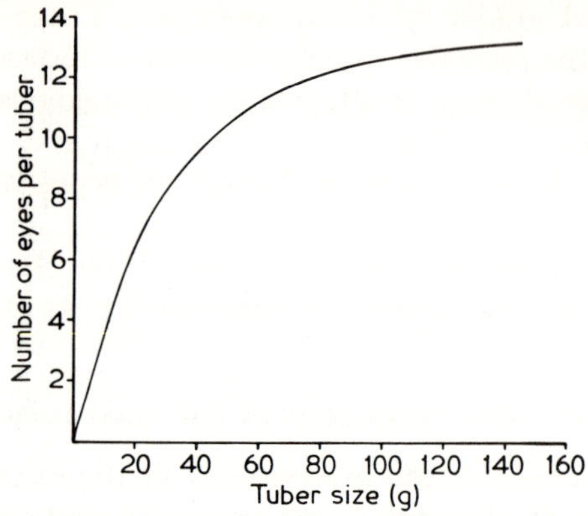

Fig. 7.2 Relationship between number of eyes per tuber and tuber size (drawn from unpublished data of Allen)

by Krijthe (1946) passes from the basal to the apical end. The number of lateral buds which separate from the apical bud and form independent eyes depends upon the degree of development of the tuber and as Fig. 7.2 shows, the number of eyes per tuber increases with tuber size, but at a decreasing rate. Each main bud subtends a number of other buds, part of the tertiary axis, so that every eye contains several buds, each of which is a potential growing point.

Apart from tuber size there is limited information on the effect of other factors on the number of eyes and buds per seed tuber. Svensson (1969) found differences in the number of eyes per seed tuber between varieties and seed lots of the same variety and grade. The latter finding suggests that the number of eyes can be affected by husbandry treatments. O'Brien (1978) found that area of seed production and date of lifting of the seed crop had small and inconsistent effects (Table 7.1).

Table 7.1 Effect of area of seed production and date of lifting of the seed crop on number of eyes per seed tuber, 38–45 mm (O'Brien, 1978)

Date of lifting seed crop (1972)	*Area of seed production*	
	Rhayader (270 m above sea level)	*Tenby (30 m above sea level)*
17 July	11.40	11.12
24 July	11.37	11.17
1 August	11.33	10.40
8 August	11.72	10.98
15 August	11.33	11.25
22 August	11.77	10.97
29 August	11.22	11.77
5 September	11.55	11.37

The effects on number of eyes of other factors known to affect tuber yield, e.g. fertilizers, appear to be unknown. Indeed, the numbers of eyes and buds present on the commonly used seed sizes do not appear to have been reported for even the most popular commercial varieties. Notwithstanding the difficulties involved in identifying eyes, especially around the apical eye, this is a serious deficiency, for the number of potential growing points is unknown.

Our knowledge of the growth of individual eyes is also poor, yet it is of importance to know whether there are differences between eyes in the growth and yield of the stems which develop. Sadler (1961) reported that in the variety, Arran Pilot, the largest bud was the apical one and the size of the other buds increased to a maximum in eyes in the centre of the tuber and then decreased again with further progression round the phyllotactic spiral. Goodwin (1967) using the same variety also found the apical bud to be the largest and the first separate secondary axis bud, the heel end bud, was the smallest. Birecki and Rostropowicz (1963) found no differences in the performance of excised eyes of uniform weight from different sizes of seed tuber. De Carvalho (1975) stated that eyes from large seed tubers were larger than eyes from small seed tubers and the differences were affected by variety. Data, however, were not presented. Allen (1978a) examined the performance of different eyes within tubers of the same size and total eye number (8) by dissecting out eyes to produce single eye tubers. As can be seen in Table 7.2, differences between eye positions were small in numbers of stems and tubers and in tuber yield. Single eye tubers did not differ significantly from intact tubers in number of stems which suggests that few or indeed only one eye developed in the intact tubers and observation of the few such tubers to survive until harvest confirmed that most stems arose from one eye, that at the apex. As Allen and Wurr (1973) have observed that most

Table 7.2 Effect of eye position on performance of single eye tubers (Allen, 1978a)

Eye position		*Number of above-ground stems* ($000\ ha^{-1}$)	*Number of tubers* ($000\ ha^{-1}$)	*Total yield* ($t\ ha^{-1}$)
(Apical)	1	103.2	1421.8	35.1
	8	86.2	1277.3	35.8
	7	87.1	1312.3	36.4
	4	102.3	1156.1	32.3
	3	92.5	1295.2	39.6
(Heel)	2	86.2	1203.7	35.0
Control	(all)	96.0	1490.0	45.5
	S.E.	±6.13	±69.49	±2.25

stems in commercial crops arise from the 'apical complex', there is evidence to suggest that the majority of eyes on seed tubers do not grow. From the very limited data available it would seem that there are only small differences between eyes which do grow. Experiments which report high proportions of eyes developing have not usually explained how the figures were obtained and have probably regarded one sprout (stem) as the growth of one eye. This indirect assessment is probably invalid in view of the evidence above and a detailed study of the growth and development of eyes is long overdue.

The seed tuber is clearly wasteful of growing points, especially when the larger sizes are used and it is surprising that little effort has been directed towards increasing the proportion of eyes which grow. In addition to seed sprouting which is discussed in the next Section, cutting of the seed tubers is one possibility. In North America this is standard practice, for seed tubers are rarely sprouted and the varieties produce relatively few, large tubers. In U.K. it has been used for early potato production (Ing, 1966) but is of no practical significance at present. The possibilities of increasing the number of eyes which develop and thereby the stem density are illustrated in Table 7.3. which summarizes

Table 7.3 Effect of seed cutting on number of above-ground stems and tubers (000 ha^{-1}) (Allen, 1978b)
(a)

		Planting material			
		Whole	*Sixteenths*	*Peel*	*S.E.*
1969	Stems	186.0	346.5	244.5	±15.54
	Tubers	784.5	938.6	823.8	±57.72
1970	Stems	151.4	207.7	—	± 7.67
	Tubers	400.6	552.0	—	±19.19
1971	Stems	144.4	191.0	—	±28.10
	Tubers	482.2	550.0	—	±80.40

(Variety–Majestic)

(b) Data for equivalent plant (whole tuber) densities.

		Whole (113 g)	*Half (56 g) cut laterally*	*Half cut longitudinally*	*Whole (56 g)*
1971	Stems	301.6	184.7	172.7	179.4
	Tubers	750.7	692.3	617.9	607.2

(Variety – Pentland Crown).

data from a series of experiments in Cambridge from 1969–71. Cutting of large seed tubers into various sized pieces increased the number of stems produced compared with whole tubers and the effect increased as the size of seed piece was reduced. Cutting of the seed tuber breaks the dormancy imposed on many eyes by the apical bud, probably aided by wound-induced gibberellins (Rappaport and Sachs, 1967). The small seed pieces were grown on normal ridge culture which retarded their emergence and usually resulted in reduced final yields; much of this deleterious effect could be removed by more appropriate planting depths. Table 7.3. also shows that a half of a large (113 g) tuber was equivalent to a whole tuber of the same size, a finding reported by several workers in the absence of disease (Stuart *et al.*, 1924, Brandreth and Bryan, 1937). This fact should allow for considerable economy in the use of larger seed, for as will be discussed later (Section 7.2.6), the seed rate required for small seed is lower than that for large seed. Machinery for cutting seed tubers is available and a detailed study of the cutting of seed tubers should be undertaken, particularly as the majority of the maincrop acreage is planted with unsprouted seed and the health of our seed is so much improved.

Eyes may also fail to grow because of the effects of pathogens. Skinspot (Oospora pustulans) has been shown to kill up to 25% of eyes on tubers of Majestic stored in bulk until planting (Hide *et al.*, 1969) and this pathogen is so widely distributed on stocks that its effect on the numbers of eyes which grow is considerable. Griffith *et al.*, (1974) showed that gangrene (*Phoma exigua*) not only killed eyes but increased the numbers of sprouts and stems which grew from the remaining eyes. Such effects of disease upon the number of sprouts and stems growing from seed stocks are likely to influence responses to plant density, especially as the stems from seriously infected tubers may be less vigorous than stems from healthy tubers.

7.1.2 Sprouts

At the end of the dormant period tubers begin to exhibit bud growth if placed in suitable temperatures. This growth leads to the development of sprouts and their appearance is well known yet definitions of what is a sprout have been few and varied. Thus, discussions of sprout growth taken from the literature are prone to misconceptions as the description of a sprout may vary. Bates (1935) defined a sprout as ‘an eye showing viable sprouts’. Although used and largely unchallenged until recently, such a definition is clearly unsatisfactory. Moorby (1967) called sprouts ‘the lateral shoots present on the tuber at planting’. Wurr (1975) defined a sprout as ‘bud development on the seed tuber which had leaf initials present and at least partly open’ and added the term sproutlet to refer to ‘bud development on the seed tuber which had leaf initials present

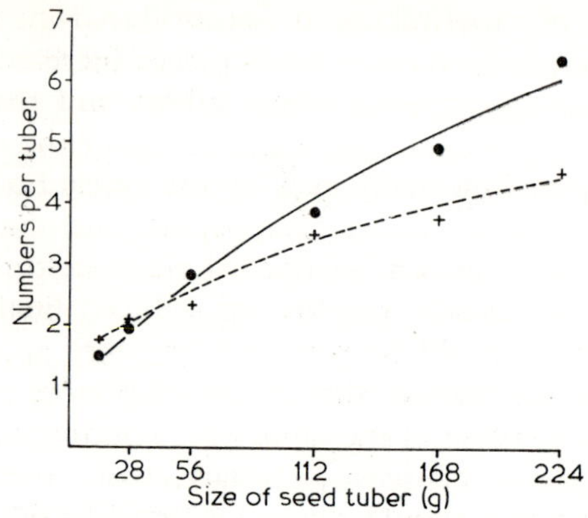

Fig. 7.3 Relationship between number of sprouts and stems per seed tuber and tuber size (after Scott and Younger, 1972). Key ●———● Sprouts + ----- + stems.

and at least partly open irrespective of whether it was derived directly from the seed tuber or whether it was a sprout branch'. Wurr's definitions do appear to be definitive (and therefore repeatable) and are the basis of comments about sprouts in this review.

The number of sprouts which develop per seed tuber is principally determined by the size of the tuber and the temperature and duration of storage. Fig. 7.3 shows that the number of sprouts per seed tuber increases with increasing seed tuber size but at a decreasing rate. The actual number of sprouts on seed tubers is much influenced by storage conditions. Tubers placed in temperatures suitable for growth at dormancy break produce few or even single sprouts as the growth of the apical bud occurs rapidly and suppresses the growth of the other buds, producing the apically dominant (or single sprout) condition. The use of low storage temperatures, later transfer of tubers to sprouting temperatures or damage to the apical bud all result in a greater number of sprouts growing (Table 7.4). Thus, prolonged, low temperature storage will result in large numbers of sprouts developing once the tubers are transferred to

Table 7.4 Numbers of sprouts per tuber $\geq$3 mm from different storage regimes. v. Home Guard (Allen *et al.*, 1978)

°*C*	*From harvest*	*Mid-November*	*Mid-January*
13	2.10	7.70	6.93
8	2.80	7.62	7.25

Data recorded 8.3.73; means of two seed sizes

temperatures suitable for growth. Tubers with many sprouts may be referred to as multi-sprouted, in comparison with the single-sprout condition.

In general, the increase in number of sprouts per seed tuber with increasing seed tuber size is restricted where the seed is sprouted for a long period for even the largest seed may produce only one or two sprouts. Early varieties such as Home Guard and Arran Comet which break dormancy in early autumn will frequently exhibit complete apical dominance in the ambient conditions of autumn. Most maincrop varieties have longer dormancy periods and do not begin to sprout until ambient temperatures are lower and usually develop several sprouts. It should not, however, be imagined that all early varieties have apically dominant seed and all maincrop varieties have seed in a multi-sprout condition for with suitable adjustments to storage conditions the converse of this may be produced.

Sprouting of seed tubers may be regarded as a means of restricting the number of potential growing points which actually grow and the more lengthy the period of sprouting the greater is the effect. As sprouting also affects the physiological age of seed tubers, which affects plant size and longevity, it has profound implications for considerations of plant density in potatoes. Unfortunately, in general, variations in the numbers of sprouts (stems) produced by seed tubers consequent upon sprouting have received limited attention in this context (see Section 7.2.6).

The length of the sprouting period also affects the morphology of the sprouts, for single sprouts often exhibit swollen bases and tend to branch, often following tip necrosis or damage. This sprout branching results in secondary stems as well as mainstems emerging above ground from this type of seed. Generally, large seed produces more branches than small and this tends to compensate for the limited number of sprouts on such seed.

Goodwin *et al.* (1969) found that the number of mainstems in the field was proportional to the number of sprouts at planting. However, not all sprouts >1 cm at planting emerged because of a re-imposition of apical dominance referred to by Morris (1967). It is probable that such effects of apical dominance occur in all sprouted tubers and result in fewer eyes developing into stems than develop sprouts. Morris (1967) suggested that only those sprouts which were 'actively growing' at planting developed into stems but it is clearly necessary to know how to recognize such sprouts and this is not known although sprout length is a useful indicator in some cases. The relationship between numbers of 'actively growing sprouts' and number of stems is variable e.g. Jarvis and Rogers-Lewis (1974), Table 7.5. That many sprouts present at planting do not develop into mainstems is illustrated by the results of

Table 7.5 Relationship between number of actively-growing sprouts and number of mainstems in two potato varieties (Jarvis and Rogers-Lewis, 1974)

	Set Size (g)	Within-row spacing (cm) 66	33	22	16.6
Pentland Ivory					
Number of actively growing	36.3	—	86.2	129.2	172.1
sprouts planted	63.5	49.4	98.8	148.1	197.7
(000's ha $^{-1}$)	101.5	54.2	108.4	162.6	216.7
No. of mainstems in June	36.3	—	78.5	109.6	137.2
(000's ha $^{-1}$)	63.5	51.7	106.2	146.9	185.5
	101.5	69.9	127.0	185.1	138.0
Record					
Number of actively growing	36.3	—	82.4	123.1	164.3
sprouts planted	63.5	48.7	97.4	146.0	194.7
(000's ha $^{-1}$)	101.5	53.7	107.3	160.9	214.6
No. of mainstems in June	36.3	—	92.7	127.7	157.2
(000's ha $^{-1}$)	63.5	68.0	124.0	168.4	215.1
	101.5	80.6	145.7	210.0	259.7

Schepers and Hoogland (1968) who found that although numbers of mainstems in the field were proportional to numbers of sprouts at planting, only 30% of sprouts had developed. This compares with 75% reported by Bleasdale (1965) for Majestic and 64% which he reported Bates (1935) to have achieved with King Edward. Schepers and Hoogland used marked sprouts and lifted all plants to determine whether sprouts observed at planting actually developed into mainstems. Such an approach is essential if the developmental pattern of sprouts is to be recorded but there are limited data of this type available. From the limited data it is clear that there is considerable variation in the proportion of sprouts which develop into stems and much of this variation cannot be partitioned. It is essential that effort is directed towards establishing reliable relationships between a sprout character recorded at planting and the stem density resulting in the field, for without such relationships it is impossible to predict the number of stems produced by seed stocks. A number of possible sprout characters have been suggested and two appear worthy of investigation. Sproutlets, which include all bud growth on tubers, were shown by Wurr (1975) to be closely related to number of above-ground stems (mainstems and secondary stems) and the validity of these relationships over a wider range of varieties and environments requires testing. Schepers and Hoogland (1968) reported that sprouts with root primordia showed an excellent linear relationship with mainstems for 98% of such sprouts developed into mainstems. No other reports of this relationship have been found but it appears worthy of testing for like sproutlets it is an easy character to record. One obvious difficulty would be the variation in sprout size at which varieties first exhibit root primordia.

Much of the maincrop acreage is planted with seed with minimal sprout growth but prediction of the number of stems resulting from this seed may be obtained by study of the storage environment (physiological age) and the method of planting.

7.1.3 Stems

There are three types of stem which may emerge above ground. Mainstems grow directly from the seed tuber and therefore may be derived from main or lateral buds (Krijthe, 1955). Secondary stems do not grow from the seed tuber and may arise either as sub-surface branches from buds on the underground part of a mainstem or where stolons emerge above ground and bear leaves. Mainstems always bear tubers but the other two types of stem may not bear tubers (Allen and Wurr, 1973).

Differences in sprouting regime cause differences in the proportions of the different types of stem produced by seed tubers and this affects the number of tubers set and the growth and longevity of the haulm. Table 7.6 shows that a short sprouting period produces almost entirely mainstem densities while a long sprouting period produces relatively few mainstems with many secondary stems resulting from lateral branch development on the sprouts. Generally, maincrop varieties produce predominantly mainstem densities as they are subjected to a short or no sprouting period while early varieties produce a mixture of main and secondary stems. Thus, sprouting affects the number and the type of stems which occur and it may be anticipated that there will be differences in the growth and yield of individual stems from different sprouting treatments.

Whatever type of stem density is involved it is essential that it is recorded accurately but data on both differences in field densities and

Table 7.6 Effect of length of sprouting period on number and types of stems produced in the field (Allen *et al.*, 1978)

	Sprouting period			
	From lifting	*From mid-November*	*From mid-January*	*S.E.*
Above-ground stems/plant	4.71	4.79	4.75	±0.441
Mainstems/plant	1.88	3.04	3.66	±0.202
Secondary stems/plant	2.83	1.75	1.08	±0.453

Variety-Home Guard. Data means of two temperatures and two seed sizes.

the life of individual stems are limited. Allen and Wurr (1973) recorded the stem densities of a sample of commercial crops in East Anglia by lifting plants in mid-season in order to determine the origin of the stems. They found that mainstems accounted for >99% of the above-ground stem density and suggested the latter as an easy way of recording the effective plant density in a growing crop. Many of the other published reports of stem densities have not indicated how or when the counts were made and as the definition of a mainstem involves the seed tuber, some assumptions may have been made in taking these counts. Without actually lifting plants or removing soil covering the seed tuber, accurate determination of mainstems is difficult, if not impossible and must involve either the acceptance of all above-ground stems as mainstems or some subjectivity in deciding upon the origin of stems. For example, Thompson and Taylor (1974) called mainstems 'those likely to bear tubers', an unsatisfactory basis for recording especially as Allen and Wurr (1973) had reported that the size of a stem was a poor guide to the number and size of tubers carried. It is likely that many published estimates of mainstems involve this subjectivity. In many cases there are no references to secondary stems which may indicate wholly mainstem densities or the dismissal of secondary stems as unimportant. It is desirable to have much more detailed information on stem density and on the factors affecting the proportions of the different types of stem. In this context it is of interest that Allen (1978b) found that secondary stems were more common on maincrop varieties grown in Pembrokeshire than he had found in East Anglia and their number increased over a long period (Fig. 7.4).

The timing of stem counts is also important for it is known that stems die and may disappear as a result of competition before crop maturity (Bleasdale, 1965; Wurr, 1971). Counts made at maturity or after defoliation are likely to be difficult and prone to underestimation, while counts taken early in the season must be taken after complete emergence. Scott and Younger (1972) reported that number of mainstems per seed tuber was determined at an early stage but Ifenkwe (1975) found that it increased slowly for almost two months after emergence. Much more information is required on the number, type, emergence pattern and longevity of stems for efficient stem recording to be practised especially where few counts are to be made. At present it appears that maincrop varieties should be counted in mid-season and in the absence of definitive evidence that only mainstems are present, all experimental counts should involve lifting of the plants.

Sometime after emergence the mainstems arising from the seed tuber assume an independent existence and the plant resulting from one seed tuber becomes a collection of competing stems. As large seed tubers usually produce more stems than small seed tubers, the spatial arrange-

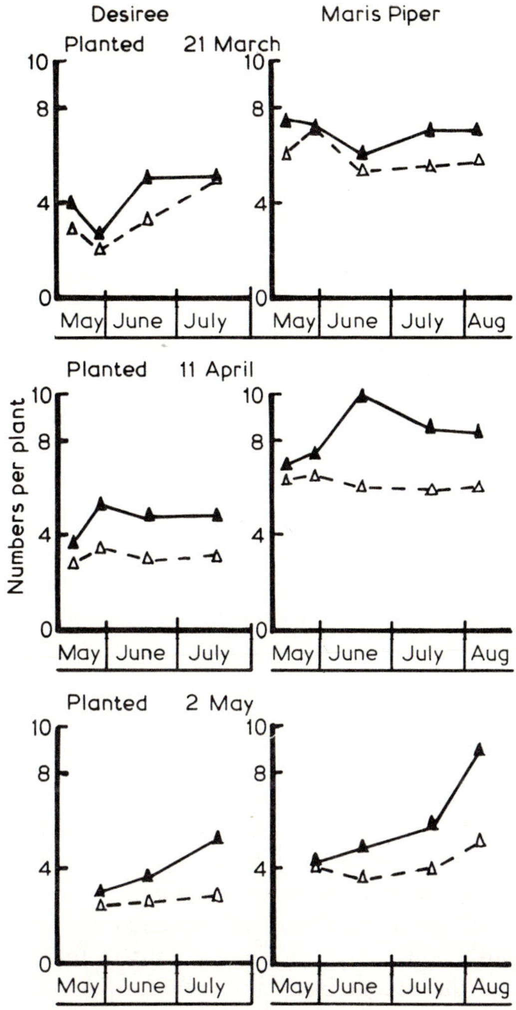

Fig. 7.4 Change in stem densities during growth for three varieties planted on dates. Key: ▲ ——— ▲ above-ground stems; Δ - - - - - - - Δ main stems; the difference between these two types being the number of secondary stems (after Allen, 1977b)

ment of a given stem density is quite different from different seed sizes. This effect is termed 'clumping' and is discussed later in Section 7.3.4.

7.1.4 Summary

This section has discussed the development of the seed tuber and the morphology and growth of the sprouts and stems. At all stages of development information is lacking and the effects of many factors together with the general growth pattern are imprecisely understood. Many of the deficiencies in information and difficulties in interpretation arise from the use of imprecise definitions of morphological characters and unjustified assumptions in their use. A detailed study of the growth and development of the seed tuber from its initiation on its parent plant to the establishment of a plant from it in the field the following year should be capable of resolving many of these points.

These deficiencies of information do not prevent the establishment of general relationships between density and yields, as the succeeding sections of this review will show. However, the variation in the number of stems produced by seed tubers is a major contributor to the variation in plant density established from year to year and severely limits the precision with which density is manipulated in this crop. This is discussed in detail in Section 7.2.6.

7.2 Units of density

The establishment of accurate relationships between tuber yield per unit area and plant density requires a definitive, repeatable unit of plant density to be identified. A number of possible units have been suggested and discussion of their merits, together with the adoption of the most suitable, must precede any discussion of the effects of plant density.

7.2.1 Number of eyes

The use of eyes as the scale of plant density was suggested by Bleasdale (1965) but, as our understanding of the development of eyes is poor and the number which develop is known to be affected by the sprouting regime, the scale has considerable deficiencies. There are also serious, practical difficulties for it is extremely difficult to count eyes because such counts depend upon the ability to recognize the lateral buds which have separated sufficiently from the apical bud to form individual eyes. For eyes close to the apical bud this is extremely difficult especially in varieties with either rather deep-seated eyes or where there is minimal separation away from the apical eye, e.g. Maris Piper and Pentland Crown. Nonetheless the number of eyes has been used as the unit of plant density but the proportion of eyes developing has not been constant over seed sizes and unequal stem densities have sometimes resulted (Thompson and Taylor, 1974). For general application, the scale is of limited usefulness.

7.2.2 Number of sprouts

The difficulties in defining sprouts have already been discussed and in view of these and the fact that the majority of the maincrop acreage is planted with unsprouted seed, the use of number of sprouts as the unit of plant density is totally invalid.

7.2.3 Number of seed tubers

The use of such a scale would assume that all sizes of seed perform similarly and as this has already been shown to be untrue the scale is unjustified.

7.2.4 Surface area of seed tuber

Reestman and de Wit (1959) found a linear relationship between the surface area of the seed tuber and the number of mainstems in the crop. They proposed the surface area of the seed tuber as a general unit of plant density. Such a scale is unlikely to be of practical value and there are reservations about the accuracy of the methods used for calculating the surface area of the tuber (Bleasdale, 1965).

7.2.5 Seed rate

This is the weight of seed planted per unit area and is the product of the number of seed tubers planted and their average weight. Growers are familiar with this unit of plant density as they buy seed by weight, appreciate variation in average tuber size and for every crop must decide upon the weight of seed to be planted. Any other unit of plant density must necessarily be achieved by the appropriate choice of seed size and number of tubers to be planted.

In view of the great practical relevance of this unit and the fact that by taking account of the size of the seed it is not subject to the criticisms levelled at other possible units, it is not surprising that the majority of reported experiments on potato densities have used seed rate as the plant density scale; those experiments which have not used the scale can be assessed by it as the number and weight of the tubers used are usually recorded. These experiments may be summarized by examining the component effects of seed size and within-row spacing (number of tubers) and then discussing the general applicability of seed rate as the plant density scale.

Effect of seed size

The results of a number of experiments over the last 40 years examining the effects of seed size on total and graded yields are summarized in Table 7.7. It may be seen that:

(a) total yield and yield over small riddle sizes increase with increase in seed tuber size; and

(b) as the riddle size over which yield is considered increases, there is a tendency for yield to increase with seed size to a maximum and then decline. Further increase in the riddle size may completely reverse the effect of seed size, so that more large ware is produced by small seed than by large seed.

Table 7.7 The effect of seed size on tuber yield (Adapted from Wurr, 1971)

Author	*Spacing and variety*	*Seed weight* (*g*)	*Yield* ($t\ ha^{-1}$)			
			Total	*>32 mm*	*>57 mm*	
Craibstone (1929)	68.6 × 30.5 cm	28	34.81	33.53	17.72	
		56	37.57	36.31	18.47	
		112	42.85	40.21	13.82	
	Great Scot	224	47.24	44.98	14.07	
			Sprouted		Unsprouted	
			Total	Ware	Total	Ware
Craibstone (1932)	68.6 × 30.5 cm	14	21.61	16.46	17.47	10.93
		28	24.75	16.71	22.99	9.80
		56	32.17	19.73	26.76	9.17
		112	34.55	19.10	30.16	9.42
	Golden Wonder	168	35.06	17.21	27.14	8.29
			Total	> 32 mm	> 44 mm	> 51 mm
Craibstone (1933)	Not stated	14	23.25	22.49	18.47	14.20
		28	30.16	28.90	24.63	15.96
		56	33.67	31.66	23.62	12.19
		112	39.20	36.82	26.01	14.20
	Golden Wonder	224	41.59	39.58	26.76	13.19
			Total	*> 42 mm*	*> 57 mm*	
Bates (1935)	71 cm × mean of 4	28	23.87	21.86	15.83	
		42	25.13	22.62	15.33	
		70	26.64	23.67	14.32	
	King Edward		S.E. ± 0.402	±0.352	±0.452	

			Total		> 42 mm	> 57 mm	
Findlay and Sykes (1938)	71 cm × mean of 4 (1st year)	28	25.26		21.66	9.80	
		42	26.76		22.87	9.60	
	73.7 cm × mean of 4 (2nd year)	70	28.85		24.80	9.80	
	King Edward		S.E. ± 0.905		± 0.829	± 0.628	
			Arran Victory			Up-to-Date	
			Total	Ware		Total	Ware
Megan and Bankhead (1938)	68.6 × 38.1 cm	28	35.06	32.04		38.07	34.18
		42	37.19	33.42			
		56	38.57	33.55		40.08	34.55
		70	40.08	35.18			
	Arran Victory and Up-to-Date	84	38.95	33.30		42.09	35.56
		98	41.21	35.31			
		112	40.58	34.30		44.23	36.06
			Total		> 30 mm		
Roer (1955)	65 × 30 cm	12.5	17.86		17.50		
		25	19.51		18.96		
		50	21.73		20.84		
		100	23.69		22.55		
	Kerr's Pink	150	24.77		22.76		
		200	26.55		24.08		
			Total		44–83 mm (less cracks, dolls, greens)		
Jarvis and Shotton (1971)	71 × 33 cm	31.4	45.06		34.48		
		62.8	47.80		35.08		
		94.2	49.43		36.74		
		125.6	49.78		35.18		
	King Edward						

Table 7.8 The effect of within-row spacing on tuber yield (adapted from Wurr, 1971)

Author	*Between-row spacing and variety*	*Seed weight or size*	*Within-row spacing (cm)*	*Yield ($t\ ha^{-1}$)*		
				Total	> 38 mm	> 51 mm
Craibstone (1930)	68.6 cm	'Large'	30.5	46.74	45.75	30.28
			45.7	43.47	42.09	32.29
	Kerr's Pink		61.0	36.06	34.93	26.14
				Total	Ware	
Craibstone (1934)	Not stated	Mean of 3	30.5	25.13	7.41	
			45.7	24.00	10.18	
	Golden Wonder		61.0	22.74	11.94	
				Total	> 32 mm	> 51 mm
Craibstone (1937)	Not stated	32–57 mm	22.9	29.90	28.90	11.69
			30.5	28.40	27.64	11.69
	Golden Wonder		45.7	27.01	26.39	14.83
				Total	> 42 mm	> 57 mm
Bates (1935)	71.1 cm					
		Mean of 3	30.5	25.63	22.87	13.82
			38.1	26.39	23.87	15.83
			45.7	25.38	23.12	16.08
	King Edward		53.4	23.37	21.36	14.83
				S.E. ± 0.603	± 0.603	± 0.452

				Total	> 42 mm	> 57 mm
Findlay and	71.1 cm 1st year		30.5	28.22	23.37	8.37
Sykes	73.7 cm 2nd year	Mean of 3	38.1	27.89	23.80	9.73
(1938)			45.7	26.14	22.79	10.23
			53.4	25.56	22.54	10.63
	King Edward			1.231	1.181	0.905
				Total	> 57 mm	
Montague and	71.1 cm	Mean of 2	30.5	61.67	13.95	
Ivins			40.7	55.14	14.93	
(1955)			50.9	51.21	15.53	
			61.1	48.68	15.88	
	King Edward		71.3	46.72	16.13	
			81.5	45.06	16.36	
				Total	> 44 mm	
Bremner and	All 71.1 cm	Mean of 2	30.5	30.66	15.58	
Taha	King Edward		45.7	30.41	18.60	
(1966)	1959		61.0	27.14	18.09	
				S.E. ± 0.829	± 0.779	
	King Edward	Mean of 2	30.5	38.45	33.17	
	1960		45.7	37.19	33.42	
			61.0	34.93	32.42	
				S.E. ± 0.955	± 0.880	
	Majestic	Mean of 2	30.5	31.66	24.63	
	1959		45.7	31.16	26.39	
			61.0	27.89	25.63	
				S.E. ± 0.829	± 0.779	
	Majestic	Mean of 2	30.5	45.23	42.97	
	1960		45.7	43.22	41.97	
			61.0	41.46	40.21	
				S.E. ± 0.955	± 0.880	

Effect of within-row spacing

The effects of within-row spacing are summarized in Table 7.8. The effect of reducing spacing is very similar to the effect of increasing seed size and similar conclusions to those made for seed size apply.

In very few experiments has a significant interaction between seed size and spacing been reported, although with large riddle sizes or small fractions of yield the tendency towards interaction increased. Thus, the merits of seed rate as the plant density scale may he assessed without consideration of possible interaction effects.

The earliest experiments often did not compare equal seed rates obtained with different seed sizes but the comparisons may be extracted from the data. Table 7.9 summarizes the relevant comparisons and shows that there was little effect of seed size at equal seed rates in the Craibstone data, nor in those of Findlay and Sykes (1938) while Bates (1935) only showed a difference for yields >57 mm. However, Singh and Wakankar (1943) showed a significant yield advantage for small seed in both comparisons of total yield and one of yield >32 mm.

Table 7.9 Effect of seed size on tuber yields at equal seed rates: older data (after Wurr, 1971). N.B. Where there is more than one comparison in each grade, the two values to be compared are numbered similarly.

Author	*Seed weight (g)*	*Total yield ($t\ ha^{-1}$)*		*Ware yield ($t\ ha^{-1}$)*	
		Spacing (cm)		Spacing (cm)	
		30.5	61.0	30.5	61.0
Craibstone (1934)	42	23.12 (1)		10.81 (1)	
	84	25.63 (2)	23.12 (1)	6.79 (2)	12.82 (1)
	168		23.87 (2)		8.04 (2)
		Spacing (cm)		Spacing (cm)	
		30.5	45.7	30.5	45.7
Bates (1935)	28	24.63		22.37	
	42		24.88		23.12
		Spacing (cm)		Spacing (cm)	
		30.5	45.7	30.5	45.7
Findlay and Sykes (1938)	28	26.89		22.37	
	42		25.38		22.1
		Spacing (cm)		Spacing (cm)	
		15.3	30.5	15.3	30.5
Singh and Wakankar (1943)	15	15.3 (1)		12.6 (1)	
	30	15.4 (2)	12.4 (1)	12.1 (2)	11.1 (1)
	60		13.6 (2)		11.7 (2)

Table 7.10 Effect of seed size on tuber yields at equal seed rates: more recent data. N.B. Where there is more than one comparison in each grade, the two values to be compared are numbered similarly

Author	*Seed weight (g)*	*Variety*	*Total yield ($t\ ha^{-1}$)*				*Yield 44–83 mm ($t\ ha^{-1}$)*			
			Spacing (cm)				Spacing (cm)			
			66	44	33	22	66	44	33	22
Jarvis and Shotton (1971)	31.4	K. Edward	–	–	–	47.2 (1)	–	–	–	34.9 (1)
	62.8		–	46.9 (1)	47.8 (2)	–	–	34.4 (1)	35.1 (2)	–
	94.2		44.1 (1)	–	49.4 (3)	–	32.5 (1)	–	36.7 (3)	–
	125.6		44.0 (2)	47.5 (3)	–	–	30.2 (2)	35.2 (3)	–	–
			Spacing (cm)				Spacing (cm)			
			66	33	22	16.5	66	33	22	16.5
Jarvis (1971)	31.4	P. Crown	–	42.9 (1)	47.6 (3)	48.9 (2)	–	36.4 (1)	40.7 (3)	42.4 (2)
	62.8		42.9 (1)	51.8 (2)	53.8 (4)	–	33.5 (1)	45.0 (2)	46.8 (4)	–
	94.2		46.8 (3)	55.9 (4)	–	–	36.9 (3)	49.1 (4)	–	–
	31.4	P. Dell	–	32.6 (1)	37.4 (3)	39.4 (2)	–	28.6 (1)	32.5 (3)	32.6 (2)
	62.8		27.9 (1)	36.9 (2)	41.5 (4)	–	25.3 (1)	32.8 (2)	35.2 (4)	–
	94.2		32.7 (3)	39.7 (4)	–	–	28.6 (3)	34.5 (4)	–	–
			Spacing (cm)				Spacing (cm)			
			33	16.5			33	16.5		
Jarvis and Shotton (1968)	2.5 cm	Majestic	–	47.4 (1)			–	39.5 (1)		
	3.2 cm		43.3 (1)	49.3 (2)			35.1 (1)	39.4 (2)		
	3.8 cm		45.2 (2)	54.1 (3)			33.6 (2)	43.5 (3)		
	5.1 cm		50.7 (3)	–			38.5 (3)	–		

In 1954 Boyd and Lessells used data from many British experiments to examine the effect of seed rate on the yield of maincrop varieties. They showed that both total and ware yield increased over the commercial seed rate range, while yield > 51 mm reached a maximum and then declined. By constructing curves showing the effect of altering seed rate by variation in seed size or within-row spacing, they found that increasing tuber size had a smaller effect on total yield but a larger effect on 'percentage ware' than reducing within-row spacing. They concluded, however, that provided the optimum seed rate was attained, the precise combination of seed size and spacing was of minor importance in determining ware yield over 38 or 45 mm riddles. This conclusion led to the view that seed rate was a reliable and accurate plant density scale for potatoes. The findings of Boyd and Lessells have profoundly influenced the consideration of potato density effects in the U.K. for the acceptance of seed rate as the plant density scale has been difficult to alter.

Since 1954 a number of workers have reported a superiority of small seed over large seed at equal seed rates, particularly at seed rates outside commercial range. The results of these experiments are summarized in Table 7.10 and show for a range of varieties that small seed outyielded large seed at the same seed rate. Holliday (1960), El Saeed (1963) and Homes (1966) explained that these effects result from the higher stem densities established from small seed at equal seed rates for as Fig. 7.3 shows the relationship between number of stems and seed size is asymptotic. Thus, with the larger seed sizes, large increases in seed tuber size produce small increases in the number of stems. As a consequence of these results and the advocacy of Holliday that the stem was the true unit of plant density for potatoes, a detailed examination of stems has been made by several workers.

7.2.6 Stems

Although the interest in stems as the unit of plant density is of recent origin it is notable that Arthur (1892) reported that increases in 'stalks' (stems) per hill gave increases in total and graded yields and Bates (1935) also appreciated the importance of stems. Bleasdale (1965) produced a detailed examination of the mainstem as the unit of plant density. Using data from several sources he plotted yields against seed rate and number of mainstems. When plotted against seed rates seed sizes appeared to give independent curves but plotted against mainstems the curves became indistinguishable. Bleasdale concluded that total yield and the proportions in the size grades were a function, independent of seed size, of the number of mainstems per unit area (Fig. 7.5). Sharpe and Dent (1968) agreed with Bleasdale's findings and found that a square root function gave the best fit to their data for both total and graded yields. Subsequently, a number of workers including Gray (1972) using mainstems and Wurr

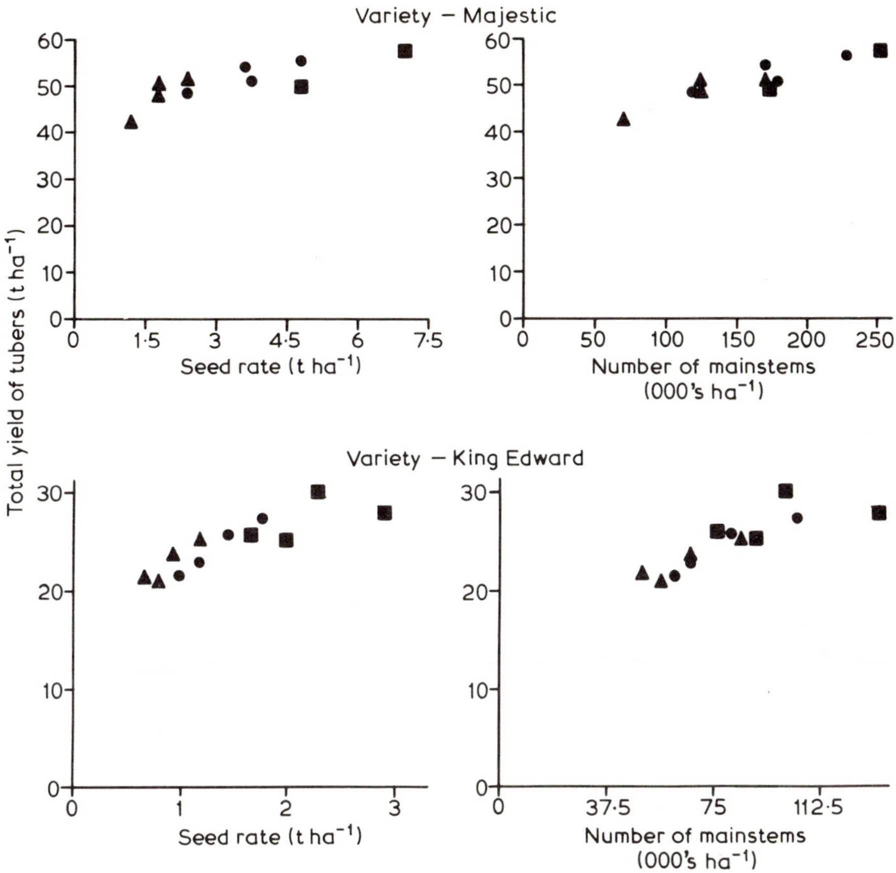

Fig. 7.5 Relationships between tuber yield, seed rate and stem density (Bleasdale, 1965). Key: ▲ Seed weight 32 g; ● Seed weight 64 g; ■ Seed weight 128 g.

(1974) using above-ground stems have found that stems are capable of relating yields by a single line for a range of seed sizes.

Good relationships between tuber yields and stems have not always been found, Goodwin *et al.* (1969) found no relationship between yield and mainstems, for at the same above-ground stem density, tubers with many mainstems and tubers with single well-branched mainstems gave similar yields. The authors considered there was confusion over the term mainstem and many other workers had included sub-surface branches in their counts. These results do not discredit the use of above-ground stems as the plant density scale. Holmes (1973) rejected stems as a useful plant density scale for in experiments where stem density was varied by sprouting regime such a scale proved quite inadequate to relate yields.

In these experiments the stem densities ranged from wholly mainstems (from short sprouting periods) to mainly secondary stems (from long sprouting periods) and in these circumstances it is likely that stems will not be capable of relating all treatments on a common scale. The performance of mainstems may be affected by the presence or absence of branches but more importantly the effects of long sprouting periods are to increase the physiological age of the seed tubers which results in earlier senescence and lower yields, compared with seed from shorter sprouting periods (Chapter 8). Thus, the performance of the stems is affected by the means of manipulating plant density. In practice crops would be grown from one seed type or the other, not a mixture of both, but it is of importance to know whether the response to increasing plant density is the same in two such contrasting types of seed. It must be emphasized that the evidence which most strongly supports the use of stems as the plant density scale is from experiments with maincrop varieties sprouted in such a way as to produce almost entirely mainstems. Even with maincrop varieties it has been found by Jarvis (1971) and Jarvis and Rogers-Lewis (1974) that the varieties, Pentland Crown and Record, show larger responses to increasing seed size than would be expected if stem density eliminated differences between seed sizes. The reasons for this are not clear and other varieties, King Edward, Pentland Dell and Pentland Ivory have been found to respond to seed rate in a way consistent with the use of stems as an overall unit of plant density. It was shown in Fig. 7.4 that maincrop varieties do produce secondary stems but little is known of the factors affecting their production and even less about the influence of such stems on plant density effects.

It is with early potato varieties that the greatest ranges of type and number of stems per seed tuber are produced, for ware crops of many varieties are produced from seed with few or single sprouts which commonly branch and late lifted ware and seed crops are produced from seed which usually has several sprouts which develop into mainstems. Unfortunately the effects of plant density in such varieties are sparsely reported in the literature. The use of higher seed rates than used for maincrop varieties is commonly noted (e.g. Cole, 1973) and the merits of large seed, usually in direct comparison with small seed, are indicated (e.g. Scott and Younger, 1972), although the data from which these comments derive are limited. Recently, Bean (1978) has studied the effects of plant density on the early variety, Home Guard in Pembrokeshire. He found that there was little difference between seed rate and above-ground stems as the plant density scale for apically dominant seed, and both scales were better than mainstems (Fig. 7.6). All seed sizes produced similar numbers of mainstems but the larger seed produced more branches and the increase was in proportion to the increase in

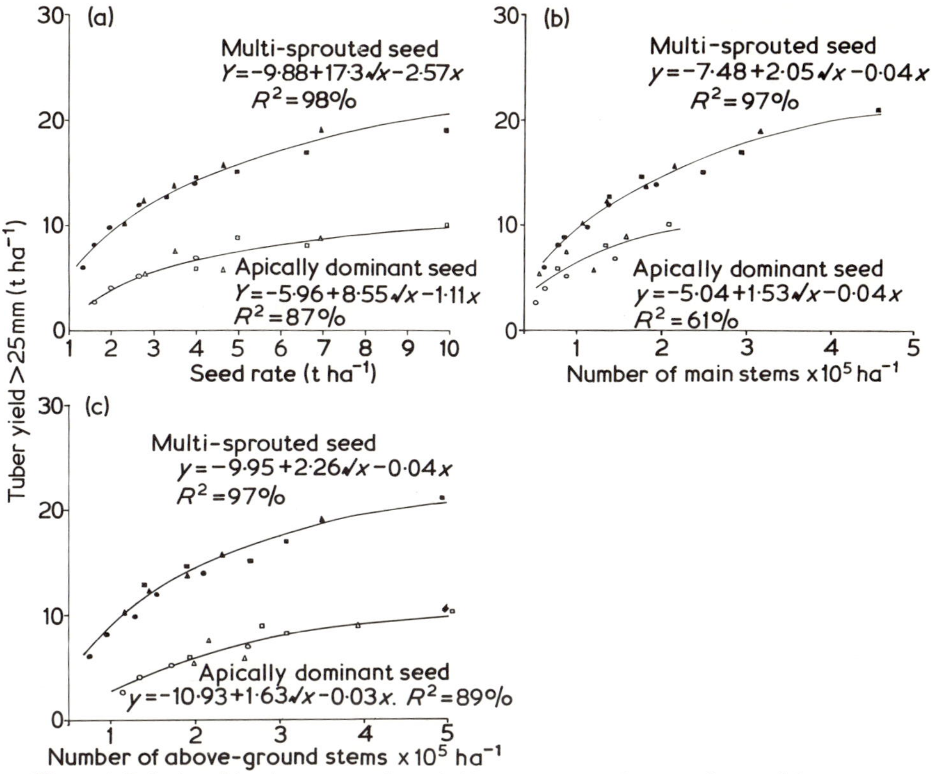

Fig. 7.6 Relationships between tuber yield > 25 mm and (a) seed rate, (b) mainstems and (c) above-ground stem densities for apically-dominant (open symbols) and multi-sprouted (closed symbols) seed of the variety, Home Guard (Bean, 1978). Key: o seed weight 40 g; Δ seed weight 70 g; □ seed weight 100 g

seed size. Consequently, the number of above-ground stems was similar at any seed rate for all seed sizes. Mainstems were a poor scale, as each seed tuber size had the same number of mainstems but widely different tuber yields. There was little evidence to support the use of large seed for early production. With seed from short sprouting periods Bean found that differences between seed rate, main stems and above-ground stems as plant density scales were very small at early harvests (June). Above-ground stem densities were again similar for all seed sizes at any seed rate although this seed produced few secondary stems. The increase in number of mainstems with increasing seed tuber size was in proportion, resulting principally from the fact that the smallest seed produced only one stem per tuber. As Fig. 7.6 shows, the relationship between total yield and any plant density scale was quite different for the two types of seed used. These data support the findings of Holmes (1973) that marked differences in the sprouting of seed may prevent tuber yields being related to stems by a single line but they also indicate that seed rate is equally inadequate in this respect. Where marked differences in sprouting occur, it is essential to restrict plant density responses to a specific type of seed. Above-ground stems and seed rate appear to be equally useful as plant density scales within a specific type of seed of Home Guard. Whether similar relationships would hold for other early and second early varieties is not known.

It is desirable to establish whether such effects of sprouting upon the response to plant density and on the most appropriate plant density scale exist in other varieties. Even if the response to plant density is not affected, large differences in numbers of stems per seed tuber may be expected to affect the combination of seed size and spacing (i.e. seed rate) required to achieve the optimum stem density. At present it may be concluded that a 'stem' scale is usually as effective as any other and in the case of the reported responses of maincrop varieties rather better.

Acceptance of the stem as the unit of plant density in the potato would allow for plant density effects to be assessed through the effective unit rather than through seed rate which has obvious limitations. The literature cited indicates that further work is required to establish the most appropriate stem parameter to be used with seed of widely different sprout characteristics. Such work should also help towards establishment of the factors which determine the number and type of stems which arise from seed tubers. Many factors such as locality, area of seed production and season which have hitherto not been considered as significantly affecting plant density merit study as to their influence on numbers of stems produced by seed tubers. The information obtained from such studies would allow the number of stems produced by seed tubers of known size and history to be predicted. Thus crops could be grown at the appropriate stem densities for specific outlets, thereby increasing the precision with which plant density is manipulated. The required within-row spacing (and hence seed rate) would be determined by the predicted capacity of the seed tubers to produce units of plant density and would vary as that productive capacity changed. At present there are no reliable means of making such alterations to seed rate although Jarvis and Shotton (1971) suggested such a correction based on the number of actively-growing sprouts per seed tuber for the variety King Edward, in which they had found large variation in the numbers of stems per seed tuber between different years. It is suggested that many factors alter the number of stems per seed tuber from crop to crop and season to season and contribute substantially to unexplained variation in the performance of particular seed rates. It follows from this that it is essential for all experimental work on density in the potato to record the numbers and types of stems present.

The adoption of a stem scale allows for easy interpretation of the effects of seed size, within-row spacing and seed rate via their effects

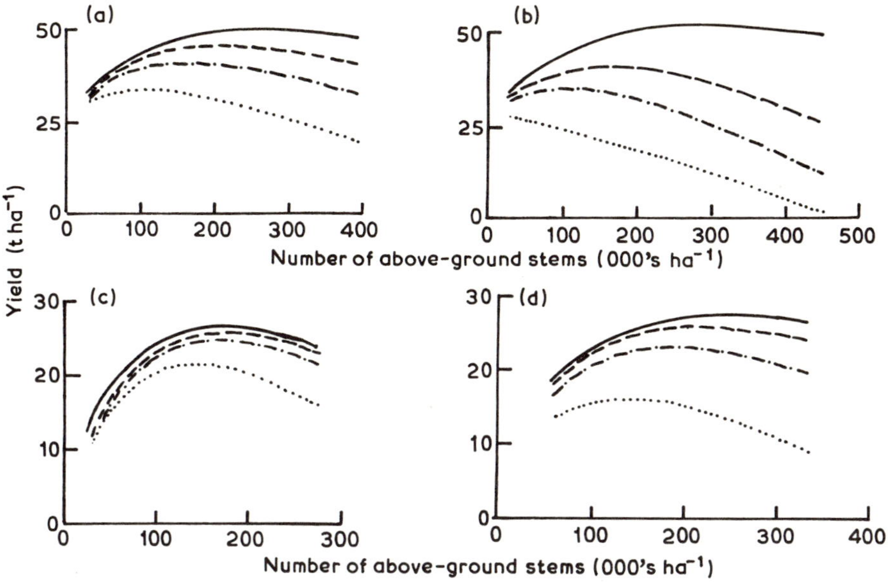

Fig. 7.7 The relationship between stem density and tuber yields for (a) Pentland Crown, 1969; (b) Maris Piper, 1969; (c) Pentland Crown, 1970; (d) Maris Piper, 1970. Key ——— total yield; – – – – – – yield > 38 mm; –·–·– yield > 44 mm; .... yield > 51 mm (Wurr, 1974).

on stem density. Fig. 7.7 shows the relationship between total and ware tuber yields and the above-ground stem densities for two maincrop varieties in two seasons (Wurr, 1974). The relationship between the total yield curve and the ware yield curves and the point of maximum ware yield varies according to variety and season. Thus, establishment of more than the general principles of plant density-yield relationships in this crop must be specific to variety and probably location. This is discussed in detail in Section 7.4.

Taking the curves shown in Fig. 7.7 the effects of seed size and within-row spacing may be explained. Increasing seed size or decreasing within-row spacing or both (and therefore increasing seed rate) result in increased numbers of stems, with reduced spacing having the greater effect. The optimum stem density for ware yield is lower than for total yield, so that increases in number of stems are more likely to exceed the optimum stem density for ware yield and reductions in ware yield from large seed sizes and close within-row spacing would be expected. Fig. 7.8 presents model responses of tuber yields to changes in seed sizes and within-row spacings. At equal within-row spacings large seed will produce a greater total yield than small seed until the spacing is reached at which the small seed achieves the optimum stem density. This within-row spacing will be much narrower than that at which large seed achieved the same

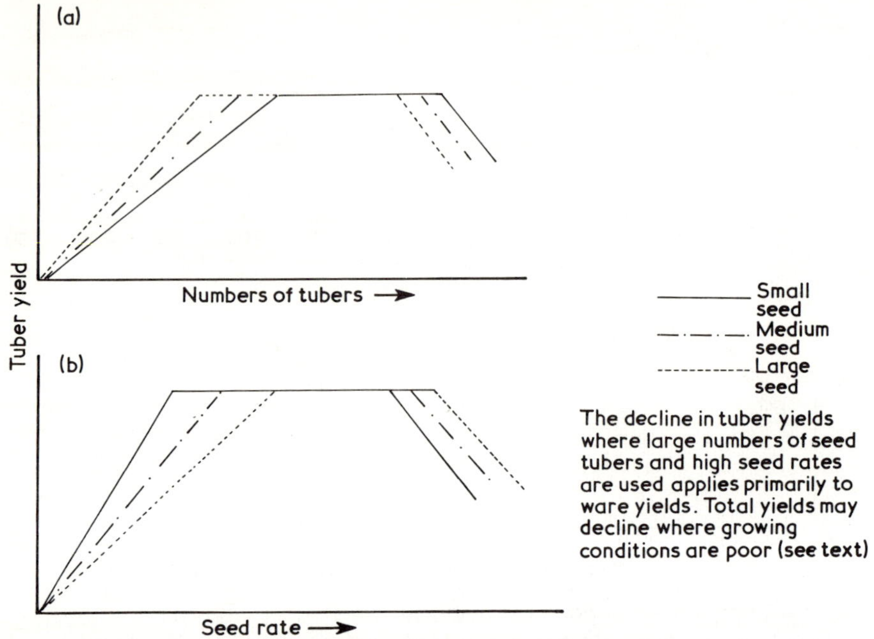

Fig. 7.8 Schematic representation of effects of seed size on relationship between tuber yield and (a) number of seed tubers planted; (b) seed rate. (Not to scale.)

stem density. At closer within-row spacings there will be no differences between seed sizes. A similar situation exists for ware yields but the actual values change. In addition, ware yields decline at close within-row spacings and large seed will decline at a wider within-row spacing than small seed. The relationship between large and small seed is therefore dependent upon the range of within-row spacings (i.e. stem densities) considered. If seed sizes are compared at equal seed rates, their relative positions are reversed, for the small seed produces more stems per unit seed weight as already described. The relationship which is observed between treatments is determined by their stem densities and the relative positions of treatments may change if they are compared at widely different parts of the yield–stem density response curve. The extensive and often confusing literature on plant density yield relationships in the potato crop is made easier to interpret when it is appreciated that the stem is the basic unit of plant density. It is no longer necessary to make the comparisons detailed above, for treatments should be compared through their stem densities.

In conclusion it may be argued that a plant density scale involving stems has been shown to relate tuber yields better than any alternative. Differences between treatments and the changes in their relationships can be explained via their stem densities and the yield–density response curve. It is intended simply to refer to stems from hereon in this review. It must be reiterated, however, that stem densities cannot be used in a predictive or advisory context without the detailed knowledge of seed tuber development already found lacking in Section 7.1. This deficiency in knowledge is capable of remedy and the continued statement of its existence ought to stimulate its solution rather than inhibit the use of the stem as the unit of plant density.

7.3 Spatial arrangement

The discussion of plant density so far has only considered the number of plants or stems per unit area, but the way in which this plant density is arranged may affect its performance. In potatoes the spatial arrangement of the density of stems may be affected by changes in row width, irregularity of within-row tuber spacing, failure of plants to emerge and by the clumping of stems consequent upon the use of seed of different sizes.

7.3.1 Row width

The traditional row widths used in the U.K. are from 61–76 cm, with the early crop being produced on the narrowest rows. However, wider rows are required to aid the mechanical culture and harvesting of the main crop (see Chapter 9) and row widths up to 90 cm are now in commercial use. At similar plant densities, changes in row widths within the range 60–90 cm have not been found to affect markedly tuber yields (Pätzold, 1964; Ziegler, 1968; North and Proctor 1971; Jarvis and Shotton, 1972) for any slight reduction in total yield has usually been compensated by improved saleability through reduced greening and cracking. Studies on quite marked deviations from square plant arrangement by Saunt (1960) and Eckersall (1965) have also shown that the potato is remarkably tolerant of changes in its plant arrangement. Attempts to grow the crop in beds (Nelson, 1967; Jarvis and Shotton, 1968) with narrow row widths have not shown any advantage and have revealed serious obstacles in this type of spatial arrangement with respect to soil cover and tuber growth.

Although the evidence is not extensive, the consistency of the results of row width experiments would suggest that for maincrops the choice of row width within the range considered is not an important factor in determining yields. There is no worthwhile information on the effects of spatial arrangement on early potatoes; as these crops are of much shorter duration and are composed of much smaller plants it is suggested that the narrower the row width the better. Information on this question is required for it is intimately connected with attempts to mechanize all aspects of early potato production.

Table 7.11 Effect of row width on tuber yields (Allen, 1972). (Data mean of three varieties)

	Seed Rate (t ha⁻¹)	*Total yield (t ha⁻¹)*		*Yield > 38 mm (t ha⁻¹)*		*Yield >51 mm (t ha⁻¹)*	
		Row widths (cm)					
		68.6	137.2	68.6	137.2	68.6	137.2
	4.06	50.1	40.9	40.0	34.3	20.0	19.1
1969	8.11	51.9	45.8	38.7	35.2	17.0	15.8
1970	4.06	40.3	34.9	37.3	32.0	28.9	24.6
	8.11	39.9	33.7	34.8	29.2	21.6	19.0

The cultural advantages which accrue from widening the rows may not be maximized at the 90 cm row width and it is relevant to ask whether even wider rows could be used. Studies on the response of potatoes to row widths beyond this level are limited. Allen (1972) compared 69 with 138 cm rows and found that although the use of the wider row width significantly reduced yields, the reduction was smaller than expected (Table 7.11). Effects were smallest on the yield of the larger sized tubers. It was suggested that higher plant densities may be justified for wider rows. Ifenkwe (1975) continued this work in the wetter environment of Pembrokeshire, comparing 66 and 132 cm row widths at a range of plant densities for two maincrop varities. The wider rows reduced numbers of stems and tubers especially at high stem densities but there were no interactions between stem density and row width in the tuber yields obtained. The wider rows reduced tuber yields early in the season but this difference became smaller as the season progressed so that differences in total and ware yields at the end of the experiments were small. Greening and growth disorders of tubers were always reduced by wider rows so that saleable yields were unaffected by the change in row width (Fig. 7.9). At the end of the experiments tuber yield could be satisfactorily related to stem density by a single line independent of row width (Fig. 7.10).

This evidence from an environment not normally associated with maincrop potatoes would suggest that wider rows than 90 cm should be seriously examined, especially where irrigation is available, for in Ifenkwe's experiments water was only seriously limiting once in three years. The adoption of even wider rows than currently being used may require changes in many other aspects of husbandry, especially in the construction of the ridges for which conventional ridging bodies would be inadequate; a discussion of which is outside the scope of this review. It is also important to examine more closely the physiological responses which allow this

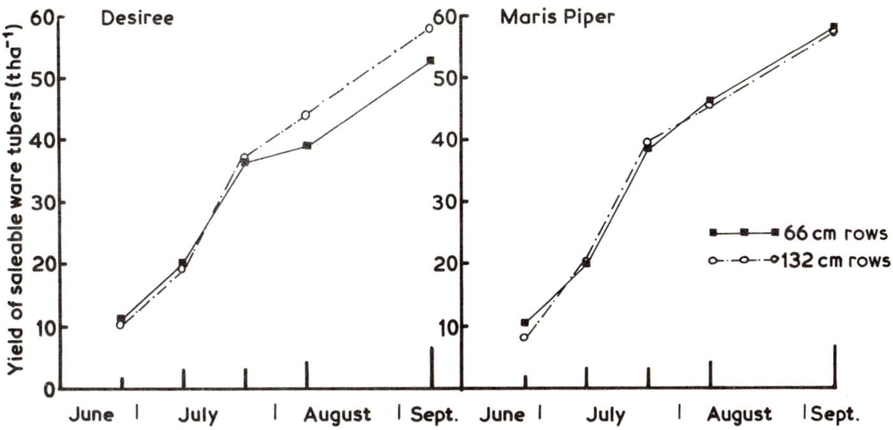

Fig. 7.9 Effect of row width on yield of saleable ware tubers (>38 mm less greened and cracked tubers) in two maincrop potato varieties. Data averaged over five planting densities (Ifenkwe, 1975). Vertical lines are LSDs.

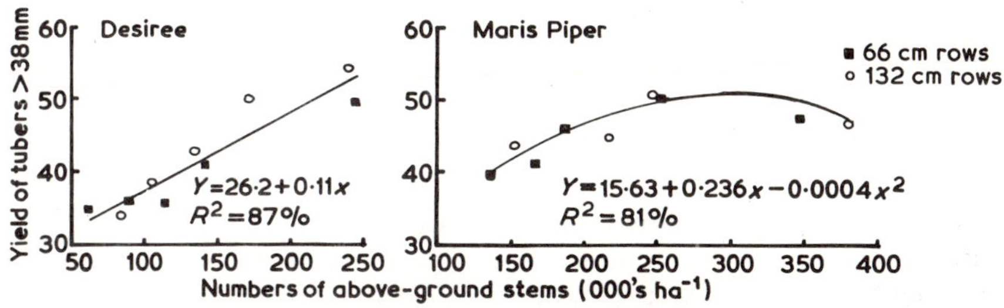

Fig. 7.10 Relationships between tuber yield and numbers of above-ground stems for combination of five planting densities and two row width of two maincrop potato varieties, Desiree and Maris Piper (Ifenkwe, 1975).

crop to tolerate such extreme spatial arrangements. Although early crop growth was markedly affected by the wider rows, this effect was compensated for by subsequent growth and development. It is possible that the restricted early growth conserved water whose later utilization was aided by changes in rooting depth stimulated by the more intense within-row spacing (for similar effect in sugar beet see Harris, 1972 and potatoes, Scott and Younger, 1972) but as no direct measurements were taken, this is speculation. What was clear in these experiments was that wider rows produced taller plants which did not lodge and which were probably much more effective in light interception and utilization later in the season than the lodged, narrow row plots. These effects deserve further close study.

Table 7.12 Effect of Irregular set spacing on total and graded tuber yields of two varieties. (Jarvis *et al.*, 1976)

Variety and grade	*Sets (ha⁻¹)*	*Coefficient of variation of set spacing*			
		0	*20*	*40*	*60*
King Edward Total yield (t ha^{-1})	32 291	43.9	44.6	45.0	42.2
	43 055	46.2	45.5	45.2	45.8
	64 582	47.8	47.1	46.3	45.9
	S.E.	± 0.86			
Saleable ware 44–83 mm (t ha^{-1})	32 291	31.4	31.3	30.7	28.7
	43 055	33.2	31.6	32.4	32.1
	64 582	33.2	32.6	31.8	31.5
	S.E.	± 0.96			
Pentland Crown Total yield (t ha^{-1})	32 291	48.5	47.4	47.1	45.8
	43 055	51.1	50.4	50.6	49.6
	64 582	55.0	53.6	52.7	52.7
	S.E.	± 0.81			
Saleable ware 44–83 mm (t ha^{-1})	32 291	36.7	35.8	35.1	34.2
	43 055	41.4	40.9	40.4	39.0
	64 582	45.1	43.7	42.7	42.9
	S.E.	± 1.48			

7.3.2 Irregularity of within-row spacing

While experimenters plant their seed tubers uniformly down the row, the adoption in commercial practice of mechanical planters has resulted in increased irregularity of spacing in farm crops. Planters of the belt-fed type are prone to extremes of irregularity unless very carefully set and operated. The general effects of this irregularity are that tuber yields are relatively unaffected by moderate irregularity but may sometimes be reduced by extreme irregularity (Table 7.12).

7.3.3 Missing plants

The failure of planted seed tubers to emerge because of disease and/or damage will affect the spatial arrangement of the remaining plants. It has been shown by Reestman (1970) that at high plant densities, missing plants have relatively little effect, a not unexpected result in view of the flat-topped nature of the yield-stem density relationship. As the proportion of missing plants increases the compensation by the remaining plants eventually becomes insufficient and yield declines. At very low plant densities there may be no compensation and missing plants will directly reduce yield.

The incidence of virus diseases and fungal diseases such as gangrene

(*Phoma exigua*) and black scurf (*Rhizoctonia solani*) affects the vigour of stems considerably and create very heterogeneous stem densities in respect of leaf production, number of tubers and longevity. The influence of these diseases upon responses to plant density is worthy of greater attention.

7.3.4 Clumping

The number of stems per seed tuber increases with increase in seed size, so that similar stem densities have quite different spatial arrangements when produced from small or large seed tubers. The existence of a relationship between yield and stem density through a single line, suggests that the effect of this stem clumping is unimportant. Nonetheless it is essential to consider whether this situation, implicit in the adoption of stems as the density scale, is justified. Bleasdale and Thompson (1965, 1966) planted one, two or three seed tubers per plant position of the variety, Majestic, at similar stem densities for two years. They found that stem clumping had no effect in one year but in the other, yields from the single seed were superior to those from the clumped seed (Table 7.13). However, in view of the good relationship between tuber yield and stems, Bleasdale has concluded that for practical purposes clumping has only a small effect. Reestman and de Wit (1959) considered that large seed were inferior to small planted at the same skin surface area because of clumping effects [although Bleasdale (1965) considered that this was caused by errors in calculating the surface areas]. In Sweden, Svensson (1966, 1972) has examined the effect of stem arrangement on yields using single sprout pieces to build up differing stem arrangements. Although he has shown that yields may be reduced by clumping, the growing period has usually been short, not more than 12 weeks and it seems likely that the greatest effects of clumping will be early in the season with compensation later. Clumping may affect the numbers of tubers set and consequently have a much greater effect on yields of specific grades. Data on this point are lacking. It should be appreciated that any marked effects of clumping in this respect would favour the use of smaller seed.

Table 7.13 Effect of seed clumping on total tuber yields (t ha^{-1}) (Bleasdale and Thompson, 1965, 1966)

	Number of seed tubers per position			
	1	*2*	*3*	*L.S.D.*
1964	42.3	45.2	41.3	4.20
1965	47.4	42.5	40.0	4.42

7.4 Effects of plant density

Following the discussion in Section 7.2 it is intended to consider stems as the unit of plant density and to discuss the effects of stems on growth, development and yield of the crop. It is accepted that these stem densities may be achieved in different ways and no mention of this is made unless essential to understanding.

7.4.1 Number of tubers

The number of tubers set by plants is determined by stem density, spatial arrangement, variety and season. Any increase in the stem density over the commercial range results in a reduction in number of tubers set per stem (Fig. 7.11). If density is increased by planting larger seed tubers then number of tubers per plant will increase despite the reduction in number of tubers per stem (Fig. 7.12). Where stem density is increased by planting more seed tubers, number of tubers per plant will decrease as number of tubers per stem decreases and number of stems per seed tuber remains unaffected (Fig. 7.13). Increasing stem density over a wide range for all varieties and the whole explored range in some varieties, results in increased number of tubers per unit area (Fig. 7.14) (Allen, 1972; Gray, 1972). As tuber yield is relatively constant over much of this stem density range the increase in number of tubers allows for manipulation of tuber size. Those varieties such as Pentland Crown (Wurr, 1971) which do not increase in number of tubers beyond a certain stem density are restricted in their control of size.

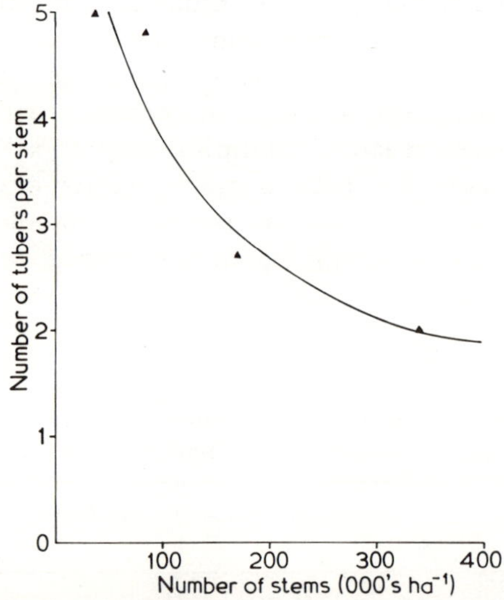

Fig. 7.11 Relationship between number of tubers per stem and number of stems per unit area (after Scott and Younger, 1972)

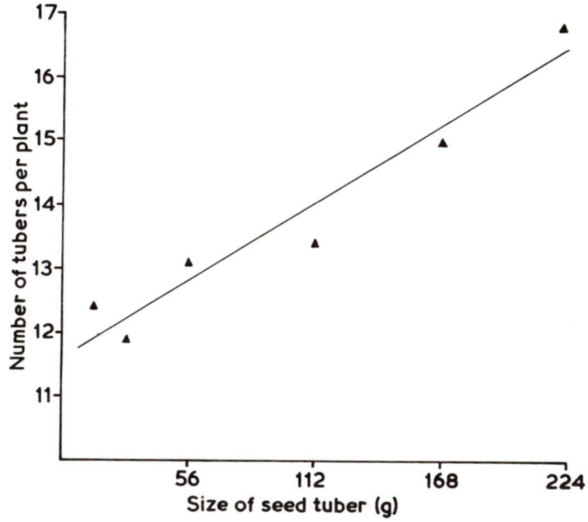

Fig. 7.12 Relationship between number of tubers per plant and size of seed tuber (after Scott and Younger, 1972)

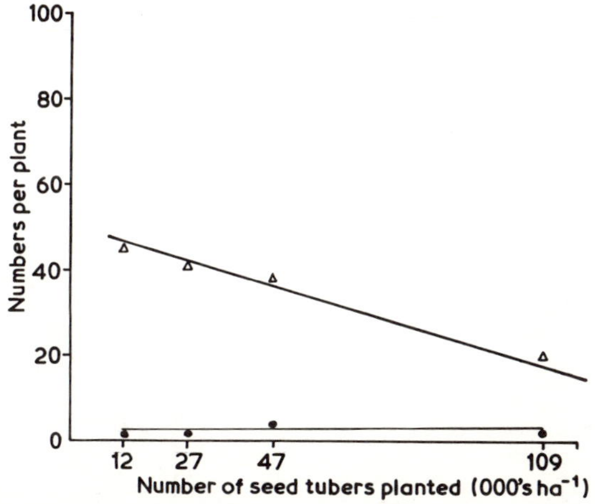

Fig. 7.13 Relationship between number of tubers per plant (Δ), number of stems per plant (●) and number of seed tubers planted (after Scott and Younger, 1972)

Spatial arrangement affects number of tubers in a similar manner to density, for increasing rectangularity (which reduces within-row spacing) reduces number of tubers set per stem. The effects of stem density and spatial arrangement on tuber numbers are well understood but the number of tubers set is also markedly affected by variety and season. Table 7.14 shows the number of tubers from several varieties

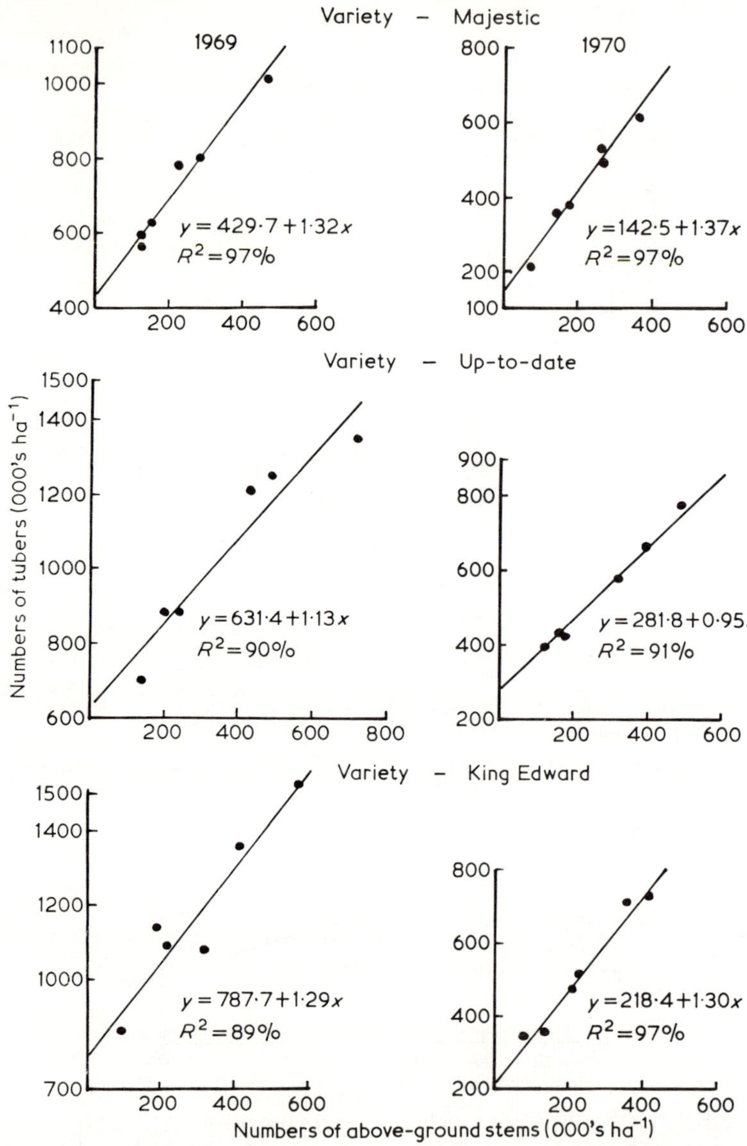

Fig. 7.14 Relationship between number of tubers per unit area and number of stems per unit area (Allen, 1972)

planted at similar stem densities in different seasons. Although the varieties show similar relative values in all seasons, the absolute values vary enormously, principally according to the extent of water supply at tuber initiation. This large variation in number of tubers from the same planting material complicates the effects of density and unless irriga-

Table 7.14 Effect of season on numbers of above-ground stems and tubers produced from similar seed rates of different varieties

Author	*Variety*	*Location*	*Season*	*Number of above-ground stems (000's ha⁻¹)*	*Number of tubers (000's ha⁻¹)*
Allen (1972)	King Edward	Cambridge	1969	299.0	1173
			1970	233.5	522
	Up-to-Date		1969	365.5	1045
			1970	275.5	544
	Majestic		1969	228.5	732
			1970	209.5	430
Jarvis and Shotton (1971)	King Edward	Terrington	1965	88.7	
			1966	148.3	
			1967	103.3	
Bremner and Taha (1966)	King Edward	Nottingham	1959	123.4	1542.5
			1960	243.5	971.6
	Majestic		1959	111.4	1136.3
			1960	141.9	691.8
Bean (1978)	Arran Comet	Trefloyne Pembs.	1975	146.7	552.9
			1976	282.7	1007.5
	King Edward		1975	146.7	1279.3
			1976	254.6	1332.4
Gray (1972)	Maris Peer	Wellesbourne	1969	590.0	2660.0
			1970	550.0	1490.0
Ifenkwe (1975)	Maris Piper	Trefloyne Pembs.	1972	230	821
			1973	235	780
	Desiree		1972	182	580
			1973	148	520

tion is available is largely uncontrollable by the grower. Doubling the number of tubers in one year compared with another with little change in number of stems is effectively doubling the stem density with consequent effects upon yields and grading. Thus, the numbers of tubers set is a modifying influence on the effects of stem density.

As increases in number of tubers continue beyond the stem densities giving maximum ware and total yield, it is possible greatly to reduce average tuber size by using stem densities in excess of those required for maximum ware yields. Crops for canning and seed which require tubers of small size with precise upper and lower riddle sizes are grown at stem densities above those for ware production. In varieties like Pentland Crown, there is limited scope for controlling tuber size, and small seed is uncommon. The control of tuber size is not very precise, for, apart from seasonal fluctuations in numbers of tubers, it seems likely that all tubers are not equally affected by increasing competition and little is

Table 7.15 Effects of very high seed rates on yields of large-sized tubers

Author	*Seed rates (t ha^{-1})*	*Yield (t ha^{-1}) > 51 mm*	*Number of tubers > 51 mm (000's ha^{-1})*
Allen, 1972	4.06	20.1	117
	8.11	18.1	115
Data mean of three varieties	16.22	13.0	85

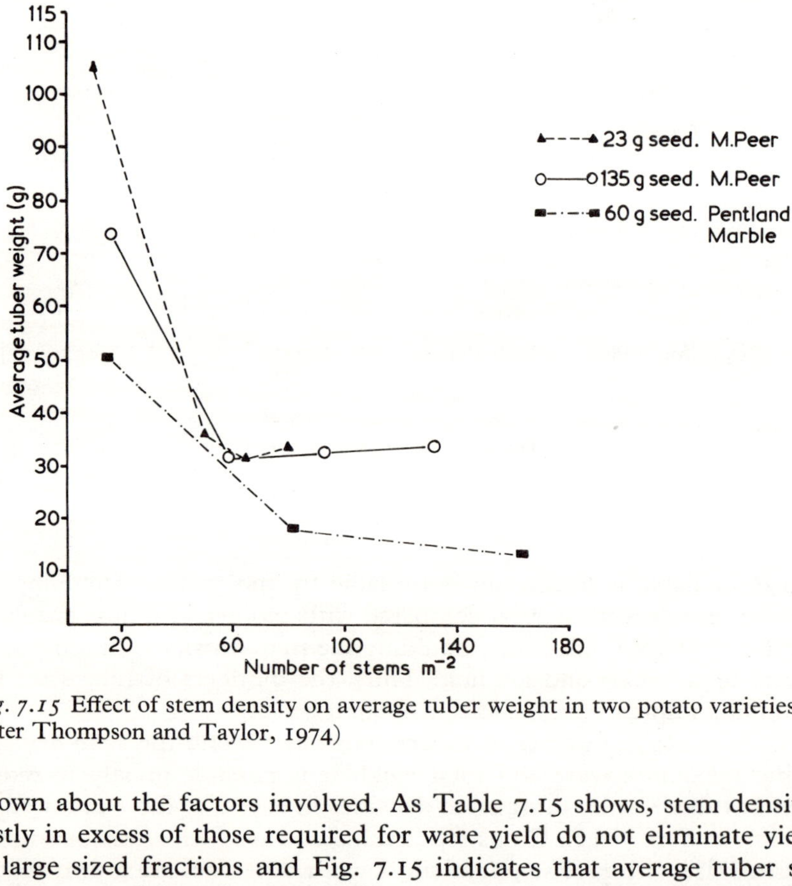

Fig. 7.15 Effect of stem density on average tuber weight in two potato varieties (after Thompson and Taylor, 1974)

known about the factors involved. As Table 7.15 shows, stem densities vastly in excess of those required for ware yield do not eliminate yields in large sized fractions and Fig. 7.15 indicates that average tuber size declines at a decreasing rate. This suggests that tubers differ in their response to competition and some may be preferentially positioned in relation to supply (Gray, 1973; Gray and Smith, 1973) so that they remain largely unaffected by competition while others are 'labile' and with increased competition may suffer and perish. These differences in tuber position may be reflected in different tuber growth rates and patterns of growth. Wurr (1973b) has reported that individual tubers do differ in their growth rates. Elucidation of the causal factors and the extent of their effects would greatly facilitate the manipulation of tuber size through number of tubers, for it may enable the range of tuber sizes per stem (plant and crop) to be reduced.

There remains a possibility that the seed sizes used to achieve required stem densities may affect the tuber size distribution and thereby the yields of finely-graded fractions of the total yield. The more regular spatial arrangements of stems resulting from the use of small seed may be reflected in slightly enhanced tuber set and more regular tuber growth which does not influence total or ware yields but affects the yield of the finer grades. Bean (1978) found that small seed increased number of tubers set compared with larger seed at similar stem densities and this affected the grading of the yield from similar total yields. The significance of this effect if substantiated, would depend upon the financial aspects of producing for outlets with apparently precise size limits.

7.4.2 Total yields

Fig. 7.7 shows the relationship between total yield and stem density for harvests at or approaching crop maturity. Total yield increases to a maximum with increasing stem density and then either remains unchanged with further increase in density or eventually begins to decline. The relationship appears to be 'flat-topped' in most reported cases (e.g. Shorpe and Dent, 1968) and any decline in yield at high stem densities seems to occur when growing conditions are poor, especially where water is lacking (Wurr, 1971; Ifenkwe, 1975). The absence of a marked point of inversion in the curve is important, for small changes about the optimum, which may result from the practical methods of establishing crops, are unlikely to affect yields markedly.

It is not intended to review the functions which have been used to relate tuber yield to plant density in the potato, but some comment is necessary. The fitting of quadratic and especially square root functions (Sharpe and Dent, 1968; Wurr, 1974) has been found satisfactory and optimum stem densities have been derived, but such functions have no biological significance. Other functions have been suggested, notably those of Bleasdale and Nelder (1960), but no definitive evidence of their biological validity has been produced. Such a function, which accounted for the components of plant density, would be beneficial in predicting effects of plant density in situations in which experiments had not been undertaken and determining the required experiments.

The general relationship between stem density and total yield would appear to hold for all varietal types for seed produced under specific sprouting regimes. Whether marked changes in sprouting and seed

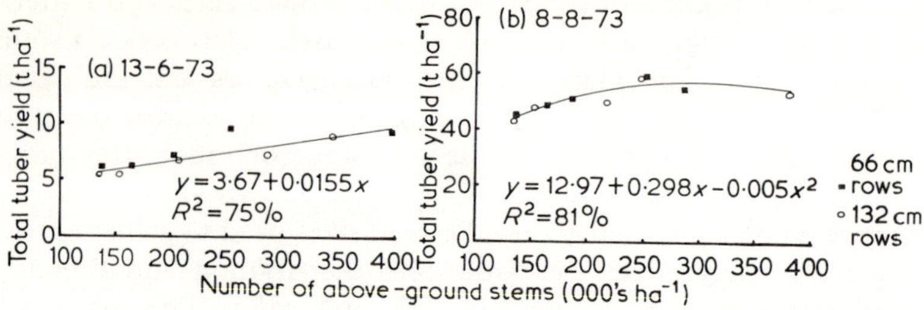

Fig. 7.16 Relationship between total tuber yield of the variety Maris Piper (a) at an early harvest, 13 June and (b) a later harvest, 8 August (Ifenkwe, 1975)

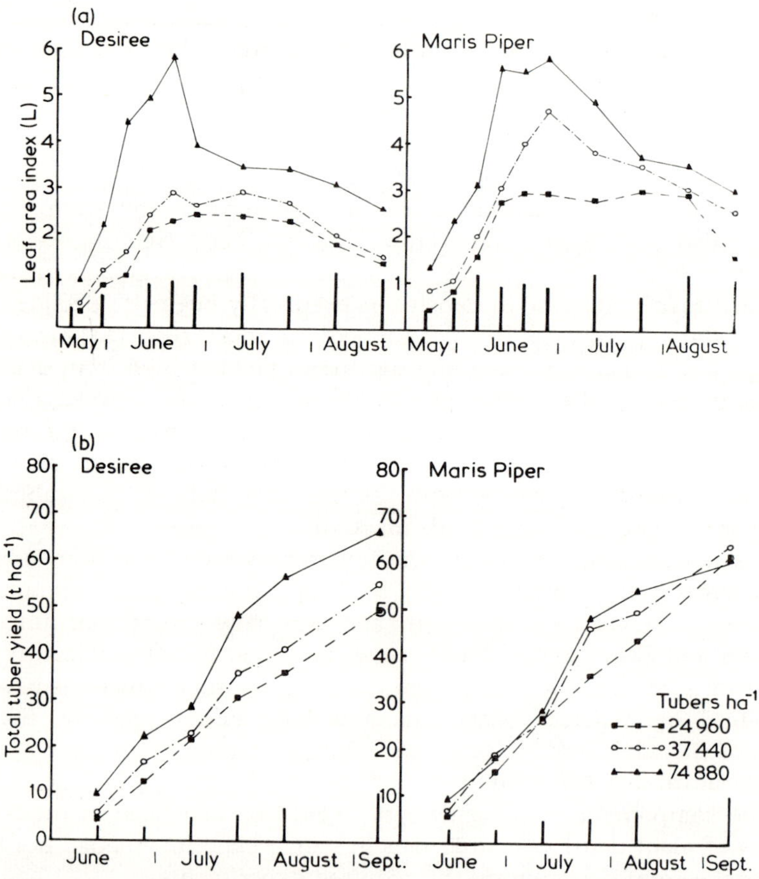

Fig. 7.17 Effect of planting density on (a) leaf area index and (b) total tuber yield of two maincrop potato varieties. Key: ▲ 74 880 seed tubers ha^{-1}; o 37 440 seed tubers ha^{-1}; ■ 24 960 seed tubers ha^{-1}. Vertical bars are LSDs and data are averaged over two row widths (Ifenkwe, 1975)

preparation affect the optimum stem density or only the seed rate required to achieve the single optimum needs to be established. Variety markedly affects the optimum stem density and although the relative positions of varieties may remain the same from year to year, the absolute values change (Wurr, 1974). These differences in optimum stem densities between varieties are of great importance and are amplified in the following section on ware yields.

Within any variety, the length of the growing season may affect the optimum stem density, for the longer the season the lower is the stem density required for maximum total yield (Fig. 7.16). Thus for the range of stem densities explored in most experiments, total yield is commonly linearly related to stem density early in tuber growth and the relationship shows a more pronounced optimum stem density as lifting is delayed. The highest stem densities increase leaf area early in the season and thereby light interception and improve early tuber growth but this advantage may subsequently be counterbalanced by increased leaf senescence, especially if water becomes limiting, or by mutual shading. The changes in leaf area and tuber yields for different stem densities of two maincrop varieties are shown in Fig. 7.17 (Ifenkwe, 1975).

7.4.3 Graded yields

The relationships between graded tuber yields and stem density (Fig. 7.7) show more pronounced optima than the curve for total yield. The larger the size grade considered, the more pronounced is the optimum and the lower is the optimum stem density (Wurr, 1974). The relationship between the stem densities for optimum total and graded yields depends upon the proportion of the total yield which falls into the grade concerned. This is determined by the number of tubers per stem; where this is low and inter-tuber competition limited, most of the yield will be in the larger sizes and differences in optimum stem densities for ware and total yields will be small. Where the number of tubers per stem is large, inter-tuber competition will reduce average tuber size and the difference in optimum stem densities will be large. Thus varieties which set many tubers per stem require very much lower stem densities, for graded yields than for total yields. Such varieties, e.g. King Edward, Maris Piper are inappropriate for the production of the largest size grades.

The large differences between varieties in their response to stem density must be appreciated if appropriate varieties are to be selected for production for specific outlets. Further, the responses to increasing stem density explain why crops for differing outlets are grown at widely different stem densities. Table 7.16 illustrates the differences in stem density required for early, canning and maincrop ware production. For the first two outlets very small tubers are marketable and high stem densities are established; in the case of the canning crop the variety,

Table. 7.16 Optimum stem densities for different types of crop

Author	*Type of crop*	*Variety*	*Optimum stem population (000's ha^{-1})*	*Comments*
Bean (1978)	Early Ware	Home Guard	400	For lifts in early June
Wurr (1974)	Maincrop Ware	Maris Piper	286	Data of year 1969
		Pentland Crown	270	
	Large Ware >51 mm	Maris Piper	10	
		Pentland Crown	102	
Jarvis and Shotton (1971)	Maincrop Ware	King Edward	150	
Sharpe and Dent (1968)	Maincrop Ware	Desiree	263	
Gray (1972)	Canning	Maris Peer	700	

Maris Peer, is used which sets many tubers per stem so that very high numbers of tubers are produced and the resulting inter-tuber competition restricts tuber size. For seed crops, stem densities intermediate between those for canning and ware are required, for seed tubers are larger than canners but smaller than ware. If the objective is to produce large ware then very low stem densities especially of varieties with relatively few tubers per stem are required. It is essential to realize that optimal yields of tubers of specific sizes can only be achieved by appropriate stem densities which will be quite different from the densities required for normal ware yields.

The length of growing season affects the optimum stem density for graded yields considerably. For early harvests of any size grade, lower densities are required which result in limited inter-tuber competition and rapid attainment by tubers of the required size. Yields are, however, low. For longer growing seasons larger and larger stem densities are justified and higher yields will result. This situation has been clearly demonstrated for canning potatoes by Gray (1972) from whose work Fig. 7.18 is taken. For production of graded tubers over a period of time, crops grown at low stem densities are harvested first and larger stem densities later. A similar situation exists in choosing varieties for very early production, for Ulster Prince which sets very few tubers per stem (sometimes only one) rapidly achieves a marketable sample of very small tonnage but does not bulk rapidly unless very high and uneconomic plant densities are established in comparison with the more favoured varieties. Consequently this variety is only used for the very earliest lifts in early areas. In earlies, the rapid fall in market price in

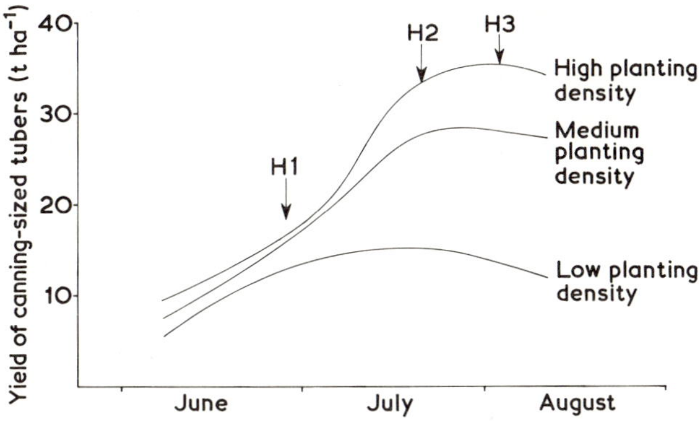

Fig. 7.18 Effect of planting density on yield of canning-sized tubers at three harvesting dates (H). H1 early; H2 mid-season and H3 a later season harvest (after Gray, 1972)

mid-season results in lower economic optimum stem densities at late lifts (Bean, 1978).

As area of production affects numbers of tubers and the extent of foliage growth, it would be expected that the required stem densities would differ in parts of the country. No evidence has been found, largely because there are no comparative data involving the wetter western part of the country, and it is usually argued that as the points of inversion in the yield–stem curves are not pronounced, any differences are unlikely to be important. This point requires clarification for differences in number of tubers may be as great between areas as between seasons and reference has already been made to the important effects of the latter on number of tubers.

7.4.4 Quality and saleability

As many of the outlets for potatoes demand tubers of specific sizes, the importance of stem density on this aspect of quality is great. It is also considerable in other aspects of quality, especially dry-matter content. High stem densities produce tubers with higher dry-matter contents than similar sized tubers from low densities (Wurr and Allen, 1974). Tuber size also affects tuber dry-matter content (Wurr and Allen, loc. cit.) and as stem density affects tuber size distribution its effect here is considerable and probably deserving of further examination. Detailed consideration of quality is given in Chapter 13.

Stem density affects the saleability of the crop by influencing the prevalence of tuber greening and growth disorders. Unfortunately there are few quantifiable separate reports of effects on these two factors as they are usually considered together. Ridges of satisfactory size and construction greatly reduce the incidence of tuber greening and this is one of the advantages of using wide rows. Large numbers of stems per plant, as from the use of large seed often increase greening, for the growth of large numbers of tubers in a restricted soil area results in tubers being forced through the upper surface of the ridge and exposed to light (Scott and Younger, 1972; Ifenkwe, 1975). Effects of stem density on tuber secondary growth are usually small and the incidence of such tubers is decreased by increasing stem density (Jarvis and Shotton, 1968; Scott and Younger, 1972). The prevalence of such disorders in the variety, Majestic, was as much a reflection of inadequate stem densities as of varietal propensity for secondary growth.

Thus, effects of stem density on saleability are composed of several opposing factors and a detailed examination of them has not been reported. Their overall effect is usually small and occasional reports to the contrary of the expected should not be viewed with concern, e.g. Gray (1972) who never had more than 6% unsaleable tubers, found that unsaleable tubers increased with increasing stem density. Of greater importance are effects of stem density and row width on the damage caused to tubers during the harvesting operation. Detailed data on this aspect are scarce and difficult to obtain but effort should be made to examine this question. The work of Jarvis (1972) has shown that the adoption of 90 as opposed to 76 cm rows allowed for faster rates of harvesting with slightly lower levels of tuber damage. The former point is important as it should allow more of the crop to be harvested in early autumn when conditions for harvesting are more favourable.

Stem density and row width also affect tuber size and as yet unpublished work by Jarvis has shown that smaller tubers are less liable to damage than larger tubers. Thus increasing stem density and widening row width exert opposing effects upon the number of tubers and therefore the tuber size distribution and possibly upon damage during harvesting. A more detailed study of these effects is required.

7.4.5 Plant morphology and growth

Increasing stem density results in taller plants with much reduced axillary branching (Ifenkwe, 1975). In the same experiment it was shown that the use of wider rows at the same stem density resulted in taller plants and restricted axillary branches. These gross changes in plant morphology have been known in general terms for some time (Taylor, 1953) but their implications in the study of stem density effects on crop growth have been neglected. These changes in density affect the standing ability of the crop and the tendency of crops grown at high stem densities to lodge deserves greater attention, for it affects the ability of the crop to intercept and utilize light. It probably represents a major factor restricting the response to increased stem density in those crops where foliage collapse occurs early in the season. No studies appear to have considered the types of stem and leaf arrangement which would fit the plant for growth at high stem densities. Discussion of desirable plant forms are conspicuously absent from the literature on this crop. It is remarkable that since Bremner and Taha (1966) reported their work on the growth and development of different stem densities, no further detailed developmental crop study of different stem densities has been published in the U.K. Much work on tuber growth has been carried out to which reference has already been made but information on the whole crop is sadly deficient. Comment has been made elsewhere (Chapter 4) as to the desirability of determining the factors affecting leaf, stem and tuber growth and their inter-relationships. As stem density affects the morphology of stems and leaves throughout the life of the crop, it should be used to examine the influence of these factors on growth and development. A better understanding of growth of this crop should benefit its production in the field. A beginning should be made by carefully recording the canopy structure of selected crops throughout their life, together with their light interception, growth and development.

7.4.6 Nutrient supply and uptake

A positive interaction between stem density and N supply was reported by van Burg (1967) and Bodlaender and Reestman (1968) but few other convincing examples of this have been reported. The optimum stem density was higher when fertilizer N was used than when none was applied but differences in required stem densities at different levels of applied N were small. N uptake has been shown to be an important factor in determining leaf longevity (Dyson and Watson, 1971) and as N concentration within the plant usually declines with increasing density (Scott and Younger, 1972), it might be expected that the larger canopies of high plant densities would justify more N, especially in long season crops. However, effects of stem density on root morphology and efficiency of nutrient uptake are largely unknown and a complete examination of the interaction between stem density and nutrition especially N, is awaited.

7.4.7 Economic and advisory considerations

As stated in the introduction, stem densities in the potato crop are manipulated in practice through the seed size and within row spacing (i.e. the seed rate) used. Thus, the optimum stem density has to be changed to a seed rate for growers to be advised of the desired stem density. In many cases at present no mention of stems is made, nor is necessary with present knowledge, for effective advice. For the silt land area of East Anglia the work of Jarvis and Shotton (1968, 1971), Jarvis (1971) and Jarvis and Rogers-Lewis (1974) has resulted in recommendations

Table 7.17 Seed rates for maximum gross margins for different seed sizes

Author	*Variety*	*Seed size (g)*	*Optimum seed rate (t ha^{-1})*	*Gross margin £ ha^{-1} at optimum seed rate*
			Seed at £30 t^{-1} Ware at £15 t^{-1}	
Jarvis (1971)	Pentland Dell	31.4	1.87	222
		62.8	3.12	200
		94.2	3.07	185
	Pentland Crown	31.4	2.09	314
		62.8	3.33	331
		94.2	4.15	324
			Seed at £25 t^{-1} Ware at £18 t^{-1}	
Jarvis and Shotton (1971)	King Edward	31.4	1.64	346
		62.8	2.55	311
		94.2	3.05	277
			Seed at £30 t^{-1} Ware at £15 t^{-1}	
Jarvis and Rogers-Lewis (1974)	Pentland Ivory	36.3	2.44	339
		63.5	3.44	324
		101.5	4.23	285
	Record	36.3	2.30	225
		63.5	2.98	230
		101.5	3.22	191

of seed rates for maximum gross margins for a number of varieties. These results are illustrated in Table 7.17. It can be seen that the optimum seed rate is dependent upon seed size, being lower for smaller seed. This would be expected from the relationships between seed size and number of stems discussed in Section 7.2 if there is one overall relationship between tuber yields and stem density. It also applies to those varieties, Pentland Crown and Record, in which there is evidence of differential seed size effects. As seed is purchased largely on a weight basis, irrespective of tuber size, and other costs are little affected by seed size, it follows that the largest financial returns will usually accrue from the use of smaller seed, for the achievement of the optimum stem density will produce similar yields. Table 7.18 illustrates this effect and shows that in Pentland Crown and Record the advantage of small seed is not apparent unless seed costs become very high in relation to ware prices. With all varieties any increases in seed costs relative to ware prices improve the relative profitability of small seed. There is very little seed as small as 31 g available of Pentland Crown so that commercially there are greater returns from the smallest readily available seed than from larger seed. Were a differential price system for seed potatoes to be introduced into this country it would be possible for growers to pay more

Table 7.18 Gross margins (£ ha^{-1}) at optimum seed rates with different seed prices

Author	*Variety*	*Seed size (g)*	*Seed price (£ t^{-1})* 20	30	40
				Ware at £15 t^{-1}	
Jarvis (1971)	Pentland Crown	31.4	336	314	294
		62.8	366	331	299
		94.2	368	324	284
	Pentland Dell	31.4	242	222	205
		62.8	235	200	170
		94.2	222	185	161
				Ware at £18 t^{-1}	
Jarvis and Shotton (1971)	King Edward	31.4	356	336	316
		62.8	321	294	269
		94.2	299	259	215
				Ware at £15 t^{-1}	
Jarvis and Rogers-Lewis (1975)	Record	36.3	255	225	201
		63.5	262	230	198
		101.5	235	191	156

for small seed than large seed and still produce the same gross margins.

These results ought to promote greater use of small seed and a greater awareness among seed producers of the value and necessity of producing small seed. The economic savings in the transport costs for the smaller total quantity of small seed were discussed by Wurr (1973a). The failure of farmers to adopt small seed in greater numbers is surprising even allowing for some deficiency in supply. A more rational use of seed will not be encouraged by the adoption of the suggestions of Evans (1975) that large seed is required for the largest tuber yields. The evidence for this advice is not known to the author and it is not a scientific basis for improvement to regard the production of high yields as justification of the correctness of the methods employed. Higher yields of tubers than are achieved in commercial practice have been a feature of potato research for the last 20 years, some yields being as high as any achieved by this 'blueprint', e.g. Gunasena and Harris (1971). This situation is not likely to change. Improvement in the economic production of ware potatoes in the U.K. would result from greater use of small seed in maincrop varieties. The results in Table 7.18 also illustrate clearly that optimum seed rates for economic returns must be specific to the seed size and the variety used. There is no general seed rate in this crop. Those varieties which produce many stems require lower seed rates for any seed size than varieties with few stems per seed tuber. There remains the important question as to whether these optimum seed rates are applicable for other areas and with different types of seed. It has been suggested that this may not be so and a study of this is required.

It must be appreciated that the use of small seed necessarily requires the use of very close within-row spacing which may be difficult to achieve with available planters. This must be accepted as a possibility but it would seem that many growers are not sufficiently aware of the value of small seed to increase their use of it and are not restricted by such considerations. It is most serious in early crop production where the extensive use of very small seed would be restricted by the narrow within-row spacings required with conventional row widths. For this reason there would be considerable benefit in a re-examination of the spatial arrangements in this crop, which may allow the small seed to be established satisfactorily.

The recommendations shown in Table 7.17 are based on the average results obtained by Jarvis and his co-workers. In any individual year the responses will vary according to the actual stem densities established and information should be sought which would help minimize this variation and improve the precision of control of stem density. In those varieties e.g. Maris Piper, Pentland Ivory and Pentland Dell in which responses to seed rates are consistent from year to year, the importance of variation in actual numbers of stems will be smaller than in varieties like King Edward in which marked changes in response occur. In such varieties, accurate prediction of numbers of stems is desired, as was proposed by Jarvis and Shotton (1971) through the use of numbers of actively growing sprouts. However, a more reliable and easily recorded sprout character than these authors used must be sought.

Seed size is also an important factor in the practicable mechanical ways of establishing modern crops. The use of mechanical planters necessitates close grading of seed for efficient operation and seed rate must be adjusted with each change in seed size and variety. If the planting is not mechanized, the ungraded 32–57 mm seed may be planted at the rate appropriate for the average tuber size. Such crops must necessarily be irregular in that small and large seed will be mixed along the row. If the planting operation damages sprouted tubers, differences in stem morphology, if not numbers of stems, may result (Jarvis *et al.*, 1973). The practical aspects of planting are dealt with in Chapter 9.

References

Allen, E.J. (1972). *J. agric. Sci., Camb.* **79**, 315–21.

Allen, E.J. (1978a). *J. agric. Sci., Camb.*

Allen, E.J. (1978b). *J. agric. Sci., Camb.* (in press).

Allen, E.J. and Wurr, D.C.E. (1973). *Potato Res.* **16**, 10–20.

Allen, E.J., Bean, J.N., Griffith, R.L. and O'Brien, P.J. (1978). *J. agric. Sci., Camb.* (in press).

Arthur, J.C. (1892). *Indiana Sta.* Bulletin **42**, 105–18.
Bates, G.H. (1935). *J. agric. Sci., Camb.* **25**, 297–313.
Bean, J.N. (1978). Density studies with early potato varieties. Ph.D. Thesis. University College of Wales Aberystwyth.
Birecki, M. and Rostropowicz, S. (1963). *Eur. Potato J.* **6**, 1–13.
Bleasdale, J.K.A. (1965). *J. agric. Sci., Camb.* **64**, 361–6.
Bleasdale, J.K.A. and Nelder, J.A. (1960). *Nature, Lond.* **188**, 342.
Bleasdale, J.K.A. and Thompson, R. (1965). *Rep. natn. Veg. Res. Station for 1964*, 37–8.
Bleasdale, J.K.A. and Thompson, R. (1966). *Rep. natn. Veg. Res. Station for 1965*, 39–40.
Bodlaender, K.B.A. and Reestman, A.J. (1968). *Neth. J. agric. Sci.* **16**, 165–76.
Boyd, D.A. and Lessells, W.J. (1954). *J. agric. Sci., Camb.* **44**, 465–76.
Brandreth, B. and Bryan, H. (1937). *J. natn. Inst. agric. Bot. for 1937.* **2**, 4.
Bremner, P.M. and Taha, M.A. (1966). *J. agric. Sci., Camb.* **66**, 241–52.
van Burg, P.F.J. (1967). *Neth. Natn. Tech.* Bulletin 4. Oct. 1967.
de Carvalho, T. (1975). Tech. Bull. 11. Part 1. 1–19. *Publ. Rhod. Ministry Agric.*
Cole, H. (1973). Early potatoes in West Cornwall 1962–1972. An economic survey, Report 19. Agric. Economics Unit University of Exeter.
Craibstone, N. Scot. Coll. Agric. (1929). *Scott. J. Agric.* **12**, 68–70.
Craibstone, N. Scot. Coll. Agric. (1930). *Scott. J. Agric.* **13**, 78–81.
Craibstone, N. Scot. Coll. Agric. (1932). *Scott. J. Agric.* **15**, 79–84.
Craibstone, N. Scot. Coll. Agric. (1933). *Scott. J. Agric.* **16**, 87–92.
Craibstone, N. Scot. Coll. Agric. (1934). *Scott. J. Agric.* **17**, 85–91.
Craibstone, N. Scot. Coll. Agric. (1937). *Scott. J. Agric.* **20**, 57–63.
Dyson, P.N. and Watson, D.J. (1971). *Ann. appl. Biol.* **69**, 47–63.
Eckersall, R.N. (1965). Implications of herbicide use on plant population, plant arrangement and cultivations in the potato crop. Ph.D. Thesis, University of Nottingham.
El Saeed, A.K. (1963). Aspects of plant population in the potato crop. Ph.D. Thesis, University of Nottingham.
Evans, S. (1975). *Outl. Agric.* **8** (4) 184–7.
Findlay, D.H. and Sykes, E.T. (1938). *Emp. J. exp. Agric.* **6**, 253–61.
Goodwin,P.B. (1967). *J. exp. Bot.* **18**, 78–86.
Goodwin, P.B., Brown, A., Lennard, J.H. and Milthorpe, F.L. (1969). *J. agric. Sci., Camb.* **73**, 167–76.
Gray, D. (1972). *J. agric. Sci., Camb.* **79**, 281–90.
Gray, D. and Bleasdale, J.K.A. (1972). The production of potatoes for canning. *Bull Potato Marketing Board.*
Gray, D. (1973). *Potato Res.* **16**, 80–4.
Gray, D. and Smith, D.J. (1973). *Potato Res.* **16**, 293–5.
Griffith, R.L., Hide, G.A., Hirst, J.M. and Stedman, O.J. (1974). *Ann. appl. Biol.* **77**, 237–50.
Gunasena, H.P.M. and Harris, P.M. (1971). *J. agric. Sci., Camb.* **76**, 33–52.
Harris, P.M. (1972). *J. agric. Sci. Camb.* **78**, 289–302.
Hide, G.A., Hirst, J.M. and Mundy, E.J. (1969). *Ann. appl. Biol.* **64**, 265–79.
Holliday, R. (1960). *Fld. Crop Abstr.* **13**, 247–54.
Holmes, J.C. (1966). *Scott. Agric.* **45**, 176–80.
Holmes, J.C. (1973). Relationship between stem number and tuber number in the potato crop. *Proc. 5th Trien. Conf. Eur. Ass. Potato Res. Norwich* (1972).
Ifenkwe, O.P. (1975). Effects of row width and plant density on growth and development of two maincrop potato varieties. Ph.D. Thesis, University College of Wales, Aberystwyth.

Ing, E.G. (1966). *Expl. Hort,* **14**, 39–42.
Jarvis, R.H. (1971). *Expl Husb.* **20**, 84–92.
Jarvis, R.H. (1972). *Expl Husb.* **21**, 85–92.
Jarvis, R.H. and Shotton, F.E. (1968). *Expl Husb.* **16**, 73–92.
Jarvis, R.H. and Shotton, F.E. (1971). *Expl Husb.* **20**, 12–29.
Jarvis, R.H. and Shotton, F.E. (1972). *Expl Husb.* **21**, 78–84.
Jarvis, R.H. and Rogers-Lewis, D.S. (1974). *Expl Husb.* **27**, 23–30.
Jarvis, R.H., Palmer, G.M. and Shotton, F.E. (1973). *Expl. Husb.* **24**, 41–7.
Jarvis, R.H. Rogers-Lewis, D.S. and Bray, W.E. (1976). *Expl Husb.* **30**, 28–41.
Krijthe, N. (1946). *Meded Landb Hoogesch. Wageningen* **6**, **47**.
Krijthe, N. (1955). *Neth. J. agric. Sci.* **3**, 291–304.
Megan, W.J. and Bankhead, J. (1938). *J. Minist. Agric. N. Ireland* **6**, 22–31.
Milthorpe, F.L. (1963). *Proc. 10th Easter School agric. Sci., University of Nottingham*, 3–16.
Montague, V.J. and Ivins, J.D. (1955). *Agriculture, Lond.* **61**, 586–7.
Moorby, J. (1967). *Eur. Potato J.* **10**, 189–205.
Morris, D.A. (1967). *Eur. Potato J.* **10**, 296–311.
Nelson, D.C. (1967). *Am. Potato J.* **44**, 17–21.
North, J.J. and Proctor, J.M. (1971). *Expl Husb.* **22**, 99–103.
O'Brien, P.J. (1978). Effects of seed crop treatments on the growth and yield of progeny tubers of early potato cultivars. Ph.D. Thesis. University College of Wales, Aberystwyth.
Pätzold, C. (1964). *Eur. Potato J.* **7**, 1–12.
Rappaport, L. and Sachs, M. (1967). *Nature, Lond.* **214**, 1149–50.
Reestman, A.J. (1970). *Potato Res.* **13**, 248–68.
Reestman, A.J. and de Wit, C.T. (1959). *Neth. J. agric. Sci.* **7**, 257–68.
Roer, L. (1955). *Forskm. Fors. Landbr.* **6**, 17–41.
Sadler, E.M. (1961). Factors influencing the development of sprouts of the potato. Ph.D. Thesis, University of Nottingham.
Saunt, J.E. (1960). Plant population studies with the potato crop. M.Sc. Thesis, University of Leeds.
Schepers, A. and Hoogland, R.F. (1968). *Jaarb. Inst. biol. Scheik. Onderz. LandbGewass.* (1968) 47–53.
Scott, R.K. and Younger, A. (1972). *Outl. Agric.* **7**, 3–9.
Sharpe, P.R. and Dent, J.B. (1968). *J. agric. Sci., Camb.* **70**, 123–9.
Singh, B.N. and Wakankar, S.M. (1943). *J. Am. Soc. Agron.* **35**, 613–16.
Stuart, W. *et al.* (1924). Size of potato sets: comparison of whole and cut seed. Bull. 1248 U.S. Dep. Agric.
Svensson, B. (1966). *Växtodling* **21**, 1–86.
Svensson, B. (1972). *Potato Res.* **15**, 346–53.
Taylor, C.E. (1953). *Ann. appl. Biol.* **40**, 778–88.
Thompson, R. and Taylor, H. (1974). *Potato Res.* **17**, 51–63.
Wurr, D.C.E. (1971). Some effects of seed size and spacing on the yield and grading of two maincrop potato varieties. Ph.D. Thesis, University of Cambridge.
Wurr, D.C.E. (1973a). Control of tuber size in the seed crop. *Proc. 5th Triennial Conf. Eur. Ass. Potato Res. Norwich* (1972).
Wurr, D.C.E. (1973b). *Rep. natn. Veg. Res. Stn.* (1972) 65.
Wurr, D.C.E. (1974). *J. agric. Sci., Camb.* **82**, 37–45.
Wurr, D.C.E. and Allen, E.J. (1974). *J. agric. Sci., Camb.* **82**, 277–82.
Wurr, D.C.E. (1975). *Potato Res.* **18**, 83–91.
Ziegler, G. (1968). *Albrecht—Thaer—Archiv.* **12**, 155–72.

Chapter 8

'Seed' tuber production and management

D.C.E WURR

8.1 Introduction

8.1.1 Market requirements

Potatoes are a basic food crop required throughout the whole year and supplied as either a first early, a second early or a maincrop. The harvesting of a first early in the U.K. begins in May and the tubers are relatively small in size, with thin skins and a limited storage potential. However, they are valued for their flavour and sell readily, rapidly replacing maincrop potatoes stored throughout the winter. Maincrops are harvested much later, generally from September onwards and their tubers are larger than earlies, with thicker, more mature skins. Second early potatoes are intermediate in character between a first early and a maincrop and are grown to meet market requirements in midsummer.

This classification usually signifies the time at which the crop is lifted to give a profitable commercial return and does not necessarily represent the actual maturity of the variety concerned. In fact the classification of a variety into one of these categories merely represents its usual use and does not mean that it cannot be used for a different purpose. Indeed, many existing varieties can be made either earlier or later than their current classification suggests by appropriate seed tuber management.

Early potatoes are harvested when yields are low but prices are high. Progressively later liftings give higher yields but lower prices and financial returns therefore depend on the combinations of these two factors. In general the first crops to market make a smaller profit than those marketed later (Cole, 1973) but growers are nevertheless keen to have varieties or technical expertise which will enable them to produce earlier crops than they can at the present time.

The production of an early crop requires the early planting of well

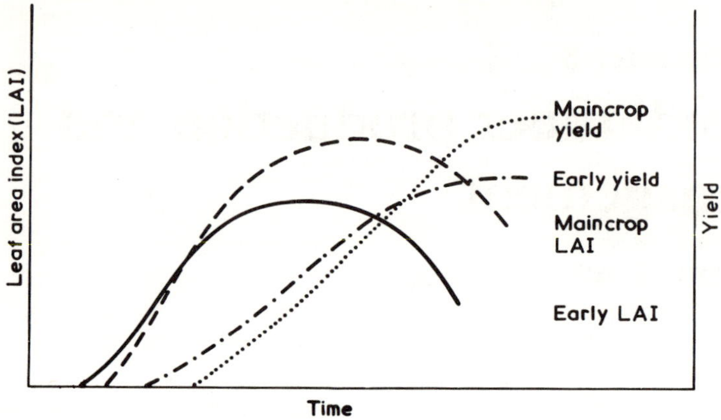

Fig. 8.1 Typical patterns of leaf and tuber development of an early crop and a maincrop.

sprouted seed in order to give rapid plant emergence and foliage development, and tuber initiation at a low leaf area index (LAI). The crop then begins to 'bulk up' at an early stage, that is it develops tuber yield early although its actual bulking rate may be relatively slow because this is proportional to the LAI shortly after tuber initiation (Ivins and Bremner, 1965).

Once tubers are initiated the products of photosynthesis are largely directed to tuber production and further leaf development is restricted (Ivins and Bremner, 1965). Thus with an early crop which initiates tubers at a low LAI, at normal plant spacings, the duration of the foliage is restricted and senescence occurs relatively early. Although this is generally true there are some varieties (e.g. Arran Comet) which initiate tubers at an early stage of haulm growth and yet still manage to produce a maximum LAI of at least five.

In contrast (see Fig. 8.1) a maincrop tends to initiate tubers at a larger LAI which is then able to supply both the needs of tuber production and those of new leaf formation so that senescence is later and the tubers bulk at a greater rate than those of an early crop and for a longer period of time. Carried to its extreme this means that the crop grows as a late maincrop, senescing naturally only in October. However, this can be a disadvantage by delaying the harvest which then takes place in cooler, wetter conditions with an increased likelihood of tuber damage, disease infection and subsequent poor storage. There are therefore practical advantages in producing a similar or slightly smaller yield for harvest in September and this can be done by growing crops which either senesce naturally or are defoliated by that time.

The sprouting of seed essentially advances its growth cycle so that it develops yield earlier than unsprouted seed. In general for crops defoliated chemically or by blight in late August or early September this advantage is maintained but thereafter a so-called 'crossover' effect occurs and if both sprouted and unsprouted crops are left to grow on, the unsprouted crop gives a higher total yield. Control over the sprout development of seed tubers is, therefore, a means of manipulating the pattern of tuber production to suit market requirements.

8.1.2 Variation

Crops grown for ware differ widely in performance from year to year, from site to site in the same season, from seedstock to seedstock at a site, and from seed tuber to seed tuber within a stock. Variation in performance due to the year and the site of production is largely attributable to environmental factors in the ware production year and these are considered elsewhere in this book. However variation attributable to the seedstock or the individual tuber may be caused by a number of factors associated with seed production and management (see Section 8.1.3).

Svensson (1966) and many other workers (see Section 8.4) have shown that there are important differences in performance between seed lots of the same variety and occasionally this may be due to the use of clones with a different yield potential as reported by Werner (1940) and Davidson and Lawley (1953). However, any research worker who has carefully selected from the same source, seed tubers of the same variety and size, with the same number of sprouts or sproutlets will confirm that there are also large differences in the performance of individual seed tubers within a given lot. Svensson showed that the coefficient of variation of the yield of plants grown from individual seed tubers, graded closely (40–50 mm), varied from about 20–50% according to the variety, the seed lot, the year and the plant population at which the crop was grown. He used seed potatoes of high quality in order to eliminate variation caused by disease and concluded that the observed variation in yield from plant to plant was largely attributable to the properties of the seed. He found evidence that the number of eyes on a seed tuber was affected by the method of cultivation or seed treatment used in the production of the seed crop and suggested that the chemical composition and the physiological age of individual tubers could also vary.

If this variation both within and between seed lots can be reduced then improvements in the level and predictability of yield must follow.

8.1.3 Seed production and management factors affecting subsequent ware yield

The major seed production and management factors which affect subsequent tuber performance are: the certification grade of the seed in so far as it relates to the health of the tuber; the physiological age of the seed; the way in which the seed crop is grown in the field and the way in which it is stored. These are discussed in the rest of this chapter.

'Seed certification' (Section 8.2) describes the aims of a certification system, the requirements for certification in the U.K. and the importance of non-certified seed. The section on physiological age (8.3) describes the concept of the seed tuber ageing physiologically and shows why chronological ageing is only a part of physiological ageing. The effects of physiological age on sprout development and crop growth are described and the mechanisms which may control its mode of action are discussed. The sections on field factors (8.4) and storage factors (8.5) show how the way that the seed crop is produced and stored can affect its subsequent performance. This is usually due to an effect on the physiological age of the seed which is represented by the level of visible sprout development on the tuber. The latter depends on two factors: the time at which the tuber breaks dormancy and the rate of sprout growth. In general the way in which the crop is grown affects the time at which dormancy is broken and the way that it is stored affects the rate of sprout growth, although varieties can differ appreciably in both respects.

The concluding section (8.6) discusses how it may be possible to measure physiological age and then put that information to practical advantage.

8.2 Seed certification

The prime aim of a seed certification system is to maintain the varietal purity of seed stocks and to control the spread of disease by prescribing tolerances and cultural conditions. To this end most countries are selective in their choice of seed area and in the U.K. great advantage is taken of the relatively cool, windy and comparatively aphid-free conditions in Northern Ireland, Scotland and Northern England to produce seed with minimal virus infection. In many countries the climate is such that this is not possible. In the Netherlands, for example, a low level of virus infection is attainable only by making early defoliation of the haulm in late July or early August obligatory, the precise date depending on the variety, the grade of seed and seasonal aphid activity. This system produces seed tubers which are relatively small in size with moderate levels of latent fungal disease infection because the skins are well set at harvest and the tubers are lifted in relatively warm dry conditions. By comparison, planting and consequently haulm destruction and harvesting in the seed areas of the U.K. takes place late in the season for climatic reasons, with attendant higher risks of tubers becoming infected with the organisms of tuber fungal diseases.

8.2.1 Seed certification in the U.K.

England and Wales, Scotland, and Northern Ireland administer their own separate certification schemes for seed potatoes but the requirements for certification are basically similar and are defined by consultation between the three Departments concerned. A unified approach to certification is clearly desirable in order that the potato directives of the E.E.C. can be interpreted on a U.K. basis.

The Directive on the Marketing of Seed Potatoes lays down minimum requirements for seed certification and marketing and these procedures will become operative in the U.K. as soon as national legislation to enact the terms of the directive has been completed.

Most existing regulations in the U.K. already more than satisfy the minimum E.E.C. requirements so any changes involved are unlikely to be drastic.

The main details of the directive are as follows:

(1) There shall be two categories of seed potatoes: basic seed which is intended for further seed production; and certified seed which is to be used for the production of ware. These categories are defined in terms of minimum quality standards but may be subdivided into grades of different standard provided that these are higher than the minimum laid down.

A protected region has been established in terms of the Prevention of Spread of Pests (Seed Potatoes) (Great Britain) Order 1974 which comprises Scotland, the English counties of Northumberland (excluding the districts of Blyth Valley and Wansbeck) and Cumbria (excluding the districts of Barrow-in-Furness and South Lakeland). Within this region certification schemes include only grades within the basic seed category and imports of any other seed from outside, whether for planting as seed or ware, are banned except under licence. Certified seed (Grade CC) will be produced only in the remainder of England and Wales.

(2) Only potatoes certified within the categories basic seed and certified seed may be marketed and this must be done in homogeneous batches in new bags, sealed and officially labelled. Re-sealing may only be done officially and the facts of re-sealing must be recorded on the label.

(3) Minimum sizes and permitted size ranges within a seed lot are defined and tolerances for out-sizes are prescribed.

(4) The treatment of seed for sale with germination inhibitors is prohibited and any chemical treatment of the seed must be declared.

In the U.K. basic seed incorporates the three grades: VTSC, (seed which is multiplied from tubers produced by virus-tested stem-cuttings), FS (Foundation Seed) and grade AA – all of which are currently produced in Scotland. Stocks of VTSC in Scotland are grown by 'VTSC growers' on land which has not produced potatoes for at least seven years and the clonal stocks are subject to a stringent scheme of testing for the presence of viruses, fungi and bacteria. In general, crops cannot be entered for certification at VTSC level before their third year of multiplication from stem cuttings nor after their fifth. Stocks of FS must be derived from VTSC grades and are eligible for certification at FS level for only

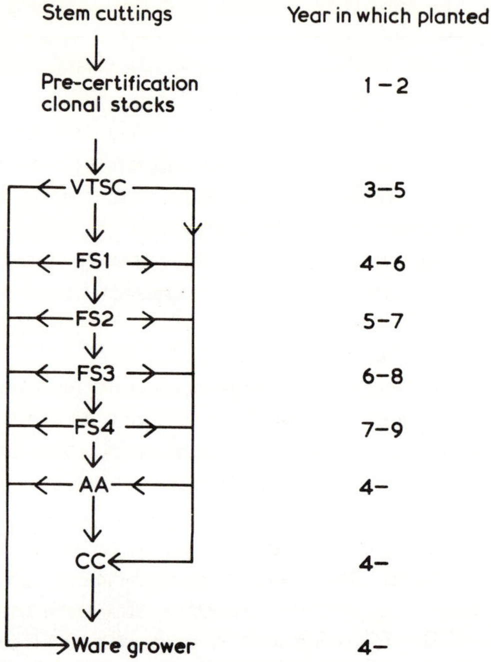

Fig. 8.2 A simple diagram illustrating possible patterns of seed movement

four years. Grade AA seed can be derived from AA or higher grades and is without time restriction.

The possible pattern of seed movement which is illustrated diagramatically in Fig. 8.2 shows that a producer of FS stocks must plant VTSC seed in order to produce FS1 stocks – the 1 designating the first year in the FS grade. FS1 stocks may then be sold to the ware grower or to produce FS2 or AA seed, or even CC seed in the certified seed area. The possible uses of FS2, FS3, FS4 and AA are similarly shown in Fig. 8.2.

Seed crops are certified following field inspections and have to meet strict regulations concerning crop separation, the length of the rotation, and the permitted levels of: groundkeepers, purity and trueness-to-type, virus diseases, blackleg and potato cyst eelworm. The permitted tolerances are progressively increased for lower grades of seed.

Crops are certified provisionally on the basis of field inspections because in terms of the new procedures following accession to the E.E.C., requirements for certification include fulfilling tuber health standards. Hence after field inspection only provisional certificates are issued and their terms will not require the issue of a separate final certificate since this will automatically depend on the consignment not being rejected at the time of preparation for sale. However, the stringent U.K. virus health standards, in conjunction with the climatic conditions in the main seed areas mean that there is no need for post-harvest virus health tests, which are an integral part of the certification schemes of existing E.E.C. member countries.

8.2.2 The production of non-certified seed

Almost half the seed planted in England and Wales is uncertified and the proportion of seed in this category has risen in recent years for a a number of reasons: increasing prices for certified seed; the availability of the varieties Pentland Crown and Pentland Ivory, which are resistant to severe mosaic and moderately resistant to leaf roll; and the increasing use of systemic insecticides. However, when the E.E.C. seed potato marketing directive is fully implemented in the U.K. the sale of uncertified seed will be banned and this could increase demand for certified stocks or alternatively it could encourage more growers to produce their own 'once grown' seed.

Certain early areas, for example Cornwall and Pembrokeshire have for many years tended to produce their own seed because of the advantages to be gained from using seed of the correct physiological age (see Section 8.3).

8.3 Physiological age

Physiological age is a much used but little defined term which Toosey (1964) described as 'the physiological state of the tuber at any given time'. It is generally regarded as illustrated by the degree of visible sprout development (Toosey, 1964) and is known to affect the subsequent growth and yield of the crop.

Kawakami (1953) measured physiological age as the time from harvesting the seed until it was planted and Krijthe (1962) found that changes in the sprouting capacity with time could be used to measure the age of tubers. She reported the sequence of ageing phenomena to be: (1) single sprout stage; (2) multiple sprout stage; (3) branching stage; and (4) small tuber formation stage. Such a sequence involves a decrease in apical dominance with older seed producing more stems in the field and indeed Iritani (1968a) using seed with a wide range of chronological ages suggested that its productivity could be measured by determining the relative speed of emergence and degree of apical dominance in a greenhouse during the winter.

8.3.1 Effect of chronological age

Kawakami (1953) recognized the importance of the physiological age of the seed in determining crop yield although he was in fact referring to chronological age, which is only a part of physiological age. Changes in chronological age result in differences in physiological age, while the reverse is not necessarily true. For example, two seed lots may have the same chronological age but if their internal 'clocks' can be made to 'tick' at different rates by storing them in different ways (see Fig. 8.8 in Section 8.6.1) then the seed lots will have different physiological ages.

The chronological age of a seed tuber is particularly important where the climate permits the growth of more than one potato crop in a year and Kawakami (1973) has described the problems in central and southern Japan where two crops a year are grown, one in the spring and one in the autumn. Seed for the spring crop is mainly produced by summer crops in northern Japan and is freely available, but for the autumn crop this seed is physiologically too old and lacking in vigour. Neither is seed from the spring crop particularly suitable because it is physiologically too young; either still being dormant or having only just broken dormancy.

A similar problem exists in many other countries and it is usually alleviated by the use of suitable varieties, dormancy breaking compounds and in some cases by earlier planting in spring in order to widen the gap between the spring and autumn crops. Nevertheless the seed used is not necessarily at the optimum age which Kawakami (1953) and other workers have shown to be important if the crop is to give maximum yield at maturity (see Fig. 8.3). Kawakami measured age from harvesting to planting and produced seed with a wide range of ages by growing seed crops at different times of the year and at different latitudes. He suggested that the optimum age of seed was approximately five months from harvest to planting and considered that the value of a seed tuber was not affected by its source other than through its effect on chronological age. He ignored any influence of the conditions under which the seed was produced and stored but there are nevertheless many reports of seed of the same chronological age from harvest to planting giving differences in performance in the field which were attributable to physiological age.

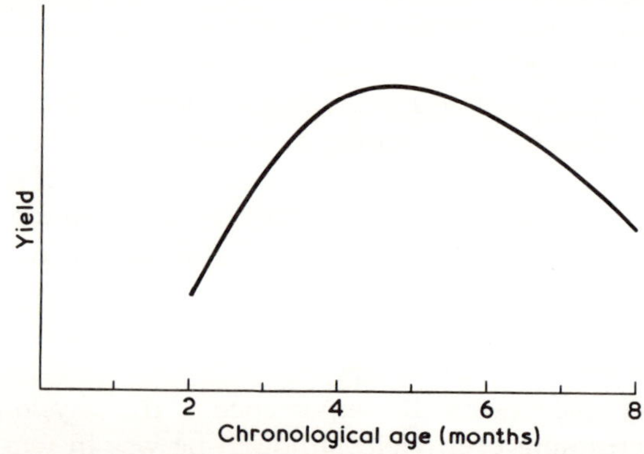

Fig. 8.3 General relationship between yield and the chronological age of the seed (after Kawakami, 1953)

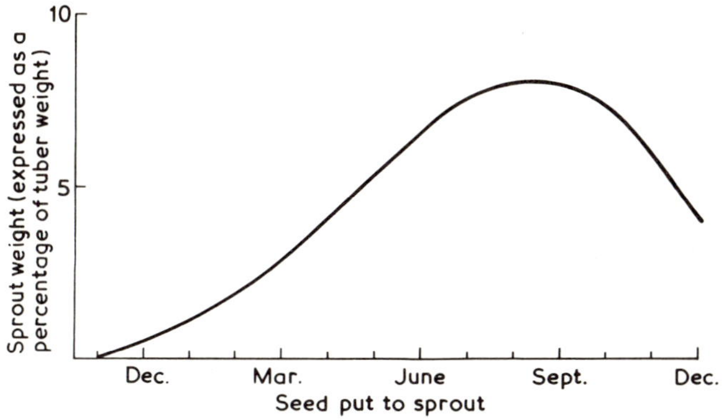

Fig. 8.4 Effect of chronological age on sprout growth (after Krijthe, 1962)

8.3.2 Effect on sprout development

The age of seed has been shown to affect sprout growth by Krijthe (1962) who kept tubers free from sprout development at 2 °C, transferring samples at monthly intervals to 20 °C where they were kept for four weeks. The sprouts were then removed and weighed and Fig. 8.4 shows that sprouting capacity measured as the weight of sprouts, expressed as a percentage of the tuber weight, increased to a maximum and then declined.

When tubers are physiologically very young (i.e. dormant) they show no bud growth but as they age bud development occurs, eventually producing sprouts which develop leaf initials and root primordia. The exact size, shape and structure of a sprout depends upon the variety and the storage conditions but in general as further ageing of the tuber occurs, the morphology of the sprout becomes modified with swelling of the base and the development of stolons and lateral branches.

8.3.3 Effect on growth and yield

Very old seed may initiate tubers prematurely while it is still in store and the same phenomenon, known as 'little potato' can also occur in the field either before or shortly after the plant has emerged. If field conditions are unsuitable for growth, tuber formation is more likely to occur before the plants emerge and may in severe cases cause a 'miss'.

Physiologically old seed emerges earlier than physiologically younger seed and initiates tubers earlier and at a lower LAI. Consequently because the products of photosynthesis are largely directed to tuber production the development of the leaf canopy is restricted and senescence tends to occur earlier than with physiologically younger seed (see Fig. 8.5). The extent to which this occurs varies according to the variety (see Section 8.1.1) and the environment in which the crop is grown but the

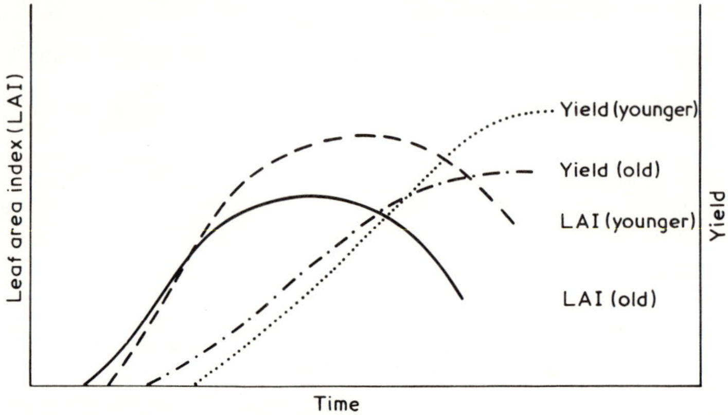

Fig. 8.5 Typical patterns of leaf and tuber development of crops grown from physiologically old and younger seed

bulking rate of a crop grown from physiologically old seed is less than that grown from physiologically younger seed (see Fig. 8.5) and although it gives a higher yield early in the season this position is eventually reversed. These two typical patterns of growth represent the requirements for producing an early crop and a maincrop respectively (see Section 8.1.1).

Physiological ageing seems to act very largely by its effect on the level of sprout development at planting. In suitable conditions this can control the rate of shoot development in relation to the time at which tubers are initiated, thus affecting the times of emergence, foliage development and senescence of the crop.

8.3.4 Mechanism of action

The controlling mechanism behind physiological age is not known. It may be a gradual build-up in the levels of growth substances which eventually trigger tuber initiation or it may be caused by some other factor such as the sugar content of the tuber or the mobilization of N within it.

The mechanism of tuber initiation is not fully understood but is probably controlled by a balance between endogenous gibberellins and some other substance (Hammes and Beyers, 1973). There are external effects of temperature and photoperiod on the growing plant, yet Madec and Perennec (1959) have shown that tuber initiation is induced by both the foliage and the mother tuber. Indeed physiologically old seed is capable of producing tubers in the complete absence of light ('little potato'). In reality there are probably several interdependant factors which are themselves affected by field and storage conditions.

For example, Burton and Wilson (1970) have shown that at maturity the total sugar content and the contents of reducing sugar and sucrose are in general directly correlated with the latitude of cultivation. Burton (1965) has also shown that the effect of the temperature of storage and the changes in sugar levels over time are complicated. He noted that there was a good correlation between the reducing sugar content and the amount of sprout growth and concluded that sweetening (see Chapter 14) may be necessary for rapid sprout development.

The level and form of tuber N may also be important. Newton (1927) found that absorption of nitrate (NO_3^-) by the tuber shortened dormancy and Headford (1962) suggested that N might be a component of the tuber reserves which was limiting to sprout growth in storage. Furthermore, Dinkel (1968) observed that following the natural dormancy of tubers there was a period when a tuber's N reserves were essentially not available to support shoot development in the light although good root and sprout growth were produced in the dark. He found that if an external inorganic source of N as NO_3^- was applied to the roots then the shoots grew well in the light and that as tubers aged they naturally became capable of supporting progressively more growth in the light, indicating a gradual release of nitrogenous compounds. The degree of unavailability of N, the period over which this occurred and the response to different sources of N were found to vary according to the variety.

8.4 The effect of field factors during the seed production year on subsequent performance

8.4.1 Location

There have been reports from many countries of seed from different sources giving different levels of sprout development, foliage vigour and yield (Kawakami, 1953; Madec and Perennec, 1956; Broadbent *et al.*, 1961; Iritani, 1967; O'Brien, 1978). These comparisons have usually been made between imported seed stocks from different origins or between imported seed and home produced seed, although in general little has been known about the way in which the imported seed has been produced and stored prior to reaching the experimental centre.

In the U.K. a comparison of the growth of plants from 'once grown' seed and from Scotch seed usually shows that 'once grown' seed emerges earlier and forms tubers earlier (Burton, 1966). It gives a greater yield than imported seed early in the season although the reverse may be true if the crops are allowed to grow on to maturity. Clearly the two types of seed stock have different physiological ages which must be attributable to environmental and cultural factors during the seed production year.

Kawakami considered differences in the performance of his seed stocks to be due entirely to differences in their chronological age, ignoring completely the effect of different environments of production on the physiology of the seed tuber, but there is good evidence that seed of the same chronological age may differ in physiological age as a result of the way in which it has been produced.

There have nevertheless been several reports that seed from different sources, produced in various ways, did not differ in performance and these show that there is nothing specific about the source of seed but that it represents the combination of many varying factors, which may or may not combine to affect seed performance. The effects of some of these factors are well known, but there are others which are less well understood.

8.4.2 Soil type and moisture content

Gray (1973) found that seed produced on a sandy loam gave more vigorous foliage than seed from a clay soil in the same area. This work did not consider yield but as Section 8.3.3 has shown, increased early vigour usually gives higher early and lower final yields.

Several workers (Bardeeva *et al.*, 1955; Filippov, 1964; Karavaeva *et al.*, 1973) have shown that seed produced on a mineral soil gave a lower final yield than seed produced on a peat soil and it is perhaps possible that this was due to differences in the level of soil temperature or to differences in the daily amplitude of soil temperature on the two soil types. Organic soils absorb heat readily owing to their dark colour but, because they often have high moisture contents, they may not warm up as much as a mineral soil. Furthermore, at the same moisture content peat soils have a lower thermal diffusivity and therefore a smaller amplitude of temperature fluctuation throughout the day than sandy soils.

There may even be a direct effect of the soil moisture content on seed performance. Certainly Yablonskii (1959) found that tubers grown in dry conditions broke dormancy earlier than those grown under wet conditions and Filippov (1964) reported that seed produced where moisture was plentiful aged more slowly than that produced where the soil was dry. However Werner (1929) found no difference in the performance of seed produced on soils of different moisture status.

8.4.3 Temperature during the growing season

Went (1959) and Iritani (1968b) showed that the performance of seed tubers depended upon the growing conditions under which the tubers were produced. They reported that tubers produced under cool conditions gave a higher final yield than those produced in warm conditions. This effect occurs because seed produced at a higher temperature tends to break dormancy early and therefore becomes physiologically older than seed produced under cooler conditions; hence it gives a lower yield at maturity.

It is not known whether this effect is due to the average temperature over the whole period of growth or whether the temperature at specific stages is critical. For example, Claver *et al.* (1957) have shown that extreme temperatures at tuber initiation can be important. A high temperature (34.0 °C) during tuber formation gave seed tubers which produced yields less than half those produced by seed formed under relatively cool (7.7 °C) conditions.

8.4.4 Altitude

There have been several reports of the altitude at which seed tubers were produced affecting subsequent ware yield. Rønsen (1971) and several workers quoted by De Bokx and Mooi (1974) found that the yield from seed grown the previous year at low altitudes was greater than that from seed produced at higher altitudes while Staĭkov and Chavdarov (1971) found the reverse to be true. However, Kozlowska (1963) showed that ware yield increased progressively as the altitude of seed production increased to 939 m but that a further increase in altitude gave a reduction in yield.

These reports vary widely but are not irreconcileable. Increasing altitude in general reduces the average temperature during the growing season so that seed grown at high altitude should be physiologically younger than that grown at a lower altitude. Fig. 8.4 has already shown that the effect of physiological age on yield depends on the stage of growth of the crop at harvest.

The effect of altitude is however inevitably more complex than this suggests since it includes other variables such as soil type, rainfall and the chilling effect of wind. Extremes of temperature are indeed important during tuber initiation (Section 8.4.3) and early storage (Section 8.5.9) when a short period of low temperature gives an earlier break of dormancy. There may be a similar effect on tubers subjected to low soil temperatures prior to harvest and indeed Wurr (unpublished data) found that seed tubers of two out of four varieties grown at 365 m sprouted earlier than tubers of the same size grown in the same way at three sites of lower altitude.

8.4.5 Time of planting

Variation of the time at which the seed crop is planted effectively produces seed of different chronological ages by moving the whole growth cycle of the crop. Late planted seed crops produce tubers which break dormancy later and sprout more slowly (see Table 8.1) but may give a higher mature yield than seed from crops which are planted earlier in the season. This technique is used to limit the physiological age of seed in areas where seed tends to get too old (Dubuffet, 1954). In fact in one area of France crops of some early varieties intended for the production of high quality basic seed cannot be planted before a specified date in order to prevent that seed becoming too old before it is used the next year.

Table 8.1 Effects of the time of planting the seed crop on the sprout development of Red Craigs Royal

	Date of planting (1974)				
	21/3	4/4	18/4	2/5	S.E.
Days from lifting to 3 mm sprout	105.5	108.5	121.9	125.8	±1.43
Length of longest sprout 28/2/75 (mm)	22.7	21.8	18.3	17.3	±0.84

(Allen, personal communication)

In U.K. conditions the maximum variation in the time of planting is likely to be found between early crops in Cornwall or Pembrokeshire which produce their own uncertified seed and the same varieties planted in Scotland to give certified seed. Scottish seed crops probably initiate tubers two months later and it is easy to see why early growers have tended to prefer their own older, 'acclimatized' seed.

8.4.6 Time of defoliation and harvest

Most seed crops are defoliated and then left for about three weeks so that the tubers can set their skins before harvest. Thus any manipulation of the time of defoliation usually, but not always involves a similar manipulation of harvest date. Seed tubers from a crop defoliated when immature and harvested 2–3 weeks later usually start sprout growth before those defoliated when more mature, but the interval between the beginning of sprout growth of the two lots is less than that between the times at which they were defoliated. This may be due to the level of some internal factor but the possible influence of different environmental conditions in the ground and in store cannot be ignored. Indeed Emilsson (1949), and Burton (1963) found that seed lots defoliated at different times but harvested at the same time began to sprout at the same time, irrespective of the date of defoliation.

However Allen (personal communication) found that there was a small but consistent effect of early defoliation in giving an earlier start to sprout growth and Holmes and Gray (1972) also reported that early defoliation affected the sprout growth of progeny tubers if sprouting was encouraged soon after harvest. They found that such effects were nevertheless too small to have any noticeable impact on the yield of the subsequent ware crop; a finding which agrees with those of Murphy *et al.* (1967) and Goodwin *et al.* (1969).

There have however been reports of important effects of the times of defoliation and harvest on yield in several references quoted by Kawakami (1953) and by Henriksen (1973), although there appears to be little consistency as to whether early or late defoliation produces the higher yield. This is not surprising, in view of the effects of physiological age on the pattern of field growth; the environment in which the ware crop is grown and the stage of growth at which it is harvested, are of prime importance. However O'Brien (1978) found that for ware crops grown to maturity, seed defoliated early gave a lower final yield than seed from crops defoliated later, and Wurr (unpublished data) also found that this was nearly always so.

In the U.K. there now appears to be little value placed on manipulating the time of defoliation of the seed crop in order to produce seed of an appropriate physiological state yet 50 years ago (Ministry of Agriculture and Fisheries, 1923) it was considered important to defoliate the crop before the haulm showed the first signs of ripening when the tubers had suitable levels of N and dry matter.

8.4.7 Stem density

The stem density at which the seed crop is grown can have a small effect on the following ware crop although this is unlikely to be important commercially. Seed grown at a high stem density tends to produce fewer stems and tubers than that which is produced at a low stem density. This is probably attributable to differences in the shape of the seed tubers. Those grown at a higher stem density tend to be more round and less elongated which explains why Walker (1968) found that high densities reduced the ratio of eyes to tuber weight. This in turn suggests why, for any seed tuber weight, the number of stems which grow is reduced by producing that seed at a high stem density.

8.4.8 Fertilizer

There have been several reports that varying the level of fertilizer applied to the seed crop has had no effect on the performance of the progeny tubers. However some workers have found that increasing the level of N applied to the seed crop, increased the N reserves in the seed tuber and produced more rapid sprout growth (Walker, 1968; Thow, 1970). Such effects have resulted in higher yields of both early and maincrop varieties during the early stages of growth but the response has been variable and in general small at maturity. Schepers *et al.* (1969) also found that high levels of applied N gave seed which tended to give fewer main stems.

Gray (1974) considered that the observed effects of the application of N to the seed crop on ware crop performance were connected with the time at which the seed crop matured and hence the physiological age of the seed tubers. Low N levels applied to the seed crop give earlier maturity and sprouting and this may produce seed which is beyond its optimum age at planting while high levels of applied N delay senescence of the seed crop and therefore produce younger seed.

There can also be other effects of fertilizer application; increasing phosphate levels advance the maturity of the crop (Harrington, 1935; Emilsson, 1961); N tends to increase the length of tubers in relation to the breadth and K to reduce it (Harrington, 1935), so affecting the number of eyes on a tuber of any one size. Early sprout growth has also been shown to be related to the K level of the tubers (Walker, 1968).

8.5 The effect of storage factors on subsequent performance

8.5.1 Sprouting

Seed which is sprouted is physiologically older than seed which is unsprouted and therefore produces plants which emerge faster, initiate tubers earlier, 'bulk' somewhat slower and senesce earlier. During most of the season the yield of a sprouted crop exceeds that of an unsprouted crop but if the latter is allowed to grow to maturity it will produce the higher yield. The relative advantage or disadvantage of sprouting thus depends on the time at which the ware crop is defoliated, either chemically or by blight or frost.

This explains why reports of the effect of sprouting on yield vary considerably. For example in the Fen area of England where crops are normally burnt off in late August or early September there is a benefit from sprouting of up to 5 t ha^{-1} (Shotton, 1964) while in areas where crops are allowed to grow on for a longer period of time, unsprouted seed can have a similar advantage.

Shotton pointed out that since the overall effect of sprouting was relatively small, differences between sprouting treatments could only be expected to be a part of this difference. Many factors affect sprout development in storage yet relatively few of them have a consistent effect from year to year and from variety to variety and this is possibly explained by two factors: (1) the difficulty of making experiments sufficiently accurate to detect small differences between treatments and (2) the fact that most practical treatments are only tested at a final harvest and accurate details of their effects on sprout growth and early plant development in the field have been ignored.

8.5.2 Length of the sprouting period

Shotton and Palmer (1972) and Jarvis *et al.* (1973) investigated the effect of the length of the sprouting period on the yield of Majestic, King Edward and Pentland Crown planted by hand. In their situation where sprouted seed outyielded unsprouted seed, an increase in the length of the sprouting period of Majestic and King Edward increased its advantage over that of a short sprouting period and this effect was closely correlated with the length of the apical sprout. However with Pentland Crown, seed sprouted for a short period of time gave a higher yield than seed

Table 8.2 Effect of length of sprouting period on the number of sprouts per tuber > 3 mm (cv. Home Guard)

Seed sprouted	*Sprouts/tuber >3 mm* 1972/73 (recorded 8/3/73)	1973/74 (recorded 28/2/74)
From lifting	2.5	2.2
15/11	7.7	8.5
15/1	7.1	8.4
S.E.	± 0.30	± 0.13

(Allen, personal communication)

sprouted more normally (3–4 months). The answer probably lies in the type of sprout development produced by the two methods; Majestic and King Edward sprouted normally, gave apical sprouts which were larger than those produced by a short sprouting period, while with Pentland Crown the reverse was true. Thus the importance of the sprouting period for hand planting, which did not damage the sprouts, appears to be related to the length of the apical sprout at planting.

However, when planted by machine the response of Pentland Crown was not the same because the ability of sprouts to withstand damage was important. Normal sprouting out-yielded a short sprouting period, probably because the short sprouting period gave much softer sprouts which were more susceptible to damage. Certainly appreciable yield reductions associated with increased levels of sprout damage have been recorded in other trials at Terrington (Jarvis and Palmer, 1973) and also on the Continent (Anonymous, 1970).

The length of the sprouting period might be expected to have an effect on the numbers of sprouts and stems produced and therefore the resulting yield (see Chapter 7 for effects of stem density on yield). The varieties used by Jarvis *et al.* (1973) showed that there was little effect of the length of the sprouting period on sprout and stem numbers but Allen (personal communication) (see Table 8.2) found that the early variety Home Guard was affected considerably. The extent to which the apical dominance mechanism (see Chapter 4) operates must therefore vary appreciably between varieties.

8.5.3 Temperature

The temperature at which seed potatoes are stored has a large effect on their subsequent performance, acting through its effect on sprout development. High temperature storage produces physiologically old tubers with well developed sprouts while low temperature storage gives physiologically younger tubers with slight sprout development. Short and

Shotton (1970) found that the mean length of the apical sprout increased with temperature to a maximum between 13–18 °C.

The temperature of storage also affects sprout morphology, higher temperatures giving an increasing tendency towards the production of sprouts with swollen bases, which form stolons and produce lateral branches as the sprout apices become necrotic (Toosey, 1964).

Fischnich and Krug (1963) have shown that increasing storage temperature reduced the time to emergence, the number of nodes and the length of the shoot axis. Storage at 12 °C produced foliage with thinner stems and smaller leaves but with a greater number of leaflets during the initial growth phase, than plants grown from tubers stored at lower temperatures. Increasing the temperature of storage gave earlier tuber initiation and development of yield but the highest final yield (if there was a long growing period) was produced by storing the seed at low temperatures.

8.5.4 Light

Appleman (1914) found that exposure to natural light prolonged the dormancy of mature tubers while possibly reducing that of immature tubers. This might perhaps have a small effect on the dormancy of seed stored in glasshouses or polythene structures but the normal use of light (either natural or artificial) nowadays is to prevent excessive sprout development once dormancy is over. Seed which is sprouted in potato stores is generally illuminated by warm-white fluorescent tubes for 8–10 h daily once sprout development has started and a single tube will be used to light 6–8 stacks of trays.

Short and Shotton (1968) reported an extensive trial which compared lighting regimes during the sprouting period and they found that there was a general trend for sprout lengths to be reduced as the duration or intensity of illumination was increased. However, they found no consistent effect of variation in the lighting regime on the yield of hand planted potatoes.

8.5.5 Humidity and ventilation

Storage at high humidity shortens the dormant period (Rosa, 1928) and gives more stems than storage at low humidity (Smith, 1937) although Workman and Twomey (1967) found that it had no effect on mature yield in the field. However Neubauer *et al.* (1965) reported that seed stored under continuous ventilation gave more large and fewer mis-shapen tubers than seed stored in still air. This is an interesting observation because nearly all the emphasis placed on seed tuber performance is related to the effect on total yield and not to the distribution of that yield. Certainly ventilation reduces sprout development (Neubauer *et al.*, 1965; Tomkins, 1966) because the accumulation of carbon dioxide encourages sprout growth but there is a confounding effect here because respiring potatoes also produce volatiles which suppress sprout growth if allowed to accumulate.

Workman and Twomey (1969) considered the effect on yield of storage atmospheres containing varying levels of oxygen and carbon dioxide because they thought that factors such as these might modify physiological age but within the normal range of storage conditions they found little consistent evidence to this effect.

8.5.6 Types of store and container

Shotton (1964) and Messer *et al.* (1972) have reported the effect of various types of seed potato store and container (tray, sack or box) on yield. Their results showed differences between treatments within a year but these differences were inconsistent from one year to another.

Seed potatoes are usually sprouted in plastic or wooden trays in glass-houses in areas which traditionally have had spare winter capacity, or in artificially lit internal stores, either purpose built or more usually converted from another use. Such systems produce adequately sprouted seed but the increasing capital cost of replacing existing old facilities has led to the investigation of cheaper alternatives such as the use of temporary plastic structures which are becoming increasingly popular.

In the U.K. there is also another problem. The timely use of fast automatic planters requires rapid handling of seed from the store to the field and into the planter and this has emphasized the need to devise systems of seed storage and handling which retain the advantages of sprouting and yet still permit fast planter filling.

One grower in the U.K. has attempted to do this by chitting seed in the traditional way in trays and then using a small gang of otherwise surplus labour to tip these trays into large pallet boxes in the seed store. These are then transferred to the field and tipped into the planter. An examination of this technique at Terrington E.H.F. (Ministry of Agriculture, Fisheries and Food, 1974) has shown that although there appeared to be a slight increase in sprout damage over the whole planting process compared with seed tipped directly from trays into the planter, yields were similar. Other techniques currently under appraisal include the use of specially constructed half-tonne chitting boxes which allow good light penetration and ventilation and seed tubers stored in them have given yields comparable with those from seed tipped directly from trays to the planter.

8.5.7 Seed size graded out

The effect of seed tuber size and weight on the number of stems produced and tuber yield is discussed at length in Chapter 7. Variation in seed tuber size can account for appreciable differences in yield and the average weight of seed tubers in the same size grade can differ from year to year, from variety to variety and from batch to batch. It is therefore essential to consider the weight of seed tubers graded out of store when comparing the effect of different seed management techniques on yield.

Potato seed sold in the U.K. at the moment must pass over a 32 mm square mesh and through a 57 mm square mesh and indeed most seed is sold graded 32–57 mm. However in recent years there has been increasing demand for smaller seed, largely because of the increasing use of automatic planters which require even-sized seed but also partly because of the financial advantages of small seed in establishing any required stem population (see Chapter 7). In fact at least 20% of seed is now sold with a top riddle size of 51 mm and ware growers pay a small premium for this. Some seed is also sold split-graded e.g. 32–44 and 44–57 mm but this is usually to suit a particular customer's requirements.

E.E.C. regulations require a minimum size of 25 or 28 mm depending on the length to greatest width ratio. There is no absolute maximum size but the variation in size between tubers in a lot must not exceed 20 mm although individual countries can vary this if they wish. In the Netherlands, for example, seed potatoes are sold in three size grades: 28–35, 35–45 and 45–55 mm.

In North America where much of the seed is cut before planting, tubers are much larger than this. For example in Canada most of the seed used is 'A' size weighing from 113–340 g and this is cut into several pieces while some 'B' size seed weighing from 43–113 g is also used whole. The general relationship between tuber size and weight is shown in Fig. 8.6.

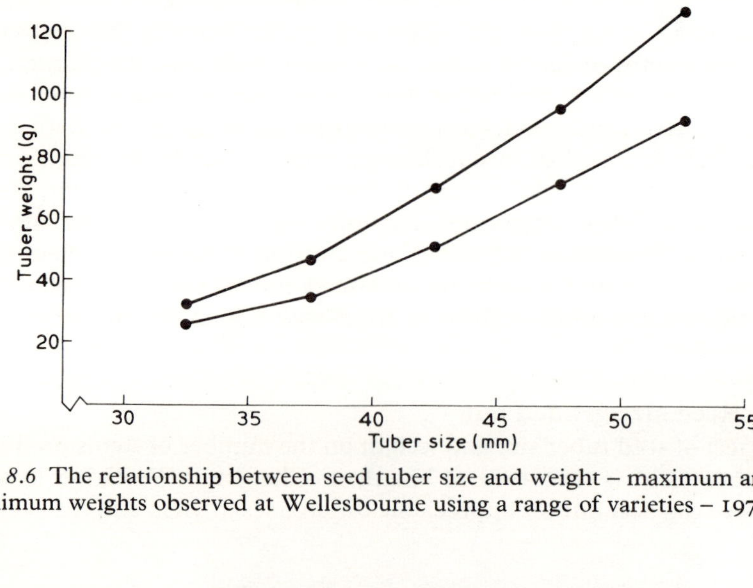

Fig. 8.6 The relationship between seed tuber size and weight – maximum and minimum weights observed at Wellesbourne using a range of varieties – 1972–5

8.5.8 Sprout damage

Sprout damage may occur at various stages in the life of a seed tuber from the time at which sprout growth begins until the tuber is planted. Seed produced in Scotland is stored in bulk prior to grading and the later in the season that 'dressing' is done, the greater the likelihood that the seed will have started to sprout. If this is so the sprouts tend to be either severely damaged or removed during grading and this may increase the number of sprouts and stems subsequently produced but is unlikely to have much effect on the physiological age of the tuber. The type and rate of sprout growth occurring following desprouting is determined by the age of the tuber when desprouted. For example, if seed is sprouted at different temperatures and then desprouted and planted, the early sprout growth increases in proportion to the original temperature of storage (Madec and Perennec, 1962).

The effect of sprout damage on yield is however largely related to its effect on the sprout development of the planted seed tuber; if this is severely reduced by desprouting during the planting operation then the whole growth cycle of the plant developing from that tuber is likely to be displaced backwards in time, while if desprouting is less severe or has occurred during grading out, so that sprout development is near optimum, then there is not likely to be much effect on yield.

It is however fair to say that the effects of desprouting on subsequent crop growth are not fully understood and that more work on this subject is needed.

8.5.9 Seed treatment

This chapter has shown that the level and type of sprout development on seed tubers is a major factor in determining the pattern of growth of the crop and the ultimate yield. In certain situations modifications of sprout development or subsequent growth are important.

Dormancy breaking

Treatments which break dormancy tend to be used in countries where two crops are grown in the same season and physiologically young seed produced by the first crop is used to plant the second. Dormancy can be broken by dipping tubers in a solution of thiourea but this gives several sprouts because it breaks apical dominance (Denny, 1926) while ethylene chlorhydrin also breaks dormancy but does not give a multi-sprout condition. Thiourea treatment improved emergence and yield (Singh and Verma, 1970) but has generally been found to be less effective than a gibberellic acid (G.A.) dip which in turn has been less effective than a mixture of thiourea and G.A. (Boiko, 1968; Voĭtekh, 1971). G.A. treatment by a dip (Van Hiele, 1961) or by a spray (Holmes *et al.*,

1970) increased the number of sprouts and stems in the field but Voĭtekh (1971) found that the most effective treatment was G.A. and Rindite. Rindite, a mixture of seven parts by volume of ethylene chlorhydrin, three parts of ethylene dichloride and one part of carbon tetrachloride, is also effective by itself. Other chemicals which have been used to break dormancy are glycol monohydrin, methyl bromide, potassium thiocyanate and ethylene, although the latter then suppresses further sprout growth.

Tedin (1938) reported that storage for a period of time at low temperature followed by storage at conventional sprouting temperature could also reduce the dormant period in some varieties when compared with continuous storage at the same sprouting temperature and Wurr and Allen (1976) have shown that such a technique increased sprout development, stem emergence and early yield in the field. Thomas and Wurr (1976) have subsequently shown that this effect is probably caused by increased levels of gibberellins in the seed tuber following cold treatment and the break of dormancy induced by mechanical vibration of the seed tubers (Birkner, 1955) may well be stimulated similarly.

Sprout suppression

The suppression of sprout growth is usually associated with the storage of ware potatoes but it can be used to prolong dormancy and therefore to prevent seed potatoes from becoming physiologically too old. Chemicals which have been used include the potassium salt of naphthalene acetic acid (Guthrie, 1938) and tecnazene (Brown and Reavill, 1954).

In the U.K. sprout suppressants may be needed in mild winters where some varieties break dormancy early and give appreciable sprout growth before they can be graded out of store, but it is currently illegal to use any compound other than tecnazene on potatoes to be sold as seed and E.E.C. regulations completely ban the use of any germination inhibitors on seed. Even tecnazene should be used very carefully because Brown and Reavill (1954) found that its use could delay emergence and reduce the final yield if the seed was not allowed sufficient ventilation before planting – an adequate period was about six weeks.

Improvement of growth and yield

Once seed becomes old it tends to be predisposed towards the ‘little potato’ condition but Langille (1969) has shown that injection of the seed tuber with GA_3 and 2-chloroethylphosphonic acid (CEPA) may help plants to outgrow this condition. CEPA releases ethylene and it is thought that this may release an axillary bud on the apical sprout which then elongates in response to the applied GA_3.

In Eastern Europe there have been many reports that the yield of the crop has been improved by soaking seed in solutions of such compounds as heteroauxin, vitamin B12, $CuSO_4$, $ZnSO_4$, $MnSO_4$, $CoSO_4$, $Co(NO_3)_2$, $(NH_4)_2MoO_4$ and H_3BO_3. Some of these salt solutions are claimed to affect such factors as the rate of shoot growth, leaf chlorophyll content and nutrient uptake but their effect is unlikely to be directly physiological; it is more likely that they act by making up a deficiency that exists in the soil where the ware crop is grown.

Other treatments affecting plant growth and yield are: the irradiation, at appropriate doses, of seed tubers which can accelerate sprout development and increase yields; a similar effect caused by subjecting tubers to electromagnetic fields which possibly act by changing the metabolism of the seed tubers so that they give earlier tuber initiation (Pittman, 1972); and the effect of centrifugal forces which gave greater sprout development in the light, probably because they made N available for growth (Anonymous, 1971; and see Dinkel, 1968).

8.6 Future developments

8.6.1 The measurement and control of physiological age

Standards of physiological age are important yet at the moment they are not easy to apply. The main reason for this is the difficulty of defining and measuring physiological age and the adoption of a single measurement system would make the application of the concept much easier.

Physiological age can be partially measured on a time scale as Kawakami has shown yet there are many factors which affect the performance of seed of the same chronological age and it is important to be able to measure their effect quantitatively. One common denominator is that of temperature both during the growing season and in store and it may be that the effects of location, soil type, soil moisture and the times of planting and harvest operate as a result of different temperature regimes.

It is possible that physiological age could be measured on a temperature based time scale and Wurr (unpublished data) accumulated the mean temperature to which the tubers were subjected (in day degrees above a base of 0° C) from the time of initiation of the seed tubers through their growth and storage until they were planted the next year. The final yield of treatments using seed with a range of ages was quadratically related to physiological age measured on an accumulated temperature scale (Fig. 8.7).

This is essentially similar to the response of yield to chronological age reported by Kawakami and suggests that there is an optimum physiological age for final yield and that if this is exceeded, total yield is reduced. This is explained quite simply by the fact that seed which is physiologically old senesces early and is outyielded by younger (normal) seed, while

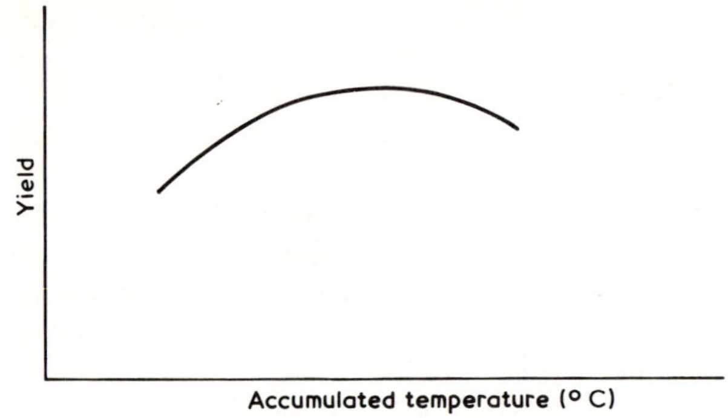

Fig. 8.7 The general relationship between final yield (cv. Désirée) and the temperature accumulated by the seed from initiation to planting.

very young seed does not have a long enough growing season to catch up normal seed.

How should such temperature records be made? The store temperature is easily recorded with a max/min thermometer or thermograph but soil temperature is more difficult to monitor. At the National Vegetable Research Station two contrasting techniques have been tried. The first involves the use of an electronic temperature integrating device which continuously monitors soil temperature with a thermistor and produces readings on a counter which are easily converted to mean or accumulated temperature. This technique uses a linear relationship between the counting rate and temperature (and therefore physiological age) but since most plant processes are the result of chemical reactions which proceed at rates that are exponentially related to temperature, the use of a system which allows for an exponential response may be more appropriate. This is indeed possible by using the inversion of sucrose (Jones, 1972) to measure accumulated temperature; a method which is cheap and suitable for use on sites that are visited infrequently.

This concept of measuring physiological age is only a first approximation but it seems to explain in general terms, many of the reported differences in performance of seed lots. However it needs refinement before it can be used more widely. For example, it may be more useful if applied over different phases of the crop's production and storage, or if a different base temperature is used and it may be necessary to allow for the amplitude of temperature fluctuation because there is limited evidence that both high and low temperatures can be important. Claver *et al.* (1957) have shown that high temperatures at tuber initiation appear to age seed while a period of low temperature can give an earlier break of dormancy than continuous storage at a higher temperature (Wurr and Allen, 1976).

There is ample evidence that variation in the physiological age of

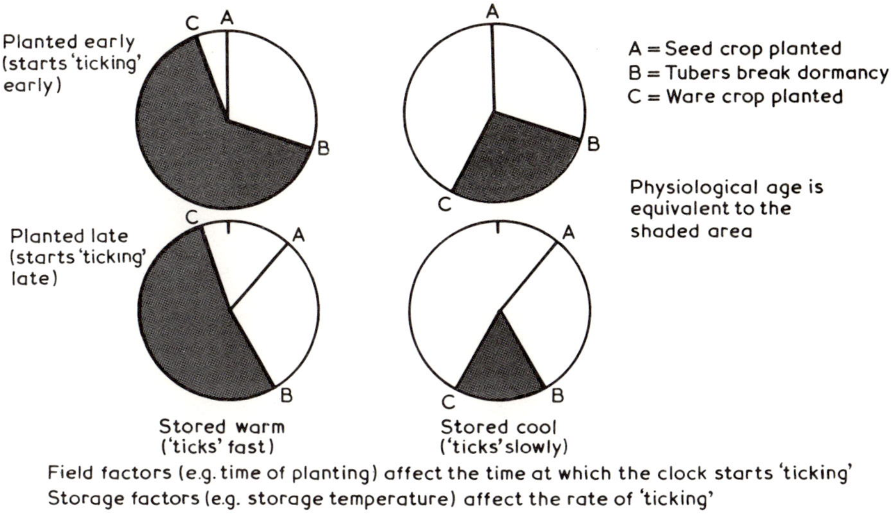

Fig. 8.8 Diagrammatic examples of the effects of field and storage factors on the physiological age of seed using the analogy of a clock.

seed lots caused by field factors has a much smaller effect than that caused by the manipulation of storage conditions but this does not mean that field conditions are unimportant. In the context of a 'ticking' clock analogy (see Fig. 8.8) field factors tend to influence the time at which the clock starts ticking (break of dormancy) while in general storage conditions affect the rate at which the clock 'ticks' (rate of sprout development). The combination of the two factors therefore affects the ultimate age of the seed. The clocks of different varieties stored in the same way may also tick at different rates which means that seed of different ages is produced.

At the moment the factors in the field which affect subsequent performance are incompletely understood but once physiological age can be measured accurately it will be possible for the ware grower to know what state seed is in when it is delivered to him and if his stores are suitably equipped he will then have the opportunity of adjusting conditions appropriately so that the seed is at optimum condition at planting. However at present relatively few growers have stores which are capable of such temperature control.

These optimum conditions will vary according to the market requirement and the variety and it may well be that varieties differ in their response of yield to accumulated temperature (Fig. 8.9). For example varieties might be of types A, B or C. A and B have different accumulated temperature requirements to give maximum yield but the shape of the two curves is similar, whereas C has the same optimum as A but is much less stable and this may give considerable yield fluctuations from year

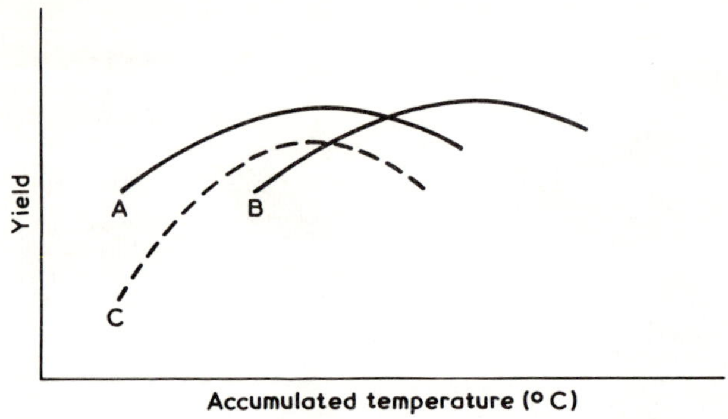

Fig. 8.9 A diagram suggesting possible yield responses to accumulated temperature of different varietal types.

to year. Yield increases rapidly to a maximum as accumulated temperature increases but beyond the optimum temperature it also drops rapidly as the variety moves into the 'little potato' condition.

Breeders might find that an assessment of the performance of a variety over a range of physiological ages would be a useful indication of the wider potential of that variety before it was released for general multiplication.

At present it is not clear precisely what measurements of sprout development at planting can be closely related to the physiological age of the seed and to the yield produced but if in the future the seed management factors affecting physiological age can be understood and controlled more accurately, it ought to be possible to produce seed lots of known ages which give particular levels of sprout development that are appropriate for specific market requirements. This is perhaps most relevant for the certified seed of early varieties which is produced in the cool conditions of Scotland or Northern Ireland, and of varieties for export, which are planted at a different time of the year to their domestic use. For example, seed for Mediterranean countries may be planted before Christmas and therefore must be physiologically mature when exported.

There is also a need for control of the age of certified and non-certified seed in the U.K. Manipulation of the physiological age of seed offers a grower the opportunity to achieve planned production of ware potatoes with crops maturing and being harvested sequentially, giving a consequent reduction in damage and its associated problems. Such control over the growth of the ware crop means that it might be possible to use fewer varieties than at the present by suitably manipulating the growth of those that are required.

References

Anonymous. (1970). *K.T.B.L. Versuchsstation Dethlingen. Jahresbericht* 1969.
Anonymous. (1971). *Agroborealis* **3**, (2) 4–7.

Appleman, C.O. (1914). *Maryland Agric. Exp. Station* Bulletin (183).

Bardeeva, A.S., Bÿchkov, A.A., Mozhaev, G.I. *et al.* (1955). *Zemledelie* (1) 122–3, abstracted in *Soils and Fertilizers* **19**, 103.

Birkner, E. (1955). *Z. landw. Vers. u. Unters Wes.* **1**, 245–54, abstracted in *Field Crop Abstracts* **10**, 118–9.

Boiko, N. (1968). *Kartofel'i ovoshchi*, (7) 8–11, abstracted in *Field Crop Abstracts* **22**, 68.

Bokx, J.A. De and Mooi, J.C. (1974). *Potato Res.* **17**, 410–33.

Broadbent, L., Brown, P.H. and Wright, R.C.M. (1961). *Expl Hort.* **5**, 37–9.

Brown, W. and Reavill, M.J. (1954). *Ann. appl. Biol.* **41**, 435–47.

Burton, W.G. (1963). In *The Growth of The Potato*, eds. Ivins, J.D. and Milthorpe, F.L., pp. 17–41. London: Butterworths.

Burton, W.G. (1965). *Eur. Potato J.* **8**, 80–91.

Burton, W.G. (1966). *The Potato*, 2nd Ed. Chapter 6, 130–1, Wageningen: European Association for Potato Research.

Burton, W.G. and Wilson, A.R. (1970). *Potato Res.* **13**, 269–83.

Claver, F.K., Tizio, R. and Montaldi, R.E. (1957). *Revta invest. agricolas. Buenos Aires* **11**, 209–14, abstracted in *Field Crop Abstracts* **12**, 48.

Cole, H.M. (1973). *Agric. Enterprise Studies in England and Wales*. Economic Report No. 19.

Davidson, T.M.W. and Lawley, D.N. (1953). *Empire J. exp. Agric.* **21**, 137–40.

Denny, F.E. (1926). *Contributions. Boyce Thompson Institute Pl. Res.* **1**, 154–68.

Dinkel, D.H. (1968). *Proc. Am. Soc. hort. Sci.* **93**, 476–84.

Dubuffet, G. (1954). *C. r. hebd. Séanc. Acad. Agric. Fr.* **40**, 84–8, abstracted in *Field Crop Abstracts* **7**, 161.

Emilsson, B. (1949). *Acta agric. suec.* **3**, 189–284.

Emilsson, B. (1961). *Växt-närings-nytt* **17**, (3) 1–4, abstracted in *Soils and Fertilizers* **24**, 486.

Fillippov, D.I. (1964). *Dokl. Akad. Nauk.* (11) 9–12, abstracted in *Soils and Fertilizers* **28**, 295.

Fischnich, O. and Krug, H. (1963). In *The Growth of The Potato*, eds. Ivins, J.D. and Milthorpe, F.L. pp. 72–8. London: Butterworths.

Goodwin, P.B., Brown, A., Lennard, J.H. and Milthorpe, F.L. (1969). *J. agric. Sci., Camb.* **73**, 167–76.

Gray, D. (1973). *Ann. appl. Biol.* **75**, 83–91.

Gray, D. (1974). *J. agric. Sci., Camb.* **82**, 363–9.

Guthrie, J.D. (1938). *Science* **88**, 86.

Hammes, P.S. and Beyers, E.A. (1973). *Potato Res.* **16**, 68–72.

Harrington, F.M. (1935). *Proc. Am. Soc. hort. Sci.* **33**, 560–2.

Headford, D.W.R. (1962). *Eur. Potato J.* **5**, 14–21.

Henriksen, J.B. (1973). *Proc. Fifth Triennial Conf. Eur. Ass. Potato Res.* 1972, 114.

Hiele, F.J.H. Van. (1961). *Eur. Potato J.* **4**, 26–39.

Holmes, J.C. and Gray, D. (1972). *Potato Res.* **15**, 220–35.

Holmes, J.C., Lang, R.W. and Singh, A.K. (1970). *Potato Res.* **13**, 342–52.

Iritani, W.M. (1967). *Am. Potato J.* **44**, 153–8.

Iritani, W.M. (1968a). *Am. Potato J.* **45**, 111–6.

Iritani, W.M. (1968b). *Am. Potato J.* **45**, 322–6.

Ivins, J.D. and Bremner, P.M. (1965). *Outlook on Agriculture* **4**, 211–17.

Jarvis, R.H. and Palmer, G.M. (1973). *Expl Husb.* **24**, 29–36.

Jarvis, R.H., Palmer, G.M. and Shotton, F.E. (1973). *Expl Husb.* **24**, 41–7.

Jones, R.J.A. (1972). *Soils and Fertilizers* **35**, 615–9.

Karavaeva, N., Esova, L. and Kholopova, É. (1973). *Kartofel' i ovoshchi* (6) 9–10, abstracted in *Field Crop Abstracts* **26**, 600.

Kawakami, K. (1953). *Mem. Hyogo Univ. Agric.* **2**, 1–114.

Kawakami, K. (1973). *Proc. Fifth Triennial Conf. Eur. Ass. Potato Res.* 1972, 151–2.
Kozlowska, A. (1963). *Eur. Potato J.* **6**, 143–59.
Krijthe, N. (1962). *Eur. Potato J.* **5**, 316–33.
Langille, A.R. (1969). *Can. J. Pl. Sci.* **49**, 800–2.
Madec, P. and Perennec, P. (1956). *Annls Amélioration des plantes* **6**, 5–26, abstracted in *Field Crop Abstracts* **10**, 118.
Madec, P. and Perennec, P. (1959). *Eur. Potato J.* **2**, 22–49.
Madec, P. and Perennec, P. (1962). *Annls physiologie végétale, Paris* **4**, 5–84.
Messer, H.J.M., Lindsay, R.T. and Neale, M.A. (1972). *Potato Res.* **15**, 246–56.
Ministry of Agriculture and Fisheries (1923). *Collected leaflets on Cultivation and Diseases of Potatoes*, 2nd Edn. 18. London: H.M.S.O.
Ministry of Agriculture, Fisheries and Food (1974). *Terrington fourteenth Annual Review 1973*, 20–2.
Murphy, H.J., Manzer, F.E., Goven, M.J. and Merriam, D.C. (1967). *Am. Potato J.* **44**, 227–31.
Neubauer, L.W., Hoyle, B.J. and Kucera, E. (1965). *Trans. Am. Soc. agric. Engineers* **8**, 138–40.
Newton, W. (1927). *J. agric. Res.* **35**, 141–6.
O'Brien, P.J. (1978). Effects of seed crop treatments on the growth and yield of progeny tubers of early potato cultivars. Ph.D. Thesis, University College of Wales, Aberystwyth.
Pittman, U.J. (1972). *Can. J. Pl. Sci.* **52**, 727–33.
Rønsen, K. (1971). *Forskning og forsøk i landbruket* **22**, 241–66.
Rosa, J.T. (1928). *Hilgardia* **3**, 99–124.
Schepers, A., Hoogland, R.F. and Krijthe, N. (1969). *Eur. Potato J.* **12**, 251–63.
Short, J.L. and Shotton, F.E. (1968). *Expl Husb.* **16**, 93–8.
Short, J.L. and Shotton, F.E. (1970). *Expl Husb.* **19**, 69–77.
Shotton, F.E. (1964). *Expl Husb.* **11**, 1–9.
Shotton, F.E. and Palmer, G.M. (1972). *Expl Husb.* **21**, 109–22.
Singh, H.P. and Verma, V.S. (1970). *Ind. J. Agron.* **15**, 183–4.
Smith, O. (1937). *Cornell Univ. agric. Experiments Station* Bulletin No. 663.
Staĭkov, G. and Chavdarov, C.H. (1971). *Rasteniev' 'dni Nauki* **8**, 45–53.
Svensson, B. (1966). *Växtodling* **21**, 1–86.
Tedin, O. (1938). *Botaniska notiser* for 1938, 257–66.
Thomas, T.H. and Wurr, D.C.E. (1976). *Ann. appl. Biol.* **83**, 317–20.
Thow, R.F. (1970). Second generation effects of NPK on the potato. Ph.D. Thesis, University of Edinburgh.
Tomkins, R.G. (1966). *Agric. Res. Council Ditton Laboratory Annual Report* 1965–6, 30–1.
Toosey, R.D. (1964). *Field Crop Abstracts* **17**, 161–8.
Voĭtekh, A.D. (1971). *Khimiya v Sel'skom Khozyaistve* **9**, 57–8, abstracted in *Field Crop Abstracts* **24**, 706.
Walker, M.G. (1968). The effects of nutrient treatments of potato plants on the performance of their progeny. Ph.D. Thesis, Bath University of Technology.
Went, F.W. (1959). *Am. J. Bot.* **46**, 277–82.
Werner, H.O. (1929). *J. agric. Res.* **38**, 405–10.
Werner, H.O. (1940). *Am. Potato J.* **17**, 66–80.
Workman, M. and Twomey, J. (1967). *Proc. Am. Soc. hort. Sci.* **90**, 268–74.
Workman, M. and Twomey, J. (1969). *J. Am. Soc. hort. Sci.* **94**, 260–2.
Wurr, D.C.E. and Allen, E.J. (1976). *J. agric. Sci., Camb.* **86**, 221–4.
Yablonskii, E.A. (1959). *Soviet Pl. Physiol.* **6**, 443–50.

Chapter 9

Mechanization and crop performance

R.H. JARVIS

9.1 Introduction: the development of mechanized production techniques

In the last 25 years techniques of potato production throughout the world have changed radically. Although developments in western Europe have not been as rapid as in the U.S.A., mechanization has nevertheless reached a more advanced level there than in some other parts of the world, and the changes which have taken place in Great Britain may therefore be taken as an example. In 1948, 65–70% of the area carrying maincrop varieties was planted by hand (Boyd and Dyke, 1950); in 1973, the last year for which adequate survey data are available, the figure had fallen to 3% (Potato Marketing Board, 1975). Complete potato harvesters were not mentioned in the report of the 1948 survey but in 1958, 3% of the crop was lifted with machines of this type (Potato Marketing Board, 1959). This figure rose to 18% in 1963, 30% in 1968 and to 57% in 1973 (Potato Marketing Board 1965, 1970, 1975).

The earlier machines, both planters and harvesters, depended heavily on hand labour to obtain satisfactory results, with a planter requiring one man for each row and a harvester a crew of at least four to separate trash, clods and stones from the tubers. Indeed on at least one complete harvester marketed in the early 1960s an alternative arrangement was provided so the potatoes could be picked out from the detritus. Recent developments have been aimed primarily at reducing the manning requirement and increasing the work rate so that the labour input per unit area can be reduced. Particularly with planters, considerable success has been achieved. For instance in a small survey carried out in 1974, only 36% of the planters required one operative per row, the remainder requiring at the most one operative per two rows. Work rates ranged from 1.74 ha day^{-1} with the older machines to 3.24 ha day^{-1} with the more modern types (Shotton, 1976). Similarly, of the harvesters covered in a survey in 1971 (Ministry of Agriculture, Fisheries and Food 1972), 16% were described as 'unmanned'.

Other changes have taken place which although of less fundamental importance are nevertheless not without significance in the present context. A notable development has been the introduction of herbicides for use in the potato crop; from the first commercial marketings in 1960, usage increased rapidly so that 44% of the crop was treated in 1968 and 67% in 1973 (Potato Marketing Board 1970, 1975). The tendency to adopt increased row widths must also be noted. While in 1968 only 1% of the crop was planted in rows wider than 76 cm, 14% of the crop fell into this category in 1973, and 9% were planted at a row width of 91 cm (Potato Marketing Board 1970, 1975). The move to wider rows has been more pronounced on farms where complete harvesters are used. In 1971 23% of the farms surveyed used a row width greater than 76 cm and 17% used rows 91 cm apart (Ministry of Agriculture, Fisheries and Food, 1972).

Although both the introduction of herbicides and the adoption of wider rows may in some cases have led to increases in ware yield (Jarvis and Shotton, 1972a, b), these measures have proved attractive to the farmer primarily because they have enabled him to reduce labour and machinery inputs. In the case of herbicides, a single pass with a sprayer can replace a number of post-planting cultivations aimed at controlling weeds. The use of wider rows may lead to reductions in labour and machinery requirements in several ways: it is not necessary to cultivate as deeply in order to obtain sufficient tilth to build an adequate ridge, thus reducing inputs at this point and also reducing the risk of producing clods which can lead to difficulties at harvest; provided a machine can travel at the same speed and deal with the same number of rows in one pass as on narrower rows, there will be a reduction in machinery time required because the length of row per unit area is less; and because there is more space between the rows, there is less likelihood of tractor wheels consolidating the ridge during inter-row cultivations or at harvest. The adoption of techniques of this type may therefore be attributed, as in the case of the widespread adoption of mechanical planters and complete harvesters, to the increasing scarcity and cost of labour. The potato crop has been particularly vulnerable to these pressures because of its traditional dependence on hand labour. Even now its labour requirement is higher than any of the other common agricultural crops grown in temperate regions. It therefore seems likely that the pressures of the past 25 years will continue in the foreseeable future. The need for the development of more effective mechanized production techniques will continue; it will be equally necessary for the particular requirements of mechanized production systems to be considered at all stages of research and development work on all aspects of the potato crop.

The reasons for this latter statement are two-fold. First, research workers need to be aware of the limitations which mechanized production techniques may impose on the particular aspect with which they are concerned; and second, there are certain problems concerned with mechanized production which require an agronomic rather than a mechanical solution.

9.2 Variation in farmers' requirements: place in the farming system

Because potatoes are grown in a number of widely differing farming systems, the degree of mechanization that a farmer requires may vary considerably. In areas where potato growing has remained linked to intensive vegetable production, as for instance on the silts of south Lincolnshire, labour input has remained at a high level on the majority of holdings. With a wide choice of high-value cash crops which can be grown, the farmer will normally aim at the highest possible return per unit area from potatoes and is in a position to use a considerable amount of labour if necessary to achieve this end. In these circumstances the grower retains the freedom to reject those aspects of mechanization which tend to reduce the overall profitability of the crop and adopt only those which are of direct benefit in his particular circumstances.

At the other extreme, further north in the same county the farmer on the Wolds is concerned mainly with cereal production and regards potatoes primarily as a 'break crop'. The labour force on the farm is very small, being limited by the requirements of the chief enterprise, and it is therefore necessary to mechanize the whole system of potato growing as fully as possible. In this case it is the return per unit of labour which must be considered first and the return per unit area from the potato crop can be maximized only within the limits imposed by the size of the labour force. In the Netherlands, where a similar situation has arisen although for different reasons, the need to economize in labour is leading to the development of an extremely sophisticated system of mechanized production, depending heavily on co-operatively or contractor owned machinery.

Although these two examples in no way cover the whole range of possibilities they serve to illustrate the principle that the extent to which potato production is mechanized may depend on a number of factors not directly connected with the crop itself. The level of mechanization may in some instances affect the general level of crop performance and in order to minimize possible adverse effects, the development of different husbandry techniques to suit different levels of mechanization becomes necessary.

9.3 General implications of mechanization

Machines have a number of general advantages compared with hand labour. Of particular importance in the present context is their ability both to run for long hours without pause and to maintain a consistent standard of work. However, with simple machines these desirable features will be realized only if working conditions and raw materials are uniform. In agriculture in general and perhaps especially with potato production, several of the factors which have to be contended with are notoriously lacking in uniformity. In particular, soil conditions may vary appreciably from hour to hour and from one part of a field to another; the size and shape of potato tubers may be extremely variable also. It is this lack of uniformity which gives rise to many of the problems associated with mechanized potato production and generally the best solutions to the various problems which have been encountered have involved both an improvement in the machine and a reduction in the variability.

Further difficulties arise in mechanized production systems because of the susceptibility of potato tubers to mechanical damage. Seed potatoes, particularly sprouted seed, require to be handled gently if their performance is not to be impaired. This is a major consideration in areas such as western Europe where soil temperatures frequently remain low for some time after planting and the early growth of the crop is slow; in such conditions damage to seed and sprouts may be of considerable significance and in many cases reduce yield (e.g. Jarvis and Palmer, 1973). In the potato-producing areas of the U.S.A., however, where growing conditions at planting time are normally very good, damage to seed is of little importance and farmers may plant freshly-cut seed-pieces with impunity.

Ware potatoes too are easily damaged by falling on to hard surfaces or as a result of blows received from machine parts during harvesting and subsequent handling. Such damage may impair the appearance of the tubers, it may lead to difficulties in storage and it may in the more extreme cases render the tubers unsaleable.

The susceptibility to mechanical damage of both seed and ware tubers places a severe restriction on the design of machines which handle potatoes at any stage of the production process. Such operations as the accurate spacing of sets along the row and the accurate size-grading of ware tubers would be much easier to mechanize if it were not for this liability to damage.

It is clear then that a successful mechanized production system must be designed in such a way that the maximum benefit is obtained from the advantages offered by the various machines while possible adverse effects must be minimized. Because of the variation in farmers' requirements, no single solution can be offered to the problems of mechanization but the same principles of problem solving need to be applied in all cases. By detailing the relative importance of the various considerations involved, it becomes possible to indicate how a system may be built up to suit a specific set of requirements.

9.4 Main problem areas

The three main areas of potato production where the greatest problems of mechanization lie are in the planting, harvesting and grading of the crop. These are the operations which traditionally required large amounts of labour and in which the replacement of hand labour by machines has been associated with the greatest difficulties. Each of these areas requires detailed consideration but because of the complicated relationships which exist between them, none can be dealt with in isolation. For the same reason, a clearer understanding of the problems which are involved can be obtained if the topics are discussed in reverse chronological order.

9.5 Detailed consideration of the main problem areas

9.5.1 Grading

Objectives

The object of grading or dressing potatoes is to separate saleable tubers from those which are unsaleable. Unsaleable potatoes normally include those which are green, cracked, mis-shapen or mechanically damaged and those which have been attacked by pests or disease organisms to an extent which has made them unacceptable to the consumer. Grading also involves sorting the tubers according to size. Usually very small tubers are not required and frequently very large tubers too are unwanted. Current (1975) standards set by the Potato Marketing Board in Great Britain for instance specify that maincrop potatoes sold for ware must be capable of passing through a 82.5 mm square mesh riddle but not through a 41 mm riddle. In some cases more than one size of saleable tubers may be riddled out. With crops grown for seed the smaller tubers are sold for this purpose while the larger are disposed of on the ware market. In the United Kingdom the traditional dividing point for the two grades is 57 mm. Similarly, in an early crop small potatoes (usually under 45 mm) may be sold for canning and the larger ones for ware use. The two aspects of grading, that is the removal of off-grades and size grading, may be considered separately.

Size grading

Sizing is a comparatively simple process to mechanize and mechanical riddles have been widely used for a number of years. The most commonly encountered difficulty arises if the tubers are long rather than round or short oval. In this case tubers tend to pass over the sieves with their long axes horizontal, which results in inaccurate grading. It can usually be prevented by agitating the riddles vertically as the tubers pass over them but this may lead to an increase in the amount of mechanical damage inflicted. This in itself is frequently a problem at grading particularly in climatic regions where the operation is usually carried out during the colder winter months. Unless care is taken to ensure that the temperature of the potato store is raised to 15–20° C for a long enough period for all tubers to reach this ambient temperature before grading, very high levels of internal bruising (blackspot) may result (Burton, 1973).

Off-grade removal

It is this aspect of grading rather than sizing which usually gives rise to much greater difficulty and it is the separation of tubers which are defective in some way which has in most cases made it necessary to retain a considerable amount of hand labour on the grading line. Little or no progress has been made in mechanizing this aspect of grading and because of the diverse nature of the defects which necessitate the removal of tubers, no straightforward mechanical solution is likely to be found. Improved labour utilization in grading can therefore be achieved only by reducing the number of sub-standard tubers to a minimum, an objective which is clearly desirable for other reasons also. The following aspects are relevant.

Seed-bed preparation

The quality of the seed-bed will influence the uptake of nutrients and water during the whole of the growing season and may therefore affect the incidence of cracking and second growth.

Variety

Greening, cracking, second growth, susceptibility to mechanical damage and to certain tuber diseases and pests are all factors which are affected by the choice of variety.

Planting

The depth of planting influences the level at which the tubers are formed and so affects the incidence of greening (Jarvis, 1973a).

Inter-row cultivation

The final ridge must be large enough to prevent greening as far as possible. Cultivations must produce sufficient tilth for the ridge without leading to excessive drying which may, in the absence of irrigation, increase the incidence of cracking, second growth and the production of badly-shaped tubers.

Blight control

Ineffective blight control measures lead to tuber infection and so increase the number of sub-standard tubers.

Pest control

Good control of pests which attack the tubers such as slugs and wireworms is necessary.

Harvesting

Tubers rejected during grading because of mechanical damage have frequently received the damage during harvesting or loading into store. In addition damage inflicted at this stage may lead to rotting during storage and again add to the number of tubers to be rejected at grading. Factors affecting the incidence of damage at harvest are dealt with in detail in Section 9.5.2 below.

Storage conditions

Unsuitable conditions may encourage the development of such diseases as gangrene (*Phoma exigua* var. *foveata*), skinspot (*Oospora pustulans*) and silver scurf (*Helminthosporium solani*); tubers blemished by these diseases must be removed during grading although the last mentioned is unlikely to seriously impair quality unless a sample is being prepared for pre-packing. Poor storage conditions may also lead to the development and spread of wet rots. In addition to the obvious loss, wet rotting slows up the grading process because the decaying tubers turn to a slurry and this, combined with soil, adheres to the healthy tubers. Apart from detracting from the appearance of the sample, it then becomes more difficult to detect other blemishes.

9.5.2 Harvesting

Objectives

A well designed and properly operated potato harvester should effectively separate tubers from the haulm and from soil, clods and stones. It should do this quickly and inflict only the unavoidable minimum of mechanical damage during the process. The first potato harvesters were required only to leave the tubers exposed on the soil surface and they were then picked by hand. A 'complete' potato harvester however is normally required to transfer the harvested crop into a cart travelling alongside or, less commonly, to carry the produce in a hopper to the end of the field before transferring it to some form of vehicle for transport to the store. Alternatively, if the crop is to sold straight from the field (a situation commonly found with early potatoes), some form of grader may be incorporated and the tubers placed in bags or sacks on the machine. Since the crop is then sold without any further inspection, it is particularly important in this case that a proper standard of separation is achieved.

Advantages and disadvantages of complete harvesters

A complete potato harvester offers a farmer the means of harvesting his crop with considerably less labour than is required by any other method. By reducing the physical effort required from the harvesting gang and by reducing the dependence on casual labour, the complete harvester enables work to continue for long hours; indeed, given a well-organized labour force, harvesting can continue for 24 h a day with breaks only for lubrication and other essential maintenance. However, many farmers have been reluctant to commit themselves to complete harvesting because of the various problems which they have experienced with these machines in practice.

It is not intended to describe the construction and development of potato harvesters in any detail. In general the objective of both the designer and the user has been to reduce to a minimum the number of operatives required on the machine. Indeed totally unmanned harvesters have been in use on the polder soils of the Netherlands and elsewhere for a number of years. These machines have a large separation area for the removal of loose soil and are capable of high outputs with low levels of mechanical damage. However, they cannot effectively separate appreciable quantities of stones or clods from the tubers. Although some progress has been made, it is probably fair to say that any device likely to be developed in the foreseeable future which will separate stones and more particularly clods from the crop will inevitably be complicated and expensive. Most machines therefore carry a number of operatives to perform this function; it should be the aim of the farmer to grow his crop in such a way that the number of workers required on the machine can be kept to a minimum and if possible the need for them removed altogether. Essentially this involves ensuring that only a very few clods require to be picked from the machine by hand. Apart from adding to the problem of separation, high clod numbers are frequently associated with high levels of mechanical damage. There appear to be two factors involved in this association. In experimental work, small increases in mechanical damage have been found to result from large increases in clod number (Ministry of Agriculture, Fisheries and Food, 1968); a more potent factor in practice may be that when many clods are present, agitation and high web speeds are used in attempts to break them and it is probable that it is these adjustments rather than the clods themselves which lead to the increased damage.

Husbandry factors affecting harvester performance

A number of aspects of harvester performance may be influenced by husbandry factors within the farmer's control. The more important of these factors include the following.

Selection of soil type

Although light sandy soils are usually very suitable for mechanical harvesting, they will produce satisfactory yields only if rainfall is adequate or irrigation available. In general, medium loams, light or medium silts and peats are most suitable for potato production. On these soils problems connected with clod formation are likely to arise, particularly in seasons following mild winters, but will not be as acute as on heavier land. On soils which have a moderately high stone content but are otherwise suitable for the crop, the use of a mechanical stone crusher or picker should be considered. It is impossible that potato harvesting will be successfully mechanized on very heavy, very strong or steeply sloping land. Even if future engineering developments lead to solutions of the problems which arise in these particularly difficult circumstances, the cost of the equipment is likely to be such that the farmer would be at a marked commercial disadvantage compared with his competitors.

Seed-bed preparation

It is particularly important to avoid producing clods in the course of pre-planting cultivations. Even on soils with only a moderately high clay content, clods formed by the injudicious use of implements in early spring usually remain in the ridge until harvest. Because of the wide range of variations in soil and climate it is unsafe to put forward detailed recommendations but nevertheless the following principles have wide practical application.

(1) Cultivations should be confined to the depth of weathered soil. Provided that autumn cultivations have been properly carried out and excessive consolidation avoided during cultivation, crop roots will penetrate freely into the lower, unworked layers but stolon growth and tuber formation will be confined to the looser parts close to the original surface and in the ridge. If unweathered pieces of soil are brought to the surface they will dry out as clods except on soils with a very light texture. In addition, if the lower layers of the soil are disturbed, tubers will tend to be formed at a lower level and the harvester share must be set to work in this partly cultivated soil to avoid slicing them; slower work and over-loading of the harvester webs will result.

(2) Implements with straight rather than curved lines should be preferred because they are less likely to bring unweathered soil to the surface.

(3) Soil compacted by tractor wheels may be disturbed by subsequent cultivations and form clods. The loading on tractor wheels can be reduced by using power-driven implements and clearly the risk of compaction is reduced if the number of cultivations is kept to a minimum.

(4) The timing of cultivations in relation to soil moisture content is critical. Although in many countries or areas experimental work has indicated the existence of an optimum planting date, in some circumstances it may be advisable to delay planting beyond this date if by so doing a better seed-bed can be prepared. Indeed if the crop is to be lifted by unmanned harvesters, no alternative choice may be available to the farmer in a wet spring.

Manuring

The susceptibility of potato tubers to mechanical damage can be affected by the supply of plant nutrients (Pätzold and Dambroth, 1970) although in many circumstances departures from the dressings recommended for maximum economic yield lead to an increase rather than a reduction in susceptibility. However, it has been shown that high levels of potassium in the tubers are associated with reduced susceptibility to internal bruising (van der Zaag and Meijers, 1970) and it is possible that in some circumstances higher rates of application would be justified for this reason.

Choice of variety

Different varieties produce tubers at different levels in the soil. If the tubers are formed relatively deeply, the depth of the harvester share will have to be increased and the separation problem increases.

Other varietal differences which are of importance in this context are susceptibility to mechanical damage and maturity date. Varieties differ in their susceptibility to both splitting and bruising (Pätzold and Dambroth, 1970; Blight and Hamilton, 1974) and low susceptibility is clearly a desirable character. With regard to maturity, a variety which reaches its full yield comparatively early may be harvested when soil conditions are relatively dry and warm (see below).

Depth of planting

Depth of planting appears to influence the depth of tuber formation (Jarvis, 1973a) and should therefore be as shallow as is consistent with the need to build a ridge to provide adequate soil cover over the sets.

Uniformity of planting depth is important also. If this is not achieved, accurate setting of the harvester share is impossible and although the machine may be overloaded with soil in some places, in others the share slices through the tubers.

Row width

Row width has an important effect on mechanical harvesting. Increasing the row width enables the crop to be harvested more rapidly by reducing the length of row per unit area; it leads to the presence of fewer clods in the ridge because it is not necessary to work so deeply to produce sufficient tilth for an adequate ridge; it reduces the risk of clod formation during post-planting cultivations by the action of tractor tyres on the side of the ridge; and it reduces the risk of damage which occurs particularly in wet seasons to the ridge and to the tubers by the tyres of the tractor drawing the harvester (Jarvis, 1972).

It has been shown that in some areas row width can be increased considerably beyond the distance traditionally used without adversely affecting the yield of maincrop varieties. For instance Jarvis and Shotton (1972a) and North and Proctor (1973) compared 75 with 91 cm rows in eastern England. They found that provided the same weight of seed was planted per unit area, similar ware yields were obtained from both row widths. In cases where greening was a problem, a reduction in its incidence on the wider rows led to a slight increase in ware yield. It is clearly desirable that advantage should be taken of the benefits of wider rows wherever possible.

Post-planting cultivations

Although post-planting cultivations are usually aimed at increasing the amount of tilth available for incorporation into the ridge as well as controlling weeds, this objective is not always achieved (Kouwenhoven, 1967). If an adequate ridge can be built at or immediately after planting, there is much to be said in favour of dispensing with further inter-row cultivations and using a residual herbicide to control weeds. In this way the frequent passage of machinery through the crop and the attendant risks of producing clods are avoided.

Time of lifting

Lifting under very dry conditions should be avoided because tubers which have lost turgor are extremely susceptible to internal bruising (Thornton *et al.*, 1973). Ideally the soil should be moist but friable and in temperate climates such conditions are likely to be found early rather than late in the traditional lifting season. Early lifting has the added advantage that soil temperatures are higher; at low temperatures susceptibility to internal bruising increases considerably (Burton, 1973).

9.5.3 Planting

Objectives

Ideally a potato planter should be capable of placing the sets accurately along the row and at constant depth and, if the seed has been chitted, without damaging the sprouts. It should be capable of a high output with a minimum of labour and be so constructed that associated activities such as transferring seed to the hopper can be carried out quickly and easily. Other desirable attributes include ease of adjustment of set spacing, depth of planting and the coverers. With most early machines the sets were placed by hand into cups or pockets situated on the periphery of a wheel which carried them to the planting shoe where they were dropped into the soil. The advantages and disadvantages of these machines are now only of historical interest; the most obvious disadvantage was their slowness, forward speed being limited by the rate at which the operators could place sets in the cups.

Factors affecting planter development

The advantage of speed of planting in as far as it affects timeliness of planting is discussed in detail below. From the farm management point of view potato planting is noteworthy for its slowness compared with the other common arable crops and increased planting rates are desirable to enable machinery and labour to be released for other tasks at a time of year when on many farms there is a strong demand for both. In the same way a reduction in the labour requirement from 5 to 2 men (including tractor driver) for a 4-row planter is a welcome improvement; indeed on some holdings the labour force has diminished to such an extent that a low requirement in this respect is essential. Faster work rates and lower labour requirements have therefore become the principal objectives in planter design. Since planters are in addition required to handle seed potatoes gently and position them accurately, it is perhaps not surprising that development has been relatively slow. This lack of progress has led to a re-assessment of the whole planting process including such aspects as seed-bed preparation and seed handling which has enabled more efficient planting systems based on existing machinery to be evolved. The topic is of particular interest because it illustrates how a combination of disciplines, in this case, work study, engineering and agronomy, can provide an acceptable solution to a practical agricultural problem. It is therefore discussed in some detail in the section below.

Improvement of planter utilization

The two most common types of planter in use in the U.K. at present are the Continental cup-fed machines, in which seed is picked up from the hopper by cups mounted on an endless chain or belt and then carried to the planting shoe; and the belt-fed type, with seed flowing from a hopper into a trough which periodically tips the sets on to moving rubber belts extending backward to the shoe. Both types are usually referred

Table 9.1 Effect of forward speed on planting activity duration

Planter type	*Hand-fed*	*Cup-fed*	*Belt-fed*
Forward speed (km h⁻¹)	*2*	*4*	*10*
Activity:		Min h^{-1}	
Planting	44	35	23
Turning	4	7	9
Filling	12	18	28

to as 'automatic' but 'semi-automatic' is a more accurate term since with all belt-fed and most cup-fed planters an operator is required to ensure that the flow of seed is continuous.

The most striking feature to be found in work study data on potato planting is almost invariably the considerable difference between 'spot' and 'overall' work rates. A typical 'spot' rate of 1.5 ha h^{-1} for a 2-row belt-fed planter working in 76 cm rows may be contrasted with the 'overall' rate of 0.38 ha h^{-1} reported by Shotton (1976) as the average of 75 machines of this type. In a number of cases part of this discrepancy has been found to be due to organizational problems in that some farmers have been slow to appreciate that the higher potential output of the modern planter demands more rapid seed-bed preparation and an improved supply of seed compared with older techniques. However even when these ancillary operations are not limiting, a considerable difference remains to be accounted for. The reason may be found in a detailed study of the planting operation itself.

Table 9.1 shows a typical breakdown of the time spent on the various activities directly connected with planting when using machines of different types. Most seed is handled in chitting trays or sacks and an average filling rate of 1 min (50 kg) $^{-1}$ has been assumed. It has also been assumed that each machine is planting 2 rows 76 cm apart in a field 300 m long at a seed rate of 2.5 t ha $^{-1}$ and that 1.5 min are required for each headland turn.

It is clear that with the higher speed of the more modern types of planter, the proportion of time spent actually planting is considerably decreased and that if planter output is to be increased the proportion of non-productive time must be reduced. There is little opportunity for improving turning time but the time spent in filling the planter hopper with seed could be reduced appreciably if a bulk handling technique were to be adopted. Projected outputs with different filling rates are shown in Table 9.2 which is based on the same premises as Table 9.1

Table 9.2 Projected effect of filling rate on activity duration and planter output

Activity	*Filling rate (min (50 kg)$^{-1}$)* 1	0.5	0.3
		Min h^{-1}	
Planting	23	29	33
Turning	9	13	15
Filling	28	18	12
Output (ha h^{-1})	0.44	0.56	0.63

except that a 25% time allowance for repairs, adjustments and personal needs has been included.

If the filling rate could be reduced from 1 to 0.3 min (50 kg)$^{-1}$, the proportion of time spent actually planting would increase by 43% and a similar percentage improvement obtained in planter output. There are no basic engineering problems in handling seed in 500 kg pallet boxes and tipping it into the planter hopper with the aid of a suitable fork-lift attachment in 3 min. Alternatively a 'high tip' trailer can be used and both methods have been employed by farmers with considerable success when the seed has not been sprouted. In this case the only difficulty is that the seed cannot be properly inspected during storage prior to planting; high quality seed and good environmental control are therefore essential. However in the case of sprouted seed there are several problems concerned with bulk storage and handling which require to be solved.

In the U.K. seed potatoes are traditionally sprouted in trays approximately 76 cm × 46 cm × 15 cm each holding 16–17 kg and similar containers are used elsewhere. With vertical corner-posts approximately 25 cm high, the trays can be stacked so that air circulates freely between them and the seed is reasonably well illuminated. In standard types of 500 kg pallet boxes it is difficult to ensure adequate ventilation and only the surface layers of seed, a small proportion of the whole, receive adequate light. One possible solution is to keep the seed cool until just before planting and then to sprout for a brief period to give short sprouts perhaps 4–5 mm long. However, this system has a number of disadvantages: in practice it is difficult to define 'just before planting' and if farm work is delayed by adverse weather, excessive sprout growth may occur; seed with short sprouts frequently gives appreciably lower yields under U.K. conditions than traditionally sprouted seed (Shotton, 1973); and contrary to a widely held belief, short sprouts are no less susceptible to damage by the planter mechanism (Jarvis, 1973b). A more useful solution appeared to be to develop a purpose-built pallet-based container in which seed could be chitted to give sprouts of the traditional type about 12–18 mm long.

At the time of writing several prototype bulk sprouting containers with a capacity of 500–600 kg have been constructed for testing and at least one is available on the U.K. market. All are constructed mainly of metal mesh with the minimum amount of heavier material required for strength and rigidity. All are based on a 'layer' or 'cell' concept; the seed is contained by the metal mesh in cells about 15 cm deep, separated from adjacent cells by an air space and arranged horizontally, vertically or diagonally according to type. Preliminary trial results suggest that seed sprouted in these containers may yield as well as seed sprouted in the traditional type of tray but the investigations have not yet been completed (Ministry of Agriculture, Fisheries and Food, 1975b). Techniques of this type are likely to provide a useful method of reducing labour requirements and improving planter performance in the short to medium term. In the longer term it is to be hoped that an improved understanding of the physiological ageing process in seed potatoes will enable methods of seed treatment to be devised so that farmers will no longer need to sprout their seed even in situations where chitting is currently recommended.

Other aspects of planter performance

There are a number of aspects of planter performance which may have a direct effect on yield and others which affect off-grade production and so influence saleable yield.

Speed of planting

In a given set of circumstances there is a simple relationship between speed of planting and time of planting. The general effect of time of planting on yield is well-known but data are not sufficiently abundant to enable accurate predictions to be made in particular situations. Normally, early planting results in earlier tuber initiation but this is sometimes accompanied by earlier leaf senescence (Ivins, 1963). In the case of potatoes grown for early ware production there is the added complication that prices decline rapidly as the season advances and normally the object is to obtain a harvestable crop as early as possible. The usual procedure therefore is to plant as soon as soil conditions permit. In the case of maincrop potatoes to be lifted at maturity the position is rather different. There is some evidence that very early planting is undesirable, particularly if the seed is well sprouted (Baldwin, 1964) and it is generally accepted that in most of the ware producing areas of the U.K. the target date should be late March or early April.

Radley (1963) established that for plantings after the second or third week of April yield declined at a rate of about 0.75 t ha^{-1} $week^{-1}$ until mid-May when the decline rate accelerated. More recent work has suggested that this figure may be an underestimate (Ministry of Agriculture, Fisheries and Food, 1975a). However, an improvement in speed of planting alone will not enable a farmer to plant the whole of his crop in early April if previously he had planted it all in late April. A more useful change to envisage is the reduction of a planting period extending throughout April to one covering only the first two weeks of that month, that is from planting half the area at a date likely to give maximum yield and half in the period of yield decline to planting the whole crop in the maximum yield period. If it is assumed that originally planting extended uniformly over the month, an increase in yield of 0.43 t ha^{-1} averaged over the whole planted area would be expected on the basis of Radley's figures. Even if this figure were doubled, the increase in yield would not be large; and in contrast, the 50% reduction in man and machine hours which would be obtained in these circumstances is likely to be of considerably more value to the grower. In many cases it is unlikely that such a marked improvement in mean planting date could be achieved but the pattern of a relatively small increase in yield and a relatively large improvement in labour and machinery utilization will be maintained.

An aspect of speed of planting which also requires consideration concerns date of planting in relation to the physical condition of the soil and clod formation. A farmer with limited planter capacity may frequently start work very early in the spring in order to avoid a yield reduction from late planting on part of the crop. In these conditions, particularly on the heavier soils, more difficulty will be experienced in obtaining sufficient tilth and there will be a much greater risk that the implements will bring to the surface pieces of wet soil which will subsequently dry to form clods. If the speed of planting can be increased, it becomes possible to avoid these very early plantings and to work only when the soil is in a suitable condition without extending the planting season any further into the period of yield decline.

Sprout damage

All automatic and semi-automatic planters damage the sprouts on chitted seed to a greater or lesser extent. Sprout damage almost invariably delays emergence and frequently reduces yield. Of the machines which have been subjected to investigation, the Continental cup-fed type appears to cause the most damage and the older hand-fed types the least; belt-fed planters have given intermediate levels (Jarvis and Palmer, 1973). The effect of sprout damage on yield has been shown to vary with variety (Winiger, 1973) and there are considerable seasonal differences; in general yield reductions are greatest in conditions when there is a good response to sprouting (Ministry of Agriculture, Fisheries and Food, 1975a). In trials carried out in west Norfolk with the variety King Edward over a 5-year period comparing a hand-fed with a Continental cup-fed planter, the increased sprout damage caused by the latter machine led to an average yield reduction of 1.6 t ha^{-1} in ware yield (Ministry of Agriculture, Fisheries and Food, 1974). Both level of damage and yield reduction may be affected by sprout type. For instance Jarvis and Palmer (1973) showed that sprouts produced during a short, warm period in an insulated store were more susceptible to damage than sprouts of a similar length produced in a longer, cooler period in a glass chitting-house; the associated yield reductions with the two types when planted with a Continental cup-fed machine were 2.4 and 0.8 t ha^{-1} of saleable ware.

The use of closely-graded seed is also an important factor in the reduction of sprout damage. Many automatic planters are fitted with an adjustable shutter to enable the flow of seed to be regulated between the main hopper and the compartment where it is either picked up by the cups or metered on to the belts. With closely-graded seed the shutter can be adjusted accurately to give the required flow of seed but with seed graded to the normal U.K. standard of 32–57 mm, the opening must be set wide enough to allow the largest sets to pass through and inevitably too many of the smaller sets pass also. This frequently leads to high levels of sprout damage, particularly on cup-fed machines as a result of the repeated passage of the cups through an excessively large volume of seed.

Most of the data discussed above relate to maincrop varieties and much of it to work carried out in the U.K. With crops grown for early ware production, sprout damage is likely to be of greater significance than with maincrops. This is reflected in the attitude of many farmers who have continued to use hand labour or hand-fed planters for the early crop or, where labour shortage has enforced change, have shown a marked preference for the less damaging belt-fed machines rather than for the cup-fed type. In the case of early maturing maincrop varieties, the subject of sprouting and sprout damage is of much less importance; it is interesting to note that in the Netherlands where the variety Bintje predominates ware producers have preferred to move away from traditional sprouting techniques and plant seed on which visible sprout development has only just started.

Irregularity of spacing

Certain types of planter, particularly those based on the flat belt system, distribute the sets much less uniformly along the row than others. A reasonably good measure of uniformity may be obtained by calculating the coefficient of variation (C.V.) for inter-set spacing; typical values are 15–25% for Continental cup-fed, 20–30% for hand-fed and 40–60% for belt-fed planters. Results from trials in which different types of planter were compared have indicated that irregular set spacing reduced yield (Jarvis and Palmer, 1973) and this has been confirmed by further experiments in which typical set distribution patterns were reproduced by hand-planting (Jarvis *et al.*, 1976). These workers reported that both total and ware yields tended to fall as irregularity of spacing increased; an average depression in ware yield of 1.7 t ha^{-1} from planting at a C.V. of 60% compared with uniform spacing along the row was recorded. The effect was more pronounced at sub-optimal set densities. There was some evidence that increasing irregularity led to increased variability in tuber size which tended to detract from the appearance of the ware sample but the differences were small and unlikely to be important in practice.

Close grading of the seed normally improves the regularity of spacing obtained with belt-fed planters because it helps to ensure uniform loading of the belts. This adds further emphasis to the need to use closely-graded seed in most types of automatic planter.

Depth of planting

Although excessively deep planting may reduce total yield, variations in depth within the normal range of 0–7.5 cm below ground level appear to have little effect. However, even within this range, saleable yield may be affected because of differences in off-grade production (Jarvis, 1973a). In general it is more difficult to build an adequate ridge over the sets if planting has been particularly shallow and a higher incidence of greening usually follows. On the other hand deeper planting may lead to higher levels of slug damage to the tubers in situations where the pest is troublesome.

Depth of planting also has an important effect on the ease of mechanical harvesting. Deeper planting leads to the formation of tubers at a lower level in the soil (although it is unlikely that the relationship between mean tuber depth and planting depth is linear) and it becomes necessary to set the harvester share to run more deeply to avoid slicing them. This will normally increase the volume of soil to be separated by the primary web and, on heavier soils, the number of clods also will be considerably greater. In this context the type of clod may vary markedly according to cultivation technique and season, from dry, compressed pieces of soil to wet semi-plastic lumps, but all are characterized by their resistance to fracture or crumbling on a normal sieving web. It is therefore extremely important that planting depth should be chosen according to the condition of the soil at planting. On the heavier soils it must frequently be accepted that not more than 10 cm of tilth reasonably free from clods will be obtained from the drier surface soil. In this case the sets should be planted close to the surface with about 5 cm of loose tilth below them. On lighter soils deeper planting is usually possible and may be desirable in order to minimize the need for subsequent ridge building cultivations.

Uniformity of planting depth is an important feature also. If the position of the tubers in the vertical plane varies along the row, the harvester share must be set to avoid slicing even the deepest tubers with the result that elsewhere the machine will be unnecessarily over-loaded. Assuming that a level seed-bed has been prepared, uniformity of planting depth depends largely on planter design. Independently-sprung planter units are to be preferred and in some circumstances less difficulty may be experienced with a 2-row rather than a 4-row machine.

9.6 Implications of mechanization in research and development work

Throughout western Europe, North America and most other parts of the world potato production has been mechanized to a greater or lesser extent. There can be little doubt that the trend towards more complete mechanization will continue, primarily as a result of rising labour costs in increasingly industrialized societies. Any attempt to improve yield or otherwise increase the efficiency of the production process must therefore be made with the requirements and limitations of mechanized systems in mind. This is not to say that fundamental research should be completely circumscribed by the constraints imposed by existing machines but rather that an awareness of the nature of the general problems associated with the mechanization of the potato crop is a necessary adjunct to the planning of any research and development project on the crop. At the same time, it is to be hoped that there will be an increasing tendency for agronomical research to be directed towards the solution of problems which have either arisen as a direct result of or have been accentuated by the increased use of mechanical equipment.

Because of the continuing development of mechanization techniques, there would be little purpose in listing the restraints and limitations of present day systems; many have been discussed as examples in Sections 9.3–9.5 above. However three general points require further emphasis and consideration; all are likely to continue to be significant features in the foreseeable future and in all three cases the prospect of a complete solution in mechanical or engineering terms appears to be remote.

The first of these concerns the lack of uniformity in the average crop of potatoes. Uniformity is a particularly desirable attribute in that uniform material can be handled by comparatively simple machines. The development of varieties or techniques which produce less variability in tuber size and shape should therefore be a long-term aim. Similarly, improved techniques for increasing the uniformity of seed-beds, depth of tuber formation and ridge size would reduce the problems of mechanical harvesting, particularly those problems concerned with the separation of tubers from clods and stones and with mechanical damage.

The second general point concerns the susceptibility of potato tubers to mechanical damage. There is a considerable need for a better understanding of the factors affecting tuber damage susceptibility and again it is possible that both plant breeding and agronomy could lead to improvements. One particular aspect of mechanical damage concerns sprouted seed potatoes. Although in the short-term the possibility exists of developing methods of producing sprouts which are more resistant to damage, a long-term solution is more likely to be found as a result of an improved understanding of the factors controlling the physiological ageing process.

The same field of work might also lead to better use being made of the comparatively short growing season available in many regions for the growth of the crop. This is the third general point, for although determined initially by latitude and climatic features, further restraints have been imposed by the increased use of heavy machinery for both planting and harvesting potatoes, principally because of the operating difficulties and risks of damage to soil structure in wet conditions. The undesirability of very early planting, particularly because of the risk of producing clods, and of late harvesting, particularly because of increased mechanical damage and storage problems, has been described above. A model solution would give more rapid haulm growth, an earlier start to bulking and in addition a faster rate of bulking than is normally found at present.

Finally, the question of choosing suitable husbandry methods in research and development work must be considered. Although fundamental research may require the use of methods which differ considerably from those used commercially, a newly developed technique needs to be tested under normal mechanized farming conditions as soon as possible and certainly before recommendations are made for its commercial adoption. For instance a research worker may reasonably test different methods of storing seed potatoes in small plots planted by hand; but he would be ill-advised to advocate the adoption of any promising technique which emerges from this work until it has been tested through a mechanical planter and shown to be successful in spite of sprout damage, uneven set spacing, uneven planting depth and other hazards which may befall. This is perhaps an obvious example, but many carefully calculated significant differences have been found to be of little consequence when subjected to commercial mechanical treatment.

References

Baldwin, J.H. (1964). *Expl Husb.* **11**, 83–7.

Blight, D.P. and Hamilton, A.J. (1974). *Potato Res.* **17**, 261–70.

Boyd, D.A. and Dyke, G.V. (1950). *N.A.A.S. Quarterly Review* (10) Winter 1950, 47–56.

Burton, W.G. (1973). *Proc. Fifth Triennial Conf. Eur. Ass. Potato Res.* (1972) 63–81.

Ivins, J.D. (1963). In *The Growth of the Potato*, eds. Ivins, J.D. and Milthorpe, F.L. pp. 303–10. London: Butterworths.
Jarvis, R.H. (1972). *Expl. Husb.* **21**, 85–92.
Jarvis, R.H. (1973a). *Expl Husb.* **24**, 37–40.
Jarvis, R.H. (1973b). *Proc. Fifth Triennial Conf. Eur. Ass. Potato Res.* (1972) 107.
Jarvis, R.H. and Palmer, G.M. (1973). *Expl Husb.* **24**, 29–36.
Jarvis, R.H. Rogers-Lewis, D.S. and Bray, W.E. (1976). *Expl Husb.* **30**.
Jarvis, R.H. and Shotton, F.E. (1972a). *Expl Husb.* **21**, 78–84.
Jarvis, R.H. and Shotton, F.E. (1972b). *Expl Husb.* **22**, 61–8.
Kouwenhoven, J.K. (1967). *Eur Potato J.* **10**, 257–71.
Ministry of Agriculture, Fisheries and Food. (1968). *Terrington Eighth Annual Review* (1967) 6–11.
Ministry of Agriculture, Fisheries and Food (1972). The utilization and performance of potato harvesters. *Farm Mechanization Studies* (24).
Ministry of Agriculture, Fisheries and Food (1974). *Terrington Fourteenth Annual Review* (1973) 17–19.
Ministry of Agriculture, Fisheries and Food (1975a). *Terrington Fifteenth Annual Review* (1974/75) 23–7.
Ministry of Agriculture, Fisheries and Food (1975b). *Terrington Fifteenth Annual Review* (1974/75) 28.
North, J.J. and Proctor, J.M. (1973). *Expl Husb.* **22**, 99–103.
Pätzold, Chr, and Dambroth, M. (1970). *Z. Acker-u PflBau* **132**, 281–94.
Potato Marketing Board (1959). Report on the survey of maincrop potatoes (1958) London: Potato Marketing Board.
Potato Marketing Board (1965). Report on the survey of maincrop potatoes (1963). London: Potato Marketing Board.
Potato Marketing Board (1970). Report on the survey of maincrop potatoes (1968). London: Potato Marketing Board.
Potato Marketing Board (1975). Maincrop potato production (1973). London: Potato Marketing Board.
Radley, R.W. (1963). In *The Growth of the Potato*, eds. Ivins, J.D. and Milthorpe, F.L., pp. 211–20. London: Butterworths.
Shotton, F.E. (1973). *Proc. Fifth Triennial Conf. Eur. Ass. Potato Res.* (1972) 114–15.
Shotton, F.E. (1976). *Expl Husb.* **30**, 1–17.
Thornton, R.E., Smittle, D.A. and Peterson, C.L. (1973). *Proc. Twelfth Ann Washington Potato Conf.* 1–10.
Winiger, F. A. (1973). *Proc. Fifth Triennial Conf. Eur. Ass. Potato Res.* (1972) 105–7.
Zaag, D.E. van der and Meijers, C.P. (1970). *Proc. Fourth Triennial Conf. Eur. Ass. Potato Res.* (1969) 93–103.

Chapter 10

Weed control

R.J. MAKEPEACE and J. HOLROYD

10.1 The problem of weeds in potatoes

10.1.1 The potato crop in the rotation

Following the introduction of the potato as a food crop to Great Britain it was grown largely as a vegetable for local consumption rather than as a saleable cash crop as it is today. Non-agricultural communities were small at this time and had ready access to holdings from which potatoes could be obtained. The need to grow potatoes on a more extensive basis emerged as cheap food sources were required for the large urban populations that appeared in the nineteenth century.

In the nineteenth century arable farming in many parts of England was dependent on the Norfolk four course rotation which consisted of each quarter of the arable land on a holding growing in sequence the following crops with all crops being grown in any single year: roots, barley, seeds (i.e. clover), and wheat. Variations of this rotation were almost infinite but they all satisfied the same principles of maintaining fertility, without fallowing, and maintaining a statisfactory distribution of labour requirements throughout the year.

The root break consisted of turnips, swedes or a similar root crop which was used for animal feed. This gradually gave way to cash crops, especially sugar beet, at the beginning of the twentieth century and potatoes towards the end of the nineteenth century. In these newer crops, instead of animals dunging direct on to the land, fertility was maintained by liberal applications of farmyard manure. The seeds break also gradually disappeared and beans or peas were grown in its place.

These changes were both gradual and sporadic, being affected by local conditions and social factors such as the cropping sequences specified in the tenant's lease. Nonetheless the areas best suited to arable farming took up these new crops and in the beginning of this century eastern England and eastern Scotland had large purely arable acreages based on the following rotations:

A	B	C	D	E
(1) Early potatoes and green manure or vegetable brassicas in the same year	(1) Potatoes (2) Wheat (3) Wheat (4) Barley	(1) Potatoes (2) Wheat (3) Sugar beet (4) Wheat	(1) Potatoes (2) Wheat (3) Barley (4) Grass ley	(1) Potatoes (2) Wheat (3) Barley (4) Barley (5) Peas or beans

There are endless variations of these rotations due to local factors. In addition the introduction of artificial fertilizers in this century removed the need to fertilize the land by rotation. This resulted in a slight increase in the number of potato crops in rotations until the occurrence of potato eelworm; a gap of at least five years between successive potato crops is needed to keep an acceptably low eelworm population. With the advent of modern nematicides and resistant varieties however the eelworm problem may now be reduced and the number of times that potatoes can be grown in the rotation may be increased (see Chapter 12).

10.1.2 The potato as a cleaning crop

One function of rotation in farming has always been that of controlling weeds. Before the advent of herbicides, cereals permitted weeds to proliferate. Hoeing was ineffective and due to their growth habit weeds brought about serious yield losses. The 'root' crops by comparison were grown on wide rows which allowed ready access for both hand and mechanical cultivation operations. Potatoes also had the additional need for deep initial cultivation in order to obtain sufficient worked soil to make the ridges in which the crop was grown. Planting would be followed by re-ridging operations and further extensive soil disturbance at harvest. The result was a large number of husbandry operations all of which would destroy weeds. In addition the potato crop produces a good leaf canopy once the foliage has met between the rows. The crop is not however a good competitor with perennial weeds although it allows ready access for hand weeding.

In the modern rotation the need to use the potato for cleaning purposes has been superseded by herbicides both in the potato crop itself and also in the other crops in the rotation. Annual broad-leaved weeds are not a problem and annual grass weeds can be controlled by several herbicides or in succeeding crops. In the case of wild oats however the potato crop can play a cleaning role by preventing seed return in that year. In longer, cereal dominated rotations, couch grass and related perennial grasses have become a problem and the potato has little beneficial effect against this type of weed.

10.1.3 Competition between weeds and potatoes

Weeds compete with the potato crop for light, nutrients and water. Yields of tubers can be severely reduced, the size of the reductions being dependent on the density and competitive ability of the particular weed population and the availability of supplies of light, nutrients and water. The yield reductions brought about in herbicide experiments, which included non-weeded and hand-weeded controls, typically range from 16–76% (Nield and Proctor, 1962).

The earlier the weeds emerge in relation to the crop the greater their competitive advantage but the easier they can be controlled with a 'contact pre-emergence' herbicide (see Section 10.4.2) before the crop emerges. Conversely potatoes 'chitted' before planting emerge earlier in relation to the weeds and thus compete more effectively with them, however such early emergence may make it impossible to use a 'contact' herbicide.

Annual broad leaved weeds because of their growth habit tend to be more competitive than annual grass weeds and because of their speed of germination and emergence, more competitive than perennial weeds, particularly in the earlier stages of crop development. Once the potato canopy has closed annual weeds are effectively suppressed, hence the main requirement of a potato herbicide is to control weed growth for the 7–12 weeks between planting and the closure of the leaf canopy.

Rapidly growing perennial weeds such as creeping thistle (*Cirsium arvense*), perennial sow-thistle (*Sonchus arvense*), and amphibious bistort (*Polygonum amphibium*) can penetrate the canopy once it has closed but rarely provide serious competition. They generally emerge later than annual weeds, after the crop has emerged and thus are difficult to control with herbicides or cultivations within the growing crop. Together with the perennial grasses – couch (*Agropyron repens* and *Agrostis gigantea*) they often make most of their growth when the potatoes begin to senesce and the canopy becomes more open.

10.1.4 Other effects of weeds

In addition to affecting yields, weeds can interfere with harvesting. This has become particularly important in recent years when the crop is largely harvested mechanically. Perennial weeds such as the couch and black bent which have tough wiry rhizome systems can cause considerable difficulties. Above-ground vegetation which can be killed off with a desiccant does not generally cause problems. In addition rhizomes may damage tubers physically by growing through them. Weeds may also act as alternate hosts for pests and diseases which can have very serious effects on the potato crops. Perhaps the best example is the peach potato aphid (*Myzus persicae*) which can live on a wide range of farm crops and weeds such as shepherd's purse (*Capsella bursa-pastoris*), black nightshade (*Solanum nigrum*), charlock (*Sinapis arvensis*) and fat-hen (*Chenopodium album*). This aphid transmits several virus diseases such as leaf roll and mosaic which have severe effects on the potato crop.

Weeds may also affect the potato crop indirectly, if herbicides applied for weed control are used incorrectly (i.e. at the wrong time or at too high a rate), and damage the crop.

10.2 Weed biology

Arable cropping in cultivated land in general favours annual and certain perennial weeds. The frequency, timing and type of cultivation are important in determining which species become prevalent. In land which has been direct-drilled for a number of years it is noticeable that whereas there is a marked decline in numbers and species of annual weeds, perennial weeds tend to increase. Annual weeds normally reproduce by seed but vegetative propagation is usually more important in perennials.

The size and significance of weed populations arising from seed in arable soils is not only dependant on the numbers of seeds present but also on such factors as their dormancy, longevity, time of emergence and depth in the soil from which they can emerge. Similar factors apply to populations arising from vegetative propagules.

10.2.1 Perennial weeds

Grass weeds

Couch grass (*Agropyron repens*) and black bent (*Agrostis gigantea*) are characterized by their possession of a tough wiry system of under-ground creeping stems or rhizomes. At intervals along these rhizomes there are nodes which bear buds in the axils of scale leaves. Left undisturbed most of these buds remain dormant, however if the rhizomes are broken up by cultivation the buds are stimulated to grow and form fresh plants. In the early stages of development food reserves are utilized and depleted by the growth of these new shoots. Thus it is possible to exhaust the plant by repeated cultivations or the use of a suitable herbicide to remove the new growth before the reserves can be replenished. On the other hand a single cultivation without follow up treatment can increase the population.

The two grasses often occur in arable fields together but they are readily distinguished by their foliage and flowering heads. Their rhizomes are superficially similar but the internodes of couch grass tend to be longer than those of black bent. Both grasses can reproduce from seed but the flowering heads of couch grass are often sterile and propagation by seed is relatively unimportant. Because of this almost complete reliance on vegetative reproduction distinctive clones often arise. The habit and vigour of growth of these clones can vary considerably and there is often a marked difference in their susceptibility to a particular herbicide. Cloning also occurs in black bent but is much less distinctive because it produces viable seed much more readily and freely. Neither grass is a strong competitor with the potato crop unless present in large quantities and when subjected to strong competition the growth of both weeds is reduced however the weeds can persist.

The deep cultivations and successive ridging operations, which were part of the normal technique for growing potatoes, generally broke up the rhizomes but kept the grasses in check until the potato canopy closed between the rows. The soil disturbance involved in harvesting moved and fragmented the rhizomes again. The potato in the past was therefore a very useful cleaning crop. However the introduction of herbicides and the reduction of labour input has meant that neither the rhizomatous grasses nor some of the broad-leaved perennials can be adequately controlled in the growing crop. Control measures now have to be taken in the autumn before the crop is planted and particular attention needs to be paid to headlands and hedgerows which often provide a reservoir for re-infestation.

Another species of bent, creeping bent (*Agrostis stolonifera*) is occasionally troublesome in potatoes. As its name implies it spreads by means of stolons which extend over the surface of the soil and thus are much more vulnerable to herbicides.

Broad-leaved weeds

Like the perennial grass weeds, several perennial broad-leaved weeds have become more troublesome as the number of cultivations involved in the growing of a potato crop has been reduced. The principal ones in the U.K. are the thistles, creeping thistle (*Cirsium arvense*) and perennial sowthistle (*Sonchus arvensis*), docks (*Rumex obtusifolius*) and of lesser importance (*R. crispus*), field bindweed (*Convolvulus arvensis*), coltsfoot (*Tussilago farfara*) and in the fenland areas the common reed (*Phragmites communis*) sometimes spreads out from the ditch banks.

Of these perennial broad-leaved weeds creeping thistle is one of the most persistent and troublesome. It occurs in both grassland and arable land. It spreads by brittle roots which break up readily on disturbance and produces fresh plants from adventitious buds. In undisturbed soil large quantities of roots can remain hidden for many years with only the occasional above-ground shoot. When the land is broken up, large numbers of plants may appear. Viable seed is rarely produced in spite of the large quantities of 'thistle down' which is apparent after flowering. Individual clones arise relatively freely as in couch grass and can be distinguished by differences in morphology. The clones also vary in their susceptibility to herbicides but large differences are uncommon. It is one of the most difficult weeds to control in the growing potato crop and cultivations are much more effective than herbicides.

The other important thistle *Sonchus arvensis* is also spread vegetatively by creeping roots but viable seed is produced and is more important in propagating the weed; clones arise less freely. It is also very difficult to control in the potato crop with herbicides unless cultivations are used frequently. However it is less widespread than creeping thistle.

Field bindweed (*Convolvulus arvensis*) is a local problem in potatoes. It is however a scrambling plant which can establish extensive mats of vegetation and interfere with harvesting. The root system is extensive and deep but regeneration occurs rarely if at all from root fragments Swan and Chancellor (1974). New shoots are generally produced in cultivated land from roots below plough or cultivation depth. When uncultivated these roots can spread widely and relatively rapidly but when cultivated they are often restricted to the area of original establishment. The above-ground parts die back in the autumn. It will flower freely in most seasons but is reputed to set viable seed in quantity only in hot summers; the seed can however remain dormant but viable for long periods in the soil.

Coltsfoot (*Tussilago farfara*) is also a relatively limited problem in potatoes and although it can grow on all soils it is most troublesome on the heavier types. It can spread rapidly by means of rhizomes which penetrate to considerable depths in the soil but is rarely a problem in grassland which indicates its susceptibility to competition. Top growth dies back during the winter but it flowers and sets seed before the leaves appear in the spring. It is distributed by seed and rhizome fragments although it is contained by frequent cultivations.

Docks (*Rumex obtusifolius* and *R. crispus*) are relatively 'spot bound' perennials which accumulate food reserves in fleshy tap roots. Cultivations will often break up and distribute the tap roots but regeneration occurs readily from the disturbed roots or root fragments whether these are buried or left on the soil surface. The broad-leaved dock (*R. obtusifolius*) regenerates more readily than curled dock (*R. crispus*) from root fragments. Both species are easily spread as seed which is produced profusely if they are allowed to flower; however they are susceptible in the seedling stage to cultivations and herbicides. Young seedlings are much more susceptible then plants arising from root fragments owing to the greater reserves of the latter.

The common reed (*Phragmites communis*) is a particular problem of the fenland areas in England where it spreads out from ditch banks on the spoil thrown out during ditch cleaning operations. The weed has an extensive rhizome system which enables it to spread considerable distances. New plants arise readily from seed or rhizome fragments.

10.2.2 Annual weeds

Annual weed populations are almost invariably composed of a mixture of species with one or two species dominant. Pure stands of one species are rare under normal conditions, but can arise where for example a herbicide has been used regularly and a particular species such as wild oat (*Avena fatua*) or cleavers (*Galium aparine*) is resistant. The composition and balance of the population is influenced by several factors such as timing, type and frequency of cultivation, fertilizer use, soil type, cropping and weather conditions. Cultivations in general stimulate germination by changing the micro-climate around the seeds for example by increasing aeration and soil water content. However, this does not modify the inherent pattern of germination of the particular species. Some species germinate at any time, e.g. chickweed (*Stellaria media*), some only in the spring e.g. knotgrass (*Polygonum aviculare*) and others only in the autumn e.g. cleavers. Similarly some species are much more common on a particular soil type where conditions favour their development e.g. cleavers and black-grass (*Alopecurus myosuroides*) on the heavier soils, corn marigold (*Chrysanthemum segetum*) on light acid soils. However, associations such as these may not always be a direct effect of soil type on weed species: the association is more often the result of the effect of a particular pattern of cultivation or cropping which is associated with a particular soil type.

Seed is all important to annual weeds, since it is their sole means of survival. Any factors which influence seed numbers, dormancy and distribution can therefore have profound effects on weed populations. Certain weed species produce large numbers of viable seeds over a short flowering period, some have an extended flowering period, others achieve the same objective with short life cycles and short intervals between succeeding generations.

Viable seed which is moist but does not germinate is said to be dormant Roberts (1972). Most weed seeds, and particularly annual weed seeds, can remain in this state for some period of time, otherwise they would be very unlikely to survive as weeds. Arable soils generally contain large quantities of seeds from preceding years and may contain up to 250 millions or more viable seeds per ha. In the absence of the addition of fresh seed the population declines at a rate which can vary enormously depending on the weed species present, the soil type and condition and the amount of cultivation Roberts (1972). Any seed requires adequate water, oxygen and a suitable temperature before it will germinate and grow. Dormancy may be broken if the seed is subjected to certain special conditions, which may involve light, freezing, rupture or breakdown of the seed coat (Roberts 1972).

Weed seed may be distributed by a variety of agencies but the most important for many species is man himself. Thus weeds are often sown with crop seed, dispersed with hay or straw used for feeding, spread on the land in farmyard manure and carried from field to field or farm to farm in machinery, merchant's lorries etc. However, natural agencies such as wind and water, and birds and animals are also important for many species.

There are only two annual grass weeds, annual meadow grass (*Poa annua*) and wild oats (*Avena fatua*) which can be serious weeds in potatoes and neither will reduce yields unless present in large numbers. The main need for their control is to reduce their potential numbers in any subsequent crops such as cereals where they could be a real problem. Among the annual broad-leaved weeds the larger growing or 'scrambling' types can threaten yields or interfere with harvesting. The following can all give rise to severe problems in potatoes: charlock (*Sinapis arvensis*), wild radish (*Raphanus raphanistrum*) and field pennycress (*Thlaspi arvense*); annual sowthistle (*Sonchus oleraceus*), mayweeds (*Matricaria* spp); knotgrass (*Polygonum aviculare*), black bindweed (*Polygonum convolvulus*), redshank (*P. persicaria*), fat hen (*Chenopodium album*), orache (*Atriplex patula*), chickweed (*Stellaria media*), fumitory (*Fumaria officinalis*), and cleavers (*Galium aparine*).

Most of the herbicides used for the potato crop must be applied before full crop emergence and thus weeds which emerge late or have a protracted period of emergence often present more of a problem than those emerging earlier: black nightshade (*Solanum nigrum*) which emerges mainly in May and June is an example.

In addition to true weeds, volunteer plants from previous crops such as cereals, or ryegrass can sometimes cause problems.

10.3 Weed control

10.3.1 Traditional methods of weed control

The cultivations carried out in the potato crop used to be similar to those carried out in other row crops. They consisted of horse drawn harrows supplemented by ample hand labour. The land to be planted to potatoes was ploughed by horse drawn units and ridged up with ridging bodies mounted on a plough frame. The seed tubers were planted by hand after which the ridges would be split over them and reformed. Subsequent cultivations consisted of harrowing down the ridges with light harrows and re-ridging perhaps two or three times before the emerged crop became too large to allow the transit of horses and equipment. This system was efficient in terms of weed control but demanding of labour.

With the advent of the tractor the power base changed and mechanized planting allowed the use of multi-row machines for all operations including weed control. The system itself did not change except that the ridges could be almost harrowed flat. There was no problem of having to re-ridge quickly as there was plenty of power available. The types of harrow used changed slightly; the horse-drawn scuffle disappeared and was replaced by light chain harrows and saddle-back harrows designed especially for the potato crops.

10.3.2 Modern cultivation techniques

The post-war years saw a dramatic increase in tractor power. This has given increased freedom to the grower to alter husbandry techniques and has led to the almost complete mechanization of the crop and to the use of row widths more suited to the crop and the machinery and not dictated by the limits of the horse. The main post-war change in weed control cultivations has been the refinement of the early cultivation techniques. Re-ridging being left as late as possible with light harrowings in between to control weeds whilst avoiding water loss due to the exposure of moist soil to the atmosphere. The system changed slightly with more emphasis on spring-tined harrows as well as chain harrows. A completely novel cultivation system has evolved in the Netherlands which consists of helical harrows which fit the shape of the ridge and revolve as they move forward. This cultivation controls weeds while maintaining a fine soil tilth, and moving the fine soil up to the ridge top. This system is also considered to be beneficial in conserving soil moisture.

The loss of soil moisture with cultivation indicates one of the shortcomings of mechanical weed control, which may lead to the formation of clods during the season that can become a nuisance at harvesting. The repeated passage of tractors may cause problems of soil compaction and hinder root development. These factors have only come to light with the advent of safe herbicides which can allow a comparison of cultivations and no cultivations. From such comparisons it has also been shown that repeated cultivation increases heat loss from the soil surface thereby increasing the amount of frost damage. The Dutch-system went far to alleviate the problems associated with cultivations but has been superceded by herbicides which have many advantages over the conventional mechanical method of weed control. Although cultivations are still extensively used in potato husbandry there are now few crops that are weeded entirely by mechanical means. For further discussion on the role of cultivations, see Pereira (1941), Sanders (1954) and Bremner (1966).

10.3.3 The use of herbicides

The use of herbicides for the control of weeds in potatoes is by comparison with their use in cereal crops a recent development. The first Weed Control Handbook (British Weed Control Council, 1957) mentions only TCA as a pre-plant treatment for grass weeds and makes a tentative recommendation for the use of MCPA in the crop. The chemical destruction of potato haulm prior to lifting was well established and the same

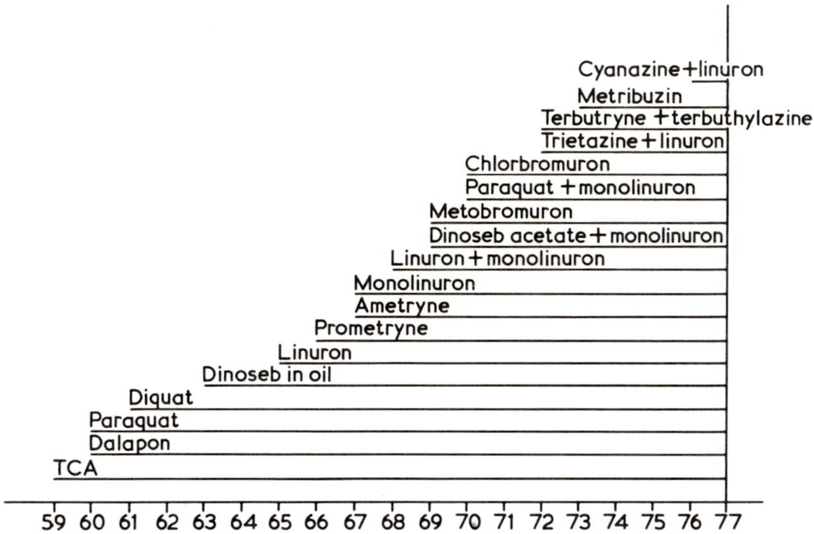

Fig. 10.1 The introduction of herbicides into the potato crop. The year of introduction is here designated as that when the recommendation first appeared in the M.A.F.F. annual '*Approved Products for Farmers and Growers*'

edition of the handbook lists the following compounds as suitable haulm desiccants: sulphuric acid; sodium arsenite; tar oil; sodium chlorate; DNOC; copper sulphate plus salt.

The first herbicides to be recommended were primarily used for other crops and tended to be recommended in potatoes as an extension of these uses; examples were TCA and dalapon. The next major development was the introduction of pentachlorophenol as a pre-emergence herbicide allowing weeds to be killed prior to the emergence of the crop. Cultivation was used for any further weed control needed. In 1960–2 diquat and paraquat were introduced as contact herbicides for pre-emergence application. However, the crop was sensitive to these herbicides and their use needed to be carefully controlled if damage to the crop was to be avoided. What was needed was a selective herbicide that did not damage the crop that could be applied at any time during its life. In 1964–5 this goal was nearly achieved with the introduction of dinoseb and linuron. While both herbicides can cause damage to the crop once it has emerged, their advantage lay in the property (particularly associated with linuron) of remaining active in the soil for several weeks after application. Thus susceptible weeds present at spraying were killed and any further germinating seeds were controlled by the residual action of the chemical. This was a considerable step forward in the technique of weed control and although the residual action is dependent on adequate soil moisture, it did enable the grower to consider for the first time weed control without any cultivations. These two compounds were followed by many similar materials such as monolinuron, prometryne, ametryne, metobromuron and chlorbromuron. Their introduction is shown in Fig. 10.1. They all have weaknesses and those that do not harm the crop are invariably poor in controlling weeds. Progress continues to be made however and with the introduction of metribuzin in 1973 for the first time there is available an effective herbicide that can be safely applied after the crop has emerged (Elliott and Cox 1964; and Elliott, 1965; Evans, 1968; Hampson and Taylor, 1974).

10.3.4 The interaction between herbicides and cultivations

From the number of products available for weed control in potatoes it might appear that few cultivations for weed control need to be carried out in the crop. This certainly appeared to be the case in the late 1960s. The fact that cultivations are still an important part of weed control in potatoes is due to the weakness of most herbicides. The most important weakness is the need for adequate moisture in the soil for herbicides to be effective and also the presence of resistant weeds. After the introduction of linuron and other reisudal herbicides, there was a trend away from cultivations until the shortcomings of such herbicides became apparent. One problem was the nature of the ridge (Shotton 1966). Residual herbicides need a fine tilth and no soil movement after application to achieve this; cultivations after planting were modified to produce a low, well rounded ridge. Where these conditions cannot be met cultivations are used for weed control and a herbicide is applied just prior to emergence to give residual action. Thus both cultivations and herbicides used alone have serious drawbacks and the disadvantages of each can be overcome by their combined use.

The general practice at present is that in rotations containing other cash crops such as sugar beet, peas, beans, carrots and vegetables the main problem is annual weeds and there is a high dependence on residual herbicides with cultivations when necessary. In cereal dominated rotations there is the complication of annual and or perennial grass weeds and here cultivations are relied on in the early part of the season followed by paraquat applied at up to 40% emergence of the crop.

10.4 Recommendations for the use of herbicides

10.4.1 The safety of herbicides to the crop and consumer

Pesticides applied to growing crops in the western world are regulated with respect to the safety of the operator applying them, the consumer and the environment. Each nation achieves regulation of pesticides by different means although the end is virtually the same. In the U.K.

these aspects of safety are controlled by the Pesticides Safety Precautions Scheme. In the case of potato herbicides the same procedure is followed as for other potato pesticides, including fungicides and insecticides. There is a slightly reduced health risk from herbicides compared with other crop protective chemicals in that they are applied very early in the life of the crop thus giving a long time interval from the time of application to the time of harvest. Haulm desiccants on the other hand are applied to the crop just before harvest and must therefore leave virtually no residue in the crop on lifting before they can be cleared for safe use. In the UK, further protection is afforded to the users of herbicides by the Agricultural Chemicals Approval Scheme.

Safety to the growing crop is an important factor examined in herbicide trials. Effects on the crop from herbicides are difficult to assess owing to the influence of the seed tuber on early growth and also to the ability of the crop to recover from temporary damage under normal growing conditions. At emergence potato shoots can be destroyed two or three times before any severe damage is done to the crop, due largely to the food reserves in the seed tuber. Weed control techniques involving cultivations rely on this effect so that they can be carried out over a prolonged period late into the life of the crop. Herbicides are similarly tolerated if of the contact type but if translocated then damage to the parent tuber can occur with long-term effects on the crop. There are in the potato crop few completely safe post-emergence herbicides and there are a number of materials discussed later that are only tolerated to a varying degree. The use of haulm desiccants is fully discussed later, but they have a very special requirement in that they must completely destroy the foliage without affecting the tuber in any way.

10.4.2 Types of herbicidal activity and application methods

No attempt is made to give a complete review of the literature, useful general references are Fryer and Makepeace (1977a, b) and Ministry of Agriculture, Fisheries and Food (1977). Where appropriate general references are given at the end of the short account of each herbicide.

Herbicides may act on plants after entry through the roots, the shoot, the foliage or all three. They may be applied before the crop is planted (pre-planting), after the crop is planted but before it emerges (pre-emergence) or after the crop has emerged (post-emergence).

Selectivity between the crop and the weeds may be dependent on a separation between weed and crop growth in time; depth protection i.e. the bulk of the herbicide is left near to the soil surface where most of the weeds are germinating and little or none reaches the roots of the more deeply planted potato tubers; or to a difference in physiology between the crop and the weeds. Herbicides used in the potato crop rely for their selectivity on one or more of these factors.

Six main types of treatment are used.

(1) Pre-planting treatments in which the herbicide is applied before the crop is planted and selectivity is dependent on its disappearance before planting. Action on the weeds may be through the foliage, the roots or both.

(2) Pre-emergence treatments in which the crop has been planted but neither it nor the weeds have emerged at the time of treatment. The herbicide is applied to the soil and either left on the surface or mixed in shallowly. Entry into the weeds may be through the roots or the emerging shoots. Selectivity may be due to physiological differences between crops and weeds or depth protection.

(3) Contact pre-emergence treatments, in which although the crop has been planted, few if any of the shoots have emerged, whereas many of the weeds have germinated at the time of treatment. Selectivity is thus due to a separation in time.

(4) Contact and residual pre-emergence treatments in which the crop has been planted but not emerged. As in (3) above there is some activity on the weeds through the foliage but the main activity is through the soil. Selectivity is due to a separation in time and a combination of depth protection and physiological resistance of the crop.

(5) Post-emergence treatments in which the crop and weeds have emerged. Entry is through the foliage and soil and selectivity is dependent on the physiological resistance of the potato. As a general rule weeds are more susceptible when young. Shielded inter-row treatments with contact non-selective herbicides are also sometimes used in which selectivity is achieved by a physical separation of the foliage of the crop from the weeds at the time of treatment.

(6) Haulm destruction or post-senescence treatments. Contact herbicides (see Section 10.5.3) are often used to kill the potato haulm before harvest. These treatments also control many weeds which may interfere with harvest. They are non-selective contact herbicides which under normal conditions do not harm the potato tubers buried below ground.

10.4.3 Pre-planting herbicides

These are usually used for the control of perennial weeds, particularly grass weeds, either in the autumn or the spring before planting.

Aminotriazole (3-amino-1,2,4-triazole)

This herbicide can enter the plant through the roots but its main route is through the foliage. It is readily translocated. When formulated with ammonium thiocyanate its activity is enhanced.

It has little or no selectivity but persistence in the soil is relatively low. Thus it can be used either in the autumn or the spring for the control of grass weeds such as couch grass and many perennial dicotyledonous weeds (e.g. docks, thistles and coltsfoot). The symptoms of treated plants are very characteristic as it prevents chlorophyl formation, and shoots which are almost white or pinkish are produced. The normal dose is 4.48 kg a.i. ha^{-1} which is applied in 220 to 440 l of water. After an autumn treatment weed control is helped by ploughing to a depth of 20 cm. After a spring treatment it is recommended that a period of 3–6 weeks should be left before ploughing to allow any residues to disappear. After this potatoes may be planted. The longer period of 6 weeks may be required on a very sandy soil. Joice and Norris (1962) have reviewed the uses of this herbicide.

Dalapon sodium (sodium 2,2-dichloropropionate)

Dalapon is used to control couch grass and other perennial grasses by treatments which can be made either in the autumn or the spring. Most of the uptake is through the foliage, whence it is translocated to the roots and rhizomes. Thus it is most effective on plants with a relatively large amount of actively growing foliage. This is encouraged by cultivation several weeks before treatments are applied. Ploughing is also normally recommended two weeks after treatment. Dalapon sodium is used at doses of 13–16 kg a.i. ha^{-1} in 330 l of water in the autumn and 10–13 kg a.i. ha^{-1} in the spring. Potatoes can be planted after ploughing in the spring.

Dalapon can also be used after the potato haulm has died back naturally or after the use of a desiccant, to control grass weeds which may grow before the crop is harvested. In addition to reducing the carry-over of weeds into subsequent crops this treatment makes harvesting easier. If there is any living potato foliage at the time of treatment red-skinned varieties are liable to lose their colour.

Glyphosate (N-(phosphonomethyl) glycine)

Glyphosate like dalapon and aminotriazole is applied to the foliage of actively growing perennial grasses. It is translocated freely and can completely kill the rhizomes of couch grass. Unlike the previous two herbicides described, it is generally most effective when there is no cultivation before treatment. There is little or no activity through the soil except at very high doses and potatoes can be planted 7–10 days after treatment. The normal dose is 1.3 kg a.i. ha^{-1} applied in 220 l of water (Evans, 1972, Hodkinson, 1974).

TCA (trichloracetic acid)

TCA is a soil acting herbicide which is absorbed mainly by the roots. It can be used in the autumn or the spring to control couch grass and other perennial grasses. Cultivation is a necessary adjunct to its use on couch grass as the rhizomes need to be actively growing at the time of treatment. At the same time the herbicide is brought into the immediate neighbourhood of the roots. Cultivation can be before and/or after application of the TCA.

Persistence of the herbicide in the soil is influenced by rainfall and the soil type, residues tend to last longer in soils of higher organic matter, but potatoes can be planted after a minimum interval of 2 months. The dose normally used in autumn is 33.6 kg a.i. ha^{-1} and in the spring 16.8–22.4 kg a.i. ha^{-1} (Nield and Proctor, 1962).

EPTC [S-ethyl N,N-dipropyl (thiocarbamate)]

EPTC is a soil acting herbicide which enters the plant through the roots and/or the shoot. As it is volatile it is essential to mix it with the soil immediately after application. Most of the annual grasses such as wild oats (*Avena fatua*) are controlled and it also gives a good suppression of the perennial grasses such as couch grass. It is normally used in the spring as soil temperatures begin to rise and growth is active. The dose is normally 4.5 kg a.i. ha^{-1} and potatoes can be sown immediately after treatment.

10.4.4 Pre-emergence herbicides

Metrobromuron (N′-(4-bromophenyl)-N-methoxy-N-methyl urea)

This potato herbicide is the only one in common use which has little or no foliar activity. Thus it must be used before the weeds have emerged or with a contact herbicide such as paraquat. A dose of 1.68 kg a.i. ha^{-1} is used on light soils and 2.24 kg a.i. ha^{-1} on medium to heavy soils to control many of the annual dicotyledonous weeds. Of the commoner dicotyledonous weeds, cleavers (*Galium aparine*) and fumitory (*Fumaria officinalis*) are resistant. Residues may cause damage if other crops are sown within 3 months of treatment. (Hein and Smith, 1966; Schuler and Ebner, 1969; Smith and Marks, 1968).

10.4.5 Contact pre-emergence herbicides

Paraquat (1,1′-dimethyl-4,4′-bipyridylium dichloride)

Paraquat has little or no activity through the soil but is rapidly absorbed by foliage which it kills relatively quickly usually within a day or two depending on temperature and light intensity. Seedling weeds are therefore killed easily but perennials generally recover. Seeds which have not germinated are not harmed and therefore a residual herbicide is often added to control subsequently germinating weeds (see Section 10.4.6).

When the weeds are in the cotyledon stage a dose of 0.56 kg a.i. ha^{-1} is sufficient but if beyond this, or if perennial grass weeds are present, a dose of 1.12 kg a.i. ha^{-1} is used in 200–300 l of water. Potato shoots should not have emerged at the time of treatment as they will be damaged however, there is relatively little permanent damage to the crop if not more than 25% of the shoots have emerged at the time of treatment (Butler, 1960; Headford, 1968).

Dimexan [*di (methoxythiocarbanyl) disulphide*]

Dimexan like paraquat, is a 'contact' herbicide which kills any seedlings which have germinated and also desiccates the aerial growth of any perennial weeds which are present. However, in addition, it has some fumigant action on the soil surface layer and kills weed seeds which are germinating near the soil surface. Perennial weeds can normally be expected to regrow. It is formulated as an emulsifiable concentrate and normally used at a dose of approximately 9.0 kg a.i. ha^{-1} in 200 l of water. Post-crop emergence treatments are also used to control the size of potato tubers (see Section 10.5.5).

10.4.6 Contact pre-emergence and residual herbicides

These herbicides are applied after sowing but before crop emergence. They have some foliar activity and so kill some seedling weeds. They also act through the soil and kill weeds as they germinate. Several commercial products are mixtures of two compounds, one with a predominately foliar action and the other effective through the soil. Most of the compounds in this category are either triazines or halogen substituted ureas. The activity of many is influenced by soil type and the dose in some situations needs to be adjusted to take account of this (Table 10.1). If early potatoes are to be followed immediately after harvest with a catch crop then consideration has also to be given to the risk of damage from persistant herbicide residues.

Ametryne (4-ethylamino-6-isopropylamino-2-methylthio-1,3,5-triazine)

Although ametryne is active through the foliage and roots it cannot be used satisfactorily on extreme soil types. Crop damage can occur on soils containing more than 80% sand and also in the higher rainfall areas of Scotland, Northern Ireland, Pembrokeshire and Cornwall. On soils of high organic matter content, however there is little or no activity through the soil. Treatments are most effective when applied to damp soil or light rain follows shortly after application. Heavy rain following application can result in some chlorosis of the foliage which is usually only

Table 10.1 Herbicide doses for particular soil types (Doses in kg a.i. ha^{-1})

Herbicide	*Soil type**									
	Very light		*Light*		*Medium*		*Heavy*		*Organic*	
	E	*M*	*E*	*M*	*E*	*M*	*E*	*M*	*E*	*M*
Ametryne	–	–	1.4	2.2	1.4	2.2	1.4	2.2	1.4	2.2
chlorbromuron	1.1–1.4	1.4	1.1–1.4	1.7	1.1–1.4	2.0	1.1–1.4	2.0	1.1–1.4	2.0
dinoseb in oil	2.5	2.5	2.5	2.5	2.5	2.5	2.5	2.5	2.5	2.5
dinoseb acetate + monolinuron	–	–	2.2	2.2	2.8	2.8	3.3	3.3	3.3	3.3
linuron	0.5–0.8	1.1–1.7	0.5–0.8	1.1–1.7	0.8–1.1	1.1–1.7	1.7–2.2	1.7–2.2	1.7–2.2	1.7–2.2
linuron + cyanazine	–	–	1.1	1.4–1.7	1.4	1.7	1.7–2.2	1.7–2.2	2.2	2.2
linuron + monolinuron	–	–	0.8–1.1	1.1	1.1	1.1–1.7	–	1.7–2.2	–	1.7–2.2
metobromuron	1.7	1.7	1.7	1.7	1.7	2.2	1.7	2.2	–	–
metribuzin	0.5	0.7	0.5	0.7	0.7	1.0	0.7	1.0	–	1.0
monolinuron	–	–	0.8–1.1	1.1–1.4	1.1	1.1–1.7	1.4–1.7	1.7	1.7	1.7
paraquat	1.4	1.4	1.4	1.4	1.4	1.4	1.4	1.4	1.4	1.4
prometryne	1.7	–	1.7	–	1.7	–	1.7	–	1.7	–
terbutryne + terbuthylazine	1.1	1.4	1.1	1.4	1.1	1.7	1.1	1.7	1.7	1.7
trietazine + linuron	1.4	1.7	1.4	1.7	1.4	1.7	1.4	2.2	–	–

* Eagle D.J. (1976).
E = Early; M = main crop.

temporary, but the varieties Pentland Dell and Pentland Crown may be more permanently affected. It is formulated as a wettable powder and the normal dose used on early potato varieties is 1.4 kg a.i. ha^{-1} and on main crop varieties 2.2 kg a.i. ha^{-1}.

Many annual broad-leaved and grass weeds are susceptible but cleavers (*Galium aparine*), speedwells (*Veronica spp*) and wild oat (*Avena fatua*) in addition to most perennial weeds are relatively resistant. Residues of the herbicide in the soil do not persist for long and any crop can be sown 8 weeks after treatment. (Nield and Proctor, 1962).

Chlorbromuron (N′-(4-bromo-3-chlorophenyl)-N-methoxy-N-methyl urea)

This herbicide, formulated as a wettable powder, is used pre-crop emergence but is most effective on many weeds at the time of emergence. Annual meadow-grass (*Poa annua*), field pansy (*Viola arvensis*) and field penny-cress (*Thlaspi arvense*) however are only susceptible pre-emergence. Black bindweed (*Polygonum convolvulus*) cleavers and speedwells are moderately susceptible and fumitory black nightshade and wild oat together with most of the perennials are resistant.

On early potato varieties doses of 1.1 to 1.4 kg a.i. ha^{-1} are used, the lower dose if most of the weeds have emerged; on main-crops doses of 1.4 to 2.0 kg a.i. ha^{-1} are used, the lower dose on light sandy soils. There is little activity in soils of high organic matter content and only treatments applied post-weed emergence are effective. Persistence in the soil is moderate and any crop can be sown 12 weeks after treatment. Umbelliferous crops such as carrot or celery can be sown after a shorter period (Smith and Marks, 1968).

Cynazine + linuron (2-chloro-4-(1-cyano-1-methylethylamine)-6-ethylamino-1,3,5-triazine + N-(3,4-dichlorophenyl)-N-methoxy-N-methyl urea)

This mixture controls most annual weeds except wild oats particularly if applied before the weeds emerge. Cleavers, corn marigold (*Chrysanthemum segetum*), knotgrass (*Polygonum aviculare*), mayweeds (*Matricaria recutita* and *Tripleurospermum maritimum* subsp. *inodorum*), groundsel (*Senecio vulgaris*) and fumitory are only moderately susceptible when they have emerged. Formulated as a wettable powder, it is used at doses of 1.1 to 2.2 kg a.i. ha^{-1} according to soil type but on very light soils the crop can be affected even at the lowest dose. On high organic matter soils the highest dose is only likely to control weeds which have emerged.

Cyanazine has a relatively short persistence but linuron can persist to affect other crops sown 2–4 months after treatment depending on the dose and soil type. Lettuce has been found to be particularly susceptible and is not sown in the same season that the herbicide is used.

Dinoseb in oil (2-(1-methylpropyl)-4,6-dinitrophenol)

This herbicide is principally a contact herbicide and is therefore used as near to crop emergence as possible when the maximum number of seedling weeds have already emerged. However it has some residual effect up to 4 weeks after treatment on all except organic soils. Many seedling weeds are susceptible but wild oats and most perennials are resistant; knotgrass corn marigold, annual meadow-grass and black nightshade (*Solanum nigrum*) are moderately resistant and black bindweed, cleavers and small nettle (*Urtica urens*) are only moderately susceptible. The normal dose is 2.5 kg a.i. ha^{-1} and there are no residue problems (Joice and Norris, 1964).

Dinoseb acetate + monolinuron (2-(1-methylpropyl)-4,6-dinitrophenyl acetate + N′-(4-chlorophenyl)-N-methoxy-N-methyl urea)

The addition of monolinuron gives a mixture which is more effective on the seedling weeds than dinoseb on its own. Fumitory is only susceptible to the contact activity, corn marigold, knotgrass and mayweeds are only susceptible at the cotyledon stage while annual meadow-grass is only susceptible when treated pre-emergence. The monolinuron provides residual activity. The mixture is not suitable for use on very light soils which are low in organic matter content. Doses of 2.2 to 3.3 kg a.i. ha^{-1} are normally used, the lower dose being used on lighter soils. Formulation is as a wettable powder. Residues are not normally a problem for subsequent crops.

Linuron (N-(3,4-dichlorophenyl)-N-methoxy-N-methyl urea)

Linuron is probably the residual herbicide which is most widely used for weed control in potatoes. It is marketed as an emulsifiable concentrate and a wettable powder. In terms of activity on the weeds there is little to choose between the two formulations. It is used pre-crop emergence but has contact activity on many seedling dicotyledonous weeds. Black nightshade, cleavers, corn marigold, groundsel, knotgrass, mayweeds, speedwells and annual meadow-grass are all resistant when treated post-emergence. Fumitory, wild oats and perennial weeds are resistant to pre- and post-emergence treatments.

It is used at dose which range from 0.5 kg a.i. ha^{-1} to 2.2 kg a.i. ha^{-1} depending on the soil type and whether the crop is early or maincrop.

On organic soils there is relatively little activity through the soil and it is therefore only effective on susceptible weeds which have emerged. Residues are not a problem for succeeding crops if these are umbelliferous e.g. carrot or parsnip, but other crops may not be sown within 2 months of using the lower doses or 3 months of the higher dose. In very dry conditions these periods are extended by 1 month. Lettuce is particularly susceptible to small residues in the soil.

Linuron is also marketed as a mixture with monolinuron formulated as an emulsifiable concentrate. The addition of the monolinuron gives an increase in effectiveness on annual grass weeds, particularly annual meadow-grass (Cowan and Joy, 1964; Hartel, 1964).

Metribuzin [*4-amino-6-t-butyl-3-(methylthio)-1,2,4-triazine-5(4H)-one*]

Metribuzin is one of the most effective herbicides which is used in potatoes. It controls most annual weeds although knotgrass and annual sowthistle (*Sonchus oleraceous*) are more susceptible to pre-emergence treatments and black bindweed and black nightshade are more susceptible to post-emergence treatments. Wild oats are only moderately well controlled.

Unlike most of the other potato herbicides metribuzin can be used on maincrop varieties, with the exception of Maris Piper and Pentland Ivory, post-emergence until the crop foliage is up to 15 cm in height. First and second early varieties are only treated pre-emergence. The doses used range from 0.5 kg 1.i. ha^{-1} to 1.05 kg a.i. ha^{-1}, the lower doses being used on early varieties and lighter soils. It is formulated as a wettable powder and applied at volume rates in excess of 200 l ha^{-1}.

Residues cause a problem for subsequent crops but ryegrass, cereals and winter beans can generally be sown in the same season providing at least 16 weeks have elapsed since treatment and the land has been ploughed to 15 cm. Dry weather conditions sometimes exacerbate the problem. (Eue *et al.*, 1969; Eue and Tietz, 1970; Mannall *et al.*, 1972).

Monolinuron (N′-(4-chlorophenyl)-N-methoxy-N-methyl urea)

This herbicide is effective on most annual weeds if used pre-emergence, the exceptions are fumitory and wild oats which are relatively resistant either pre- or post-emergence. A number of other weed species become relatively resistant when they have emerged and for this reason in addition to being used on its own, monolinuron is also formulated as a mixture with the contact herbicide paraquat.

Monolinuron alone is not normally used on very light soils because of the risk of crop damage and on other soils it is used at doses ranging from 0.84 to 1.68 kg a.i. ha^{-1} according to soil type. The mixture with paraquat is used at 1.4 kg a.i. ha^{-1} on all soil types.

There are no residue risks for subsequent crops after use of the mixture but it is recommended that other crops should not be sown within two months of using monolinuron at 0.84 kg a.i. ha^{-1} or within three months of a dose of 1.4 kg a.i. ha^{-1} (Hartel, 1964).

Prometryne (4,6-bisisopropylamino-2-methylthio-1,3,5-triazine)

The use of prometryne is restricted to early varieties of potatoes in which it is used at a dose of 1.75 kg a.i. ha^{-1} to control germinating and emerged weeds before 10% of the potatoes have emerged. Cleavers and wild oats and perennial weeds are resistant. Residual herbicidal activity is also very low on soils with a high organic matter. Crops other than carrots, parsley, celery and peas are not normally sown within eight weeks of treatment (Pfeiffer and Phillips, 1964).

Terbutryne + terbutylazine (4-ethylamino-2-methylthio-6-t-butylamino-1,3,5-triazine + 2-chloro-4-ethylamino-6-t-butylamino-1,3,5-triazine)

This mixture of triazine herbicides is used to control annual weeds at doses of 1.15 to 1.7 kg a.i. ha^{-1} in early potatoes and at doses of 1.4 to 1.7 kg a.i. ha^{-1} in main crop varieties, the dose being varied according to soil type. Wild oats and cleavers are again relatively resistant. The crop should not have emerged at the time of treatment.

The control of annual and perennial grasses can be improved by the addition of paraquat. Residues can be a problem for following crops and the land should be ploughed at least 15 cm deep, and at least 12 weeks allowed to elapse before sowing another crop.

Trietazine + linuron (2-chloro-4-diethylamino-6-ethylamine-1,3,5-triazine + linuron)

This mixture of 50% trietazine and 50% linuron is used to control annual weeds pre-crop emergence (up to 10%) on mineral soils. It is not normally used on very light or highly organic soils.

A dose of not more than 1.4 kg a.i. ha^{-1} is used on early varieties but doses of 1.7 – 2.25 kg a.i. ha^{-1} are used on main crop varieties, according to soil type. Wild oats, cleavers, black nightshade and perennial weeds are all resistant; speedwells and fumitory are only moderately susceptible. Control of these weeds is improved if the contact herbicide paraquat is added to the mixture. Residues can be a problem for subsequent crops and land should be ploughed to 15 cm and 10–14 weeks left before planting (Nield and Proctor, 1962, Pfeiffer and Phillips, 1964).

10.4.7 Post-crop emergence herbicides

Currently there are almost no herbicides which are used post-emergence, in the accepted sense in the potato crop. Some, as mentioned in the foregoing section, are used when a considerable proportion of the potato shoots have emerged, but with the possible exception of metribuzin, many are liable to cause at least temporary crop damage.

Barban (4-chlorobut-2-ynyl 3-chlorophenyl carbamate)

Barban does not cause crop damage but its use is limited to the control of wild oats which have not more than 2.5 main stem leaves. A dose of 0.35–0.70 kg a.i. ha^{-1} is applied in 110–220 l of water and gives effective control if the crop is competitive.

MCPA (4-chloro-2-methyl phenoxyacetic acid)

MCPA is sometimes used as an emergency treatment to control late germinating weeds including fat-hen and creeping-thistle (*Cirsium arvense*) at a dose of 0.70 kg a.i. ha - in 150–300 l of water. Potato varieties differ in their susceptibility and Majestic, for example, is more tolerant than King Edward. The potato foliage is generally deformed by the treatment and this may cause confusion with virus symptoms where the crop is being grown for seed.

Dalapon is also used after crop senescence to reduce late developing grass weed infestations see Section 10.4.3.

10.4.8 Application

Even the most effective herbicide must be applied accurately if it is to give good weed control and avoid damage to existing or future crops.

Most herbicides applied to the potato crop have both foliar and soil activity and must adequately cover both the emerged weeds and the soil surface. Ridging of the soil increases the effective surface area and there is a risk of shadowing, particularly if clods are present. Relatively high volume rates are generally used in application.

Sprayer types

Most sprayers are of the standard farm type. Nozzles may be angled to treat each side of the ridge to give a more even spray distribution.

Experimentally, attempts have been made to overcome the shortage of a truly post-emergence herbicide by applying a non-selective contact herbicide to the bottom of the ridge with nozzles on 'drop legs'. Contamination of the potato foliage is prevented by shielding. In practice the risks of contamination of the potato foliage and effects on yield have proved to be too great.

Nozzles and pressures

Standard fan or cone nozzles are used at moderate to high pressures in order to give adequate soil and foliage cover.

10.4.9 Special requirements

The ridging of the soil for potato culture requires that the tractors on which the sprayers are mounted should have a relatively high clearance and are fitted with narrow wheels and tyres. These latter run in the furrows without disturbing the sides of the ridges.

Soil type

As indicated previously (Table 10.1) the dose of most soil acting herbicides has to be modified according to soil type. Lower rates are used on lighter sandy soils and higher rates on heavier or organic soils. As a general rule herbicidal activity is higher on lighter soils and the crop is consequently more at risk. On heavier soils activity is lower and weeds are more difficult to kill.

Soil conditions

Soils should be relatively free of clods and the ridges rounded so that after herbicide application the soil surface is stable and there is relatively little erosion of the surface soil and herbicide into the furrows. Should this happen or the soil surface be disturbed by cultivation, weeds may germinate in herbicide free conditions on the ridges.

Weather

If the soil surface is dry and no rain follows shortly after herbicide application the activity of soil-acting herbicides may be reduced and weed control consequently be poor. Moisture is necessary to carry the herbicide to weed seeds which germinate below the soil surface. Conversely heavy rain soon after treatment may carry herbicide down to the potato tubers and cause damage.

Residues and succeeding crops

Most of the soil-acting herbicides have some residual activity but the persistence of this activity varies from herbicide to herbicide. As indicated in Section 10.4.8 application needs to be even with no over-dosing or overlapping as this will increase the risk of damage to succeeding crops.

The likelihood of persistence and the choice of succeeding crops have been indicated under the respective herbicides in Section 10.4.3–10.4.8.

10.5 Potato haulm destruction and the regulation of tuber size

10.5.1 Ease of harvesting

The ease with which a potato crop is harvested is dependent on a number of factors, all of which need to be optimal if the operation of lifting is not to be seriously hampered and the quality of the crop reduced. On maincrop potatoes natural senescence will occur in the late summer, accompanied by changes in the tuber which give the tuber natural protection against damage and disease entry. This process will be advanced in dry years or where the foliage has been seriously damaged by potato blight or insect damage. In wet summers, and in particularly wet autumns, the potato foliage continues to grow. The haulm becomes increasingly difficult to remove and will seriously impede harvesting machinery. There is also in the latter situation the problem of excessive weed growth which can be a greater nuisance than the potato haulm itself. In general neither haulm nor weeds are a problem in early potatoes but in second earlies the haulm is invariably very vigorous at lifting and needs to be removed before harvesting can begin.

10.5.2 Mechanical haulm destruction

Interference at harvesting from haulm and weeds pre-1950 was considerably less because of the use of single row spinners for lifting and hand labour for picking up the tubers. With increasing mechanization of the crop, changes in husbandry, the introduction of new varieties and more effective control of blight and pests, means had to be found of removing the haulm so that the grower could time his harvesting operation better and also speed up the process. There is also a need to be able to kill off the haulm in order to prevent blight spores from infecting tubers and to advance curing in the soil in order to reduce damage from mechanical harvesters. An added advantage is the reduction of dirt on the tubers.

The mechanical removal of potato haulm can partly be achieved by the lifter itself. One virtue of the spinner was that it threw the haulm further than the tubers thus separating the two. This only works when harvesting one row at a time. Elevators and mobile harvesters do not cope with a large amount of haulm and harvesters in particular can be seriously impeded by any quantity. Machinery built solely for the destruction of haulm did not become widely available until the early 1960s. Forage harvesters were and may still be used to remove potato haulm. They suffer from the disadvantage of blowing the haulm on to other ridges making it necessary to lift as the haulm is removed, and also they can only be used at ridge height leaving the haulm and weed in the furrow. The most effective haulm destroyers have been the pulverizer types which use rubber or metal flails revolving at high speeds. The flails are set to follow the profile of the ridges giving good results in the furrows. They leave behind pulverized trash that readily dies and which falls through the elevator chains on the harvester. These machines have the advantage of being able to operate on the same day as the harvester.

10.5.3 Chemical haulm destruction

The use of chemicals for the destruction of potato haulm was very quickly adopted by growers as a quick and effective means of cleaning the land prior to harvest and also as a means of controlling weed growth. The adoption of the technique was aided by the need to stop blight spores from reaching the soil and causing blight in the tubers. It proved a happy marriage to burn off the haulm when blight became established. The crop could then be left so that when lifting began there had been ample time for the vegetation to die down completely.

Chemicals were used before 1939. By 1958 the practice of chemical haulm destruction was well established. Sulphuric acid, sodium arsenite tar oil, sodium chlorate, DNOC and copper sulphate plus salt were some of the compounds used. Tar oil, DNOC and copper sulphate ceased to be used by 1960 due largely to their unpleasantness. Sodium chlorate is still used in some places, notably Scotland, where high rainfall can ensure that it is washed from the soil and harmful residues are not left for succeeding crops. Sodium arsenite was withdrawn from use in 1961 due to its toxic nature.

Sulphuric acid has remained in use as a haulm desiccant and is still one of the most reliable materials. At the height of its popularity, sulphuric acid was used on a number of agricultural crops and was applied by contractors with special sprayers. Its use declined with the increase in farm sprayers as these machines had metal parts which were attacked by the acid. The farm sprayers of today are made from plastics and fibreglass and can be more easily modified and as a result the use of sulphuric acid sprays has increased in recent years. It is particularly useful in dry years when diquat and dinoseb reduce the keeping qualities of the tubers. The acid is used as Brown Oil of Vitriol (BOV) either direct at 225 litres ha^{-1} or as a 12% solution applied at 1125 litres ha^{-1}. When diluting the sulphuric acid it should always be carefully added to the water and not vice versa. If the acid or diluted acid get on to the skin it must be immediately washed with large amounts of water and neutralized with washing soda. Stock should be kept out of treated crops for at least 3 days after spraying and it is possible for acid on the foliage to damage clothes and the skin for up to a week after spraying.

The withdrawal of sodium arsenite was followed by the introduction of first dinoseb and then diquat. Although dinoseb is a scheduled poison it was rapidly adopted as it can be grower applied and does not need special spray machinery. Dinoseb is formulated for haulm destruction as an oil based emulsion and there are some 20 products on the market all containing 0.9%. These are applied at 22 l ha^{-1} to 33 l ha^{-1} depending on the density of the haulm. The amount of water used is quite important as good coverage of stems and leaves is necessary. Dinoseb is both a good haulm and weed killer although it is ineffective on grass. Under very dry conditions it can be translocated into the tuber causing internal vascular browning. These effects are not as serious as those produced by diquat. It is necessary to wait 10 days between spraying dinoseb and lifting the crop.

Diquat is the most widely used haulm killer and has the advantage of being neither a scheduled poison nor corrosive. It is easily applied and does not need particularly high volume rates of water to be effective. It is recommended at 0.77 kg a.i. ha^{-1} in 200–550 l ha^{-1} of water. It gives a quick kill but is not as effective as sulphuric acid or dinoseb on weeds. The major drawback to the use of diquat is that when applied under very dry conditions it is translocated to the tubers and causes internal vascular browning. This is quite pale at the time of lifting but intensifies in stored potatoes leading to heel- or stem-end rot. Tissues outside the vascular ring become necrotic, and eventually decay from the invasion of other secondary rotting organisms. Potatoes showing slight vascular marking can be marketed for ware but any further degeneration makes them unsaleable. The internal rotting also prevents their use as seed. The conditions causing internal vascular browning are related to soil type, occurring more frequently on light rather than on heavy soils. It is also more likely to occur when there is high humidity, e.g. after light showers or at the beginning or end of the day. When conditions are clearly unsuitable for the use of diquat or dinoseb, sulphuric acid or mechanical haulm destruction should be used and the crop lifted as soon as possible after haulm removal.

10.5.4 Regulation of tuber size

The regulation of tuber size has many potential applications in potato growing but to date there have arisen no reliable methods of achieving a predictable control of tuber size. In the post-war years most attempts to alter the size of tubers were aimed at enlargement as a means of boosting yield. A variety of chemicals were sprayed on to the crop foliage but the only compounds to produce any response were the hormone herbicides applied at low doses. A formulation of 2,4-D was test marketed for this use but there were difficulties in timing the spray, damage from overdosing and taint problems, and the project was droped.

In recent years, efforts have been aimed at producing the maximum number of tubers of a specified size. Potatoes grown for seed have very specific size limits. All attempts to control the size of seed produced have been confined to cultural methods, e.g. higher plant populations associated with timely haulm destruction. Attention has been given to the canned potato market. This needs a smaller range of tubers from certain varieties, especially King Edward, Désiree and Maris Peer. Various growth regulator chemicals have been used in trials but none have proved commercially acceptable.

The only chemicals to be recommended commercially for the control of tuber size are the contact herbicide and desiccant dimexan, and the growth regulator *N*-dimethylaminosuccinamic acid. Dimexan acts by restricting foliar growth thereby affecting tuber growth. It is applied at 7.5 kg a.i. ha $^{-1}$ to the crop when the potato plants are about 6–8 cm high. This has increased the yield of tubers in the medium and small range, even though the total yield of the crop has been reduced. It can also be of use to prevent splitting of large tubers. *N*-dimethylaminosuccinamic acid is applied later in the life of the crop when tubers are from 6–12 mm in diameter. The potatoes should be growing well at the time of application and as the uptake of the chemical is slow, application should be followed by a minimum period of 12 h dry weather. *N*-dimethylaminosuccinamic acid is applied at the rate of 1.9 kg ha $^{-1}$ in 450–700 l ha $^{-1}$ of water. A minimum period of at least 3 weeks must be allowed between application and harvest.

10.6 Potatoes as weeds

The potato as a crop is restricted in the frequency of its occurrence in normal rotations because of the serious danger of encouraging eelworm in the soil. There has been a tendency however, over recent years, for the number of years between successive potato crops to be reduced. The development of modern pesticides to control eelworm problems has removed some of the inherent risks on the continent. In Europe potatoes are sometimes grown in alternate years between cereal crops.

In the United Kingdom it is not uncommon for a potato crop to occur in some rotations once in three or four years. This has resulted in the build-up of a large population of potato tubers in the soil as volunteers which present quite serious problems. In addition to presenting a new weed problem volunteer potatoes are also important host plants for diseases and viruses. This problem is particularly important in the vicinity of potato clamps and also where potatoes are riddled out of store and unwanted tubers disposed in dumps around the farm giving rise to foci for the dissemination of virus diseases and potato blight. As a hygiene measure on the farm it is of the utmost importance that potatoes on clamps and around farm stores should be destroyed as soon as they make growth the following year. There are a variety of means of doing this but, when choosing the method, consideration must be given to the use of that land for subsequent cropping or other storage that may occur in that area.

10.6.1 Destruction of potatoes on clamp sites

Only one or two materials are currently recommended for the destruction of potatoes from potato clamps. The most useful to date have been the granular herbicides dichlobenil and chlorthiamid. These are normally recommended for use in fruit and certain ornamental situations but are very destructive to the potato. They also have the advantage of being in a granule form which can be applied over a small area by hand. Treatment must be carried out in the winter and the rate of use should be in the order of 200 kg ha^{-1} of granules. They leave very high soil residues that will inhibit any further plant growth during the year of treatment and therefore where potato clamps are used on the headland this must be borne in mind when planting subsequent crops. Where it is important that there is no residue, a foliar herbicide should be used and for this glyphosate or aminotriazole would appear to be the most satisfactory materials. They have the advantage that they can be sprayed on to the foliage in the following spring and with glyphosate there will be no residues in the soil to affect any planting subsequent to treatment; with aminotriazole an interval must be left in order to allow residues of the herbicide to disappear from the soil. The use of foliar weedkillers for destruction of volunteer potatoes in this situation gives rise to danger because sufficient growth must be present for the chemical to be properly taken up and there is then a chance that vectors of virus diseases could breed on these plants.

10.6.2 Volunteer potatoes

The problem

The reduction in hand labour available to farmers and the consequent increasing mechanization of potato harvesting has meant that the volunteer potato, as a weed, has assumed increasing importance in recent years. All the tubers of a potato crop cannot be harvested mechanically and many, particularly the smaller ones, are left in the field after harvest. Lutman (1977) found that between 120 000 and 370 000 tubers ha^{-1} were left in the soil after harvest. Most of these tubers were less than 4 cm in diameter and had fallen between the bars of the elevators during harvest.

Potato tubers are susceptible to frost and if left on the soil surface are generally killed during the winter. Burial in the soil gives protection and potatoes burried below 2 cm are not generally killed in a mild winter. Tubers buried 10–15 cm deep may survive even a hard winter. Lumkes and Sijtsma (1972) in Holland showed that tubers must be exposed to at least 50 frost h with temperatures below $-2\,^{\circ}C$ (e.g. 25 h at $-2\,^{\circ}C$ or 10 h at $-5\,^{\circ}C$) before they are killed.

Surviving tubers produce plants the following season which, depending on the competitiveness of the crops in which they are growing, in turn produce new tubers. Lutman (1977) found that in winter wheat an average of 1–2 tubers per plant were produced whereas in less competitive spring barley there were 2–4 tubers per plant. Thus, particularly in mild winters, volunteer potatoes can survive for at least 4–5 years in most arable crops.

Occurrence

In cereals, volunteer potatoes are rarely competitive as they emerge relatively late in the spring. However, should the crop become laid or the foliage sparse due to pests or disease, they can become dominant relatively quickly and interfere with harvest. As already mentioned even in a competitive cereal crop, daughter tubers are produced which will perpetuate the problem.

Slower growing and less competitive crops such as sugar beet and row crops such as onions, carrots and brassicae can be severely reduced by competition from volunteer potatoes. If crops of peas and beans grown for processing contain volunteer potatoes, pieces of haulm or potato berries may contaminate the harvested product. These are poisonous and can be a hazard to consumers.

If volunteer potatoes persist they can constitute a hazard to following potato crops even if they are grown in a 4–5 year rotation. Populations of potato cyst eelworm (*Globodera rostochiensis* Wall) can be maintained and some virus, bacterial and fungal diseases e.g. blackleg (*Erwinia carotovora* var. *atroseptica* (van Hall Dye), gangrene (*Phoma exigua* var. *foveata* (Foister Boerema) and rugose mosaic leaf roll viruses (Doncaster and Gregory, 1948; Perambelon and Lowe, 1971; Adam and Todd, 1974) may also be maintained.

Nilsson (1975) suggested that volunteers may also make the control of Colorado beetle (*Leptinotorsa decemlineata* Say) more difficult by providing hidden breeding sites.

Control

Cultural

Little control has been achieved by cultural methods although the use of tine cultivation rather than the plough in the autumn can help to reduce the numbers of tubers which survive. The plough buries tubers whereas tines leave them on or near the surface where they are more vulnerable to frost.

Cropping

Potatoes can survive in most arable crops as already indicated although highly competitive crops such as oil seed rape should reduce them to a minimum. Relatively few plants will persist in a two year ley although haulm may be a hazard to grazing animals or in silage or hay in the first year.

Herbicide

There are few herbicides which will control potatoes. Due to their relatively large food reserves regrowth can occur even when a large proportion of the haulm has been killed. However both glyphosate and aminotriazole have controlled large potato plants in June and early July and prevented the production of viable daughter tubers (Lutman 1976a). Neither herbicide is however selective in emerged annual crops although a shielded sprayer may allow the use of glyphosate in some row crops. The most promising possibility is to use either herbicide in cereal stubbles on potato regrowth after early harvest of an arable silage crop or winter barley. However not all the potato plants regrow and there is always a risk of some survival.

The growth regulator herbicides used in cereals will distort the haulm but rarely give the complete kill necessary to prevent the formation of daughter tubers. In any case the emergence of the potato shoots is so extended that many do not emerge until the cereals are past the stage of growth at which they can be safely treated.

Propyzamide and trifluralin mixed into the soil in March have reduced and delayed the emergence of potato shoots and there are possibilities of following these treatments with late sown horticultural crops such as culture and brassicae. Field reports have indicated that metoxuron can give good control in carrots but in experiments it has proved less effective (Lutman 1976b).

On stone free soils possibly one of the most promising developments which originated in Holland is the fitting of a crusher to the potato harvester so that tubers not harvested are crushed before being returned to the field.

References

Adam, J.W. and Todd, J.M. (1974). *Potato Res.* **17**, 349–50.

Bremner, P.M. (1966). *Proc. 8th British Weed Control Conf.* 1–7.

British Weed Control Council (1957) *Weed Control Handbook*, 1957.

Butler, A.J. (1960). *Proc. 4th British Weed Control Conf.* 254–9.

Cowan, W.T. and Joy, N.B. (1964). *Proc. 7th British Weed Control Conf.* 430–40.

Cox, T.I. and Elliott, J.G. (1965). *Weed Research* **5**, 158–68.

Doncaster, J.P. and Gregory, P.H. (1948). *Agric. Res. Council Ref.* 7.

Eagle, D.J. (1976). Soil texture classification for the adjustment of herbicide dose *Proc 1976 British Crop Protection Conf.—Weeds*, 981–8.

Elliott, J.G. and Cox, T.I. (1964). *Proc. 7th British Weed Control Conf.* 401–10.

Eue, L. and Tietz, H. (170). *Pflanzenschutz-Nachrichten Bayer* **23** (3) 208–18.

Eue, L., Westphal, K., Dickore, K. and Meiser, W. (1969). *Proc. Eur. Weed Research Council 3rd Symposium New Herbicides* **1**, 125–32.

Evans, D.M. (1972). *Proc. 11th British Weed Control Conf.* 64–70.
Evans, S.A. (1968). *J.R. agric. Soc.* **129**, 149–68.
Fryer, J.D. and Makepeace, R.J. (1977a). *Weed Control Handbook, Vol. I Principles*, 6th Edn. Oxford: Blackwell's Scientific Publications.
Fryer, J.D. and Makepeace, R.J. (1977b) *Weed Control Handbook, Vol II Recommendations*, 8th Edn. Oxford: Blackwell's Scientific Publications.
Hampson, C.P. and Taylor, J.A.H. (1974). *Proc. 12th British Weed Control Conf.* 1153–8.
Hartel, K. (1964). *Proc. 7th British Weed Control Conf.* 441–9.
Headford, D.W.R. (1968). *Proc. 9th British Weed Control Conf.* 547–52.
Hein, G. and Smith, J.M. (1966). *Proc. 8th British Weed Control Conf.* 8–12.
Hodkinson, H.D. (1974). *Proc. 12th British Weed Control Conf.* 1003–13.
Joice, R. and Norris, J. (1962). *Proc. 6th British Weed Control Conf.* 427–34.
Joice, R. and Norris, J. (1964). *Proc. 7th British Weed Control Conf.* 423–9.
Lumkes, L.M. and Sijtsma, R. (1972). *Landbouw en Plantenziekten* (1) 17–36.
Lutman, P.J.W. (1976a). *The Seed Potato* (**16**) 18–19.
Lutman, P.J.W. (1976b). *Proc. 1976 British Crop Protection Conf.—Weeds*, 211–6.
Lutman, P.J.W. (1977). *Weed Research* **17**.
Mannal, H.G., Davis, M.E. and Whitworth, S.H. (1972). *Proceedings 11th British Weed Control Conf.* 519–27.
Nilsson, I. (1975). *16th Swedish Weed Conf.* 1. *Reports* D8–10.
Nield, J.R.A. and Proctor, J.M. (1962). *Proc. 6th British Weed Control Conf.* 697–712.
Perambelon, M.C.M. and Lowe, R. (1971). *Ann. Ref. Scot. hort. Res. Inst.* 50–1.
Pereira, H.C. (1941). *J. Agric. Sci.* **31**, 212.
Pfeiffer, R.K. and Phillips, J.D. (1964). *Proc. 7th British Weed Control Conf.* 411–22.
Roberts, E.H. (1972). *Seed Viability*, ed. Roberts, E.H. London: Chapman and Hall.
Sanders, H.G. (1954). *British Weed Control Conf.* 15–18.
Schuler, J. and Ebner, L. (1964). C3126 *Proc. 7th British Weed Control Conf.* 450–60.
Shotton, F.E. (1966). *Proc. 8th British Weed Control Conf.* 21–7.
Smith, J.M. and Marks, T.G. (1968). *Proc. 9th British Weed Control Conf.* 533–8.
Swan, D.G. and Chancellor, R.J. (1974). *Proc. Western Soc. Weed Sci.* **27**, 20–1.
Ministry of Agriculture, Fisheries and Food (1977).
Ministry of Agriculture, Fisheries and Food, London. S.T.L. 167.

Chapter 11

Disease aspects of potato production

G.A. HIDE and D.H. LAPWOOD

11.1 Introduction

Disease control is a prerequisite for improving and maintaining yield and quality of the potato crop and since the potato became widely grown serious outbreaks of disease and crop failures and consequent social and economic effects have repeatedly provided incentive for improvement. For example, in the mid-eighteenth century the widespread occurrence of leaf roll in Germany and the U.K. led to poor yields, and in the mid-nineteenth century late blight devastated crops in the New York and New England states of the U.S.A. A little later, the catastrophic blight epidemics ravaged the Irish crop, resulting in famine and mass emigration of the population. Towards the end of the nineteenth century, wart disease, found first in Czechoslovakia and later in other parts of Europe, threatened the survival of the cultivars of the day. When the causes of these diseases were discovered, early investigators formulated what have become established as the basic methods of disease control. For example, studies of viruses and wart led to the establishment of seed production in areas less favourable to the spread of disease, isolated from the main potato growing regions and where close control by inspection for freedom from diseases as well as for purity of cultivar could be maintained.

The potato is prone to more than a hundred diseases caused by bacteria, fungi, viruses or mycoplasmas but fortunately relatively few reach serious proportions in any one growing area. Whereas late blight is generally the most important disease wherever potatoes are grown, some diseases are serious only in cool temperate regions (e.g. skin spot and gangrene) and others only in warmer climates (e.g. early blight, bacterial wilt or brown rot).

Table 11.1 Disease of the potato

Causal agent and common disease name	*Distribution*	*Symptoms*	
		Foliage	*Tubers*
VIRUSES AND OTHER AGENTS			
Virus X, mild mosaic	World-wide	Mottling	–
Leaf roll virus	World-wide	Rolling	–
Virus Y, severe mosaic	World-wide	Necrosis, stunting, mottling	–
Virus A, mild mosaic	World-wide	Mosaic	–
Paracrinkle virus	Europe, N. America	Crinkling	–
Virus S	World-wide	Mosaic	–
Tobacco rattle virus	Europe, N. America, Brazil, Japan	Stunting, distortion	Internal necrosis
Mop top virus	W. Europe, Peru	Chlorotic markings	Internal necrosis
Yellow dwarf virus	N. America	Chlorosis, stunting	Internal necrosis
Spindle tuber 'viroid'	N. America, U.S.S.R., S. Africa	Stunting	Malformation
Witches' broom (mycoplasma-like bodies)	Europe, N. America, Australasia, China	Chlorosis, stunting	–
BACTERIA			
Corynebacterium sepedonicum (Spieck. and Kotth.) Skaptason and Burkholder, Ring rot	Europe, N. America	Wilting	Rot in vascular region
Erwinia carotovora (Jones) Bergey *et al.*			
var. *atroseptica* (van Hall) Dye, Blackleg, Soft rot	Europe, N. America	Wilting	Soft rot
var. *carotovora*, Soft rot	Europe, N. America	Wilting (Arizona USA)	Soft rot
Pseudomonas solanacearum (E.F.Sm.) E.F.Sm., Brown rot	Tropics, Warm-temperate zones	Wilting	Soft rot
FUNGI			
Alternaria solani (Ell. and Mart.) Sol., Early blight	World-wide	Progressive necrosis, defoliation	Rot
Botrytis cinerea Fr., Grey mould	World-wide	Decay of senescing stems	Rot
Collectotrichum coccodes (Wallr.) Hughes, ● ●	World-wide	Premature senescence	Superficial lesions
Fusarium spp., Fusarium wilt	N. America	Wilting	Rot
Fusarium solani var. *caeruleum* (Sacc.) Booth Dry rot	Europe	–	Rot

Helicobasidium purpureum (Tul.) Pat., Violet root rot	World-wide	Premature senescence	Superficial lesions and rot
Helminthosporium solani Dur. and Mont., Silver scurf	Europe, N. America	–	Blemish
Macrophomina phaseolina (Tassi) Goid., Charcoal rot	N. America, India	Stem rot	Rot
Oospora pustulans Owen and Wakef.*, Skin spot	N. Europe, N. America, Australasia	Non-emergence	Blemish, kills eyes
Phoma exigua Desm. var. *exigua* } Gangrene	Europe, N. America, Australasia	Non-emergence	Rot
Phoma exigua Desm. var. *foveata* (Foister) Boerema } Gangrene	Europe	Non-emergence	Rot
Phytophthora erythroseptica Pethybr., Pink rot	Europe, N. America, Australasia	Wilting, premature senescence	Rot
Phytophthora infestans (Mont.) de Bary., Late blight	World-wide	Progressive necrosis, defoliation	Rot
Pythium ultimum Trow., Watery wound rot.	World-wide	–	Wet rot
Rhizoctonia solani Kuhn., Black scurf	World-wide	Non-emergence	Superficial sclerotia
Sclerotinia sclerotiorum (Lib.) de Bary, Stalk break	Europe, N. America	Stem rot	–
Spongospora subterranea (Wallr.) Lagerh., Powdery scab	World-wide	Galls on roots	Superficial pustules
Streptomyces scabies (Thaxt.) Waksman and Henrici., Common scab	World-wide	–	Superficial lesions
Synchytrium endobioticum (Schilb.) Perc., Wart	Europe, N. and S. America, S. Africa, Asia	Deformation of petioles and leaves	Deformation
Verticillium albo-atrum R & B } Verticillium wilt, Early death	World-wide	Chlorosis, premature senescence	Vascular browning
Verticillium dahliae Kleb. } Verticillium wilt, Early death	World-wide	Chlorosis, premature senescence	Vascular browning

*Now *Polyscytalum pustulans* (Owen and Wakef.) Ellis

Although it cannot be stated with authority it seems very probable that most potato pathogens have been transported from country to country in the normal course of international trade in potatoes over the centuries. Speedier transport may well have increased their chances of survival on tubers; certainly expanding trade has provided greater opportunities in this century. The absence of a disease from a particular country may result from failure of introductions to become established because of unfavourable climate, but in more recent times import restrictions could have prevented the initial introduction of some diseases into some countries whose climate is favourable for their development. For example, importation of tubers affected with bacterial ring rot is prohibited into the U.K. where the disease is unknown, although it is prevalent in many parts of Europe.

Pathogens may attack foliage, root systems and tubers, so disease can be important throughout the crop cycle. Seed tubers may deteriorate before or soon after planting especially when seed is cut; the growing plant may be weakened by direct attack or systemic infection and thereby less productive, or the roots or foliage may be attacked directly causing their premature death. Tubers may become infected during growth or, often more importantly, through wounds made at harvest or subsequent handling. Diseases therefore are capable not only of decreasing production but also of blemishing and rotting tubers so that the quality of the crop is adversely affected.

Since potatoes are propagated vegetatively, the seed tuber often provides the major source of inoculum, ensuring the continued survival of some diseases (e.g. late blight), but symptoms may not always be present, for example, when infection is 'latent' (e.g. blackleg) or when inoculum is carried as a surface 'contamination' as often occurs with dry rot and gangrene. Groundkeeper or 'volunteer' plants surviving from previous potato crops may also help perpetuate diseases (e.g. viruses) as do some solanaceous and other weed hosts. Sometimes susceptible agricultural crops provide a reservoir of inoculum (e.g. bacterial brown rot) whereas the pathogens causing powdery and common scab, pink rot and wart can survive considerable periods in soil in the absence of known hosts.

World-wide losses of potato yield through disease have been estimated at 30% (Cramer, 1967). In the U.S.A., average losses for the period 1951–60 were 19%, with blight, leaf roll, X and Y viruses and verticillium wilt accounting for at least 2% each. However, these figures almost certainly exclude losses occurring after lifting from damage and disease which in the U.K. can account for 30% of the harvested yield (Twiss and Jones, 1965).

By contrast, estimates of costs to the potato industry of preventing disease are more difficult to make. In any calculation, the cost of producing 'disease-free' seed, manufacturing, developing and applying plant protective chemicals, developing specialized machinery for potato handling and storage, breeding disease resistant cultivars and of the necessary research, advisory and disease forecasting services are among many factors to be considered.

Expenditure on disease control will differ in different areas, depending on the value of the crop and the number of diseases of economic importance. In the U.K., growers acknowledge the need to buy certified seed to ensure freedom from virus diseases and wart and to apply fungicides to control late blight. Only recently, however, have diseases such as gangrene, blackleg, skin spot and silver scurf received the attention they warrant as major causes of losses in production, in storage or in presentation quality. In the latter respect the sale of washed, pre-packed potatoes has elevated some hitherto unimportant blemishing diseases such as silver scurf to a completely new status.

This chapter describes how the main potato diseases listed in Table 11.1 affect production during the sequence from seed to stored crop, how they are spread and perpetuated, the methods developed for their control and the possibilities for improvement.

11.2 Effects of diseases

11.2.1 Damage to seed and early plant growth

Seed crops can carry a variety of diseases, and in many countries are inspected and certified during growth to ensure freedom from viruses. Such certification rarely guarantees freedom from tuber-borne bacterial and fungal disease since these seldom cause above-ground symptoms and mostly develop after the tubers have been stored. Thus, Pérombelon (1972) found all Scottish seed stocks to be contaminated with blackleg bacteria and when Hirst *et al.* (1970) surveyed stocks of seed of cv. King Edward received in England and Wales in 1963–70 from Scottish sources, they found an average of 56% tubers affected with skin spot, 9% with gangrene, 3% with dry rot and 2% with late blight. Stocks showed wide variation in the proportion of diseased tubers and distribution patterns of different diseases between different stocks were of two basic types. Gangrene, dry rot, late blight, powdery and common scab always showed skew distributions, as shown in Fig. 11.1(a), indicating that most stocks had few diseased tubers, and very few stocks had many tubers affected, probably resulting from much mechanical injury (gangrene, dry rot), ineffective chemical control (late blight) or much soil inoculum (common and powdery scab). By contrast, skin spot and silver scurf had rectangular distributions, that is with small differences in the proportion of stocks affected to different degrees by these diseases [Fig. 11.1. (b)]. However, the disease susceptibility of cultivars affected the distribution pattern, e.g. that of skin spot in the more resistant Pentland Crown tended to skew.

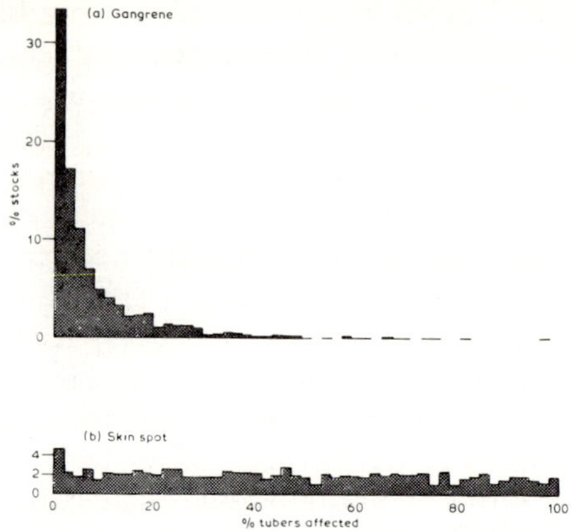

Fig. 11.1 Contrasted distribution patterns of gangrene and skin spot in King Edward seed planted in 1963–74. (a) gangrene; (b) skin spot

The main causes of seed tuber decay before planting are gangrene (*Phoma exigua* var. *foveata*; sometimes var. *exigua*) dry rot (mainly *Fusarium solani* var. *caeruleum*), late blight (*Phytophthora infestans*) and soft rot (*Erwinia carotovora* var. *carotovora* and *E. carotovora* var. *atroseptica*) either alone or in combination (Bokx and Mooi, 1974).

Cutting seed, as widely practised in the U.S.A., increases the likelihood of *Fusarium* rots, *Erwinia* soft rots, ring rot (*Corynebacterium sepedonicum*) and, in the warmer states of the U.S.A., brown rot (*Pseudomonas solanacearum*). Sprout growth from seed tubers during storage and after planting can be adversely affected by diseases. Black scurf (*Rhizoctonia solani*) can kill tips of developing sprouts and roots when seed is sprouted in high humidity (Mordue, 1964) and, in cooler potato growing regions, skin spot (*Oospora pustulans*) kills buds in the eyes of unsprouted seed (Hide *et al.*, 1969b as also can wart (*Synchytrium endobioticum*) at high storage humidity (Noble and Glynne, 1970). Silver scurf (*Helminthosporium solani*) can decrease sprout vigour (Mooi, 1968) but usually, despite its prevalence, it does little damage.

Soil conditions after planting should favour rapid growth so that shoots emerge quickly and give full plant populations. Gaps (misses) and delay in emergence are usually caused by rotting of the seed tubers or by failure or delay in sprouting in cold and dry or waterlogged soils, sometimes by planting errors and inadvertent use of sprout inhibitors or by bird or animal damage. Rapid and even emergence is of general importance in potato production and it is essential to growers of early maturing cultivars to ensure the early yields that command premium prices. Therefore, in the U.K., the physiological disorders 'little potato' (van Schreven, 1956), which may be associated with *R. solani*, and 'coiled sprout' sometimes associated with *Verticillium nubilum* (Moorby and McGee, 1966), can delay emergence of shoots and production of tubers.

Most of the diseases of stored seed also affect the establishment and growth of plants. Besides the rapid decay of seed tubers or seed pieces by *Fusarium* spp. and bacterial rots after planting, shoots may be killed before or soon after emergence by blackleg or *R. solani*. Even if plants are not killed, growth of shoots and stolons may be modified and affect the number and size of tubers that form. When the stem number is decreased, as happens when seed is severely affected with skin spot, fewer but larger tubers develop (Boyd and Lennard, 1961), whereas other diseases, for example, gangrene, stimulate sprouting of seed and so increase the number of stems and the number of small tubers (Griffith *et al.*, 1974).

Infection of stolons and tuber initials also affects the number and size of tubers, for example, *R. solani* can prune stolons causing them to branch and form more but smaller tubers (Hide *et al.*, 1973b) whereas *O. pustulans* usually decreases tuber numbers (Hirst and Salt, 1959) and thereby increases tuber size.

Compared with many annual crops, potatoes are grown wide apart within and between rows to give a relatively low plant population, so failure of plants to emerge causes relatively large and easily observed gaps. However, losses in yield due to gaps are not proportional to the number of missing plants because plants bordering gaps are able to benefit from decreased competition and can compensate to some extent for the 'miss'. With a single gap, the yield of each of the two adjacent plants within the row and in neighbouring rows may be increased respectively by 25 and 5% so that overall, the compensatory growth from these four plants can account for 60% of the potential yield of the missing plant. With larger gaps (two or more adjacent plants missing) production by neighbouring plants may be larger than for single plant gaps but the amount of compensation will be proportionally less.

In experiments to measure compensation in which plants were removed randomly soon after emergence to simulate crops with a range of 0–24% of plants missing, crop yields were decreased 0.3% for every 1% of plants removed (70% compensation) but statistically significant yield decreases were only reached when 15% of plants were missing (Hirst *et al.*, 1973). When plants were removed at flowering to simulate gaps caused by death of plants, for example, by blackleg, then adjacent plants compensated less and yields were decreased 0.8% for every 1% of plants removed (20% compensation), the differences becoming significant only when 6% or more plants were removed. These results were derived from crops in which the surviving plants were healthy and able to benefit from decreased competition; in crops which are gappy because of disease this may not be true so that in practice, compensation may not always

Table 11.2 Effects of planting differently diseased King Edward seed tubers on growth, total yield and size distribution of tubers

Symptoms on seed tubers	*Plants emerged after 6 weeks (%)*	*Final plant population (%)*	*Number of stems/plant*	*Total yield ($t\ ha^{-1}$)*	*% of yield*			
					Undersize	*Small*	*Large*	*Oversize*
Skin spot								
None	91	98	4.0	49.3	7	40	52	1
Severe	44	80	2.9	39.5	5	27	63	5
Random selection of seed	76	94	3.7	48.7	8	36	54	1
Skin spot (dead eyes)								
None	90	98	3.5	51.0	7	37	55	1
Severe	12	52	2.4	26.5	6	27	61	5
Random selection of seed	90	94	3.4	49.1	7	35	57	1
Black scurf								
None	89	98	3.7	50.4	9	37	53	1
Severe	79	96	3.7	46.7	12	39	48	1
Random selection of seed	88	98	3.8	49.7	11	40	48	1
Gangrene								
None	89	97	4.2	49.7	11	44	44	1
Severe	64	82	4.9	39.6	13	47	40	1
Random selection of seed	73	89	4.6	43.3	13	40	45	1

Data from Hide *et al.* (1973a,b) and Griffith *et al.* (1974).

be as great as these results indicate. Table 11.2 shows how plant populations and growth were affected by three diseases and illustrates this point clearly. For example, seed tubers with eyes severely affected with skin spot produced half the plant population of unaffected seed and only half the yield. However, this is an over-simplification because tubers within a seed stock vary in disease severity. So much so that when tubers without symptoms selected from stocks of seed with skin spot or black scurf were compared with a random selection of seed from those stocks, then the latter often yielded as much as the former presumably because vigorous plants compensated for weak or missing neighbours. By contrast, random selections of seed from stocks with gangrene yielded less than seed without lesions, probably because affected seed produced plants with many stems, so limiting the compensatory growth and yield of surviving neighbouring plants.

11.2.2 Stunting and premature death of foliage

Infection with viruses

Plants can become systemically infected with viruses following transmission either mechanically or through vectors and, unlike diseases caused by bacteria and fungi, once the infection has become established development of disease within the plant is not greatly affected by weather conditions. Usually viruses decrease plant vigour by shortening internodes, and cause colour changes in foliage which becomes mottled or chlorotic. Yields are decreased and some viruses cause internal tuber symptoms which cannot be seen until the affected tuber is peeled. Crops affected in this way may be rendered unsaleable even though a relatively small proportion of tubers are affected. Some viruses may rapidly kill plants whereas others cause mild or no symptoms and the reaction of different cultivars to the same virus can vary so that symptoms alone may not be a reliable guide to the virus involved.

With some virus diseases, symptoms can be quite different in the year when plants become infected (primary symptoms) from those showing in plants derived from infected seed (secondary symptoms). For example virus Y, transmitted by *Myzus persicae* and other aphids, may cause near complete defoliation (leaf drop streak) within 3–4 weeks of infection and yields are decreased according to the time and rate of defoliation. In the following season plants from infected seed are severely stunted, mottled (severe or rugose mosaic) and die early often giving only half the yield of uninfected plants and this composed of a high proportion of small tubers.

The primary symptoms of some virus infections are mild or, if infection occurs late in the season, plants show no symptoms, but even so a proportion of tubers from them when planted as seed will produce plants with severe symptoms in the following season. Plants infected during the growing season with leaf roll virus transmitted by aphids, and potato yellow dwarf virus transmitted by leaf hoppers, may show merely slight rolling of upper leaves, often only on one stem. In the next season however, leaf roll infected tubers produce stunted plants with rolled lower leaves. These are thickened and brittle with accumulation of carbohydrates and are characteristically harsh and rattle if shaken. Aerial tubers may form in leaf axils and yield is halved. Secondary symptoms of leaf roll and virus Y also develop on plants following previous infestation by viruliferous aphids of their sprouts before planting. Seed infected with yellow dwarf virus produces chlorotic and stunted plants and sometimes the growing point is killed; in warm soil tubers may fail to sprout or plants die soon after emergence. Any tubers that are produced are small and misshapen and, because rust coloured lesions develop in the pith, quality is also affected (Walker and Larson, 1939).

Two other diseases that cause severe secondary symptoms are witches' broom and spindle tuber. Plants produced from tubers infected with witches' broom, caused by mycoplasma-like bodies (Harrison and Roberts, 1969), are slow to emerge and have many weak stems bearing simplified pale-edged leaves (Todd, 1954). Spindle tuber is caused by a virus-like particle comprising only RNA ('viroid') and infected plants remain erect but with rugose leaves and the few tubers produced are cylindrical or irregular in shape. The viroid can be transmitted in true seed (Fernow *et al.*, 1970) and therefore can be a problem to the potato breeder.

Some virus diseases produce only mild symptoms or none at all, in both seasons. Virus X, transmitted mechanically by leaf contact, machinery or on clothing, may produce no symptoms in some cultivars whereas in others it causes an interveinal mottle but without visibly affecting plant vigour, although yield may be decreased. Plants infected with virus A, transmitted by aphids, show mild symptoms but in association with virus X or Y cause crinkle symptoms. Paracrinkle and virus S, both transmitted by aphids, produce such slight symptoms in plants that their prevalence in some cultivars remained unrecognized until plants were freed from them (Bawden and Kassanis, 1965).

Virus diseases decrease yield but as infected plants are usually scattered through the crop, neighbouring healthy plants can compensate in growth, as discussed in Section 11.2.1. For example, in three crops with 5–15, 15–25 and 25–50% of plants affected with secondary symptoms of leaf roll and virus Y, yields were decreased respectively by 2, 7.5 and 16% (Schick, 1952). However, the amount of compensation may be less than expected if infection spreads from diseased to neighbouring healthy plants especially if the spread occurs early in the season. Also, when several diseased plants are together, they too can benefit from decreased competition, so calculation of losses in yield from crops with known proportions of diseased plants is complex (Reestman, 1970).

Destruction of foliage

Early blight (*Alternaria solani*) and late blight (*Phytophthora infestans*) are probably the most important fungal foliage diseases of potatoes. Both are dependent on weather conditions for sporulation and spread and the amount of yield loss they cause depends on how early and quickly they destroy the foliage. Late blight can attack and destroy foliage of developing plants at any stage of growth given suitable weather (Cox and Large, 1960). In Mexico where crops are planted in anticipation of the rainy season, susceptible European cultivars are attacked soon after emergence, yielding little or nothing (Lapwood, 1971). At the other extreme, as in much of the U.K., blight does not become epidemic until the canopy closes across rows allowing the development of a suitable micro-climate for the disease to spread (Hirst and Stedman, 1960). The fungus sporulates mainly on the undersides of leaves in a band around the edge of the necrotic lesions and the sporangia may be released into the air or splashed in water droplets during rain to other leaves. Persistent leaf wetness is needed for successful infection and epidemics develop only in wet weather and progress most rapidly when rain is frequent. Because it is a disease which mostly develops from discrete lesions, the rate of foliage destruction depends on how soon every leaf becomes infected, although in some cultivars stem infections can be common, resulting in death of the stems distal to the girdling lesions.

Young foliage seems temporarily resistant to early blight. In Colorado, U.S.A., first (primary) infections produced from spores in soil are few and develop on the lower leaves in June and provide the spores for secondary spread which may be delayed, however, for as long as 2 months. When spread does occur it can destroy the foliage very rapidly (Harrison *et al.*, 1965). Yields may be decreased by up to 40% (Harrison and Venette, 1970), although in areas where the disease is severe, crops may further be affected by wilting and early death caused by *Verticillium* spp.

The growth of tubers is dependent on the functioning leaf area, so diseases that defoliate plants have serious effects on yield. The progress curve for a late blight epidemic, expressed as the proportion of leaf area destroyed, is essentially of the sigmoid type, and Large (1952) found in England and Wales that tuber growth stops when 75% of the leaf area is destroyed. From data on the progress of late blight in fungicide-sprayed and unsprayed crops, reasonably reliable estimates of the percentage loss in yield can be made by reference to their bulking curves.

In Canada where the growing season is 40 days shorter than in the U.K., this method seriously under-estimated losses (James *et al.*, 1971), and a method based on weekly increments of foliage destruction, with epidemics classified as 'early' or 'late' depending on when they occurred in relation to crop development, was found more reliable (James *et al.*, 1972).

Wilting and premature death of plants

Several pathogens may cause wilting symptoms because they infect roots and invade the vascular system. Water supply to the foliage is restricted and typically leaves become chlorotic and sometimes roll before dying. If infection occurs late, symptoms may be confused with and by senescence.

Of the bacteria attacking potato plants, three cause diseases which are of major economic importance; *Erwinia carotovora* var. *atroseptica* the cause of blackleg, *Corynebacterium sepedonicum* causing ring rot and *Pseudomonas solanacearum* causing brown rot. All are tuber-borne and some affected tubers give rise to diseased plants. *E. carotovora* var. *atroseptica* produces a necrotic rot on stems extending from the seed tuber up to and, in very wet weather, above soil level. Leaves turn yellow and leaflets roll upwards but not all stems on a plant may be affected. In Arizona *E. carotovora* var. *carotovora* has been suspected as a cause of blackleg (Stanghellini and Meneley, 1975). Infection with *C. sepedonicum* also causes yellowing and rolling of leaves early in the season in warm areas, although in Europe symptoms develop later and can be confused with late blight or senescence (Dounine, 1961). Similarly infection with *P. solanacearum* results in wilting of plants (Conroy, 1969) but like *C. sepedonicum* does not cause a 'blackleg' symptom.

Infection by fungi, including species of *Verticillium* and *Fusarium* and *Phytophthora erythroseptica* can also cause wilting and early death. *Fusarium* spp. mostly cause wilting in warm climates following infection of roots or decay of seed tubers before plants are fully established. *Verticillium* spp. infecting through roots grow into the vascular system and decrease leaf area (Harrison and Isaac, 1969). Later, lower leaves become unilaterally chlorotic and symptoms progress up the stem until the plant becomes desiccated and dies. Crop yields can be halved (Guthrie, 1960). Severity of the disease is increased by infestation of roots with nematodes including *Pratylenchus penetrans* and *Heterodera rostochiensis* which seems to assist infection of roots (Morsink and Rich, 1968; Corbett and Hide, 1971). In N. America the disease can be caused by *V. albo-atrum* and *V. dahliae* (Busch, 1967) whereas in the U.K., *V. dahliae* is the predominant species (Isaac and Harrison, 1968).

Several other pathogens can cause wilting and death of foliage. In wet areas, stems may be infected by *Sclerotinia sclerotiorum* (stalk break) causing leaf chlorosis and death of stems, which bear sclerotia in the stem cavity (Partyka and Mai, 1962). *Botrytis cinerea* (grey mould) usually infects leaves but may progress along petioles to infect stems late in the growing season, causing a soft wet rot. Occasionally injured tubers are attacked. In warm climates *Colletotrichum coccodes* (black dot) infects roots and stems, causing premature death in dry weather. The fungus may also prune stolons and blemish tubers (Thirumalachar, 1967).

11.2.3 Infection of tubers during growth

Progeny tubers may be attacked during growth by bacteria and fungi which gain entry through lenticels and growth cracks, unsuberized tissue around eyes, insect wounds, intact skin or via the stolons. After infection, tubers may rot in soil, become deformed or blemished or occasionally the pathogens remain latent to cause disease later. Inoculum may be present at the tuber surface or in the surrounding soil and infect only when tubers are stored in conditions favourable for infection or are injured at or after lifting.

Rots

Tubers that are rotted by fungi usually remain intact unless they are invaded by secondary organisms. Sporangia of *Phytophthora infestans* washed down from diseased foliage can infect tubers through lenticels, eyes or growth cracks to produce the typically hard brown marbled progressive rot. *Macrophomina phaseolina* infects through lenticels, eyes, stolons and wounds made by larvae of the tuber moth (*Phthorimaea operculella*) to cause black sunken lesions and later blackening of internal tissues (Paharia, 1960; Thirumalachar, 1955). *Phytophthora erythroseptica* gains entry through stolons causing a wet but firm rot progressing from the heel towards the rose end of the tuber. Sometimes the infection may occur at the 'eyes'. When lifted, infected tubers frequently have small clumps of soil adhering to the skin where moisture has exuded from lenticels. The cut surface characteristically turns a bright salmon pink when first exposed to the air, giving the disease its common name, pink rot. Occasionally the reddish brown mycelium of *Helicobasidium purpureum* (violet root rot) is found on tubers. The fungus penetrates the skin and causes a soft rot.

Bacteria infect tubers through lenticels and stolons generally causing more disintegration of tissues than do fungi, resulting in soft mushy rots. *Erwinia carotovora* var. *atroseptica* and *E. carotovora* var. *carotovora* cause a general disintegration of tissues which turn purple/black when cut. *Corynebacterium sepedonicum* attacks the tissues of the vascular region, in which cavities develop causing typical 'ring rot' symptoms. These may later be masked by secondary bacteria invading other tissues and causing complete disintegration of tubers. *Pseudomonas solanacearum* also infects the vascular elements which are stained brown (brown rot) and liquid can be squeezed from infected tubers.

Blemishes

Several diseases deform or blemish the tuber skin or flesh and so decrease quality and saleability of crops. In wet soil, *Synchytrium endobioticum* penetrates eyes to form warty outgrowths that disfigure tubers. *Spongospora subterranea* infects lenticels in similar conditions and development of powdery scabs containing resting spores (spore balls) occurs when soil gradually dries (Kole, 1954). In continuing wet conditions resting spores may germinate, the fungus penetrates deeper into tuber tissue resulting in the development of tumour-like outgrowths. The actinomycete *Streptomyces scabies* (common scab) causes little malformation. Infection of the lenticels of developing tubers gives rise to surface scabs which enlarge so long as the tubers swell. Dry weather favours infection and, when prolonged during early tuber formation, encourages severe attacks that can affect a large proportion of the tuber surface area (Lapwood and Hering, 1970) and so decrease saleability. Tubers can be deformed by *Rhizoctonia solani* and the sclerotia (black scurf) that develop on the tuber surface towards the end of the growing season may affect the marketing quality of tubers.

Tubers become systemically infected with the viruses that infect foliage but usually show no symptoms. A few viruses, however, do cause internal tuber symptoms which can be particularly insidious because they are not seen until the tubers are prepared for cooking. Consequently a small proportion of affected tubers can render a whole crop unacceptable. The viruses capable of tuber damage include potato yellow dwarf (Section 11.2.2), potato mop top, transmitted by the zoospores of the powdery scab fungus *Spongospora subterranea* (Jones and Harrison, 1969) and tobacco rattle, transmitted by ectoparasitic nematodes of the genus *Trichodorus* (van Hoof, 1968). The last two viruses cause arc-shaped rusty-brown necroses of the tuber flesh (spraing) which are occasionally visible on the surface of the tuber; yields can be decreased by 25% or more.

'*Latent diseases*'

During growth, tubers can be infected or contaminated by bacteria and fungi that may not cause disease until after crops are lifted. Bacteria may penetrate lenticels or through stolons but cause rots only when tubers are stored or planted. Spores of *Helminthosporium solani* (silver scurf) produced on seed tubers cause symptomless infection of young tubers, firstly at the stolon end; and *Oospora pustulans* (skin spot) from lesions on seed tubers, roots, stolons and stem bases infects scale leaves and buds of tuber initials and young tubers. Sometimes silver scurf and skin spot symptoms are seen on tubers at lifting (Kharkova, 1961; Mooi, 1968) but mostly they develop in store. *Phoma exigua* (gangrene) from soil, decaying seed tubers or from pycnidia on dead stems may occasionally invade tubers before lifting (Todd and Adam, 1967), but more often, like *Fusarium* spp, spores contaminate the tuber surface and surrounding soil and do not infect until tubers are injured at or after lifting.

11.2.4 Infection of tubers at lifting

When potato crops are lifted, tubers are particularly at risk from the bacteria and fungi present at the tuber surface, in the surrounding soil, or harboured in the decaying foliage, seed or infected progeny tubers. Wounds inflicted by stones and machinery may make tubers unsaleable but more importantly may provide sites for infection by pathogens. In the U.K. almost one-quarter of all tubers from 600 farms surveyed in 1965 and 1966 had wounds penetrating a few mm into the flesh (Church *et al.*, 1970). Some wounds, such as cuts, heal quickly and completely with the formation of a cork barrier, but crush wounds are more slowly and often less completely healed and are therefore more likely to become infected. Wounds are most severe when lifting from dry stony soils and in dry seasons, especially if tubers are immature. Much fertilizer can delay maturity and increase the incidence of mechanical damage (Paterson and Gray, 1972).

Immature tubers are prone to infection by *Fusarium* spp., *Pythium ultimum* (watery wound rot) and *Alternaria solani* (Boyd, 1967a; Venette and Harrison, 1973) but their resistance later increases and, in Scotland, wounded tubers are most resistant to infection by *F. solani* var. *caeruleum* when fully mature. *Phoma exigua* and *Macrophomina phaseolina* (charcoal rot) can also infect wounds made during lifting, and tubers can be infected by *Phytophthora infestans* from inoculum in soil and on foliage but, because sporangia are short-lived, delaying harvest for 2 weeks after foliage destruction can minimize this risk.

11.2.5 Post-lifting disease

Deterioration can start soon after crops are lifted, during transport or at the retail outlets. Early marketing can avoid the risk of loss in store but some crops must be stored to ensure continuity of supply.

During transport and at retail outlets

Most loss follows warm and humid conditions at or soon after lifting, because they favour rapid rotting by the soft rot bacteria and the fungi *Pythium ultimum* and *Phytophthora erythroseptica*. These problems are usually more acute in immature crops and are often associated with poor ventilation in transit (Cromarty and Easton, 1973). Secondary rotting by bacteria can also affect tubers infected with late blight. At retail outlets tubers offered for sale are also at risk especially if they have been cleaned by brushing or washing and packed in poorly ventilated plastic bags. These conditions encourage bacterial rotting especially of immature tubers (Scholey *et al.*, 1968). Susceptibility is decreased by storage before washing.

In store

Most mature crops are kept in store and the storage conditions should encourage wound healing initially and thereafter should be adjusted to minimize water loss and sprouting. The method of storage varies, according to the available facilities and to the degree of environmental control required (Chapter 14), from clamps in the field insulated with straw and soil, to improvised buildings or purpose-built stores with facilities to control the environment. Disease development is greatly influenced by storage conditions (Boyd, 1972) and, because crops for processing or for sale as table potatoes or as seed may require different environments, so different diseases may become dominant.

In stores kept cool (2–5 °C) a major rotting disease is gangrene (*Phoma exigua*) which develops on wounds that have not healed (Malcolmson and Gray, 1968). Externally, lesions show as dark, dry spreading depressions and cavities may form within tubers often much larger than the visible symptoms would suggest. Dry rot (*Fusarium* spp.) can be arrested in cool (5 °C) stores but re-activated on return to warm conditions. Under drying conditions rots gradually lose moisture and become shrunken, developing concentric wrinkles round the site of infection on which cushions of spore-bearing mycelium develop, while internally the affected tissue is usually light brown. In damp conditions the rots advance more rapidly and, as with tubers affected by late blight, secondary organisms can hasten the decay and sometimes initiate massive soft rotting. Susceptibility of tubers to infection increases during storage. Like gangrene, dry rot can develop from wounds inflicted when tubers are sorted after storage.

In warm stores bacteria, either in wounds or lenticels or following fungal infections, are the main cause of rapid deterioration, which is most likely to occur if crops are loaded, or become wet. Rotting of individual tubers may begin soon after loading and, if conditions are favourable, moisture exuded from these will affect neighbouring tubers leading to local pockets of rotting tubers in which temperatures rise, accelerating the rate of spread through the stack. If conditions become cooler and drier, infections developing at lenticels may be arrested and affected tissues collapse causing spots, a form of 'pit rot', or larger superficial 'hard rot' lesions (Logan, 1964).

Of the blemishing diseases, skin spot can become particularly severe in cool (2 °C) humid stores. Symptoms usually develop 2–3 months after lifting and subsequently the number of spots increases and buds in the 'eyes' of affected tubers may be killed (Hide *et al.*, 1969b). Infection occurs through 'eyes', lenticels or skin abrasions and the disease is often prevalent in crops lifted in wet cool conditions (Boyd and Lennard, 1962). Humid stores encourage the sporulation and spread of *O. pustulans*, and also the formation and spread of wart and the development of tumours on tubers with powdery scab. In warm storage, silver scurf spreads over the tuber from the stolon end and the fungus sporulates on the lesions to spread to neighbouring tubers. Affected tubers are unattractive when washed before sale and may shrivel.

11.3 Survival and spread of pathogens

There are a number of ways in which potato pathogens survive but all require dispersal in space and time to infect succeeding crops. Some are associated with soil and can be regarded as soil-borne. Usually these spread only slowly, unless introduced with infected seed tubers. However, they may survive for long periods by forming thick-walled resting structures, e.g. *Spongospora subterranea* whose spore balls may remain viable in soil for more than 10 years, and *Synchytrium endobioticum* resting sporangia for at least 25 years. Also found in soil are the oospores of *Phytophthora erythroseptica* and *P. infestans* (the latter probably only in Mexico), the chlamydospores of *Fusarium* spp, and sclerotia of *Rhizoctonia solani*, *Sclerotinia sclerotiorum* and *Verticillium dahliae*. A few pathogens survive saprophytically in soil in the absence of potatoes or potato crop residues. For example, *Streptomyces scabies* is commonly found in soils under grass while *Rhizoctonia solani* and *Helicobasidium purpureum* are also soil inhabitants, although they can infect crops other than potatoes. *Colletotrichum coccodes* can survive in soil for at least a year after decomposition of crop residues and *Phoma exigua* for at least 7 years after potato crops but in what form is not known. Phytopathogenic bacteria do not form resting spores and seldom survive for long in soil. An exception is *Pseudomonas solanacearum* which can survive for 4–6 years in bare fallow and for up to 10 years in soils with non-susceptible plants (Schuster and Coyne, 1975). The viruses causing spraing can survive in the soil in their vectors, the fungus *S. subterranea* and nematodes of the genus *Trichodorus*.

Other pathogens survive only in infected plant remains, and so the length of survival of *Alternaria solani*, *Oospora pustulans* and *Erwinia carotovora* varieties depends on the rate of decomposition of the debris. Many of the above pathogens and also *Corynebacterium sepedonicum* can be found contaminating planting and harvesting implements as well as sacks and storage equipment.

Many pathogens survive from season to season in infected tubers, which may be lifted and used as seed, or remain in soil unharvested and perpetuate infection on volunteer plants. Most viruses survive in tubers which produce systemically infected foliage; an exception is tobacco rattle virus which is seldom transmitted from infected seed tubers. Lesions on seed tubers or soil adhering to them can be sources of bacterial and fungal inoculum for infecting the resulting crop and for introducing diseases into new areas.

Most bacterial diseases originate from seed tubers as do skin spot and silver scurf, and discarded tubers affected with late blight may produce infected stems and foliage, which constitute primary sources of infection each year (Boyd, 1967b). Plants other than potatoes may also become infected with potato pathogens but it is not known how important these are as sources of inoculum or as means of survival.

For the continuance of diseases, pathogens require not only means of survival but also effective methods of spread. Pathogens on seed tubers may be released into soil as infected tubers rot (*Phoma exigua*, *Fusarium* spp. and the bacterial pathogens), as spores produced on tuber lesions (*Helminthosporium solani*, *Oospora pustulans*), or mycelium may grow out into the soil, e.g. from sclerotia (*Rhizoctonia solani*). Dissemination may be assisted by water, by soil fauna and the movement of soil by implements and wind (Easton *et al.*, 1969). The presence of growing roots or tubers can stimulate the release of motile zoospores which are able to move in soil water films from the resting spores of *S. subterranea* (Kole, 1954) and probably of *S. endobioticum.*

Over-ground, pathogens are spread by airborne spores, in insect vectors or on implements. Spread of airborne fungal spores and subsequent infection requires specific weather conditions, for example, sporangia of *Phytophthora infestans* are washed or splashed in rain on to neighbouring plants or down to the tubers where they require a water film in which to germinate. Spores of *Alternaria solani* in soil are blown or splashed in heavy rain on to leaves to initiate primary lesions, which produce spores in dew or following rain or irrigation (Rotem and Reichert, 1964). The spores are released and spread by wind when leaves dry (Ohms and Fenwick, 1961). Rain is also needed to release spores from pycnidia of *Phoma exigua* on dead stems and to wash *Erwinia carotovora* var. *atroseptica* from blackleg stems to tubers in soil. These bacteria are also spread by insects attracted to blackleg stems (Graham and Hardie, 1971) and by tractors and implements contaminated by passing through infected crops.

The amount and rate of spread of aphid transmitted viruses depends on the build-up of the vector populations, the stage of development of the crop when first infested and vector movement within and between crops (Broadbent, 1953). For example, in the south of England early infestations of *Myzus persicae* carrying virus Y can almost destroy crops, and early infection with leaf roll may result in the appearance of secondary symptoms in the same season. There can be a 50-fold increase in the number of infected plants in one season. Older plants resist infection so that the later the aphids arrive in crops, the less damaging are these infestations. Usually the number of plants infected with leaf roll and virus Y increases by up to 10-fold in one season, for example in a stock of Ulster Prince grown for three successive seasons, the incidence of leaf roll was respectively, 0.3, 2.1, 13.3% and of virus Y, 0.3, 0.1 and 5.1% (Broadbent *et al.*, 1960). The amount of spread may also be affected by the behaviour of virus in aphids. For example, virus Y is readily transmitted soon after acquisition but is non-persistent, the ability to transmit being restricted to a small number of feeding probes and therefore spread is limited. However, leaf roll virus requires an incubation period in the aphid before transmission can be effected, but is then persistent so that it can be spread widely following long flights between crops.

Seed tubers infected with viruses usually produce diseased plants, but not all the progeny tubers of newly infected plants are necessarily diseased because, perhaps, only one stem of the several independent stems comprising a plant, and the tubers attached to it, are infected (Doncaster and Gregory, 1948). Usually infected plants produce a larger proportion of small tubers. If these are planted as seed the proportion of seed sized tubers infected increases in successive seasons, providing more initial sources of the virus and leading to increased spread. Crops become less vigorous and yields decline progressively (often termed 'degeneration' or 'running out'). However, some viruses infecting very susceptible cultivars kill plants before tubers form and therefore tend to be self-eliminating (e.g. potato yellow dwarf).

11.4 Control of diseases

A knowledge of the etiology of a disease and of the methods of crop production are essential if the disease is to be prevented or its severity and effects minimized. The control measures employed will often depend on eliminating, or severely restricting, the most important sources of inoculum. Some diseases can be avoided by planting disease-free seed tubers, planting into pathogen-free soil or by legislation preventing introduction of new diseases. Eradication of pathogens may be possible by treating seed tubers or soil, by removing alternate hosts and infected host tissues, or by a rotation of crops. Crops may be protected by modifying their environment, e.g. with irrigation, or by the application of chemicals that prevent infection or control vectors. Development of disease in tubers may be prevented or restricted by modifying the storage environment. Probably the easiest method in practice is to use resistant cultivars and thereby avoid the necessity of other measures.

11.4.1 Cultural (non-chemical) control

Non-chemical control usually means modifying cultural practices to make the whole environment as unsuitable as possible for the disease in all its phases, by decreasing inoculum available to the crop, e.g. by cultivations to eliminate volunteer plants, as well as restricting spread during growth or storage by various means. Diseases can sometimes be avoided by wise choice of site, by the rotation of crops or by improved soil drainage. Severity may be lessened by changing planting or lifting dates or depth of planting and adjusting temperature and humidity in store to encourage the healing of wounds and thus decreasing the rate of disease development.

Growing conditions

Diseases resulting from soil-borne inoculum are among the most difficult to control and complete prevention is seldom achieved. Rotation with non-susceptible crops can decrease early death of plants caused by *Verticillium albo-atrum* (Emmond and Ledingham, 1972) but not when it is caused by *V. dahliae* (Powelson, 1964). Soil-inhabiting *Rhizoctonia solani*, including strains non-pathogenic to potatoes but able to form sclerotia on the tuber surface, increases with the frequency of potato crops (Afanasiev and Morris, 1948). Soil infested with oospores of *Phytophthora erythroseptica* requires a rest from susceptible crops of 4 years to ensure comparative freedom from pink rot, whereas with powdery scab and wart, practicable rotations are unlikely to be effective.

Flaming infected stems before lifting decreases *Verticillium* inoculum in soil and increases yield of subsequent potato crops (Easton *et al.*, 1972) but, like rotations, the method is unlikely to be successful when the soil population is already large, even if it proved practicable. Improving drainage promotes rapid growth after planting and thereby decreases the chance of pink rot and the bacterial diseases that are encouraged by waterlogged soil; and eliminating potato volunteers could decrease inoculum of viruses, bacteria and fungi that are harboured in tubers.

Liberal applications of inorganic manures, especially N, delay the development of symptoms of *Verticillium* spp. and *Alternaria solani* so that growth of the crop is prolonged and yields are improved (Easton, 1970; Barclay *et al.*, 1973). However, an imbalance in fertilizer can affect the severity of foliar diseases and render the tubers more liable to mechanical injury. Excessive N increases the speed of foliage destruction by late blight although this can be partly offset by increasing phosphate (Awan and Struchtemeyer, 1957). With early blight both the efficiency of control by chemicals applied to foliage and yields are decreased when excess N is given (Soltanpour and Harrison, 1974). Organic manuring can decrease pink rot (van Haeringen, 1938) and also common scab (Weinhold *et al.*, 1964) which can be severe in the first crop after long-term grass and especially on light sandy or gravelly soils and on land recently limed.

Common scab is also effectively controlled by irrigation provided this achieves a soil water potential of not less than − 0.4 bar (pF 2.6) for at least 6 weeks after tuber initiation (Lapwood *et al.*, 1973). When soil populations of *S. scabies* are large or watering impracticable, the disease may be controlled by chemicals or combination of both treatments (Davis *et al.*, 1974). Irrigation decreases charcoal rot indirectly by lowering soil temperature and *Verticillium* by encouraging anaerobic decomposition of organic residues containing the pathogen (Watson and Huber, 1971). It may, however, increase the incidence of late blight and blackleg (Rotem and Palti, 1969).

Delaying planting until the soil becomes warm and planting shallowly with sprouted seed tubers, decreases the risk of seed tuber decay by bacteria and *Fusarium* spp., and encourages development of shoots from tubers affected with skin spot. Also, damage by *Rhizoctonia solani* on stems (stem canker) is largely prevented and coiled sprout decreased. Deep planting may increase stem canker (van Emden, 1965), but a tall pointed ridge is best for preventing greening of tubers and infection by late blight (Lacey, 1966). Coiled sprout is associated with the planting of seed with over-long sprouts and little potato with planting into cold soil immediately after desprouting. Avoidance of these practices will largely overcome the problems.

The amount of tuber infection is sometimes affected by lifting date, but different diseases are affected in different ways. For example, *R. solani* and *Oospora pustulans* infect tubers increasingly during the growing season and the diseases they cause, black scurf and skin spot, are decreased by early lifting. Lifting soon after foliage destruction can decrease the prevalence of gangrene (Logan *et al.*, 1975) probably by avoiding the accumulation of *P. exigua* in soil. In India, lifting early before the ambient temperature rises at the end of the growing season decreases charcoal rot. There may be less silver scurf when crops are lifted early but in store the disease can develop more on early-lifted crops than on those crops lifted later (Mooi, 1968). With *Alternaria solani*, although the amount of inoculum on foliage increases towards the end of the growing season, this is accompanied by increasing tuber resistance, so that crops lifted late may develop least disease. Similarly as tubers mature, their resistance to infection by *Fusarium solani* var. *caeruleum* increases and can be induced by destroying foliage prematurely (Boyd, 1967a). Provided lifting is delayed to allow tuber skin to 'set', this practice offers a means of control, but sometimes at the expense of yield.

Much mechanical damage on tubers is caused by lifting from cold, dry and stony soils, providing opportunity for wound infections to occur. Damage also occurs later when crops are removed from store and much can be prevented by adjustment and protection of implements and by avoiding handling tubers when they are cold (Malcolmson and Gray, 1968).

Management of the storage environment

Modifying the conditions in which tubers are stored can prevent diseases developing, although with some diseases, control by this method is only palliative, the disease subsequently resuming its progress if the environment becomes suitable. Minimizing mechanical injury, apart from decreasing direct losses, is an important measure of control for some diseases but, since injuries can never be avoided entirely, the provision of conditions that favour rapid wound healing offers a useful additional means of control. In the U.K., 'curing' for 10–14 days at 15 °C immediately after loading into store decreases the amount of gangrene that develops later in cooler conditions. Cut wounds heal quicker than more complex injuries (e.g. crush wounds).

The development of bacterial rotting soon after tubers are put into store may be favoured by warm 'curing' conditions, especially if tubers carry surface moisture from soil, or from rain during lifting or transport to the store. Condensation on exposed tubers, which occurs when ambient temperatures fall below that of the stack, can be avoided by a thick straw covering by way of insulation, any condensation then forms in the straw layer, keeping the tubers dry. Curing can decrease skin spot but increase silver scurf if tubers are not subsequently held cool (e.g. 2 °C). However, curing is unlikely to affect the survival of most pathogens present on the tuber surface so that tubers injured later, as during quality or size grading before sale, may become infected unless they are cured again.

Low storage temperatures, imposed to avoid sprouting and water loss, encourage gangrene and skin spot. Development of severe silver scurf, charcoal rot and *Fusarium* rots is largely arrested by these conditions but may recommence when tubers are removed to warmer conditions; consequently in India charcoal rot is encouraged to develop by storing seed crops warm for 2–3 weeks so that affected tubers can be seen and removed when the crop is size graded.

When seed tubers are dried after lifting and stored in boxes or trays, sporulation and spread of pathogens is prevented and, if the temperature is raised, sprouting of the seed is encouraged. With tubers removed from bulk stores and stored dry and warm, skin spot and damage to eyes from *O. pustulans* will be minimized. Dry conditions may also minimize infection of wounds but are unlikely to prevent gangrene once established. Heating can rid tubers of late blight, skin spot, gangrene and silver scurf (Førsund, 1960; Fox *et al.*, 1970; Hide, 1975), but the limited safety margin between killing the pathogens and damaging buds in tuber eyes makes the method unsuitable for general use. Leaf roll virus also can be eliminated by heating tubers (Upreti and Nagaich, 1968) but, except in special circumstances as perhaps for valuable nuclear stocks that are heavily infected, the method is seldom used in practice.

Other cultural methods

Some viruses including X, Y, A, S and paracrinkle can be eliminated by growing plantlets from excised apical meristems on nutrient agar (Kassanis and Varma, 1967). Callus tissue from meristems may also be used although there is a risk of chromosome abnormalities. These methods are useful when all tubers of a stock are infected and in the U.K., the cultivar King Edward, totally infected with paracrinkle, was almost completely replaced with virus-free seed stocks in Scotland 11 years after the first meristem culture (Bawden and Kassanis, 1965).

Larger portions of stems can be rooted in standard compost to produce plants free from some bacteria and fungi (Hirst *et al.*, 1970). This method is useful for producing nuclear clones free from diseases that are tuber-borne but strict hygiene is necessary to maintain this level of health since diseases caused by pathogens surviving in soil or plant debris are quickly re-acquired when stocks are multiplied in farm conditions.

11.4.2 Chemical control

Soil

Some diseases can be controlled by application of chemicals to soil, although nematode control by soil-applied chemicals has generally met with more success. Extension of the practice for disease control awaits the development of more effective compounds for soil-borne pathogens at economic cost. Systemically-acting soil-applied materials are becoming available and the prospect of phloem-translocated chemicals effective against root and tuber diseases could lead to a rapid change in disease control practice. Currently, however, control by soil applied chemicals finds but limited application for potato diseases.

Several materials have been used to prolong growth and increase yields of crops infected with *Verticillium* spp., including nematicides which are effective when the disease is associated with nematodes in soil. Fumigation with methyl bromide, a very toxic material with a wide spectrum of activity against arthropods, nematodes, fungi and weed seeds, decreases the severity of disease but involves the use of gas-tight sheets making it cumbersome and costly for large scale application. Pre-planting treatments with dazomet or chloropicrin are also effective as are the nematicides, aldicarb, DD, and telone, even when nematodes known to increase the disease are apparently not present in soil (Easton, 1970). Benomyl incorporated in soil delays the appearance of symptoms, and is beneficial to subsequent potato crops, although the major effect seems to be by controlling nematodes (Hide and Corbett, 1974).

Quintozene (PCNB) incorporated in soil decreases common scab, stem canker and black scurf so that saleable tuber yields are improved, but the possibility of phytotoxicity and taint in tubers limits its general acceptance. Common scab may also be decreased by soil applications of a range of chemicals including urea formaldehyde, sulphur, gypsum, copper salts and magnesium sulphate but whether through effects on the pathogen or on the host is not understood. Sulphur has also been reported to decrease pink rot, powdery scab and blight.

Seed tubers

Treating seed tubers with chemicals offers a method of eliminating the relatively small but important amounts of inoculum they carry. The most common fungicides previously used on seed tubers contained mercury and acted as eradicants. *Fusarium* rots, gangrene, skin spot, silver scurf and black scurf are controlled by washing and dipping in solutions or suspensions of organic mercury compounds (Ayers and Robinson, 1956; Boyd, 1960; Graham, 1960). The treatment encourages sprouting and early even emergence and increases the proportion of small tubers in the crop, although it may increase the incidence of blackleg.

Treating seed pieces with organo-mercury solutions also controls bacterial rotting (Robinson *et al.*, 1960). Antibiotics perform similarly but can predispose them to infection by *Fusarium* spp. and *Phoma* spp. (Bonde and Malcolmson, 1956). Recent awareness of the toxic hazards of mercury compounds and associated problems in the disposal of effluent has stimulated a search for alternative and less noxious materials.

In Scotland, fumigating potatoes with 2-aminobutane in gas-tight chambers decreases gangrene and skin spot in stored tubers (Graham *et al.*, 1973) and, like treatment with mercury compounds, best control is achieved by fumigating soon after lifting. Benzimidazole fungicides (benomyl, thiabendazole) applied as dusts, dips or sprays can be used as seed dressings to control black scurf, gangrene, skin spot, *Fusarium* rots and tuber-borne *Verticillium* spp. (Hide *et al.*, 1969a; Biehn, 1970). Although these fungicides may become systemic when applied to roots, they penetrate little beneath the periderm of treated seed tubers (Tisdale and Lord, 1973). Sporulation of *Oospora pustulans* and *Helminthosporium solani* on treated seed after planting is prevented and infection of progeny tubers greatly decreased. Gangrene is also controlled on seed, but these fungicides fail to prevent infection of progeny tubers.

Foliage

Chemicals may be dusted, or more commonly sprayed, on to foliage either to kill insects which transmit virus diseases (see Section 11.2.2 and Chapter 12), or as protectants against early or late blight. Systemically-acting chemicals which can be applied to foliage to protect the whole plant, including the underground parts, are actively being sought but at present none is near commercial application.

Both early and late blight can become epidemic but the timely application of fungicide slows or prevents their progress and so prolongs the life of the foliage, allowing the bulking of the crop to continue. For late blight, the timing and frequency of spray applications, fungicides and formulations and the weather conditions favouring infection, spread and epidemic development have been much studied (Cox and Large, 1960; Hirst and Stedman, 1960; Evans *et al.*, 1965). Such information has enabled many countries to operate a forecasting service to warn farmers of outbreaks of this disease and when conditions are suitable for its spread. Inorganic chemicals based on copper (which can be phytotoxic) were used for many years but have been largely replaced by organic dithiocarbamates (mancozeb, zineb, maneb etc.) and organo-tin compounds (fentin hydroxide, fentin acetate). High volume sprays applied to run-off or medium volume sprays (200–350 l ha^{-1}) designed to wet the foliage short of run-off, are applied by tractor-mounted sprayers. Dusting is sometimes preferred as a quicker method using light-weight equipment, especially in poor ground conditions. More recently, highly concentrated formulations suitable for low or ultra-low volume application from aircraft have been introduced. After evaporation, the concentrated droplets leave 'spots' of chemical which is then released in rain at rates determined by the retention qualities of the formulation (Hislop, 1967).

Early blight is becoming increasingly important in the U.S.A., especially in areas where overhead irrigation is more widely practised. First applications are usually of dithiocarbamates or captafol, timed according to plant size and maturity or when primary infection of lower leaves has occurred. The number and timing of spray applications vary widely in practice, although research has shown that three well-timed sprays can be very effective (Harrison and Venette, 1970; Douglas and Groskopp, 1974).

Tractors used for spraying crops can damage foliage and compact soil so that yields from damaged rows are decreased by up to 30%. Loss of crop has been estimated at 5–7% (Cox and Large, 1960; Struchtemeyer *et al.*, 1963), but depends mainly on the proportion of rows damaged and varies greatly with soil type and season.

11.4.3 Legislation and certification of seed

The purpose of legislation and certification schemes is to prevent introduction of alien diseases and decrease the spread of endemic ones. Before it was realised that introduction of new diseases could be prevented or initial introductions eradicated, diseases became established in new areas that provided a favourable climate. Nowadays each country has regulations prohibiting import of tubers carrying certain diseases, or of all tubers from areas where certain diseases are known to be endemic. Such procedures are more likely to be effective against pathogens that have no airborne spores, for example *Synchytrium endobioticum* and *Corynebacterium sepedonicum*, than those having airborne spores or vectors which can travel long distances (for example *Alternaria solani*, *Phytophthora infestans*, insect transmitted viruses). In the past, quarantine restrictions have usually only delayed the introduction of diseases but this delay could provide a valuable opportunity to develop control methods, or perhaps breed resistant cultivars in anticipation of eventual introduction. However, to embark on such a programme would demand the knowledge, or at least the conviction, that the disease would, if introduced, pose a serious threat, and this is not always so. Moreover, it would involve the introduction of the pathogen concerned for experimental purposes, which is always attended by some risk even with the constraints of rigid quarantine regulations.

Spread of endemic diseases can be restricted by legislation prohibiting production in, or movement away from, areas known to be infested with serious pathogens, and by voluntary certification schemes controlling health of seed tubers. Seed crops are often produced in areas isolated from the main growing regions where aphids are scarce and where crops can be maintained free of virus diseases. In the U.K. the importance of wart disease and the persistence of the pathogen in soil were realized in the early 1900s, but no method of control was known. Although some cultivars did not become infected ('immune'), seed stocks often contained rogue tubers of non-immune cultivars, so schemes were introduced to ensure trueness to type. Roguing (negative selection) and later multiplication from disease-free plants(positive selection) were used to improve the health of seed and tolerances for diseases and abnormalities in field inspections were gradually decreased (Chapter 8). Recently in Scotland and N. Ireland propagation by stem cuttings has been introduced to produce nuclear stocks free from pathogenic bacteria and fungi.

Certification of seed can only control those diseases that are tuber-borne and, even if stocks are multiplied from healthy seed, they are not protected from diseases that are acquired from soil. Healthy seed can be successfully produced if multiplication is done hygienically and supplemented with chemicals that can keep inoculum at a very low level, and the regular and continual replacement of diseased seed with recently propagated healthy stocks should itself help in this direction.

11.4.4 Resistant cultivars

Replacing susceptible cultivars with resistant ones is the most convenient, and for the grower, the most economical and practical method of controlling diseases, and even in the early nineteenth century differences in susceptibility to disease were noted between cultivars. Genetic sources of resistance, as well as of many other desirable characters, found in existing cultivars or in wild or cultivated *Solanum* spp., have to be incorporated into new cultivars.

Immunity to disease is seldom achieved and may not be long-lasting because uncommon pathogenic races can be selected, or the pathogens themselves adapt and become able to infect previously immune plants. Cultivars exhibiting tolerance were once thought valuable for controlling virus diseases. Now they are considered unsatisfactory as their use restricts the growing of non-tolerant cultivars, which may be preferred in other important respects, by maintaining perpetual reservoirs of inoculum. At present, two types of resistance are exploited. Hypersensitivity relies on single genes and with late blight infected cells die rapidly and prevent further spread of the disease, but with some virus diseases, systemic spread is rapid and the whole plant dies. However, reliance on single major genes conferring hypersensitivity is now generally recognized as unsatisfactory at least for airborne pathogens capable of epidemic diseases (e.g. *P. infestans*, causing late blight). Experience suggests that such pathogens exist as mixed populations from which new races are likely to emerge able to overcome this type of resistance in response to the extreme selection pressure it imposes. Multigenic (field) resistance is more broadly based and seldom eliminates pathogens but allows a low level of disease to develop; less selection pressure is imposed and the resistance is likely to be more durable. Most breeding of potato cultivars has concentrated on resistance to wart, late blight and virus diseases.

Immunity to wart has long been recognized, although improved methods of testing showed that 'immune' cultivars gave differing amounts of hypersensitive reaction (field immunity). Characters governing resistance are dominant and readily inherited and variability of the pathogen is low, so the chances of selecting new, virulent strains (or races) seem remote, although races with the ability to infect cultivars immune to U.K. strains of the pathogen, have been found in N. America, Central Europe and India (Noble and Glynne, 1970; Black, 1971).

Hypersensitivity and multigenic resistance have been used in breeding cultivars resistant to late blight. Hybrids derived from the Mexican *Solanum demissum* showed hypersensitivity but later succumbed to new races of *P. infestans*. As a result, a genetic basis for classifying resistance (R genes) was developed and at least ten genes have been discovered, all of which are complemented by virulence genes in the pathogen able to overcome the resistance they confer. Variability in the pathogen dictates that cultivars depending on hypersensitivity for their resistance to late blight are unlikely to maintain their resistance for very long and this approach has been abandoned in most breeding programmes in favour of multigenic resistance. This is likely to prove more durable and seems equally effective against all known races of the fungus. Progress of the disease is restricted by effects on different components of its development, including resistance to infection, slow growth of the pathogen through tissues, an extended incubation period and a reduction in the number of spores produced, resulting overall in a prolongation of the life of the foliage.

Many genes govern the reaction of *Solanum* spp. to viruses and in some established cultivars field immunity is a dominant character. Single gene hypersensitivity to strains of virus X is found in *S. acaule* and to Y and A in *S. stoloniferum*. Multigenic resistance to virus Y and leaf roll is known in *S. phureja*, *S. acaule* and *S. demissum* and these sources are preferred since, as with late blight, single major gene resistance is likely to be less successful than multigenic resistance. Characters that discourage insect vectors may also be exploited to decrease disease (Chapter 12).

11.5 Future prospects

Potato production in a disease-free environment is as yet unknown and will be difficult to achieve, but must be the aim if the costs of preventing diseases and the losses they cause are to be avoided. At present, the use of resistant cultivars is probably the easiest and cheapest method of disease control for the farmer. However, the case for incorporating resistance to more than a few major diseases in a breeding programme, supposing sources were available, is not strong because each additional demand lengthens the time required to achieve a successful cultivar, and this is already a slow and expensive process.

Late blight has remained the most important disease in world potato production for more than 100 years and any major improvement in control must still come either from resistant cultivars or from new chemicals, possibly downward translocated, and improved methods of application. Chemicals applied to soil may help control diseases that cause premature death of plants, but present materials seldom greatly affect amounts of inoculum in soil and it may prove more profitable to explore methods of prolonging crop growth (e.g. by supplementary fertilizer) to provide the grower with economic yields. Alternatively, losses from soil-borne disease may be decreased by avoiding production on fields known to harbour inoculum.

The prospects seem brighter for diseases that are mostly tuber-borne, provided there are no major alternative sources of infection, and chemicals are available to prevent re-contamination of 'clean' stocks. However, the extra effort and costs of improving and maintaining health may only be justified if benefits accrue at all stages of the multiplication programme from the first tubers freed from disease to the crop destined for food.

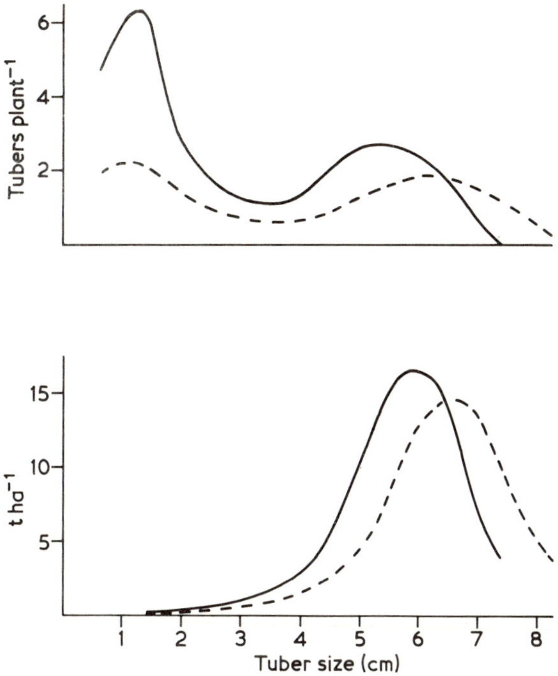

Fig. 11.2. Trends of tuber size distribution (numbers plant $^{-1}$ or t ha $^{-1}$) in crops grown from healthy (complete line) and commercial seed (broken line)

Also vigilance will be needed to ensure that other pathogens do not fill the ecological niches vacated by the present major pathogens. So far in the U.K., seed, free from tuber-borne pathogens, has yielded an extra 10% (Hide and Bell, 1974) but benefits may be larger than this as the crops also developed less disease in store. However, freedom from diseases also has its disadvantages, for example, in the increased yields from healthy seed there is often a larger proportion of small tubers (Fig. 11.2). Consequently, methods of producing the crop to counteract, or take advantage, of this tendency need to be investigated.

Storing potatoes always involves risks and improvement in storage technology is needed to avoid infrequent catastrophes as well as the smaller chronic losses which amount annually to a significant proportion of world production. However, high costs will prohibit the installation of fully controlled stores for as long as the potato remains a cheap staple food. The major problems are the presence of pathogens and mechanical damage on tubers as they are loaded into store. If crops likely to deteriorate (in quality or quantity) could be distinguished from those which have a potential for long storage, decisions on which crops to store and subsequent management of stores could be modified to allow appropriate orderly marketing with minimal losses. This might be done by simple assessments of disease potential during growth (Hide *et al.*, 1974) and assessments of injuries at lifting. If sources of infection can be decreased by using healthy seed tubers or applying chemicals at loading or later in store, the storage environment could be adjusted to provide conditions that suit the production of the best product rather than the present compromise imposed by the need for disease suppression. Mechanical damage is one of the most important causes of direct and indirect loss of crop and its avoidance, of all the factors discussed, would undoubtedly make the most significant single contribution to production. It is a problem worthy of considerable effort since the benefits could be very great indeed: it has been estimated that on average more than 20% of the crop in U.K. sustains largely avoidable damage at harvest.

Besides agronomic aspects of diseases, there are many facets of the biology of pathogens that need to be investigated before effective and continuing control can be assured. Much is still to be learned about the relative importance of different sources of inoculum, its spread and the time and methods of infection. The role of volunteer plants and of perhaps hitherto unknown sexual stages of fungi need to be evaluated. Resistant cultivars may be the best method of decreasing effects of disease; meanwhile, other control measures can be made more effective if sources of inoculum can be decreased.

References

Afanasiev, M.M. and Morris, H.E. (1948). *Am. Potato J.* **25**, 17–23.
Awan, A.B. and Struchtemeyer, R.A. (1957). *Am. Potato J.* **34**, 315–9.
Ayers, G.W. and Robinson, D.B. (1956). *Am. Potato J.* **33**, 1–5.
Barclay, G.M., Murphy, H.J., Manzer, F.E. and Hutchinson, F.E. (1973). *Am. Potato J.* **50**, 42–8.
Bawden, F.C. and Kassanis, B. (1965). *Rothamsted Report for 1964*, 282–90.
Biehn, W.L. (1970). *Pl. Dis. Reptr.* **54**, 254–5.
Black, W. (1971). *Scottish Plant Breeding Station Report for 1970–1*, 52–60.
Bokx, J.A. de and Mooi, J.C. (1974). *Potato Res.* **17**, 410–33.
Bonde, R. and Malcolmson, J.F. (1956). *Pl. Dis. Reptr.* **40**, 615–9.
Boyd, A.E.W. (1960). *Eur. Potato J.* **3**, 137–54.
Boyd, A.E.W. (1967a). *Ann. appl. Biol.* **60**, 231–40.
Boyd, A.E.W. (1967b). *Scott. Agric.* **46**, 78–82.
Boyd, A.E.W. (1972). *Rev. Pl. Path.* **51**, 297–321.
Boyd, A.E.W. and Lennard, J.H. (1961). *Eur. Potato J.* **4**, 361–77.
Boyd, A.E.W. and Lennard, J.H. (1962). *Pl. Path.* **11**, 161–6.
Broadbent, L. (1953). *Biological Reviews* **28**, 350–80.
Broadbent, L., Heathcote, G.D. and Burt, P.E. (1960). *Eur. Potato J.* **3**, 251–62.
Busch, L.V. (1967). *Can. Pl. Disease Survey*, **47**, 76–8.
Church, B.M., Hampson, C.P. and Fox, W.R. (1970). *Potato Res.* **13**, 41–58.
Conroy, R.J. (1969). *Agricultural Gazette of New South Wales*, **80**, 175–8.
Corbett, D.C.M. and Hide, G.A. (1971). *Ann. appl. Biol.* **68**, 71–80.
Cox, A.E. and Large, E.C. (1960). *U.S. Dept Agric. Hb* 174.

Cramer, H.H. (1967). *Plant Protection and World Crop Production*. Leverkusen: Farbenfabriken Bayer AG.
Cromarty, R.W. and Easton, G.D. (1973). *Am. Potato J.* **50**, 398–407.
Davis, J.R., McMaster, G.M., Callihan, R.H., Garner, J.G. and McDole, R.E. (1974). *Phytopathology* **64**, 1404–10.
Doncaster, J.P. and Gregory, P.H. (1948). *The Spread of Virus Diseases in the Potato Crop*. London: H.M.S.O.
Douglas, D.R. and Groskopp, M.D. (1974). *Am. Potato J.* **51**, 361–8.
Dounine, M.S. (1961). *Corynebacterium sepedonicum* in the USSR. *Report of the International Conference on Corynebacterium sepedonicum and Pseudomonas solanacearum*. European Mediterranean Plant Protection Organization.
Easton, G.D. (1970). *Am. Potato J.* **47**, 419–26.
Easton, G.D., Nagle, M.E. and Bailey, D.L. (1969). *Phytopathology* **59**, 1171–2.
Easton, G.D., Nagle, M.E. and Bailey, D.L. (1972). *Phytopathology* **62**, 520–4.
Emmond, G.S. and Ledingham, R.J. (1972). *Can. J. Pl. Sci.* **52**, 605–11.
Evans, E., Couzens, B.J. and Griffiths, W. (1965). *Wld Rev. Pest Control* **4**, 84–92.
Fernow, K.H., Peterson, L.C. and Plaisted, R.L. (1970) *Am. Potato J.* **47**, 75–80.
Førsund, E. (1960). *Nordisk Jordbrugsforskning*, Suppl. 1, 298–301.
Fox, R.A., Dashwood, E.P. and Wilson, H.M. (1970). *Scottish Horticultural Research Institute Report for 1969*, 30–1.
Graham, D.C. (1960). *Eur. Potato J.* **3**, 80–9.
Graham, D.C. and Hardie, J.L. (1971). *Proc. 6th British Insecticide and Fungicide Conf.* 219–24.
Graham, D.C., Hamilton, G.A., Quinn, C.E. and Ruthven, A.D. (1973). *Potato Res.* **16**, 109–25.
Griffith, R.L., Hide, G.A., Hirst, J.M. and Stedman, O.J. (1974). *Ann. appl. Biol.* **77**, 237–50.
Guthrie, J.W. (1960). Early dying (Verticillium wilt) of potatoes in Idaho. *Idaho Agric. Exp. Station Res. Bulletin*, 45.
Harrison, B.D. and Roberts, I.M. (1969). *Ann. appl. Biol.* **63**, 347–9.
Harrison, J.A.C. and Isaac, I. (1969). *Ann. appl. Biol.* **64**, 469–82.
Harrison, M.D. and Venette, J.R. (1970). *Am. Potato J.* **47**, 81–6.
Harrison, M.D., Livingston, C.H. and Oshima, N. (1965). *Am. Potato J.* **42**, 279–91.
Hide, G.A. (1975). *Pl. Path.* **24**, 233–6.
Hide, G.A. and Bell, F. (1974). Agronomic effects of healthier seed. *Rothamsted Experimental Station Report for 1973*, 145.
Hide, G.A. and Corbett, D.C.M. (1974). *Ann. appl. Biol.* **78**, 295–307.
Hide, G.A., Hirst, J.M. and Griffith, R.L. (1969a). *Proc. 5th British Insecticide and Fungicide Conf.* 310–4.
Hide, G.A., Hirst, J.M. and Mundy, E.J. (1969b). *Ann. appl. Biol.* **64**, 265–79.
Hide, G.A., Hirst, J.M. and Stedman, O.J. (1973a). *Ann. appl. Biol.* **73**, 151–62.
Hide, G.A., Hirst, J.M. and Stedman, O.J. (1973b). *Ann. appl. Biol.* **74**, 139–48.
Hide, G.A., Lapwood, D.H., Legg, P.R., Griffith, R.L. and Bell, F. (1974). *Rothamsted Experimental Station Report for 1973*, 144–5.
Hirst, J.M. and Salt, G.A. (1959). *Trans. Br. mycol. Soc.* **42**, 59–66.
Hirst, J.M. and Stedman, O.J. (1960). *Ann. appl. Biol.* **48**, 471–88.
Hirst, J.M., Hide, G.A., Griffith, R.L. and Stedman, O.J. (1970). *J.R. agric. Soc.* **131**, 87–106.
Hirst, J.M., Hide, G.A., Stedman, O.J. and Griffith, R.L. (1973). *Ann. appl. Biol.* **73**, 143–50.
Hislop, E.C. (1967). *Ann. appl. Biol.* **60**, 265–79.
Isaac, I. and Harrison, J.A.C. (1968). *Ann. appl. Biol.* **61**, 231–44.

James, W.C., Callbeck, L.C., Hodgson, W.A. and Shih, C.S. (1971). *Phytopathology* **61**, 1471–6.
James, W.C., Shih, C.S., Hodgson, W.A. and Callbeck, L.C. (1972). *Phytopathology* **62**, 92–6.
Jones, R.A.C. and Harrison, B.D. (1969). *Ann. appl. Biol.* **63**, 1–17.
Kassanis, B. and Varma, A. (1967). *Ann. appl. Biol.* **59**, 447–50.
Kharkova, A.P. (1961). *Botanicheskii Zhurnal, SSSR*, **46**, 399–407.
Kole, A.P. (1954). *Tijdschr. PlZiekt.* **60**, 1–65.
Lacey, J. (1966). *Eur. Potato J.* **9**, 86–98.
Lapwood, D.H. (1971). *Ann. appl. Biol.* **68**, 41–53.
Lapwood, D.H. and Hering, T.F. (1970). *Potato Res.* **13**, 296–304.
Lapwood, D.H., Wellings, L.W. and Hawkins, J.H. (1973). *Pl. Path.* **22**, 35–41.
Large, E.C. (1952). *Pl. Path.* **1**, 109–17.
Logan, C. (1964). *Eur. Potato J.* **7**, 45–56.
Logan, C., Copeland, R.B. and Little, G. (1975). *Ann. appl. Biol.* **80**, 199–204.
Malcolmson, J.F. and Gray, E.G. (1968). *Ann. appl. Biol.* **62**, 89–101.
Mooi, J.C. (1968). *Instituut voor Plantenziektenkundig Onderzoek, Mededeling* 482.
Moorby, J. and McGee, S. (1966). *Ann. appl. Biol.* **58**, 159–70.
Mordue, J.E.M. (1964). *School of Agriculture Report, University of Nottingham for 1963*, 42–8.
Morsink, F. and Rich, A.E. (1968). *Phytopathology* **58**, 401.
Noble, M. and Glynne, M.D. (1970). *FAO Pl. Protection Bulletin*, **18**, 125–35.
Ohms, R.E. and Fenwick, H.S. (1961). *Idaho Agric. Extension Service Bulletin*, 346.
Paharia, K.D. (1960). *Indian Potato J.* **2**, 1–11.
Partyka, R.E. and Mai, W.F. (1962). *Phytopathology* **52**, 766–70.
Paterson, M.I. and Gray, E.G. (1972). *Potato Res.* **15**, 1–11.
Pérombelon, M.C.M. (1972). *Ann. appl. Biol.* **71**, 111–7.
Powelson, R.L. (1964). *Am. Potato J.* **41**, 303.
Reestman, A.J. (1970). *Potato Res.* **13**, 248–68.
Robinson, D.B., Ayers, G.W. and Campbell, J.E. (1960). *Am. Potato J.* **37**, 203–12.
Rotem, J. and Palti, J. (1969). *Ann. Rev. Phytopathology* **7**, 267–88.
Rotem, J. and Reichert, I. (1964). *Pl. Dis. Reptr.* **48**, 211–5.
Schick, R. (1952). *D. Landiwirtschaft*, **3**, 618–27.
Scholey, J., Marshall, C. and Whitbread, R. (1968). *Pl. Path.* **17**, 135–9.
Schuster, M.L. and Coyne, D.P. (1975). *Nebraska Agric. Exp. Station Res. Bulletin*, 268.
Soltanpour, P.N. and Harrison, M.D. (1974). *Am. Potato J.* **51**, 1–7.
Stanghellini, M.E. and Meneley, J.C. (1975). *Phytopathology* **65**, 86–7.
Struchtemeyer, R.A., Epstein, E. and Grant, W.J. (1963). *Am. Potato J.* **40**, 266–70.
Thirumalachar, M.J. (1955). *Phytopathology* **45**, 91–3.
Thirumalachar, M.J. (1967). *Am. Potato J.* **44**, 241–4.
Tisdale, M.J. and Lord, K.A. (1973). *Pesticide Sci.* **4**, 121–30.
Todd, J.M. (1954). *Pl. Path.* **3**, 17–20.
Todd, J.M. and Adam, J.W. (1967). *Proc. 4th British Insecticide and Fungicide Conf.* 276–84.
Twiss, P.T.G. and Jones, M.P. (1965). *Eur. Potato J.* **8**, 154–72.
Upreti, G.C. and Nagaich, B.B. (1968). *Am. Potato J.* **45**, 373–7.
van Emden, J.H. (1965). *Eur. Potato J.* **8**, 188–9.
van Haeringen, G.H. (1938). *Tijdschr. PlZiekt.* **44**, 247–56.

van Hoof, H.A. (1968). *Nematologica* **14**, 20–4.
van Schreven, D.A. (1956). *Pl. Soil*, **8**, 49–55.
Venette, J.R. and Harrison, M.D. (1973). *Am. Potato J.* **50**, 283–92.
Walker, J.C. and Larson, R.H. (1939). *J. agric. Res.* **59**, 259–80.
Watson, R.D. and Huber, D.M. (1971). *Proc. Int.* Verticillium *Symp.* Wye College, 32.
Weinhold, A.R., Oswald, J.W., Bowman, T., Bishop, J. and Wright, D. (1964). *Am. Potato J.* **41**, 265–73.

Chapter 12

Pest aspects of potato production

Part 1. Nematode pests of potatoes

K. EVANS and D.L. TRUDGILL

12.1 Introduction to their biology

Plant-parasitic nematodes are small (0.2–10 mm long) worm-like animals, able to move between soil particles, between folded leaves of plant buds, in the air spaces of leaves and stems or in plant tissues themselves. Their movement and activity in soil is influenced by the thickness of the water films surrounding soil particles and they are inactive in dry soils. Agricultural soils usually contain many species of plant-parasitic nematodes, as well as predatory and microphagous types, and the species composition is affected by many factors including climate, soil type and cropping patterns. Species parasitic on plants all possess a mouth stylet which they use to puncture plant cells, and in all except the Trichodoridae this stylet is hollow and used to extract cell contents. Many feed on the cytoplasm which accumulates in the region of the stylet following the injection of saliva, although some actually modify the cells they feed upon so that the supply of cytoplasm is increased and they are able to become sedentary feeders and lose their vermiform shape (*Heterodera*, *Globodera*, *Meloidogyne*, *Nacobbus*). Such nematodes usually produce large numbers of eggs and include the most harmful species.

Nematodes attacking potatoes can be grouped according to whether they attack stems and leaves, tubers or roots (see Table 12.1).

12.1.1 Nematodes infesting stems and leaves

The stem nematode, *Ditylenchus dipsaci*, is thought to be adapted to be able to attack stem tissues by virtue of the pectolytic enzymes it produces (Riedel and Mai, 1971). The enzymes dissolve the middle lamella between adjacent cells and enable the nematode to move through the tissue

Table 12.1 The main nematode pests of potatoes

Latin name	*Common name*	*Part of plant attacked*	*Known geographical distribution*
Globodera rostochiensis	Golden nematode of potato	Roots	World-wide
Globodera pallida	White potato cyst-nematode	Roots	World-wide
Meloidogyne hapla	Northern root-knot nematode	Roots	N. America and temperate areas in general
Nacobbus aberrans	False root-knot nematode	Roots	Peru and Bolivia
Pratylenchus penetrans	Root lesion nematode	Roots	N. America and Europe
Ditylenchus destructor	Tuber-rot nematode	Tubers	N. America, W. Europe and U.S.S.R.
Ditylenchus dipsaci	Stem nematode	Stems, leaves, tubers	W. Europe
Trichodorus and Paratrichodorus spp.	Stubby root nematodes	Transmit tobacco rattle virus (spraing) to tubers	N. America and Europe

but some of the enzymes causing this dissolution may be of host origin. *Ditylenchus dipsaci* has been reported causing distortion of potato haulms in Germany and The Netherlands (Seinhorst, 1957) and in England (Newton and Duthoit, 1954). The aerial parts of plants are invaded from the soil only during wet weather, but once within the protected environment of the plant multiplication is rapid and haulms may be severely distorted and stunted.

12.1.2 Nematodes infesting tubers

Ditylenchus dipsaci may also attack tubers, eventually causing tuber rot, but does so less frequently than the tuber nematode *Ditylenchus destructor*. Both species invade the tubers via eyes or lenticels, but *D. dipsaci* lesions may extend throughout the tuber whereas *D. destructor* lesions are more superficial. Damage caused by both species is accentuated by secondary fungal or bacterial infections. Weeds provide alternative hosts for *D. destructor* but damage to tubers occurs only rarely in Europe because of the relatively high standards of weed control, the use of good quality seed and crop rotation. Severe damage has been reported from Canada and the U.S.A. where infested land is now avoided, and also from Russia. The ecology, biology and control of these nematodes is reviewed by Winslow and Willis (1972).

Tuber quality may be spoilt by spraing disease caused by infection with tobacco rattle virus (TRV) transmitted during the feeding of *Trichodorus* and *Paratrichodorus* spp. Typical symptoms of spraing are necrotic arcs within the tuber, though these may also be caused by potato mop top virus transmitted by the fungus *Spongospora*. Nine species of *Trichodorus* and *Paratrichodorus* in Europe and three of *Paratrichodorus* in the U.S.A. are vectors of TRV (van Hoof, 1968). The Trichodoridae are all ectoparasitic nematodes with wide host ranges, but most species are confined to lighter sandy soils (Winfield and Cooke, 1975) so, although spraing is widespread in western Europe, significant amounts of infection only occur if the soil is moist enough for nematode activity in the period following tuber initiation (van Hoof, 1964). When they are abundant *Trichodorus* spp. sometimes cause stunting of roots, and therefore tops, and abrasion of the surface of tubers due to mass feeding.

The root-knot nematodes (*Meloidogyne* spp.) are endoparasitic with sedentary females. Besides producing galls on infested plant roots they also infest tubers, producing gall-like protuberances which render affected tubers unsaleable. The soil temperature requirements of *Meloidogyne* spp. vary, *M. hapla* having an optimum development temperature of 25° C (Griffin and Jorgenson, 1969) so that it is mainly a pest of temperate and Mediterranean regions, whereas *M. javanica* and *M. incognita* have even higher optimum development temperatures and their effective geographical range is limited. *Meloidogyne hapla* is the dominant species attacking potatoes in Europe, North America and Japan, but in Australia, Africa, South America and parts of Asia, *M. javanica* and *M. incognita* cause most damage.

12.1.3 Nematodes attacking roots

The nematodes attacking roots may be classified as those indigenous to countries where potatoes have been introduced and those indigenous only to the countries of origin of the potato. The false root-knot nematode *Nacobbus serendipiticus bolivianus* (synonymized with *N. aberrans*, Sher, 1970) and the cyst-nematodes, *Globodera rostochiensis* and *G. pallida*, are probably indigenous only to Peru and Bolivia, but *N. serendipiticus* has not been widely distributed with the potato crop whereas the two cyst-nematodes have, and are now the main nematode pests of potatoes in many countries. For this reason the cyst-nematodes will be considered in detail.

Nematodes indigenous to countries where potatoes have been introduced

The list of nematodes occasionally associated with potato growing and perhaps causing poor growth includes species of *Belonolaimus*, *Helicotylenchus*, *Hemicycliophora*, *Hexatylus*, *Paratylenchus*, *Rotylenchulus*, *Rotylenchus*, *Tylenchorhynchus* and *Xiphinema* (Winslow and Willis, 1972). Most are ectoparasites with wide host ranges and when large populations are produced by preceding crops they may be somewhat injurious to potatoes.

Species of *Meloidogyne*, *Longidorus*, *Trichodorus* and *Pratylenchus* have been associated with crop losses in several countries. *Meloidogyne* spp. are probably most important for their effect on tuber quality but they also produce galls on the roots and interfere with root growth.

Unlike *Meloidogyne* spp., which are sedentary endoparasites, *Pratylenchus* spp. are migratory endoparasites, and the vermiform juveniles and adults move freely into and out of roots. They are equipped with a powerful stylet which they use to break down cell walls as they move within the roots and their host ranges are wide although some hosts enable them to multiply much more than others (Oostenbrink *et al.*, 1957). *Pratylenchus penetrans* is one of the most harmful species in temperate regions but is uncommon in warmer climates where *P. brachyurus* (Africa) and *P. coffeae* (Japan) replace it. *Pratylenchus penetrans* damage occurs in patches,in which the plants are stunted and chlorotic and yield poorly (Dickerson *et al.*, 1964).

Longidorus spp. are migratory ectoparasites but are able to feed deep inside roots with their long (up to 130 μm) mouth stylets. Most species feed on root tips where they induce galls. Reports of damage are few, but as with many other ectoparasitic species this may be because of inadequate observations. *Paralongidorus maximus* caused typical galling of root tips and depressed growth in patches in potato crops in Germany (Sprau, 1960). In a sandy soil in England populations of *L. leptocephalus* below cultivation depth may have prevented roots entering the sub-soil, so depriving the crop of moisture in mid-season (Evans, 1971). Large numbers of *Trichodorus* spp. feeding on root tips may also stunt root growth.

Nematodes indigenous to the potato-growing areas of South America

Nacobbus serendipiticus bolivianus

Three species of nematodes indigenous to South America are severe pathogens of potatoes and of these the false root-knot nematode, *N. serendipiticus bolivianus*, is still restricted to South America, mainly in the altiplano regions of Peru and Bolivia. This restricted distribution may be due to poor persistence in the absence of a host crop so that only on the altiplano, where almost all the commonly grown crops (potatoes, quinoa, mashuar) are alternative hosts for this nematode, does it survive. Potatoes in this area are frequently severely damaged and *N. serendipiticus* is often the dominant nematode pest (Franco and de Scurrah, 1972), out-competing cyst-nematodes which also occur. In nearby areas where a greater range of crops including non-hosts (barley, beans, onions) are grown, cyst-nematodes, which are able to persist in the absence of a host crop, are the dominant nematode pests of potatoes.

Globodera rostochiensis and Globodera pallida

The two species of cyst-nematodes are dominant in different areas of South America; in Peru *G. Pallida* is the dominant species with *G. rostochiensis* only in the south and in Bolivia. Populations of *G. rostochiensis* in Panama, Venezuela and Colombia probably developed from cysts imported with seed potatoes from Europe (Fig. 12.1).

Potato cyst-nematodes have few hosts outside the genus *Solanum*, but are well adapted to the potato crop. As they are the most important nematode pests of potatoes (causing annual losses of some £7–10 million in the U.K., Jones, 1970) they have been the subject of much research and will figure largely in this review. Although they are highly specialized plant parasites it seems likely that the damage they cause is similar to that caused by other root-parasitic nematodes.

Life cycle of potato cyst-nematodes

The discovery that some potato cultivars were resistant to potato cyst-nematodes (Ellenby, 1952) was followed soon afterwards by the discovery of nematode populations able to overcome this resistance (Dunnett, 1957; Jones, 1957a, 1958). The different strains of nematodes were

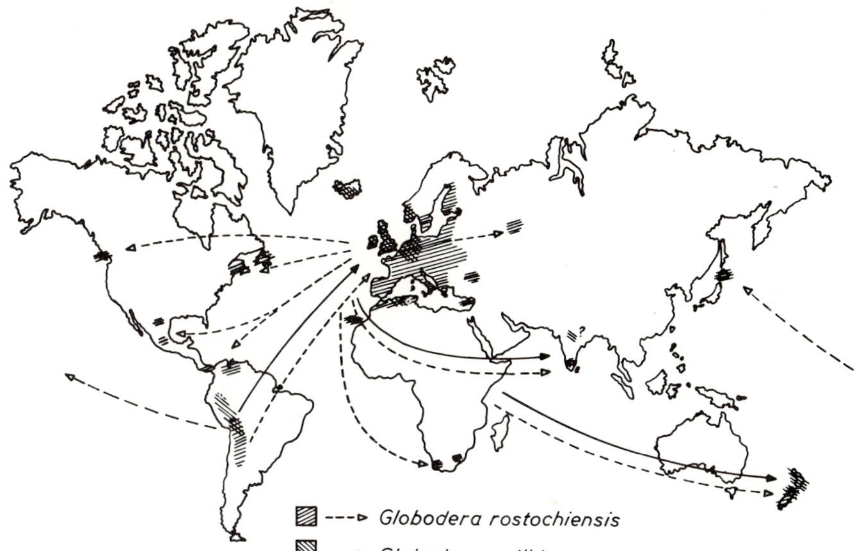

Fig. 12.1 The known world distribution of the two species of potato cyst-nematodes; the arrows indicate their probable movements

called 'pathotypes' by Cole and Howard (1966) and until recently all pathotypes were thought to belong to one species. Recent research, however, has shown that there are two distinct species. Guile (1966) noticed that pathotypes able to reproduce on varieties of potato containing the H_1 major gene for resistance derived from *Solanum andigena* had females that remained white up to the stage of cyst formation instead of passing through the commonly-described golden-yellow phase. Webley (1970) showed that populations with yellow females had shorter larval stylets than those with white females, and Parrott (1972) that the two types did not interbreed readily. Trudgill and Parrott (1972) showed that the patterns of water-soluble protein bands produced by electrophoresis in polyacrylamide gels differed greatly for yellow and white females and the two types were described as separate species by Stone (1973). Each species has pathotypes which differ in their ability to reproduce on potatoes with genes for resistance (see Stone, 1976). For instance pathotypes of *G. rostochiensis* able to reproduce on plants containing the H_1 gene for resistance have been reported from the Netherlands and Bolivia. There are small differences in the life cycles of the two species (McKenna and Winslow, 1972) but it is thought that their biology is essentially similar.

The dormant eggs of the nematodes are contained within a cyst formed from the toughened integument of the female. In the absence of a host crop only a small proportion of the eggs hatch each year but when a host

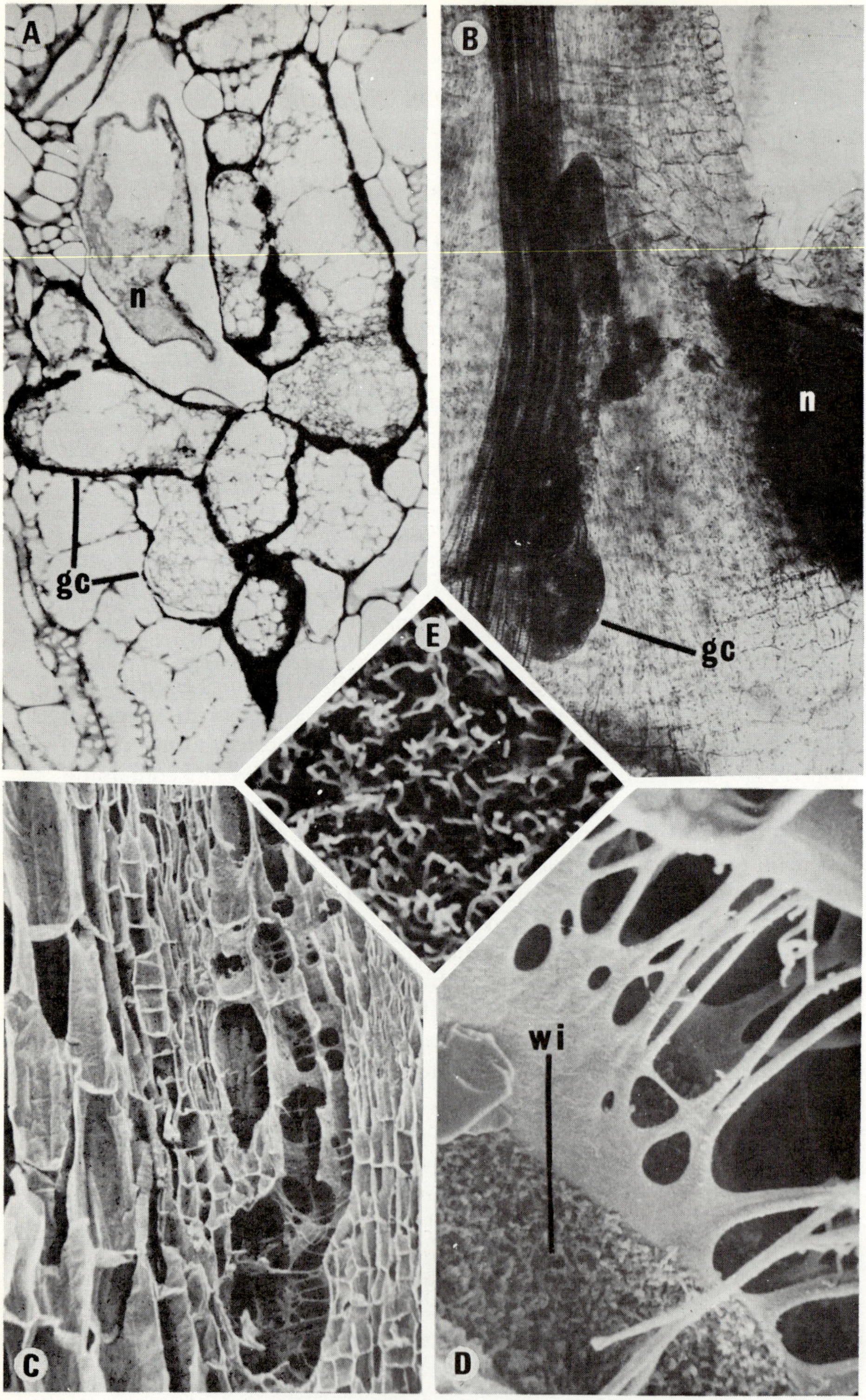
A
n
gc
B
n
gc
E
C
wi
D

crop is grown most of the larvae are stimulated to hatch by a specific chemical 'hatching factor' diffusing from the host roots. Second stage larvae hatch and invade the roots, cutting a path through the cells until they reach the stele where they induce formation of specialized transfer cells (syncytia) on which the now sedentary larvae feed. The syncytia are formed mainly from phloem parenchyma cells by the breakdown of cell walls and enlargement of the cells. Special wall ingrowths develop on the walls adjacent to vessels to increase the surface area of the plasmalemma (Fig. 12.2) and so increase the rate of transport of solutes (Jones and Northcote, 1972).

Provided a larva induces a syncytium of sufficient size it receives all the food necessary to develop into an adult female, but when the syncytium is inadequate for this purpose the larva either dies or becomes a male. Experimental evidence suggests that the availability of food affects the sex of the developing larvae, poorly nourished larvae becoming males and well-nourished larvae becoming females (Trudgill, 1967). After fertilization the female deposits eggs in her own body, her cuticle tans and when she dies the dead 'cyst' containing hundreds of eggs is left in the soil. In the absence of host crops soil infestations may persist for 20–30 years, and it is in this dormant condition that potato cyst-nematodes have been spread, either in infested soil or adhering to tubers harvested for use elsewhere as seed. The large reproductive capacity of these nematodes (up to 70-fold and more) and their slow rate of decline in the absence of a host crop (about 30% per annum) make them a persistent and serious pest of potatoes.

12.1.4 Future changes in world distribution of potato cyst-nematodes

Potato sickness due to cyst-nematodes was not noticed in Europe until the late 1800s, suggesting that they were introduced to Europe with breeding material imported in the mid-1800s to improve resistance to potato late blight. The pathways of their distribution are largely a matter of speculation but the known distribution of the two species is shown in Fig. 12.1. Major potato-growing areas from which both species are

---

Fig. 12.2 (a) Transverse section of potato root showing giant cell complex around head of female potato cyst-nematode. x 240 (b) Whole mount of female potato cyst-nematode in potato root showing giant cell complex associated with the stele. x 200 (c) Scanning electron micrograph of a longitudinal section of a soybean cyst-nematode-infested root with cell contents digested away to show dissolution of cell walls and formation of a syncytium. x 200 (d) Scanning electron micrograph of a longitudinal section of a tobacco cyst-nematode-infested root showing dissolved cell walls and wall ingrowths. x 1000 (e) Wall ingrowths in giant cells induced by root-knot nematodes in roots of balsam. x 3000
(n) nematode; (gc) giant cells; (wi) wall ingrowths.
(a) Preparation by R.A. Rohde, photograph by C.C. Doncaster;
(b) Preparation by D.L. Trudgill, photograph by Rothamsted Experimental Station; (c–e) by M.G.K. Jones.

absent include the bulk of North America and Australia, whilst much of India, Russia and possibly China are also still free.

In spite of quarantine measures further spread to uninfested areas is likely, and as varieties with the H_1 (ex. *andigena*) gene for resistance, such as Maris Piper which now represents over 10% of the maincrop acreage in the U.K., are increasingly grown, *G. pallida* or the aggressive pathotypes of *G. rostochiensis* will become more widespread and increasingly dominant in mixed populations. Spread of *G. pallida* will be especially difficult to prevent due to the lack of good sources of resistance. Pure populations of *G. rostochiensis* pathotype Ro1, such as comprise most North American infestations, failed to reproduce when potatoes with the H_1 gene for resistance were grown for several seasons in pot and field tests, and it seems possible that such populations could be eliminated by continuous growing of varieties with gene H_1 (Trudgill and Parrott, 1973). Varieties with a good measure of resistance to both species may be developed and work to identify suitable sources of resistance in tuber-forming species of *Solanum* from the Andes is continuing. Eradication using chemicals is difficult but appears to have been achieved in Israel (Anonymous, 1974).

12.2 Effects of nematodes on growth

Oostenbrink (1972) lists potatoes as being moderately susceptible to *D. dipsaci*, *M. hapla*, *P. penetrans* and *T. teres*, but the lack of published information prevents meaningful conclusions about their relative importance. *Nacobbus serendipiticus* is locally important in Peru and Bolivia (Franco and de Scurrah, 1972) but again there is a lack of quantitative information. The general symptoms in crops infested with root-feeding nematodes are patches of stunted plants. Although much information on the relationship between nematode numbers and yield is available for cyst-nematodes, information on the fundamental mechanisms of damage has only recently been obtained and much still remains to be determined. Therefore the following sections are concerned with work on mechanisms of cyst-nematode damage in resistant and susceptible potato varieties aimed at increasing our understanding and providing a basis for future improvements in crop management. The similarity of above-ground symptoms caused by all types of root-feeding nematodes suggests that models developed from work on cyst-nematodes may be applied to infestations by other root-pathogenic species.

12.2.1 Effects on yield

Because potato cyst-nematodes are largely immobile and pass only one main generation per year, there is a good correlation between pre-planting nematode density and damage. From the results of 30 field

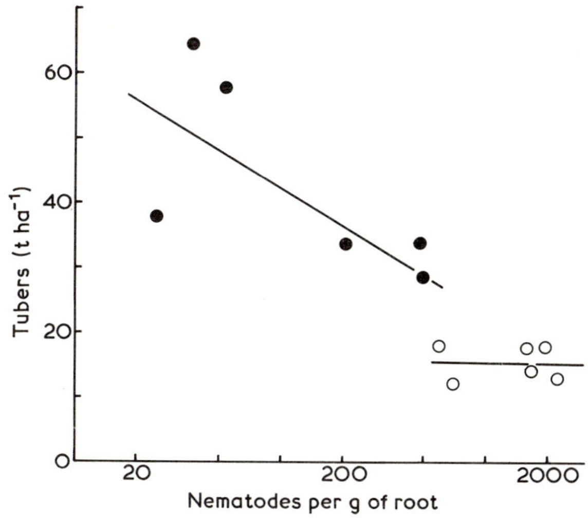

Fig. 12.3 The tuber yield of Pentland Dell potato plants 20 weeks after planting related to numbers of nematodes found in the roots 8 weeks after planting (from Trudgill *et al.*, 1975b). (● Fumigated with dazomet; o unfumigated.)

trials, Brown (1969), using untransformed straight line regressions, found that in most trials there was an average loss of 0.85 ton acre $^{-1}$ (2.13 t ha $^{-1}$) for each 20 eggs g $^{-1}$ of soil before planting. Linear regressions for the different trials tended to be parallel over a wide range of potential yields suggesting that yield loss was independent of potential yield and that the effects of nematodes were dominant.

Other models show the relationship between nematode numbers and yield as a sigmoidal one (Seinhorst, 1965). At low nematode densities the plant is able to tolerate or repair the damage sustained, and at very high densities the sites available for nematode feeding become saturated and a certain minimum yield remains unaffected. Seinhorst's model describes a wide range of nematode/yield relationships but it is empirical, and apart from the concepts of tolerance and minimum yield does not further our understanding of how nematodes damage plants. Experience has shown that the maximum pre-planting population of potato cyst-nematodes that potatoes will tolerate without loss of yield is about 10 eggs g $^{-1}$ on a sandy loam soil but 10–20 eggs g $^{-1}$ on a black fen soil (Jones, 1966).

The primary effects of cyst-nematode attack are on root growth and function with secondary effects on the rates of haulm growth and senescence, which in turn affect tuber yield. Trudgill *et al.* (1975a,b) describe an experiment on sandy soil in which the effects of different numbers of potato cyst-nematodes on the growth, nutrient uptake and yield of nematode resistant (Maris Piper) and nematode susceptible (Pentland

Dell) varieties were observed. Their results will be discussed in detail because they provide a more comprehensive picture of the effects of potato cyst-nematodes on potato growth, although the repeated growing of potatoes on a sandy soil may have led to attacks by secondary pathogens.

Repeated fumigation with dazomet greatly increased final tuber yield of the susceptible variety Pentland Dell even when the nematode infestation after fumigation was moderately large, whereas yields from unfumigated plots were uniformly poor and only about a third of yields from the best fumigated plots (Fig. 12.3). The difference in yields between fumigated and unfumigated plots with moderate numbers of nematodes may have been due to the control, by dazomet, of secondary root-rot pathogens.

In the same experiment the resistant variety Maris Piper yielded significantly more than Pentland Dell, even when infested with many *G. pallida* to which it is susceptible, suggesting that Maris Piper may be more tolerant of potato cyst-nematode attack or have greater resistance to secondary root pathogens.

12.2.2 Effects on root morphology and growth

Trudgill (1968) showed, by adding different numbers of potato cyst-nematode larvae to pots in which potatoes were growing, that root growth was greatly slowed when nematode attack was severe. Since nematodes reduce the rate of root expansion in this way the volume of soil which the plant is able to exploit and therefore the available reservoirs of water and nutrients are decreased.

Evans *et al.* (1973) observed the distribution of roots of the potato varieties Pentland Dell and Maris Piper in soil cores taken to a depth of 70 cm 12 weeks after planting, and found that when many nematodes were present root systems of both varieties were small and did not penetrate much below plough depth (50 cm below the crests of the ridges). The availability of nutrients to plant roots varies greatly and is affected by many factors. N is very mobile in the soil but K is more strongly adsorbed by soil minerals and so can only be efficiently extracted from those parts of the soil intensely ramified by roots. P is even less mobile. Assuming various radii of influence for roots Evans *et al.* calculated the volumes of soil from which nutrients were available to their plants, and found the best relationship between yield and volume of soil explored when the radius was 0.5 cm (Table 12.2), approximately the distance over which plant roots are able effectively to extract potassium (Nye, 1968).

Although a number of assumptions were made in the calculations the results in Table 12.2 provide circumstantial evidence that cyst-nematodes may limit growth of potato plants because insufficient amounts of the less mobile nutrients (such as K) are available to the plants. However, the information given is not sufficient to conclude whether or not nematodes interfere with the efficiency of root function or movement of solutes within the root system, but Price *et al.* (1976) concluded that

Table 12.2 The effect of nematode populations on the estimated volumes of soil ($m^3 \times 10^{-2}$) exploited by potato plants* (from Evans *et al.*, 1973)

	Pentland Dell			*Maris Piper*	
Nematode population size	*Small*	*Large*	*Small*	*Large*	*Large*†
Depth					
Ridge soil	1.8	0.7	2.7	2.7	1.1
Ploughed layer	2.1	0.1	3.0	1.5	1.3
Sub-soil	0.6	0.0	1.0	0.1	0.3
Total	4.5	0.8	6.7	4.3	2.7
Yield of tubers/plant (g)	1071	200	1424	977	815
Total K in tops and tubers (g)	7.8	1.7	9.4	5.8	2.9

*N.B. It is assumed that nutrients move to roots over a distance of 0.5 cm.
†*G. pallida*, all other populations *G. rostochiensis*

Heterodera avenae did not decrease the efficiency per unit length with which oat seedling roots took up and translocated labelled P and K.

12.2.3 Effects on nutrient uptake

Most evidence suggests that nematodes do affect total amounts of minerals taken up by plants and their translocation to the tops. The elements whose uptake is most affected are P, K, magnesium (Mg) and calcium (Ca) (Parris and Jehle, 1943; Lownsbery, 1956; Sher, 1957; Kirkpatrick *et al.*, 1959; Oteifa and Elgindi, 1962). The concentration of N is usually unaffected, but concentrations of sodium (Na), Mg and Ca may be increased.

In the experiments of Trudgill *et al.* (1975a, b) nematodes significantly decreased the amount of K in the haulm dry matter (Fig. 12.4) and comparing these results to those of Carpenter (1957) suggests that heavily-infested plants were chronically K-deficient and that the rate of K uptake limited haulm expansion.

Interpretation of Trudgill *et al.*'s results is not straight-forward because the percentage dry matter is greater in the haulms of plants dwarfed by nematodes than in the haulms of lightly-infested plants. This is probably because nematode-infested plants are chronically water-stressed, which would reduce cell expansion rather than cell division (Slatyer, 1967). Individual leaves are smaller in infested plants, presumably because individual cells are smaller. Infested plants therefore probably have a greater ratio of cell wall material to cell content and hence a greater dry matter content than uninfested plants. The increase in percentage dry matter in infested plants recorded by Trudgill *et al.* is

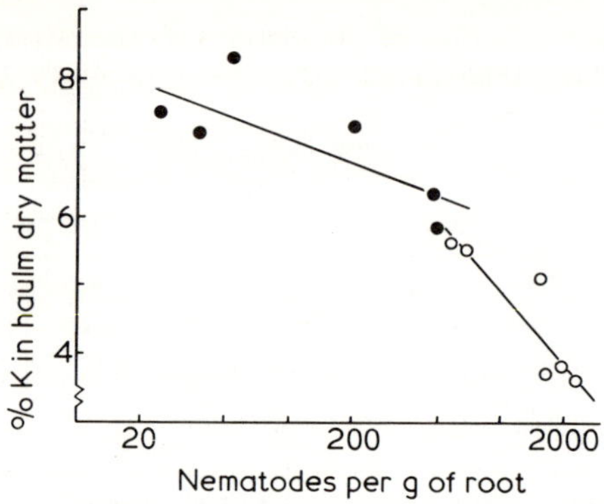

Fig. 12.4 K content of haulm dry matter of Pentland Dell potato plants 11 weeks after planting related to numbers of nematodes found in the roots 8 weeks after planting (from Trudgill *et al.*, 1975b). (● Fumigated with dazomet; o unfumigated)

Table 12.3 Plant nutrients (mg g^{-1} of fresh haulm) in the potato variety Pentland Dell 11 weeks after planting related to numbers of potato cyst-nematodes g^{-1} of root (from Trudgill *et al.*, 1975b)

	mg g^{-1} of fresh haulm of:					
Nematodes/g root	*K*	*N*	*P*	*Mg*	*Ca*	*Na*
25	6.01	3.53	0.429	0.352	1.49	0.012
38	6.29	3.40	0.442	0.319	1.18	0.017
55	6.39	3.39	0.435	0.305	1.12	0.011
205	5.74	3.30	0.413	0.252	1.42	0.015
485	5.85	3.35	0.413	0.279	1.63	0.018
498	5.03	3.12	0.386	0.238	1.91	0.014
588	4.94	3.08	0.322	0.242	2.02	0.019
694	5.46	3.18	0.323	0.258	1.74	0.020
1560	5.09	3.69	0.334	0.324	2.50	0.022
1688	3.85	3.54	0.302	0.244	2.70	0.028
1903	4.03	3.60	0.334	0.286	2.76	0.024
2235	3.38	3.28	0.314	0.260	2.75	0.025

considerably greater than that which would be expected from the changes in leaf water potential alone that they measured.

In Table 12.3 the amounts of each nutrient are expressed in mg per g of fresh weight of haulm for Pentland Dell (measured 11 weeks after planting) infested with various numbers of nematodes. As nematode numbers increased, the amount of K g^{-1} fresh weight of haulm decreased greatly. Amounts of Mg and P also decreased, although the amounts of Mg were still probably adequate for normal plant growth. Amounts of N were unaffected but those of Ca and Na increased two-fold. This increase in concentration of Ca and Na probably maintained cation concentration, which would otherwise have been excessively low due to shortage of K, and may have been due to a breakdown in the calcium exclusion mechanism. These changes in concentration probably explain the results of Stelter and Meinl (1969) who measured the total ash content of cyst-nematode infected potato plants and concluded that nematodes affected uptake of neither nutrients nor water. A few other workers have reached similar conclusions for different nematode/crop associations (Hunter, 1958; Jenkins and Malek, 1966; Bergeson, 1966, 1968).

Nematodes decreased the P content of plants (Trudgill *et al.* 1975b) but the P content of plants in unfumigated plots was also much less than that of plants in fumigated plots containing similar numbers of nematodes. Fumigation can increase the amounts of mycorrhizal fungi associated with roots (Bird *et al.*, 1974) which may in turn increase P uptake and so explain the results observed.

The poor growth of nematode-infested plants observed by Evans *et al.* (1973), was probably due to inadequate K uptake resulting from poor root growth (Table 12.2, total K in tops and tubers). K, however, may not always be the nutrient limiting growth of nematode-infested potato plants. In an experiment on a site adjacent to the one used by Trudgill *et al.* two nematode susceptible varieties of potato (Record and Pentland Crown) received four rates of NPK fertilizer. The soil was deficient in Mg and Trudgill *et al.* (1975), comparing plants receiving the higher fertilizer rates with those receiving lower rates, found that nematodes decreased plant growth and the amounts of Mg taken up by the plants. Yields and Mg content of both varieties were decreased by increasing numbers of nematodes, and there were good correlations between these factors (Fig. 12.5).

Potassium content of plants in this experiment was unaffected except in the variety Record, growing in plots with the lower rates of fertilizer, but the results clearly show how potato cyst-nematode damage to roots decreases uptake of the less available cations.

As heavily-infested plants may support large numbers of nematodes, significant amounts of N are incorporated in the nematodes themselves. Abstraction of nutrients from the plant by nematode feeding has been suggested as a mechanism of damage (Hollis, 1963), but analyses of the haulm dry matter of nematode-infested plants by Trudgill *et al.* (1975a) showed that in plants dwarfed by chronic shortages of other nutrients the N content was unaffected. It is possible, however, that N concentration may be one of the primary factors limiting the growth of nematode-infested plants. Potato plants can take up and store K and some other nutrients in amounts in excess of their minimum requirements. The decreased amounts of K, Mg or P in the haulms of nematode-infested

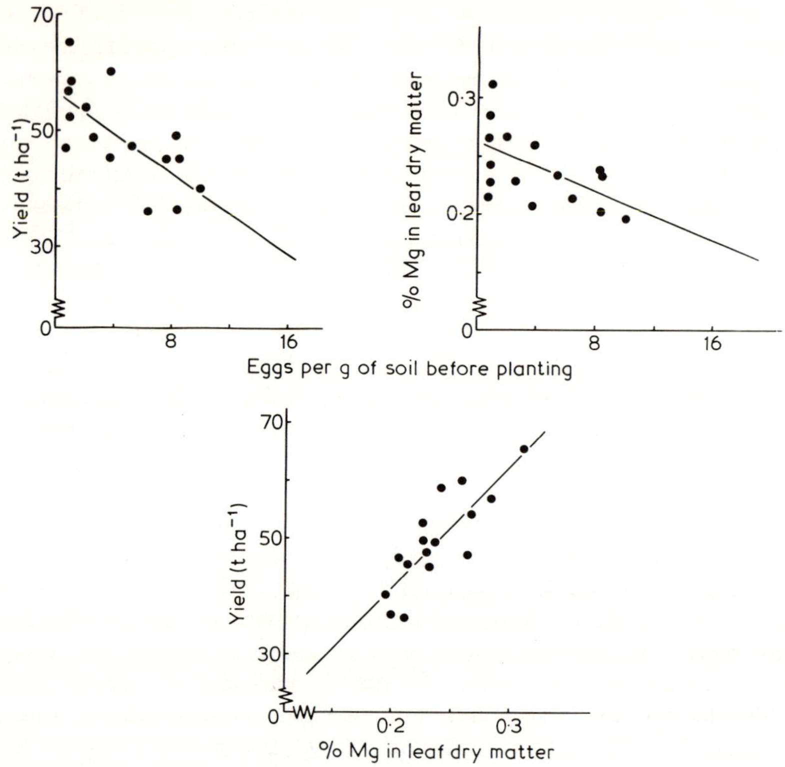

Fig. 12.5 Relationships between yield, Mg content of the leaf dry matter and pre-planting nematode population, for Pentland Crown potato plants (modified from Trudgill *et al.*, 1975 – high fertilizer rates only)

plants may still be adequate for plant growth, whereas the relatively small differences in N concentration found between nematode-infested and healthy plants may be due to growth limitation by decreased N uptake in the infested plants. Pot experiments have suggested that N and P may be more important than K in limiting the growth of infested plants and that during the early stages of growth auxins or growth inhibitors may be involved (Trudgill, loc. cit.).

12.2.4 Effects on haulm growth

The typical above-ground symptoms of potato cyst-nematode damage are stunting of the haulm and early senescence (Grainger, 1951). Measurement of ground cover at weekly intervals showed that nematodes had decreased the rate of haulm growth 6 weeks from planting (Fig. 12.6), and that infested plants senesced considerably earlier than uninfested ones (Evans *et al.*, 1975).

Regression of leaf area 11 weeks after planting (when plants were at

Fig. 12.6 Changes, with time, in ground cover of Pentland Dell potato plants infested with few, intermediate numbers or many potato cyst-nematodes (from Evans *et al.*, 1975)

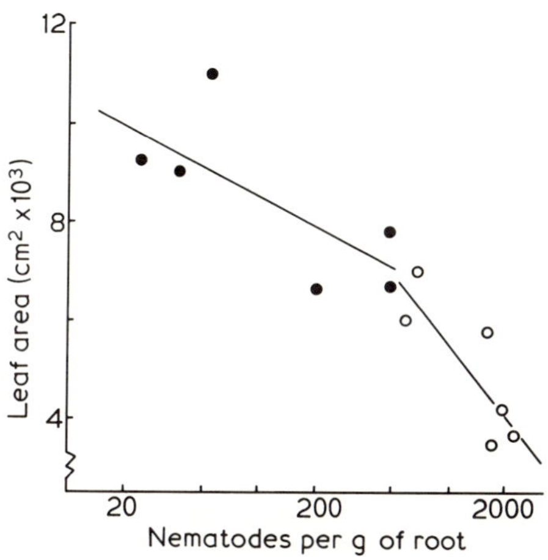

Fig. 12.7 Leaf area of Pentland Dell potato plants measured 11 weeks after planting related to numbers of nematodes found in the roots 8 weeks after planting (from Trudgill *et al.*, 1975b). (● Fumigated with dazomet; o unfumigated)

their maximum size) against nematode numbers showed that up to this time leaf area was largely independent of fumigation treatment. The number of nematodes present was the dominant factor, with the form of the curve much the same as that in Fig. 12.3 for K content of the haulm dry matters against numbers of nematodes (Fig. 12.7). Trudgill *et al.* (1975a) also made detailed growth analyses, 10 weeks after planting, of plants growing poorly in a heavily-infested, unfumigated plot. They showed that these plants were smaller than those in an adjacent, lightly-infested, fumigated plot because the numbers of stems were less and internode lengths and individual leaf areas were smaller, although

Table 12.4 Growth analysis and N and K contents of heavily- and lightly-infested Pentland Dell potato plants, 10 weeks after planting (from Trudgill *et al.*, 1975a)

	Heavily-infested unfumigated	*Lightly-infested fumigated*
Haulm		
Fresh weight (g)	199	538
% Dry matter	9.2	8.1
Height (cm)	34.2	43.4
Number of stems/plants	4.2	8.6
Number of nodes/stem	13.0	11.2
Number of leaves/stem	11.6	10.6
Leaf area/plant (cm^2)	1598	5882
Area each leaf (cm^2)	32.8	64.5
% N in dry matter	4.5	4.8
% K in dry matter	4.0	6.8
Tubers		
Fresh weight (g)	49	77
Number of tubers	15.4	19.4
% dry matter	14.6	13.6
% N in dry matter	2.3	2.1
% K in dry matter	2.3	2.5
Roots		
Fresh weight (g)	48.6	104.2
% N in dry matter	2.8	2.5

numbers of leaves per stem were unchanged. The full results are given in Table 12.4.

12.2.5 Effects on water uptake

As nematodes affect both root growth and nutrient uptake it is not surprising that they also affect water uptake. O'Bannon and Reynolds (1965) found that cotton plants infested with *Meloidogyne* spp. grew as well as uninfested plants when the soil was kept at field capacity, but that they used more water per unit weight. When the soil was allowed to dry to 50% of field capacity between waterings infested plants grew more slowly than uninfested ones. Odihirin (1971) found that at first *M. incognita* greatly increased the rate of transpiration of infested tobacco plants but that eventually the plants wilted even when abundant water was available, apparently because the roots became too severely damaged to meet the demand for water. Wallace (1974) evaluated the effects of root-knot nematodes on photosynthesis and nutrient demand by roots of tomato plants, and concluded that the nematode's most important effect was probably on water uptake and thereby on photosynthesis.

Field observations suggest that the early senescence of potato plants infested with many cyst-nematodes is associated with water stress, and by observing the water relations of infested plants Evans *et al.* (1975) were able to show that the onset of senescence 12–13 weeks after planting was associated with increased internal water deficit. Earlier in the season young plants responded to water stress by closing their stomata to conserve water, but eventually they adapted to the continuing water stress and opened their stomata. The decreasing water uptake of heavily-infested plants due to failing root systems was partly balanced by the progressive shedding of leaves, but death eventually occurred when leaf water potentials continued to decrease in spite of large increases in stomatal resistance.

12.2.6 Interactions with other organisms

In the experiment described by Trudgill *et al.* (1975b) fumigation with dazomet delayed senescence, shown by comparing leaf area after 14 weeks with that after 11 weeks (Fig. 12.8). Plants in unfumigated plots senesced much earlier than those in fumigated plots and this is best explained by suggesting that secondary root-rot pathogens are involved. The variety Maris Piper stayed green longer than Pentland Dell and had whiter roots, perhaps indicating that it has greater resistance to root pathogenic fungi. Grainger and Clark (1963) reported that potatoes suffer greater loss of yield when grown in soil containing both *Rhizoctonia*

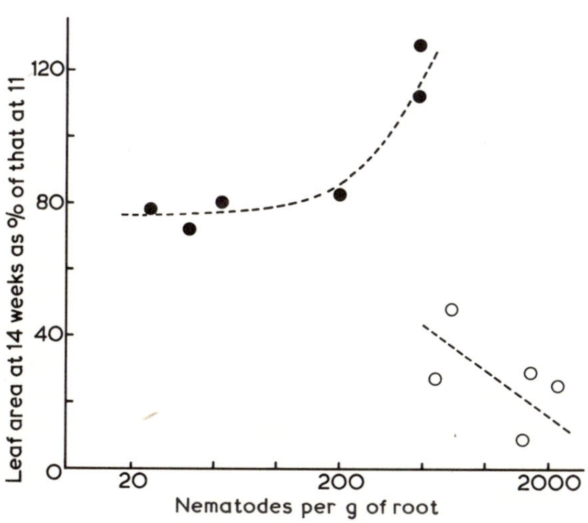

Fig. 12.8 Leaf area of Pentland Dell potato plants 14 weeks after planting as a percentage of that 11 weeks after planting related to numbers of nematodes found in the roots. (from Trudgill *et al.*, 1975b). (● Fumigated with dazomet; o unfumigated.)

solani and potato cyst-nematodes than in soil containing either organism alone.

Corbett and Hide (1971) showed that *G. rostochiensis* increased the severity of the disease caused by *Verticillium dahliae* in potatoes, but the interaction between *P. penetrans* and Verticillium wilt of potatoes reported by Morsink and Rich (1968) seems less clearly defined. Yield was even noted to increase with increasing fungal inoculum at constant nematode levels. Other interactions of economic importance are those involving species of *Trichodorus* and *Paratrichodorus* with strains of tobacco rattle virus, already mentioned.

12.2.7 Conclusions

Invasion of plant roots by small numbers of nematode larvae sometimes stimulates root growth (Jones, 1957b), but invasion by large numbers slows root growth and increases root proliferation, so changing the form of the root system and decreasing considerably the volume of soil the plant is able to exploit. Rates of uptake of nutrients are lowered and those least mobile in the soil (K, Mg and P) are those which may then limit haulm growth. Poor root growth and nutrient uptake cause haulms to be dwarfed; leaves are smaller, fewer stems develop, and yield is reduced because tubers are fewer and smaller. Infested plants probably initiate almost as many tubers as uninfested plants but fewer survive to harvest, possibly because they are resorbed. Fewer, smaller tubers are harvested from infested plants because leaf area duration is less as a result of smaller leaf area and early senescence.

Although the greatest effect of cyst-nematodes is to limit the effective size of the root system, the extraction of nutrients during nematode feeding is a far from trivial effect. The specialized syncytial transfer cells (Fig. 12.2) have increased surface areas available for movement of solutes, and the partial recovery of moderately infested plants which is sometimes noticed in late summer may be due to cessation of the drain of nutrients from the plant as the female nematodes reach maturity and die. Estimations of the N content of female nematodes (Trudgill, unpublished data) suggest that if they used all the N in ingested cytoplasm they would remove about 5% of the total plant N. However, if they used only a quarter then they would remove about 20% of the total plant N. Abstraction of similar amounts of other mineral nutrients or carbohydrates would have a significant debilitating effect on the plant. However, Trudgill *et al.* (1975a) detected no changes due to nematode infestation in the net assimilation rates of plants. Net assimilation rate might be expected either to increase because the nematodes act as a sink for assimilate or to decrease because of physiological effects of the nematodes on the plant.

Other ways in which nematodes may affect plant growth include disturbing the balance of plant growth substances. Setty and Wheeler (1968) found that roots of tomato plants infected with *Meloidogyne* spp. contained more auxin than uninfected roots, and Bowman and Bloom (1966) showed that infection with *M. incognita* significantly lowered the resistance of tomato plants to Fusarium wilt.

However, the general principles of the effects of nematodes on plants are probably those outlined by Trudgill *et al.* (1975a, b) and Evans *et al.* (1975) for potato cyst-nematodes and potato plants, i.e. stunting of roots, reduced nutrient uptake and poor top growth, water stress and a sudden increase in water stress after invasion of roots by secondary pathogens, followed by death. More work is required to investigate the value of 'foliar-feeding' of nematode-infested plants with mineral nutrients, the role of secondary root-rot pathogens and the effect of nematode invasion on root growth, particularly in relation to tolerance.

12.3 Methods of decreasing nematode damage

Damage due to nematodes is avoided if they are absent from the environment in which a crop is grown. If a pest is present, however, its numbers can be decreased to levels that the crop will tolerate by chemicals, by a variety of biological means such as growing resistant or non-host varieties or crops, by planting the crop in cool soils where its roots grow but the nematodes are not very active, by harvesting the crop before the pest has completed its life cycle, and possibly by the use of natural pathogens or predators for biological control.

12.3.1 Legislative control

Potato cyst-nematode is recognized as the most serious pest threatening the American potato industry and as it is one of the most difficult of all crop pests to control most countries from which it is thought to be absent have strict regulations governing the importation of any produce that might carry it. This includes seed potatoes, transplants, nursery stock, bulbs and corms. Quarantine regulations may also be imposed within a country to prevent further spread once the pest is known to have entered. For example, in Steuben county on mainland U.S.A. and on Long Island large areas are quarantined and, in addition to stringent regulations concerning movement from these areas, of hay, straw, propagating material and farm machinery, all potatoes grown must be packaged in non-reusable paper bags. Potatoes may not be grown on or near land known to be infested (Spears, 1968). However, it is difficult to detect light infestations of potato cyst-nematode and it may be introduced and spread over a wide area before its presence is noticed, so quarantine measures are difficult to apply effectively (Jones, 1969).

In order to minimize losses when the pest is known to be present, infested land may be put under 'notice' and the frequency of potato growing severely restricted. In Eire farmland found to be infested with the nematode must be sown with grass immediately and in Northern Ireland planting of potatoes, tomatoes or plant propagating material is prohibited on any land where the nematode is discovered. Within the European Economic Community member countries are required to issue regulations which prevent production of seed potatoes on land where the nematode has been detected and in ware potato growing areas only resistant varieties may be grown on land officially declared to be infested, unless the land is treated with an approved nematicide or the potatoes are to be lifted before the life cycle of the nematode is complete. In Scotland seed potatoes may only be grown on a particular field one year in seven, and only on land which is free of potato cyst-nematode. The growing crop is also inspected before a certificate is granted. Measures such as these all help to decrease losses by preventing spread of the nematode, but even when legislation does not enforce control measures there are a number of other precautions which the sensible farmer may use in order to reduce his risks.

12.3.2 Damage avoidance

By using suitable rotations losses due to potato cyst-nematode may be decreased, as the nematode population declines by about 30% per year in the absence of potatoes, although groundkeeper potatoes may slow the rate of decline at low population levels (den Ouden, 1967). Eventually populations decline to a level at which no significant damage is caused when a potato crop is grown, and in practice in England potatoes can be grown one year in four or five on heavy soils, but only once in six or seven years on light soils or peats (Cooper, 1953; Winfield, 1965). However, some farmers in the past have adopted shorter rotations and accepted some loss as this can be more profitable than having longer rotations. Cyst-nematode populations can be estimated from soil samples and the service provided in England and Wales by the Agricultural Development and Advisory Service (A.D.A.S.) has been important in decreasing crop losses in infested areas.

When potatoes are lifted early the females may be killed before they have produced many eggs (van den Brande and d'Herde, 1964) which means that early potatoes can be grown year after year on the same land without much yield loss. Such areas are usually relatively warm and the absence of frosts allows very early planting and growth of potatoes before soil temperatures are high enough for nematode activity (e.g. the Ayrshire coast of Scotland, Pembrokeshire, Cornwall).

In some areas damage is avoided because soil temperatures seem to be too great for potato cyst-nematodes. On the coast of Peru potatoes are grown under irrigation and inspection of the growing crop has revealed cyst-nematodes, presumably introduced with the seed. However, the nematodes seem to die out during the non-growing season when no irrigation water is supplied and soil temperatures become high (Simon, 1955). Also in Peru the distribution of the false root-knot nematode *Nacobbus serendipiticus* is limited by unknown biological factors to the area of the altiplano around Lake Titicaca. This restricted distribution may be due to a slow rate of artificial spread, in turn due to the restricted commerce of the region and failure of the nematode to survive in dried soil (the factor which makes possible the rapid spread of cyst-nematodes), or because it requires a host crop every year to survive in soil. Whatever the reason most potato producing areas of the world are free of this nematode, potentially able to cause total crop loss. *Meloidogyne* spp. although at present not causing large yield losses in potatoes, pose a serious potential threat to potatoes adapted for growing in sub-tropical and tropical areas.

12.3.3 Use of resistant varieties

About half the potato cyst-nematode infestations in England are of *G. rostochiensis* pathotype Ro1 and so are unable to multiply on varieties containing the major gene H_1. Commercially acceptable varieties of this type, such as Maris Piper, therefore have a useful role to play, although only in areas in which *G. pallida* is scarce or absent. Dunnett (1961) found that the wild potato *Solanum multidissectum* had a major gene H_2 which gave resistance to some populations able to reproduce on H_1 hybrids, and more recently Howard *et al.* (1970) and Howard and Fuller (1971) reported yet another major gene H_3, which gives partial resistance to British populations of *G. pallida*. Eventually these genes will be incorporated in commercially acceptable varieties and, if no more pathotypes are discovered in Britain, it will be possible to decrease potato cyst-nematode populations by growing resistant varieties. However, as more genes for resistance are discovered so pathotypes able to overcome that resistance are identified. In the Netherlands and Bolivia pathotypes of *G. rostochiensis* able to reproduce on H_1 hybrids have been discovered, and more pathotypes of *G. pallida* exist than are presently found in Britain (Howard, 1972). Growing an H_1 hybrid would favour increase of the aggressive part of a mixed population and no potato line, wild or otherwise, has yet been found with complete resistance to all populations of potato cyst-nematodes. Some lines of *S. vernei* have a wide spectrum of polygenically-based resistance and although it is difficult to breed from such parent material a number of varieties with resistance derived from *S. vernei* are now grown for starch production in the Netherlands. The search for further sources of resistance is continuing, especially in Peru.

Breeding for resistance to *Meloidogyne* spp. is less advanced, but some sources of resistance have been identified. However, resistant varieties for this nematode may have a shorter life than those for cyst-nematodes as *Meloidogyne* spp. are parthenogenetic and have enormous reproductive capacities, which might enable resistance-breaking pathotypes to arise and spread more quickly.

Potato cyst-nematode numbers are decreased more by a suitable resistant variety each year (by 75–80%) than when a non-host is grown (30%) because of the specific hatching factor. With *Pratylenchus* spp., which have very wide host ranges and against which no resistance has been reported, excessive damage may be prevented by avoiding planting potatoes after a good host such as barley.

12.3.4 Use of tolerant varieties

Potato varieties differ considerably in their tolerance to nematode attack, if tolerant varieties are defined as those which suffer a smaller reduction in yield at high nematode densities. Huijsman *et al.* (1969) tested the tolerance to cyst-nematode attack of 118 potato varieties and found that the variety Multa significantly out-yielded all others. When Multa was compared in a pot test with the intolerant variety Libertas the total dry matter produced by Multa with an initial nematode density of 1600 eggs g^{-1} of soil was 50% of that produced in the absence of nematodes. The same percentage reduction occurred in Libertas at an initial density of only 300 eggs g^{-1} of soil. Huijsman *et al.* also equated resistance, tolerance and susceptibility to cyst-nematode attack with different cell reactions in the roots.

Incorporation of cyst-nematode tolerance characteristics into potato varieties would be most useful if the variety were also resistant. Growing a resistant variety usually reduces the nematode population in the soil by more than 70%. Nevertheless much root damage and yield loss may occur if the initial population is dense. This would be largely avoided if the resistant variety were also tolerant. However, it is difficult to screen seedlings for tolerance to nematode attack as the nature of tolerance is unclear and no suitable screening technique is available.

12.3.5 Chemical control

Treatment of soil with chemicals on a field scale to control soil-inhabiting nematodes is costly but many thousands of acres are treated annually in the U.S.A., as are an increasing number in Europe. The chemicals used may be soil fumigants, organophosphates or oximecarbamates. Although 11 kg or less a.i. ha^{-1} of a granular oximecarbamate may prevent cyst-nematodes from feeding and multiplying (Whitehead, 1973) large yield increases are necessary because the chemicals are expensive and require special precautions during application because of their poisonous nature. Oximecarbamate nematicides, such as oxamyl and aldicarb, are applied as granules and work best when well incorporated into the top-soil before a susceptible crop is planted. Unlike fumigant nematicides, which kill nematodes, the oximecarbamate nematicides seem to disorientate the nematode larvae and prevent them from invading the plant roots (Boparai and Hague, 1974). The cost of such treatment is presently about £100 ha $^{-1}$ in the U.K. Brown and Sykes (1973) tested oximecarbamate and organophosphate nematicides and found that both groups of compounds were effective in decreasing the transmission of tobacco rattle virus to potato tubers by *Trichodorus* spp. when applied in the ridge at planting time.

12.3.6 Integrated control

No single method of controlling potato cyst-nematode is ideal; control by rotating potatoes with non-host crops means that potatoes cannot be grown as frequently on a field as many farmers would wish; resistant varieties are not yet available for all pathotypes of the nematode and their use encourages multiplication of 'aggressive' portions of a population; chemical control is both expensive and sometimes of uncertain outcome. Jones (1970) outlines various rotations which incorporate the use of a resistant variety with and without nematicides and shows that very large reductions in the nematode populations may be obtained, even with quite short rotations, so that a susceptible crop might then be grown. Including a susceptible crop would have the advantage of slowing selective multiplication of aggressive pathotypes, so prolonging the useful life of the resistant variety. The use of a nematicide may have additional benefits on land where harmful nematodes such as *Longidorus* and *Trichodorus* spp. are troublesome. Other nematodes may also be killed and in the Netherlands, where the use of nematicides has been compulsory for a number of years when potatoes are grown frequently on potato cyst-nematode infested land, considerable benefits to cereal crops have been obtained through the control of *H. avenae*.

12.3.7 Cost effectiveness of control methods

In order to assess the cost effectiveness of a control measure we need to know the potential of the nematodes for damage, the yield increases obtained and the cost of the control measure. The costs of the various control measures available are difficult to assess but there can be little doubt that sufficient damage is caused by potato cyst-nematodes to justify considerable effort to control them. If a nematode pest is thought to be absent from a country it is worth trying to avoid or delay its introduction, but once an area is infested the potato-growing farmer is faced with the alternatives of control by rotation, nematicides or resistant varieties. Providing resistant varieties yield well and are of good quality they are much the best method of control as there is little extra cost to the farmer, such as that of chemicals and their application, or loss of income due to widened rotations and the enforced growing of crops less profitable than potatoes. However, resistant varieties are costly to develop and many thousands of pounds may have to be spent over the 10 or 20 years before successful ones are marketed.

Appropriate resistant varieties are not always available and the only alternative open to many farmers is to employ a rotation designed to keep nematode numbers below damaging levels. Empson and James (1966, 1967) examined the economics of enforced changes in rotation and found that lengthening the period between potato crops from three to five years was profitable on farms where all fields were infested, provided the yield potential was 10 ton acre^{-1}, but less so when it was 15 ton acre^{-1}. In order to establish suitable rotations, the distribution and numbers of the nematode have to be estimated, and such a service was, until recently in Great Britain, provided free by the National Agricultural Advisory Service (N.A.A.S.) and the Department of Agriculture for Scotland (D.A.F.S.). Adamson (1972) estimated that the value of the advice given after each sample is processed might be as much as £87 and this service is still available for a nominal charge from the Agricultural Development and Advisory Service (A.D.A.S.) and (D.A.F.S.). However, the cost of maintaining this service should be considered when costs of control are assessed on a national scale. More obscure costs are also involved such as not growing potatoes on the best potato land as often as would be desirable, the cost of nematode-free certification schemes for seed potatoes and propagating material such as bulbs, and the potential harm to a country's export trade in seed potatoes and other planting material. If a pest can be totally excluded from a country many of these costs are avoided, but they are replaced by those of monitoring land for appearance of the pest and maintaining quarantine measures.

12.3.8 Conclusions

The extensive movement of men, machinery and planting material which occurs nowadays means that nematodes are continually being introduced to new areas, and that eventually all species will have access to all areas which provide a suitable environment and suitable hosts. However, quarantine controls, such as the United States Department of Agriculture's campaign against the golden nematode (*G. rostochiensis*), can be effective in delaying introduction of pests. The golden potato cyst-nematode was first identified on Long Island in 1941, and in 1944 New York State instituted quarantine procedures. Systematic surveys showed that many properties on the island were infested but it was not until 1967 that the nematode was discovered on the mainland in Steuben County (Spears, 1968), so the enforcement of quarantine undoubtedly slowed the spread of the nematode.

If a nematode pest is very widespread strict legislative control cannot be effectively applied but if it is relatively sparsely distributed legislation can slow further spread and minimize losses. Perhaps more importantly it means that seed may be produced on uncontaminated land so that spread via seed is eliminated. Operation of reliable clean seed schemes is a prerequisite for any control programme. Pure *G. rostochiensis* pathotype Ro1, populations may be eliminated by repeated growing of H_1 ex-*andigena* varieties, but in general where potato cyst-nematode is endemic farmers will have to learn to live with it by using rotations, nematicides and resistant varieties in an integrated policy designed to keep populations below the threshold for damage.

12.4 Future developments in nematode research

Because many plant parasitic nematodes live in the soil it is often difficult to determine the amount of damage they cause, especially as nematodes are only one of several agents which may impair root function. There are usually obvious associations between the nematodes and disease symptoms in the plant when *Globodera* spp., *Meloidogyne* spp. or *Nacobbus* spp. attack potatoes but the effects of ectoparasites such as *Longidorus* spp. and endoparasites such as *Pratylenchus* spp. are more insidious and difficult to diagnose. A good example of the difficulties of diagnosis is the story of the so-called 'Docking Disorder' of sugar beet, which was variously ascribed to fungal pathogens and unusual chemical and physical conditions of the soil before Gibbs and Harrison (1963) associated the damage with *Longidorus* spp. and Whitehead *et al.* (1966) associated it with *Trichodorus* spp. More work is required to assess the effects of these nematodes on potatoes.

Increased knowledge of the mechanisms of nematode damage to plants may enable us to apply remedial treatments such as sprays containing plant nutrients applied to the foliage. N, P, K and Mg are the nutrients most likely to improve the growth of nematode-infested plants. Research is also required to determine the optimum rates of fertilizer for plants infested with small to moderate numbers of potato cyst-nematodes.

Increased knowledge of the mechanisms of nematode damage to plants might also provide plant breeders with a more rational approach to the problem of testing progeny for resistance and tolerance. Giebel (1970) attempted to develop a chemical test to determine the degree of plant resistance to *G. rostochiensis* and found that resistant and susceptible varieties differed in the ratio of mono- to polyphenols found in their roots. Current potato breeding programmes do not seek to exploit tolerance to potato cyst-nematode, largely because of the lack of a simple technique for evaluating tolerance. Huijsman *et al.* (1969) described the different anatomical reactions in the roots of susceptible, resistant and tolerant varieties but the techniques involved in assessing the reactions of root cells to nematode feeding are too difficult to apply on a routine basis. As a primary effect of potato cyst-nematode is slowed root growth, measurements of the effect of invading larvae on root growth may provide one measure of tolerance. Tolerance to larval invasion damage would be useful in both resistant and susceptible varieties, but would be especially useful when combined with resistance as a tolerant susceptible variety is likely to allow build-up of very large nematode populations. Tolerance to secondary root pathogens may also be worth seeking but as yet there is insufficient information about their role in hastening senescence of nematode-infested plants, except possibly that of *Verticillium dahliae* (Corbett and Hide, 1971).

There is also a need to understand better the ways in which nematodes interact with the plant environment to cause damage. Small numbers of potato cyst-nematodes caused increased damage to potatoes grown in soil poor in Mg (Trudgill *et al.*, 1975), and Evans (1971) suggested that *L. leptocephalus* caused most damage to potatoes when water was scarce.

There is still scope for improvement in control of nematodes and the losses they cause in potatoes. Biological control of potato cyst-nematodes seems unlikely as they are as devastating in the area from which they originated as they are in the areas to which they have been introduced. However, biological agents are probably important in limiting populations of other cyst-nematodes (Kerry, 1974). The use of potato varieties resistant to *G. rostochiensis* pathotype Ro1, may diminish the occurrence of non-aggressive biotypes but will increase the occurrence of aggressive biotypes, making the breeding of varieties with resistance to these strains a priority. Howard (1972) describes resistance found in *S. andigena* which gives good resistance to British populations of *G. pallida*, and this will eventually be combined with genes H_1 (already widely used, e.g. in Maris Piper) and H_2 (from *S. multidissectum*) to give varieties with resistance to most British populations of potato cyst-nematodes (Howard and Fuller, 1971). In order to provide varieties resistant to all pathotypes, such as occur in South America, a wider spectrum of resistance is required. This is probably to be found in wild potatoes such as *S. vernei*, *S. oplocense* and *S. spegazzinii*, but seems to be polygenically inherited and this coupled with the differences in ploidy level between the wild species and *S. tuberosum* makes incorporating this resistance in commercially acceptable varieties difficult.

Improvements in the use of nematicides may also be possible. Spraying cabbage plants with oxamyl decreased invasion of the roots by *H. schachtii* larvae (Potter and Marks, 1971) but Whitehead *et al.* (1973) reported that there was no evidence to suggest that potato cyst-nematodes could be controlled in this way. However, this approach may be effective in protecting potatoes from ectoparasitic nematodes such as *Longidorus* spp. and preventing infection of tubers with tobacco rattle virus transmitted by *Trichodorus* spp. For the future, however, integration of control measures as suggested by Jones (1970) and applied on a whole farm basis, using estimates of soil nematode populations by extension services, seems likely to provide the best control of cyst- and other nematode pests of potatoes.

Some nematode problems of potato growing may be solved by agronomic changes. For instance, it may be possible to eliminate spraing in potatoes by growing barley, which is a non-host for most strains of TRV, prior to growing potatoes, although good weed control would be essential. However, the spread of potato cyst-nematodes (especially *G. pallida*) and an increased awareness of the effects of other nematodes will mean that nematode problems in potatoes in temperate regions will continue to occur for many years to come. And in the same way that cyst-nematodes have spread *N. serendipiticus* may spread beyond its present fairly limited distribution in Peru and Bolivia. More research is required to understand what factors control the distribution of such a potentially serious pest, especially as it has been reported in a glasshouse in England (Graham, 1959). It seems likely too that potatoes will be grown more in lowland tropical areas and root-knot nematodes will undoubtedly prove troublesome in these localities, making research into root-knot nematode resistant lines another priority. However, the rapid progress made in recent years in understanding and controlling nematode problems of potatoes suggests that control measures will usually keep abreast of them.

References

Adamson, A. (1972). *A.D.A.S. Q. Rev.* **6**, 1–11.

Anonymous (1974). *European and Mediterranean Plant Protection Organisation Publications*, Series B. **77**.

O'Bannon, J.H. and Reynolds, H.W. (1965). *Soil Sci.* **99**, 251–5.

Bergeson, G.B. (1966). *Phytopathology* **56**, 1287–9.

Bergeson, G.B. (1968). *Phytopathology* **58**, 49–53.

Bird, G.W., Rich, J.R. and Glover, Sue U. (1974). *Phytopathology* **64**, 48–51.

Boparai, J.S. and Hague, N.G.M. (1974). *Meded. Faculteit Landbouwwet enschappen Rijksuniversiteit, Gent* **39**, 719–25.

Bowman, P. and Bloom, J.R. (1966). *Phytopathology* **56**, 871.

Brande, J. van den and d'Herde, J. (1964). *Nematologica* **10**, 25–8.

Brown, E.B. (1969). *Ann. appl. Biol.* **63**, 493–502.

Brown, E.B. and Sykes, G.B. (1973). *Ann. appl. Biol.* **75**, 462–4.

Carpenter, P.N. (1957). *Bull. Maine Agric. Exp. Station, University of Maine, Orono* **562**, 1–23.

Cole, C.S. and Howard, H.W. (1966). *Ann. appl. Biol.* **58**, 487–95.

Cooper, B.A. (1953). *Hort. Educ. Ass. Ann. Report*, 106–15.

Corbett, D.C.M. and Hide, G.A. (1971). *Ann. appl. Biol.* **68**, 71–80.
Dickerson, O.J., Darling, H.M. and Griffin, G.D. (1964). *Phytopathology* **54**, 317–22.
Dunnett, J.M. (1957). *Euphytica* **6**, 77–89.
Dunnett, J.M. (1961). *Scottish Plant Breeding Station Report for 1961*, 39–46.
Ellenby, C. (1952). *Nature, Lond.* **170**, 1016.
Empson, D.W. and James, P.J. (1966). *N.A.A.S. Q. Rev.* **73**, 22–9.
Empson, D.W. and James, P.J. (1967). *N.A.A.S. Q. Rev.* **76**, 160–5.
Evans, K. (1971). *Report of Rothamsted Experimental Station for 1970*, 155–6.
Evans, K., Parkinson, K.J. and Trudgill, D.L. (1975). *Nematologica* **21**, 273–80.
Evans, K., Trudgill, D.L., Thompson, Alison J. and Pandé, Lalita, U. (1973). *Report of Rothamsted Experimental Station for 1972*, 154–5.
Franco, J. and Scurrah, Maria M. (1972). In *Enfermedades de la Papa en el Peru. Boletin Tecnico* (77) *Estacion Experimental Agricola La Molina.* Ch. VI. DGIA: Lima.
Gibbs, A.J. and Harrison, B.D. (1963). *Report of Rothamsted Experimental Station for 1962*, 113.
Giebel, J. (1970). *Nematologica* **16**, 22–32.
Graham, C.W. (1959). *Pl. Path.* **7**, 114.
Grainger, J. (1951). The Golden Eelworm. *Res. Bull. 10 West of Scotland Agric. Coll., Auchincruive, Ayr.*
Grainger, J. and Clark, Marjorie, R.M. (1963). *Eur. Potato J.* **6**, 131–2.
Griffin, G.D. and Jorgenson, E.C. (1969). *Pl. Dis. Reptr* **53**, 259–61.
Guile, C.T. (1966). *Pl. Path.* **15**, 125–8.
Hollis, J.P. (1963). *Nematologica* **9**, 475–94.
Hoof, H.A. van (1964). *Meded. LandbHoogesch. opzoekingsstations van de staat te Gent.* Deel XXIX **3**, 944–55.
Hoof, H.A. van (1968). *Nematologica* **14**, 20–4.
Howard, H.W. (1972). *Ann. appl. Biol.* **71**, 263–6.
Howard, H.W., Cole, C.S. and Fuller, J.M. (1970). *Euphytica* **19**, 210–16.
Howard, H.W. and Fuller, J.M. (1971). *Pl. Path.* **20**, 32–5.
Huijsman, C.A., Klinkenberg, Caroline H. and Ouden, H. den (1969). *Eur. Potato J.* **12**, 134–47.
Hunter, A.H. (1958). *Soil Sci.* **86**, 245–50.
Jenkins, W.R. and Malek, R.B. (1966). *Soil Sci.* **101**, 46–9.
Jones, F.G.W. (1957a) *Nematologica* **2**, 185–92.
Jones, F.G.W. (1957b). *Nematologica* **2**, 257–72.
Jones, F.G.W. (1958). *Pl. Path.* **7**, 24–5.
Jones, F.G.W. (1966). *Report of Rothamsted Experimental Station for 1965*, 301–16.
Jones, F.G.W. (1969). In *Nematodes of Tropical Crops*, ed. Peachey, J.E. pp. 67–80. Farnham Royal: Commonwealth Agricultural Bureaux.
Jones, F.G.W. (1970). *Jl R. Soc. Arts* **118**, 179–99.
Jones, M.G.K. and Northcote, D.H. (1972). *J. Cell Sci.* **10**, 789–809.
Kerry, B.R. (1974). *Nematologica* **20**, 259–60.
Kirkpatrick, J.D., Mai W.F., Fisher, E.G. and Parker, K.G. (1959). *Phytopathology* **49**, 543 (Abstract).
Lownsbery, B.F. (1956). *Phytopathology* **46**, 376–9.
McKenna, L.A. and Winslow, R.D. (1972). *Ann. appl. Biol.* **71**, 274–8.
Morsink, F. and Rich, A.E. (1968). *Phytopathology* **58**, 401 (Abstract).
Newton, H.C.F. and Duthoit, Cecily M.G. (1954). *Pl. Path.* **3**, 139–40.
Nye, P.H. (1968). *J. Soil Sci.* **19**, 205–15.
Odihirin, R.A. (1971). *J. Nematology* **3**, 321–2.

Oostenbrink, M. (1972). In *Economic Nematology*, ed. Webster, J.M. Ch. 20, pp. 497–514. London: Academic Press.
Oostenbrink, M., s'Jacob, J.J. and Kuiper, K. (1957). *Tijdschr. PlZiekt.* **63**, 345–60.
Oteifa, B.A. and Elgindi, D.M. (1962). *Nematologica* **8**, 216–20.
den Ouden, H. (1967). *Nematologica* **13**, 325–35.
Parris, G.K. and Jehle, R.A. (1943). *Pl. Dis. Reptr* **27**, 235.
Parrott, Diana M. (1972). *Ann. appl. Biol.* **71**, 271–3.
Potter, J.W. and Marks, C.F. (1971). *J. Nematology* **3**, 325 (Abstract).
Price, N.S., Clarkson, D.T. and Hague, N.G.M. (1976). *Proc. Thirteenth Int. Symp. Nematology, Dublin*, p. 76.
Riedel, R.M. and Mai, W.F. (1971). *J. Nematology* **3**, 28–38.
Seinhorst, J.W. (1957). Phytonematology in Western Europe, *Technical Committee, Southern Regional Nematology Project* (S-19), Auburn, Alabama.
Seinhorst, J.W. (1965). *Nematologica*, **11**, 137–54.
Setty, K.G.H. and Wheeler, A.W. (1968). *Ann. appl. Biol.* **61**, 495–501.
Sher, S.A. (1957). *Phytopathology* **47**, 703–6.
Sher, S.A. (1970). *J. Nematology* **2**, 228–35.
Simon, J.E. (1955). *Informe mensual Estacion experimental agricola La Molina* **29**, 8–14.
Slatyer, R.O. (1967). *Plant Water Relationships*. London: Academic Press.
Spears, J.F. (1968). The Golden Nematode Handbook. *U.S. Dept Agric., Agric. Handbook* **353**.
Sprau F. (1960). *Nematologica Suppl.* **2**, 49–55.
Stelter, H. and Meinl, G. (1969). *Biol. Zbl.* **88**, 575–82.
Stone, A.R. (1973). *Nematologica* **18**, 591–606.
Stone, A.R. (1976). *Report of Rothamsted Experimental Station for 1975*, 195–6.
Trudgill, D.L. (1967). *Nematologica* **13**, 263–72.
Trudgill, D.L. (1968). The effect of environment on sex determination in *Heterodera rostochiensis*. Ph.D. Thesis, University of London.
Trudgill, D.L., Evans, K. and Parrott, Diana M. (1975a). *Nematologica* **21**, 169–82.
Trudgill, D.L., Evans, K. and Parrott, Diana M. (1975b). *Nematologica* **21**, 183–91.
Trudgill, D.L. and Parrott, Diana M. (1972). *Nematologica* **18**, 141–8.
Trudgill, D.L. and Parrott, Diana M. (1973). *Ann. appl. Biol.* **73**, 67–75.
Trudgill, D.L., Parrott, Diana M., Evans, K. and Widdowson, F.V. (1975). *Nematologica* **21**, 281–6.
Wallace, H.R. (1974). *Nematologica* **20**, 27–33.
Webley, D. (1970). *Nematologica* **16**, 107–12.
Whitehead, A.G. (1973). *Ann. appl. Biol.* **75**, 439–53.
Whitehead, A.G., Greet, D.N. and Fraser, Janet E. (1966). *Proc. Third Br. Insecticide and Fungicide Conf., Brighton, 1965*, pp. 100–5.
Whitehead, A.G., Tite, D.J. and Fraser, Janet E. (1973). *Ann. appl. Biol.* **73**, 325–8.
Winfield, A.L. (1965). *N.A.A.S. Q. Rev.* **67**, 110–17.
Winfield, A.L. and Cooke, D.A. (1975). The ecology of *Trichodorus*. In *Nematode Vectors of Plant Viruses*, eds. Lamberti, F., Taylor, C.E. and Seinhorst, J.W. pp. 309–41. London: Plenum.
Winslow, R.D. and Willis, R.J. (1972). In *Economic Nematology*, ed. Webster, J.M. pp. 17–48. London: Academic Press.

Chapter 12

Pest aspects of potato production

Part 2. Pests other than nematodes

R.W. GIBSON

12.5 Introduction

Many pests damage the cultivated potato; Dudley *et al.* (1952) list over a hundred species in N. America alone. Most are either nematodes (discussed previously), insects or molluscs, although there are a few within other taxonomic groups. The major pests are insects especially of the orders Hemiptera, Coleoptera and Lepidoptera and are listed in Table 12.5.

Several major pests species including the peach-potato aphid *Myzus persicae* (Sulzer), the potato aphid *Macrosiphum euphorbiae* (Thomas), and the potato tuber-worm *Phthorimaea* (= *Gnorimoschema*) *operculella* (Zeller) are world-wide in distribution. Others are pests on at least two continents, for example, the aphid *Aphis nasturtii* Kaltenbach, the Colorado beetle *Leptinotarsa decemlineata* Say, flea-beetles (*Empoasca* spp.) and the lady-beetles (*Epilachna* spp.). The wireworms (Elateridae), as a group, are important world-wide but different genera are present in different regions.

It is obviously difficult to estimate yield losses caused by pests, but Cramer (1967) has estimated that they directly cause losses of 24.2×10^6 t of potatoes, amounting to 9.7% of the world crop, and, as vectors, they contribute to the estimated 90.3×10^6 t lost through disease. Potato pests cause damage in a number of ways. Some, such as Colorado beetles and lady-beetles defoliate the plant, whereas the hemipteran pests (mainly aphids and leaf hoppers) remove sap and may inject toxic saliva, but both lessen tuber yield. Others, such as wire-worms, eat into tubers considerably decreasing the value of the crop but perhaps only slightly decreasing yield. Some pests, notably Hemiptera, also transmit viruses which cause disease in the year of infection and in subsequent crops

Table 12.5 Major pests of the potato crop

Common name	*Scientific name*	*Main area where it is a pest*	*Whether a pest on non-solanaceous crops*	*Whether common on non-solanaceous wild plants*
Peach-potato aphid	*Myzus persicae*	World-wide	Yes	Yes
Potato aphid	*Macrosiphum euphorbiae*	World-wide	Yes	Yes
Buckthorn aphid	*Aphis nasturtii*	Europe and N. America	No	Yes
Glasshouse-potato aphid	*Aulacorthum solani*	World-wide		Yes
Potato flea-beetle	*Epitrix cucumeris*	America	Yes	Yes
Potato leaf hopper	*Empoasca fabae*	America	Yes	Yes
Colorado beetle	*Leptinotarsa decemlineata*	America and Europe	No	No
Potato tuberworm	*Phthorimaea operculella*	World-wide warm-temperate and tropical regions	No	?
Lady-beetle	*Epilachna* spp. especially *E. vigintioctopunctata*	Australasia	Yes	?
Wireworms	(Family Elateridae)	World-wide group but geographically-restricted species.	Yes	Yes
Slugs	(Several genera)	World-wide group but geographically-restricted species.	Yes	Yes

grown from tubers of infected plants (see Chapter 11). Such diseases are costly since they either necessitate frequent purchase of uninfected seed tubers or the perennial acceptance of smaller yields. Although the potato is a cosmopolitan crop, about 87% of world production is concentrated in Europe and the U.S.S.R., especially in the belt comprising Britain, France, Germany, Czechoslovakia, Poland and eastwards into the U.S.S.R. In this region, aphids, Colorado beetles, wireworms and, in the U.S.S.R., lady-beetles are important pests.

In England and Wales, where neither Colorado beetles nor lady-beetles are pests, Strickland (1965) estimated yield loss in maincrop and second-early potatoes, in terms of extra land required to compensate for pest damage, as:

Wireworms		648 ha
Slugs		1 336 ha
Aphids:	direct effect of virus diseases	1 500 ha
	extra cost of fresh seed	12 432 ha
Total hectare equivalent loss		15 916 ha
Area grown in 1963		180 630 ha

These estimates do not include the cost of measures to exclude Colorado beetles but they demonstrate the cost of the production of seed tubers in areas separate from the main crop and where aphid-borne viruses spread slowly. Thus, in Britain, over 80% of ware crops are grown in England and Wales but less than 20% of certified seed is grown there; seed potato being produced mainly in Scotland and N. Ireland (Potato Marketing Board, 1974). Cramer (1967) estimated that excluding losses caused by virus diseases, insect pests directly cause a loss of about 3% of production in Britain, and about 5% in the rest of Europe, where other pests are present.

The potato crop was introduced to Europe from S. America during the sixteenth century and was grown quite extensively by the seventeenth century. Perhaps surprisingly, potato pests have not spread with the potato; the majority were already present in the habitats into which man has introduced the potato (the nematodes, see Chapter 12, Part 1 appear to be the chief exception) although a few pest species, such as the Colorado beetle (Section 12.7.3), have then spread via the agency of the crop. As a result, pest species usually have at least one indigenous host and are not entirely dependent for survival on the potato crop. The wide geographical range of such pests as the potato leaf hopper, the flea-beetle and the aphids *M. persicae* and *M. euphorbiae* is not related to the spread of the potato crop; these insects are all pests on several other crop plants.

12.6 Feeding habits

Pest species are generally polyphagous or, if restricted to solanaceous plants, feed on species in at least two genera. Polyphagous species usually do not require 'secondary plant substances' to stimulate feeding but instead host plants may be accepted on the basis of the presence of nutritive substances and the absence of toxins or feeding deterrents (Fraenkel, 1969). *M. persicae*, *M. euphorbiae* and even the less polyphagous Colorado beetle and tuberworm all feed readily on diets containing only simple nutritive chemicals such as sugars and amino-acids (Mittler and Dadd, 1962; Cartier and Morin, 1965; Ritter, 1967; Meisner *et al.*, 1974).

12.6.1 Leaf-eating pests

Within the Coleoptera, the Colorado beetle, flea-beetle and lady-beetle are important leaf-eating pests. Both adult and larval Colorado beetles and lady-beetles feed on the foliage, those of the latter skeletonizing the leaflets (Anonymous, 1964). Colorado beetles will eat the entire leaf, even biting into the petiole and stem and a few larvae eat large amounts of plant tissue (Fig. 12.9). At Ottawa, Canada, it was found that a Colorado

Fig. 12.9 2 potato plants, 9 days after 20 Colorado beetle larvae had hatched on each plant.

beetle eats in its larval and adult stages about 4.23 g of leaf per season (Gibson *et al.*, 1925), assuming a normal uncontrolled population of 2.5×10^6 beetles ha^{-1}, Hirst (1975) calculated that 1700 kg of leaves ha^{-1} would be eaten by each generation. 20 to 30 adult lady-beetles or 30 to 40 larvae per plant halved tuber yield of infested plants (Shabliorskii and Chebanov, 1962). Lady-beetles pupate on the foliage but Colorado beetles pupate underground so pupae may be unaffected by insecticide applied to the leaves. The adults also over-winter underground in a state of diapause; in northern latitudes of Europe and America, there is only one generation each year but in warmer, southern latitudes, several generations occur.

Only adult flea-beetles feed on potato foliage, making small round holes in the leaves. The eggs are laid in cracks in the soil and the larvae feed on roots and (especially those of *E. tuberis* Gent.) tubers. There may be more than one generation per year in warm climates and adults overwinter either underground or under herbage. This insect is widespread in N. America and is a pest on several crops (general references: Dudley *et al.*, 1952; Morrison, *et al.*, 1967).

The potato tuberworm and cutworms are important lepidopteran potato pests and, as with almost all the Lepidoptera, only the larvae cause damage. First instar tuberworms (*P. operculella*) feed within leaves, forming mines between the upper and lower epidermis, but later instars may tunnel into the petiole and stem. Larvae also tunnel into tubers exposed above ground, causing much damage (Graf, 1917), and infestations on poorly-stored or un-harvested tubers provide larval overwintering sites (Bacon, 1960; Broodryk, 1971).

Cutworms are larval noctuid moths and their name is derived from their habit of biting through the stems of young plants near ground level, so a few larvae can do much damage. There are several species: *Agrotis spinnifera* Hb. feeds only on the foliage, others, such as *A. ypsilon* Rott. also feed on the tubers (general references: Dudley *et al.*, 1952; Srivastava and Khan, 1962; Hodgeson *et al.*, 1974).

There are several N. American species of grasshopper which eat the leaves of potato crops (Dudley *et al.*, 1952) and leaf-mining flies (*Liriomyza* spp.) are occasional pests in both N. and S. America (Arellano, 1963; Landis *et al.*, 1967).

Leaf-eating pests have occasionally been implicated as vectors of viruses and bacteria. Thus Goss (1931) considered that flea-beetles, Colorado beetles and grasshoppers may transmit spindle tuber viroid and unmottled curly dwarf virus. Colorado beetles and flea-beetles may also be vectors of the bacterium causing brown rot (Dudley *et al.*, 1952). In addition to decreasing the leaf area available for photosynthesis, leaf-eating pests may upset the balanced development of the plant, for example, by preventing the re-cycling of nutrients from eaten leaves.

Surprisingly, Ladd (1963) had to remove more than half the leaf area of a plant before yield was significantly restricted and Prasad (1960) also found that the removal of foliage by increasing numbers of flea-beetles had little effect until more than 50% of foliage was dead or devoured, after which yield declined sharply. Dyson (1965) also found that decreasing the leaf area of plants by 40% by removing leaflets from each leaf at the time of tuber initiation did not decrease tuber yield and suggested that the sink strength of the tubers and not the supply of photosynthate may be limiting tuber growth.

12.6.2 Sap-feeding pests

All the major sap-feeding pests are in the insect sub-order, Homoptera. They have elongated mouthparts forming a tube composed of paired mandibular and maxillary stylets with a central food canal and a separate salivary canal (Figs. 12.10–12.12). This tube is inserted into the plant, usually to reach and withdraw sap from the phloem. Sap removal is, by itself, damaging to plants, but toxic saliva and infectious virus particles may also be injected during feeding.

Of the Homoptera, the aphids (Aphididae) are by far the most damaging. The cosmopolitan aphid, *Myzus persicae*, is sufficiently important a pest on potatoes and other crops to warrant its own review (van Emden, *et al.*, 1969) but *Macrosiphum euphorbiae*, *Aphis nasturtii*, and *Aulacorthum solani* are also major pests. These species all feed on the foliage but *Rhopalosiphoninus latysiphon* feeds on the underground portions of the stem and roots (Gibson, 1971a). In the complete, or holocyclic lifecycle, eggs are laid in autumn on the 'primary', often a woody, host plant.

Fig. 12.10 The aphid *M. euphorbiae* feeding in a potato leaf vein

Fig. 12.11 The stylets of *M. euphorbiae* protruding from the rostrum and penetrating a leaf vein

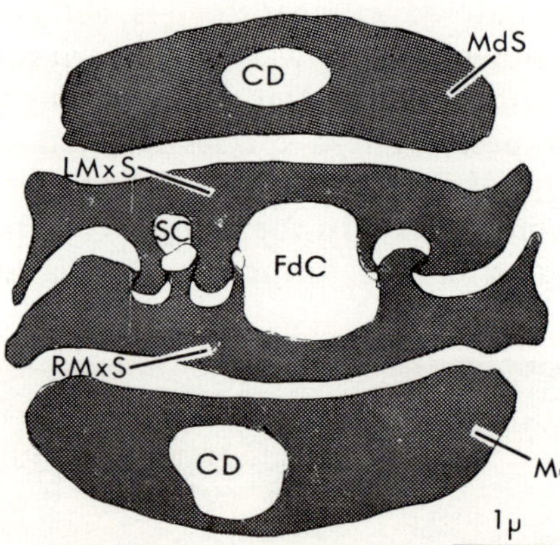

Fig. 12.12 A cross-section of the stylet of the aphid *M. persicae*. CD, central duct; FdC, food canal; SC, salivary canal; L Mx S and R Mx S, left and right maxillary stylets; Md S, mandibular styles. (From Forbes, 1969)

These eggs are produced by sexual reproduction whereas all subsequent reproduction is viviparous and parthenogenetic. Larvae hatch from the eggs in spring, feed on the primary host, mature and produce offspring asexually. There may be several generations on the primary host, but eventually winged adults develop which fly away to colonize 'secondary', often herbaceous, host plants. Parthenogenetic reproduction may continue for several generations on the secondary host plants until the autumn, when winged adults fly back to the primary host. On the secondary host plants, adults may be either winged (alate) or wingless (apterous); a greater proportion of alates usually develop in crowded conditions and these may fly away, sometimes being blown several hundred kilometres at high altitudes, to colonize more secondary host plants. *Rhamnus* spp. (buckthorn) *Rosa* spp. and *Prunus* spp. (e.g. peach trees) are respective primary host plants of *A. nasturtii*, *M. euphorbiae* and *M. persicae* and, as well as potatoes, other crop plants and a variety of annual weeds, are their common secondary host plants. Sometimes, the life cycle is incomplete (anholocyclic); for example, there are biotypes of *M. persicae* which breed parthenogenetically throughout the year overwintering on stored, sprouting potatoes, on hardy herbaceous plants or on glasshouse crops.

The non-requirement of a mate aids the rapid increase of low density populations such as might be present after alates colonize a crop, and the absence of the egg stage may accelerate population growth by shortening generation time and avoiding egg mortality. Dispersal to fresh secondary host plants, either of the same or different species, continues throughout summer, providing the new colonies with a temporary respite from disease, parasites and predators (general reference: van Emden, 1972).

Leaf hoppers (Jassidae) are another important group of hemipteran pests, most reports of damage coming from N. America, although there are reports from Egypt (Abdel-Salam *et al.*, 1972). As with aphids, long-distance wind-borne dispersal may be an important feature of their phenology. Thus, the potato leaf hopper *Empoasca fabae* cannot survive the winter in the northern and central states of N. America (Decker and Cunningham, 1967) but these areas are recolonized each spring by adults blown, sometimes at altitudes of 0.8 km or more, from warmer southern states (Pienkowski and Medler, 1964). The six-spotted leaf hopper, *Macrosteles fascifrons* is another pest which overwinters in southern states of N. America (Simpson *et al.*, 1964). Leaf hoppers are oviparous, *E. fabae* usually laying its eggs in the stem or leaf veins of potato plants. The eggs hatch into larvae resembling adults except in being wingless; there are several generations each year. *E. fabae* is polyphagous and is also a pest on clover and lucerne (general references: Painter, 1951; Dudley *et al.*, 1952; Landis *et al.*, 1967; Delong, 1971).

The potato psyllid *Paratrioza cockerelli* (Sulc.) is another homopteran pest of potato plants in N. America. Again, this pest is polyphagous and its population may build up in spring on wild plants before moving to the potato crop. The winged adults are the dispersive instar and have been trapped at altitudes of over 1 km; the larvae are sedentary (general references: Daniels, 1937; Wallis, 1946; Pletsch, 1947). Whitefly (Aleyrodidae) are occasionally pests on potato plants and they have a similar life cycle to the potato psyllid, eggs hatching into sedentary larvae and finally developing into delicate, white-winged adults.

An occasional non-homopteran pest of potatoes is the cosmopolitan and polyphagous spider mite, *Tetranychus urticae* (L). Like Homoptera, this mite has a tube, called the stylophore for imbibing plant sap. Although wingless, the mites can be dispersed by wind. This species may become a pest after insecticides have selectively killed their predators (Landis *et al.*, 1967) (general reference: Evans *et al.*, 1961).

Damage caused by sap-feeding pests

Sap-feeding pests damage plants by removing sap, by injecting toxic saliva and by infecting with viruses. Some physical damage may also be caused by stylet penetration, and honeydew deposits may become colonized with sooty moulds. It is usually impossible to distinguish between the damage caused by each of these means, although the effect of virus infection may be distinguished since symptoms are not dependent on the vector remaining on the plant.

Bradley and Ganong (1951) have reported peak populations in excess of 10 000 *M. persicae*, *M. euphorbiae* plus *A. nasturtii* per plant on potato crops in New Brunswick, Canada. In the neighbouring state of Maine, yields, were lessened when each plant was infested for a total of 20 400 aphid-days, equivalent to a steady aphid population of 272 aphids for the 75 days that crops were usually infested. For every subsequent 10-fold increase in aphid infestation, yield was lessened by rather more than 10% (Shands *et al.*, 1972b). In Britain, Taylor (1957) recorded populations of several hundred *M. persicae* and *M. euphorbiae* per plant and, in 1952, more than 1×10^4 *A. nasturtii*. Carden (1965) considered that maximum aphid populations of 5 per leaf lessen yield of maincrop potatoes by about 1.7 t ha^{-1} and 10 per leaf lessen it by about 2.5 t ha^{-1}. However, yields of infested field plants may have been curtailed by aphid-borne virus diseases as well as by aphid feeding.

The phloem sap, which is the usual food source of aphids, contains large concentrations of sugars, amino-acids and other synthesized plant products. A minimum estimate of the volume of phloem sap extracted can be obtained by measuring the frequency of aphid excretion and size of the droplets (Banks and Macaulay, 1964); Table 12.6 shows rates of excretion measured using adult *M. persicae* and *M. euphorbiae* feeding on top and bottom leaves. The results from Table 12.6 suggest that populations of several thousand aphids may, each day, be removing several millilitres of sap.

Aphids, especially *A. solani*, may also cause necrotic spots to develop at feeding sites and young infested leaves may fail to expand properly, becoming crinkled and curled. Plants on which *M. euphorbiae* have fed may develop a symptom known as top-roll, the upper leaflets becoming rolled (Fig. 12.13); in one experiment, 90% of stems in caged, infested

Table 12.6 Volume (μl) of honeydew excreted in 2 h by single adults

	Top leaves	*Bottom leaves*	*Least significant difference ($P = 0.05$)*
M. persicae	0.12	0.07	0.021
M. euphorbiae	0.50	0.16	0.122

Fig. 12.13 Potato plant (cv. Majestic) with top-roll

plots developed top-roll. These plants yielded 45% less than un-infested plants (Gibson, 1974a), and the rolled leaves of naturally-affected field plants covered less ground and photosynthesized less efficiently than those of unaffected plants (Table 12.7). Although several active chemicals have been isolated from aphid saliva, none have been proved to cause these morphogenic effects (general reference: Pollard, 1973).

The potato leaf hopper *E. fabae*, causes its own characteristic feeding damage known as hopperburn. In affected plants, the tips and margins of leaflets curl upwards, turning yellow and brown (Dudley *et al.*, 1952). Insecticides applied to control the leaf hopper increased yield over that of untreated controls by between 22 and 47% in one experiment (Wressell and Driscoll, 1964). The smaller yield of infested plants may partly be

Table 12.7 The rate of photosynthesis and the leaf cover of normal and top-roll affected plants

	With top-roll	*Without top-roll*	*Standard error of difference*
Photosynthesis ($g\ CO_2\ m^{-2}$ leaf h^{-1})	0.80	1.69	± 0.17
Leaf cover (%)	67.5	81.5	± 4.5

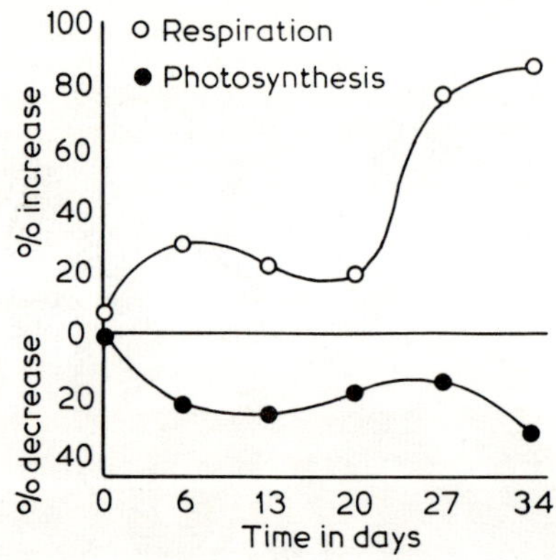

Fig. 12.14 Percentage change from the control of photosynthetic and respiratory rates of potato leaves taken after plants had each been infested with 50 adult potato leaf hoppers (from Ladd and Rawlins, 1965)

caused by phloem sap removal, but feeding by hoppers both increases respiration and decreases photosynthesis, the latter effect continuing after the hoppers have gone (Fig. 12.14). Sugar concentration is generally greater in infested leaves, perhaps because the saliva of *E. fabae* contains amylase and invertase although toxins as well as phloem disruption could be involved (general reference: Delong, 1971).

Mild symptoms of psyllid yellows, the feeding damage caused by the potato psyllid *P. cockerelli*, include leaf curling and yellowing but, when more psyllids are present, the leaves may become purple and the plant stunted. Also, small aerial tubers may be produced in addition to small subterranean ones. In 1938, in Wyoming, it was estimated that the potato crop was halved by psyllid damage (Edson and Wood, 1939). Surprisingly, although adults feed on potato plants, only the nymphs cause psyllid yellows (general reference: Pletsch, 1947; Staples, 1968).

Virus transmission by sap-feeding insects

Many potato viruses are transmitted by homopteran, especially aphid pests. They may cause some loss of yield during the first season a plant is infected, but more costly is the transference of virus to subsequent generations via infected tubers and the need to grow seed tubers separately from the main crop in areas where the insect vectors are rare (see Chapters 8 and 11). Several aphid- and hopper-borne viruses cause damage to potato crops. Potato virus Y (PVY) (general reference: Delgado-Sanchez and Grogan, 1970), and leaf roll (general reference: Peters, 1970)are common virus diseases which illustrate many of the properties of aphid-transmitted viruses.

Although there are some intermediate (semi-persistent) forms, most insect-transmitted viruses can be divided into: persistently-transmitted viruses, e.g. leaf roll and non-persistently-transmitted viruses, e.g. PVY.

As the name suggests, persistently-transmitted viruses can be transmitted for several days after the vector has fed on an infected plant and often, as with leaf roll, for the remainder of the vector's life. Vector infectivity is not lost during a moult and it may be possible to make individuals of the vector species viruliferous by injecting them with haemolymph taken from an infected specimen; leaf roll virus has been transferred from one *M. persicae* to another in this manner (Stegwee and Ponsen, 1958). Persistently-transmitted viruses have been described as circulative (Kennedy *et al.*, 1962); they are usually not readily transmitted immediately after the vector has fed on an infected plant but appear to pass through the gut wall into the haemolymph and may infect plants by being injected with the saliva. For some persistent viruses, such as potato yellow dwarf, there is evidence that the virus multiplies within its hopper vector (Chiu *et al.*, 1970) but it is not clear whether this occurs with leaf roll (Peters, 1970). Leaf roll is transmitted by at least eleven aphid species including *M. persicae*, *M. euphorbiae*, *A. nasturtii*, *A. solani* and *latysiphon* (Kennedy *et al.*, 1962).

Non-persistent viruses can be transmitted immediately after feeding but infectivity is quickly lost. The optimum feeding period for aphids to acquire PVY is from 30 s to 5 min and the aphid is immediately infectious requiring less than 60 s feeding to infect, suggesting that the virus is carried on or near the mouthparts. Infectivity is lost during moulting when the mouthparts are cast off along with the rest of the cuticle and applying virus-inactivating treatments to the tips of the stylets renders most aphids non-viruliferous although such treatments may affect their feeding (Bradley, 1964). At least sixteen aphid species, including *M. persicae*, *M. euphorbiae*, *A. gossypii*, *A. nasturtii* and *A. solani* transmit PVY (Kennedy *et al.*, 1962). Particles of non-persistent viruses may not just adhere to insect mouthparts as a contaminant; the virus/vector relationship seems to be much closer than this. Thus, not all aphids can transmit a particular virus, both vector and non-vector species even being in the same genus (Bradley, 1964). There is now evidence that a 'helper' agent is needed for virus particles to be transferred by a vector from one plant to another. One of the first indications of its presence

Table 12.8 Transmission of PVY by aphids probing supernatant component before or after virus

*First membrane**	*Second membrane*	*Experiment*	
		1	*2*
Supernatant	Virus	7/8†	8/8
Virus	Supernatant	0/8	0/8
Supernatant + virus	—	8/8	7/8

*Aphids were allowed to probe for 5–10 min on each membrane.
†Number of plants infected/number tested; ten aphids were used for each test plant.
From Govier and Kassanis (1974).

was supplied by observations on the transmission of potato aucuba mosaic virus (PAMV). This virus is transmitted non-persistently by *M. persicae*, but only when 'aided' by PVY-infected plants. Aphids transmitted PAMV if the plant was also infected with PVY or if the aphids had recently fed on a plant infected with PVY, but not by feeding only on a PAMV-infected plant (Kassanis and Govier, 1971). Later, it was discovered that *M. persicae* could not transmit PVY from sachets containing a synthetic diet together with PVY purified by centrifuging sap of infected plants but, if some of the supernatant was included or if the aphids first fed on a satchet containing supernatant, PVY could be acquired and transmitted (Table 12.8). The component in the supernatant necessary for transmission of PVY is presumably the 'helper' agent in PVY-infected plants which aids transmission of PAMV.

The behaviour and abundance of aphid vectors are, of course, important in determining the amount of virus spread but the transmission charactersitics of the virus are also important. Only alates are able to travel from field to field but long-distance migrants carrying non-persistent viruses may become non-infective by the time they alight on another potato plant; only persistent viruses such as leaf roll can be transferred long distances. Alates, especially those of *M. persicae* are probably more effective than apterae at transmitting viruses from plant to plant within a field because alates, especially the younger ones, can fly to other plants but the apterae (often more numerous) may be important in spreading virus to all the stems growing from a tuber and to adjacent plants. Apterous forms may also spread virus in stored seed potatoes, since the tubers are close together. Different species and stages of aphids may differ in their abilities to transmit a virus (Robert, 1971); in Britain, *M. euphorbiae* and especially *M. persicae* are the major vectors of leaf roll and PVY. Early virus spread is generally from infected plants within the crop, as incoming aphids are mostly from hosts other than potatoes. Weather may influence spread via aphid numbers and behaviour; warmer weather and mild winters are often associated with subsequent aphid abundance and warm calm conditions increase aphid activity. However, warm weather also directly increases virus spread since both *M. euphorbiae* and *M. persicae* are more effective vectors of leaf roll at 24 than at 6 or 12 °C (Robert and Rouzé-Jouan, 1971) (general references: Broadbent, 1953; Watson and Plumb, 1972).

12.6.3 Root- and tuber-eating pests

These are usually the larval stages of insects, the adult living above-ground, often doing no damage. Important examples are wireworms which are the larvae of click-beetles (Elateridae), larval flea-beetles (*Epitrix* spp.), tuberworms which are larval tubermoths (*P. operculella*) and cutworms (*Agrotis* spp.) which are larval noctuid moths. The other major pests are slugs (gastropod molluscs) of which both adult and immature forms are generally subterranean and feed on the potato crop.

Wireworms, as a group, cause much damage. Species of several genera are involved, including *Limonius*, *Conoderus*, *Ctenicera*, and *Agriotes*. Wireworms are polyphagous feeders, damaging roots of several crop plants, and in fields of potatoes, also eating into the tubers. Roots exude attractants to the larvae and feeding is stimulated by such substances as sugars, fat and fatty acids, peptones and trypsin-digested potato protein (Crombie and Darrah, 1947; Thorpe *et al.*, 1947). Wireworms may take several years to become adults and so are quite frequently pests 'out of place', in that the eggs may not have been laid in a potato crop, but earlier when another crop was present, Thus damage to potato crops may be related to the attractiveness of previous crops to egg-laying adults; it is often most serious when they are planted immediately after pasture (Strickland *et al.*, 1962). The Great Basin wireworm, *Ctenicera pruinina* (Horn.) of N. America provides an extreme example since it only breeds in dry soils; when this habitat is irrigated and used for growing arable crops, the larvae are pests for up to three years, ceasing to be a problem when all the larvae have pupated (Landis *et al.*, 1967).

Adult flea-beetles have already been mentioned as leaf-eating pests but the feeding of their larvae also damages roots and tubers. Larvae of *Epitrix cucumeris* feed mainly on the roots of potato plants whereas the larvae found eating into tubers are usually those of the tuber flea-beetle, *E. tuberis* (Morrison *et al.*, 1967). The cutworms have already been mentioned as pests of potato foliage but species such as *Agrotis ypsilon*, *A. interacta* Wlk. and *A. segetis* Schiff. also feed on the tubers (Dudley *et al.*, 1952; Srivastava and Khan, 1962). The tuberworm *P. operculella* is another leaf-eating pest but the larvae also tunnel into tubers incompletely covered by soil and into unprotected stored tubers.

Several species of slugs are pests and amongst those encountered in Britain are *Arion hortensis* Ferussac, *A. fasciatus* Nilsson, and *Agriolimax reticulatus* Müller (Gould, 1965). The globular eggs may be several millimetres in diameter, are gelatinous, relatively unprotected and usually laid in groups underground. Development to the adult is direct, young slugs being similar in appearance to adults but smaller and sexually immature (general reference: Janus, 1965). Slugs do not have a water-impermeable cuticle but are covered in mucus; they are restricted to humid environments such as the soil and under herbage. Slugs can burrow through soil to feed on the potato tubers before they are harvested. They are polyphagous, feeding even on dead material and they remain as a perennial population in the soil. They are not very mobile, so the amount of damage they cause is largely dependent on the size of the population in the soil before planting and the rate at which that population increases (Symonds, 1973).

Damage caused by root- and tuber-eating pests

Direct damage

Damage to the tubers usually has little effect on total yield but instead decreases saleable value. If only a small percentage of tubers is damaged, it may be worth-while removing them by hand. If many are damaged, this becomes uneconomic and the crop may have to be sold cheaply for animal food or to industry. Pests of tubers can be a serious problem; Lilly (1973), states that 'wireworms have long been a limiting factor in the commercial production of quality potatoes in irrigated districts of S. Alberta', and Carpenter and Scott (1974) reported a similar situation in Idaho. In Britain, damage by slugs and wireworms is equivalent to about 1% of production (Strickland, 1965), but in 1965, 40% of tubers of slug-susceptible varieties were damaged in a naturally-infested field experiment (Winfield *et al.*, 1967) so damage can be severe.

The tuberworm is the only important pest of stored tubers and it also damages tubers exposed above-ground in the field (Graf, 1917; Kasargode and Nagpurkar, 1920; Shorey *et al.*, 1967). However, this insect is a pest only in warm areas of the world such as California, India and parts of Australasia, which is not where the bulk of the world crop is grown.

Disease spread

Potato blackleg disease is caused by the bacterium *Erwinia carotovora* var. *atroseptica*. This disease is widespread and economically important (see Chapter 11) causing foliage death of field plants and tuber soft-rotting. The bacterium can over-winter on tubers and be spread by contact to other stored tubers. In 1926, Leach found evidence that in N. America, the seed corn maggot *Hylemyia cilicrura* Rond also spreads this disease. Eggs are laid near planted seed tubers into which the larvae tunnel. The incidence of blackleg disease was associated with seed-tuber infestation and the bacterium could be isolated from egg batches and from within surface-sterilized adults and puparia. The seed corn maggot is not a pest on potatoes in Britain and here blackleg had been assumed to result from planting infected tubers. The recent use of stem-cuttings eliminated blackleg from treated stock (see Chapter 11), but, in 1971, blackleg occasionally reappeared. Flies (genus *Drosophila*) found in the crop were contaminated with the blackleg bacterium (Graham and Hardie, 1971) and it has now been shown that *D. melanogaster* Meig. can carry the bacterium both internally and externally and can infect wounded plants (Molina *et al.*, 1974). As the use of stocks derived from stem-cuttings lessen the chance of transmission through the tuber, so the significance and importance of dipteran vectors of *E. carotovora* var. *atroseptica*, and perhaps of other bacteria, may increase.

Aphids may be pests on stored, sprouting, seed potatoes, and can spread virus from tuber to tuber; virus spread is dealt with in Section 12.6.2.

12.7 Means of decreasing pest damage

12.7.1 Chemical insecticides

Modern organic insecticides have revolutionized pest control and complemented the increased use of fungicides and fertilizers. McNew (1963) considered that the use of DDT and a more effective fungicide (NABAM) played a major role in enabling yield ha^{-1} in the U.S.A. to nearly double in the nine years after their introduction. Although the use of DDT is now restricted, the percentage of crops treated with insecticide is still increasing (Fig. 12.15) and new, sometimes more specific, insecticides have been developed, especially in the carbamate and organophosphorous groups.

Molluscides are one of the few chemical treatments usually formulated with a food bait. In contrast most insecticides are applied to the crop either as granules, sprays or dusts usually making the crop itself poisonous. Insecticides may be subdivided into systemic insecticides which are

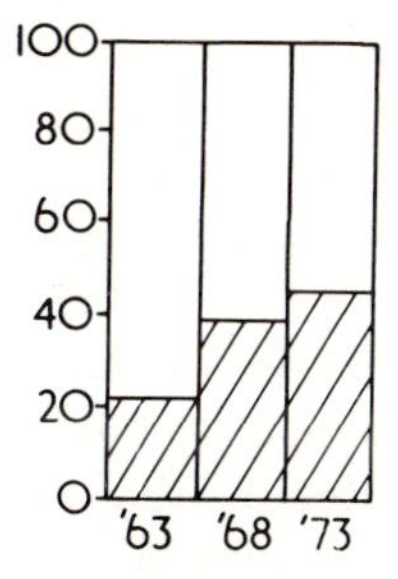

Fig. 12.15 The percentage of British crops treated with insecticides (Potato Marketing Board, 1975)

absorbed into the plant and non-systemic insecticides which remain on the plant surface. The systemic insecticides are selected for short persistence, because they must not remain to contaminate the tubers after harvest, non-systemic insecticides may be more persistent. Some systemic insecticides (e.g. phorate) are formulated as granules which can be incorporated into the soil during planting and, after their absorption by the crop, can kill phytophagous insects for perhaps two or more months. This may avoid or delay the need to treat growing crops, saving labour and avoiding crop damage. The systemic action of insecticides protects new growth and can compensate for incomplete cover by sprays whereas non-systemic insecticides do not protect new growth and repeated applications may be required to maintain pest control. Soil applied systemic insecticides may have little effect on predators and parasites of foliage-infesting pests (Reynolds and Metcalf, 1962), but they can have a deleterious effect on beneficial soil-inhabiting animals such as earthworms (Edwards *et al.*, 1967). Applications of contact insecticides to the foliage may kill predators and result in spectacular increases in pest populations. Carbaryl (Sevin) is so effective at killing predators that it has been used to increase aphid numbers in field experiments (Radcliffe and Lauer, 1971).

Some insecticides can control a range of pests. Thus, granules of phorate applied to the soil have, for example, lessened damage by aphids, wireworms, leaf hoppers, Colorado beetles, flea-beetles, psyllids and thrips. This insecticide may also limit virus spread since some pests controlled are also vectors.

The effectiveness of an insecticide in limiting virus spread is much affected by the transmission characteristics of the virus and whether the source of infection is within or outside the treated crop. This is illustrated in results by Hille Ris Lambers *et al.* (1953). The systemic aphicide Systox had little effect on the spread of the non-persistent virus, potato virus Y, which requires only short acquisition and inoculation feeds and has no intervening latent period, the aphids surviving long enough to spread the virus. Thus, spraying with Systox did not prevent PVY spreading from infected plants to previously uninfected plants (Table 12.9, treatment 2). In contrast, persistent viruses such as leaf roll require long acquisition feeds by aphids and have a latent period before they are infective. Spraying infected plants with Systox can kill aphids before they are infective lessening the spread of leaf roll to adjacent plants (Table 12.9, treatment 5), whereas in the absence of insecticide, leaf roll quickly spread to adjacent plants (Table 12.9, treatment 6). Systox did not completely prevent infection by either leaf roll or PVY even when there were no adjacent infector plants (Table 12.9, treatments 1 and 4), presumably because viruliferous aphids migrating into the crop were not killed before they had transmitted. However, applying oil to

Table 12.9 Effect of an insecticide (Systox) and proximity to a virus source on virus spread

Treatment number	*Chemical treatment (every 10 days)*	*Middle 'infector' row*	*Test rows listed in order of proximity to infector rows*									
			1	*2*	*3*	*4*	*5*	*6*	*7*	*8*	*9*	
			(a) % virus Y infected plant									
1	Systox spray	Uninfected	20	19	11	14	19	15	16	25	14	16.9
2	Systox spray	100% virus Y	87	83	79	58	56	48	49	42	42	60.4
3	Water spray	Uninfected	7	13	21	18	14	15	13	12	11	13.8
			(b) % leaf-roll infected plants									
4	Systox spray	A few leaf-roll infected plants	11	14	17	20	18	18	21	20	21	17.8
5	Systox spray	100% leaf-roll infected	18	25	18	25	25	21	20	27	22	22.2
6	Water spray	100% leaf-roll infected	74	49	30	28	21	24	25	19	37	34.1

Adapted from Hille Ris Lambers *et al.* (1953).

the foliage of potato plants has impeded the transmission and spread of PVY in glasshouse and field experiments with little sign of phytotoxicity and only slight losses in yield (Bradley, 1963; Bradley *et al.*, 1966).

The use of insecticides has led to the development of tolerant strains of insects requiring more insecticide than usual to be lethal. Insecticide tolerance, for example, has been found in Colorado beetles (Hurkova, 1968), flea-beetles (Kring, 1958), tuberworms (Champ and Shepherd, 1965), *M. persicae* (Fellowes and Ferguson, 1974), and the southern potato wireworm *Conodera falli* (Workman, 1964). Tolerance might be combatted by adding inhibitors of insect detoxification systems (Wilkinson, 1973) or by using other insecticides such as the synthetic pyrethroids (Elliott *et al.*, 1974). Insecticides in common usage exert at least part of their effect via the nervous system. However, insecticides not relying on this mode of action are being developed and tested, for example, chitin synthesis inhibitors (Turnipseed *et al.*, 1974), chemicals with insect juvenile hormone activity can upset development (Benskin and Perron, 1973; Tamaki, 1973), and chemosterilants which, although not killing the adult, lessen their fecundity (Chawla *et al.*, 1973). It is important to keep the problem of pest tolerance of insecticides in perspective; pests can also surmount other practices used to suppress their populations. Insect strains which can colonize 'resistant' varieties have evolved (Nielson *et al.*, 1970) and the co-evolution of pest and introduced parasite can result in the pest being less effectively controlled by the parasite than when it was initially introduced (Olson and Pimentel, 1974).

12.7.2 Resistance and tolerance

Resistance to, or tolerance of, pests can provide one of the cheapest answers to preventing damage. The large number of wild tuberous *Solanum* species and cultivars of *S. tuberosum tuberosum* and *S. tuberosum andigena*, provide a diverse gene pool containing sources of resistance to many potato pests (Painter, 1951). Despite this, only a few widely-grown cultivars (e.g. Sequoia) are particularly resistant to even one of the common potato pests, although some are resistant to certain insect-transmitted viruses (see Chapter 15). The availability of insecticides has decreased the need for resistant cultivars but there are other probable reasons for their absence:

(1) Resistance or tolerance to one pest may achieve little, if insecticide has to be applied to prevent other species causing damage; resistance against several pests may be required to achieve a reduction in insecticide usage.

(2) Resistance or tolerance to one pest may be inherited independently of resistance to another, hampering breeding for a cultivar with resistance to several pests.

(3) Susceptibility may be linked with agronomically valuable crop characteristics.

(4) Potato breeding is usually done where insects which may transmit viruses to valuable plant stocks are few or absent, so pest susceptibility may be less obvious to the plant breeder.

Resistance to leaf hoppers and flea-beetles has been found in various wild *Solanum* spp. (Sleesman, 1940; Ephraim and Radcliffe, 1968) and also in *S. tuberosum* cultivars, but at least in the latter resistance to the one pest has been associated with susceptibility to the other (Sanford and Sleesman, 1970). Certain *S. tuberosum* cultivars are more tolerant than others of large leaf hopper populations and Hibbs *et al.* (1964) have shown that the increase in plant sap sugar concentration caused by the feeding of the potato leaf hopper is less in a tolerant cultivar than in a susceptible one.

Several alkaloids including solanine, solanidine, demissidine and leptine I deter the leaf hopper from feeding on artificial diets (Dahlman and Hibbs, 1967) but in the Solanaceae, alkaloids may be synthesized in the roots and transported to the foliage via the xylem (Bollard, 1960) little or none being present in the phloem sap (Eschrich, 1970), the main food supply of the leaf hopper (Smith and Poos, 1931). Phloem-feeding aphids may similarly avoid alkaloids (Guthrie *et al.*, 1962), but the Colorado beetle, which eats the whole leaf, cannot. Not surprisingly, the type and concentration of alkaloid may determine whether a plant is a suitable host for the latter. The cultivated potato contains the alkaloid solanine which has little adverse effect on the Colorado beetle even up to 50 times normal leaf concentration but *Lycopersicon esculentum* (tomato) and *S. demissum* contain large concentrations of alkaloids of a type (respectively tomatine and demissidine) which deter feeding and confer resistance (Kuhn and Low, 1955). Hsiao and Fraenkel (1968) list many plants containing alkaloids or other secondary plant substances confering resistance. However, large concentrations of alkaloids in the foliage may be correlated with large concentrations in the tubers (Deahl *et al.*, 1973), making the tubers unsuitable for human consumption.

Aphid abundance may be affected by the availability of such food substances as sugars and amino-acids, and susceptibility of *S. tuberosum* cultivars has often been associated with such factors as age, maturity date of the variety, and soil fertility (Bald *et al.*, 1946; Taylor, 1957; Harrewijn, 1970). In wild potato species, resistance to *M. persicae* is more common in members of the series Pinnatisecta and Bulbocastana and resistance to *M. euphorbiae* is more common in the series Longipedicellata and Bulbocastana, although several species with resistance to one or both aphid species occur in other *Solanum* series. This aphid resistance is often retained to a considerable extent in hybrids with *S. tuberosum*. Often, *Solanum* species resistant to *M. persicae* are also resistant to *M. euphorbiae*, but the presence of exceptions suggests there is more than one basis for resistance. Resistance to the leaf hopper *E. fabae* is negatively correlated with resistance to *M. euphorbiae* and not significantly correlated with resistance to *M. persicae* (Radcliffe and Lauer, 1970, 1971). The resistance of at least three wild *Solanum* spp. to *M. persicae* and *M. euphorbiae* was lessened by detaching leaves (Gibson, 1971c) so resistance in at least these species is affected by the physiological state of the plant. Resistance to aphids seems not always to depend on the suitability of the phloem sap as food. Adams (personal communication) using an electronic method of monitoring aphid feeding (McLean and Weight, 1968), has been able to select plants which are generally rejected by *M. persicae* after only superficial probes. What causes rejection is unknown but, as well as conferring resistance, it may restrict the spread of persistent viruses which require long acquisition and inoculation feeds. However, it is possible that making aphids more restless might increase the spread of non-persistent viruses unless aphid activity is considerably lessened by the suppression of aphid populations.

A further source of resistance is that based on four-lobed glandular hairs which, when touched, rupture and release a sticky substance which gradually sets (Gibson, 1971d). Although hairs with this ability do not seem to occur on the cultivated potato, they are present on leaves and stems of *S. polyadenium*, *S. berthaultii* and *S. tarijense*. *S. polyadenium* is resistant to the aphid, *M. persicae* (Adams, 1946), to the Colorado beetle (Torka, 1950) and to the potato flea-beetle and leaf hopper (Sleesman, 1940). The sticky material released from the four-lobed hairs on *S. polyadenium* glues aphids to the plant (Gibson, 1971d) (Figs. 12.16

Fig. 12.16 The foot of an aphid stuck to glandular hairs of *S. tuberosum* × *S. berthaultii*. There is an intact four-lobed hair to the left of the aphid's foot

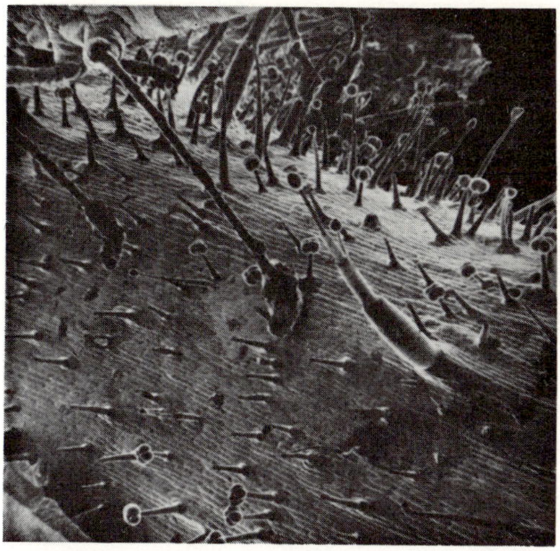

Fig. 12.17 An aphid stuck to the stem of *S. tuberosum* × *berthaultii*. x 50.

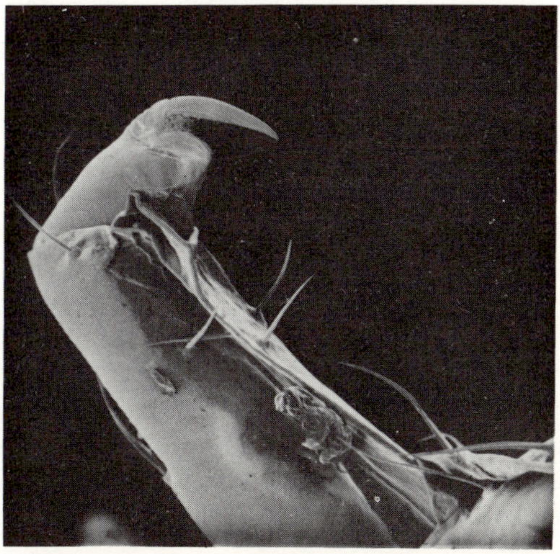

Fig. 12.18 The foot of a Colorado beetle larva which had fed on *S. tuberosum* Note the large terminal claw used for gripping the plant. x 40.

and 12.17) and either glues larval Colorado beetles to the plant or so encases their feet that, unable to grip, they fall off (Gibson, 1976) (Figs. 12.18 and 12.19). This provided considerable protection against both insects, 33% of colonizing aphids being stuck in 24 h and all larval Colorado beetles being stuck or falling off within 8 days; in both insects, affected specimens quickly died. Although there is no direct evidence that glandular hairs provide resistance to leaf hoppers and flea-beetles,

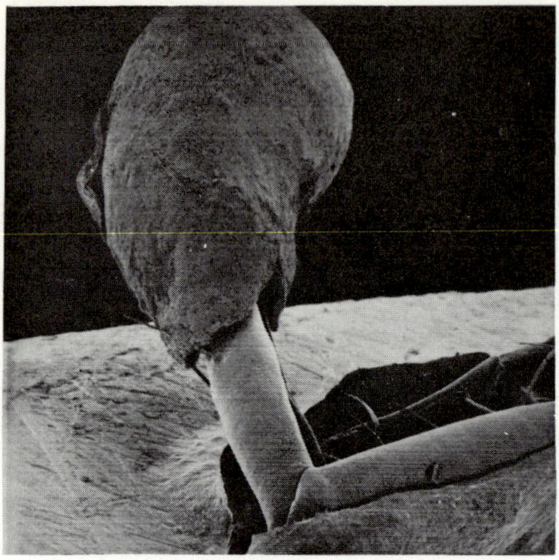

Fig. 12.19 The foot of a Colorado beetle larva which had fed on *S. polyadenium* Note the large enveloping mass of glandular material. x 70

susceptibility of *S. polyadenium* to the latter was increased by washing the leaves (Sleesman, 1940); this may have discharged most of the hairs. *S. berthaultii* has been crossed with the cultivated potato producing hybrid progeny with aphid-trapping glandular hairs (Gibson, 1974b); also, hybrids of *S. polyadenium* with the cultivated potato are maintained in the U.S.A. (Ross and Rowe, 1966). If this resistance was bred into the cultivated potato, it might restrict feeding damage by a variety of pests and also lessen virus spread, since trapped vectors cannot move to another plant.

Breeding may also contribute to protecting the tubers. For example, certain cultivars have been found to be less likely than others to be eaten by wireworms (Telford, 1942) and ones with tubers which develop deep underground are protected from colonization of the tubers by the tuberworm (Bacon, 1960).

12.7.3 Eradication and exclusion

Several potato pests, notably the tuberworm and most aphids, are already widely distributed and probably occupy much of their potential range. However, others are geographically restricted and, to combat their spread, many countries have regulations governing the import of insects and insect-infested cargoes. The Colorado beetle provides an example of the danger that exotic species may constitute and the measures that can be taken to prevent spread.

The original habitat of the beetle was the foothills of the Rocky Mountains in Colorado, where the buffalo bur, *Solanum rostratum*

Dunol is still a favourite host plant and on which the beetle gave no indication of its potential as a pest. It became a pest on potatoes when the area occupied by this crop was extended by European settlers to this region. In 1850 it was found in crops in Nebraska, by 1865 it had crossed the Mississippi and by 1874 it had reached the Atlantic and was spreading north into Canada. By 1877 it was found in Germany and Holland, having crossed the Atlantic on ships, but it was eradicated. However, in 1922, it colonized the Bordeaux area of France and eradication measures failed. By 1940 it had spread south into Spain and north-east into Germany, by 1945 it had reached Czechoslovakia and by 1949 Russia although it did not become well-established until the late 1950s. It had also reached Yugoslavia by 1946 and has now spread into Asiatic Turkey (general references: McIntosh, 1927; Thomas, 1952; Janson, 1961, 1971).

Despite spreading several thousand kilometres across Europe and into Asia, the Colorado beetle has failed to become established in the British Isles. It seems to be unable to fly across the English Channel (34 km wide) but it can be carried with cross-Channel traffic. There are restrictions on the import of potatoes and other agricultural and horticultural products, especially during the spring, summer and autumn and infested cargoes are refused entry unless treated with insecticides. However, the large amount of cross-Channel traffic makes it inevitable that some beetles, including fertilized females, enter Britain undetected and the aid of the public to find such beetles has been sought by advertising campaigns, especially by posters (Fig. 12.20). Even so, breeding colonies have been found on British crops in several years including 1901, 1933, and there was a serious outbreak directly after World War II which was finally eradicated in 1952.

In the last outbreak (1946–1952), all beetles, larvae and eggs were hand-collected, the haulm treated with insecticide, and the soil injected with carbon disulphide and dusted with DDT to destroy pupae. Nearby areas were also searched and, if nothing found, treated with DDT. In 1947, 2000 ha of potato crops were searched (at an average of 1.4 ha per man-day) and 4000 ha were treated with insecticide as a precaution (Gimingham and Thomas, 1948). In the subsequent 3 years, 10 000 ha or more were treated annually with insecticide as a precaution against infestation, some being areas previously found to be infested and then replanted with potatoes to lessen the risk of dispersal of any overwintered adults. The absence since 1952 of recolonization from the Continent has probably been aided by the effective use of insecticides to control the pest on the Continent and the widespread use in Britain of aphicides, many of which also kill the beetle.

It is difficult to assess how large a threat the Colorado beetle is to potato crops in Britain. The success of the eradication campaign suggests

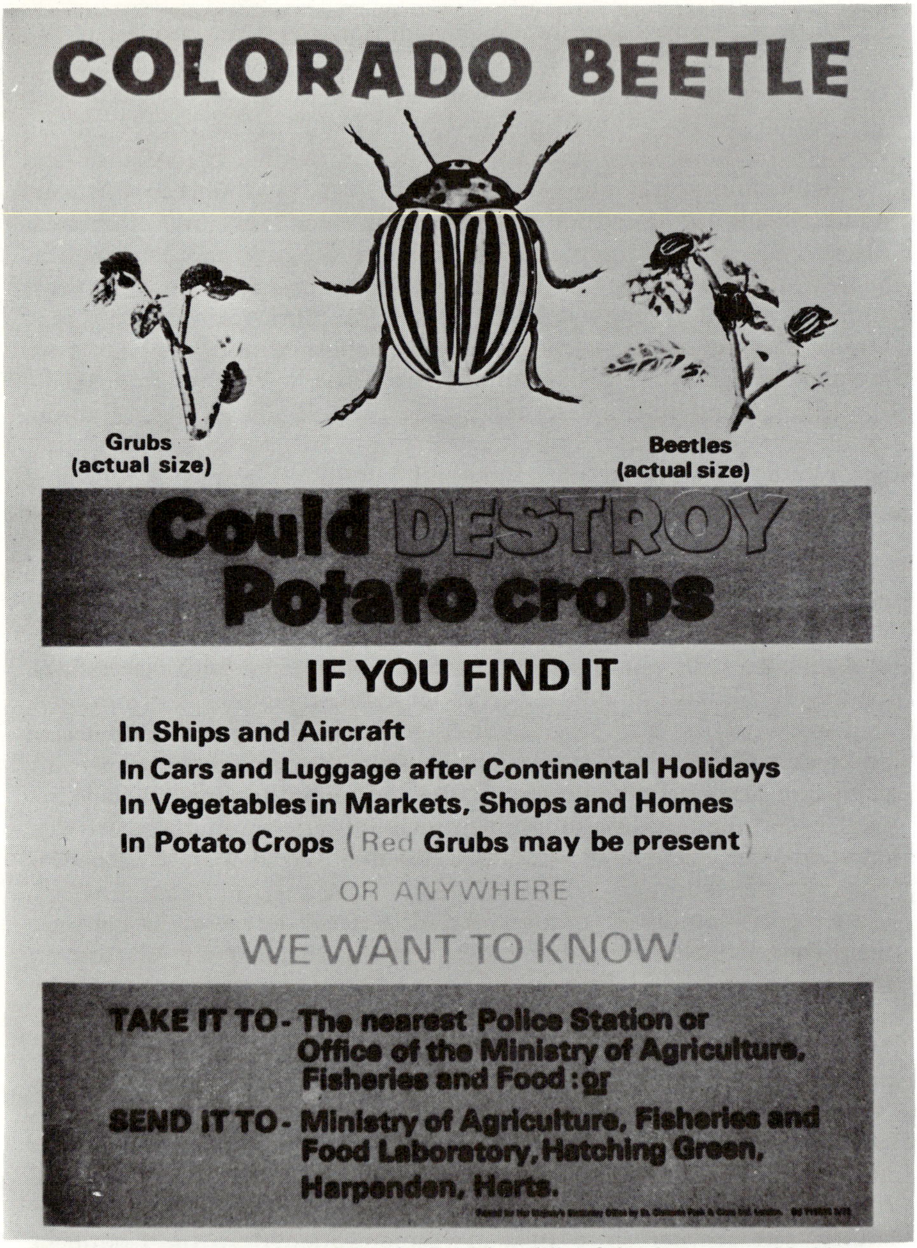

Fig. 12.20 A poster used to advise the British public of the danger provided by the Colorado beetle

that British conditions may not be ideal for it, but clearly the beetle can breed here. Modern insecticides can control this pest, but the measures taken to prevent its establishment are probably much cheaper than the cost of insecticide treatment if it ever became established. Even during the last outbreak, the annual cost of exclusion and eradication measures has been estimated as being £55 000 which is small compared to the value of the crop at that time (about £69 million) (Ordish, 1952).

12.7.4 Cultural control measures

Cultural control measures have usually been of value only against soil-infesting pests and are completed before the potatoes are planted. Such measures cannot be taken against foliage-infesting pests, all of which colonize the growing crop. Time of planting and ridging can affect aphid abundance (Shands *et al.*, 1972a) but, at this stage, crop husbandry is primarily concerned with the direct requirements of the crop and pest damage is usually restricted by the use of insecticides. Controlling aphids and leaf hoppers before they colonize the potato crop is difficult because their winter host plants are usually widespread and abundant, and the insects can disperse several hundred kilometres. However, despite the mobility of vectors, virus spread has been successfully limited by crop hygiene co-ordinated over a large area (Simpson *et al.*, 1964).

Damage to tubers is usually caused by a pest present at the time of planting. There may be little augmentation of the population from outside the crop, so damage can be lessened by decreasing the pest population before planting or by avoiding planting a crop where the pest is abundant such as after a previous potato crop. Where crops are rotated, potatoes may still need to be avoided after certain other crops; thus wireworms are often damaging in crops planted after grass (Strickland *et al.*, 1962), although they are usually rare following crops of lucerne *Medicago sativa* L. (Gibson, *et al.*, 1958). Kring (1963) has shown that if green rye is ploughed under, this can drastically decrease wireworm populations, apparently because the wireworms become infected with *Pseudomonas fluorescens* Migula, a bacterium associated with rotting rye. In irrigated areas, allowing the soil to dry out may decrease wireworm populations and flooding may decrease centipede populations (Dudley *et al.*, 1952). There are simple and cheap methods of determining some pest populations and these can help a farmer to arrange his crops to minimize damage. Thus, wireworm populations can be determined by sieving soil samples (Onsager, 1974) and slug activity can be assessed by observing the amount of damage to tubers planted out early in spring before the crop itself is planted (Symonds, 1973).

Damage to foliage by the tuberworm cannot easily be controlled by cultural methods, although general crop hygiene is helpful, but deep-planting of the seed tubers followed by ridging can ensure that developing tubers remain buried and protected from infestation by larvae (Graf, 1917).

12.7.5 Biological and behavioural control

The control of pest populations by other organisms (biological control) in the environment is common in nature. The magnitude of control has occasionally been demonstrated by insecticides such as Carbaryl (Sevin) which, by selectively killing aphid predators, allowed an aphid population to increase to 28 times that on un-treated plants (Peterson, 1963). A wide range of organisms kill potato pests; they include viruses, bacteria, fungi, nematodes and members of several insect orders. When this natural control is insufficient to prevent damage, measures to increase biological control have sometimes been tried. A few individuals of a potential controlling organism may be introduced, usually an exotic predator or parasite, and permanent control of the pest is provided by the otherwise unaided survival and multiplication of the introduced organism. Alternatively, large numbers of the organism may be repeatedly introduced to maintain control as in spraying crops with suspensions of pest-attacking viruses or fungi.

Several hymenopterous parasites have been introduced to control potato tuberworm. The braconid *Orgilus lepidus* Muesebeck, initially found parasitizing tuberworms in Argentina and Uruguay (Muesebeck, 1967), has been introduced into Australia, Bermuda, Cyprus, India, Mauritius, New Zealand, U.S.A. and S. Africa (Oatman *et al.*, 1969). Oatman and Platner (1974) name several other hymenopterous parasites that have been introduced into California and, in their survey, the exotic, together with the indigenous species, parasitized more than 50% of tuberworms. Despite this, they still recommend cultural methods preventing exposure of potato tubers as being necessary to avoid tuber infestation. Reed (1971) working in Australia has shown that a spray containing a granulosis virus which attacks tuberworms may be at least as effective in controlling this pest as commonly-used insecticides. The virus can spread to nearby fields, perhaps via the insectivorous bird, the silver-eye (*Zosterop gouldi*), since the virus remains pathogenic even after passage through its gut (Matthiessen and Springett, 1973). Tuberworms can be intensively reared on tubers, facilitating large-scale production of virus. A granulosis virus attacking tuberworms has also been found in S. Africa (Broodryk and Pretorius, 1974).

The colonization of Europe by the Colorado beetle instigated the introduction of the predatory bug *Perillus bioculatus* Fabr. (Heteroptera: Pentatomidae) from N. America. In order to maintain it in parts of Europe, Szmidt and Wegorek (1967) suggested that perhaps there should be an annual mass release of artificially reared specimens. The fly, *Doryphorophaga doryphoryae* (Riley) is the predominant parasitoid of Colorado beetle larvae in parts of Canada (Kelleher, 1966) and has been tried in the U.S.S.R. for control of the beetle (Shutova, 1961). A parasitic nematode has experimentally been sprayed on crops to control Colorado beetles but although the nematode can be reared in several insect hosts, stored for more than a month and easily applied to crops, its effectiveness was limited by its susceptibility to desiccation (Welch and Briand, 1961).

Fig. 12.21 A Colorado beetle infected with the fungus *B. bassiana*. (Fungus identified by N. Wilding.)

Sprays of suspensions of spores of the entomophagous fungus *Beauveria bassiana* (Bals.) (Fig. 12.21), have controlled field populations of the beetle in the U.S.S.R; a sub-lethal amount of DDT has been included with the spores in the spray formulation to improve control (Dyadechko and Sikura, 1963).

Aphid populations are often severely depleted by parasitoids, predators and entomophagous fungi, but the time lag between aphid colonization of the crop and colonization of the aphid population by parasites etc., allows the aphid population to increase rapidly early in the season; Shands *et al.* (1965), found that parasites had no appreciable adverse effect on field populations containing *M. persicae*, *M. euphorbiae*, and *A. solani* and *A. nasturtii* and predators such as cocinellids (ladybirds) only had an effect late in the season, usually after July. The colonization of potato crops by predators can to some extent be hastened either by introducing their larvae (Shands *et al.*, 1972c) or by spraying on their eggs (Shands and Simpson, 1972a). This requires a method of rearing many predators, which although it can be fairly intensive still seems to necessitate a diet of live aphids (Shands *et al.*, 1970). The entomophagous fungus *Entomophthora thaxteriana* can be propagated on an artificial medium and the spores refrigerated until needed (Soper, *et al.*, 1975); such an organism may be cheap enough to use in sprays against aphids.

So far, there seem to be no examples in which a potato pest has been sufficiently suppressed by introduced predators or parasites so as to eliminate damage, although pest populations have been diminished. However, the large-scale culture of virus and fungal control agents for routine spraying on crops is attracting much attention. Since their effects are similar to those of chemical insecticides, they must be shown to be economically better, more convenient to handle, able to be applied nearer the time of harvest, or to be without the undesirable side-effects of chemical insecticides. However, the use of biological control agents may not be without hazard; FAO/WHO (1973) have, for example, suggested a series of safety tests for insect viruses (general reference on use of micro-organisms: Burgess and Hussey, 1971).

Another approach to controlling pest populations is either to utilize their behaviour to achieve mass removal of adults by trapping, killing or rendering sterile, or to affect disruptively their behaviour. Light-traps have been tested as a means of trapping adult wireworms, but insufficient were caught to suppress significantly the population (Day *et al.*, 1973). Valeric acid, the sex pheromone released by females of the wireworm *Conoderus falli* to attract males, may be a useful trap bait for surveying wireworm populations and for wireworm control (Jacobson *et al.*, 1968). Pheromone trapping may be especially effective against such wireworm species as *Limonius canus* in which a majority of the males emerge before the females and may be trapped without competition from natural sources (Onsager *et al.*, 1968). There is also evidence of a sex pheromone released by female tubermoths (Rothschild, 1969/70) and Colorado beetles (Wilde *et al.*, 1969) so synthetic baits may also be used for trapping males of these pests. Aphids become agitated by 'alarm' pheromones released by disturbed or injured individuals, so there are pheromones which can affect the behaviour even of parthenogenetic generations and it may be possible to use these to affect them adversely (Bowers *et al.*, 1972). Light reflected from aluminium foil has successfully repelled alate aphids from certain crops but, although its use with potato crops has occasionally been associated with smaller aphid populations, the effect was not consistent (Shands and Simpson, 1972b).

I thank staff of the photography and stereoscan sections of Rothamsted Experimental Station for preparing all the illustrations except Fig. 12.10 and 12.11, for which I gratefully acknowledge the assistance of members of Long Ashton Research Station.

References

Abdel-Salam, A.M., Assem, M.A., Hammad, S.M. and Eid, G.H. (1972). *Z. angew. Ent.* **70**, 76–86.

Adams, J.B. (1946). *Am. Potato J.* **23**, 1–22.

Anonymous (1964). *Agric. Gaz. N. S. W.* **75**, (9).

Arellano, U.M. (1963). *Agronomia* (La Molina, Peru) **30**, 62–76.

Bacon, O.G. (1960). *J. econ. Ent.* **53**, 868–71.

Bald, J.G., Norris, D.D., and Helson, G.A.H. (1946). *Council Sci. Ind. Res., Bull.* **196**.

Banks, C.J. and Macaulay, E.D.M. (1964). *Ann. appl. Biol.* **53**, 229–42.

Benskin J. and Perron, J.M. (1973). *Can. Ent.* **105**, 619–22.

Bollard, E.G. (1960). *Ann. Rev. Pl. Physiol.* **11**, 141–66.
Bowers, S.W., Nault, L.R., Webb, R.E., and Dukty, S.R. (1972). *Science* **177**, 1121–2.
Bradley, R.H.E. (1963). *Can. J. Microbiol.* **9**, 369–80.
Bradley, R.H.E. (1964). In *Plant Virology*, eds. Corbett, M.K. and Sisler, H.D. Ch. 7. Gainesville: University of Florida Press.
Bradley, R.H.E. and Ganong, R.Y. (1951). *Can. J. Zool.* **29**, 329–38.
Bradley, R.H.E., Moore, C.A. and Pond, D.D. (1966). *Nature* **209**, 1370–1.
Broadbent, L. (1953). *Biol. Rev.* **28**, 350–80.
Broodryk, S.W. (1971). *Phytophylactica* **3**, 73–83.
Broodryk, S.W. and Pretorius, L.M. (1974). *J. ent. Soc. sth. Afr.* **37**, 125–8.
Burgess, H.D. and Hussey, N.W. (Ed.) (1971). *Microbial control of Insects and Mites*, London: Academic Press.
Carden, P.W. (1965). *Proc. Third Br. Insecticide Fungicide Conf.* 308–16.
Carpenter, G.P. and Scott, D.R. (1974). *J. econ. Ent.* **67**, 665–7.
Cartier, J.J. and Morin, St. G. (1965). *Ann. Ent. Soc. Quebec* **10**, 69–79.
Champ, B.R. and Shepherd, R.C.A. (1965). *J. Agric. Animal Sci.* **22**, 69–81.
Chawla, S.S., Perron, J.M. and Huot, L. (1973). *J. econ. Ent.* **66**, 963–5.
Chiu, R., Liu, H., MacLeod, R. and Black, L.M. (1970). *Virology* **40**, 387–96.
Cramer, H.H. (1967). *Plant Protection and World Crop Production.* Farbenfabriken Bayer Ag-Leverkusen, Germany.
Crombie, A.C. and Darrah, J.H. (1947). *J. exp. Biol.* **24**, 95–109.
Daniels, C.B. (1937). Controlling Colorado potato pests. *Colorado Exp. Station Bull* (437).
Dahlman, D.L. and Hibbs, E.T. (1967). *Ann. ent. Soc. Am.* **60**, 732–40.
Day, A., Stanley, J.M., Webb, J.C. and Hartsock, J.G. (1973). *J. econ. Ent.* **66**, 757–60.
Deahl, K.L., Young, R.J. and Sinden, S.L. (1973). *Am. Potato J.* **50**, 248–52.
Decker, G.C. and Cunningham, H.B. (1967). *J. econ. Ent.* **60**, 373–9.
Delgado-Sanchez, S. and Grogan, R.G. (1970). Potato virus Y. *C.M.I./A.A.B. Description of Pl. Viruses*, (37).
Delong, D.M. (1971). *Ann. Rev. Ent.* **16**, 179–210.
Dudley, J.E., Landis, B.J. and Shands, W.A. (1952). Control of potato insects, *U.S.D.A. Farmers Bull.* 2040.
Dyadechko, N.P. and Sikura, A.I. (1963). Summary in *Biol. Abst.* **43**, (20907).
Dyson, P.W. (1965). An analysis of growth and yield of the potato crop with particular reference to nutrient supply. Ph.D. Thesis, University of London.
Edson, H.A. and Wood, J.I. (1939). *Pl. Dis. Repr, Suppl.* **118**, 85–118.
Edwards, C.A., Thompson, A.R. and Lofty, J.R. (1967). *Proc. Fourth Br. Insecticide and Fungicide Conf.* 48–54.
Elliott, M., Jones, N.F., Pulman, D.A., Farnham, A.W. and Needham, P.H. (1974). *Rothamsted Experimental Station Report for 1973*, 168–70.
Van Emden, H.F. (Ed.) (1972). *Aphid Technology*, London and New York: Academic Press.
Van Emden, H.F., Eastop, V.F., Hughes, R.D. and Way, M.J. (1969). *Ann. Rev. Ent.* **14**, 197–270.
Ephraim, J.P. and Radcliffe, E.B. (1968). *Proc. North Central Branch, Am. Ass. Econ. Ent.* **23**, 148–52.
Eschrich, W. (1970). *Ann. Rev. Pl. Physiol.* **21**, 193–214.
Evans, G.O., Sheals, J.G. and Macfarlane, D. (1961). *The Terrestrial Acari of the British Isles. An Introduction to their Morphology, Biology and Classification* Vol. 1. London: British Museum.
FAO/WHO (1973). *The use of Viruses for the Control of Insects, Pests and Disease*

Vectors. Food and Agriculture Overseas/World Health Organization.

Fellowes, R.W. and Ferguson, A.M. (1974). *N. Z. J. exp. Agric.* **2**, 83–8.

Forbes, A.R. (1969). *Can. Ent.* **101**, 31–41.

Fraenkel, G. (1969). *Entomologia exp. appl.* **12**, 473–86.

Gibson, K.E., Lane, M.C., Cook, W.C. and Jones, E.W. (1958). Effects of some crop rotations on wireworm populations on irrigated lands. *U.S. Depart. Agric. Technical Bull.* (1172).

Gibson, R.W. (1971a). *Ann. appl. Biol.* **69**, 89–96.

Gibson, R.W. (1971b). A study of the distributions of five aphid species on potato plants. Ph.D. Thesis, University of Bristol.

Gibson, R.W. (1971c). *Ann. appl. Biol.* **68**, 245–51.

Gibson, R.W. (1971d). *Ann. appl. Biol.* **68**, 113–19.

Gibson, R.W. (1974a). *Ann. appl. Biol.* **76**, 19–26.

Gibson, R.W. (1974b). *Potato Res.* **17**, 152–4.

Gibson, R.W. (1976). *Ann. appl. Biol.* **82**, 147–50.

Gibson, R.W., Whitehead, D., Austin, D.J. and Simkins, J. (1976). *Ann. appl. Biol.* **82**, 151–3.

Gibson, A., Gorham, R.P., Hudson, H.F. and Flock, J.A. (1925). The Colorado Potato Beetle in Canada. *Department of Agic. Dominion of Canada, Bull.* (52).

Gimingham, C.T. and Thomas, I. (1948). *J. Minist. Agric.* **55**, 55–63.

Goss, R.W. (1931). Infection experiments with spindle tuber and unmottled curly dwarf. *Ncbraska Agric. Exp. Station Res. Bull.* (53).

Gould, H.J. (1965). *Pl. Path.* **14**, 109–11.

Govier, D.A. and Kassanis, B. (1974). *Virology* **61**, 420–6.

Graf, J.E. (1917). The potato tuber moth. *U.S.D.A. Bull.* (427).

Graham, D.C. and Hardie, J.L. (1971). *Proc. Sixth. Br. Insecticide Fungicide Conf.* 219–24.

Guthrie, F.E., Campbell, W.V. and Baron, R.L. (1962). *Ann. ent. Soc. Am.* **55**, 42–6.

Harrewijn, P. (1970). *Entomologia exp. Appl.* **5**, 114–18.

Hibbs, E.T., Dhalman, D.L. and Rice, R.L. (1964). *Ann. ent. Soc. Am.* **57**, 517–21.

Hille Ris Lambers, D., Reestman, A.J. and Schepers, A. (1953). *Netherlands J. agric. Sci.* **1**, 188–201.

Hirst, G.W. (1975). Meteorology and the Colorado potato beetle. *World Meteorological Organization Tech. Note* (137).

Hodgeson, W.A., Pond, D.D. and Munro, J. (1974). *Can. Department Agric. Publ. 1492.*

Hsiao, T.H. and Fraenkel, G. (1968). *Ann. ent. Soc. Am.* **61**, 485–92.

Hurkova, J. (1968). *Acta entomol. Bohemoslavac* **65**, 188–207.

Jacobson, M., Lilly, C.E. and Harding, C (1968). *Science* **159**, 208–9.

Janson, H.W. (1961). *Agric. Lond.* **67**, 522–3.

Janson, H.W. (1971). *Agric. Lond.* **78**, 161.

Janus, H. (1965). *The Young Specialist looks at Land and Freshwater Molluscs* London: Burke.

Kasargode, R.S. and Nagpurkar, S.D. (1920). *Bombay Department Agric. Bull.* (102) 26–37.

Kassanis, B. and Govier, D.A. (1971). *J. gen. Virol.* **13**, 221–8.

Kelleher, J.S. (1966). *Ann. ent. Soc. Am.* **59**, 1059–61.

Kennedy, J.S., Day, M.F. and Eastop, V.F. (1962). *A Conspectus of Aphids as Vectors of Plant Viruses.* London: Commonwealth Institute of Entomology.

Kring, J.B. (1958). *J. econ. Ent.* **51**, 823–8.

Kring, J.B. (1963). *Frontiers Pl. Sci.* **15**, 6–7.

Kuhn, R. and Low, I. (1955). In *Origin of Resistance to Toxic Agents*, ed. Sevag, M.G., Reid, R.D. and Reynolds, O.E., pp. 122–32. New York: Academic Press.

Ladd, T.L. (1963). *Dissertation Abst.* **23**, 4787–8.

Ladd, T.L. and Rawlins, W.A. (1965). *J. econ. Ent.* **58**, 623–8.

Landis, B.J., Semel, M. and Simpson, G.W. (1967). *Potato Handbook for 1967*, 12–29.

Leach, J.G. (1926). *Phytopathology* **16**, 149–76.

Lilly, C.E. (1973). *J. econ. Ent.* **66**, 1205–7.

Matthiessen, J.N. and Springett, B.P. (1973). *Aust. J. Zool.* **21**, 533–40.

Meisner, J., Ascher, K.R.S. and Lavie, D. (1974). *Z. angew. Ent.* **77**, 77–106.

Mittler, T.E. and Dadd, R.H.E. (1962). *Nature, Lond.* **195**, 404.

Molina, J.J., Harrison, M.D. and Brewer, J.W. (1974). *Am. Potato J.* **51**, 245–50.

Morrison, H.E., Gentner, L.G., Koontz, R.F. and Every, R.W. (1967). *Am. Potato J.* **44**, 137–44.

Muesebeck, C.F.W. (1967). *Proc. ent. Soc. Washington*, **69**, 177–8.

McIntosh, T.P. (1927). *The potato; its History, Varieties, Culture and Disease.* Edinburgh: Oliver and Boyd.

McLean, D.L. and Weight, W.A. (1968). *Ann. ent. Soc. Am.* **61**, 180–5.

McNew, G.L. (1963). Pest control in relation to human society. In *New Developments and Problems in the Use of Pesticides.* Washington: National Academy of Science.

Nielson, M.W., Lehman, W.F. and Marble, V.L. (1970). *J. econ. Ent.* **63**, 1489.

Oatman, E.R. and Platner, G.R. (1974). *Environ. Ent.* **3**, 262–4.

Oatman, E.R., Platner, G.R., and Greary, P.D. (1969). *Ann. ent. Soc. Am.* **62**, 1407–14.

Olson, D. and Pimentel, D. (1974). *Environ. Ent.* **3**, 621–4.

Onsager, J.A. (1974). *Am. Potato J.* **51**, 313–17.

Onsager, J.A., McDonough, L.M. and George, D.A. (1968). *J. econ. Ent.* **61**, 691–3.

Ordish, G. (1952). *Untaken Harvest*, London: Constable et Comp., Ltd.

Painter, R.H. (1951). *Insect Resistance in Crop Plants.* New York: Macmillan Co.

Peters, D. (1970). Potato leafroll virus. *C.M.I./A.A.B. Description of Plant Viruses.* (36).

Peterson, A.G. (1963). *Am. Potato J.* **40**, 121–9.

Pienkowski, R.L. and Medler, J.T. (1964). *Ann. ent. Soc. Am.* **57**, 588–91.

Pletsch, D.J. (1947). The potato psyllid, *Paratrioza cockerelli* (Sulc.), its biology and control. *Montana Agric. Exp. Station Technical Bull.* (446).

Potato Marketing Board (1974). *Potato Statistics Bulletin December 1974*, London: Potato Marketing Board.

Potato Marketing Board (1975). Report on the survey of maincrop potato production 1973. London: Potato Marketing Board Report.

Pollard, D.G. (1973). *Bull. ent. Res.* **62**, 631–714.

Prasad, S.K. (1960). *Indian J. Ent.* **22**, 231–2.

Radcliffe, E.B. and Lauer, F.I. (1970). *J. econ. Ent.* **63**, 110–14.

Radcliffe, E.B. and Lauer, F.I. (1971). *J. econ. Ent.* **64**, 1260–6.

Reed, E.M. (1971). *Bull. Ent. Res.* **61**, 207–22.

Reynolds, H.T. and Metcalf, R.L. (1962). *J. econ. Ent.* **55**, 2–5.

Ritter, F.J. (1967). *Meded. Rijkfaculteit Landbouwwetenschappen Gent*, **32**, 291–305.

Robert, Y. (1971). *Potato Res.* **14**, 130–9.

Robert, Y. and Rouzé-Jouan, J. (1971). *Potato Res.* **14**, 154–7.

Ross, R.W. and Rowe, P.R. (1966). Supplement to the inventory of tuber-

bearing *Solanum* species. *Wisconsin Agric. Exp. Station Bull. (533) (Suppl.).*
Rothschild, G.H.L. (1969/70). Sex pheromone of potato moth. *Ann. Report of C.S.I.R.O. Division of Ent.* **33**.
Sanford, L.L. and Sleesman, J.P. (1970). *Am. Potato J.* **47**, 19–34.
Shabliorskii, V.V. and Chebanov, G.E. (1962). *Biological Abst.* (1964) **45**, (3720).
Shands, W.A. and Simpson, G.W. (1972a). *J. econ. Ent.* **65**, 1383–7.
Shands, W.A. and Simpson, G.W. (1972b). *J. econ. Ent.* **65**, 507–10.
Shands, W.A., Holmes, R.L. and Simpson, G.W. (1970). *J. econ. Ent.* **63**, 315–17.
Shands, W.A., Simpson, G.W. and Murphy, H.J. (1972a). Effects of cultural methods for controlling aphids on potatoes in Northeastern Maine. *Maine Life Sci. Agric. Exp. Station Technical Bull. (57).*
Shands, W.A., Simpson, G.W. and Storch, R.H. (1972c). *J. econ. Ent.* **65**, 799–809.
Shands, W.A., Simpson, G.W., Muesebeck, C.F.W. and Wave, H.E. (1965). Parasites of potato infesting aphids in Northeastern Maine. *Maine Agric. Exp. Station Technical Bull. (T19).*
Shands, W.A., Simpson, G.W., Seaman, B.A., Roberts, F.S. and Flynn, C.M. (1972b). Effects of differing abundance levels of aphids and of certain virus diseases upon yield and virus disease spread in potatoes. *Maine Life Sci. Agric. Exp. Station Technical Bull. (56).*
Shorey, H.H., Deal, A.S., Hale, R.L. and Snyder, M.J. (1967). *J. econ. Ent.* **60**, 892–3.
Shutova, N.N. (1961). *Biological Abst.* **40** (25196).
Sleesman, J.P. (1940). *Am. Potato J.* **17**, 9–12.
Simpson, G.W., Landis, B.J. and Shands, W.A. (1964). *Potato Handbook*, **9**, 5–21.
Smith, F.F. and Poos, F.W. (1931). *J. agric. Res.* **43**, 267–85.
Soper, R.S., Holbrook, F.R., Majchrowicz, I. and Gordon, C.C. (1975). Production of *Entomophthora* resting spores for biological control of aphids. *Maine Life Sci. Agric. Exp. Station Technical Bull. (76).*
Srivastava, B.K. and Khan, R.M. (1962). *Indian Potato J.* **4**, 88–92.
Staples, R. (1968). *J. econ. Ent.* **61**, 1378–80.
Stegwee, D. and Ponsen, M.B. (1958). *Entomologia exp. appl.* **1**, 291–300.
Strickland, A.H. (1965). *J. R. Soc. Arts* **83**, 62–81.
Strickland, A.H., Bardner, H.M. and Waine, R.A. (1962). *Pl. Path.* **11**, 93–105.
Symonds, B.V. (1973). *Pl. Path.* **22**, 30–4.
Szmidt, A. and Wegorek, W. (1967). *Entomophaga* **12**, 403–8.
Tamaki, G. (1973). *Can. Ent.* **105**, 761–5.
Taylor, C.E. (1957). *Nat. Ass. Seed Potato Merchants Year Book* 1955–6, 18–26.
Telford, H.S. (1942). *North Dakota Agric. Exp. Station Bull.* (4), (5) 7–8.
Thomas, I. (1952). *Pl. Path.* 1, 103–8.
Thorpe, W.H., Crombie, A.C., Hill, R. and Darrah, J.H. (1947). *J. exp. Biol.* **23**, 234–66.
Torka, M. (1950). *Am. Potato J.* **27**, 263–71.
Turnipseed, S.G., Heinrichs, E.A., Da Silva, R.F.P. and Todd, J.W. (1974). *J. econ. Ent.* **67**, 760–2.
Wallis, R.L. (1946). *J. econ. Ent.* **39**, 689–94.
Watson, M.A. and Plumb, R.T. (1972). *A. Rev. Ent.* **17**, 425–52.
Welch, H.E. and Briand, L.J. (1961). *Can. Ent.* **93**, 759–63.
Wilde, J. De, Lambers-Suverkropp, K.H.R. and Van Tol, A. (1969). *Netherlands J. Pl. Path.* **75**, 53–9.

Wilkinson, C.F. (1973). *Chemtech* August, 492–7.
Winfield, A.L., Wardlow, L.R. and Smith, B.F. (1967). *Pl. Path.* **16**, 136–8.
Workman, R.B. (1964). *Proc. Florida St. hort. Soc.* **77**, 210–12.
Wressell, H.B. and Driscoll, G.R. (1964). *J. econ. Ent.* **57**, 992–3.

Chapter 13

Tuber quality

D. GRAY and J.C. HUGHES

13.1 Introduction

The potato is one of the major staple food crops and appears to be particularly well adapted to growth in Northern latitudes but attempts are now being made to introduce it into the lowland areas of the tropics and to improve its protein quality and content (International Potato Centre, 1973). Its major centres of production are N. Europe and the U.S.S.R. which account for nearly 70% of the world acreage and over 75% of total output. In contrast, the U.S.A. and Canada together only account for 3% of world acreage but 6% of production. Of the European production 52% is grown in Poland and the Federal and Democratic Republics of Germany (Anonymous, 1972a). Although in these cases a proportion is used for industrial starch and alcohol production, consumption is generally higher than elsewhere and up to 175 kg/person/annum are reputedly consumed in Poland (see Burton, 1974).

On a world-wide basis the crop produces more dry matter (DM) and protein per hectare than the major cereal crops (Table 13.1) although because of its relatively low tuber DM content about three times as much of the raw product has to be eaten to provide an equivalent energy intake to that of cereals. In the past potatoes have been a significant energy source for industrial communities and have also provided significant quantities of protein. Groot (1947) estimated that in the Netherlands up to 10–25% of total protein requirement was met by potato consumption. However, in most developed communities its contribution to this part of the diet is now small but even at the relatively low level of present day consumption it is still a major source of vitamin C. Burton (1974) has calculated that in the U.K., which has a current intake of about

Table 13.1 Total DM production, DM production per unit area, total protein production and production per unit area of the major subsistence crops

	Total DM production ($t \times 10^7$)	*Total protein production* ($t \times 10^6$)	*DM production* ($t\ ha^{-1}$)	*Protein production* ($t\ ha^{-1}$)
Wheat	27.5	32.9	1.30	0.156
Rice	26.7	23.2	1.97	0.172
Maize	23.5	24.7	2.13	0.224
Barley	11.4	11.6	1.46	0.148
Sorghums/millet	8.2	7.4	0.73	0.066
Potatoes	6.6	6.0	2.93	0.266
Sweet potatoes, yams	3.9	2.9	3.82	0.280
Cassava	3.4	0.8	4.92	0.115
Soybeans	4.2	16.7	2.62	1.043

(Source F.A.O. Production Yearbook, 1972).

200 g day^{-1}, over 30% of the intake of Vitamin C from fruit and vegetables comes from potatoes but this figure is considerably higher in other European countries. In addition, the potato provides minerals principally iron, thiamine, nicotinic acid and riboflavin and small amounts of vitamin A.

Despite this, the consumer in the developed parts of the world pays little attention to its nutritive value compared with that placed on aesthetic requirements. Indeed, quality in potatoes is usually taken to mean those components such as its appearance and those which contribute to culinary quality such as flesh colour before and after cooking and the texture of the cooked product. These aspects have a very marked effect upon the acceptability of potatoes as a food leading to preferences for different qualities even on a local basis, for example, in the preference for mealy potatoes in the north of England as opposed to waxy ones in the south.

Increasingly a major proportion of the crop is processed before sale, particularly in North America but to a lesser extent in Europe (Hampson, 1972), as chips (crisps in the U.K.) frozen French fries (chips in the U.K.), dehydrated flake or powder and canned potatoes of various forms. In these cases not only is the size, depending on the particular use, shape and freedom from blemishes and damage important but other aspects of quality assume greater significance than is usually the case for freshly harvested potatoes destined for domestic use. This is exemplified by the DM contents required for both chip-making and dehydration, and the low reducing sugar contents required for chip-making. How these particular factors affect chip quality is fairly well understood but nevertheless, it is not always possible to grow crops having the desired quality. For other quality attributes such as tuber damage or texture the degree of control is even more limited because the underlying basis of differences in texture and damage between varieties and from different sites is not well understood. Whereas poor quality in these respects has not been a major constraint on consumption in the past there is increasing evidence that potatoes with undesirable textural properties and damaged tubers within a sample could adversely affect it in the future (Potato Marketing Board, 1964, 1974a, 1974b).

Numerous factors are concerned with potato quality and relate to the morphology, structure and chemical composition of the tuber all of which can be affected by the environment during growth. As a result it is often difficult in commerce to produce tubers having consistent quality. It is thus important to know not only the biochemical and morphological basis of differences in quality but also to understand the part played by variety, cultural practice and environment and their relative importance. In this chapter we consider those aspects of quality primarily associated with external morphology and those primarily associated with the internal structure and composition such as nutritional quality in relation to the provision of protein, vitamins and carbohydrates and culinary quality including texture, internal and external damage, after-cooking blackening and enzymic and non-enzymic browning. The improvement of quality attributes by breeding techniques is not discussed and has been reviewed elsewhere (Howard, 1974; Chapter 15).

13.2 Quality associated with morphology and external appearance

The morphological features such as the shape of the tuber including internal and external deformities, its size, depth of eyes, the colour of the skin and flesh together with surface blemishes due to disease, superficial damage or a combination of both will initially determine the acceptability of the product to the consumer. For potatoes destined for processing or direct for the table use these quality requirements are of very considerable importance since poor quality can lead to considerable wastage so increasing the costs of preparation. In this section we consider these qualities with the exception of susceptibility to damage since this depends in part on the mechanical and rheological properties of the tuber which are more closely related to the internal structure and composition of the tuber.

13.2.1 Size of tubers

The size of tubers required by the consumer varies considerably and will depend upon the ease of handling for household purposes and also upon the acceptable levels of peeling loss since this will be proportionally greater for small than large tubers. Generally, tubers within the size range 4.0–7.5 cm riddle size are preferred but smaller tubers 2.0–4.0 cm are used for canning whole. Tuber size is determined in part by variety and tuber set per stem but in commerce a large measure of control over size is achieved by varying the plant population (see Chapter 7).

13.2.2 Depth and appearance of skin

This may be important in determining susceptibility to scuffing damage (see Section 13.3.5) but it is also important because varieties are recognized by the appearance of the skin, for example, in netted skinned potatoes such as Netted Gem.

The thickness of the periderm is usually a varietal feature (Artschwager, 1924) but it is influenced also by cultural factors. For example, high levels of N-fertilizer and deep planting produce a thin periderm but phosphate and irrigation produce a thick skin. High soil temperatures can induce the formation of a rough or scaled skin (Yamaguchi *et al.*, 1964a).

13.2.3 Depth of eyes, normal pigmentation of the skin and flesh

The basis of skin pigmentation in potatoes is the presence of anthocyanins dissolved in the cell sap of the periderm or peripheral cortical cells (Burton, 1966). Flesh colour, with some notable exceptions, is usually white or yellow and up to a dozen different carotenoids have been identified in potato tubers. Brunstetter and Wiseman (1947) and Caldwell *et al.* (1945) have demonstrated a close relationship between the content of these and flesh colour. These attributes and that of depth of eyes are largely genetically controlled (Salaman, 1926; Howard, 1970).

13.2.4 Greening

Greening of tubers as a result of chlorophyll formation in the periderm layers on exposure to light reduces the acceptability of the product and because of the parallel development of solanidine greened tubers are sometimes bitter when cooked. Greening may occur in the field but it is a more troublesome problem during storage in retail outlets. Methods of reducing its development so increasing shelf life have been reviewed by Smith (1968).

Light intensities as low as 3–11 W m^{-2} for as short a period as 24 h induce greening (Larsen, 1949; Gull and Isenberg, 1958) but its development is affected by variety (Larsen, 1949; Akeley *et al.*, 1962), by maturity (Yamaguchi *et al.*, 1960) and by temperature (Larsen, 1949). For example no greening is found at 5 °C but it is extensive at 20 °C and the effect is greater in immature tubers. In the field the major cause of greening is insufficient cover over the tubers at planting (Lewis and Rowberry, 1973) but the effect can be aggravated by varieties whose tubers form near the surface and by the clumping of stems which results from the use of large seed planted at wide spacings (Bleasdale and Thompson 1965). This occurs because competition is less severe at wide spacings allowing more tubers to set, some of which are forced to the surface where they green.

Associated with the development of a green pigment is the formation of steroid alkaloids, mainly, solanidine which is of importance because it can cause off-flavours on cooking at concentrations of 15–20 mg/100 g (Zitnak and Johnston, 1970) but in normal potatoes its concentration is 5–10 times lower than this (Lampitt *et al.*, 1943) and because most of it is in the skin the concentration in prepared potatoes is usually low and only in exceptional cases has it been responsible for outbreaks of poisoning (see Burton, 1966, 1974). The amount of solanidine formed is also characteristic of the variety and concentrations as high as 35 mg/100 g have been recorded for cv. Lenape (Zitnak and Johnston, 1970) a variety now withdrawn from commerce for this reason. Concentrations are also higher in immature than mature tubers, but the concentration does not appear to be influenced by fertilizer regime (Pallman and Schindler, 1942).

13.2.5 Shape of tubers, second growth and hollow heart

Shape

Tuber shape is largely a varietal characteristic (Salaman, 1926) and Howard (1970) has shown that long is dominant to round. Burton (1966) recognizes four distinct shapes; round, oval, pointed oval and kidney shaped but others (Whitehead *et al.*, 1953) recognize seven. Shape is also influenced in some varieties by the cultural and environmental conditions which favour rapid and sustained growth such as wide spacing and ample supply of nutrients. High levels of applied-N (Neumann, 1925; Westover, 1941) and irrigation but low levels of K (Wohrmann, 1966) increase the length relative to the width of the tuber.

Uniformity of tuber shape within a population is also an important characteristic particularly for some forms of processing where peeling and slicing are carried out. Tuber shape is principally affected by variety and plant spacing; close spacing giving a more uniform sample. It is also controlled by temperature, for example, those in the range 12–20° C produce a more even shape than temperatures above or below this range (Yamaguchi *et al.*, 1964a).

Second growth

A number of types of tuber deformity are found relatively frequently when stress occurs during the period of tuber growth. These deformities, including cracking, twisting of the tubers, 'knobbly' or 'dolled' tubers and chain tubers, are collectively known by the term 'second growth'

(Lugt, 1960). The development of these tubers is often associated with periods of drought followed by rainfall or irrigation resulting in a rapid recommencement of growth (Bodlaender *et al.*, 1964). In severe cases the secondary tuber inhibits the growth of the primary tuber and movement of materials from the primary tuber can be so extensive as to lead to a reduction in the quality of fried products (Iritani and Weller, 1973). However, the usual effect of an incidence of second growth, which can reach 20% of the crop, is to reduce the external quality and appearance of a sample of tubers.

Although these deformities are associated with periods of drought, cessation of tuber growth is not a prerequisite for the induction of second growth since continuous irrigation can increase its incidence (Corey and Meyers, 1955). Bodlaender *et al.* (1964) have now shown that second growth can be induced simply by exposing plants to short periods of high temperature. It cannot be induced to occur in newly detached tubers previously given a heat treatment, but in those attached to the parent plant second growth occurs whether the heat stimulus is given to the foliage or the tubers. This suggests that the tubers revert to a non-tuberizing stage under high temperatures in which rapid growth of the stolons, stems and leaves occurs but, when the temperature falls, the plant reverts to a tuberizing condition. Exposing plants to different temperatures can induce different types of second growth (Lugt *et al.*, 1964). However, it is not known exactly how temperature acts to give either release of a single bud from dormancy in the case of chain tuberization or several as in the case of knobbly tubers.

Treatments such as irrigation which cool the soil directly or close spacing (Robins and Domingo, 1956) which produce a closed canopy more rapidly than wide spacing can reduce the incidence of secondary growth but equally other reports (Kraus, 1943, 1945; Corey and Meyers, 1955) suggest that these treatments have no effect. Irrigation after the formation of the second growth will usually increase its severity. However, it is difficult to compare results because the timing of irrigation in relation to the stage of growth of the plant or the plant population used is not given. Presumably, in reality there is an interplay of factors which are finely balanced and their effect may depend on the time of application of water in relation to foliage and tuber development. Varietal differences in susceptibility to this disorder have been noted (Bodlaender *et al.*, 1964) although no varieties are completely free from its effects.

Hollow heart and external cracking

In contrast to second growth, marked varietal differences in susceptibility to another disorder, hollow heart, have been found (Clark and Lombard, 1951) and there are reports that varieties prone to this disorder rarely show extensive occurrence of second growth (Edmundson, 1935). The disorder starts with the death of a small area of the pith cells after the cell contents have disappeared (Levitt, 1942). This then widens as the tuber grows forming a series of large cracks producing the hollow centre. The disorder appears to be associated with rapid foliage growth promoted by high temperatures and soil moisture (Crumbly *et al.*, 1973) at the time of tuber initiation because the initial stages of the disorder are not usually found in large tubers. In support of this, defoliation in the early stages of growth and shading reduce it but wide spacing and high rates of applied-N, which both stimulate haulm growth increase its occurrence. Crumbly *et al.* (1973) have shown using ^{14}C that materials move out of the cells under conditions of inadequate water supply. Perhaps redistribution of assimilates or the lack of certain micronutrients (Krantz and Lana, 1942; Kallio, 1960) are responsible for the death of the central cells of the pith.

External cracks can develop in tubers associated with periods of rapid growth. Although the occurrence of these is known to be largely associated with particular varieties, e.g. Majestic, and with climatic conditions favouring rapid changes in the rate of tuber growth relatively little is known about it although Whitehead *et al.* (1953) suggest its occurrence is associated with a low level of boron in the affected tissue.

13.3 Quality associated with internal factors

The internal structure and composition of the tuber determines its nutritional quality, the amount of damage produced during handling, the colour of the cooked product (after-cooking blackening, enzymic browning, non-enzymic browning), texture and yield of processed products. The range of intrinsic factors involved is very large, for example, texture is controlled by cell wall structure and composition, cell size, starch and the pectic substances of the middle lamella all of which vary to differing extents from tissue to tissue and tuber to tuber depending upon the environment during growth and the stage of maturity when the tubers are harvested. Thus many quality defects are more pronounced in some tissues and tubers than others. It is also important to realize that not all tubers initiate at the same time nor do they grow at the same or even constant rate (see Chapter 4). Thus at any given stage of development tubers of the same size may be of different ages and as a result will show marked variation in composition.

Clearly, account needs to be taken of this variation and of interactions between varietal, environmental, and cultural factors in an experimental analysis of differences in quality, but often methods of sampling, preparation and cooking methods are not always clearly reported and analytical procedures vary widely thus making comparisons between

Table 13.2 Nutritionally Important constituents of the potato.

	*Amount provided by 100 g FW**	*U.K. daily requirement*†
Protein (N × 6.25)	2.1 g	45–75 g
Energy	0.3 MJ	11 MJ
Vitamin C	25 mg	30 mg
Thiamine	0.1 mg	1.4 mg
Riboflavin	0.02 mg	1.7 mg
Nicotinic acid	0.5 mg	18 mg
Iron	1.0 mg	10 mg

*From Burton, 1966; Lampitt and Goldenberg, 1940
†From Anonymous, 1972b.

the results of different workers difficult. An analysis of the environmental and cultural factors affecting a particular quality are further complicated because assessments of quality are often subjective and although a terminology and nomenclature have been defined (Lugt *et al.*, 1962) many descriptions in experiments are ambiguous. Furthermore, the expression of differences in quality often depend on the way in which the tubers are prepared or cooked. In view of this it is perhaps not surprising that results of the effects of different cultural practice and of different soil and aerial environments on quality often appear conflicting.

13.3.1 Nutritive value

The potato provides significant quantities of protein, vitamin C, carbohydrates, iron and to a lesser extent vitamins of the B complex and vitamin A (Table 13.2). On average in the U.K. potatoes probably provide about 4–4.5% of the daily energy and protein intake, about 7–10% of the iron and riboflavin intake and over 25% of the daily vitamin C requirement (Anonymous, 1972b).

Energy

The contribution of the potato to energy intake varies considerably depending on whether it is cooked in water or fat (oil). For example, whereas boiled potatoes provide about 0.3 MJ per 100 g FW, chips (French fries) provide up to 0.6 MJ (Burton, 1974) the former providing at U.K. levels of consumption about 5% and the latter 15% of the daily energy requirement. Carbohydrates provide most of the energy source and these constitute about 75% (with a range from 61.5–91.5%) of the total solids (Lampitt and Goldenberg, 1940). As a source of energy its major disadvantage is the considerable bulk (approximately 3.5 kg) that would have to be consumed to meet daily requirements.

Protein

In recent years there has been considerable interest in potato proteins because despite their low absolute content and content relative to carbohydrate, they have a high biological value (Woodward and Talley, 1953; Schuphan, 1959) equalling soybean protein (Kaldy, 1972). This is because the soluble non-protein N complements the protein (Chick and Cutting, 1943; Chick, 1951). However, potato protein is low in the S containing compound methionine (Hughes, 1948) so, whilst at U.K. levels of consumption up to 1/5 of the adult requirement could be met for most of the other essential amino acids, only about 1/10 of the methionine requirement would be supplied (Burton, 1966).

Much of the published information on the distribution of nitrogenous constituents in the tuber is found in the older literature (for a review see Lampitt and Goldenberg, 1940) and it is often difficult to compare results with more recent work because methods of analysis and extraction have been improved in recent years. In general the N-content of the tuber ranges from about 0.11 to 0.58 g per 100 g FW and about 30–50% of this is soluble protein and another 10% is insoluble, proteinaceous and largely located in the skin (Neuberger and Sanger, 1942). The soluble protein is of two types, 70% tuberin and 30% tuberinin which have similar amino acid contents but differ in their iso-electric points (Groot *et al.*, 1947). The remainder of the nitrogenous material is largely present as the amides, glutamine and asparagine which account for over 50% of the soluble non-protein N (Neuberger and Sanger, 1942; Mulder and Bakema, 1956). The latter is concentrated largely in the skin and the former in the pith whereas the soluble protein is distributed evenly throughout the tuber.

The nutritive value of potato protein varies considerably between lots of the same variety (Chick and Slack, 1951) but little is known about the influence of varietal, cultural and environmental factors. The effects of preparation, cooking and storage on the nutritional value of potato protein is described by Hughes (1948) and Chick and Slack (1951).

Protein content (g/100 g DM) appears to be inversely related to DM content (Talley *et al.*, 1961) where this is varied by the amount of applied N or by variety. This may frustrate attempts to increase per capita protein consumption using cooked potato in underdeveloped countries where there is also a low calorie intake. The relationship is less clear where DM content is varied by differences in tuber size, for example, Bleasdale and Thompson (1969) found that total N concentration was greater in small than large tubers, which have a higher DM content but Desborough and Weiser (1974) could find no consistent effects of tuber size on protein concentration. Such comparisons based on direct and indirect (N x 6.25) measures of protein highlight the difficulty of comparing results of different workers, particularly as the N-concentration of the protein can vary between 13.5 and 14.8% (Mulder and Bakema, 1956; Desborough and Weiser, 1974) depending upon location and fertilizer usage.

Differences in protein concentration (often based on total N) between commercial varieties can vary greatly between sites and seasons (Mulder and Bakema, 1956; Talley *et al.*, 1961; Desborough and Weiser, 1974) probably as a result of differences in soil fertility (Varis, 1973), particularly N. The effect of site on the relative times of maturity may also be important because predominantly more N is moved to the tubers during the later stages of growth (Carpenter, 1949). The effects of applied N on protein concentration are, however, small compared with the effects of N on total yield of tubers and yield of protein ha^{-1}. Recent work by Desborough and Weiser (1974) has demonstrated large differences in tuber protein concentration between Solanum species which suggests that protein content and particularly that of methionine could be improved by breeding methods. However, there is a need in the development of this work to adopt standardized techniques, to extend chemical analysis of the protein to its constituent amino acids and to develop rapid methods for the assessment of its biological quality perhaps in combination and competition with other currently used foodstuffs.

Vitamin C

Vitamin C (Rosenberg, 1942) is present in the tuber in both the reduced (ascorbic acid) and oxidized state (dehydroascorbic acid) but the content of the latter is usually considerably lower (0–14%). Dehydroascorbic acid is readily converted into diketogulonic acid on cooking and as this is not capable of being reduced to ascorbic acid its formation represents a loss of vitamin C. The concentration of vitamin C in the potato tuber ranges from 3–40 mg/100 g FW (Lampitt and Goldenberg, 1940). It is not influenced by tuber DM content or by size (Noonan *et al.*, 1951) and it appears to be evenly distributed throughout the tuber (Lampitt *et al.*, 1945a). Although higher contents have been recorded in the bud than the stem end of the tuber (Smith and Gillies, 1940) and in the skin than the cortex (Mudambi and Hanning, 1961) it is difficult to verify this because peeling the tissues damages cells, leading to a loss of ascorbic acid which will consequently be greater in the thinner tissue layers.

During the growth of the crop the concentration of vitamin C rises reaching a maximum at haulm senescence and then falls if harvest is delayed (Lampitt *et al.*, 1945a). Differences in vitamin C concentration between varieties have been recorded (Karikka *et al.*, 1944; Lampitt *et al.*, 1945b) but these and the effects of fertilizers, site and even the stage of maturity when the crop is harvested are small compared with the effects of the temperature and length of period of storage (Barker and

Mapson, 1950). The rate of loss in store and the final level reached depends to some extent on the variety (Smith and Gillies, 1940) and also on crop maturity (Barker and Mapson, 1950). Considerable quantities are lost during cooking and processing and this aspect is dealt with in detail by Leichsenring *et al.* (1951) and Burton (1966).

13.3.2 Specific gravity (starch or DM content)

The specific gravity of tubers is an important measure of quality used extensively by processors to assess the suitability for the production of French fries, chips and dehydrated products since the yield of product is greater per unit fresh weight from tubers with a high solids content.

It is also often used as an indicator of mealiness (Warren and Woodman, 1974), breakdown of tissues on cooking (Talburt and Smith, 1967) and susceptibility to internal bruising (Ophuis *et al.*, 1958). However, the relationships between DM and these attributes of quality frequently do not hold true and some of the effects may be indirect (see sections 13.3.4 and 13.3.5).

Starch is by far the largest component of the solids content ranging from 60–80%. The relationship between starch and total solids is relatively little affected by environmental conditions and so DM content is usually equated with the content of starch (Von Scheele *et al.*, 1937). This relationship is affected by variety as a result of differences in air spaces between cells in the tuber (Nissen, 1967) and by internal disorders such as hollow heart (Drew and Deasy, 1941). Nevertheless, as a general indicator of DM content the determination of SG has much merit and because of its ease of measurement it is widely used by the processing industry to assess quality for particular uses.

The selection of suitable varieties is probably the most reliable way at present to control tuber DM content in commerce (see Chapter 15). Tubers of early maturing varieties usually but not invariably have lower DM contents than those of late maincrop varieties (see Burton, 1966 for a review) and so during late spring and most of the summer months in the northern hemisphere the supply of tubers having a high DM content which are preferred for dehydration and chip manufacture is limited.

Genotypic control of tuber DM content is however modified by cultural and environmental factors although interactions between these and variety are usually small (Killick and Simmonds, 1974). In general, cool dull years and short growing seasons reduce DM production and % DM content whilst the reverse is true in warm sunny years and long growing seasons. However, the precise effects of water supply, temperature and radiation on starch and DM accumulation are not clearly understood (see Chapter 4). Applied N and K usually reduce tuber DM content (Kunkel and Holstad, 1972) but the effects are small over the range of levels of fertilizer giving optimum yields. With K fertilizers the chloride usually reduces DM content to a greater extent than the sulphate (Dickens *et al.*, 1962) although the effects are usually small and may depend on the soil moisture (Harward *et al.*, 1956). The effects of applied P on % DM content are usually small (see review by Kunkel and Holstad, 1972) but they vary, some authors reporting a decrease and others an increase for increasing amounts of P.

The effects of haulm destruction (Akeley *et al.*, 1955; Murphy and Goven, 1959) and irrigation (Smith & Nash, 1941; Pratt *et al.*, 1952) on % DM are also variable. In the former the precise effects are likely to depend upon the time of harvest in relation to the stage of maturity when defoliation occurs and also on the weather conditions after defoliation (Talburt and Smith, 1967). Irrigation during the early stages of growth increases DM content because starch production is increased but irrigation later in the season or continuous irrigation depress DM contents.

However, these responses are only generalizations and considerable variability in the effect of agronomic factors on % DM is reported in the literature, for example in the level of N and K giving maximum % DM contents. It is probable that the variable effects that cultural and environmental factors have on % DM contents can only be fully understood by relating the effects of the treatments on DM content to the growth and activity of the foliage and tubers and to crop maturity.

As tubers increase in size during the growing season the % DM content of the population increases from about 15–16% soon after tuber initiation to a maximum which varies between 18–28% (Appleman and Miller, 1926). In the later stages of growth DM contents can fluctuate considerably or, after haulm senescence they may fall. These changes in DM content reflect changes in the rate of increase in starch and cell wall material compared with water uptake. In the early stages of growth, deposition of starch is rapid but the rate declines as growth proceeds (see Chapter 4) despite continued rapid uptake of water. The rate of uptake of water may in certain situations e.g. rainfall or irrigation reduce % DM content for a short period of time. Changes in this fine balance between water uptake and starch synthesis and deposition may also account for the reduction in DM content during haulm senescence but losses due to respiration may also be important. So, defoliation of crops during the main stage of growth will almost invariably result in tubers of a lower DM content than those allowed to continue growth. But the results from later defoliation will usually be variable. Similarly increasing levels of applied N, which delay tuber initiation and prolong foliage growth may only markedly reduce % DM content if the growth of the foliage is curtailed.

The effects of cultural and environmental factors on the tuber size distribution may also indirectly affect % tuber DM contents. Usually

% DM contents increase with an increase in tuber size to a maximum and then fall with a further increase in size (Wurr and Allen, 1974), but for any tuber size they are higher the higher the plant population used to produce the tubers (Bleasdale and Thompson, 1969; Gray, 1972). Since changes in plant population change the tuber size distribution and the effects can vary from site to site the different responses of tuber DM to changes in population reported in the literature (see Hughes, 1974) may, in part, depend upon its effects on the size distribution of tubers. Fertilizers also effect the size distribution of tubers but it is not known if the variability in tuber DM produced in response to fertilizers is also related to differences in tuber size distribution because frequently only a restricted size range of the tuber population is assessed for DM content at harvest.

How differences in DM content associated with differences in tuber size arise is not well understood but they may be related to differences in the rates of growth of the medulla and cortex as tubers grow since these regions differ considerably in DM content (Glynne and Jackson, 1919). However, little is known about the deposition of starch, rates of cell division and expansion in different tissue layers and in different sizes of tuber, the relative growth rates of different tissue layers and how these are affected by varietal, cultural and environmental conditions.

Variability in DM content between tubers of the same size and in the distribution of DM within the tuber are also important aspects of quality since these can affect the yield and uniformity of the processed product from different batches of tubers. The effects of variety, culture and environment on the former are not well documented although Cole (1975) has shown that small tubers show greater tuber to tuber variability in DM content than large ones. It is well established that the DM content is highest in the vascular layers and declines towards the skin and medulla and from the stem to the bud end (Glynne and Jackson, 1919; Baijal and Van Vliet, 1966; Cole, 1975). Although Cole (1975) reported differences in DM content of different tissue layers between varieties it is not known whether this is affected by cultural and environmental factors.

13.3.3 Non-enzymic browning

Although sugars only constitute a small fraction usually up to 3% of the total DM they are particularly important because even at these concentrations they cause browning of chips and French fries during the frying process. Hodge (1953) has described two types of browning reaction, (1) caramelization and (2) those based on carbonyl-amino reactions or the so-called Maillard reaction (Maillard, 1912). Thornton (1940) ascribes browning in potatoes to the former but they are unlikely to be of great importance because a greater energy is required for the reaction than the Maillard reaction and furthermore, at temperatures usually used for frying (165–170°C) browning of sugars only takes place rapidly in the presence of amino acids (Schallenberger *et al.*, 1959). The first formed products in the Maillard reaction have now been shown to be α-γ unsaturated carbonyl compounds (Burton and McWeeny, 1963; Burton *et al.*, 1962, 1963). The development of the melanoidins from these compounds takes place through a series of complex reactions involving conjugation followed by polymerization. The depth of colour produced depends not only on the amount of sugars and amino acids present but also upon the type of chemical reactions leading to chromophore development which are affected by pH and temperature (Burton and McWeeny, 1964). In model systems the rate of browning is dependent upon α-amino N and not sugars but in practice α-amino N is rarely limiting and so does not usually determine the rate of reaction or the colour developed. As a result the amount of colour is largely dependent on the amount of reducing sugars present (Sweetman, 1930; Schwimmer *et al.*, 1957; Schallenberger *et al.*, 1959) except at high temperatures when hydrolysis of sucrose occurs (Denny and Thornton, 1940) and the disaccharide content is then important. Nevertheless, the correlations between browning and reducing sugar content vary considerably, from 0.32 to 0.99. (See Carlsson, 1970; Habib and Brown, 1956, 1957; Hope *et al.*, 1960; Schwimmer *et al.*, 1957; Schallenberger *et al.*, 1959). Habib and Brown (1956) and Hope *et al.* (1960) ascribe these variations to amino N irrespective of changes in reducing sugar content. It is difficult, however, to compare the results of these authors since the thickness of slice fried and the frying temperatures differed. Ascorbic acid is also thought to play a role in browning since paper discs soaked in it and then fried will brown but the concentration in the tuber is probably never high enough to cause an unacceptable level of browning (Talburt and Smith, 1967).

Because of the importance of obtaining samples which are free from browning on processing considerable efforts have been made to specify an acceptable upper limit of reducing sugar. This is usually about 0.10% on a FW basis but levels up to 0.25% may sometimes be accepted (Burton and Wilson, 1970). Higher levels can be accepted depending on the cooking method (Carlsson, 1970) and this aspect is fully discussed by Talburt and Smith (1967).

At tuber initiation both reducing sugars and sucrose are at high levels (0.75% and 1.5% on a FW basis, respectively) and then fall, more rapidly for sucrose, as the tubers grow (Burton and Wilson, 1970). Towards harvest these rise again but the amount of reducing sugar probably increases more than for sucrose possibly as a result of a decrease in soil temperature (Carlsson, 1970) promoting invertase activity (Sowokinos, 1973). The content of reducing sugars at harvest is affected by environmental conditions particularly temperature but varieties differ considerably in both initial contents of reducing sugars and also the level at harvest (Carlsson, 1970). Sowokinos (1973) showed that varieties with a consistently good processing performance after a period of storage depleted their reducing sugar content more rapidly during growth of the crop than those of poor processing performance. Good processing performance after a period of storage was also related to low-levels of sucrose at harvest. However, it is still not clear from this or other work (Vliet and Schreimer, 1960) precisely how levels of sucrose and reducing sugar at harvest and after a period of storage are related (Isherwood, 1973) because although invertase activity is important in causing increased reducing sugar levels at harvest (Sowokinos, 1973) it does not appear to be so important during storage (Pressey, 1969).

In the north of the U.K. (56–58° N) sugar levels in tubers are always higher throughout growth than in those grown in the south (52° N) and usually the tubers brown more frequently on frying (Burton and Wilson 1970). Low temperatures just before harvest also increase the incidence of browning (Kissmeyer-Nielsen and Weckel, 1967) and probably account for high sugar levels and greater browning in crops with delayed maturity either brought about by late planting or high levels of N (Kunkel and Holstad, 1972). Although irrigation may prolong foliage growth (Llewellyn, 1967) its effect on browning are variable depending on the stage of irrigation. Applications of water late in the season give considerably higher levels of browning than early applications (Talburt and Smith, 1967), who also concluded that the effects of P and K fertilizers on chip colour are small although Welte and Müller (1966) have shown that increasing levels of K can lead to lower reducing sugar levels.

In practice the levels of sugars are controlled by the temperature of storage. By increasing the temperature tubers with high levels of sugar can be reconditioned (see Chapter 14) but not all varieties react in the same way (Burton, 1966) and there is often considerable variability from tuber to tuber.

13.3.4 Texture of the cooked potato

Many field experiments and investigations of the chemical composition of the tuber have been carried out on various aspects of cooked texture since the turn of the century (see reviews by Burton, 1966; Talburt and Smith, 1967; Reeve, 1967; Linehan and Hughes, 1969a; Buren, 1970; Warren and Woodman, 1974). Although it is known that texture is dependent on variety, maturity and environment, it is still a matter of some controversy which intrinsic factors play a role and it is not always possible to produce tubers with the correct textural properties. This problem is particularly acute when potatoes are grown for processing. The reasons for this are the complexity of the intrinsic factors involved in controlling texture, the interaction of variety, maturity and environment and the definition and measurement of texture.

Texture of foodstuffs depends on the interaction of a number of physical characteristics – hardness, cohesiveness, viscosity, elasticity and adhesiveness (Szczesniak, 1963). It is suggested that in any particular food, texture would be the result of the interaction of a number of these characteristics.

Various terms have been used to describe the texture of cooked potatoes – mealy, floury, smooth, grainy, coarse, waxy, soapy, soggy. Mealiness is a term that is frequently used in American literature and is synonymous with 'floury'. Mealiness (and flouriness), by most definitions used in the literature, appears to have three main components: friability, (disintegration, crumbliness), mouth feel, and dryness (from appearance). However, it is not always possible to assess the relative weighting that various workers have given to disintegration and palatability in the subjective assessment of mealiness. Because of these difficulties a more detailed and a better defined system was introduced by the European Association of Potato Research (Lugt *et al.*, 1962) which has the following categories: consistency, mealiness, dryness and structure.

Many objective methods have been used in an attempt to measure the texture of the cooked potato (see Linehan and Hughes, 1969d; Warren and Woodman, 1974). However, it is not always clear which characteristics are being measured. These methods are based on weight loss after cooking, compressive, puncture and tensile measurements and extrusive methods.

The tendency for potatoes to break down on cooking (slough) which is generally agreed to be due to cell separation, plays an important role in determining the suitability of potatoes for particular processed products or for domestic use as salad, boiled or baked potatoes. However, there is still some controversy as to the relative role that starch and pectic substances play in cell separation (see Hoff, 1972; Warren *et al.*, 1975) and it is still not clear why certain samples of potatoes disintegrate or slough more than others. Specific gravity has often been shown to be correlated with mealiness and breakdown, particularly within a variety, although there are many instances where a relationship has not been found (see Linehan and Hughes, 1969a; Gray, 1972). The hypothesis that cells separate and rupture on cooking due to the exertion of a swelling pressure produced by gelling of starch granules has become firmly established in the literature. However, there appears to be no experimental evidence for starch swelling pressure (see Hoff, 1972) and, moreover, certain workers have found the reverse effect; increased firmness (Rathsack, 1935; Barmore, 1937; Linehan and Hughes, 1969b) or retained weight (Bretzloff, 1968) being associated with higher amounts of starch or amylose. However, Warren and Woodman (1974) believe that increased firmness found with higher levels of starch is a reflection of increased viscosity and not cell adhesion.

Although it is generally accepted that the solubilization during cooking of pectic substances in the middle lamella will play a role in determining cell adhesion of cooked potatoes, it has proved extremely difficult to demonstrate. The evidence has often been circumstantial and until recently (Hughes 1971; Hughes and Faulks, 1973; Hughes *et al.*, 1975a, b,c; Warren *et al.*, 1975) a direct relationship has only been found in a few instances (see Linehan & Hughes, 1969a,b). There are possibly a number of reasons why it has proved difficult to demonstrate. Firstly, the cooked potato is very high in gelled starch which may mask the effect of pectic substances and also give difficulty in analysis. Secondly, the large variations which occur in cell size in potatoes (Barrios *et al.*, 1963; Linehan *et al.*, 1968; Gray, 1972) may also conceal the effect of pectic substances, and thirdly, extraction procedures for pectic substances are often empirically derived and are not based on knowledge of the organization of the cell wall. Some of these difficulties were overcome in experiments carried out by Hughes and co-workers (Hughes, 1971; Hughes *et al.*, 1975a). It was shown that when starch and cell size were kept constant in the starting material (by using discs) compressive strength (altered by varying cooking time and addition of ions) was closely related to the degradation of pectic substances as reflected by their release into the cooking medium. Furthermore, a relationship between compressive strength and release of pectic substances on cooking could even be found in material which differed in cell size (tubers of different maturity) provided the pectic substances released on cooking were expressed on a cell wall basis (Hughes and Faulks, 1973; Hughes *et al.*, 1975c).

Pectic substances in plant cell walls are a mixture of polymers of partially esterified galacturonic acid and neutral sugars which may be broken down during cooking, to lower molecular weight products by a number of mechanisms, for example, removal of Ca and Mg bridges between galacturonic acid chains by chelation or ion exchange, cleavage, within a galacturonic acid polymer, of glycosidic bonds adjacent to methoxyl groups by transelimination and hydrogen bond breakage. However, there is no general agreement at present as to the relative role these mechanisms play in the breakdown of pectic substances in the potato during cooking. It has been generally attributed to the chelation of Ca by the naturally occurring chelating agent phytin. This theory has principally gained support from the effect of chelating agents on raw tissue and from the firming effect obtained by the addition of Ca salts prior or during cooking (see review Linehan and Hughes, 1969a; Hughes *et al.*, 1975b). However it has been shown recently that solubilization of pectic substances in potato tissue on cooking can occur by transelimination (Keijbets, 1974) and ion exchange and hydrogen bond breakage (effect of monovalent ions) (Hughes and Faulks, 1973, Hughes *et al.*, 1975c).

The conditions prior to and during cooking and composition of the tissue, for example, methoxylation of pectic substances, pH and ions, have a considerable effect on the extent and type of degradation that the pectic substances undergo. For example, during the boiling of potatoes, the high temperature and the near neutral pH of the tissue favour breakage of methoxylated pectic substances by transelimination (Albersheim, 1959; Albersheim *et al.*, 1960). If tubers are held at moderate temperatures (50–80°C) prior to boiling sloughing is reduced. This effect was originally attributed to starch retrogradation and a concomittant reduction in starch swelling pressure (Reeve, 1967), but Bartolme and Hoff (1972) believe that it is due to the demethoxylation of polyuronide units by pectin methylesterase at moderate temperatures resulting in an increased number of carboxyl groups becoming available for Ca and Mg bridge formation. The possibility exists that the increased firming could also have resulted from a reduction in transelimination due to a loss of methoxyl groups. Pre-soaking of potatoes at room temperature or the addition of Ca salts can also reduce sloughing on cooking.

The former effect is thought to be due to the leaching, prior to boiling, of monovalent ions (Zaehringer and Cunningham, 1971) which have been shown, during cooking, to be as effective as the chelating agent sodium hexametaphosphate in solubilizing pectic substances and reducing intercellular adhesion (Hughes *et al.*, 1975c). Although the effect of $CaCl_2$ has been attributed to formation of Ca bridges between polyuronide chains, it may be partly due to a reduction in transelimination due to a lowering of the pH of the cooking liquor which occurs when potato tissue is boiled with $CaCl_2$ (Hughes, 1971; Hughes *et al.*, 1975a, b). It would appear from the above and circumstantial evidence in the literature (Wager, 1963; Davies and Le Tourneau, 1967; Bretzloff and McMenamin, 1971; Hughes *et al.*, 1975c) that chelation of Ca by naturally occurring phytin does not necessarily play the key role in the solubilization of pectic substances during cooking.

It has been shown that the composition of the polymers – the relative amount of galacturonic acid and neutral sugars – released from the potato tissue on cooking is altered during the course of cooking (Hughes and Faulks, 1973) and also possibly by the presence of monovalent ions (Hughes *et al.*, 1975c). This work indicates that before the complex relationship between pectic substances and intercellular adhesion can be evaluated it may be necessary to investigate the changes in polymer structure that occur on cooking in relation to the molecular structure of the pectic substances in the raw potato cell wall. Although some very elegant work on the detailed molecular structure of pectic material has been carried out on other plant material (Bauer *et al.*, 1973; Talmadge *et al.*, 1973; Keegstra *et al.*, 1973) no such detailed work has yet been carried out on potato cell walls.

In other plant material, a protein component has been shown to be important in the structure of cell walls (Lamport, 1965). A protein component has been found in potato cell walls (Hoff and Castro, 1969) but work to the present indicates that it does not play a key role in cell adhesion of cooked mature potatoes (McClendon, 1964; Linehan and Hughes, 1969c). It may, however, be important in immature tubers.

It is now doubtful whether starch can directly increase cell separation. The relationship found between specific gravity and increased disintegration as measured by sloughing or mealiness is probably indirect and possibly reflects differences in cell wall composition or cell size. Starch, however, may directly affect the palatability aspect of mealiness (see Warren and Woodman, 1974). Although the release of starch from cells can play a major role in the texture of reconstituted potato products, this will not only be dependent on the amount and quality of starch and the presence of ions, which can affect the swelling of starch, but also on the composition of the cell walls and the changes which take place in the cell walls on cooking (see review by Reeve, 1967).

Increased cell size has been shown to be related to reduction of cell adhesion on cooking (Barrios *et al.*, 1963; Linehan *et al.*, 1968; Gray, 1972). Cell size is affected by variety (Linehan *et al.*, 1968), maturity (Plaisted, 1957; Gray, 1972) and environmental conditions (Lehmann, 1926; Reeve, 1971). It is not clear whether this is a direct effect of smaller cells giving a greater surface area per unit volume of tissue and greater strength or whether it reflects differences in cell wall composition between small and large cells.

Possibly all the above factors play some part in influencing the texture of the cooked potato. Many of these factors are determined genetically and by the state of development and type of growth pattern (relative amount of cell division compared to cell expansion) of the tuber, consequently the most consistent differences in texture are produced by variety (see Howard, 1974), maturity (see reviews) and plant spacing (Gray, 1972). Mature tubers and those grown at close spacing generally slough more than immature tubers and those from low populations (Gray, 1972). Although the effect of maturity has been correlated with specific gravity and cell size (Gray, 1972) differences in cell wall composition may also play a role (see Hughes *et al.*, 1975c).

The effect of climate, site and fertilizers on texture are often less consistent and more difficult to predict possibly because of the complex interaction of environmental components on the growth pattern of the tuber. Although potassium sometimes reduces mealiness (see reviews), year, site and even field may have as great an effect as fertilizers (Le Tourneau and Zaehringer, 1965; Schippers, 1968). Generally hot years produce tubers which disintegrate more readily on cooking.

The precise control of texture by manipulation of environmental factors may thus be dependent on attaining a greater understanding of the intrinsic factors involved (cell size and cell wall composition and of the effect of the environment on the growth pattern of the tuber with which these compositional factors are closely related.

13.3.5 Mechanical damage including internal bruising

Mechanical damage is a major problem in most potato growing countries. In a recent survey carried out in the U.K. (Potato Marketing Board, 1974b) a third of the tubers showed some form of damage by the time they reached the store. Apart from the losses incurred due to rejected tubers additional losses to the producer result from the higher labour cost of sorting out damaged tubers and increased wastage in store due to increased moisture loss and to the higher incidence of secondary infections.

Mechanical damage can be broadly divided into two groups, firstly external damage including skin damage, in the form of scuffing, splits and crushing injury and secondly internal bruising (blackspot). Although scuffing damage does not usually include damage to the flesh this aspect of damage is often considered under splits. All these forms of impact damage can occur during harvesting or further handling, as a result of tubers striking moving or stationary parts of the equipment, striking stones, clods or one another. Internal bruising can also be affected by the development of pressure spots as a result of prolonged storage. All these aspects of damage are affected by variety, maturity and growing and storage conditions. Cracking and internal bruising due to impact are also markedly increased by handling at low temperatures.

The amount of damage occurring on impact will be determined by the mechanical and rheological properties of the tuber (its susceptibility to damage) and the external forces exerted upon it. The external forces exerted on the tuber will be determined by the height of fall and weight of tuber (velocity and kinetic energy at impact), the shape of the tuber, surface area of contact, the speed of moving parts and the cushioning effect of soil. Damage can therefore be controlled by manipulation of two groups of factors. Firstly by selecting tubers with appropriate mechanical and rheological properties. Whilst these tuber properties will be determined by variety, maturity and environmental factors during growth and storage, the rheological properties are also affected by temperature. Secondly by reducing external forces imposed on the tubers during lifting and various stages of handling by proper design and use of mechanical lifting and handling machinery (see Chapter 9).

An understanding of the rheological properties of the potato is therefore fundamental for the rational assessment of the way variety and environmental conditions affect the susceptibility of tubers to damage, and also the amount and type of external forces that tubers can withstand without damage, which is important for the design and operation of mechanical equipment. Investigations of the rheological properties of potato tissue e.g. tissue strength and elasticity or firmness have been carried out in relation to external cracking and internal bruising (blackspot), using a range of techniques (Finney, 1969, 1972). Tissue strength is usually evaluated as the total energy required to rupture material (force deformation curves) or the force to rupture point. The factors that may determine the force required to rupture the tissue are the strength of cell walls and strength of the intercellular adhesion (Finney, 1969). Elastic properties and firmness of the tuber have been evaluated by determining the force to achieve a given deformation or deformation at a standard force or the force deformation ratio. One characteristic they may all measure to some degree is the modulus of elasticity (Finney, 1969), which can also be measured by resonance frequency techniques (Falk *et al.*, 1958). Finney (1969) considered that the modulus of elasticity could be affected by at least three factors; rigidity of walls, stiffness of intercellular bonding and turgor pressure within the cells.

The direct assessment of various types of damage is either carried out on material as harvested or on carefully harvested material which has been damaged by laboratory tests. These tests may closely simulate certain harvesting conditions as in the case of shaker tests or they may reproduce in other ways the type of forces the tuber is likely to receive during harvesting and handling. The latter tests consist of dropping tubers from given heights on to different surfaces or dropping weights on to tubers either directly or using a pendulum system. Systems using the tuber or a weight as part of the pendulum have the advantage that it is possible to inflict damage and obtain information about the rheological properties of the tuber during impact.

External damage

Although external damage is controlled to a large extent by variety it can also be markedly affected by environmental conditions particularly site and season. However the effect of individual environmental factors is often difficult to evaluate because of the interaction of growing conditions and also their diverse effects on the rheological properties of tubers and on tuber size. Moreover, the effect of growing conditions and even variety on the potential damage behaviour of tubers may be masked by harvesting conditions (soil structure, moisture, temperature and equipment used) which are known to influence damage greatly (Potato Marketing Board, 1974b). An understanding of the relationship between damage and the external forces exerted on the tuber during harvesting is necessary in the design and operation of harvesting and handling equipment and is discussed in Chapter 9, but see also Tabachuk, 1953; Timofeev, 1956; Volbracht and Kuhnke, 1956; Hesen and Kroesbergen, 1960; Parke, 1963; Meinl, 1972.

In order to assess the influence of variety and environmental conditions on damage many workers have attempted to overcome the influence of harvesting conditions by using rheological tests (tissue strength and aspects of elasticity and firmness) to predict damage (see Lampe, 1960; Meinl and Effmert, 1966; Gall *et al.*, 1967; Robertson, 1970; Hunnius and Munzert, 1972; Hunnius *et al.*, 1972; Meinl, 1972; Blight and Hamilton, 1974). One of the important properties of the tuber in relation to splitting and cracking is possibly the strength of the skin and underlying tissue. The increased strength of potato tissue, as measured by penetrometer methods (Lampe, 1960; Robertson, 1970) or more sophisticated force deformation curves (Hunnius and Munzert, 1972), has been shown to be related to a reduction in field damage. However, tissue strength, which appears to be largely a varietal character although also influenced by growing conditions, does not appear to explain all the variation found in field damage. Indeed, when a very wide range of varieties were investigated by Blight and Hamilton (1974) some of the 62 varieties gave anomalous results; for instance Desiree had within its maturity group the lowest damage index although the low tissue strength readings indicated otherwise.

There may be a number of reasons why penetrometer readings are an unreliable test for predicting damage. It may be that static tests do not necessarily reflect how tubers behave under impact, or that other rheological properties of the tuber, possibly elasticity or firmness, may also be important factors in determining damage susceptibility. The elasticity of tubers, measured from the rebound of a pendulum, has been shown to be related to mechanical damage but again not all the variation in damage could be accounted for in this way (Gall *et al.*, 1967). Although certain workers (Meinl and Effmert, 1966) consider that internal pressure on the skin or flesh cannot be used as a means of predicting damage, tuber firmness nevertheless has been shown consistently to influence the type of damage produced on impact. It has been shown directly and indirectly within a variety, that excessively firm tubers are more prone to external cracking whilst flaccid tubers are more prone to internal bruising (Werner and Dutt, 1941; Hesen and Kroesbergen, 1960; Sawyer and Collin, 1960; Finney and Findlen, 1967). A more foolproof rheological test for predicting mechanical damage may be dependent therefore on developing methods which take into consideration the interaction of a range of rheological properties, e.g. tissue strength and 'elastic' properties, under dynamic conditions.

Finney *et al.* (1964) showed that rheological properties, which may be associated with damage, altered towards the end of the growing season and during curing after harvest. They suggested that the tubers would become more susceptible to external damage as the tubers matured shortly before harvesting, but much less susceptible after curing when the energy they could withstand increased by 30–80%. Hesen and

Kroesbergen (1960) also showed a reduction in susceptibility to external damage during storage but an increase in internal bruising.

Field damage has been shown to be affected by site, season, fertilizers and cultural practices. However, for reasons previously stated it is not always clear why differences occur. It has often been shown that soil conditions affect damage. More damage is found generally in tubers harvested from sandy soils than in tubers harvested from clay soils (Hesen and Kroesbergen, 1960; Potato Marketing Board, 1974b). Rainfall and irrigation have been found to increase damage. Lampe (1960) found that the centre with the highest rainfall produced tubers which damaged with the lowest rupture force, and Finney and Findlen (1967) showed that tubers from irrigated plots damaged more than tubers from unirrigated plots. Generally cultural conditions which reduce the turgor or firmness of the tuber, such as root cutting and cessation of irrigation well in advance of harvesting, reduce susceptibility to mechanical damage (Werner and Dutt, 1941; Finney and Findlen, 1967). Increased tuber weight has been shown to increase damage (Parke, 1963; Meinl, 1972; Hunnius *et al.*, 1972). Environmental conditions, which increase tuber size, may therefore increase the number of damaged tubers. However, when tuber size is taken into consideration, there may nevertheless be an effect of environment on the rheological properties of tubers and their susceptibility to damage. For example, it has been shown that N fertilizers can affect the rheological properties of tubers (Meinl and Effmert, 1966) and reduce damage (Hunnius *et al.*, 1972). Hunnius and co-workers showed that when tubers from various N treatments were compared with tubers of a similar size from the treatment receiving no N, the percentage of severely damaged or multiple damaged tubers was sometimes significantly reduced. These workers suggested that the increased ability of these tubers to withstand damage was due to a reduction in cell size which is known to occur with N fertilizers (Lehmann, 1926).

Internal bruising (blackspot)

Although a considerable amount of work has been carried out by means of field experiments and storage experiments, it is not clear why certain tubers, particularly of different varieties, are more prone to blackspot than others (see reviews by Jacob, 1959; Burton, 1970; Zaag and Meijers, 1970). Blackspot, which is more prevalent at the stem end of the tuber, is typically a blue–grey (but sometimes brown) spherical zone in the region of the vascular tissue found one to three days after impact damage. The skin often shows no visible signs of damage. The pigments responsible for blackspot are produced by the oxidation of phenolic substrates (tyrosine and possibly chlorogenic acid) by phenolases. During the course

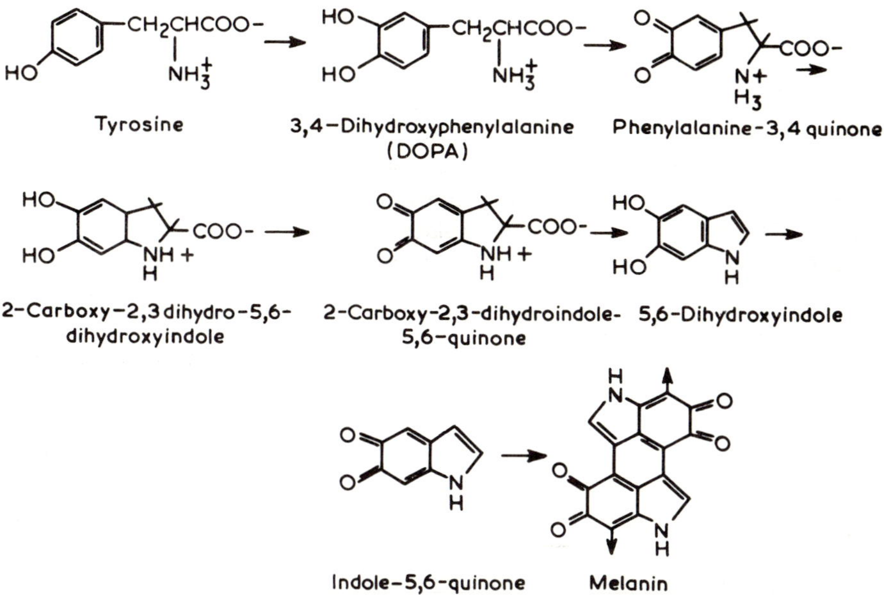

Fig. 13.1 The formation of melanin from tyrosine. The first two reactions are catalysed by potato phenolase. The others are or can be non-enzymic

of the reaction, brown, red and finally black (melanin) pigments are formed by the oxidation of tyrosine and lighter coloured products by the oxidation of chlorogenic acid (see Fig. 13.1). Under conditions of stress, breakage of the cells or membranes occurs, the cell contents become disorganized and enzymes and substrates react. Although it is known that the level of enzymes and substrates can increase as a result of stress, such as impact damage and radiation (Cheung and Henderson, 1972) it is not clear what role these increased levels play in the development of blackspot. It has been shown that the intensity of blackspot in impacted tubers of a number of varieties was similar, whether the blackspot was allowed to develop naturally over three days or rapidly (2 h) using oxygen under pressure (Duncan, 1973). In these experiments it would appear that the time needed to develop blackspot was dependent on the availability of oxygen.

There are, therefore, a number of intrinsic factors which determine the susceptibility of a tuber to blackspot, which can be put into two broad categories. Firstly the potential of the tissue to produce coloured oxidation products which depends on the presence of sufficient enzymes, substrates and oxygen since these can limit either the rate or amount of pigment formed. Secondly, the susceptibility of the tissue to damage since unless the cells are damaged sufficiently for the enzyme and substrate to react no blackspot will develop.

The former will be determined by the biochemistry of the tuber and the latter by mechanical and rheological properties of the tuber and the externally applied stress. It is therefore not surprising that biochemical properties of the tuber alone do not explain blackspot (Vertregt, 1968). Nevertheless, until recently (Es and Hartmans, 1975; Hughes *et al.*, 1975d; Olsson and Fridell, 1975) the susceptibility of the internal tissue to cell damage had received less attention than the biochemical aspects. The oxidation of tyrosine by phenolase to produce melanin is well documented and the effect of variety and environmental conditions on both enzyme and substrate have been investigated by a number of workers (see Section 13.3.6 and Burton, 1970).

The susceptibility of the internal tissue to damage will be dependent, as in the case of splitting and skin breakage, on the mechanical and rheological properties of the skin and underlying tissue. However, the relative role and the rheological properties of the two tissues involved may be different in these two problems. Robertson (1970) suggested that blackspot takes place in tubers with a strong skin, but a weak underlying tissue. Firmness, turgor pressure and specific gravity have been consistently shown to be related to blackspot.

It has been shown, by laboratory and field experiments, that, over a certain range of turgidity, blackspot susceptibility is reduced by an increase in turgidity and that the converse applies when turgor is reduced by water loss using sugar solutions or drying conditions in the field or store (see Kunkel and Gardner, 1959, 1965). Sawyer and Collin (1960) used a durometer for measuring firmness and showed a direct relationship between loss of firmness (turgidity) and increased blackspot within a variety. Turgor, however, does not always explain the differences in susceptibility between varieties and moreover extremely low or high turgor levels can have inconsistent effects. For example, when the tubers become extremely flaccid due to sprouting or drying conditions in the field while the plants are still green, the tubers become less prone to blackspot (Kunkel and Gardner, 1959), whilst extremely turgid tubers may be prone to shattering (Kunkel *et al.*, 1970).

The role of turgor, within a variety, in controlling blackspot may possibly be explained by its effect on the mechanical properties of the tissue. It is known that cells with a low turgor pressure deform to a greater extent under a given force than cells with a high turgor pressure (Nilsson *et al.*, 1958), consequently under impact these cells may be more likely to reach the critical deformation point at which damage to walls and membranes occurs. However, between varieties, it has been shown that as well as modulus of elasticity which is determined to a large extent by turgor pressure other rheological properties such as force to rupture point and deformation to rupture point also vary (Finney *et al.*, 1964). Two groups of factors may therefore be important in determining differences in internal tissue damage between varieties. Firstly the amount of deformation that takes place for a given force which will depend on the stress in the walls (turgor pressure), and secondly the amount of deformation the tissue can withstand before damage to membranes and walls occur (see Hughes *et al.*, 1975d). Both these factors may be influenced by cell size and in certain instances increased cell size has been shown to increase susceptibility to internal damage (Es and Hartmans, 1975; Olsson and Fridell, 1975) and bruising (Hudson, 1975). The above factors together with variations in the biochemical properties of the tuber could explain why firmness or turgor do not always account for varietal differences in blackspot susceptibility.

A direct relationship has been found between specific gravity and blackspot within varieties, within groups of plants under irrigation and vine-killing operations, K experiments and in storage (see Ophuis *et al.*, 1958; Jacob, 1959; Kunkel and Gardner, 1965). However, between varieties the relationship has been inconsistent, e.g. Ophuis *et al.* (1958) found a relationship whilst Jacob (1959) and Sawyer and Collin (1960) failed to do so. Jacob stated that some of the more susceptible varieties had a lower specific gravity than some of the more resistant varieties. There is, however, still some disagreement whether the effect of specific gravity is direct or whether it reflects changes in water content with its concomitant effect on turgor (Kunkel and Gardner, 1965; Kunkel *et al.*, 1973).

In stored potatoes a very good relation exists, within varieties, between specific gravity and firmness as measured by a durometer (Sawyer and Collin, 1960) and water loss and tuber firmness as measured by compression (Sharrock, 1968). The relationship found between increased blackspot and increased specific gravity in storage experiments, and possibly also in irrigation and vine-killing experiments, may therefore be indirect and reflect changes in turgor due to water loss. However, it cannot be excluded that specific gravity, in certain material, is indirectly related to blackspot in other ways. For example, specific gravity may reflect changes in cell size or properties of the cell walls. On the other hand specific gravity may even have a direct effect on damage because of its close relationship with the amount of starch in the tuber which may damage cell membranes when the cells are rapidly deformed under high impact velocities.

Although increasing the level of applied K or N reduces specific gravity of tubers the effect of these two fertilizers on blackspot is quite different. K generally reduces blackspot whilst N sometimes has the opposite effect. Specific gravity therefore cannot be directly or indirectly the major factor in determining blackspot in both instances. Within a K trial a reduction in blackspot may be associated with tubers with a lower specific gravity and higher K content (see Zaag and Meijers, 1970), however the reduced susceptibility is probably due to a combination of a number of compositional factors all of which alter in a favourable direction as a result of high applications of K. It has been shown for instance that both the potential of the tissue to produce coloured oxidation products (see section 13.3.6) and the susceptibility of the tissue to damage (Hughes *et al.*, 1975d) are lower in tubers from plants fed high levels of K. It has been suggested that turgor may play a role within K treatments (Kunkel *et al.*, 1973). However Hughes and co-workers (1975d) showed that, although less internal damage occurred in tubers from plants fed high levels of K, tuber firmness and the amount of dynamic deformation on impact was similar in tubers from all K treatments. This reduced susceptibility may be due to differences in the properties of cell walls, cell size and specific gravity which make the tissue resistant to damage even at dynamic deformations which cause cell damage in tubers from low K treatments. It is possible that K itself may even have a direct effect on the extensibility of the walls as has been shown for other plants (Probine and Preston, 1962; Rayle and Cleland, 1970). It is not clear why high levels of applied N sometimes increase susceptibility to blackspot but it may be due to an interaction of a number of factors including increased tuber size and possibly differences in tuber turgor (Kunkel *et al.*, 1973).

Low temperature during lifting (Johnson and Wilson, 1969) or handling during storage (Waint *et al.*, 1951; Ophuis *et al.*, 1958) increases blackspot, possibly through its effect on the rheological properties of the cells. The injury may be reduced by raising the temperature of susceptible tubers before handling (Fig. 13.2) (Waint *et al.*, 1951; Ophuis *et al.*, 1958). However, this effect of temperature has recently been shown to be modified by tuber turgidity (Smittle and Thornton, 1972). Carbon dioxide content of the soil has been suggested, by certain workers (Howard *et al.*, 1961; Yamaguchi *et al.*, 1964b), to be a contributing factor in increasing susceptibility of tubers to blackspot in wet soils. Kunkel *et al.* (1970), however, found completely conflicting results; blackspot decreased in tubers from moist soils and CO_2 reduced or had little effect on blackspot. They suggested that the differing results could be due to the occurrence of different types of blackspot (Weaver and Steen, 1966).

The development of pressure spots after extended storage may also influence the amount of blackspot which occurs when tubers are impacted during unloading of stores. The pressure areas on tubers (softened, flattened or indented areas occurring as a result of the pressure of adjacent tubers) are more susceptible to the development of blackspot, as a result of impact, than firmer regions of the tubers (Waint *et al.*, 1951). However the increased susceptibility of 'pressure spot' tubers to blackspot may be due to both the deformed tissue being more susceptible to damage, and also to these tubers being generally 'softer' than 'non-pressure spot' tubers and thus more susceptible to damage. In carefully handled material

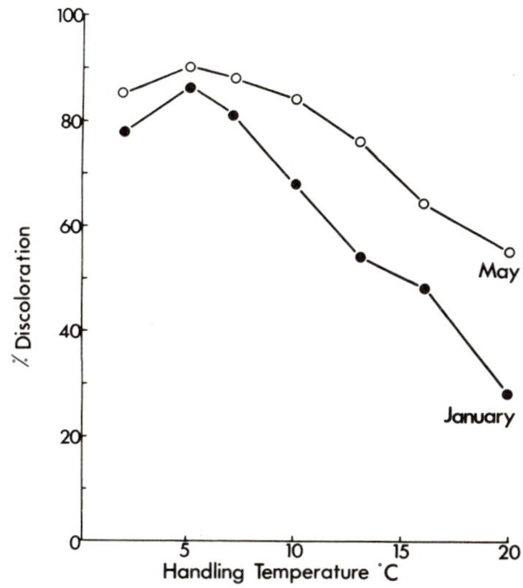

Fig. 13.2 Influence of length of storage period and temperature of handling on blackspot of cv. Eigenheimer (after Ophuis *et al.*, 1958)

(Waint *et al.*, 1951) blackspot only occurred after the 'pressure spot' tubers had been impacted.

Careful handling of tubers at suitable temperatures and the production of tubers which are not excessively large are key factors in controlling all forms of damage. Although it is known that both external and internal damage are affected by variety and environmental conditions the reasons for these differences are not well understood and the control of damage by these means is often imprecise. An explanation for these differences is possibly dependent firstly on more being known of the role that a range of rheological factors play in both aspects of damage and of how they are related to the composition of the tuber and secondly on the development of more foolproof tests for predicting damage so as to overcome the masking affect of harvesting conditions on damage. It should be stressed that both 'tissue strength' and elastic properties possibly play a critical although different role in external cracking and internal bruising and that care should thus be taken in predicting either type of damage on the basis of only one type of measurement. Furthermore the potential of the tissue to produce coloured oxidation products will also play an important role in blackspot. The most satisfactory test for this latter defect thus has been to damage tubers and assess the amount of blackspot produced.

As an oversimplification the role of the rheological properties of the tuber could be described as follows: an extremely turgid tuber with a 'weak skin' may be susceptible to external cracking whilst a flaccid tuber with a 'strong skin' but weak underlying tissue would be susceptible to blackspot providing it had a high potential for producing coloured oxidation products. From the above it can be seen that control of external damage either by variety selection or manipulation of environmental factors should be considered in close conjunction with the control of internal bruising otherwise there may be the danger of decreasing the susceptibility to one defect at the expense of increasing susceptibility to the other defect. Nevertheless despite our lack of knowledge some *ad hoc* control is possible. Variety, as in the case of most defects, plays a major role. However, since external cracking and internal bruising appear to be independent varietal characteristics, it may not always be possible to choose a variety which has the correct composition for a particular purpose, and which is also resistant to both defects. For example, the high tuber dry matter variety Record is extremely resistant to external cracking but very susceptible to internal bruising.

The manipulation of environmental conditions which affect tuber firmness may also help in the control of damage. Soil moisture and condition of the haulm around harvesting time and length and type of storage are all factors which are known to affect tuber firmness and consequently could markedly affect the type and amount of damage. However, it may be difficult to obtain the precise degree of firmness that is necessary to obtain the minimal amount of damage (Thornton *et al.*, 1973) since if the tubers are too firm cracking may occur whilst if they are too soft internal bruising may occur. Although the effect of environmental conditions on aspects of tissue strength are less well understood the reduction in blackspot susceptibility resulting from high applications of potassium on certain soils may be due to an effect on internal tissue strength.

13.3.6 Enzymic browning

The prevention of enzymic browning, which occurs when potatoes are cut or peeled, is important in the preparation of pre-peeled potatoes for domestic purposes, larger catering establishments and during the processing of potatoes when the holding time of pre-peeled or sliced potatoes may have to be extended.

The cause of enzymic browning is well understood. It is due to the oxidation of phenolic compounds by phenolase (Mapson and Swain, 1961). The two major phenolic substrates in the potato are chlorogenic acid and the amino acid tyrosine. It has been found that, although chlorogenic acid is readily oxidized in the presence of the enzyme, products of its oxidation are not deeply coloured, whereas tyrosine gives dark brown products and subsequently melanin pigments (see Fig. 13.1). Chlorogenic acid can also act as an activator for the enzyme, but as in the case of the enzyme, its concentration is not limiting (Mapson *et al.*,

1963). Consequently, of these factors, tyrosine is generally considered to be the most important in controlling the amount of enzymic browning.

As in the case of after-cooking blackening, a clear distinction is usually found between high and low browning varieties in material grown both in the field and under controlled conditions (Mapson *et al.*, 1963; Hughes and Mapson, 1967). The behaviour of intermediate varieties is, however, more variable, and they fall into either category depending upon the environmental conditions. This ranking is related to the amount of tyrosine they contain. However, in the case of certain varieties, seasons and growing conditions, the amount of browning produced by a given amount of tyrosine can be quite different (Hughes and Mapson, 1967; Umaerus and Olsson, 1974). Nevertheless, Umaerus and Olsson found that there were no cases where high tyrosine content gave low discoloration, although sometimes low tyrosine varieties gave fairly high discoloration. At a given site the ranking of extreme varieties is similar, but the behaviour of any individual variety can also be markedly affected by site and season (Mapson *et al.*, 1963; Hughes and Mapson, 1967). For example, the amount of browning and tyrosine was always higher in Majestic tubers grown at a site with acid peat soil and a high rainfall (Lancashire) than at a site with a sandy loam soil and a lower rainfall (Norfolk) (see Fig. 13.3).

The relative part that soil and climate played in controlling browning was investigated by Mapson *et al.* (1963). Their results indicated that soil has a smaller effect than climate on enzymic browning and tyrosine content. In certain cases, a direct relationship was found between the

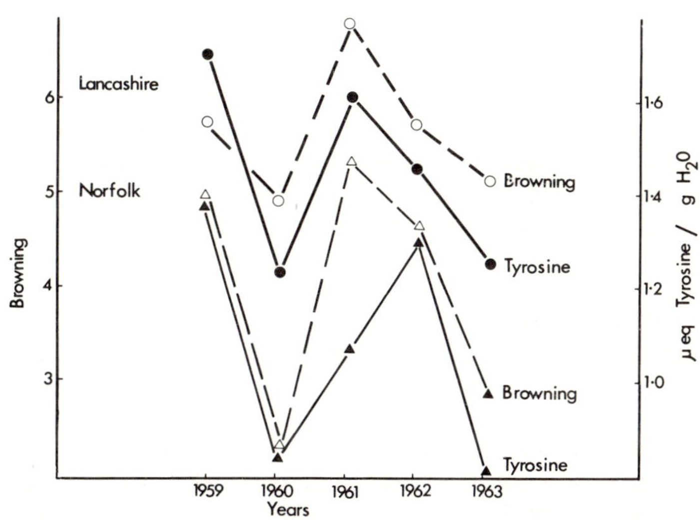

Fig. 13.3 Effect of site and season on the level of tyrosine and enzymic browning cv. Majestic (after Hughes and Mapson, 1967)

rainfall at a given centre and the amount of tyrosine in the tubers. These workers showed, in controlled water regime experiments, that the amount of browning and tyrosine increased as the amount of water given to the plants increased. The time of application of water also affected enzymic browning. For example, in experiments where one half of the total water was given either early, middle or late in the season, the amount of tyrosine was highest in the plants given extra water in the middle of the season. These results suggest that rainfall itself, and not reduction in sunlight and temperature, can affect enzymic browning.

Although the above field experiments indicate that climatic factors can often play a larger role than edaphic factors, mineral nutrition is known to affect enzymic browning. It has been shown by numerous workers that the levels of enzymic browning, tyrosine and sometimes phenolase are reduced when higher levels of K are fed to plants (Mulder, 1949; Welte and Muller, 1966; Mondy *et al.*, 1967; Rinno *et al.*, 1972). Differences have also been found between KCl and K_2SO_4; lower amounts of enzymic browning and tyrosine being produced when plants were fed large amounts of KCl compared with equivalent amounts of potassium as K_2SO_4 (Hughes and Mapson, 1967).

Although N fertilizers have been shown to increase considerably (sometimes 2-fold) the total soluble amino acid pool (see Hoff *et al.*, 1971), the level of tyrosine is only slightly reduced (Mulder and Bakema, 1956; Hoff *et al.*, 1971). N has therefore rarely been found to increase the level of enzymic browning, except under conditions of very low K (Rinno *et al.*, 1972).

As has been mentioned earlier, a direct relationship has been found between tyrosine and browning. However, this has been found to vary, depending on variety and environmental conditions (Mapson *et al.*, 1963; Hughes and Mapson, 1967; Umaerus and Olsson, 1974), suggesting that other factors play a part. Enzymic browning cannot occur unless the cells are cut or bruised, and the ease with which the cells under the cut layer may be damaged may, therefore, be a contributory factor, but it cannot be excluded that some other factors (e.g. ascorbic acid) may also be important.

13.3.7 After-cooking blackening

As its name implies, this defect shows up only after the tuber has been cooked, and appears at its worst when tubers of susceptible varieties are boiled or steamed. The discoloration also shows up in frozen French fries, canned and dehydrated potatoes. It is not confined to any one variety or cultural condition, although both were early shown to be important (see review by Hughes and Mapson, 1967; bibliography by Talley *et al.*, 1969). This discoloration is found in tubers grown in most parts of the world.

The pigment responsible for the discoloration is a complex of chlorogenic acid and iron, which is formed on cooking, and oxidized on cooling to give a coloured complex of ferri-dichlorogenic acid (Hughes and Swain, 1962a,b). However, the formation of this pigment can be affected by pH and by the presence of naturally occurring chelating agents (citric and malic acid and phosphate) in the tubers. It was found that of the substances which compete with the chlorogenic acid for the iron to form a colourless complex, citric acid was by far the most effective. The distribution of blackening within tubers was found to be dependent on the ratio of chlorogenic acid to citric acid; the latter acid always being higher at the bud end of the tuber, where little blackening takes place. Hughes and Swain (1962a) in contrast to Wurster and Smith (1963), found little variation in iron within the tuber. Although iron may play a part in determining the amount of blackening in different tubers, the ratio of chlorogenic acid to citric acid in the tuber has been shown to be a major factor accounting for differences in blackening within individual tubers and between varieties at different sites and in different years (Hughes and Evans, 1967).

It has long been known that after-cooking blackening was affected by variety, soils, fertilizers and season (Fig. 13.4). However, due to the interaction of these factors, the results often appeared to be conflicting,

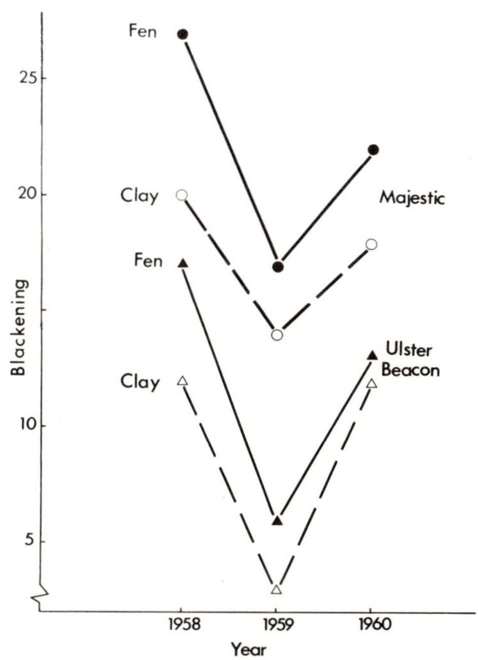

Fig. 13.4 Effect of variety soil and season on after-cooking blackening (after Hughes and Mapson, 1967)

although certain factors were usually found to make the tubers more susceptible to after-cooking blackening, namely high ratio of N to K and cool, wet seasons. Since the underlying factors controlling blackening have been known, it has been possible to assess the relative role that variety and environmental factors play in controlling the level of the substances involved and, to a large extent, explain the interactions and thus suggest logical ways of controlling this defect in the field.

Hughes and Evans (1967) showed that differences between varieties were largely due to varying amounts of chlorogenic acid and, to a lesser extent, citric acid. Although in any given variety the absolute amount of blackening and the levels of chlorogenic acid and citric acid were dependent on soil and season, the ranking of the extreme varieties for blackening in any given soil or year was usually similar and dependent on the amount of chlorogenic acid they contained and, to a lesser extent, citric acid. Thus, those varieties high in chlorogenic acid blackened the most, whilst those low in chlorogenic acid blackened the least. Varieties intermediate in chlorogenic acid varied in blackening according to the environmental conditions (Hughes and Evans, 1967). Whilst the level of these two acids can be used as an indicator of blackening behaviour, it has been shown that the relationship does not hold true for all varieties (Anonymous, 1969).

The effect of edaphic factors, unlike the varietal effect, is principally on citric acid and, to a lesser extent, chlorogenic acid (Hughes and Evans, 1967). These workers showed that potatoes grown on the fen soil (high organic matter with low K) blackened more than tubers grown in the clay soil (low organic matter and high K) and were low in citric acid and sometimes higher in chlorogenic acid. It is known that K affects the synthesis of organic acids in plants (Jacobsen and Ordin, 1954) and very good correlations were found between the K and citric acid content of the varieties examined by Hughes and Evans (1967). The lower amount of K and thus citric acid in these tubers is possibly due to the fen soil being lower in K and higher in N, which has an antagonistic effect on the uptake of K. Heisler *et al.* (1962) have also shown that blackening is related to the amount of K in the tubers but only in certain samples. However, better correlations were found between blackening and organic acids (Heisler *et al.*, 1964). In certain soils, the uptake of iron (Fe) may also be affected by soil conditions such as pH and high levels of Ca, and under these conditions Fe may play a more important role in controlling blackening. Fe has, in fact, been shown by Heisler *et al.* (1963) to play a part in controlling blackening but significant correlations were again only found in certain samples. Thus, soil will influence blackening by the way it affects the uptake of Fe and the minerals which affect the synthesis of the two organic acids.

The role that K and N play in controlling the chemistry and discoloration of potatoes has been substantiated by experiments carried out on material grown under controlled mineral conditions (Hughes and Mapson, 1967). It was shown as in field experiments, that the amount of citrate in the tuber was directly proportional to the amount of K in the tuber except when chloride was present in the nutrient solution fed to the plants. Due to the antagonistic effect of ammonium ions on the uptake of K the level of citric acid was reduced when ammonium ions or urea were present in the nutrient solution. Apart from the effect of high levels of nitrate and urea, mineral nutrition had little effect on the synthesis of chlorogenic acid. However, the effect of high N might explain the higher amounts of chlorogenic acid sometimes found in the tubers grown on the fen soil. Again, in these experiments good correlations were found between blackening and the ratio of chlorogenic acid to citric acid (Hughes and Mapson, 1967). The level of water soluble iron was found to be markedly altered only when extremely high levels of chloride were fed to the plants; these plants were chlorotic and had greatly reduced yields. There are, however, indications that, with certain phosphate treatments, the level of Fe can be reduced whilst the plants appear quite normal (Evans, private communication).

Potatoes grown in Great Britain and some other countries frequently blacken more in cool, wet years. The results of Hughes and Evans (1967) clearly indicated that this was due to the higher amounts of chlorogenic acid (and sometimes lower amounts of citric acid) found in these varieties in such years.

It is known that immature tubers sometimes blacken more than mature tubers (Wager, 1947; Hughes and Evans, 1969). It was shown by Hughes and Evans that the intensity of blackening decreased and that the distribution changed from bud to stem end during the season. These changes were associated with changes in the amount and distribution of citric and chlorogenic acid. Heisler *et al.* (1969) also showed that there was a difference in distribution of citric acid and K in larger tubers at maturity. Large tubers had less citric acid and K at the stem end than the smaller tubers. The changes in distribution of citric acid and K, therefore, may be a reflection of the growth pattern of the tuber and not of maturity *per se*.

An understanding of the underlying factors – the role of chlorogenic acid and citric acid and the extent to which they are affected by variety and environmental conditions – has made it possible to throw light on why K and N affect blackening, and also why some fertilizer experiments have given apparently inconsistent results or only slight effects (see reviews mentioned). Choice of variety can be critical in such experiments; low chlorogenic acid varieties will blacken little even with very low levels of K, since the former acid is the limiting factor. Thus control in the field can often be carried out by considering fertilizer application in relation to each site so as to avoid producing high ratios of N/K in the soil and also by choosing varieties carefully for each site. For example, at sites known to produce blackening, where ratios of N/K may be high because of soil structure and/or the climate cool and wet, low chlorogenic acid varieties should be chosen. At sites known to give little blackening where levels of K are high and levels of N are low and/or with a warm sunny climate, high chlorogenic acid varieties can be used.

Although consideration of these factors can often be used to control blackening, anomalous results are sometimes found (Anonymous, 1969). Possibly Fe may play a greater role in certain instances and investigation of the role of Fe and factors which affect its uptake should be considered.

13.4 General conclusions

The 'quality' and nutritional value of a tuber at harvest is a product of the effects of varietal, cultural and environmental factors on the growth of the crop but the extent to which anyone or number contributes to differences in quality depends upon the attribute being considered. For example, tuber shape, skin and flesh colour and depth of eyes are largely determined by variety, cultural and environmental factors being relatively unimportant. Because of this it is relatively easy to breed for improved shape or for reductions in the depth of eyes and exercise control over these attributes in the field. In some instances where environmental factors alone control a quality, for example, greening, an adequate soil cover during 'ridging up' will reduce or even eliminate it.

However, for other quality attributes, such as second growth and hollow heart, DM content, texture, enzymic and non-enzymic browning, external damage, internal bruising, after-cooking blackening and nutritional quality, both environmental and cultural factors have a more variable effect and this is often complicated by interactions between these and with variety. Consequently it is often difficult, particularly when using susceptible varieties, to specify cultural practices which will give a consistent quality.

Despite the complexity of the effects of the environment on quality studies of the basis of differences, for example, in after-cooking blackening and how these are affected by environmental and cultural factors has placed methods of controlling this disorder on a more logical basis. As a result it has also been possible to outline successful selection procedures for breeding programmes (Howard, 1974) aimed at producing varieties with consistently low after-cooking blackening indices. In other cases e.g. nutritional quality, damage and texture a logical basis of control of these is not yet possible. The importance of DM and reducing sugar levels as determinants of chip yield and colour is well established but it is not yet possible to produce consistent quality because of the variable effects of cultural and environmental factors. The limitation to a greater understanding and therefore better basis for control of these two aspects lies in the consideration of the interaction of environmental factors on the carbohydrate balance in relation to the growth and maturity of the leaf canopy and tuber.

In the case of damage and texture, two increasingly important aspects of quality, control measures are as yet entirely empirical. The understanding of the effects of varietal, cultural and environmental factors on these is limited by an inadequate knowledge of the intrinsic factors involved. Both of these attributes appear to be determined to a considerable extent by the physical and chemical properties of the cell walls. However, further definition and understanding of the problem will depend upon the development of objective methods of measuring both damage and texture and the relation of the results from these tests to the rheological properties and chemical composition of the cell and anatomy of the tuber.

References

Akeley, R.V., Houghland, G.V.C. and Schark, A.E. (1962). *Am. Potato J.* **39**, 409–17.

Akeley, R.V., Stevenson, F. and Cunningham, C.E. (1955). *Am. Potato J.* **32**, 304–13.

Albersheim, P. (1959). *Biochem. biophys. Res. Commun.* **1**, 253–6.

Albersheim, P., Neukom, H. and Deuel, H. (1960). *Archs Biochem. Biophys* **90**, 46–51.

Anonymous (1969). *50th Report Accounts National Institute agric. Botany*, England, Cambridge. p. 48.

Anonymous (1972a). *F.A.O. Production Yearbook*. Rome.

Anonymous (1972b). Household Food Consumption and Expenditure. *Annual Report of the National Food Survey Committee*. London: H.M.S.O.

Appleman, C.O. and Miller, E.V. (1926). *J. agric. Res.* **33**, 569–77.

Artschwager, E. (1924). *J. agric. Res.* **27**, 809–35.

Baijal, B.D. and Van Vliet, W.F. (1966). *Eur. Potato J.* **9**, 179–92.

Barker, J. and Mapson, L.W. (1950). *New Phytol.* **49**, 283–303.

Barmore, M.A. (1937). *Fd Res.* **2**, 377–83.

Barrios, E.P., Newson, D.W. and Miller, J.C. (1963). *Am. Potato J.* **40**, 200–8.

Bartolme, C.G. and Hoff, J.E. (1972). *J. agric. Fd Chem.* **20**, 266–70.

Bauer, D.W., Talmadge, K.W., Keegstra, K. and Albersheim, P. (1973). *Pl. Physiol. Lancaster* **51**, 174–87.

Bleasdale, J.K.A. and Thompson, R. (1965). Potatoes. *Annual Report of National Vegetable Research Station, Wellesbourne for 1964*, 37.

Bleasdale, J.K.A. and Thompson, R. (1969). *Eur. Potato J.* **12**, 173–87.

Blight, D.P. and Hamilton, A.J. (1974). *Potato Res.* **17**, 261–70.

Bodlaender, K.B.A., Lugt, C. and Marinus, J. (1964). *Eur. Potato J.* 7, 57–71.

Bretzloff, C.W. (1968). *Am. Potato J.* **45**, 17–32.

Bretzloff, C.W. and McMenamin, J. (1971). *Am. Potato J.* **48**, 246–53.

Brunstetter, B.C. and Wiseman, H.G. (1947). *Pl. Physiol. Lancaster* **22**, 421–37.

Buren, J. van, (1970). *Critical Revs Fd Technol.* **1**, 5–24.

Burton, H.S. and McWeeny, D.J. (1963). *J. Sci. Fd Agric.* **14**, 291–302.

Burton H.S. and McWeeny, D.J. (1964). *Chem. Ind.* 462–3.
Burton, H.S., McWeeny, D.J. and Biltcliffe, D.O. (1962). *Nature* **196**, 40–1.
Burton, H.S., McWeeny, D.J. and Biltcliffe, D.O. (1963). *J. Sci. Fd Agric.* **14**, 911–20.
Burton, W.G. (1966). *The Potato : A survey of its history and of the factors influencing its yield, nutritive value, quality and storage.* Wageningen: European Association of Potato Research, H. Veenman and Zonen N.V.
Burton, W.G. (1970). *Proc. Fourth Triennial Conference Eur. Ass. Potato Res.* (Brest, 1969). pp. 79–92.
Burton, W.G. (1974). *Potato Res.* **17**, 374–409.
Burton, W.G. and Wilson, A.R. (1970). *Potato Res.* **13**, 269–83.
Caldwell, J.C., Brunstetter, B.C., Culpepper, C.W. and Ezell, B.D. (1945). *Canner* **100** (13), 35–122, **100** (14), 15–34, **100** (15), 14–27.
Carlsson, H. (1970). *Vaxtodling* **26**, 9–64.
Carpenter, P.N. (1949). Mineral accumulation in potato plants. *Bull. Maine Agric. Exp. Station*, 562.
Cheung, K.W.K. and Henderson, H.M. (1972). *Phytochemistry* **11**, 1255–60.
Chick, H. (1951). *Br. J. Nutr.* **5**, 261–5.
Chick, H. and Cutting, M.E.M. (1943). *Lancet* **245** (2) 667–69.
Clark, C.F. and Lombard, P.M. (1951). Description of and key to American potato varieties. *U.S.D.A. Circular* 741.
Chick, H. and Slack, E.B. (1951). *Biochem. J.* **45**, 211–21.
Cole, S.C. (1975). *Potato Res.* **18**, 28–37.
Corey, G.L. and Meyers, V.J. (1955) Irrigation of Russet Burbank potatoes in Idaho. *Idaho Agric. Exp. Station Bull.* 246.
Crumbly, I.J., Nelson, D.C. and Duysen, M.E. (1973). *Am. Potato J.* **50**, 266–74.
Davies, W.C. and Le Tourneau, D. (1967). *Am. Potato J.* **44**, 355–62.
Denny, F.E. and Thornton, N.C. (1940). *Contr. Boyce Thompson Inst. Pl. Res.* **11**, 291–303.
Desborough, S.L. and Weiser, C.J. (1974). *Am. Potato J.* **5**, 185–96.
Dickens, J.C., Harrap, F.E.G. and Holmes, M.R.J. (1962). *J. agric. Sci., Camb.* **59**, 319–26.
Drew, J.P. and Deasy, D. (1941). *J. Dep. Agric. Repub. Irl.* **38** (2), 220–38.
Duncan, H.J. (1973). *Potato Res.* **16**, 306–10.
Edmundson, W.C. (1935). Distance of planting Rural New Yorker no. 2 and Triumph potatoes as affecting yield, hollow heart, growth cracks and second-growth tubers. *U.S.D.A. Circular* 338.
Es, A. van, and Hartmans, K.J. (1975). *Potato Res.* 18, 338.
Falk, S., Hertz, C.H. and Virgin, H.I. (1958). *Physiologia Pl.* **11**, 802–17.
Finney, E.E. (1969). *Agric. Eng.* (Aug) 462–5.
Finney, E.E. (1972). *Fd Technol.* February 1972, 68–77.
Finney, E.E. and Findlen, H. (1967). *Am. Potato J.* **44**, 383–6.
Finney, E.E., Hall, C.W. and Thompson, N.R. (1964). *Am. J. Bot.* **41**, 178–86.
Gall, H., Lamprecht, P. and Fechter, E. (1967). *Eur. Potato J.* **10**, 272–85.
Glynne, M.D. and Jackson, V.G. (1919). *J. agric. Sci., Camb.* **9**, 237–58.
Gray, D. (1972). *Potato Res.* **15**, 317–34.
Groot, E.H. (1947). *Archs néerl. Physiol.* **28**, 277–361.
Groot, E.H., Janssen, L.W., Kentie, A., Oosterhius, H.K. and Trap, H.J.L. (1947). *Biochim. biophys. Acta* **1**, 410–14.
Gull, D.D. and Isenberg, F.M. (1958). *Proc. Am. Soc. hort. Sci.* **71**, 446–54.
Habib, A.T. and Brown, H.D. (1956). *Fd Technol.* **10**, 332–6.
Habib, A.T. and Brown, H.D. (1957). *Fd Technol.* **11**, 85–9.
Hampson, C.P. (1972). *J. R. agric. Soc.* **133**, 19–26.

Harward, M.E. Jackson, W.A., Nielson, L.W., Pilanol, J.R. and Mason, D.D. (1956). *Proc. Soil Sci. Soc. Am.* **20**, 59–65.

Heisler, E.G., Siciliano, J. and Porter, W.L. (1969). *Am. Potato J.* **46**, 98–107.

Heisler, E.G., Siciliano, J. and Treadway, R.H. (1962). *Fd Technol.* **16**, 120–4.

Heisler, E.G., Siciliano, J., Woodward, C.F. and Porter, W.L. (1964). *Fd Res.* **29**, 555–64.

Heisler, E.G., Siciliano, J., Treadway, R.H. and Woodward, C.F. (1963). *J. Fd Sci.* **28**, 453–9.

Hesen, J.C. and Kroesbergen, E. (1960). *Eur. Potato J.* **3**, 40–6.

Hodge, J.E. (1953). *J. agric. Fd Chem.* **1**, 928–43.

Hoff, J.E. (1972). *J. agric. Fd Chem.* **20**, 1283–4.

Hoff, J.E. and Castro, M.D. (1969). *J. agric. Fd Chem.* **17**, 1328–31.

Hoff, J.E., Jones, C.M., Wilcox, G.E. and Castro, M.D. (1971). *Am. Potato J.* **48**, 390–4.

Hope, G.W., MacKay, D.C. and Townsend, L.R. (1960). *Am. Potato J.* **37**, 28–33.

Howard, F.D., Laborde, J.F., Yamaguchi, M. and Knott, J.E. (1961). *Proc. Am. Soc. hort. Sci.* **78**, 406–12.

Howard H.W. (1970). *Genetics of the Potato*. London: Logos Press.

Howard, H.W. (1974). *Potato Res.* **17**, 490–511.

Hudson, D.E. (1975). *Am. Potato J.* **52**, 9–14.

Hughes, B.P. (1948). *Br. J. Nutr.* **12**, 188–95.

Hughes, J.C. (1971). *Potato Res.* **14**, 326.

Hughes, J.C. (1974). *Potato Res.* **17**, 512–47.

Hughes, J.C. and Evans, J.L. (1967). *Eur. Potato J.* **10**, 16–36.

Hughes, J.C. and Evans, J.L. (1969). *Eur. Potato J.* **12**, 26–40.

Hughes, J.C. and Faulks, R.M. (1973). Texture of cooked potatoes of different maturity in relation to pectic substances and cell size (Abstract). *Proc. Fifth Triennial Conf. Eur. Ass. Potato Res.* (Norwich, 1972) 163.

Hughes, J.C. and Mapson, L.W. (1967). *Proc. 2nd Int. Congress Fd Sci. Tech.* (Warsaw, 1966) 119–35.

Hughes, J.C. and Swain, T. (1962a). *J. Sci. Fd Agric* **13**, 229–36.

Hughes, J.C. and Swain, T. (1962b). *J. Sci. Fd Agric.* **13**, 358–63.

Hughes, J.C., Faulks, R.M. and Grant, A. (1975a). *J. Sci. Fd Agric.* **26**, 731–8.

Hughes, J.C., Grant, A. and Faulks, R.M. (1975b). *J. Sci. Fd Agric.* **26**, 739–48.

Hughes, J.C., Faulks, R.M. and Grant, A. (1975c). *Potato Res.* **18**, 495–514.

Hughes, J.C., Grant, A. and Faulks, R.M. (1975d). *Potato Res.* **18**, 338.

Hunnius, W. and Munzert, M. (1972). *Z. Acker-u PflBau* **136**, 245–60.

Hunnius, W., Bachthaler, G. and Munzert, M. (1972). *Potato Res.* **15**, 54–66.

International Potato Centre (1973). *Annual Report for 1973*. Lima, Peru.

Iritani, W.M. and Weller, L. (1973). *Am. Potato J.* **50**, 223–33.

Isherwood, E.A. (1973). *Phytochemistry* **12**, 2579–91.

Jacob, W.C. (1959). *Mem. Cornell Univ. agric. Exp. Stn* **368**, 1–86.

Jacobsen, L. and Ordin, L. (1954). *Pl. Physiol. Lancaster* **29**, 70–5.

Johnston, E.F. and Wilson, J.B. (1969). *Am. Potato J.* **46**, 75–82.

Kaldy, M.S. (1972). *Econ. Bot.* **26**, 142–4.

Kallio, A. (1960). *Am. Potato J.* **37**, 338–43.

Karikka, K.J., Dudgeon, L.T. and Hauck, H.M. (1944). *J. agric. Res.* **68**, 49–63.

Keegstra, K., Talmadge, K.W., Bauer, W.D. and Albersheim, P. (1973). *Pl. Physiol Lancaster* **51**, 188–96.

Keijbets, M.J.H. (1974). Pectic substances in the cell wall and the intercellular cohesion of potato tissue during cooking. *Agric. Res. Reports 827 :* Institute for Storage and Processing of Agriculture Products (I.B.V.L.) Wageningen, The Netherlands.

Killick, R.J. and Simmonds, N.W. (1974). *Heredity* **32**, 109–12.
Kissmeyer-Nielsen, E. and Weckel, K.G. (1967). *Eur. Potato J.* **10**, 312–26.
Krantz, F.A. and Lana, E.P. (1942). *Am. Potato J.* **19**, 144–9.
Kraus, J.E. (1943). Progress Report of the Idaho Potato Research Programme. *Idaho agric. Exp. Stn Mimeo Leaflet 217.*
Kraus, J.E. (1945). *Am. Potato J.* **22**, 134–42.
Kunkel, R. and Gardner, W.H. (1959). *Proc. Am. Soc. hort. Sci.* **73**, 436–44.
Kunkel, R. and Gardner, W.H. (1965). *Am. Potato J.* **42**, 109–24.
Kunkel, R. and Holstad, N. (1972). *Am. Potato J.* **49**, 43–62.
Kunkel, R. Holstad, N. and Butala, H. (1973). *Am. Potato J.* **50**, 339.
Kunkel, R., Weaver, M.L. and Holstad, N.M. (1970). *Am. Potato J.* **47**, 105–17.
Lampe, K. (1960). *Eur. Potato J.* **3**, 13–29.
Lampitt, L.H. and Goldenberg, N. (1940). *Chem. Ind.* **18**, 748–61.
Lampitt, L.H., Baker, L.C. and Parkinson, T.L. (1945a). *J. Soc. Chem. Ind.* **64**, 18–22.
Lampitt, L.H., Baker, L.C. and Parkinson, T.L. (1945b). *J. Soc. Chem. Ind.* **64**, 22–6.
Lampitt, L.H., Bushill, J.H., Rooke, H.S. and Jackson, E.M. (1943). *J. Soc. Chem. Ind.* **62**, 48–51.
Lamport, D.T.A. (1965). *Adv. Bot. Res.* **2**, 151–218.
Larsen, E.C. (1949). Investigations on cause and prevention of greening in potato tubers. *Idaho agric. Exp. Stn Res. Bull.* 16.
Lehmann, R. (1926). *Planta* **2**, 87–131.
Leichsenring, J.M., *et al.* (1951). Factors influencing the nutritive value of potatoes. *Univ. Minnesota agric. Exp. Stn Tech. Bull.* 196.
Le Tourneau D. and Zaehringer, M.V. (1965). *Res. Bull. agric. Exp. Stn, Univ. Idaho,* **64**, 1–32.
Levitt, J. (1942). *Am. Potato J.* **19**, 134–43.
Lewis, W.C. and Rowberry, R.G. (1973). *Am. Potato J.* **50**, 301–10.
Linehan, D.J. and Hughes, J.C. (1969a). *J. Sci. Fd Agric.* **20**, 110–12.
Linehan, D.J. and Hughes, J.C. (1969b). *J. Sci. Fd Agric.* **20**, 113–19.
Linehan, D.J. and Hughes, J.C. (1969c). *J. Sci. Fd Agric.* **20**, 119–23.
Linehan, D.J. and Hughes, J.C. (1969d). *Eur. Potato J.* **12**, 41–8.
Linehan, D.J., Stooke, C.E. and Hughes, J.C. (1968). *Eur. Potato J.* **11**, 221–5.
Llewellyn, J.C. (1967). The influence of water regimes on the growth and development of the potato crop. Ph.D. Thesis, University of Nottingham, England.
Lugt, C. (1960). *Eur. Potato J.* **3**, 307–24.
Lugt, C., Bodlaender, K.B.A. and Goodijk, G. (1964). *Eur. Potato J.* **7**, 219–27.
Lugt, C., Goodijk, G. and Gaastra-Ubbels, D. (1962). *Potato Newslett. Neth.* 3–15.
Maillard, L.C. (1912). *C. r. hebd. Séanc. Acad. Sci. Paris* **154**, 66–8.
Mapson, L.W. and Swain, T. (1961). In *Production and Application of Enzyme Preparations in Food Manufacture. SCI Monograph* **11**, 121–130, *Soc. Chem. Ind., London.*
Mapson, L.W., Swain, T. and Tomalin, A.W. (1963). *J. Sci. Fd Agric.* **14**, 673–84.
McClendon, J.H. (1964). *Am. J. Bot.* **51**, 628–33.
Meinl, G. (1972). *Ziemniak (J. Instijtut Ziemniaka, Bonin, Poland)* 1972, 173–202.
Meinl, G. and Effmert, B. (1966). *Züchter* **36**, 263–72.
Mondy, N.I., Mobley, E.O. and Gedde-Dahl, S.B. (1967). *J. Fd Sci.* **32**, 378–81.
Mudambi, S.R. and Hanning, F. (1961). *J. Fd Sci.* **26**, 492–4.
Mulder, E.G. (1949). *Pl. Soil* **2**, 59–121.

Mulder, E.G. and Bakema, K. (1956). *Pl. Soil* 7, 135–66.

Murphy, H.J. and Goven, M.J. (1959). Factors affecting the specific gravity of the white potato in Maine. *Bull Maine agric. Exp. Stn* (583).

Neuberger, A. and Sanger, F. (1942). *Biochem. J.* **36**, 662–71.

Neumann, H. (1925). *J. Landwirtschaft* **73**, 7–38.

Nilsson, S.B., Hertz, C.B. and Falk, S. (1958). *Physiologia Pl.* **11**, 818–37.

Nissen, M. (1967). *Eur. Potato J.* **10**, 85–99.

Noonan, J.C., Webb, R.E., Attaya, R.B. and Miller, J.C. (1951). *Am. Potato J.* **28**, 521–4.

Olsson, K. and Fridell, J.E. (1975). *Potato Res.* **18**, 338.

Ophuis, B.G., Hesen, J.C. and Kroesbergen, E. (1958). *Eur. Potato J.* **1**, 48–65.

Pallman, H. and Schindler, K. (1942). *Schweiz Landwirtschaftliche Monatschefte* **20**, 21–7.

Parke, D. (1962). *J. agric. Engng Res.* **8**, 173–7.

Plaisted, P.H. (1957). *Pl. Physiol. Lancaster* **32**, 445–53.

Potato Marketing Board, (1964). *Report on a consumer survey*. London.

Potato Marketing Board (1974a). *Potatoes—Consumer Buying behaviour and attitudes during 1972–3*. London.

Potato Marketing Board (1974b). *Report on a National damage survey 1973*. London.

Pratt, A.J., Lamb, J., Wright, J.D. and Bradley, G. (1952). Yield, tuber set and quality of potatoes. *Cornell Univ. agric. Exp. Stn Bull.* 876.

Pressey, R. (1969). *Am. Potato J.* **46**, 291–7.

Probine, M.C. and Preston, R.D. (1962). *J. exp. Bot.* **13**, 111–27.

Rathsack, K. (1935). *Der Speisewert der Kartoffel*. Berlin.

Rayle, D.L. and Cleland, R. (1970) *Pl. Physiol. Lancaster* **46**, 250–63.

Reeve, R.M. (1967). *Am. Potato J.* **44**, 41–50.

Reeve, R.M., Timm, H. and Weaver, M.L. (1971). *Am. Potato J.* **48**, 450–6.

Rinno, G., Goerlitz, H., Wahl, H. and Ebert, K. (1972). *Arch. Acker Pflanzbau Bodenkunde*, **16**, 551–60.

Robertson, I.M. (1970). Prediction of susceptibility to bruising. (Abstract). *Proc. Fourth Triennial Conf. Eur. Ass. Potato Res.* (Brest 1969) 245.

Robins, J.S. and Domingo, C.E. (1956). *Agron. J.* **48**, 488–92.

Rosenberg, H.R. (1942). In *Chemistry and Physiology of the Vitamins*. New York: Interscience Publishers.

Salaman, R.N. (1926). *Potato varieties* Cambridge: Cambridge University Press.

Sawyer, R.L. and Collin, G.H. (1960). *Am. Potato J.* **37**, 115–26.

Schallenberger, R.S., Smith, O. and Treadway, R.H. (1959). *J. agric. Fd Chem.* 7, 274–77.

Scheele, C. Von, Svensson, G. and Rasmusson, J. (1937). *Landw. Versuchs Stationen, Berlin*, **127**, 67–96.

Schippers, P.A. (1968). *Eur. Potato J.* **11**, 266–76.

Schuphan, von W. (1959). *Z. Pflanzenernährung, Dungung Bodenkunde* **86**, 1–14.

Schwimmer, S., Hendel, C.E., Harrington, W.O. and Olson, R.L. (1957). *Am. Potato J.* **34**, 119–32.

Sharrock, J.T.R. (1968). *Eur. Potato J.* **11**, 266–76.

Smith, A.M. and Gillies, J. (1940). *Biochem. J.* **34**, 1312–20.

Smith, O. (1968). *Potatoes: Production, storing and processing*. Avi Publishing Corporation: Westport, Connecticut.

Smith, O. and Nash, L.B. (1941). *Proc. Am. Soc. hort. Sci.* **38**, 507–12.

Smittle, D.A. and Thornton, R.E. (1972). *Am. Potato J.* **49**, 360–1.

Sowokinos, J.R. (1973). *Am. Potato J.* **50**, 234–47.

Sweetman, M.D. (1930). *J. agric. Res.* **41**, 479–90.

Szczesniak, A.S. (1963). *J. Fd Sci.* **28**, 385–9.
Tabachuk, V.I. (1953). *Leningradskii sel' Skokhozyaistvennyi Institut Zapiskii* **7**, 90–9 (N.I.A.E. Bedford, England. Translation 130).
Talburt, W.F. and Smith, O. (1967). *Potato processing.* The Avi Publishing Corporation: Westport, Connecticut.
Talley, E.A., Fitzpatrick, T.H., Porter, W.L. and Murphy, H.J. (1961). *J. Fd Sci.* **26**, 351–5.
Talley, E.A., Zaehringer, M.V., Reeve, R.M., Hyde, R.B., Dinkel, D.H., Heisler, E.G. and Pressey, R. (1969). *Am. Potato J.* **46**, 302–12.
Talmadge, K.W., Keegstra, K., Bauer, W.D. and Albersheim, P. (1973). *Pl. Physiol. Lancaster* **51**, 158–73.
Thornton, N.C. (1940). *Am. Chem. Soc. Div. Agric. Fd Chemy*, Abstract Papers. 99th Meeting Cincinatti, Ohio, 18–19.
Thornton, R.E., Smittle, D.A. and Peterson, C.L. (1973). Reducing potato harvester damage. *Proc. 12th Ann. Washington Potato Conf. Trade Fair.*
Timofeev, A.N. (1956). *Sb. Trudov po Zemledel' cheskoi Mekhanike vsesoyuznoi Leningradskoi sel' skokhozyaistvennykh Nauk*, 3, 406–16. (N.I.A.E. Bedford, England. Translation 131)
Umaerus, M. and Olsson, K. (1974). *Potato Res.* **17**, 157.
Varis, E. (1973). *Acta Agralia fennica* **128**, 1–11.
Vertregt, N. (1968). *Eur. Potato J.* **11**, 34–44.
Vleit, W.F. van, and Schriemer, W.H. (1960). *Eur. Potato J.* **3**, 263–71.
Volbracht, O. and Kuhnke, U. (1956). *Kartoffelbau* **7**, 74–7 and 95–8.
Wager, H.G. (1947). *J. agric. Sci., Camb.* **37**, 270–4.
Wager, H.G. (1963). *J. Sci. Fd Agric.* **14**, 583–6.
Waint, J.S., Findlen, H. and Kaufman, J. (1951). *Am. Potato J.* **28**, 753–65.
Warren, D.S. and Woodman, J.S. (1974). *J. Sci. Fd Agric.* **25**, 129–38.
Warren, D.S., Gray, D. and Woodman, J.S. (1975). *J. Sci. Fd Agric.* **26**, 1689–97.
Weaver, M.L. and Steen, H.A. (1966). *Proc. Am. Soc. hort. Sci* **56**, 343–8.
Welte, E. and Müller, K. (1966). *Eur. Potato J.* **9**, 36–45.
Werner, H.O. and Dutt, J.O. (1941). *Am. Potato J.* **18**, 189–208.
Westover, K.C. (1941). *Am. Potato J.* **18**, 155–63.
Whitehead, T., McIntosh, T.P. and Findlay, W.M. (1953). *The Potato in Health and Disease.* (3rd Edn) Edinburgh: Oliver and Boyd.
Wohrmann, K. (1966). *Züchter* **36**, 90–5.
Woodward, C.F. and Talley, E.A. (1953). *Am. Potato J.* **30**, 205–12.
Wurr, D.C.E. and Allen, E.J. (1974). *J. agric. Sci. Camb.* **82**, 277–82.
Wurster, R.T. and Smith, O. (1963). *Am. Potato J.* **40**, 415–20.
Yamaguchi, M., Hughes, D.L. and Howard, F.D. (1960). *Proc. Am. Soc. hort. Sci.* **75**, 529–36.
Yamaguchi, M., Timm, H. and Spurr, A.R. (1964a). *Proc. Am. Soc. hort. Sci.* **84**, 412–23.
Yamaguchi, M., Flocker, W.J., Howard, F.D. and Timm, H. (1964b). *Proc. Am. Soc. hort. Sci.* **85**, 446–56.
Zaag, D.E. van der, and Meijers, C.P. (1970). *Proc. Fourth Triennial Conf. Eur. Ass. Potato Res.* (Brest, 1969) 93–103.
Zaehringer, M.V. and Cunningham, H.H. (1971). *Am. Potato J.* **48**, 385–9.
Zitnak, A. and Johnston, G.R. (1970). *Am. Potato J.* **47**, 256–60.

Chapter 14

The physics and physiology of storage

W.G. BURTON

14.1 The response of the tuber to its environment

14.1.1 Physical adjustment to the environment

Temperature

In a tuber which is in thermal equilibrium with its surroundings, metabolic heat production raises the temperature slightly above the environment – the specific heat of the tissue is about 3.6 J $(g°C)^{-1}$ (Mann, reported in Burton, 1966) and a typical heat production of, say, 50 mJ g^{-1} h^{-1} (Section 14.1.4) could thus raise the temperature 0.014°C h^{-1} were it not transferred to the surrounding air. This transfer involves a temperature gradient of about 0.2°C from the centre to the periphery, and about 0.1°C from just below the skin to the surrounding air (Burton, 1969). These are the approximate equilibrium values.

A tuber placed in an environment differing in temperature, equilibrates at a rate depending on the temperature difference, the thermal conductivity of the tissue, the surface area and the rate of air movement. Re-equilibration of ware tubers exposed to an 8°C temperature differential takes about 12 h with free air circulation (Burton, 1969). Farrimond (1972) found a heat transfer of about 0.8 W from 200 g tubers in an air current of 1 m s^{-1} with a mean surface temperature differential of about 3°C. Theoretically, the rate of heat transfer from a spherical body to an air stream is approximately proportional to $V^{0.6}$, where V is the linear velocity of the stream, but, as Mann (1963) pointed out, heat transfer from the centre of a potato is limited by the thermal conductivity of the tissue. He found the overall heat transfer from potatoes of about 7 cm diameter, weighing about 215 g to be proportional to $V^{0.31}$. The index may vary with the size, shape and moisture content of the tuber, and in practice proportionality to the cube root of the velocity is a convenient approximation.

Internal atmosphere

The main barrier to the aeration of the potato tuber is provided by the skin, through which gaseous exchange occurs only through the lenticels (Wigginton, 1973). The average distribution of these varies from about 0.5 to 3cm^{-2} of surface in the fully grown tuber (Burton, 1965a; Wigginton, 1973) and Wigginton measured the oxygen diffusion through individual lenticels of mature tubers (cv. King Edward) to average about 1×10^{-4} ml h^{-1} mb^{-1} O_2 deficit, with a range of from 2.4×10^{-5} to 3×10^{-4} ml h^{-1} mb^{-1}.

The overall permeability to O_2 of the skin of mature stored tubers, averaged over the whole surface, lies in the range 7×10^{-5} to 2×10^{-4} ml cm^{-2} h^{-1} mb^{-1} (Burton, 1950, 1965a; Mapson and Burton, 1962). Freshly harvested tubers may show a rather higher permeability (about 2.5×10^{-4}) and freshly harvested immature tubers higher still (about 1×10^{-3}), although this decreases during storage to a value equal to that of mature stored tubers (Burton, 1965a). CO_2 diffuses in the gas phase at about 80% the rate of O_2, and the permeability of the skin to gaseous diffusion of CO_2 may thus be taken to be 80% of the above figures.

The above periderm permeabilities permit diffusion, in response to respiration at normal temperatures, without the development of a large deficit of O_2 or much accumulation of CO_2, in the atmosphere of the peripheral intercellular space. At a temperature of 10°C it could contain about 2–3% CO_2 and 18–20% O_2. There is some gradient from periphery to centre, calculated to be about $br^2/6$ % in a spherical tuber, where b is

Table 14.1 Composition of internal atmosphere of potato tubers, cv. Arran Consul (based on Burton, 1950)

Temperature (°C)	*% composition measured directly at periphery*		*% composition calculated from degree of O_2 saturation measured at centre : CO_2 value assumes RQ = 1*	
	O_2	CO_2	O_2	CO_2
5	18.1	2.6	19.4	1.9
10	18.2	1.8	15.6	6.6
15	16.8	3.9	16.9	5.0
20	16.1	4.7	14.2	8.4
25	13.0	7.0	11.5	11.7
28	—	—	10.2	13.4
37	—	—	2.1	23.5

the rate of O_2 uptake in ml ml^{-1} tissue h^{-1} and *r* the radius in mm (Burton 1950). At normal rates of respiration and tuber sizes this gradient is unlikely to exceed about 0.5%.

The good aeration indicated above is achieved with intercellular spaces totalling only about 1–2% of the tissue volume (Gorter and Nadort, 1941; Burton and Spragg, 1950; Burton, 1950), the distribution of the spaces being such that the cell sap is practically equilibrated with the intercellular atmosphere (Burton, 1950).

At temperatures leading to higher rates of respiration than at 10 °C, the degree of O_2 depletion and CO_2 accumulation is greater (Table 14.1), and at very high temperatures can be harmful. At 37 °C 'blackheart' can develop in a few days.

Effect of a water film on the internal atmosphere

Because O_2 enters the tuber, and CO_2 leaves, by gaseous diffusion through the lenticels, the internal atmosphere responds to conditions which interfere with this. For example, if the surface of a tuber is completely covered by a film of water, diffusion must occur, much more slowly, in solution. As a result of this, Burton and Wigginton (1970) found a persistent film of water to lead to tubers becoming anaerobic in about 6.5 h at 10 °C and in about 2.5 h at 21 °C.

14.1.2 Wound healing

Apart from the small wound where the tuber is detached from the stolon, commercial harvesting results in some abrasions and splits. The former in particular can result in enhanced evaporative loss and in the ingress of wound parasites such as *Phoma*. The wounds, however, do not remain open but heal, provided the environmental conditions permit. Priestley and Woffenden (1923) described the first step in healing as a 'blocking' of the wounded surface by layer of suberin which prevented invasion by pathogens and reduced evaporation. A phellogen was then initiated from which layers of suberized periderm cells arose. The formation of this wound-cork is influenced by the factors listed below.

Type of wound

According to Artschwager (1927), abrasion results in a deep seated periderm, whereas the wound-periderm which forms after a smooth clean cut may be within one layer of cells of the surface. If a tuber is bruised and the periderm broken by impact, there is, typically, irregular development of wound-periderm below the margin of the injury, or beneath breaks in the original periderm, but only slight or no development beneath tissues which are merely crushed. In general, healing is more rapid and more uniform over clean cuts than over other types of injury.

Physiological state of the tuber

Wigginton (1974) reported considerable variation, which he could only ascribe to unexplained differences in physiological state, in the rates of both suberization and wound-periderm formation in individual tubers of apparently uniform samples. Apart from this individual variation there is a continuous reduction in the rapidity of wound-periderm formation throughout development and storage, from young tubers 5 mm in diameter to mature tubers stored for 12 months after harvest (Lange and Rosenstock, 1964), the reduction being more marked the higher the storage temperature (Lange and Rosenstock, 1965). There is also a decline in the number of layers of periderm cells in the wound cork (Lange and Rosenstock, 1964).

Variety

Many workers have reported varietal differences in the rate of wound healing (Priestley and Woffenden, 1923; Artschwager, 1927; Smith and Smart, 1959; Nielsen, 1968; Wigginton, 1974) although Wigginton pointed out that the varietal differences he observed may not have been significant because of the variation between individual tubers. Differences may not be confined to wound healing under constant conditions, but may extend to the degree of response to changes in temperature (Artschwager, 1927) or oxygen pressure (cf. the results of Lange *et al.*, 1970, on cv. Saskia, and of Lipton, 1967, on cv. White Rose).

Temperature

Wound healing is more rapid the higher the temperature over the range 2.5 to 20 °C (Appel, 1906; Priestley and Woffenden, 1923; Artschwager, 1927; Radatz, 1967; Wigginton, 1974). On average, the wounded surface becomes lightly suberized in about 1–2 weeks at 2.5–5 °C, in about 4 days at 10 °C, and in 1–2 days at 20 °C. Complete suberization of the superficial layer of cells could take 3–6 weeks at 5 °C, 1–2 weeks at 10 °C and 3–6 days at 20 °C. Wigginton (1974) could detect no formation of wound periderm until 3–5 days after wounding at 20 °C, 1–2 weeks at 10 °C, and up to 4 weeks at 5 °C. The development of two layers of wound-periderm required about 5–7 days at 20 °C; 9–16 days at 10 °C and 4–9 weeks at 5 °C. Experience suggests that the timescale of wound healing in freshly harvested tubers at 10 °C is adequate to prevent serious invasion by wound parasites, indicating that suberization of the surfaces within 4 days, with subsequent wound cork formation, may normally be sufficiently rapid.

O_2 concentration

Wigginton (1974) found anaerobic conditions prevented both suberization and periderm formation at 20 °C. Slight suberization occurred in 1% O_2, and increasingly in higher concentrations up to 10% and sometimes 21%. Periderm formation did not occur below 3–5% and also increased with higher concentrations up to 21%. The foregoing is in agreement with the findings of Lipton (1967), but Lange *et al.* (1970) stated that in cv. Saskia some periderm formation occurred even in 1% O_2 and was as effective as in air in 5% O_2.

CO_2 concentration

Wigginton (1974) found 5, 10 and 15% CO_2, in the presence of 21% O_2, prevented periderm formation, and progressively reduced suberization at 10 °C.

Water vapour pressure deficit

With a high water vapour pressure deficit (wvpd), drying occurs at the surface of the wound and the resulting crust inhibits or greatly delays suberization. On the other hand, a water-saturated atmosphere hinders wound healing by causing proliferation of cells at the wounded surface (Olufsen, 1903; Lange *et al.*, 1970). Description of the effects of humidity on wound healing will be restricted here to those reported by Wigginton (1974), as he appears to have been the only worker to have adequately controlled and monitored the humidities in the experimental environment. He found little effect upon the rate of healing at 10 °C (cv. King Edward) until the wvpd exceeded 2.4 mb, although there was some slight delay even with a wvpd of 0.8 mb compared with the optimal 0.2 mb (98% r.h.). Delay was more serious with higher deficits, but with a wvpd approaching 5 mb there was still complete suberization of the surface and the formation of one layer of periderm. At 20 °C there was little effect until the wvpd reached nearly 7 mb and the effect of 11 mb was comparable with that of 5 mb at 10 °C. With a wvpd of 16 mb at 20 °C (30% r.h.) neither suberization nor periderm formation occurred. The less harmful effects of an increased wvpd at 20 °C, compared with those at 10 °C, were stated by Wigginton to be due to the more rapid healing at the higher temperature, enabling adequate suberization to occur before desiccation proceeded sufficiently to depress the process. In summary, and allowing for the different temperatures at which wound healing occurs in practice, some wvpd is necessary to prevent proliferation, but it should not exceed about 2 mb and is optimally about 0.2–0.5 mb.

Sprout suppressants

The various chemicals or radiation treatments which can be used to suppress sprout growth will also suppress the formation of wound periderm, with consequent enhanced potential for attack by rotting organisms and for excessive water loss – see Ellison and Cunningham (1953) and Cunningham (1953) with respect to the methyl ester of α-naphthaleneacetic acid; Ives (1955) and Castello Branco (1957) with respect to iso-propyl-*n*-phenylcarbamate; Burton (1968) with respect to 3-5-5 trimethylhexan-l-ol; and Waggoner (1955) and Sawyer (1956) with respect to γ-irradiation. As a general rule sprout suppressants should be applied only after wound healing is complete, although tetrachlornitrobenzene appears to provide an exception to this principle.

14.1.3 Water loss

Water vapour pressure deficit

Water loss exemplifies a move of a dynamic system towards equilibrium – in this case between a tuber which is in equilibrium with an approximately water saturated atmosphere, and a surrounding atmosphere which usually is not saturated. The rate at which water will evaporate from a tuber in response to the above tendency is directly proportional to the difference between the equilibrium water vapour pressure of the tuber (i.e. of a saturated atmosphere at tuber temperature) and the water vapour pressure of the air in contact with it. If the tuber and the air are at the same temperature this difference approximates closely to the water vapour pressure deficit of the air.

Temperature

If the temperature of the tuber is changed, then the water vapour pressure in air saturated at tuber temperature is changed, to an extent which is shown in Table 14.2. If the temperature of the air is changed, over the range encountered in potato storage, the vapour pressure of the water in it, if the quantity remains unchanged, is unaltered. For example, air saturated with water vapour at 10 °C contains 9.33 g of water vapour m^{-3}, exerting a pressure of 12.16 mb. If we raise the temperature to 20 °C, at constant atmospheric pressure, the cubic metre will expand to 1.035 m^3, still containing 9.33 g water vapour – i.e. 9.01 g m^{-3}, exerting a similar pressure (Table 14.2). By contrast, the effective vapour pressure of a tuber raised from 10 to 20°C will be increased from 12.16 to 23.09 mb.

The quantity of water does not always remain unchanged, however, and a fall in air temperature can result in a lower water vapour pressure if the temperature drops to below dew-point. If we lowered the temperature of the m^3 of air, saturated at 10, to 5 °C, then its volume would contract to 0.98 m^3, which if it still held the original 9.33 g of water vapour would contain 9.50 g m^{-3}. Air at 5 °C however can only hold 6.76 g m^{-3}, exerting a pressure of 8.65 mb, and the excess water therefore condenses out.

Sufficient has been said here to enable the initial effects of any changes in air or tuber temperature upon the rate of water loss to be calculated. These initial effects are transitory, and alter in predictable fashion as the temperatures of the tuber, or of the air, or of both, change in the direction of thermodynamic equilibrium between tuber and air; and also as, if the volume of air is limited, the water evaporated into it, from the tuber, increases its water vapour pressure (see below).

Table 14.2 The content and pressure of the water vapour in 1 m^3 of saturated air at a total barometric pressure of a standard atmosphere (= 1013.25 mbar) (from Burton, 1973)

Temperature (°C)	*Weight of water vapour (g)*	*Volume of water vapour (l) (= % by vol. x 10)*	*Pressure of water vapour (mbar)*	*Weight of water vapour $mbar^{-1}$ pressure (g)*
0	4.84	6.00	6.08	0.80
1	5.18	6.44	6.53	0.81
2	5.54	6.92	7.01	0.80
3	5.92	7.43	7.53	0.79
4	6.33	7.96	8.07	0.79
5	6.76	8.54	8.65	0.78
6	7.22	9.15	9.27	0.78
7	7.70	9.80	9.93	0.78
8	8.21	10.49	10.63	0.77
9	8.76	11.22	11.37	0.77
10	9.33	12.00	12.16	0.77
11	9.93	12.82	12.99	0.77
12	10.57	13.69	13.87	0.76
13	11.25	14.62	14.81	0.76
14	11.96	15.60	15.81	0.76
15	12.71	16.64	16.86	0.76
16	13.50	17.74	17.97	0.75
17	14.34	18.90	19.15	0.75
18	15.22	20.12	20.39	0.75
19	16.14	21.42	21.70	0.74
20	17.12	22.79	23.09	0.74
25	22.80	30.87	31.28	0.73
30	30.04	41.88	42.43	0.71

Table 14.3 Calculated rates of evaporation from potato tubers at different rates of ventilation

Characteristics of air:	*10°C; water holding capacity, 9.33 mg l^{-1}; saturated water vapour pressure, 12.16 mbar; wvpd, 1 mbar (= a deficit of 0.77 mg water l^{-1})*			
Characteristics of tuber:	*(1) Mature stored tuber; 10°C; surface area, 100 cm^2; wt., 100 g; evaporative potential, 10 µg cm^{-2} h^{-1} $mbar^{-1}$*		*(2) Immature tuber 1 day after harvest; 10°C; surface area, 100 cm^2; wt., 100 g; evaporative potential, 400 µg cm^{-2} h^{-1} $mbar^{-1}$*	
Rate of ventilation	*Evaporation from tuber*			
lh^{-1}	*mg h^{-1}*	*% of potential*	*mg h^{-1}*	*% of potential*
1	0.58	58	0.77*	2
2	0.75	75	1.54*	4
4	0.87	87	3.08*	8
10	0.94	94	7.69	19
20	0.97	97	14.6	36
100	0.99	99	31.9	80
200	1.00	100	35.6	89
1000	1.00	100	39.0	98

*Loss limited by water-holding capacity of air.

Ventilation

The effect of changing the rate of ventilation is not marked except at very low rates, or except in the case of tubers – for example, immature tubers – which can lose water readily. Evaporation from a tuber into the air raises the amount of water in the latter and hence its water vapour pressure, thus reducing the wvpd. If the amount of air concerned is small the proportionate effect of evaporation of a given amount of water into it is great. Increasing the amount of air by increased ventilation can considerably reduce this effect. Conversely, if the amount of air is great, the effect is small and still further increasing the amount of air can have little influence. This is illustrated in Table 14.3 using typical values for the characteristics of the tubers.

Limitation by the periderm

The periderm limits the rate of evaporation from a tuber to much below that from a free water surface, exposed to the same wvpd. The dead cork cells are not freely permeable to water which migrates slowly through them to the surface where it evaporates in response to the wvpd. About 98% of the water loss appears to occur in this way and only about 2% by direct diffusion in the gas phase through the lenticels (Burton, 1973). Limitation by the periderm results in rates of water loss, from mature undamaged tubers exposed to free air circulation, of the order of 6–10 μg cm^{-2} h^{-1} mb^{-1} (Burton, 1973) whereas in the absence of the skin, Burton (1955) found a rate prior to wound healing, of about 3.5 mg cm^{-2} h^{-1} mb^{-1}, several hundred times as great.*

Handling wounds

The figures above suggest that breaks in the skin caused by harvesting or grading, could have a marked effect upon the rate of water loss. An abrasion of 1 cm^2, on the surface of a tuber of surface area 100 cm^2, could theoretically increase the loss in the first hour after handling from, say, 1 to 4.5 mg mb^{-1}. This is not unrealistic. Burton and Hannan (1957) found the loss from cv. Dr McIntosh to average about 30 μg cm^{-2} h^{-1} mb^{-1} during the first week of storage at 10°C subsequent to harvest. During this week wound healing was occurring, as shown by the fact that the average loss, after 2 weeks, had settled down to a value of about 10 μg cm^{-2} h^{-1} mb^{-1}, and thus during the first days of storage the loss must have been appreciably greater than 30 μg cm^{-2} h^{-1} mb^{-1}. Wound healing during the early stages of storage, facilitated by providing the favourable conditions indicated in Section 14.1.2, is thus important. In its absence the potential evaporative loss from tubers, damaged no more than is usual in commercial handling, could well be double or treble that from undamaged tubers.

Maturity

Burton (1973) gave values for loss from immature tubers ranging in weight from 0.49 to 135.57 g. At the stage at which they could be harvested as 'earlies' they could, after one day, be losing water at a rate of 100–600 μg cm^{-2} h^{-2} mb^{-1} – about 15–100 times as fast as mature stored tubers. The rate of loss dwindles if such tubers are stored under conditions favourable for suberization, and approaches that from mature tubers after about 2 weeks (Appleman *et al.*, 1928).

Sprout growth

Evaporation from the surface of sprouts can be about 1 mg cm^{-2} h^{-1} mb^{-1} (Burton, 1955, 1966). The potential water loss from sprouted tubers is

*This figure has been recalculated, from the original data, in different units from those given in the publication cited.

thus much greater than from unsprouted tubers. Burton and Hannan (1957) demonstrated a linear relationship between the percentage by weight of sprout growth and the rate of water loss from tubers exposed to free air movement, every 1% by weight of sprouts increasing evaporative loss by 0.075% of the initial tuber weight week $^{-1}$ mb $^{-1}$ wvpd.

Variety

Appleman *et al.* (1928) noticed marked varietal differences in the rate of evaporation from potatoes immediately after harvest and for a few weeks thereafter. These could have been due to differences in maturity at harvest and in the rate of wound healing (see Section 14.1.2). Later in the storage season, varietal differences in the rate of loss were insignificant in the varieties they studied, although with other varieties it has been possible to demonstrate different rates of loss (Burton, 1973), perhaps related to the thickness of the periderm. Later still in the storage season, varietal differences in the rate of sprouting may lead to marked varietal differences in evaporation (Appleman *et al.*, 1928).

Cooling effect of evaporation

At potato storage temperatures, say 5°C, the latent heat of vaporization of water is about 2.49 kJ g^{-1}. The specific heat of potatoes at the same temperature is about 3.6 J°C^{-1}. The evaporation of 1 mg of water will thus remove an amount of heat sufficient to lower the temperature of 1 g of potato tissue by 0.69°C.

14.1.4 Respiration

General

The respiration of a potato tuber releases bound energy by the breakdown of starch molecules, the energy so released being in part transferred into other constituents of the tuber and in part lost as heat. Isherwood (1973) suggested that the basal respiration at 10°C – about 4 mg CO_2 kg^{-1} h^{-1} – was equivalent to 1.4 mmol ATP 100 g^{-1} 24 h^{-1}. From the pathways he suggested for carbohydrate interconversion this would be equivalent to a continuous turn-over of carbohydrate, expressed as anhydrohexose units, at a rate of 0.5 mmol 100 g^{-1} 24h^{-1}. The cycles of reactions involved include the sequential oxidation – by dissolved O_2 which is itself reduced to water – and reduction of a terminal oxidase; and decarboxylation steps which result in the evolution of CO_2. The volumes of O_2 combined and CO_2 evolved, measured over a period of some months from harvest, are about the same; but within this period there appears to be a consistent pattern of change in the respiratory quotient. Burton (1974) found a tendency for a respiratory quotient less than 1.0 (about 0.8–0.9) during the early storage period and also later, during sprout growth; and a respiratory quotient greater than 1.0 (up to about 1.3) during the period of minimum respiration, in the middle of the storage period, during which sprout growth started. Apart from these trends, O_2 intake and CO_2 output become temporarily out of balance if, for instance, the temperature is changed (Burton, 1974) or if the potatoes are handled (Burton, 1974; Isherwood and Burton, 1975); and there appears to be a shift towards a permanent excess of CO_2 evolved from senescent tubers (Isherwood and Burton, 1975).

As described in Section 14.1.1, the cells of the potato tuber, under normal storage conditions, are well oxygenated. The main terminal oxidase of the tuber, cytochrome-*c* oxidase (Levy *et al.*, 1948; Schade *et al.*, 1949; Thimann *et al.*, 1954), has a high affinity for O_2 – Burton (1974) gave a value of 7×10^{-8} M for the Michaelis constant of the combination of O_2 in potato tissue – and is thus fully saturated. Other oxidases with lower affinities are only partially saturated, but may be responsible for 20–30% of the O_2 uptake of mature freshly harvested tubers (Mapson and Burton, 1962; Burton, 1974). Doubts as to the functioning of low-affinity oxidases in intact tubers (Schade *et al.*, 1949; Ducet and Chevillotte, 1970) have been based on a misapprehension of the degree of aeration of the tuber.

Two pathways of respiration have been suggested, prior to the final transfer of electrons by the terminal oxidase system. Both involve the initial mobilization of the starch as reactive hexose phosphate esters, which are then respired either by the pentose phosphate pathway (Romberger and Norton, 1961) or, via pyruvate, by the tricarboxylic acid cycle (Anderson *et al.*, 1952; Barker and Mapson, 1955). In that several intermediates and enzymes are common to both pathways they may well be interrelated (Norton, 1963). The evidence, up to the mid-1960s, for the two pathways was summarized by Norton (1963) and Burton (1966). It is quite possible that both operate to variable extents, depending upon such factors as storage conditions, maturity, state of dormancy, damage and infection by disease.

Maturity

The rate of respiration of potato tubers, measured after removal from the plant, is high soon after tuber initiation and decreases throughout the growing season, reaching a fairly steady, though slowly declining value, at maturity (Appleman and Miller, 1926; Singh and Mathur, 1937; Burton, 1964, 1974). Burton (1964) reported an O_2 uptake of 50 ml kg^{-1} h^{-1} at 10°C by tubers, cv. Majestic, average weight 5.4 g, at the end of June, falling to about 4–5 ml kg^{-1} h^{-1} by mature tubers in mid-September. Potatoes harvested immature will thus be respiring more rapidly than mature tubers, to an extent related to the degree of immaturity. This high rate of respiration is not maintained after harvest but falls off rapidly, approximating to that characteristic of mature tubers after 2–4 weeks (Burton 1963a, 1974).

Period of storage

Mature tubers show no rapid fall after harvest – their rate of respiration is already low – but there is a downward drift until the start of sprout growth. Fairly typical values, for both O_2 uptake and CO_2 output at 10°C, would be 4 ml kg^{-1} h^{-1} at harvest falling to 2 ml kg^{-1} h^{-1} before the start of sprout growth and then rising progressively (see below under 'sprout growth'). The pattern for immature tubers, after the initial fall, is similar.

Variety

Varietal differences in the rate of respiration at any given time might be expected to result from differences in maturity or state of dormancy. The basal rate of mature dormant tubers may also differ. Schulz (1927), however, who found some evidence of this, did not consider it to be fully conclusive.

Temperature

The short-term effect of temperature upon the rate of respiration is well illustrated by the results of Müller-Thurgau (1882), showing a two-fold increase for a 10 °C rise in temperature over the range 0–20 °C. This short-term effect is not maintained during storage at different temperatures and after the re-establishment of all the dynamic equilibria between temperature-sensitive reactions which may influence the overt symptoms of respiration, the Q_{10} may be very different from 2. Burton (1974) found a mid-storage temperature coefficient, 20/10 °C, of about 1.2 – 1.3, but respiration at 2 °C (sweetened tubers) could be higher than at 20 °C. Later, increased respiration accompanying sprouting – most vigorous at 20 °C and non-existent at 2 °C – changed this pattern.

O_2 concentration

The major part of the respiration of intact tubers is mediated by a high oxygen-affinity terminal oxidase, presumed to be cytochrome-*c* oxidase, O_2 combination by which is unaffected by O_2 concentration except at low levels (Craft, 1963, studied 1–6% O_2; Mapson and Burton, 1962, 1.4–100% O_2; and Burton, 1974, air, 5% and 100% O_2). Mapson and Burton (1962), however, found O_2 uptake at 20°C to be increased by 16% above the air value in pure O_2, which in view of the concentrations of O_2 found in the centre of tubers in air (see Section 14.1.1) can only be explained on the basis of increased uptake by a low-affinity oxidase (e.g. polyphenolase) and not by increased combination with cytochrome-*c* oxidase. This conclusion is supported by the comparatively slight effects of reducing the concentration of O_2. Table 14.4 shows results of this, typical of tubers during the first few weeks after harvest. Burton (1974) found low-affinity uptake to be up to about 30% of the whole at the time of harvest, but to fall almost to zero over a period of about $1\frac{1}{2}$–3 months. Thereafter, it reappeared in the sprouting tuber, although it is quite possible that this reappearance was in the respiration of the sprouts and not of the tubers upon which they were growing. The consequent change in response to changed O_2 concentrations is illustrated in Table 14.5.

Consideration of the effects of O_2 concentration on CO_2 output is

Table 14.4 Typical relation, within a few weeks of harvest, between ambient O_2 concentration and O_2 uptake by potato tubers at 20°C (cv. King Edward, O_2 uptake in air, 5–8 ml kg^{-1} h^{-1}) (based on Mapson and Burton, 1962)

O_2%	*O_2 uptake as % uptake in air*
14.7	92
9.8	81
7.0	73
5.0	70
2.5	59
1.4	55

Table 14.5 Change, during storage at 10°C, in relation between ambient O_2 concentration and O_2 uptake by potato tubers at 10°C (cv. Majestic). Uptake relative to that in 100% O_2 (values in ml kg^{-1} h^{-1} in parentheses) (based on Burton, 1974)

Weeks after harvest	*2–3*	*4–5*	*6–7*	*8–9*	*10–11*	*13–14*	*15–16*
100% O_2	100	100	100	100	100	100	100
	(3.8)	(2.8)	(2.2)	(2.3)	(2.4)	(2.3)	(2.9)
Air	74	95	109	99	75	89	75
5% O_2	51	79	99	81	56	57	62

complicated by a delayed response of decarboxylation (Burton, 1974). Despite this, exposure for only a few hours to 5% O_2 leads to some reduction in CO_2 production compared with production in air (Burton, 1974), and the effect becomes more marked as the O_2 concentration is further reduced (Craft, 1963). Even in N_2, however, Craft found the production of CO_2 over 24 h to be about 60% of that in air; although some of this could have resulted from the delayed decarboxylation of intermediates present at a level characteristic of the previous aerobic state. Exposure for some hours to pure O_2 has little effect on CO_2 output. Prolonged exposure leads at first to an increase in output, over a period of two weeks or so, but thereafter the effects of 'O_2 poisoning' become apparent and after 5–6 weeks CO_2 output may be negligible (Barker and Mapson, 1955).

CO_2 concentration

Burton (1952) measured O_2 uptake and CO_2 output in the presence of 5–7% CO_2 over periods of 11–14 weeks. If comparison is restricted to experiments in which accumulation of volatiles other than CO_2 was at least qualitatively the same, it would appear that the increased CO_2 reduced both O_2 uptake and CO_2 output by about 25–30%, the respiratory quotient being practically unaffected. O_2 concentrations were not quite the same in these experiments, but the differences were unlikely to have had a marked effect upon the rate of respiration.

Metabolic inhibitors and stimulants

Hanes and Barker (1931) exposed tubers to HCN, at concentrations ranging from 0.14 to 1.45 ml l^{-1}, for periods of several weeks. Both O_2 uptake and CO_2 output were stimulated equally to an extent directly related to HCN concentration. Typically the stimulation rose to a peak – at 0.47 ml l^{-1} and above, this could be eight times the air value – and then fell off, although never down to the air value. At the lowest concentration no peak was discernible, there being a steady increase to about 30% above the air value. At 0.47 ml l^{-1} and above, injury occurred, the more rapidly the higher the concentration, indicated by an increased respiratory quotient and in some cases by internal discoloration. Craft (1963), in short-term experiments, found stimulation of O_2 uptake and CO_2 output at 25°C (tubers previously stored at 12.8°C) to be caused by a much higher concentration of HCN (1%). In the case of tubers previously stored at 0°C, however, both O_2 uptake and CO_2 output were depressed by this level of cyanide. Craft also found 95% CO 5% O_2, compared in effect to 95% N_2 5% O_2, to effect respiration in a way similar to 1% cyanide.

Herklots (1928) and Huelin (1932) found continuous exposure to low concentrations of ethylene caused a marked temporary rise in the rate of CO_2 production by potato tubers. Using 1000 parts 10^{-6} ethylene a peak value of 2–2.5 times the control was reached in 2–3 days and was followed by a fall to a value rather more than the control. Huelin (1933) found the maximum response to be produced by 100 parts 10^{-6} ethylene. Above this concentration there was evidence of a fall in response. Below it there was a definite fall, but even at 1 part 10^{-6} there was a 38% stimulation of CO_2 output.

Most attempts, other than the above, to elucidate the pathways of respiration by using metabolic inhibitors and stimulants have employed discs of tuber tissue, because only by such methods can inhibitors or stimulants, other than gaseous ones, be introduced. The pathways so elucidated, however, are those occurring in discs of tuber tissue, and not necessarily in intact tubers. Thus malonate (5×10^{-2} M, pH 5.0) inhibits the O_2 uptake of freshly cut discs by only 10% (Romberger and Norton, 1961) suggesting that the tricarboxylic acid cycle may be comparatively unimportant in their respiration but this respiration is proceeding at a multiple of the rate of that of the intact tuber. No deduction can be made on these grounds as to the extent of the participation of the TCA cycle in the respiration of the latter (Norton, 1963). Bearing this in mind, the following effects upon the respiration of freshly cut discs have been observed (Hackett *et al.*, 1960):

HCN (4.6×10^{-4} M), 70% inhibition; antimycin, 41% inhibition at 0.1 mg l^{-1} and 80% inhibition at 10.0 mg l^{-1}; azide (10^{-3} M, pH 5.0), 65% inhibition; dinitrophenol (5×10^{-5} M) 133% stimulation.

If discs are 'aged' for 24 h or more in running tap water their respiration again increases, to a multiple of that of freshly-cut discs, and the response to inhibitors and stimulants also changes. There is greatly increased sensitivity to malonate (Norton, 1963), and a reduction in the inhibition by other agents listed above, in some cases to the extent of replacement by stimulation (Hackett *et al.*, 1960).

Irradiation

γ-irradiation causes a marked temporary increase in respiration, rising to a peak in a week or so and then falling to about the control level (Sussman, 1953).

Respiration in relation to composition

The rate of respiration is a measure of the rate of turn-over of the respiratory cycles and of the rate at which substrate is being mobilized and fed into these cycles. It need bear no relation to the concentration of any intermediate or substrate unless the latter is so depleted as to limit the rate. There may be apparent correlations between, say, sucrose content and the rate of respiration, as suggested by Barker (1936a). This is compatible with the high rates at low temperatures, which cause tubers to sweeten; of immature tubers, which contain a high concentration of sucrose; and following γ-irradiation, which causes a marked temporary increase in sucrose concentration. No consistent relationship can be demonstrated over a wide range of conditions and treatments, however. The cases just mentioned are all observed at times when rapid interconversion of carbohydrate is occurring, and Isherwood (1973) calculated that the increased respiration was closely equivalent to the increased formation of ATP, involved in the carbohydrate conversions.

Sprout growth

Sprout growth, which involves mobilization and transfer of material from the tuber to the sprout and its synthesis into the sprout structure, is accompanied by increased respiration. The respiration of a sprouting tuber includes both that of the tuber and of the sprouts, and Burton *et al.* (1955) found an approximately 50% increase in respiration to have occurred by the time sprout growth had reached about 1% by weight. The increase continues with increasing growth, and Isherwood and Burton (1975) recorded a 4–5-fold increase in respiration by the beginning of June in considerably sprouted tubers. Much of this may be accounted for by the respiration of the sprouts themselves. In desprouted tubers Isherwood and Burton found the increase to be not more than 30% above the basal level.

Handling

Barker (1935) found that turgid tubers could be handled without serious disturbance of their respiration, but that if the tubers were soft, gentle squeezing could increase the rate of respiration by about 30%. More vigorous treatment, such as that involved in washing and drying the tubers, was found by Burton (1974) to more than double the O_2 uptake by slightly wilted tubers and to cause a 30% increase in that by freshly harvested tubers. In both cases the effect was temporary and disappeared in the course of a day. Isherwood and Burton (1975) found handling to lead to variable respiratory quotients, an effect which could last for several weeks.

Heat production

For every g of carbohydrate consumed in normal respiration about 16 kJ of energy are released. In part this is transferred into energy-rich phosphate molecules and utilized in metabolism; in part it is lost as heat. There have not been many measurements of the proportion so lost. Green *et al.* (1941) suggested it was as much as about 90%. If we accept this for tubers in which no marked change is occurring, then for every g of CO_2 produced in respiration about 10 kJ of energy are released in the form of heat. The amount could differ during periods of rapid change but evidence for this is lacking.

14.1.5 Break of dormancy and sprout growth

General

There has been argument as to whether the buds on a potato tuber are ever dormant, in that some change and development is always occurring, albeit slowly (Davidson, 1958). Certainly it would be difficult, if not impossible, to pinpoint a time before which no change occurred and after which growth started. As a practical concept, however, dormancy, or the resting state, has a real meaning. It is common experience, in the case of many varieties, that soon after harvest a tuber may be placed at, say, 20°C for 3 weeks without any noticeable bud-growth occurring; while a few months later a tuber, which had been held at 2°C to prevent growth, would, when placed at 20°C, immediately start to sprout, and growth in 3 weeks would be considerable. Admittedly cell division could probably have been detected, by microscopic examination, to be occurring in the first tuber; and the difference in behaviour may be regarded as quantitative rather than qualitative; but this quantitative difference is very great and in practical terms there is a difference in the state of the buds. In the case of the first tuber they could be described as 'dormant', as preferred by Burton (1963a), or 'resting', as preferred by Emilsson (1949). Their different behaviour, compared with that of those on the second tuber, is a manifestation of a different biochemical balance in the buds, and is subject to manipulation by biochemical means, either by direct treatment with substances thought to contribute to the balance, or with substances having similar biological effects; or by placing the tuber under conditions such that the balance changes naturally.

A bud which is defined as dormant or resting, as above, is normally drifting the whole time during storage towards a state of growth. Other influences may intervene to prevent growth commencing – an example is the suppression of lateral buds by a vigorously growing apical bud – but apart from this the drift is towards growth. Because it is a drift, and not a sharp change, and because some of the changes may be very local, it is difficult to be precise as to the moment when the biochemical balance becomes permissive of consistent growth in the bud as a whole, and to correlate changes in the balance with the start of growth. The substances and reactions concerned in the balance have also not been fully identified. Gibberellins are known to be present (Okazawa, 1960) and to increase considerably at about the time of sprout growth (Smith and Rappaport,

1961). GA_3 has also been shown to accelerate RNA and DNA syntheses and mitotic activity in plugs with buds from initially dormant tubers (Rappaport and Wolf, 1969), and stimulates break of dormancy (Rappaport *et al.*, 1957). On the other side of the balance, gibberellin inhibitors are known to be present in the tuber (Okazawa, 1960).

Hemberg (1947) first suggested that the rest period of potato buds was controlled by inhibitors, later equated with the inhibitor-β complex of Bennett-Clark and Kefford (1953). Hemberg was unable, however, to inhibit the growth of potato buds by these substances, extracted from the potato, which inhibited the elongation of *Avena* coleoptiles. Burton (1956a) emphasized that if an inhibiting substance were to be postulated as controlling dormancy, then it must be shown to inhibit, not the elongation of *Avena* coleoptiles, but the growth of potato sprouts. He failed to show any effects on the growth of potato sprouts on an agar medium *in vitro*. Blumenthal-Goldschmidt and Rappaport (1965), however, succeeded, by repeated application, in inhibiting the growth of potato buds by the inhibitor-β complex. The inhibition was reversed by GA_3. They speculated that abscisic acid ('abscisin II'), shown by them to inhibit potato bud growth, and suggested by Wareing *et al.* (1966) to inhibit gibberellin synthesis, might be the inhibiting component of the β complex. Abscisic acid was later isolated from potatoes (Cornforth *et al.*, 1966). Rappaport and Wolf (1969) reviewed the control of dormancy in potato buds and concluded that the substances that appeared to be most involved were gibberellins and abscisic acid. The correlation of changes in chemically identified and estimated gibberellins and abscisic acid in potato buds with the dormant or active state of the buds is, however, still lacking.

Burton (1957, 1961, 1963a) considered that it was incorrect to measure the dormant period from the date of harvest, as was usual in published work, in that the date of harvest had no physiological significance – for example the foliage at the time could be still functional, or senescent, or could even have been dead for some weeks. He suggested that the most logical physiological or biochemical period from which to measure the beginning of dormancy, was the period during which tuber initiation was occurring. In other words a period during which the biochemical balance in the stolon tip was swinging away from a state permissive of extension growth towards that favouring lateral expansion, cell enlargement, and ultimately the practical cessation of cell division. He later developed this theme further (Burton, 1972).

Variety

Extensive information on the dormant or resting periods of different varieties after harvest was published by Emilsson (1949; 51 varieties) and by Schippers (1956b; 41 varieties). Burton (1963a) gave durations of the dormant periods of eleven varieties, measured both from the time of tuber initiation and from the time of harvest. The average duration after harvest found by Schippers at 10°C, for tubers harvested immature in mid-July, was 19.5 weeks and the range 17–27 weeks. Durations after harvest found by Burton at 10°C for tubers harvested mature in September averaged 8 weeks, range 5–14 weeks, and after tuber initiation averaged 24 weeks, range 20–33 weeks. There is no evidence in any of the above work that the duration of the dormant period bears any relation to the maturity class of the variety.

There does not appear to be any necessary connection between the length of the dormant period and the subsequent rate of growth (Burton, 1963a) but, like the dormant period, the rate of sprout growth shows marked, and in general consistent, varietal differences. These are illustrated by some of the results of Emilsson (1949), although in considering these we must remember to compare only varieties with the same dormant period after harvest and which therefore commenced growth at the same time. Burton (1966) quantified the differences between the rates of growth exhibited by four varieties by comparing the constants in the differentiated form of the sigmoid growth curves, demonstrating a five-fold difference between Craig's Defiance and Home Guard at one extreme and Arran Consul at the other. There is no obligatory relationship between maturity class and the rate of sprout growth, although in fact the maincrop varieties which have survived in commerce tend to exhibit a slower rate of growth than do commercial early varieties.

Season

Burton (1966) in discussing the effect of season upon the length of the dormant period, concluded that this was normally comparatively slight uless the season was either abnormally hot and dry, or cold and wet, shortly before the time of harvest, but when the foliage was still alive. He quoted his own results (Burton 1963a) and those of Krijthe (1962) to demonstrate, respectively, an increase of about a month in the duration of dormancy after a cold wet growing season and a decrease of as much as 9 weeks after a markedly hot and dry season. This need not always be the case, however. Burton (1963a) had also shown that samples from serial liftings during a variable season could differ considerably in bud dormancy.

Temperature

In general, over the range normally encountered in practice, the higher the storage temperature the shorter the dormant period (Wright and Peacock, 1934; Schippers, 1956a; Burton, 1963a). Schippers' results showed an 18% reduction in the period of dormancy after harvest if the storage temperature was raised from a constant 10 °C to a constant 20 °C. Lowering the storage temperature from 10 to 5 °C lengthened the period by 67%, and from 10 to 3 °C by over 150%. This marked effect of small reductions in temperature at the lower end of the range is noteworthy, and suggests an accumulation of the inhibiting side of the biochemical balance at low temperatures, analogous to the accumulation of sugars, similarly caused. Storage for about a month at a temperature of −1 °C, just above the freezing point of the potato, may lead to a very much prolonged dormant period or may permanently impair the ability to grow.

The above results relate to constant storage temperatures. Variable temperature regimes have effects which are not predictable from our present knowledge, ranging from a reduction in overall dormant period if storage includes a period of some weeks at 0–5 °C (Emilsson, 1949; Schippers, 1956a; Burton, 1973), through samples upon which temperature regime appears to have no effect, to samples which exhibit extended dormancy if the temperature of storage falls even briefly below 20 °C. (Schippers, 1956a).

When sprouting has started, the rate of growth on any one sample of tubers is related to temperature, normally exhibiting an apparent maximum at about 20 °C, although the true optimum temperature may be about 25 °C if growth is not limited by nutritional deficiency (Dyson and Digby, 1975a, b). Growth is usually very slow at temperatures as low as 5 °C although some varieties may show slight growth even at 2 °C (Krijthe, 1946). Sprout growth on the variety Majestic at different temperatures, related to that at 20 °C (= 100), was found by Burton (1958) to be 27 at 10 °C, 62 at 15 °C and 56 at 25 °C. Diurnal fluctuation of temperature between 2 and 10 °C, the period at each temperature being between 8 and 16 h, prevents or considerably reduces growth (Burton, 1969).

Burton (1966) pointed out that in addition to the direct effects of temperature upon sprout growth there would be indirect effects attributable to the effects of temperature upon the composition of the tubers upon which the sprouts are growing. He illustrated this by reference to the results of Barker (1937) and Schippers (1956a).

Period of storage

The pattern of dormancy break and subsequent growth differs with the length of the period which has elapsed before environmental conditions permit their occurrence. Dormancy or otherwise is a property of each individual bud and there is a gradation in the depth of dormancy from the oldest, most deeply dormant, to the youngest, least dormant bud. If freshly harvested tubers are stored at a temperature at which vigorous growth is possible the youngest (distal) bud normally grows first, and products of its growth inhibit the other buds, leading to apically dominant growth. If the tubers had been stored at a temperature too low for growth the biochemical balance in successively older buds would have completed, in turn, the swing to a state permissive of growth. If samples were transferred at intervals to a temperature favourable for growth, an increasing number of buds would grow, with roughly equal vigour, on successive samples, there being no opportunity for the establishment of apical dominance (Appleman, 1918; Bushnell, 1928, 1929; Kawakami, 1952; Toosey, 1959). Apart from the number of buds growing, the vigour of growth on successive samples would show a change, increasing to a maximum and then decreasing (Schippers, 1956b; Krijthe, 1962) until finally the potential for growth was practically non-existent. The time scale of this change in vigour is shorter the higher the storage temperature (Krijthe, 1962) and is a varietal characteristic, the fall in potential being sufficiently early in some varieties, e.g. Ulster Premier, at moderate storage temperatures, to be of commercial importance in reducing emergence in the field (Sharrock and Burton, reported in Burton, 1963b).

O_2 concentration

O_2 concentration affects both the break of dormancy, as indicated by the number of sprouts per tuber which start to grow, and the subsequent growth of sprouts as indicated by the size they attain.

The concentration in which the greatest number of buds start to grow at 10–20°C is about 2–4% ($2–3 \times 10^{-5}$ M in the sap) at the beginning of the storage season, but rises to 14–23% ($2–3 \times 10^{-1}$ M in the sap) by June (Burton, 1968). There is a similar rise in the concentration which is optimal for sustained growth, from 4–5% to 17–20%. Late in the storage season, low concentrations of O_2, which stimulate growth earlier, cause marked suppression of growth. Burton suggested that in these effects of O_2 concentration we may be concerned with the optimally anaerobic reversible metabolism of a growth inhibitor, the enzyme system concerned being possibly sulphydryl dependent. Rakitin and Suvorov (1935) stated that exposure for a week to anaerobic conditions destroyed the dormancy of freshly harvested tubers. The conditions in their experiments may have been doubtful, but additional evidence is afforded by the results of storage in 1% O_2 for a few days upon subsequent sprout growth in air at 20°C. Early in the storage season this can stimulate growth. Later, when we may postulate the growth inhibiting system to be at a minimum, it has no effect (Burton, 1973).

The fact that 2% O_2 may be optimal for the start of growth, but does not thereafter permit vigorous growth, has been tentatively ascribed to an inability to mobilize stored carbohydrate in 2% O_2 (Burton, 1971).

CO_2 concentration

Sprout growth at 10 °C is stimulated by increasing the concentration of CO_2 in the atmosphere up to an optimal value of about 2–4%, giving a concentration of free CO_2 in solution in the cell sap of about 2×10^{-3} M (Burton, 1958). Concentrations above this have a less and less stimulatory effect, until at about 7–10% ($4–5 \times 10^{-3}$ M in the sap) growth is similar to that in air (10^{-3} M in the sap). Still higher concentrations reduce growth and a concentration as high as 15% at 10 °C, giving a concentration of about 8×10^{-3} M in the sap, will inhibit it. Stimulation of sprout growth by increased CO_2 results in part from an increase in the number of cells produced, but also from a marked increase in the length of the fully elongated cells.

Other metabolic volatiles

The stored potato tuber evolves, in addition to CO_2, very small quantities of a number of volatile substances (Burton, 1952, 1965c; Burton and Meigh, 1971). If these accumulate in the storage atmosphere they can completely inhibit sprout growth at normal atmospheric concentrations of O_2 and CO_2, and inhibit or considerably reduce it in stimulatory concentrations – for example in 5% O_2 or 5% CO_2 (Burton, 1965c). The active substances have not been identified, but Burton and Meigh (1971) suggested that their molecular structure probably included olefinic, ethereal and/or S containing groups.

Humidity

It has been reported (Emilsson, 1949) that the dormant period after harvest is shorter if the humidity of the storage atmosphere is high. This is based on a stated difference between 100% r.h. and drier conditions, but the original papers may not refer to 100% r.h., but to moist conditions, i.e. covering the tubers with moist sphagnum moss (e.g. Smith, 1929; Thornton, 1939). This could give a film of moisture on the tuber, and reduce its content of O_2 (Burton and Wigginton, 1970) with consequent effects on dormancy (Burton, 1968). Genuine atmospheric humidity has not been shown to have any effect on the length of the dormant period (Davidson, 1958), but when sprout growth has started, it may influence the form of the sprouts, branching being greater under dry conditions, and production of adventitious roots greater under humid conditions.

Light

Sprout length is considerably reduced by exposure of the sprouting tuber to light, the most active wavelengths being in the ranges 350–450 nm and 650–950 nm, which even at low intensities can reduce sprout length to only about 4% of that of sprouts grown in the dark (Wassink *et al.*, 1950).

Chemical suppressants and stimulants

Very many substances have some effect on growth. In so far as any generalization may be attempted, many S compounds, particularly those containing sulphydryl groups, stimulate sprouting (Miller, 1931; and see Hammett, 1929); as do many oxidizing agents (Appleman, 1916; Rosa, 1923), a number of derivatives of ethylene (Denny, 1926a,b) and urethane (Dettweiler, 1952). The vapours of several alcohols will reduce or inhibit sprout growth (Burton, 1956b) as will the methyl ester of α-naphthalene acetic acid (Denny *et al.*, 1942; van Stuivenberg and Veldstra, 1942; Luckwill, 1948), various chlorinated nitrobenzenes (Brown, 1947; Luckwill, 1949; Brown and Mellor, 1952; Brown and Reavill, 1954), phenyl urethane and a number of related compounds (Dettweiler, 1952) including iso-propyl-*n*-phenyl carbamate (Rhodes *et al.*, 1950; Dettweiler, 1952; Downie, 1952; Fischnich, 1956), and its chloro derivative (Marth and Schultz, 1950; Sawyer and Dallyn, 1957). Maleic hydrazide, applied to the foliage, inhibits subsequent sprout growth in the crop (Marshall and Smith, 1951; Paterson *et al.*, 1952; Kennedy and Smith, 1951, 1953; Franklin and Thompson 1953; Fischnich and Pätzold, 1954).

Irradiation

Sprout growth may be prevented for a considerable period, in some cases indefinitely, by irradiating the tubers with γ-rays at 5 000–20 000 rad (Sparrow and Christensen, 1954; Brownell *et al.*, 1954, Burton and Hannan, 1957; Pätzold and Weiss, 1957) electrons (Hannan, 1954) or X-rays (Hagberg and Nylsom, 1954; Pätzold and Kolb, 1957).

Wounding

Scrubbing off the periderm (Schmid, 1901), cutting or peeling (Appleman, 1914, 1916) or superficial burning (Dostal, 1942) can all lead to the break of dormancy. Wounding can also of course lead to fungal or bacterial rotting and, although cut seed pieces are used in some regions, this is more to economize in the weight of seed tubers than to promote sprout growth. Wounding (in the form of excising plugs of tissue with buds) is however useful when potatoes are used experimentally for the bioassay of growth inhibitors.

Internal sprouting

The growth of sprouts into the tuber instead of normal growth was described many years ago (Müller, 1846; Gager, 1912). It results from the growth of axillary buds at the base of external sprouts of which the growing point has been destroyed. Davis (1961) described the inducing factors as (a) injury and loss of dominance of the apical meristem without either inhibition of the development of unexposed buds or complete inhibition of basal portions of the sprout, followed by (b) circumstances which allow the full force of the expanding sprout to be exerted on the periderm – by, for example, the pressure of another tuber against the eye of origin, or by the mechanical components of a basipetally expanding cluster of buds. Both predisposing factors can follow the use of inadequate concentrations of sprout inhibitors giving initial suppression but later a rosette of buds (see e.g. Twiss, 1963). It is however the rosette following apical damage, not the subsequent presence of the low concentration of inhibitor, which induces the complaint (Sawyer, 1961; Sawyer and Dallyn, 1964).

14.1.6 Changes in composition

General

The potato tuber contains very many potentially reactive systems of substrates and enzymes – in fact the presence of complex molecules which have been elaborated in the tuber during growth suggests a potentiality for continued reactions, and for reverse reactions. When a mature tuber is harvested its content of the major known constituents has reached a fairly steady level. Change is still occurring – we know that there is a perpetual cycling of some constituents and may suppose that there is cycling of others – but the production and breakdown of many of them is nearly in balance: not quite, as some change can be detected, and of course there is a perpetual loss of carbon from the system in the CO_2 evolved in respiration.

Every reaction is influenced by its immediate physical and chemical environment. The latter is affected by the products and demands of every neighbouring reaction; which reactions are themselves subject to similar influences. The whole tuber complex is thus very labile, and if the storage conditions, or the stimuli to which the tuber is exposed, are very different from those under which it has been growing, then the state of near-balance is lost, and readjustments, which may be considerable, occur. If the fresh stimuli include mechanical damage or infection by pathogens, then the complexity of the response can be much increased and may involve reactions, following membrane disruption or the introduction of different reactants or enzymes, for which there is normally little or no opportunity.

The complete response of the tuber to the time and conditions of storage, and to other stimuli, as shown by changes in composition, is unknown. We do, however, have some knowledge of a number of changes, mainly those in the content of constituents of known or suggested importance to the quality or dietary value of the tuber.

Percentage dry matter

Taking starch to be the primary respiratory substrate and assuming an average evolution of CO_2 during storage of rather more than 3 ml kg^{-1} h^{-1} (see Section 14.1.4), then in one month a tuber will have lost about 0.3% of its weight in the form of dry matter and gained about 0.15% of its weight in water produced during respiration. If respiration were the only source of net weight loss, a tuber with an initial percentage dry matter of 22.0% would, after one month, have a dry matter of 21.7% and after six months, 20.4%. Water loss is also occurring during storage however (Section 14.1.3) and the percentage of dry matter need not therefore decrease, although it does unless the potatoes are stored under conditions such that water loss is excessive. Under good storage conditions a drop of about 0.1% a month might be expected.

Carbohydrates

Normally, about 70–75% of the dry matter of the potato tuber consists of starch, synthesized from sucrose translocated to the tuber during growth, and potentially liable to be broken down again to sucrose and reducing sugars. Of these the concentration is low in the mature tuber at the time of harvest, the ratio of starch to total sugar being of the order of 30–50 : 1. There is, however, a perpetual turnover in the system: starch–sucrose–reducing sugars–starch; and changed environmental conditions may well upset the balance and lead to a change in the starch: sugar ratio. If the potatoes were maintained in store at the same temperature as they were at the time of harvest, say 15 °C, there might be a slight rise or fall in sugar content, but no marked change until after the lapse of some months, when the tubers would start to sweeten, slowly at first but with increasing rapidity (Burton *et al.*, 1959; van Vliet and Schriemer, 1963; Burton, 1965b; Fig. 14.1). Burton (1966) gave the term 'senescent sweetening' to this phenomenon, which is irreversible and linked to some change in the membranes surrounding the starch grains (Isherwood, 1976).

The timing and degree of senescent sweetening are influenced by variety and exhibit parallelism with potential sprout growth, there being some evidence that this is more vigorous the more marked the sweetening (Burton, 1965b). If sprout growth is prevented by chemical or physical means the sweetening is even greater, possibly because of the lack of demand from growing sprouts (Burton, 1965b; Isherwood

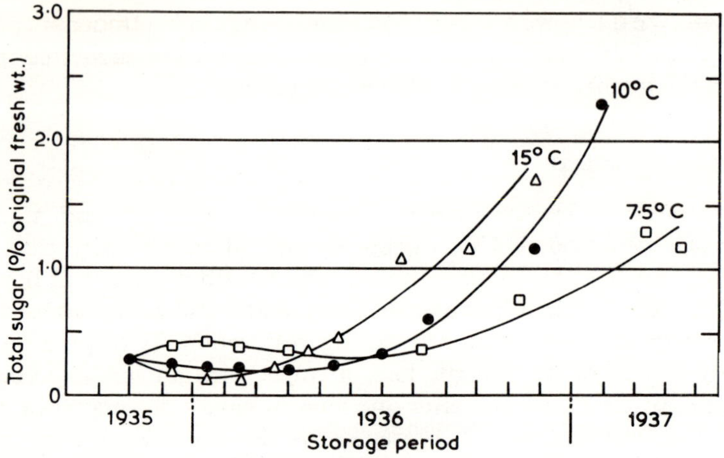

Fig. 14.1 Senescent sweetening of potato tubers (cv. King Edward) after prolonged storage at 7.5°C and less prolonged storage at 10 and 15°C (adapted from Barker, 1938)

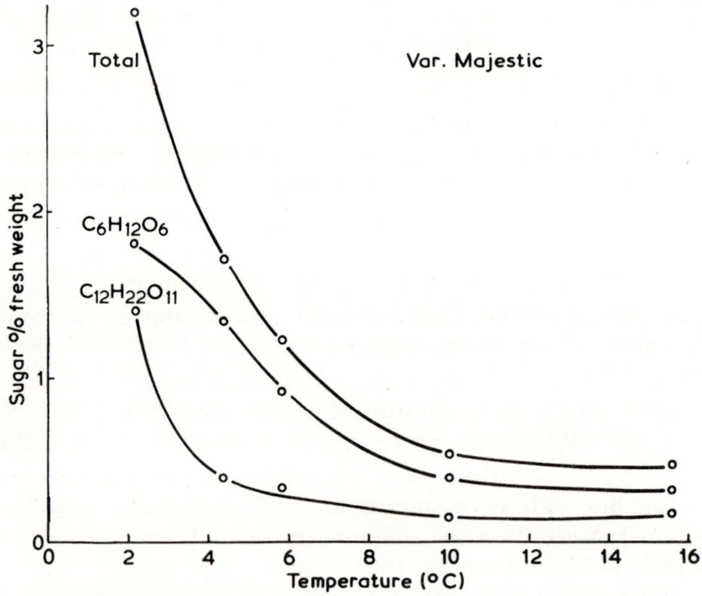

Fig. 14.2 Low temperature sweetening: sugar contents of potato tubers (cv. Majestic) after storage for 4 weeks (17th Dec.–14th Jan.) at different temperatures (from Burton, 1965b)

and Burton, 1975). Senescent sweetening starts later the lower the storage temperature (Fig. 14.1), but a low temperature does not give a means of avoiding sweetening, because marked sugar accumulation occurs in tubers stored at low temperatures (Fig. 14.2). Low temperature sweetening differs from senescent sweetening in being to a large extent reversible if the temperature is raised, but it too may involve changes in the permeability of the membrane surrounding the starch grains, in this case reversible. There have been many attempts to equate it with temperature-induced changes in the relative activities of enzymes or enzyme inhibitors. Pressey (1970) considered sucrose phosphate synthetase to be the enzyme responsible for sucrose formation in cold-stored potatoes. Subsequent hydrolysis of sucrose to glucose and fructose may be regulated by the comparative levels of invertase and an invertase inhibitor (Pressey and Shaw, 1966; Pressey, 1966), although Pressey (1970) suggested that the availability of sucrose at the site of invertase action could be a controlling factor.

The reversibility of temperature-induced carbohydrate changes is not complete, and the residual effects of previous storage temperature can be detected for a long time. Burton (1975) found that after 90 days at 10 °C the sugar content tended to be higher in potatoes previously cool stored for 4 weeks than in those stored throughout at 10 °C. Conversely the sugar content of tubers previously stored at 15.5 °C was lower.

γ-irradiation of normal unsweetened tubers causes marked temporary sucrose accumulation, reaching a maximum in 3–7 days, and then falling to a little above the original content during about 3 weeks (Burton *et al.*, 1959; Kubicki, 1961; Burton, 1975). The effect appears to be independent of the time which has elapsed after harvest and before irradiation (Burton *et al.*, 1959), but if this prior storage has been at a temperature which has already induced considerable sweetening, irradiation may give no further increase, or even a decrease (Pedersen, 1956; Rubin and Metlitsky, 1958; Burton, 1975). Reducing sugars are also affected, not so markedly as sucrose soon after harvest, but increasingly seriously the later the date of irradiation, the effect being also more prolonged than in the case of sucrose (Burton *et al.*, 1959). Senescent sweetening is considerably accelerated by irradiation, but, as mentioned above, this effect may be common to all methods of sprout suppression.

The foregoing description of carbohydrate changes relates to tubers harvested mature. Similar changes occur also in tubers harvested immature, but in addition these show marked changes in the few days following harvest. At the time of harvest, immature tubers have a high content of sucrose. After harvest, during storage at temperatures of 10°C and above, this falls rapidly, in the course of a few days, to a low level similar to that in mature tubers. The fall is accompanied by an equivalent, or even more marked, rise in reducing sugars which, although they later decrease slowly, may remain at a high level during subsequent storage (Burton, 1965b).

N fraction

There are no major changes in the crude composition of the major components of the N fraction during most of the storage period, but changes occur during sprout growth. These are mainly in amide and free amino acid fractions (Tagawa and Okazawa, 1953; Szalai, 1957a, b,c, 1959 a,b; Talley *et al.*, 1964). The largest and most consistent changes occur in the content of proline, which increases, particularly towards the end of storage, and is translocated to the sprouts (Heilinger and Breyhan, 1959; Breyhan *et al.*, 1959; Heilinger, 1961; Talley *et al.*, 1964). Harvesting wounds or disease lesions could lead to a local increase in the content of tyrosine, necessarily at the expense of some other nitrogenous constituent. Johnson and Schaal (1957) found a doubling in the content in cells immediately adjacent to the wound, but the effect on the overall composition of the N fraction of the whole tuber would be slight.

Phenolic compounds

Phenolic compounds of known importance to the cooking quality of the tuber include tyrosine, already mentioned above; chlorogenic acid and perhaps caffeic acid. Of the two last, chlorogenic acid increases in concentration in the buds during storage (Amberger and Schaller, 1973); is synthesized and increases very markedly in the cells adjacent to wounds (Politis, 1948; Johnson and Schaal, 1957; Ozeretskovskaya and Vasyukova, 1965); and shows a temporary rise, followed by a fall, after γ-irradiation (Penner and Fromm, 1972).

Chlorophyll

Chlorophyll is synthesized, particularly in the peripheral layers, if potato tubers are exposed to light, the synthesis increasing with increasing light intensity over the range 140–3300 lx (Liljemark and Widoff, 1960). Liljemark and Widoff found daylight-type fluorescent (350–750 mm) to be effective in promoting the synthesis. Within this range of wavelength, green light, 500–600 nm, was least, and blue, 400–500 nm, and red, 600–700 nm, increasingly more effective. The degree of chlorophyll synthesis which occurs under given conditions appears to be an heritable varietal characteristic (Akeley *et al.*, 1962). Harkett (1975) found the rate of synthesis to be markedly temperature dependent, a maximum rate being reached at 15°C and negligible synthesis occurring in 16 days under 100–1100 lx at 2°C. Once formed, chlorophyll is not lost, at least over a period of 33 days, during subsequent storage in the dark (Harkett, 1975).

Melanin and other pigmented oxidation products

Reactions between enzymes and substrates in the potato are regulated by the presence of functional membranes around the organelles in which either or both may occur. This can impose a form of physical separation between an enzyme and a potential substrate, as a result of which no reaction is normally possible. If the membranes are damaged, either physically or chemically, then aberrant reactions can take place. The most important of these is the formation of melanin as the end-product of a chain of reactions ensuing from the polyphenolase-mediated formation of 3–4 dioxyphenylalanine from tyrosine by hydroxylation and oxidation. Physical membrane damage may occur during storage as a result of pressure, more particularly in the case of wilted tubers which distort readily; or as a result of breakage by ice crystals if freezing of the cell sap occurs – the freezing point is -1 to $-2°C$. We then get melanin formation in the damaged tissue, giving the characteristic pressure bruises or symptoms of frost injury. If tubers are subject for a few days to anaerobic conditions, or high ($>15\%$) CO_2, or to high temperatures which result in an O_2-deficient and CO_2-rich internal atmosphere (see Section 14.1.1), loss of structure or complete breakdown, possibly local, occurs, accompanied by melanin formation – in the case of anaerobic conditions only after the access of O_2. In these cases the cause of membrane disruption must be chemical. The high-temperature injury is known as 'black-heart'.

Although melanin formation is the most important source of discoloration in the flesh, there are other potential causes which also involve polyphenolase-catalysed oxidations. Both chlorogenic and caffeic acids are potential substrates, the former giving a greenish-brown end-product (Metlitskii and Tsekhomskaya, 1964).

Glycoalkaloids

The potato tuber normally contains small quantities (about 2–10 mg/100 g) of solanidine and its glycoalkaloids. Exposure to light induces their further synthesis in the peripheral tissue, red light being the most effective (Conner, 1937). Zitnak (1955) found exposure to sunlight for 6–8 h at near-freezing temperatures to result in a 3-fold increase. In view of the effect of temperature, mentioned below, the increase could often be more serious. McKee (1961) reported a considerable increase to follow wounding, the concentration rising within 4 days (temperature unspecified) from undetectable to 1 mg g^{-1} fresh weight in the tissue to a depth of 0.3 mm below the wounded surface. There was little increase 3.2 mm or more below the wound. Salunkhe *et al.* (1972) found the increase in wounded tissue to be greater in the light than in the dark. As in the case of chlorophyll formation, temperature has a marked effect: Salunkhe *et al.* finding little synthesis of glycoalkaloids in slices held, even in the light, for 48 h at 0 °C; some increase at 8 °C; and marked synthesis at 15 and 24 °C. Accumulation also follows infection by pathogens (Locci and Kúc, 1967).

Terpenoids

Changes in composition following infection by pathogens are not strictly relevant to a consideration of the storage of healthy tubers. They are mentioned here because of the widely publicized (Renwick, 1972) but unsubstantiated (Elwood and MacKenzie, 1973; Smith *et al.*, 1973; Emanuel and Sever, 1973; Chaube *et al.*, 1973) suggestion of the teratogenic effects of blighted potatoes. Among the more marked changes resulting from infection, apart from the glycoalkaloid synthesis mentioned above, is a great increase in the content of terpenoids, particularly of rishitin and phytuberin, which are normally present in negligible amounts and which may accumulate to over 1 mg g^{-1} fresh weight in rotted tissue (Lyon, 1972). Increases have been reported after inoculation with *Phytophthora infestans* (Tomiyama *et al.*, 1968), various other fungi (*loc. cit.*; Varns *et al.*, 1971), and *Erwinia caratovora* var. *atroseptica* (Lyon, 1972). The increases are local and marked accumulation does not occur far beyond the edge of the rot (Lyon, 1972).

Vitamins

The potato makes an appreciable contribution to the dietary intake of thiamin, nicotinic acid and vitamin C; and, in the case of very yellow fleshed varieties, to vitamin A. With a mixed diet it is only the provision of vitamin C which is important – the potato is the major source in the northern European diet (see Burton, 1974) – and losses during storage are correspondingly of interest. Of the other vitamins mentioned, Yamaguchi *et al.* (1960) found only slight loss of thiamin after prolonged storage at 10 °C and none at 5 °C; and negligible change in nicotinic acid at either 5 or 10 °C. Meiklejohn (1943), however, found up to 50% loss of thiamin in sprouted tubers.

Vitamin C is taken here to include both the reduced (ascorbic acid) and oxidized form (dehydroascorbic acid). There are several changes which can occur during storage, including synthesis, interconversion of the two forms, conversion of dehydroascorbic acid to diketo-gulonic acid (which represents a loss of vitamin C), and other reactions leading to loss. The net result which has usually been reported is a marked loss of vitamin C during storage, from about 30 mg/100 g at harvest to about

10 mg/100 g by May being typical (see Burton, 1966, for references to the older literature), about 2/3 of the loss occurring during the first 3 or 4 weeks. Recent extensive results by Müller (1975), on samples of 28 varieties stored at 8°C, seem to contradict this. The content at harvest was comparatively low, averaging about 19 mg/100 g, but during the first 4 weeks of storage there was, on average, little loss; nor was it marked after 10 weeks – in fact in many varieties it was negligible. It may be that the storage temperature used by Müller was optimal for preventing loss of vitamin C.

There was evidence of a temperature optimum in the work of Effmert *et al.* (1961) but marked loss was found by Zilva and Barker (1939) at 10°C, not greatly different from 8°C, and the question of the extent of storage losses of vitamin C appears to be unsettled. Barker and Mapson (1950) found much evidence of effects of temperature, there being, in freshly harvested tubers and after prolonged storage, an increase in ascorbic acid during low temperature sweetening. Different results were observed in the middle of the storage season, and clarification of the age-response and of the reasons for this is needed. Barker and Mapson (1952) found anaerobic conditions prevented change in ascorbic acid – either synthesis or loss – but they stated this inhibition was not permanent, there being a slow decrease in content after exposure to anaerobic conditions for 3–4 weeks. They stated that there was no sign of tissue injury in the experimental tubers. This is unusual after anaerobisis for 3–4 weeks, however, and a more detailed study of the effects of oxygen concentration, in relation to temperature of storage, is required. Johnson and Schaal (1957) found an appreciable, although temporary, increase in ascorbic acid in damaged tuber tissue.

Other constituents

There must be many unknown changes in quantitatively minor constituents of the tuber, which have not been studied because their importance in tuber metabolism, or to the quality or nutritive value of the tuber, has not been established. Changes in gibberellins, of probable importance in sprout growth; of inhibitors, possibly related to dormancy; and of citric acid, of known importance to the quality of cooked potatoes (Hughes, 1963) have been followed. Gibberellins increase considerably about the time of sprout growth (Smith and Rappaport, 1961). There is also a loss of inhibiting substances – as measured by inhibition of *Avena* coleoptile growth – during storage (Hemberg, 1943, 1947), though there may also be a subsequent increase (Burton, 1956a). Citric acid content may fluctuate during storage at 8°C, but there is little consistent change in content after 10 weeks of storage (Müller, 1975).

14.2 The environment within a stored bulk of potatoes

14.2.1 General

The foregoing sections have been concerned with the response of a single tuber to the environment immediately surrounding it. They can be applied to a stored bulk of potatoes only to the extent that we know the environments immediately surrounding the individual tubers which make up the bulk. When potatoes are heaped together, the outer surfaces of the top tubers will be in contact with the outside air or air in the building, but even in these tubers the environment of the under surfaces will be affected by heat and moisture from the potatoes below. In the centre of the heap the environment may be very different from that at the surface. Before we can make use of most of the information in Section 14.1 therefore, for anything other than single tubers, we need to have a fairly precise idea of the environment in a stored bulk.

14.2.2 Temperature

Convection and thermodynamic equilibrium

We have seen (Section 14.1.1) that a tuber reaches thermodynamic equilibrium with its environment with the establishment of a small temperature differential. This means that the whole of the metabolic heat production is then being transferred to the surrounding air as fast as it is being produced. The rate of heat production depends on the rate of respiration, but for a mature 200 g tuber at normal storage temperatures, it could be expected to be within the range 8–16 J h^{-1} (Section 14.1.4). Transfer of this heat to the surrounding air raises the temperature of the latter – the specific heat of air is about 1.0 J g^{-1}°C, and 10 J would thus raise the temperature of 1 g (i.e. about 800 cm^3) of air by 10 °C, or of 1 l of air by 8 °C.

In the case of a single tuber exposed to free air circulation in a room, the ratio of the weight of air to tuber is so great that the effect of the transfer of metabolic heat upon air temperature is negligible. 1 ton of fairly clean potatoes heaped together will, however, have only about 0.5 m^3 of air in the spaces between them. The heat production of the ton of potatoes (40–80 kJ h^{-1}) would be sufficient to raise the temperature of this amount of air by about 65–130 °C in 1 h. The air is not trapped, however, but can escape from the heap. Within the temperature range encountered in a potato stack, every 1 °C rise in air temperature leads to a drop of about 0.36% in its density (there is also a small additional drop due to humidification of the air in a stack), and this warmed lighter air will rise out of the heap at a rate which would be proportional to the drop in density (or the rise in temperature) were it not that the resistance to air flow through heaped potatoes increases in proportion to $V^{1.8}$ (Burton *et al.*, 1955). The net result is that the air rises out of the heap at a rate proportional to $T_D^{0.55}$ where T_D is the temperature difference activating the movement (Burton *et al.*, 1955). The more the temperature rises, therefore, the faster the air rises out of the heap, cooler air being drawn in from the surroundings to replace it and being in turn warmed and replaced by cooler air. A convective movement of air into and out of the heap has thus been activated. Every m^3 of air leaving the heap at a temperature 1 °C above that at which it entered would remove about 1.25 kJ. Also, as pointed out by Hylmö *et al.* (1975a), since the convective air can be assumed to be leaving the stack saturated, it will cause evaporative cooling. The saturation is at ambient temperature, not stack temperature, for reasons given in Section 14.2.4 below, and thus at usual potato store temperatures, for every mbar wvpd in the ambient air 1 m^3 convection will remove about 0.78 g water (Table 14.2) and thus about 1.95 kJ.

The wvpd of the ambient air varies. Burton (1966) gave rough average values for eastern England ranging from a minimum of 0.8 mbar in the winter months to a maximum of 2.7 mbar in May and June, the average over the whole storage season, October to May, being about 1.4 mbar. If we use this value, with the realization that it is a local overall average, and not a precise figure, then every m^3 of convective air will remove about 2.75 kJ by evaporation, averaged over the season. Burton *et al.* (1955) derived a relationship between the approach velocity of convective air movement and the temperature difference activating it. They also found that in a stack of potatoes in equilibrium, the maximum temperature was about 1.5 times the average temperature. The precise temperature difference activating convection and over what distance may be open to question. Burton *et al.* (1955) used the average stack temperature, which gives a value 20% less than that calculated using the maximum temperature. The average will be used here, with the above caveat. On this basis, with a stack of fairly clean potatoes 3 m high, comparable with commercial practice, it can be calculated from the relationship mentioned above that 1 °C temperature difference would activate convection at a rate of 4.6 $m^3 t^{-1} h^{-1}$, capable of removing 5.75 kJ h^{-1} plus about 12.6 kJ h^{-1} by evaporation. A rise of the temperature difference to 3 °C would increase the convection to 8 $m^3 t^{-1} h^{-1}$ (i.e. $\alpha T_D^{0.55}$), capable of removing 30 kJ h^{-1} plus about 22 kJ h^{-1} by evaporation, about equivalent in total to the normal average heat production.

In general terms, the effectiveness of convection in removing heat increases in proportion to $T_D^{1.55}$ so far as direct cooling is concerned (Burton *et al.*, 1955), but in proportion to $T_D^{0.55}$ so far as evaporative cooling is concerned. Respiration, and heat production, may also increase with increasing temperature, but at normal storage temperatures, to a much less extent than does the effectiveness of convection (Section 14.1.4); and, except under exceptional circumstances, which will be mentioned below, thermodynamic equilibrium is always established. Burton *et al.* (1955) found it to be reached after a period of about 2 weeks without marked fluctuation in ambient temperature. In stacks in such a state the temperature is of course not uniform – if the average temperature were 3.5 °C above ambient there would be gradation from little above ambient to perhaps 5 °C above (see 'Dimensions of the stored mass' below), and the rate of convection, and its effectiveness, would show local variation and local transfer of heat. Burton *et al.* (1955), in bins of potatoes 1.9–3.7 m high, found average temperatures to differ from ambient by from 1.8 to 3.0 °C and maximum temperatures by 2.8 to 5.8 °C. They derived a general equation for the calculation of average temperatures in stacks in thermodynamic equilibrium with the ambient air, based on continuous records of temperature measurements, at up to 73 points in any one stack, on 12 stacks ranging from about 10–100 t. Modified to SI units and with a correction to allow for evaporative heat loss (Hylmö *et al.*, 1975a), disregarded in the original equation, it is:

$$(T_p - T_a)^{2.8} = \frac{K}{1.17 \times 10^{-2}}\left(\frac{Q_m - C - E}{A}\right)^{1.8} \qquad (14.1)$$

Where T_p = average temperature of the potatoes (°C)
T_a = average temperature of the ambient air (°C)
K = coefficient of resistance of air flow of the potatoes
Q = metabolic heat production (kJ t^{-1} h^{-1})
m = weight of potatoes (t)
C = heat loss by conduction (kJ)
E = heat loss by evaporation (kJ)
A = area of top surface of stack (m^2).

The value for K varies with the sample of potatoes. Burton *et al.* derived a value of 7.7×10^{-5} for a commercial sample with little adherent earth. The data of Ophuis (1957) give a value of 5.3×10^{-5} for cleaned potatoes. With 20% adherent earth, Wager *et al.* (1952) found 2×10^{-4} and the same potatoes, sprouted to the extent of 5% by weight gave 8×10^{-4}. There is thus a 15-fold difference between extremes, but for good unsprouted commercial samples a value of 8×10^{-5} may be a reasonable estimate. Q, as we have seen from Section 14.1.4, is likely to be in the range 40–80 kJ $t^{-1}h^{-1}$. C, in the case of large commercial stacks, is small in comparison with convective loss and may be disregarded. Measurements of weight loss under commercial conditions suggest that water evaporates at a rate, averaged over the storage season, of about 10 g t^{-1} h^{-1}, representing (Section 14.1.3) a heat loss of about 25 kJ t^{-1} h^{-1}, of the order of rather less than half the normal total loss. The importance of this in the overall establishment of heat balance was first realized by Hylmö (see Hylmö *et al.*, 1975a) although he may have somewhat overestimated the proportion of heat normally so lost. Dry ambient air would give excessive evaporation and correspondingly greater cooling. On average, in England, we could probably use a figure of about 35 kJ t^{-1} h^{-1} for $(Q_m - C - E)$.

Dimensions of the stored mass

Burton *et al.* (1955) found the temperature in an unventilated mass of potatoes, averaged over the whole height, to increase from the periphery inwards and to reach a maximum value at a distance from the sides which was about equal to the height of the stack. Thereafter there was no further increase. The cooler periphery is due to cool air being drawn in by convection at the sides of the stack, an effect which is noticeable over the distance given. Over the rest of the stack the air is presumably drawn in mainly over the top surface. It follows from the foregoing that the lateral dimensions of potato stacks of commercial size have little effect upon the temperature of the potatoes, although in small bins peripheral cooling would have an appreciable effect. This expected result has been confirmed in commercial stacks of 500 t and upwards.

In contrast to the above, the height of a stack influences its temperature markedly. Table 14.6 gives the maximum temperatures reached at different levels in two unventilated stacks of potatoes. A simple useful rule is that the maximum temperature in an unventilated stack will approximate to about $1.8h$ °C above average ambient temperature, where h is the height of the stack in m, and the average temperature to about $1.2h$ °C above. The maximum temperature is not reached at the top of the stack, where cool air is being drawn in by convection, but usually about 0.5–1 m below the top surface. The minimum is usually about 1 °C above average ambient and is normally situated at the most exposed top corner.

Table 14.6 Approximate maximum temperatures reached at different levels in unventilated bins of fairly clean, mature unsprouted potatoes (based on Burton *et al.*, 1955)

Total height of potatoes (m)	*Level at which temperature was taken (m above floor)*	*Approximate maximum temperature (°C above average ambient)*
1.85	0.25	1.5
	1.0	3
	1.7	3
3.7	0.25	2.5
	1.0	4
	1.75	5.5
	2.5	7
	3.55	5

Ambient temperature

The thermodynamic equilibrium values discussed above are all related to ambient temperature. The temperature of a stack of potatoes, if it is out of equilibrium with the surrounding air, will, as outlined above, always move in the direction of that equilibrium. There will be no tendency to equate with the actual ambient air temperature, but with a higher temperature according to the height of the stack. Thus the average temperature of a stack 3 m high will follow the general trend of a temperature about 3.5 °C higher than the average daily ambient, although with an amplitude of fluctuation normally less than 1 °C per day. The thermal inertia of a stack of potatoes is too great for it to follow rapid ambient fluctuations. Similarly the maximum temperature of the stack would follow the general trend of a temperature about 5.5 °C higher than average ambient, again with damped fluctuation. It was stated above that the minimum temperature was usually that of the potatoes in the most exposed top corner. This is true of the state of thermodynamic equilibrium; and true also of the long-term average. These potatoes however follow a temperature about 1 °C above average daily ambient, with only such damping of fluctuation as results from their limited heat conductivity (Section 14.1.1). If there was a sudden marked rise in ambient temperature they could, transiently, be the warmest potatoes in the stack – see, for example the temperatures given by Wager *et al.* (1952, Fig. 6).

The response to ambient temperature, described above, is illustrated in Fig. 14.3, in which the average temperature of potatoes stored to a depth of 2.5 m is shown in relation to a temperature 3 °C above daily average ambient, over a period of 80 days.

Inclusion of earth

The effect of earth included in the stack is 2-fold. A coating of earth on the tubers may influence the rate of transfer of heat from tubers to air, which was given in Section 14.1.1; and earth in the spaces between the tubers will increase the resistance of the stack to air flow. Little information exists on the first effect, which could be expected to be much influenced by the moisture content of the earth. The effect of increased resistance to air flow can be assessed from Equation 14.1. As we have seen above, inclusion of excessive earth has been observed in one case to increase the value of K in this equation by a factor of about 3, which means that $(T_p - T_a)^{2.8}$ would have been increased 3-fold and hence $(T_p - T_a)$ by a factor of about 1.5. Potatoes stored to a depth of 3 m would thus reach thermodynamic equilibrium at an average temperature 5–5.5°C above ambient instead of 3.5°C above. The magnitude of the effect obviously depends upon the amount and nature of the earth included,

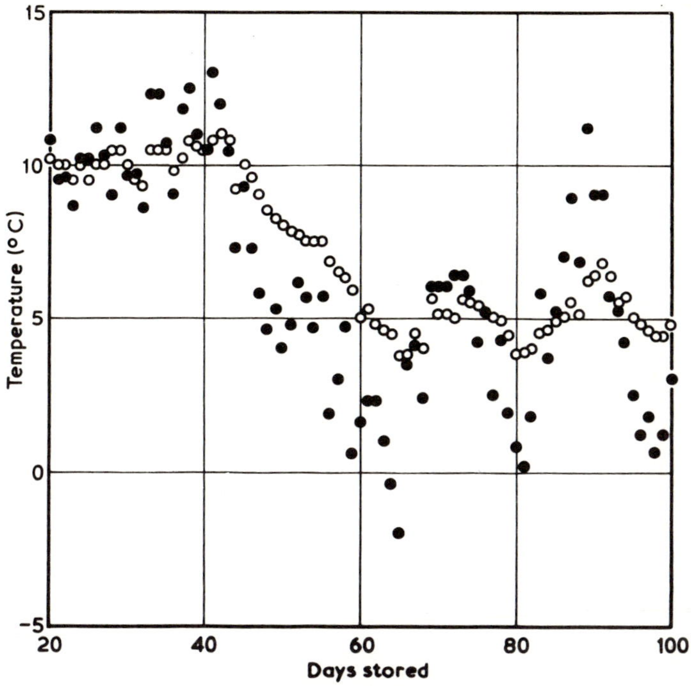

Fig. 14.3 Daily average temperatures, over a period of 80 days, of potatoes (cv. Majestic) stored in an unventilated bin to a height of 2.5 m, related to a temperature 3°C higher than daily average store air temperature (Data from Wager *et al.*, 1952) O: potato temperature; ●: store air temperature plus 3 °C

Note: (1) Approximate thermodynamic equilibrium during periods of minor fluctuation in store temperature (days 20–43)

(2) Damped response to major consistent trends in store temperature, tending to re-establishment of thermodynamic equilibrium with temperature 3°C above ambient (days 44–100)

and the consequent influence upon K, which will vary from case to case. The above factor of 1.5 is quoted merely to give an idea of the likely upper limit of the effect.

Sprout growth

The effect of sprout growth is also 2-fold. It is accompanied by increased respiration (Section 14.1.4) and consequent heat output; and the sprouts growing in the spaces between the tubers increase the resistance of the stack to air flow. Again we can make use of Equation 14.1. Sprout growth to an average extent greater than 5% by weight is rarely encountered in practice. Tubers sprouted to this considerable extent could be respiring at 4–5 times the rate of unsprouted tubers, which means that Q_m in the equation could be 5 times greater and $Q_m - C - E$ possibly as much as 8 times greater – although E is potentially much greater in sprouted than in unsprouted potatoes (Section 14.1.3) it is, in unventilated storage, limited by the water-holding capacity of the convective air, and thus by the rate of convection, and could be perhaps doubled in the case under consideration. $(T_p - T_a)^{2.8}$ could thus be increased by a factor of at least $5^{1.8}$ and possibly $8^{1.8}$ and hence $T_p - T_a$ by a factor of 2.8–4.2. In addition there is the effect upon resistance to flow, i.e. upon K, which could be increased by a factor of 4, as we have seen above (admittedly in the presence of considerable earth). This would increase $T_p - T_a$ by a further factor of 1.6. The overall effect could well be a 5-fold increase in $T_p - T_a$. The average temperature of a 3 m high stack, instead of being 3.5°C above ambient would tend to be 17–18°C above. Again, as in the case of inclusion of earth, this figure is not precise, but it gives an idea of the potentially serious effect of considerable sprout growth in unventilated stacks.

Overheating

We have discussed above the establishment of thermodynamic equilibrium. With normal potatoes in stacks of normal height and under normal ambient conditions this is established with no very great temperature rise. On the other hand, we have seen that, with considerably sprouted potatoes, the rise necessary to activate sufficient convection is quite considerable. As the temperature of the potatoes rises in the move towards the establishment of equilibrium it would be possible in such potatoes to reach a situation in which the extra heat generated as a result of a further rise in temperature, with increased respiration, was greater than could be removed by the increase in convection brought about by that rise in temperature. The production of heat would thus outstrip its removal and overheating would occur. This is imminent (Burton *et al.*, 1955) when

$$\left(\frac{T_p - T_a + \mathrm{d}T}{T_p - T_a}\right)^{1.55} = \frac{Q + \mathrm{d}Q}{Q} \qquad (14.2)$$

where T_p, T_a and Q are as defined for Equation 14.1 and:

$\mathrm{d}T$ = a small rise in temperatures, say 0.5–1°C

$\mathrm{d}Q$ = increase in production of metabolic heat (kJ t^{-1} h^{-1}) caused by an increase of temperature $\mathrm{d}T$.

It is possible to derive the conditions under which overheating could occur from Equation 14.1 (Table 14.7).

Table 14.7 Conditions under which uncontrollable overheating could occur in unventilated stacks of potatoes (approximate: based on Burton *et al.*, 1955)

	Mean potato temperature (°C) at which uncontrollable overheating could start		*Mean ambient temperature (°C) at which foregoing dangerous potato temperature could be reached*	
	Height of storage (m)			
	1.85	*3.7*	*1.85*	*3.7*
Characteristics of potatoes				
(a) Freshly harvested				
Immature, fairly clean	27	22	13	6
Immature, much earth	23	18	8	0
Mature, fairly clean	30	27	19	13
Mature, much earth	27	23	14	8
(b) Stored, unsprouted				
Fairly clean	36	30	24	19
Much earth	30	27	20	14
(c) Stored, considerably sprouted				
Fairly clean	27	22	13	6
Much earth	23	18	8	0

14.2.3 Composition of the atmosphere

O_2 and CO_2

We have seen above that air may be moving into a potato stack by convection at a rate of about 10 $m^3\ t^{-1}\ h^{-1}$. CO_2 is being produced, and O_2 absorbed, at a rate of only 2–4 $l\ t^{-1}\ h^{-1}$. We would thus expect depletion of O_2 and accumulation of CO_2 to be negligible, differing by about 0.04% from ambient. The same would be essentially true even with the much lower convection rates suggested by Hylmö *et al.* (1975a,b). Mann (unpublished work at the Ditton Laboratory) inserted copper sampling tubes into heaps during loading and could detect only slight differences ($< 0.1\%$) from ambient air during subsequent storage. Crook and Watson (1950) found that prior to the earth cover being placed on a field clamp, the composition of its atmosphere did not differ noticeably from that of the surrounding air.

Other metabolic volatiles

That potato tubers produce some volatile substances, other than CO_2, is shown by the fact that they have a detectable odour, which Buttery and Ling (1973) attributed to methoxy-3-isopropylpyrazine. The production by the potato of volatile substances which will inhibit sprout growth was mentioned in Section 14.1.5, although there is no evidence that the sprout-suppressing volatiles and the odorous volatiles are identical. The concentration of such substances in the atmosphere of potato stacks has not been measured, but there is evidence that it reaches levels active in suppressing sprout growth. For example, it is a matter of common observation, first recorded by Butler (1919), that ventilation may increase the amount of sprout growth. In unventilated stacks, when ambient temperatures, and hence peripheral stack temperatures, are favourable to growth in Spring, it is noticeable that growth is much more vigorous round the periphery of the stack than elsewhere. Temperatures round the periphery, although favourable, are not so favourable to growth as the higher temperatures in the centre (Section 14.1.5) and it seems reasonable to attribute the difference to the sweeping away of suppressant volatiles by the rapid peripheral air movement. Similarly enhanced sprout growth may be observed over open ducts beneath a potato stack, again, of course, only if ambient temperatures are high enough for growth to occur readily. Conversely, sprout growth in the centre of an unventilated stack may often be slight despite favourable temperatures (see e.g. Wilson *et al.*, 1962).

14.2.4 Evaporative capacity of convection

Upper limit to the evaporative capacity

The air which enters a potato stack by convection is usually not fully saturated with water, and thus has a capacity to take up more, and a consequent wvpd in response to which evaporation occurs from the potatoes. As the air moves through the stack it is warmed and can thus take up still more water (Table 14.2), but when the warm air rises out of the stack it is cooled by admixture with incoming air, and by the potatoes in equilibrium with this mixture. Ultimately the air is cooled to ambient temperature and cannot then hold any more water vapour than the water-holding capacity of air at ambient temperature. If in fact it has taken up more than this in the warm state then the excess will condense out. In a stack with no such covering as a layer of straw, this condensation will occur on the top potatoes. If there is a sufficiently thick straw covering, the top potatoes will be warmer and the air will have moved above them before being cooled down to dewpoint, and condensation will occur in the straw. In either case, the air which leaves the stack, plus its covering, cannot on average be more than saturated at store air temperature. The upper limit to the evaporative capacity of convection is thus provided by the quantity of water which the limited volume of convective air can take up. Any excess above this which condensed on the top potatoes or the straw would humidify subsequent convective air entering through the top of the stack and reduce subsequent evaporative loss. The long-term average would be as above. If we take the normal rate of convection to be about 10 m^3 t^{-1} h^{-1}, then for every mbar wvpd in the ambient air the upper limit of evaporative loss would be about 7.8 g t^{-1} h^{-1} (Table 14.2) – about 0.13% $week^{-1}$ $mbar^{-1}$.

Evaporation from an unventilated bulk of potatoes

We can compare the above estimate of the maximum possible loss from unventilated potatoes with some results of Wager *et al.* (1952) relating to a small bin (9.5 t) of Majestic potatoes. These were weighed in and out at the beginning and end of storage. The amount of earth was excessive (about 18%). From the average temperature, 6.6°C, 3.8°C above average ambient, and the measured resistance to air flow, we can calculate an average rate of convection of about 7.7 m^3 t^{-1} h^{-1}, less than would be expected with cleaner potatoes. The average temperature of the store air was 2.8°C and its average wvpd 1.5 mbar. It thus held 4.66 g of water m^{-3} and had a saturated capacity of 5.84 g m^{-3} (Table 14.2). Hence it was capable of removing 9.1 g of water t^{-1} h^{-1} – about 0.15% $week^{-1}$. The potatoes actually lost 3.8% by respiration and evaporation over a period of 23 weeks. Of this, about 0.4–0.8% might have been due to respiration (Section 14.1.6), leaving an evaporative loss of 3–3.4%, an average of 0.13–0.15% $week^{-1}$, approximating to the suggested possible maximum. We can thus probably regard the actual evaporation from unventilated potatoes as being equal to the capacity of the convective air on entry i.e. for fairly clean potatoes, about 0.13% per week per mbar ambient wvpd.

14.3 Modification of the environment and response of stored potatoes, according to requirements

14.3.1 Requirements

This chapter is not concerned with the requirements of potatoes for different uses, but a brief indication of the limits of these requirements will be given here, to provide a context for the discussion of suitable modifications of storage conditions.

Wound healing

For all uses, wound healing is essential immediately after harvest, or, where relevant, after grading. This involves a period of about 2 weeks at 10 °C or above (Section 14.1.2). It is occasionally possible to take a calculated risk in not allowing wound healing, where it is known that storage is to be so short that wound pathogen development is unlikely to be serious – in buffer stocks for canning for example. It must be realized, however, that there is some risk.

Seed

Points to be considered are: method of planting and the importance placed upon being able to plant rapidly when conditions are right; size of handling and storage units to suit the method of planting and other requirements; importance placed upon avoiding sprout damage – rapid mechanical planting could damage sprouts which had grown and perhaps delay emergence, but slower planting could lead to later planting and even longer delay; length of sprout which can be tolerated in the light of the foregoing; importance or otherwise of apical dominance. We assume that exposure to light is not deleterious, nor advantageous except for its effect in giving stronger sprouts; that sweetening is not a disadvantage; and that it is desirable for the dormant period to be over – even if the sprouts are not growing they should be on the verge of so doing. Uniformity of behaviour is important, which means that storage conditions, in particular temperature, should be uniform.

In accordance with the above it could be suggested that the minimum storage temperature should be well above any danger of low temperature injury (Section 14.1.5), say 2 °C; that the long-term maximum should not be such that undesirable growth occurred in the variety stored i.e. usually 5 °C or below; and that the temperature range throughout any batch of seed should not exceed 2 °C at the most. There could be a short period at, say 10 °C to encourage growth if and when required.

The main requirement for seed storage therefore is close temperature control within the range 2–5 °C after wound healing.

Ware for domestic and catering consumption

Potatoes for consumption must be protected from light because of the undesirable changes induced by exposure to it (Section 14.1.6). Wilting and sprout growth should be avoided as far as economically possible. Apart from these, storage impinges on domestic and catering requirements by affecting the degree of internal bruising during grading. If it is feasible, susceptible stocks should be allowed to rise to 15 °C or so before grading. Sugar content, affected by storage temperature (Section 14.1.6), is not critical provided that potatoes which have sweetened at a low storage temperature are allowed to de-sweeten, at 10 °C or above, for about 2 weeks during distribution and in the home. Senescent sweetening is not normally a serious problem, although it can become objectionable in very late stored potatoes. Vitamin C content should be of importance to the consumer, but we cannot yet specify optimum conditions for its retention.

From the above we could conclude that a temperature of 5 °C or so, after wound healing, could be optimal. This would avoid serious sprout growth (Section 14.1.5) and senescent sweetening (Section 14.1.6). Low temperature sweetening could be acceptably reversed during distribution. The temperature range would need to be kept within narrow limits if we depended upon temperature to avoid sprout growth – say 4–6 °C. An alternative method would be to hold the temperature roughly in the range 5–10 °C and use chemical sprout suppressants for late storage – it is of course a waste of money to use these on potatoes which are only to be stored for a period during which sprouting could be prevented by a simple storage regime such as night ventilation. Whatever method of storage is adopted the avoidance of wilting is important – water loss should be kept, if possible, below 5%.

Ware for processing

'Processing' usually means either 'crisp' or 'chip' manufacture or dehydration in the form of flakes or granules. The storage requirements for all these are compatible and they can be considered together. Avoidance of internal bruising is important but is not really a storage problem except for raising the temperature, if necessary, before grading. The main requirement of the manufacturers is a very low and uniform content of reducing sugars – preferably about 0.1% of the fresh weight. For this, in the early part of the storage season, the higher the temperature the better, up to about 20 °C (Section 14.1.6). Storage of potatoes in any quantity at a temperature of 20 °C is hazardous, however, and a temperature of about 10 °C is the best commercial compromise. Because of the requirement for uniformity a range of ± 1 °C is desirable. A manufacturer usually requires supplies which last until freshly harvested potatoes are available at an economic price. This entails late storage, and late storage at 10 °C results in a degree of senescent sweetening which is unacceptable to the processor. It would be avoided by storage at 5 °C, but this gives low temperature sweetening which, although reversible, is incompletely so, and not uniform (Section 14.1.6).

A compromise solution is to store the potatoes required for late use at about 7–8° C after wound healing. Senescent sweetening is sufficiently delayed but there is some low temperature sweetening (Fig. 14.2). This is reversed by brief storage, batch by batch, at a high temperature, up to 20°C, and the residual rather higher sugar and lack of uniformity accepted as the lesser of two evils. Storage at the temperatures mentioned would result in vigorous sprout growth, and chemical suppression is essential except for the stocks stored only briefly. Wilted potatoes present processing difficulties, and water loss should, if possible, be kept below 5%.

Canning potatoes

Canned potatoes are typically small immature tubers. These normally have a high content of sugar and the sweetening which results from storage at about 4–5°C, after wound healing, is apparently acceptable. This avoids sprout growth. Immature potatoes are very susceptible to water loss (Section 14.1.3) and have a high initial rate of respiration (Section 14.1.4) and consequent heat production. Avoidance of drying conditions in store is thus essential, and storage in bulk to depths greater than about 2 m is inadvisable.

Industrial potatoes

The main industrial use of potatoes is for starch manufacture. Sweetening represents a loss of starch – the extra sugar accumulation at a temperature of 4°C, above that at 10°C, illustrated in Fig. 14.2, represents a loss of about 1.3% by weight of starch, probably about 8–9% of that originally present. Sweetening must thus be avoided. Unventilated storage in bulk, although with ducts under the stack for the introduction of a sprout inhibitor if desired, at a temperature within the approximate range 8–12°C, is probably the best method. Sprouting represents a loss of starch, but 1% by weight of sprouts only represents a loss of about 0.1% of starch, which together with the extra respiration accompanying sprouting may be about 1% of the original content. The cost of sprout suppression must thus be set against the realizable advantage. Wilting is not important.

14.3.2 Environment: temperature

General

It will have been seen from the above that fairly close control of temperature is desirable for the efficient storage of seed, ware for processing, and canning potatoes. The financial advantages to be derived from this control must be balanced against the cost of obtaining it. In the case of ware for domestic or catering consumption, and industrial potatoes, close control of temperature is not necessary provided the period of storage, or the other measures taken, are such that sprouting is not excessive and wilting, in potatoes for consumption, preferably does not exceed about 5%. Bulk storage with convective circulation, supplemented, if the storage is to a depth of over 2.5 m, by the facility to ventilate at about 70 $m^3 t^{-1} h^{-1}$ through ducts under the stack when the temperature of the outside air is suitable, is usually adequate for ordinary ware storage. Sprout suppression would be necessary if storage were prolonged. The extra expense of any more sophisticated method of storage would need to be justified by an increased monetary return from such storage.

Subdivision of the mass

A small unit, such as a 0.5–1 t crate of potatoes, behaves as a miniature unventilated bulk. The dimensions are such that the maximum temperature, controlled by convection (Section 14.2.2), approximates to 1 °C above ambient, and the minimum temperature to ambient. We thus have fairly uniform temperature conditions in each crate. It remains to ensure that the crates do not differ from each other, which means that the air throughout the store must be at about the same temperature. This is done by circulating the store air. The circulating air, moving over the potatoes is warmed by the metabolic heat – 30 m^3 passing over a ton, producing 35 kJ net after evaporative cooling (Section 14.2.2), would be raised nearly 1 °C in temperature, which means that circulation at 30 $m^3 t^{-1} h^{-1}$ would result in a 1 °C gradient in the store air. Allowing also for the gradient in each crate, circulation at this rate is sufficient to hold the potatoes within a range of about ± 1 °C of mean store temperature. It remains to control the latter at the required temperature – ideally about 3–4 °C for seed; about 10 °C for short-term storage for processing; about 7 °C for long-term storage for processing, and about 4–5 °C for canning.

Potatoes for such uses can best justify the expense of crate storage, for which the crates represent a sizeable capital investment. In the main potato growing areas of England it is possible, on average, to obtain sufficient control for every purpose, until about March, by an adequately insulated store and judicious use of ventilation with outside air during the night, or whenever the temperature is suitable, using the same fans as for recirculation, i.e. with a capacity of 30 $m^3 t^{-1} h^{-1}$. This suffices for seed storage. For storage during the succeeding months for other purposes, in some cases until July or August, refrigeration is necessary. Ideally, refrigeration would also be desirable for seed storage in years when the winter was very warm, but the economic justification for it would have to be based on the frequency of such years and the monetary loss from not having such a facility.

Ventilation of the stack

Instead of subdividing the crop into crates and circulating the store air to reduce temperature scatter, the same result can be achieved by storing the potatoes in bulk and continuously ventilating them by means of ducts in or on the floor. Continuous ventilation at 30 $m^3\ t^{-1}\ h^{-1}$ would give a difference of about 1 °C between the minimum and maximum temperature in the stack, assuming a net heat production, after evaporative cooling, of 35 kJ $t^{-1}\ h^{-1}$. Ventilation ensures a reasonably uniform temperature but does not control the level of that temperature, and this must be done by controlling the temperature of the ventilating air by recirculation, the air intake being above the stack, coupled with admixture with outside air of suitable temperature, or with refrigeration.

Two rates of ventilation have been mentioned above. 30 $m^3\ t^{-1}\ h^{-1}$ was suggested for air circulation, and also for ventilation with suitable outside air, in cases such as crate stores and bulk stores with recirculation where air movement is virtually continuous and there is no build-up of temperature among the potatoes. A rate of 70 $m^3\ t^{-1}\ h^{-1}$ was mentioned in connection with the intermittent ventilation of stacks through which air was not continuously circulated. These stacks of potatoes, for domestic or catering use, may not usually justify expensive temperature control, as by refrigeration. Warm spells of weather can lead to undesirably high temperatures – as we have seen (Section 14.2.2), an ambient average of 10 °C would result in an average temperature of 13–14 °C, and a maximum of 15–16 °C, in a 3 m high stack. It is therefore necessary to take full advantage of cooler spells of weather to cool the stack by ventilation.

For this purpose the faster the rate of ventilation the better, up to the point where frictional heating of the ventilating air, and also increased power requirements, render further increase in rate either undesirable or uneconomic. The fact that cooling is proportional only to about the cube root of the rate of ventilation (Section 14.1.1) is relevant here. Recommended rates vary from 70 (Mann, 1963) to 160 $m^3\ t^{-1}\ h^{-1}$ (Ophuis, 1957). These higher rates of ventilation would also be desirable for crate stores and bulk stores with recirculation if it were attempted to store potatoes late without provision for refrigeration. By ventilating whenever the air was cooler than the potatoes, but above freezing point, and using 160 $m^3\ t^{-1}\ h^{-1}$, Ophuis calculated that it was possible on average in the Netherlands, to hold potatoes at mean temperatures of about 10°C in September, 7°C in October, 4°C from November to April, 8°C in May, 11°C in June and 13°C in July and August. In S.E. England, Burton and Mann (1954) calculated that potatoes could on the average be held, by a similar rate of ventilation, at 5–6°C until the end of April and perhaps at 4°C until the end of March. In Edinburgh in the 1968/9 season, Nash and Lennard (1973) using a rate of about 100 $m^3\ t^{-1}\ h^{-1}$, held the maximum temperature of 30–40 t lots within a range of about 4–6°C until the end of April. Thereafter ventilation when the air was cool became increasingly ineffective, and towards the end of May the maximum temperature in the stacks had risen to about 13 °C.

The cooling achieved by intermittent ventilation has been considered theoretically by Ophuis (1957), Burton (1966) and, in more detail, with experimental confirmation, by Hylmö *et al.* (1975b).

Refrigeration

The above sections have shown that for most purposes the temperature of stored potatoes can be controlled acceptably until March or April by using intermittent ventilation with cool air, coupled, where uniformity of temperature is important, with recirculation of the store air. It has also been shown that this method cannot be relied upon later in the storage season, and the temperature can then only be controlled by refrigeration. This can be justified for uses for which temperature control is essential, and it is usually applied by passing the circulating air over cooling surfaces, such that the inlet temperature of the air is about 0.5 °C below the desired mean potato temperature. Such a method of temperature control is efficient. Burton and Mann (1954), for example, stored 80 t of potatoes at a mean temperature of 5.2 °C with a difference, between the minimum and maximum temperatures in the mass, of about 0.8 °C.

There is one essential proviso, however, relating to the size of cooling surfaces. Ventilating air removes water from the potatoes at a rate proportional to the wvpd of the air (Section 14.1.3). In the case of refrigeration the air passing over the cooling surface deposits moisture or ice on it, if the surface temperature is below the dew-point of the air, until the moisture content of the air falls to that equivalent to saturation at the cooler temperature. The lower this is the less the moisture retained in the air leaving the coolers, consequently the higher its wvpd, and the greater the evaporation from the potatoes; the resultant increased moisture content of the air leaving the potatoes being again reduced on its next passage over the cooling surface. Efficient transfer of heat, and hence temperature control, can be achieved either by a large cooling surface at a temperature not much below the desired air temperature, or by a smaller surface operating at a lower temperature. The latter, however, could cause serious wilting of the potatoes. Burton and Mann (1954) obtained satisfactory results using finned pipe coolers capable of removing approximately 26 kJ h^{-1} (≡ 7.2 W), per °C temperature differential, for every ton of potatoes they served. The whole of the normal heat production could thus be removed by a 2 °C differential, even disregarding evaporative cooling, the wvpd of the air entering the potatoes being therefore only about 1 mb in the absence of heat leakage into the store – the latter would necessitate an increased differential.

14.3.3 Environment: evaporative capacity

Ventilation

Convection at a rate of 10 $m^3\ t^{-1}\ h^{-1}$ is equivalent to the lowest rate of ventilation given in Table 14.3, at which humidification of the air is such that evaporation from a tuber is only about 60% of the potential evaporation into air of any given inlet wvpd. Evaporation from typical mature stored tubers, capable of losing 0.17% $week^{-1}\ mbar^{-1}$ would thus be restricted to about 0.1% $week^{-1}\ mbar^{-1}$ were it not that, as we have seen (Section 14.2.4), the convective air is warmed as it passes through the stack, its evaporative capacity being thus enhanced.

Replacement of convective air movement by continuous positive ventilation at a higher rate, through ducts under the potatoes, affects both the temperature and the humidification of the air. The humidification is less because of the greater volume of air, and the consequent restriction on evaporative loss is progressively removed, as illustrated in Table 14.3, as the rate of ventilation is increased. Metabolic warming of the air is decreased, again because of its greater volume. There is also cooling of the air because of evaporation from the potatoes. In the case of convection, metabolic warming exceeds evaporative cooling and there is net warming, but in the case of positive ventilation it would be theoretically possible for cooling to exceed warming. This has in fact been observed by Hylmö *et al.* 1975a). With a heat production of 50 kJ $t^{-1}\ h^{-1}$ and a potential for water loss of 10 g $t^{-1}\ h^{-1}\ mbar^{-1}$ it can be calculated that cooling and warming are in balance with an average wvpd in the stack ranging from about 2.1 mbar (ventilation rate 200 $m^3\ t^{-1}\ h^{-1}$) to 3.5 mbar (ventilation rate 10 $m^3\ t^{-1}\ h^{-1}$). With increase of wvpd above the balance point, cooling progressively exceeds warming; with decrease below the balance point, warming progressively exceeds cooling. The resultant temperature changes affect the operative wvpd in the stack, cooling leading to a decrease, over and above that due to humidification; and warming leading to an increase which can more than offset humidification. Calculated values for the average wvpd which is operative in potato stacks ventilated at different rates with air of stated wvpd are given in Table 14.8. Rather different values would be obtained with potatoes differing in heat production and water loss, and with different storage temperatures – this last because the increase in wvpd resulting from a 1 °C rise in temperature rises progressively from 0.45 mbar at 0–1 °C to 1.39 mbar at 19–20 °C (Table 14.2).

The main point of interest which emerges from Table 14.8 is that with humid air, the faster the rate of continuous ventilation the less the wvpd operative in the stack, and the lower would be the evaporation. This is to be expected, in that, with humid inlet air, the main factor influencing wvpd is temperature rise, and this is less the higher the rate of ventilation.

Table 14.8 Calculated average wvpd operative in a continuously ventilated potato stack in a state of balance with air of constant temperature and humidity*

	Rate of ventilation ($m^3\ t^{-1}\ h^{-1}$)						
	10	*20*	*40*	*60*	*80*	*100*	*200*
Inlet wvpd of ventilating air (mbar)	*Average wvpd operative in stack (mbar)*						
0	1.1	0.6	0.4	0.3	0.2	0.2	0.1
0.5	1.5	1.0	0.8	0.7	0.6	0.6	0.5
1.0	1.8	1.4	1.2	1.2	1.1	1.1	1.0
2.0	2.5	2.2	2.0	2.0	2.0	2.0	2.0
3.0	3.2	2.9	2.9	2.9	2.9	2.9	2.9
4.0	3.8	3.8	3.8	3.8	3.8	3.8	3.9

*The following assumptions are made: heat output, 50 kJ $t^{-1}\ h^{-1}$; potential water loss, 10 g $t^{-1}\ h^{-1}\ mbar^{-1}$ ($\equiv$ 0.17% $week^{-1}\ mbar^{-1}$); inlet temperature of ventilating air, 10°C; relationship between rate of ventilation and water loss as in Table 14.3.

With drier inlet air, on the other hand, the rate of ventilation, at least within the range of normal commercial rates, has little or no effect upon water loss. The foregoing applies only to continuous ventilation when a state of heat balance has been reached and the temperature gradient is solely due to the interaction between heat production, water loss and the rate of air flow. The evaporative capacity of ventilating air during out-of-balance periods, as for example when potatoes are being cooled by ventilation, is altogether different and is discussed below. It will have been noticed that the conclusions above, on the effect of the rate of ventilation upon evaporative capacity and water loss, appear at first sight to contradict the previous conclusions reached when discussing convection. This is only because, as mentioned in Section 14.2.4 condensation at the cool top of a convectively cooled stack humidifies the ingoing air. Replacement of convection by positive ventilation from below at a similar rate removes this source of humidification and can, as shown in the first column in Table 14.8, appreciably increase water loss if the ambient air is humid.

If ventilation is used for cooling a stack of potatoes it is frequently in the form of intermittent ventilation when suitable cool air is available. The effect of this upon water loss can be illustrated by an example based upon temperature measurements on a 4000 t stack, by Hylmö *et al.* (1975b). The stack, 4.5 m high, was continuously ventilated with humid air to maintain uniformity of temperature and minimal evaporation. The fans were then switched off for 3 days, thus allowing the stack to rise to a temperature ranging from 7.8–8.8 °C. It was then cooled by ventilation at 30 $m^3\ t^{-1}\ h^{-1}$ with air averaging 6.1 °C and 97% r.h. From the humidity

Table 14.9 Calculated water loss, and equivalent average operative wvpd, in a potato stack during cooling by ventilation (temperatures during cooling from Hylmö *et al.*, 1975)

Time after start of ventilation (h)	*Air entering stack*		*Air leaving stack*		*Water loss* $(g\ t^{-1}\ h^{-1})$	*Operative wvpd equivalent to water loss*† *mbar*
	Temperature °C	*Water content* $g\ 30\ m^{-3}$	*Temperature °C*	*Water content* $g\ 30\ m^{-3}$		
0–1	6.1	210	8.8	260	50	5
10–11	6.1	210	8.2	250	40	4
20–21	6.3	215	7.8	240	25	2.5
50–51	6.1	210	6.6	225	15	1.5

*See Table 14.2. r.h. at entry 97%; on leaving 99.5%.
†Assuming a water loss of 10 $g\ t^{-1}\ h^{-1}\ mbar^{-1}$.

at entry and the figures in Table 14.3, it can be estimated that the air would probably leave the stack at about 99.5% r. h. The calculated water loss during the period of cooling is given in Table 14.9, from which it will be seen that the rate of water loss at the beginning of the cooling period would have been about 6 times as great as given in Table 14.8 for continuous ventilation at the same rate with air of similar humidity, and that the weighted mean rate of loss over a 50 h cooling period would have been over 3 times as great – 27 g t^{-1} h^{-1}, which, on the assumption of a water loss of 10 g t^{-1} h^{-1} $mbar^{-1}$ would be equivalent to a mean operative wvpd in the stack of 2.7 mbar. Losses from stacks differing more from the temperature of the ventilating air would be even greater during cooling by ventilation.

Recirculation

Recirculation is the term applied to continuous ventilation with store air, in the case of bulk storage being introduced through ducts under the potatoes and the air intake usually in the roof space above the potatoes. If the store is sealed, depending for temperature control upon heat leakage through the structure, the air is rapidly humidified by its passage through the potatoes – the volume of air in a filled bulk store could be some 2 m^3 t^{-1}, capable of holding at a temperature of, say 7 °C, only about 15 g of water vapour per ton stored (Table 14.2). There will be some loss of this water by condensation on cold parts of the structure and by air leakage from the store; there may also be dilution when it is necessary to introduce outside air, for cooling purposes, but nevertheless the store air can often be maintained with a very low wvpd, of the order of 0.5 mbar. In such a case, recirculation is equivalent to continuous ventilation with air with an inlet wvpd of 0.5 mbar. At a rate of 30 m^3 t^{-1} h^{-1} this would give an operative wvpd in the stack of about 0.9 mbar (Table 14.8) and an evaporative loss of perhaps 0.15% $week^{-1}$.

Refrigeration

The effect upon humidity of the temperature differential between coolers and air has already been discussed (Section 14.3.2). An efficiently designed refrigeration plant could ensure that the ventilating air entering the potatoes had a wvpd not exceeding about 1 mbar, the wvpd in the stack, again with ventilation at a rate of 30 m^3 t^{-1} h^{-1}, being about 1.3 mbar (Table 14.8) and evaporative loss perhaps rather more than 0.2% $week^{-1}$. If refrigeration is employed, it is often intermittent and coupled with continuous recirculation, the overall average loss being then somewhat less. It would be pointless to have the expense of refrigeration without virtually continuous ventilation, as temperature gradients in the stack would override any benefit of a low inlet wvpd (Table 14.9).

Humidification

We have seen that ventilation takes two forms: continuous, which may be in the form of recirculation, to maintain a uniform temperature; and intermittent, often with outside air, to cool a stack which, between periods of ventilation, is unventilated except by convection and thus develops a temperature gradient (Section 14.2.2). In the case of continuous ventilation the wvpd which is operative in the stack is governed by the inlet humidity and the rate of ventilation, particularly the former (Table 14.8). In the case of intermittent ventilation, the most important factor influencing the operative wvpd is the temperature gradient which has developed in the stack (Table 14.9). As in the case of refrigeration, humidification of the inlet air using, for example, a water spray in the air stream, can only really be effective in the absence of a large temperature gradient – that is with continuous ventilation. With this, humidification, as used by Hylmö in Sweden, will give an entry wvpd of about 0.3 mbar and an operative wvpd, in a stack ventilated at 30 m^3 t^{-1} h^{-1}, of perhaps 0.7 mbar. Evaporative losses can thus be kept down to about 0.12% week $^{-1}$.

14.3.4 Response: sweetening

Low temperature sweetening

Low temperature sweetening is mainly a disadvantage to the processor – for other users, subsequent de-sweetening, even though incomplete, is adequate. Any practical means of avoiding sweetening must thus satisfy the requirements of the processor, and give results better than those obtained by de-sweetening sweetened potatoes. As yet this is not possible, but some relevant work is cited here.

The effect of low O_2 concentrations has been studied. Nelson and Auchincloss (1933) found the sucrose synthesizing system to be inoperative in the absence of O_2; and anaerobic conditions, obtained either by storage in N_2 (Samotus and Schwimmer, 1963; Harkett, 1971) or under water (Samotus, 1971), reduce or prevent sugar accumulation at low temperatures. 1% O_2 is almost as effective as completely anaerobic conditions (Harkett, 1971) but higher concentrations lead to more sweetening and show little practical promise (Workman and Twomey, 1970; Harkett, 1971).

Barker (1936b) and Denny and Thornton (1941, 1942) found 5–20% CO_2 decreased the rate of low temperature accumulation of sugar, although the amount finally accumulated after some months could be the same or much more than in air. The effect of CO_2 appears to exhibit complex time/temperature/concentration relationships would militate against commercial application until they were fully understood, but, apart from this, concentrations of CO_2 above 10% can lead to tissue damage and 5% can lead to increased sucrose accumulation despite a lower accumulation of reducing sugars. All in all, the use of high concentrations of CO_2 seems to offer little prospect of practical use.

Barker (1939) found that gradual stepwise reduction of temperature from 10 to 3 or 1°C, over a period of 10–21 days, led to a lower rate of sweetening at the low temperature but the final degree of sugar accumulation after prolonged storage was unaffected. Coversely, prior storage at 20°C resulted in more rapid sweetening. Although the rate of sweetening was reduced by gradual lowering of temperature, the reduction was of no practical importance – from the processors' point of view it might merely mean that the potatoes were unusable after 2 weeks, instead of after 1 week, at a low temperature, and this provides no solution for long-term storage.

Burton (1969) limited both sweetening and sprouting to low levels by diurnal fluctuation of storage temperature between 2 and 10°C. If practicable this would be an adequate solution to the problem but there are objections on the grounds of the cost of the refrigeration necessary and more particularly, because of water loss during cooling from 10 to 2°C (see Section 14.3.3).

It is possible to breed potato varieties which do not sweeten undesirably during storage at 4°C (Lauer and Shaw, 1970) but this property has not yet been introduced into a successful commercial variety. It may however be a hopeful approach to the problem of low-temperature sweetening.

Senescent sweetening

The practical means of avoiding serious senescent sweetening have already been mentioned above (Section 14.3.1 'Ware for processing'). As indicated, it does not provide a totally satisfactory answer to the processors' requirements. In this property, too, varieties differ, which gives the possibility of avoiding the problem by the choice of a suitable variety, provided that its other properties are acceptable. Burton (1965b) illustrated very marked differences between the sweetening of seven British commercial varieties. Three of these would have been too sweet for any use after 5–6 months at 10°C. The sugar content of one of them, King Edward, would have been marginally acceptable even for crisp manufacture after 7 months. It is not a good crisping variety on other grounds, however.

14.3.5 Response: evaporation

Four characteristics of the tuber influence evaporation: its maturity, possibly the thickness of its periderm, the extent to which handling wounds have healed, and sprout growth. All these characteristics have been discussed in some detail in Section 14.1.3 and it is unnecessary to add more here, except to point out that the suggestion that periderm thickness plays a part, although reasonable, is based on very little evidence.

14.3.6 Response: sprouting

Section 14.1.5 described the conditions which were favourable or otherwise to the break of dormancy and sprouting. It also mentioned varietal differences and methods of stimulating or preventing growth. This section is concerned with the practical means of altering the natural response of the tuber to its environment. It will not be concerned with altering the environment. This, in the form of refrigeration at 2°C, can completely prevent sprout growth and, as mentioned above, can give adequate control for commercial purposes at 4°C. Refrigeration has been adequately discussed, however, in the foregoing sections. In addition there are uses for which it is not suitable and we must then store the tubers at temperatures at which they could sprout readily. Apart from sprout suppression there are commercial situations where stimulation of sprout growth is desirable. This can be the case where two potato crops can be grown in the season and it is desired to plant seed tubers when they would still be dormant, immediately after harvest. There is a similar requirement where seed tubers, harvested in one locality, are exported to another locality for immediate planting. There can also be a need for stimulation of growth in samples of seed, in connection with schemes for the certification of the crop, and this, so far, has been the major application.

Chemical stimulation

Substances which will stimulate sprout growth were mentioned briefly in Section 14.1.5. Three have given promise of commercial application – carbon disulphide, ethylene chlorhydrin and gibberellic acid. Of these ethylene chlorhydrin is most used, mixed with 1–2 dichloroethanol and carbon tetrachloride (7:3:1, respectively, by volume), the mixture, introduced by Denny (1945), being known as 'rindite'. The method of application is typically as described by Keller and Berces (1966) for tubers from 4 weeks after haulm killing onwards. A container, which is capable of being tightly closed, is filled with potatoes surrounding a tube, made of wire netting, running from bottom to top, which contains a coil or strips of filter paper. The rindite is poured on the filter paper, to the amount of 10 ml per 10 l of empty container volume, and the container closed and left at 24°C for 48 h. The tubers are then trayed up and kept under diffuse lighting at 24°C to sprout.

Alternatives to rindite have not so far proved as effective, but an alternative is desirable because of the toxicity of rindite vapour.

Chemical suppression

Of the substances mentioned in Section 14.1.5 as suppressing sprout growth, a number have been tried on the commercial scale and five have been used with success commercially – 3-5-5-trimethylhexan-1-ol, isopropyl-*n*-phenylcarbamate, isopropyl-*n*-chlorophenylcarbamate, tetrachlornitrobenzene and maleic hydrazide. Sprout suppressants leave residues in the tubers, the level of which may be subject to regulation, and this, and the acceptability of the particular substance to the responsible authority must be borne in mind.

Maleic hydrazide is not a storage treatment, being applied as a foliar spray in the form of an aqueous solution of the diethanolamine- or sodium salt, at a rate of 2.5 kg of the chemical ha^{-1}, 3–5 weeks before the death or destruction of the foliage. As a sprout suppressant it is very effective provided application can be made at the right time, but there is the possibility of reduction in yield and of misshapen tubers, particularly with early application. Applied too late it is ineffective, and successful use of maleic hydrazide requires long periods of predictable weather. It must be remembered, of course, that seed tubers cannot be taken from a maleic hydrazide-treated crop.

Tetrachlornitrobenzene ('technazene') is applied as a dust, containing about 5% of the active ingredient in an inert filler, at a rate of about 100 g of the chemical t^{-1}, as the potatoes are loaded into store. It gives reasonable control of sprout growth, although not for prolonged storage. According to Brown and Reavill (1954), free aeration for several weeks, as occurs if seed tubers are trayed up, removes all trace of tetrachlornitrobenzene from the tubers, and subsequent growth is normal. Seed tubers may thus be taken from a treated store, provided they are graded out several weeks before they are required and trayed up or otherwise exposed to free air circulation. Tetrachlornitrobenzene is the only sprout suppressant, applied post-harvest, which does not lead to enhanced rotting by wound pathogens if applied before wounds have healed.

Iso-propyl-*n*-phenylcarbamate ('propham') and iso-propyl-*n*-chlorophenylcarbamate ('chlorpropham') are normally applied together as a mixture at a rate of about 10 g of the chemicals ton^{-1}. It is possible to apply them as a dust (2% active ingredient in an inert filler) but because of adverse effects upon wound healing (see e.g. Ives, 1955) this must be distributed by blowing it through the main ventilating system after wound-healing is complete. Such a method does not give adequate distribution in box stores. In these the chemical can be applied at the time of harvest in the form of 'delayed release granules' from which an inhibitory concentration does not build up around the tubers until after the wounds have healed. If this method is adopted it is of course essential that the tubers are stored for the first 2–3 weeks under conditions favourable for wound healing, otherwise a damaging concentration can build up before healing is complete. An alternative method of distribution in box stores is to vaporize the chemicals at about 200°C and introduce the vapour into the recirculating system. The store is then closed for 24–48 h with the circulating fans running continuously, after which there is a fairly uniformly distributed deposit on the tubers. In bulk stores a similar method may be used, or application may be in the form of an aerosol of a methanol solution of the chemicals introduced into the main ventilating system with the fan running. The aerosol is deposited on the tubers and the methanol evaporates leaving a deposit of the sprout suppressant. A dose rate of 10 g t^{-1} is inadequate for prolonged storage, but a higher rate could give residue problems, and it is usual to make three applications during a long period of storage. Application is nearly always done by a contractor.

3-5-5-trimethylhexan-1-ol (nonanol) is a liquid and is introduced as a vapour, at a concentration of 0.1 mg l^{-1} of air, into an airstream blown through the main ventilating system of bulk stores at a rate of 10 m^3 t^{-1} h^{-1}. In box stores the vapour is introduced into the recirculating system. The simplest method of introducing the vapour into the air is to blow the latter over towelling dipping into a container of the liquid, the area of towelling being adjusted to give roughly the correct concentration. This is determined with sufficient accuracy by measuring the quantity of nonanol required to replenish the container, the rate of ventilating being known. Applicators incorporating a drip feed on to a hot plate have been used, but under commercial conditions have suffered from considerable variation in rate of feed. Nonanol, used as above, gives complete sprout suppression, but it is not persistent and growth can occur within 2–3 weeks of stopping the ventilation. The consumption of the chemical can be halved by alternating 2-week periods of nonanol application with 2-week periods of convection or ordinary ventilation, consumption over a 2-week period thus being about 35 kg/100 t. Nonanol provides the cheapest method of sprout suppression but involves continuous supervision over the 2-week period and is now little used.

Irradiation

It was stated in Section 14.1.5 that irradiation will completely prevent sprout growth. It has other effects: there is marked temporary sweetening, and senescent sweetening is hastened, so that very prolonged storage of irradiated tubers is not possible (Burton *et al.*, 1959). For processing use the content of reducing sugars may be too high after 3 months or so, but for domestic consumption an acceptable sugar content could be retained for perhaps 6 months or longer (*loc. cit.*). With irradiation at, say

10 krad, we can thus obtain complete sprout suppression for medium-term storage. The method involves the provision of a facility for exposing a shallow layer of the tubers to a fully shielded radiation source, usually cobalt-60, for a time sufficient for adequate dosage. The provision of such a facility could scarcely be envisaged on the farm. If it is provided at a central storage depot it could not irradiate the whole tonnage of the catchment area at the time of harvest unless the weather were so uniformly good that a steady intake was possible over a fairly long predetermined period. In other cases double handling of a considerable tonnage might be necessary.

There is however, recurrent interest in the commercial use of irradiation, and a brief description of a recent installation, first operational in the 1974–5 season, may place the method in perspective. It is situated in Shihoro, Japan, and can irradiate 15–17 t h^{-1}, the intention being to treat about 30 000 t over a 3 month period of continuous operation. It employs a 300 kCi cobalt-60 source to give a mean dose of 10 krad. The capital cost of this facility to irradiate 15–17 ton h^{-1} was 1.5 M U.S. $ (then 0.7 M £). Part of this can be offset against alternative uses in the remaining 9 months of the year – irradiation of onions, packed meals etc. – and allowing for this the cost of treating the potatoes has been estimated at about 10–11 U.S. $ t^{-1} (then 4.5–5 £).

References

Akeley, R.V., Houghland, G.V.C. and Schark, A.E. (1962). *Am. Potato J.* **39**, 409–17.

Amberger, A. and Schaller, K. (1973). *Chemie Mikrobiologie Technologie Lebensmittel*, **2**, 39–41 (Quoted from Chem. Abs. **79** (1973) 124883r).

Anderson, D.G., Stafford, H.A. Conn, E.E. and Vennesland, B. (1952). *Pl. Physiol.* **27**, 1179–80.

Appel, O. (1906). *Ber. dt bot. Ges.* **26**, 118–22.

Appleman, C.O. (1914). *Bull. Maryland agric. Exp. Stn* **183**, 181–226.

Appleman, C.O. (1916). *Bot. Gaz.* **61**, 265–94.

Appleman, C.O. (1918). *Bull. Maryland agric. Exp. Stn* **212**, 79–102.

Appleman, C.O. and Miller, E.V. (1926). *J. agric. Res.* **33**, 569–77.

Appleman, C.O., Kimbrough, W.D. and Smith, C.L. (1928). *Bull. Maryland agric. Exp. Stn* **303**, 159–75.

Artschwager, E.F. (1927). *J. agric. Res.* **35**, 995–1000.

Barker, J. (1935). *New Phytol* **34**, 407–8.

Barker, J. (1936a). *Proc. R. Soc. B*, **119**, 453–73.

Barker, J. (1936b). *Report Fd Investigation Board 1935*, 118–23.

Barker, J. (1937). *Report Fd Investigation Board 1936*, 179–80.

Barker, J. (1939). *Report Fd Investigation Board 1938*, 193–5.

Barker, J. and Mapson, L.W. (1950). *New Phytol.* **49**, 283–303.

Barker, J. and Mapson, L.W. (1952). *New Phytol.* **51**, 90–115.

Barker, J. and Mapson, L.W. (1955). *Proc. R. Soc. B*, **143**, 523–49.

Bennett-Clark, T. and Kefford, N.P. (1953). *Nature* **171**, 645.

Blumenthal-Goldschmidt, S. and Rappaport, L. (1965). *Pl. Cell Physiol.* **6**, 601–8.

Breyhan, Th., Heilinger, F. and Fischnich, O. (1959). *Landw. Forsch.* **12**, 293–5.

Brown, W. (1947). *Ann. appl. Biol.* **34**, 422–9.

Brown W. and Mellor, H.C. (1952). U.S. Patent 2 615 801, Oct. 28, 1952 *Chem. Abs.* (1953) **47**, 2927h.

Brown, W. and Reavill, M.J. (1954). *Ann appl. Biol.* **41**, 435–47.

Brownell, L.E. *et al.* (1954). Utilization of the gross fission products. *Univ. Michigan Engng Res. Institute*, Progress Report **7**.

Burton, W.G. (1950). *New Phytol.* **49**, 121–34.

Burton, W.G. (1952). *New Phytol.* **51**, 154–61.

Burton, W.G. (1955). *Proc. Institute Refrigeration*, **51**, 168–72.

Burton, W.G. (1956a). *Physiologia Pl.* **9**, 567–87.

Burton, W.G. (1956b). *Nature* **178**, 218.

Burton, W.G. (1957). *Fd Sci. Abs.* **29**, 1–12.

Burton, W.G. (1958). *Eur. Potato J.* **1** (2) 47–57.

Burton, W.G. (1961). *Proc. 1st Triennial Conf. Eur. Ass. Potato Res. Braunschweig, 1960*, 79–117.

Burton, W.G. (1963a). In *The Growth of the Potato*, eds. Ivins, J.D. and Milthorpe, F.L. pp. 17–41. London: Butterworth.

Burton, W.G. (1963b). *Ann. Report Ditton Covent Garden Labs Jan. 1962–May 1963*, 25.

Burton, W.G. (1964). *Eur. Potato J.* **7**, 90–101.

Burton, W.G. (1965a). *J. exp. Bot.* **16**, 16–23.

Burton, W.G. (1965b). *Eur. Potato J.* **8**, 80–91.

Burton, W.G. (1965c). *Eur. Potato J.* **8**, 245.

Burton, W.G. (1966). *The Potato*. Wageningen: Veenman.

Burton, W.G. (1968). *Eur. Potato J.* **11**, 249–65.

Burton, W.G. (1969). *Eur. Potato J.* **12**, 81–95.

Burton, W.G. (1971). Postscript to Harkett, P.J. (1971).

Burton, W.G. (1972). In *Crop Processes in Controlled Environments*, ed. Rees, A.R., Cockshull, K.E., Hand, D.W. and Hurd, R.G. pp. 217–33. London: Academic Press.

Burton, W.G. (1973). *Proc. 5th Triennial Conference Eur. Ass. Potato Res. Norwich, 1972*, 63–81.

Burton, W.G. (1974). *Potato Res.* **17**, 113–37.

Burton, W.G. (1975). *Potato Res.* **18**, 109–115.

Burton, W.G. and Hannan, R.S. (1957). *J. Sci. Fd Agric.* **12**, 707–15.

Burton, W.G. and Mann, G. (1954). *Agriculture* **60**, 466–72.

Burton, W.G. and Meigh, D.F. (1971). *Potato Res.* **14**, 96–101.

Burton, W.G. and Spragg, W.T. (1950). *New Phytol.* **49**, 8–10.

Burton, W.G. and Wigginton, M.J. (1970). *Potato Res.* **13**, 180–6.

Burton, W.G., Mann, G. and Wager, H.G. (1955). *J. agric. Sci.* **46**, 150–63.

Burton, W.G. Horne, T. and Powell, D.B. (1959). *Eur. Potato J.* **2**, 105–16.

Bushnell, J. (1928). *Proc. Potato Ass. Am.* **15**, 15–18.

Bushnell, J. (1929). The normal multiple sprouting of seed potatoes. *Bull. Ohio agric. Exp. Stn*, **430**.

Butler, O. (1919). Storage of potatoes. *Circular New Hampshire agric. Exp. Stn*, **20**.

Buttery, R.G. and Ling, L.C. (1973). *J. agric. Fd Chem.* **21**, 745–6.

Castello Branco, A.C. de S. (1957). Discussion on paper by Masten, V.J. and Hočevar, J. In *Proc. 2nd Int. Pl. Protection Conference, Fernhurst, England, 1956*, 169.

Chaube, S., Swinyard, C.A. and Daines, R.H. (1973). *Lancet* (1) 329–30.
Conner, H.W. (1937). *Pl. Physiol.* **12**, 79–98.
Cornforth, J.W., Milborrow, B.V. and Ryback, G. (1966). *Nature* **210**, 627–8.
Craft, C.C. (1963). *Am. Potato J.* **40**, 289–98.
Crook, E.M. and Watson, D.J. (1950). *J. agric. Sci.* **40**, 227–32.
Cunningham, H.S. (1953). *Phytopathology* **43**, 95–8.
Davidson, T.M.W. (1958). *Am. Potato J.* **35**, 451–65.
Davis, R.M. (1961). *Am. Potato J.* **38**, 411–13.
Denny, F.E. (1926a). *Am. J. Bot.* **13**, 118–25.
Denny, F.E. (1926b). *Am. J. Bot.* **13**, 386–96.
Denny, F.E. (1945). *Contr. Boyce Thompson Inst. Pl. Res.* **14**, 1–14.
Denny, F.E. and Thornton, N.C. (1941). *Contr. Boyce Thompson Inst. Pl. Res.* **12**, 79–84.
Denny, F.E. and Thornton, N.C. (1942). *Contr. Boyce Thompson Inst. Pl. Res.* **12**, 405–29.
Denny, F.E., Guthrie, J.D. and Thornton, N.C. (1942). *Contr. Boyce Thompson Inst. Pl. Res.* **12**, 253–68.
Dettweiler, C. (1952). *Planta* **41**, 214–39.
Dostal, R. (1942). *Gartenbauwissenschaft* **16**, 195.
Downie, W.A. (1952). *J. Department Agric. Victoria, Australia*, **50**, 61–6.
Ducet, G. and Chevillotte, P. (1970). (Quoted from *Chem. Abs.* **74** (1971) 72836c)
Dyson, P.W. and Digby, J. (1975a). *Potato Res.* **18**, 290–305.
Dyson, P.W. and Digby, J. (1975b). *Potato Res.* **18**, 363–77.
Effmert, B., Meinl, G. and Vogel, J. (1961). *Zuchter* **31**, 23–32.
Ellison, J.H. and Cunningham, H.S. (1953). *Am. Potato J.* **30**, 10–14.
Elwood, J.H. and MacKenzie, G. (1973). *Nature* **243**, 476–7.
Emanuel, I. and Sever, L.E. (1973). *Teratology* **8**, 325–31.
Emilsson, B. (1949). *Acta agric. Suec.* **3**, 189–284.
Farrimond, A. (1972). (Paper presented to the Engineering Section at the *5th Triennial Conference Eur. Ass. Potato Res.* Summary only published in the *Proceedings*, 1973, p. 135).
Fischnich, O. (1956). *Landbouwkundig tijdschrift* **68**, 743–55.
Fischnich, O. and Patzold, C. (1954). *Angew. Bot.* **28**, 41–52.
Franklin, E.W. and Thompson, N.R. (1953). *Am. Potato J.* **30**, 289–95.
Gager, C.S. (1912). *Bot. Gaz.* **54**, 515–24.
Gorter, A. and Nadort, W. (1941). *Proc. Acad. Sci. Amsterdam*, **44**, 1112.
Green, W.P., Hukill, W.V. and Rose, D.H. (1941). Calorimetric measurements of the heat of respiration of fruits and vegetables. *Technical Bull. U.S. Department Agric.* 771.
Hackett, D.P., Haas, D.W., Griffiths, S.K. and Niederpruem, D.J. (1960). *Pl. Physiol.* **35**, 8–19.
Hagberg, A. and Nybom, N. (1954). *Acta Agric. scand.* **4**, 578–84.
Hammett, F.S. (1929). *Protoplasma* **7**, 297.
Hanes, C.S. and Barker, J. (1931). *Proc. R. Soc. B*, **108**, 95–118.
Hannan, R.S. (1954). *Fd Sci. Abs.* **26**, 121.
Harkett, P.J. (1971). *Potato Res.* **14**, 305–11.
Harkett, P.J. (1975). *Abs. Conference Papers 6th Triennial Conf. Eur. Ass. Potato Res., Wageningen, 1975*, 179–80.
Heilinger, F. (1961). *Proc. 1st Triennial Conf. Eur. Ass. Potato Res. Braunschweig, 1960*, 241–2.
Heilinger, F. and Breyhan, Th. (1959). *Landbauforschung Volkenrode* **9**, 17–18.

Hemberg, T. (1943). *Ark. Bot.* **30 B** (7).
Hemberg, T. (1947). *Acta Horti Bergiani* **14**, 133–220.
Herklots, G.A.C. (1928). The effect of ethylene in the respiration of plant organisms. PhD. Thesis, University of Cambridge.
Huelin, F.E. (1932). *Report Fd Investigation Board 1931*, 90–2.
Huelin, F.E. (1933). *Report Fd Investigation Board 1932*, 53–4.
Hughes, J.C. (1963). In the section on 'Blackening of plant tissues' In *Report Low Temperature Res. Stn, Cambridge 1962*, 23–5.
Hylmö, B., Persson, T., Wikberg, C. and Sparks, W.C. (1975a). *Acta agric. scand.* **25**, 81–7.
Hylmö, B., Persson, T., Wikberg, C. and Sparks, W.C. (1975b). *Acta agric. scand.* **25**, 88–91.
Isherwood, F.A. (1973). *Proc. 5th Triennial Conf. Eur. Ass. Potato Res. Norwich, 1972*, 156–7.
Isherwood, F.A. (1976). *Phytochemistry* **15**, 33–41.
Isherwood, F.A. and Burton, W.G. (1975). *Potato Res.* **18**, 98–104.
Ives, J.V. (1955). *Pl. Path.* **4**, 17–21.
Johnson, G. and Schaal, L.A. (1957). *Phytopathology* **47**, 253–5.
Kawakami, K. (1952). *Mem. Hyogo Agric. College, Sassayama, Japan*, 1–114.
Keller, E.R. and Berces, S. (1966). *Eur. Potato J.* **9**, 1–14.
Kennedy, E.J. and Smith, O. (1951). *Am. Potato J.* **28**, 701–12.
Kennedy, E.J. and Smith, O. (1953). *Proc. Am. Soc. hort. Sci.* **61**, 395–403.
Krijthe, N. (1946). *Meded. LandbHoogesch. Wageningen* **47** (6).
Krijthe, N. (1962). *Eur. Potato J.* **5**, 316–33.
Kubicki, K. (1961). *Proc. 1st Triennial Conference Eur. Ass. Potato Res. Braunschweig, 1960*, 245–7.
Lange, H. and Rosenstock, G. (1964). *Phytopath. Z.* **51**, 136–52.
Lange, H. and Rosenstock, G. (1965). *Phytopath. Z.* **52**, 188–201.
Lange, H. Rosenstock, G. and Kahl, G. (1970). *Planta* **90**, 109–18.
Lauer, F. and Shaw, R. (1970). *Am. Potato J.* **47**, 275–8.
Levy, H., Schade, A.L., Bergman, L. and Harris, S. (1948). *Archs Biochem* **19**, 273–86.
Liljemark, A. and Widoff, E. (1960). *Am. Potato J.* **37**, 379–89.
Lipton, W.J. (1967). *Am. Potato J.* **44**, 292–9.
Locci, R. and Kuć, J.A. (1967). *Phytopathology* **57**, 1272.
Luckwill, L.C. (1948). *Annual Report Agric. Hort. Res. Station, Long Ashton, Bristol, 1947*, 137–42.
Luckwill, L.C. (1949). *Annual Report Agric. Hort. Res. Station, Long Ashton, Bristol, 1948*, 124–9.
Lyon, G.D. (1972). *Physiol. Pl. Path.* **2**, 411–16.
Mann, G. (1963). *Ann. Report Ditton Covent Garden Laboratories, 1962–3*, 30–1.
Mapson, L.W. and Burton, W.G. (1962). *Biochem. J.* **82**, 19–25.
Marshall, E.R. and Smith, O. (1951). *Bot. Gaz.* **112**, 329–30.
Marth, P.C. and Schultz, E.S. (1950). *Am. Potato J.* **27**, 23–32.
McKee, R.K. (1961). *Akademie Landwirtschaftswissenschaften Berlin: Tagungsberichte Nr. 27*, 277–89.
Meiklejohn, J. (1943). *Biochem. J.* **37**, 349–54.
Metlitskii, L.V. and Tsekhomskaya, V.M. (1964). *Biokhimiya* 151–62.
Miller, L.P. (1931). *Contr. Boyce Thompson Inst. Pl. Res.* **3**, 309–12.
Muller, K. (1846). *Bot. Zg* **4**, 769–72.
Müller, K. (1975). Veränderungen wertgebender Inhaltsstoffe in der Kartoffelpflanze und -knolle im Verlauf von Vegetation und Lagerung,

ihre Beteutung für die Qualität der Knolle. *Hefte für den Kartoffelbau, 17: Schriftenreihe Förderungsgemeinschaft Kartoffelwirtschaft e.V., Hamburg.*

Müller-Thurgau, H. (1882). *Landw. Jbr* **11**, 751–828.

Nash, M.J. and Lennard, J.H. (1973). *Potato Res.* **16**, 47–52.

Nelson, J.M. and Auchincloss, R. (1933). *J. Am. Chem. Soc.* **55**, 3769–72.

Nielsen, N.K. (1968). *Acta Agric. scand.* **18**, 113–20.

Norton, G. (1963). In *The Growth of the Potato*, ed. Ivins, J.D. and Milthorpe, F.L., pp. 148–59. London: Butterworths.

Okazawa, Y. (1960). *Proc. Crop Sci. Soc. Japan* **29**, 121.

Olufsen, L. (1903). *Bot. Zentralblatt* **15**, 269–308.

Ozeretskovskaya, O.L and Vasyukova, N.I. (1965). *Dokl. Akad. Nauk SSSR* **161**, 968–70.

Ophuis, B.G. (1957). *Neth. J. Agric. Sci.* **5**, 180–94.

Paterson, D.R. Whittwer, S.H. Weller, L.E. and Sell, H.M. (1952). *Pl. Physiol* **27**, 135–42.

Pätzold, C. and Kolb, W. (1957). *Beit. Biol. Pfl.* **33**, 437–58.

Pätzold, C. and Weiss, H.M. (1957). *Angew. Bot.* **31**, 93–105.

Pedersen, S. (1956). *Fd Technol.* **10**, 532–6.

Penner, H. and Fromm, H. (1972). *Z. Lebensmittelunters. u-Forsch.* **150**, 84–7.

Politis, J. (1948). *Revue cytol. cytophys. veget.* **10**, 232–3.

Pressey, R. (1966). *Archs Biochem. Biophys.* **113**, 667–74.

Pressey, R. (1970). *Am. Potato J.* **47**, 245–51.

Pressey, R. and Shaw, R. (1966). *Pl. Physiol.* **41**, 1657–61.

Priestley, J.H. and Woffenden, L.M. (1923). *Ann. appl. Biol.* **10**, 96–115.

Radatz, W. (1967). *Landbauforsch. Völkenrode* **17**, 153–8.

Rakitin, J.V. and Suvorov, N.N. (1935). *C. r. (Dokl.) Acad. Sci. URSS*, **IV (IX)** 295–7.

Rappaport, L. and Wolf, N. (1969). In *Dormancy and Survival*, ed. Woolhouse, H.W. pp. 219–40. Cambridge: Cambridge University Press.

Rappaport, L., Lippert, L.F. and Timm, H. (1957). *Am. Potato J.* **34**, 254–60.

Renwick, J.H. (1972). *Br. J. Social Preventive Med.* **26**, 67–89.

Rhodes, A., Sexton, W.A. Spencer, L.G. and Templeman, W.G. (1950). *Research* **3**, 189–90.

Romberger, J.A. and Norton, G. (1961). *Pl. Physiol.* **36**, 20–9.

Rosa, J.T. (1923). *Proc. Am. Soc. hort. Sci.* **20**, 180–7.

Rubin, B.A. and Metlitsky, L.V. (1958). *2nd Int. Conf. Peaceful Uses Atomic Energy, Geneva, 1958.*

Salunkhe, D.K., Wu, M.T. and Jadhav, S.J. (1972). *J. Fd Sci.* **37**, 969–70.

Samotus, B. (1971). *Potato Res.* **14**, 145–9.

Samotus, B. and Schwimmer, S. (1963). *J. Fd Sci* **28**, 163–7.

Sawyer, R.L. (1956). *Proc. 7th Am. Potato Utilization Conf.* 3–5.

Sawyer, R.L. (1961). *Am. Potato J.* **38**, 203–7.

Sawyer, R.L. and Dallyn, S.L. (1957). *Farm Res.* **23** (6) 6.

Sawyer, R.L. and Dallyn, S.L. (1964). *Am. Potato J.* **41**, 59–69.

Schade, A.L., Levy, H., Bergmann, L. and Harris, S. (1949). *Archs Biochem.* **20**, 211–19.

Schmid B. (1901). *Ber. dt. bot. Ges.* **19**, 76.

Schippers, P.A. (1956a). *Sticht. Aardappelbewaring, Wageningen: Publikatie* **108**, *Serie A.*

Schippers, P.A. (1956b). *Sticht. Aardappelbewaring, Wageningen: Publikatie* **112**, *Serie A.*

Schulz, K.G. (1927). *Landwirt. Versuchsstationen* **105**, 23–73

Singh, B.N. and Mathur, P.B. (1937). *Ann. appl. Biol.* **24**, 469–74.
Smith, C., Watt, M., Boyd, A.E.W. and Holmes, J.C. (1973). *Lancet* **269** (1).
Smith, O. (1929). *Hilgardia* **4**, 273–306.
Smith, O.E. and Rappaport, L. (1961). *Adv. Chem.* **28**, 42–8.
Smith, W.L. and Smart, H.F. (1959). *Phytopathology* **45**, 649–54.
Sparrow, A.H. and Christensen, E. (1954). *Nucleonics* **12** (8) 16–17.
Stuivenberg, J.H.M. van and Veldstra, H. (1942). *Landbouwkundig tijdschrift*, **54**, 611–32.
Sussman, A.S. (1953). *J. cell. comp. Physiol.* **42**, 273–84.
Szalai, I. (1957a). *Acta biol. (Acta Univ. szegediensis)* **3**, 33–40.
Szalai, I. (1957b). *Acta biol. (Acta Univ. szegediensis)* **3**, 41–9.
Szalai, I. (1957c). *Acta biol. (Acta Univ. szegediensis)* **3**, 51–4.
Szalai, I. (1959a). *Acta agron. Acad. sci. hung.* **9**, 253–64.
Szalai, I. (1959b). *Physologia Pl.* **12**, 155–61.
Tagawa, T. and Okazawa, Y. (1953). *J. Fac. Agric. Hokkaido (imp.) Univ.* **50**, 65–73.
Talley, E.A., Fitzpatrick, T.J. and Porter, W.L. (1964). *Am. Potato J.* **41**, 357–66.
Thimann, K.V., Yocum, C.S. and Hackett, D.P. (1954). *Archs Biochem. Biophys.* **53**, 239–57.
Thornton, N.C. (1939). *Contr. Boyce Thompson Inst. Pl. Res.* **10**, 339–61.
Tomiyama, K., Ishizaku, N., Sato, N., Masamune, T. and Katsui, N. (1968). In *Biochemical Regulation in Diseased Plants of Injury* pp. 287–92. Tokyo: Phytopathological Society of Japan.
Toosey, R.D. (1959). *Agriculture J. Minist. Agric.* **66**, 346–50.
Twiss, P.T.G. (1963). *Ann. Rep. Ditton Covent Garden Laboratories, Jan. 1962–May 1963*, 21.
Varns, J., Currier, W. and Kuc, J. (1971). *Phytopathology* **61**, 968–71.
Vliet, W.F. van and Schriemer, W.H. (1963). *Eur. Potato J.* **6**, 201–17.
Wager, H.G., Burton, W.G. and Mann, G. (1952). *J. agric. Sci.* **42**, 266–75.
Waggoner, P.E. (1955). *Am. Potato J.* **32**, 448–50.
Wareing, P.F., El Antably H.M.M., Good, J. and Manuel, J. (1966). *Wiss. Z. Univ. Rostock. Reihe Mathematik Naturwissenschaften* **16**, 667–72.
Wassink, E.C. Krijthe, N. and Scheer, C. van der (1950). *Proc. K. ned. Akad. Wet.* **53**, 1228–39.
Wigginton, M.J. (1973). *Potato Res.* **16**, 85–7.
Wigginton, M.J. (1974). *Potato Res.* **17**, 200–14.
Wilson, A.R. Twiss, P.T.G. and Lessells, W.J. (1962). *Eur. Potato J.* **5**, 147–65.
Workman, M. and Twomey, J. (1970). *Am. Potato J.* **47**, 372–8.
Wright, R.C. and Peacock, W.M. (1934). Influence of storage temperature on the rest period and dormancy of potatoes. *U.S. Department Agric. Technical Bull.* 424.
Yamaguchi, M., Perdue, J.W. and MacGillivray, J.H. (1960). *Am. Potato J.* **37**, 73–6.
Zilva, S.S. and Barker, J. (1939). *Report Fd Investigation Board 1938*, 199–201.
Zitnak, A. (1955). Factors influencing the initiation and rate of solanine synthesis in tubers of *Solanum tuberosum L.* PhD. Thesis University of Alberta, Canada.

Chapter 15

The production of new varieties

H.W. HOWARD

In any crop, improvements in the phenotype may be brought about either by changes in the environment (e.g. better applications of fertilizers; irrigation in dry seasons; disease and pest control) or by improvements in the genotype (i.e. new varieties). Many improvements in phenotype may depend upon genotype × environment interactions, a classical example of such an interaction being that between heavy applications of fertilizers and yield in cereals. Heavy applications of fertilizers will usually only give increased yields if the variety has a high resistance to lodging. Conversely the newer lodging-resistant varieties will only give higher yields than the older varieties at the highest levels of fertilizer application.

15.1 History of potato breeding

Unconscious selection for higher yields must presumbaly have occurred as soon as potatoes became cultivated plants. Just how soon conscious selection was made by cultivators in prehistoric times in South America will never be known. That there may have been selection for several characters seems obvious and, as Dodds and Long (1955) point out: 'it is of interest to observe that the greatest variability of types of anthocyanins seems to occur in the tuber, the organ with a selection value to early agriculturalists'.

As pointed out by Hawkes (Chapter 1), the next stage in potato breeding occurred after tetraploid cultivated potatoes of the Andigena group (i.e. *S. tuberosum* ssp. *andigena*) were introduced at the end of the sixteenth century in Europe. The introductions were apparently short-day types for tuberization and it was necessary to select for ability to produce high yields of tubers under the long days of a European summer before the potato could become a farm crop in Europe.

From Europe the potato was taken to many parts of the world. It was introduced in the early eighteenth century by settlers into North America, South Africa and Australia. It was, for example, brought to North America by a group of Irish presbyterians in 1719 (Stevenson and Clark, 1937; Salaman, 1949) and was first grown at Londonderry, New Hampshire. In the southern states of the U.S.A. the potato *(S. tuberosum)* is still often called the Irish potato to distinguish it from the sweet potato *(Ipomoea batatas)*.

There are doubts as to when and how the potato was introduced into some countries. It was, for instance, probably introduced into Mexico directly from South America by the Spaniards at the same time as it was introduced into Europe (Ugent, 1968). Similarly very little is known about the early introduction of the potato into India; it was apparently in the early seventeenth century, only about 40 years after its first introduction into Europe, and it was probably brought to India by the Portuguese when they opened up trades routes from Europe to India (Puskarnath, 1969). There were then at a later date further introductions from Europe.

Once introduced into a country, the development of new varieties tended to be done separately in each country. There was, however, some interchange of material. One variety was of special importance. In 1851 the Reverend Chauncey Goodrich of New York obtained via the U.S. Consul in Panama a small collection of tubers from South America. One of these potatoes Goodrich named 'Rough Purple Chile'; he also raised a number of selfed-seedlings from it, one of which, Garnet Chile, was a widely grown variety for many years. Later breeders used Goodrich's stocks for hybridization with stocks of European origin, and many North American varieties can be traced back to Goodrich's Garnet Chile. They include Beauty of Hebron, Russet Burbank, Early Rose, Green Mountain, Prolific and Triumph (Stevenson and Clark, 1937). Early Rose has been an important parent in the breeding of European as well as North American varieties of early maturity (Hawkes, 1956).

In the U.K. the first known reference to distinct varieties was in 1730 when five types of potato were recognized (Davidson, 1935, 1939). In Germany, according to Danert (1956), 5 different varieties were described in 1747, 40 in 1777 and as many as 186 in 1854 (not all of these, however, had names, some were identified only with a number). In the U.S.A. Stuart (1937) quotes Bailey as stating that 'only two varieties were listed in 1771, yet by the end of the eighteenth century they were numerous'.

As Stuart (1937) and others have pointed out, the severe epidemics of blight *(Phytophthora infestans)* that swept over America and Europe in the years 1845–7 gave a marked stimulus to the breeding of new varieties of potatoes. At a later time, about 1910, the recognition and alarming spread of wart disease *(Synchytrium endobioticum)* in Europe, but not in other parts of the world, led to a new era in potato breeding (Salaman, 1926). 'Immune' varieties were found and it proved possible to transmit this 'immunity' to new varieties. The next change in potato breeding was more recent. Following the Russian potato collecting expeditions in Latin America of 1925 and later, there was a renewed interest in broadening the types of parents used in breeding to include both S. American cultivated types and wild species (Hudson, 1936). This has led to many new potato varieties being bred from crosses involving either wild species or new introductions of Andigena and therefore having genes or even whole chromosome segments not found originally in Tuberosum potatoes (Howard, 1970).

15.2 Breeding methods

15.2.1 Somatic mutations

Somatic mutations occur in potatoes with a frequency of about one plant per 200 000–500 000 (Heiken, 1958, 1960). They produce chimeras (Howard, 1970, Chapter 9). Most somatic mutations are deleterious and have to be rogued out, although very rarely they may give rise to new varieties (Howard, 1970, Table 33). This applies especially to tuber colour mutations from white or white-splashed pink to full pink. On the other hand the production of new potato varieties nearly always involves sexual reproduction, i.e. the transference of pollen from the anthers of flowers of one variety to the stigmas of flowers of another variety in order to produce berries containing true seeds, and not the chance discovery of somatic mutations.

15.2.2 Consequences of vegetative reproduction

The fact that potatoes are propagated normally by vegetative reproduction makes their breeding easier than that of crops with sexual reproduction in one important aspect. Any selection will keep true to type, except for rare mutations (see above), so long as vegetative reproduction is followed. New varieties may be selected therefore as plants from crosses in the F_1 generation. Similarly it is relatively easy to use wide crosses, induced mutations and artificially produced polyploids in the production of new varieties because of the absence of the necessity of coming true-to-type after sexual reproduction.

Probably the biggest disadvantage to the breeder that is associated with vegetative reproduction is the care that must be taken to keep plants free of virus diseases at all stages of selection and of multiplication of clones. Nearly all potato viruses (spindle tuber virus is an important exception) are not transmitted through true seed and hence every seedling is in the beginning virus-free. It is the breeders' task to ensure that they remain so. Varieties in which every plant was virus-infected were common in the past; they can now be freed of virus by meristem culture (Kassanis and Varma, 1967). It should also be noted that the development of seed certification and of virus-testing schemes during the past 50 or so years has made the breeders' task harder – varieties now have potentially an unlimited life and do not degenerate so that they have to be replaced by new varieties.

Other disadvantages to the breeder following from having vegetative reproduction include (a) many established varieties are extremely heterozygous and transmit in crosses their good qualities to a very small proportion of their progeny; (b) varieties may not flower or be pollen sterile (ovule sterility is rare) – pollen sterility is very common and imposes a limit on the number of desirable crosses that can be made; (c) selection in the seedling year may be very inefficient because of the different method of propagation and in the first clonal year because of the size of set; (d) material cannot be stored easily and clones must be grown each year; (e) the rate of increase is much slower than by seeds; (f) large amounts of material have to be stored under conditions which are neither too hot nor too cold; and (g) transport of tubers is much more difficult than that of seeds.

15.2.3 Induction of flowering

Although wild potato species, most cultivated diploid potatoes and most Andigena clones flower profusely, many Tuberosum varieties produce only a few flowers and in several important varieties, e.g. Bintje and King Edward, no flowers are normally found. Inflorescences are formed, but after a short time the flower buds cease to grow and the buds fall off at the abscission layer.

There are two very useful methods for preventing bud abscission and obtaining flowers. These are grafting on to tomatoes and 'growing on a brick'. Grafting scions of potatoes on tomato stocks is successful for producing flowers with many varieties, e.g. King Edward (Carson and Howard, 1945). In 'growing on a brick' (Thijn, 1954), sprouted tubers are planted on a brick or tile above soil level and covered with sand. When the roots have grown through the sand into the soil, the sand is washed away; this allows the new tubers to be removed when very small. Both tomato grafting and removal of young tubers result in no sinks for carbohydrates and other assimilates, and hence the flower buds receive more nourishment. With King Edward and some other varieties it is also advisable to take out the axillary buds below the inflorescences. If these axillary buds develop into shoots, they tend to compete with the inflorescences for nutrients.

Long-days (Patterson, 1953; Steineck, 1957) and late planting (Rothacker, 1957) are also conducive to flowering. In addition spraying with auxin analogues can be useful in preventing flower bud abscission (Wiersema, 1950).

15.2.4 Flower retention and berry formation

Even when flowers have been produced and pollinated by a variety known to be pollen-fertile, there is a strong tendency in many varieties for the flowers to drop off and not to produce berries. The use of cut stems with flowers from field-grown plants placed in jars of water, sometimes containing an antibiotic to control bacteria, can be very successful in producing a good set of berries (Peloquin and Hougas, 1959). Spraying with 2-4D or other auxin analogues two or three days after pollination may prevent the abscission of flowers and allow the ovaries to start to develop into berries with seeds (Wiersema, 1950). Spraying at an earlier date often leads to parthenocarpic berries with no seeds.

15.2.5 Pollen sterility: storage of pollen

Many Tuberosum varieties are pollen sterile. This can be a serious nuisance in breeding programmes in that it may be impossible to make certain desired crosses. There is, as yet in potatoes, no method of making pollen-sterile varieties pollen fertile. In breeding programmes involving several crosses or back-crosses it is often possible to arrange to retain pollen fertility in the offspring until the final generation.

The inheritance of pollen fertility and sterility has been studied by several authors. Sterility has been suggested to be due to a single dominant gene, or to recessive genes, or to gene–cytoplasm interactions. These different causes of pollen sterility are not mutually exclusive, as in a study of diploid species it has been found by Grun *et al.* (1962) and by Grun and Aubertin (1965) that different types of pollen sterility may be due to either dominant or recessive genes in sensitive cytoplasms. In non-sensitive cytoplasms the various dominant and recessive genes have no effect.

Storage of pollen can be very useful in breeding programmes as it obviates the necessity of having both parents flowering at the same time. Deterioration of pollen is slower at low temperatures and at + 2.5 °C potato pollen will remain viable for about a month (Blomquist and Lauer, 1962). By drying anthers and then storing them at – 20 °C, pollen can be kept viable for as long as two years (King, 1955; Howard, 1958).

15.2.6 Seeds: dormancy and storage

Most freshly harvested potato seeds are dormant but this dormancy disappears after storage for several weeks (Simmonds, 1963). It can also be removed by treatment with gibberellic acid. Seed of Tuberosum potatoes retains a certain amount of dormancy even after storage for

9 months and seed stored for 21 or so months often germinates much better than seed stored for only 9 months. As seed and tuber dormancy appear to be correlated or even to be conditioned by a common biochemical control (Simmonds, 1964), there is a danger that, if only those seeds which germinate rapidly are pricked out and late germinating seeds are discarded, there will be an unconscious selection for low tuber dormancy.

Stored dry and under normal laboratory conditions, most samples of potato seeds retain their viability for at least 10 years. Cold temperature ($-20\,^{\circ}C$) can result in seeds remaining viable for at least 20 years (Howard, 1969, 1975).

15.2.7 Raising and selecting seedlings

It is usual now in most countries for potato seedlings to be grown in pots in aphid-proof glasshouses and not in the field. The seeds are sown in pans and the seedlings pricked out into the pots when large enough. A common size of pot (which may be plastic) is 10 cm diameter. Smaller pots allow more seedlings to be grown but the tubers are rather small. Larger pots will permit larger tubers to be obtained but will reduce the number of seedlings per unit area. In order to make maximum use of glasshouses, two crops of seedlings may be grown each year.

Most potato virus diseases, an important exception is spindle tuber virus (Diener, 1971), are not transmitted through true seed and hence every seedling is initially free of virus infection. Glasshouses screened against aphids, or regular fumigations to kill aphids, or both, enable the seedlings to be kept virus-free.

Because of the heterozygous nature of potato varieties and the small chance of combining in any one seedling the many desirable characters required in a new variety, it is usual in potato breeding to grow many thousand seedlings each year (see also Section 15.10.1). This large number of seedlings usually necessitates keeping only one tuber of each selected seedling because of the labour which would be required in the following clonal year in planting different numbers of sets of different selections.

Performance in the seedling year may not obviously be closely related to performance in subsequent clonal years, firstly because the plants are grown from seed and not tubers and secondly because the plants are grown in small pots in a glasshouse.

Selection criteria in the seedling generation have been discussed by Maris (1962) and Howard (1963a). Although discarding seedlings because they are either too 'wild' (i.e. with very long stolons and very late maturity; types needing short days for tuber production) or are too weak growing may be almost an 100% efficient, selection for many desirable characters such as high yield and good shape appears to be inefficient and seedlings discarded as low yielding and having bad shaped tubers may produce high yields of good shaped tubers in the later clonal generations. Hence both authors suggest that selection of seedlings should not be too stringent.

15.2.8 Selection in the first clonal year

It is normal practice to plant the tubers of the selected seedlings in the field in a good 'seed' district where spread of virus diseases is minimal. Selecting the best clones in this first clonal year may be difficult for two reasons. Firstly there is only one plant per clone and, as Engel (1957) pointed out, there are very big errors when attempts are made to estimate yields from observations on either single plants or small clones. Secondly the tubers planted in this first clonal year may vary considerably in size and some may have been stored for a long period. On the other hand it would appear to be possible to select the first clonal year plants efficiently for characters such as tuber shape and depth of eye.

15.2.9. Selection in later clonal generations

As the number of selections retained decreases, so the number of plants grown per selection is increased (a scheme used at the Plant Breeding Institute, Cambridge as shown in Table 15.1). The larger numbers of plants per clone both facilitates selection in the good 'seed' district and allows yield trials to be grown in ware districts and tests made for disease and pest resistance and for quality. By about the tenth year it should be possible to have selected clones with a sufficient number of tubers for submission for official trials. A similar scheme which was used at the Scottish Plant Breeding Station was described by Black (1953).

Table 15.1 Scheme used at the Plant Breeding Institute, Cambridge, for selecting and multiplying potato clones

Year	
1	25 000 seedlings, of which about 8000 selected and retained as single tubers.
2	8000 clones at one plant per clone; about 1000 selected
3	1000 clones at 4 plants per clone; about 120 selected
4	120 clones at 10 plants per clone
5	About 40 clones at 30 plants per clone
6	About 8 clones at 100 plants per clone
7	About 4 clones at 350 plants per clone
8	About 2 clones at 600 plants per clone
9	Probably one clone only at 1400 plants per clone
10 and 11	First and second years of statutory performance trials

15.2.10 Yield trials

Small scale yield trials are possible in years 8 and 9 in ware districts. In the previous years the amount of seed tubers available is too small for replicated trials and reliance has to be put on 'observation' plots. Although these observation plots are mainly for comparing different selections from the same cross or similar crosses in order to select or discard clones, it is advisable to include plots of standard varieties so that some idea can be obtained as to how the yield of selected clones compares with that of the standard varieties.

15.2.11 'Seed' multiplication

As soon as a clone is entered in official trials, or perhaps a year or so previously, plans must be made for the multiplication of a 'seed' stock for release to seed growers for general growing. It is also advisable to make a number of single plant selections and to propagate these each year as single plant progenies. If the main stock should unfortunately pick up some virus infection, then it will usually be possible to start a new stock from one or more of the single plant progenies.

15.3 Determination of breeding objectives

Much of plant breeding consists of crossing variety A which has a high yield but medium quality or low resistance to a particular disease or pest with variety B of medium yield but good quality or high resistance to the disease or pest in the hope of producing variety C of yield at least as high as that of A and with the good qualities of B. Potato breeding is no exception to this generalization.

15.3.1 Survey of present varieties

A survey of the varieties most widely grown should give information on which are the highest yielding and what their major defects are. It then remains to be decided as to how an increased yield might be obtained and whether there are any obvious parents which may have the characters necessary to counteract the major defects of the high yielding varieties.

15.3.2 Types of parent

Four main types of parent are available to the potato breeder: Tuberosum varieties, Andigena varieties, varieties of the cultivated diploid species, and wild species. It is now generally agreed that for most parts of the world a Tuberosum variety, i.e. belonging to *S. tuberosum* ssp. *tuberosum* and able to produce a high yield under long-day conditions, should always be one parent of a cross and that where a desired character cannot be found in Tuberosum varieties, then the next source should be an Andigena variety if possible. If not available in Andigena, then the cultivated diploid potatoes should be considered and use only made of wild species as parents as a last resort.

15.4 Breeding objectives: yield

No new variety has any chance of succeeding unless it is at least as high yielding as the present varieties which it will have to replace. It should on the contrary yield more than the existing varieties and it is necessary for the breeder to consider how this can be achieved.

15.4.1 Date of maturity and yield

Yield in potatoes can depend very much on the time of harvest, the later a variety can be dug the higher the yield usually. It is therefore necessary before yield trials can be done to decide into which 'maturity' group a variety belongs: early (first early), mid-season (second early), early maincrop or late maincrop. Although referred to as a 'maturity' classification, it is important to note that not all early varieties mature early and that the criterion of an early variety is the ability to produce a profitable yield early in the growing season. On the other hand late maincrops do not produce an early crop.

Investigations on the frequency of early maturing seedlings under long-day conditions have been made by Möller (1956) – see also Howard (1970, Table 11). The results suggest that time of maturity depends upon a number of genes and that varieties are heterozygous for most of these. Thus early × early does not give more than about 60% of earlies, and early × late only about 20% of earlies.

15.4.2 Rate of establishment

The rate at which potato plants make early top growth and give a good cover of the ground is obviously of importance in weed control. A rapid early growth of tops also appears to be necessary for high yields, particularly for early potatoes. This poses a problem in selection of new varieties. In selecting for lateness in sprouting so as to improve keeping quality, the breeder may at the same time select types which emerge late and make slow growth in the first few weeks after planting.

15.4.3 Heterosis

The diploid cultivated potatoes are all out-breeders with a well developed gametophytic type of self-incompatibility controlled by a series of *S* alleles. Although pollen-fertile tetraploid cultivated potatoes are self-compatible, they still have the *S* alleles (Cipar *et al.*, 1967) and appear to be out-breeders showing a depression in yield when inbred (Krantz, 1946). The small amount of basic material from which most Tuberosum varieties have been bred until recently has been stressed by Simmonds (1962). There were only a few introductions from S. America; in addition, the varieties bred from this small number of introductions were reduced by the blight epidemics of the 1840s, and the new varieties produced with a certain degree of resistance to blight must have contained an even smaller store of variability. Assuming that Tuberosum potatoes are out-breeders and that there is a lack of genetic variability, then it should be possible by using new introductions of Andigena potatoes, or possibly of other species such as the cultivated diploids, to breed varieties which will show heterosis and have very high yields.

There is some indication that this is so. It was, for example, found by Glendinning (1969) that Andigena × Tuberosum progenies outyielded Tuberosum × Tuberosum progenies by an average of 13%, the two types of progenies having similar mean maturities. Both types had similar mean tuber weights and the increase in yield was due to a higher mean number of tubers per plant. It may also be significant that the two highest yielding maincrop varieties in the U.K. in 1974 both trace back to parents not within the narrow European Tuberosum germplasm, Pentland Crown having USDA 41956 as one grand-parent and Maris Piper the Andigena stock CPC 1673. Similarly Toxopeus (1952) suggested that, even if the advantages due to resistance to the common races of blight are excluded, certain clones obtained by back-crossing from hybrids of *Solanum demissum* × *tuberosum* are very high yielding.

15.4.4 Number of tubers and tuber size

Total yield of a potato crop obviously depends upon number of plants per unit area, number of tubers per plant and the weights of individual tubers. These three variables are not independent and in certain circumstances even halving the number of plants per unit area may not decrease yield, there being more and bigger tubers at the lower plant density. Number of tubers per plant is also dependent on size of set planted and of the treatment of the sets before planting (maintenance of apical dominance to give a single sprout or conditions favourable for multiple sprouting). These two factors make work on the inheritance of tuber number and tuber size rather difficult. They also complicate selection of families in breeding programmes. In the first clonal year a rather small tuber is normally planted because the seedlings have been grown in relatively small pots (Section 15.2.7). Plants would not therefore be expected to have too many tubers. On the other hand because of prolonged storage of tubers produced from glasshouse-grown seedlings the sets may be multi-sprouted and hence likely to produce numerous tubers. Selection should obviously not be too severe for tuber number in the first clonal year and steps should be taken to ensure that sets as normal as possible are used in the second clonal year.

Usually it is not total yield which is important but yield of saleable tubers, i.e. tubers neither too large nor too small. On the whole the breeder selects for a yield of tubers of a uniform size and all of the optimum saleable size. This will usually be for a maincrop in the U.K. between 12 and 18 tubers per plant for a high yield.

Tuber numbers, size and yield have been investigated by Engel (1957) in several crosses and also in the first inbred generation from selfing four varieties. He found that 75% of plants from crosses had from 6 to 12 tubers per plant and that there were very few with less than 6 tubers. A very much larger range in tuber numbers was found for first-generation inbred clones from Voldagsen, there being about 50% of clones with more than 12 tubers. In breeding it may be useful never to use in a cross two varieties which both have numerous tubers as this can lead to many descendants being discarded for being too numerous. If a variety with otherwise useful characters has too many tubers per plant, then it may be advantageous to use for the other parent a variety with a relatively small number of rather large tubers per plant.

15.4.5 Disease resistance and yield

Breeding varieties resistant to diseases and pests (Sections 15.8 and 15.9) may often be the easiest way of breeding for higher yields. In localities where cyst nematodes (Section 15.9.1) are common, resistant varieties not only yield higher than susceptibles but also, because of the type of resistance which decreases nematode populations, lead to susceptible varieties grown later on the same land also having increased yields.

15.5 Breeding objectives – quality for domestic use

Quality in potatoes usually refers to the appearance and edibility of the cooked tubers or to the suitability of the tubers for a particular manufacturing process. Obviously what is good quality for one type of cooking or manufacture is not necessarily good quality for another type of cooking or manufacture. Although it is usual to consider quality under headings such as dry matter content, discoloration and flavour, these are only some aspects of quality and others are also of importance to the consumer. These other components of quality include tuber shape and size, depth of eyes, resistance to diseases, mechanical damage and greening, rate of sprouting, and skin and flesh colour (Howard, 1963b, 1974).

15.5.1 Tuber shape and depth of eye

In most countries the most acceptable shape of tuber is the long oval, although a few widely grown varieties have long tubers and although it has been suggested that round tubers could suffer less damage during harvesting. The genetics of tuber shape are probably not too complicated, long being dominant to round but there being at least four genes with cumulative effects (Howard, 1970). So far as the breeder is concerned, it is not usual to make crosses of 'long' with 'long' as the progeny from such a type of cross very often contains a high proportion of seedlings with too long, thin and useless tubers.

There is disagreement on the inheritance of depth of eyes, some authors suggesting deep eyes to be dominant to shallow and others the reverse. On the whole it would appear that crosses of 'shallow' with 'shallow' do produce some offspring which have eyes of medium depth. Such offspring, however, do not occur with a high frequency whereas the use of only one parent with deep eyed tubers can result in many seedlings having to be discarded because their tubers have eyes which are too deep.

15.5.2 Disease resistance

It is not usual to consider disease resistance as an integral part of good quality but in certain places resistance to a disease may be of great importance for good quality. For example, in areas where the nematode-transmitted tobacco rattle virus causes spraing, varieties such as Bintje and Record, which appear to be immune to the disease, are the only varieties which can produce good quality tubers without internal blemishes (Richardson, 1970). Similarly, now that so many potatoes are sold as washed tubers in cellophane bags, diseases which cause surface blemishes, such as common scab (*Streptomyces scabies*), are of much more importance than they were previously. On certain soils in dry years it is impossible to produce tubers of some varieties without a heavy scab incidence and scab resistance is therefore one part of good quality in these circumstances.

15.5.3 Mechanical damage

The initial criterion of many domestic consumers in judging quality is a low frequency of greened, blighted or damaged tubers rather than those of certain cooking qualities. This is understandable, as all three defects are easily seen and produce both a waste of time and of potatoes when the tubers are peeled [mechanical damage is also of considerable importance to the processor (Section 15.6) and may necessitate extra labour for inspection of tubers before they are processed as well as loss of material].

Mechanical damage may be superficial ('scuffing') or lead to cracks, deep bruises or even cuts. Superficial damage depends to a considerable extent on the tubers being fully mature when harvested. Breeding to produce varieties which mature relatively early, i.e. early maincrops rather than late maincrops, can obviate scuffing. There are varietal differences in susceptibility to bruising, and there is some evidence that tubers of high specific gravity and high dry matter are more susceptible than those of low degrees of both. Laboratory tests of susceptibilities to bruising can be carried out on small numbers of tubers by methods such as dropping a plunger of known weight from a set height on to the tuber or placing them in a shaking machine for a given time. Bruises take three or four days to show up. Staining with *p*-cresol (Robertson, 1957) may help in showing the damaged areas. While laboratory tests may help breeders in selecting for tubers with some resistance to mechanical damage, particularly when breeding for those of high dry matter content, it can be suggested that avoidance of mechanical damage is a problem for agricultural engineers in designing better machinery rather than for breeders in developing varieties with tubers resistant to mechanical damage.

15.5.4 Flesh colour

Domestic consumers in the U.K. and the U.S.A. prefer varieties with white flesh whereas those in most European countries prefer yellow flesh. Yellow is dominant to white and determined by a single gene but there are in addition modifying genes for different degrees of yellowness (references in Howard, 1970). Thus breeders in the U.K. and the U.S.A. will not have to discard any selections because of yellow flesh so long as they are employing white × white crosses but would have to discard about 50% of selections in yellow × white crosses because of yellow flesh (assuming the yellow-flesh parent to be heterozygous for yellow).

15.5 Dry matter content and texture

Texture of the cooked tubers, which is split into five characters in Table 15.2, has often been assumed to be associated with starch content and, because starch is the major solid component of the tuber, with high dry matter which is usually measured as specific gravity. However, as Linehan and Hughes (1969) point out, variations in the texture of the cooked potato can be explained only partly by variations in the starch content of the tuber and it is possible that the categories 'disintegration' and 'consistency' of Table 15.2 might depend on the cohesion of the tuber cells while 'dryness' might refer to the surface properties of the individual

Table 15.2 Characters used to assess cooking quality (after 'Potato Quality Research', working party of the European Association for Potato Research; see Lugt, 1961)

(a) Texture–this is split into 5 sub-characters, each with four degrees of expression, as shown below:

Character	*Degree of expression*			
Disintegration	None	Slight	Moderate	Complete
Consistency	Firm	Fairly firm	Soft	Soft; uneven
Mealiness	Not	Slightly	Mealy	Very mealy
Dryness	Humid	Slightly humid	Slightly dry	Dry
Structure	Fine	Fairly fine	Slightly coarse	Coarse

(b) Colour – white, cream, yellow, deep yellow
(c) Discoloration – none, slight, moderate, pronounced
(d) Flavour – neutral, slight, some, strong

Table 15.3 Inheritance of mealiness of cooked tubers (from Moller and Vogel 1961)

Cross number	*Parents*	*% of progeny with tubers which are* *not mealy*	*medium mealy*	*very mealy*
2	Lusew54.881/34 × Schwalbe	4	69	27
3	Ontario × Schwalbe	68	32	0
4	Jubel × Schwalbe	67	33	0
8	Saskia × Hochprozentge	100	0	0

cells and cell aggregates. 'Mealiness could result from an interaction between intercellular adhesion and 'dryness.' Warren and Woodman (1974) suggest that mealiness is controlled by the solids content of the tuber but that disintegration on cooking does not depend on solids content.

Although texture of the cooked tuber does depend to some extent on environmental factors, it is well established that genotype plays the most important role in determining texture and varieties have been classified for many years according to degree of mealiness. Möller and Vogel (1961) (see Table 15.3) found progenies in which no seedlings had mealy tubers when cooked, others in which there was a mixture of non-mealy and medium mealy, and still others in which there were very few non-mealy and most seedlings had tubers which cooked either medium mealy or very mealy. So far as texture of cooked tubers is concerned, potato breeders usually ensure that at least one parent of a cross has tubers which cook to give the desired texture and the main difficulty is selecting clones with tubers of the wanted cooking type. It is not usually difficult to discard clones which are obviously either too mealy or too waxy: the difficulties are in selecting for the usually required degree of slight mealiness. Tubers which are slightly mealy can be used for most purposes – chips (pommes frites), mashed, baking and salad – whereas a waxy (non-mealy) tubers are only suitable for salad use and very mealy potatoes only for mashing (Lugt, 1961).

15.5.6 After-cooking blackening

Stem-end blackening of tubers after cooking depends upon both genotype and environmental conditions, particularly soil type. The easiest way for the breeder to test selections for non-blackening is to grow them on a soil known to be conducive to produce tubers which blacken after cooking, e.g. Soil B of Table 15.4. It is now known (Hughes and Evans, 1967) that the ratio of chlorogenic and citric acids correlates well with differences in the intensity of blackening over a wide range of varieties. Potatoes with low contents of chlorogenic acid do not blacken and high contents of citric acid reduce blackening when moderate amounts of chlorogenic acid are present. After-cooking blackening is a character

Table 15.4 After-cooking blacking of tubers in selections from the cross Pentland Crown × Maris Piper – unpublished results of C.S. Cole. Maris Piper shows no blacking on soil B and Pentland Crown slight to moderate

	Number of selections in which after-cooking blackening was			
Soil	*absent*	*slight*	*moderate*	*severe*
A	11	15	7	0
B	6	13	3	6

Soil A – sandy loam at Plant Breeding Institute, Trumpington, Cambridge.
Soil B – Peat soil at Arthur Rickwood Experimental Husbandry Farm, Mepal, Cambridgeshire.

with high heritability and from using one variety as parent with good resistance to blackening it is usual to find at least 20% of the progeny with good resistance (Table 15.4).

15.5.7 Flavour

Flavour, the fourth character, in Table 15.2, is the most difficult character of the four to assess. Pronounced bad flavours are usually only found when wild species have been used in breeding programmes to introduce disease and pest resistance, and Howard (1973) therefore suggested that in choosing parents for such work the order of priority should be: first cultivated tetraploid potatoes, secondly cultivated diploids and lastly wild species. An extreme example of the production of a variety with bad quality was that of the variety Lenape, bred from the wild species *S. chacoense*, which had under certain conditions poisonous amounts of glucoalkaloids in its tubers (Zitnak and Johnston, 1970). On the other hand, with sufficient back-crosses to cultivated varieties it is possible to obtain good quality in breeding programmes in which a wild species has been used.

Assessment for the more subtle types of flavour (neutral, slight and some) involves the use of organoleptic tests and the results can be difficult to interpret as different observers have different preferences. One special difficulty with potatoes is that texture differences tend to confuse flavour differences.

15.5.8 Consumer recognition

It is very useful if the tubers of a good quality variety can be recognized easily by the consumer. This applies, for example, to King Edward VII (white, splashed pink) and Désirée (full pink) in England, Golden Wonder (long and russet skin) in Scotland and Arran Victory (purple) in Northern Ireland. Varieties with white tubers and smooth skins tend to be lumped together, even though they include Maris Piper (good quality) and Pentland Crown (quality sometimes indifferent).

15.6 Breeding objectives: quality for processing

There has been, in the last 30 years or so, a very big increase in the proportion of the potato crop which is processed for domestic consumption and not sold fresh. According to Talburt and Smith (1967), the U.S.A. potato crop in 1964 was disposed of as follows: 54% fresh food, 28% processed for food, 10% used for seed, 4% starch and miscellaneous products, and 4% loss. The potatoes processed were divided into 42% for chips (i.e. crisps in the U.K.), 36% for frozen French fried (chips in the U.K.), 16% for dehydration, 4% for miscellaneous uses and only 2% for canned 'new' potatoes.

In other countries the proportion of potatoes sold fresh for domestic use is higher in relation to that sold to processors. The amount processed in the U.K., has, however, seen a rapid growth, from 9.7% of the total crop in 1968–9 to 16.7% in 1972–3. Ten years ago the biggest outlet for processed potatoes was as chips (i.e. crisps) but there has been a big increase in the amount of potatoes sold as frozen chips (i.e. French fries, pommes frites).

Potatoes for the various types of processing may need to have quality characters not found in potatoes for the fresh domestic market. This may lead to varieties being grown solely for processing, for example the Dutch variety Record, which has yellow flesh, in the U.K. for crisps (chips) manufacture.

15.6.1 Dry matter (DM) content

Potatoes for the manufacture of chips (crisps in the U.K.), frozen French fries and dehydrated products should all have a high DM content. DM content is determined by both variety and environment. Provided that suitable control varieties are included, it is not difficult to select for high DM content by measuring it as specific gravity. It appears to be controlled by many genes and the best chance of producing a new high DM variety is to select in crosses between high DM parents. Potatoes for canning need low DMs and should be waxy rather than mealy.

15.6.2 Reducing sugar content

A low content of reducing sugars is required in potatoes for the manufacture of chips, frozen French fries and dehydrated products. A rough idea of the content of reducing sugars in tubers can be obtained by using diabetes-test papers. Later selection is usually based on degree of brownness of crisps made experimentally. This is not difficult for breeders to carry out, but care has to be taken of the temperature of storage as all tubers have higher reducing sugar contents when kept at low temperatures (see Sections 13.4.1 and 14.1.6).

15.6.3 Discoloration of raw flesh

Discoloration of peeled or cut raw tubers is a result of enzymatic oxidation in injured cells when they are exposed to air or O_2. It is of commercial importance primarily in the pre-peeling business. The discoloration, which appears to be correlated with the concentration of tyrosine in the tuber, is highly heritable; in some crosses very few selections have tubers showing discoloration whereas in other crosses very few selections have tubers with no discoloration.

15.6.4 Other characters

Just as is required in potatoes for domestic table use, tubers for processing should be resistant to damage. There appears to be a negative correlation between dry matter content and resistance to damage. It is thus very important for breeders to pay attention to resistance to damage when assessing selections with tubers of high dry matter content. It may be that there has to be a compromise between dry matter content and resistance to damage, the final selection having tubers of medium dry matter content plus some resistance to damage.

15.7 Breeding objectives – quality in industrial potatoes

In some countries potatoes are an important crop for the production of starch. The main requirement of industrial varieties is a very high starch content and a high yield of starch per acre. There are also differences between varieties in the composition of the starch, there being significant differences between varieties for the amylase (long, relatively unbranched chains) to amylopectin (highly, branched chains) ratio and for phosphoric acid content of the starch. Selection for high amylase content is possible.

15.8 Breeding objectives – disease resistance

Like most crop plants the potato suffers from many diseases and pests and the chance of breeding the 'perfect' variety with high yield, good quality and resistance to all pathogens is very small. The breeder has to decide, therefore, which are the most important pathogens to which a new variety should be resistant. Although the use of a resistant variety will usually be the cheapest method of control and may also be the simplest way of increasing yields, there are cases where reliance should be put on other control methods and the breeder not given the very hard task of producing a resistant variety.

In assessing whether a breeding programme for resistance to pathogens is realistic, it is necessary to consider first the importance of the pathogens, secondly the type and degree of resistance available, thirdly the inheritance of resistance, and fourthly tests for resistance. It is obviously of no importance to breed for resistance to a pathogen of minor importance and any breeding programme must begin with a source of resistance which can be hybridized with the standard varieties. It is also obviously of the greatest importance to know whether resistance is dominant to susceptibility and how many genes are involved. Satisfactory tests for resistance and which are not too time-consuming are of paramount importance.

Assessments of the relative amounts of damage caused by different pathogens may be available, but it may be necessary to carry out surveys and/or field experiments to determine the importance of different diseases and pests (see below: common scab; foliage blight). The relative importance of pathogens may alter with changes in agricultural systems or with the increase in popularity of new varieties (see below: wart; dry rot).

The modern tendency is to concentrate on non-race-specific resistance rather than race-specific because the former is expected to be more durable than the latter. Race-specific resistance is often of the hypersensitive type and due to dominant genes (see Section 15.8.1 for gene *Nx* giving immunity to virus X and Section 15.8.4 for *R* genes giving resistance to blight). It is a very easy type of resistance for the breeder to use both because of its simple inheritance and because of easy tests for resistance. Unfortunately it is also the type of resistance easily overcome by different races of the pathogens. It may still be useful for soil-borne pathogens where the dispersal of races is not so rapid as for pathogens producing millions of airborne spores.

Non-race-specific resistance (horizontal resistance, 'field' resistance, etc., see Robinson, 1969) is nearly always due to polygenes and is therefore often a difficult type of resistance to handle, particularly in a back-crossing programme, because all the genes are transmitted to only a low proportion of progeny. It is also very often a relatively difficult type of resistance to detect.

A relatively low degree of resistance which is transmitted to a high percentage of progeny may offer a better chance of success than a very high degree transmitted to a low percentage of progeny because there will be more resistant progeny in which to select for agronomic characters, quality and resistance to other pathogens. On the other hand it will probably be more difficult to test for a low degree of resistance rather than a high degree. Breeding from a cultivated source of resistance will usually be easier than breeding from a wild source, particularly as it will be easier to get good quality (see Section 15.5.7).

The stage at which testing for resistance to pathogens takes place needs consideration. For example, if breeding for resistance to two pathogens is being undertaken, then tests for resistance to pathogen A which are easy to carry out and require little material for testing would normally be done before tests for resistance to B which are more complicated and need more material. Also, if tests are of about equal difficulty, then tests for resistance to A which is only transmitted to a small proportion of progeny should take precedence over tests for resistance to B which is transmitted to a high proportion of progeny. Similarly, if there is a severe screening for agronomic characters other than disease resistance in the early stages of selection, then it will usually be advantageous to carry out tests for resistance at a later stage when the number of selections is relatively small.

Simple and at the same time efficient tests for resistance are of the utmost importance. It is also necessary to monitor the tests so as to be absolutely certain that they are neither too severe nor too lax. In many cases this can best be done by including three controls: susceptible, moderately resistant, and highly resistant. It is not uncommon to find, however, at the start of an investigation that there is inadequate knowledge as to the degrees of resistance and susceptibility of the standard, widely grown varieties.

15.8.1 Viruses

Several types of virus resistance can be distinguished: tolerance, 'infection resistance', hypersensitivity (usually giving 'field-immunity'), and extreme resistance (or immunity). Tolerance of a virus, i.e. showing no obvious symptoms and little reduction in yield, is usually considered to be a dangerous type of resistance in potatoes because it causes difficulties in seed certification schemes and because a tolerant variety may act as a reservoir of virus from which non-tolerant varieties can be infected. In countries in which there is no seed certification scheme or where infection of initially virus-free stocks takes place quickly, it may on the other hand be valuable.

Infection resistance can be defined as the type of resistance in which in the field only a small percentage of plants becomes infected. It often appears to be due to polygenes, crosses of moderately resistant × moderately resistant giving some plants with a resistance higher than that of either parent and crosses of highly resistant × susceptible giving usually most progeny with a moderate resistance.

Hypersensitivity and immunity on the other hand are due to single, dominant genes. These genes are widespread, being found in Tuberosum and Andigena potatoes, in the cultivated diploids and in several wild species. There are, however, several important viruses to which neither hypersensitivity nor immunity have been found.

Because most viruses can be controlled by using good 'seed' districts and certification schemes, it can be argued that resistance to virus diseases should be given a low priority in most breeding programmes and that attention should be concentrated on those diseases and pests which can only be controlled by the use of resistant varieties. This ignores that control of viruses by seed certification schemes and growing seed in special districts may be costly.

Virus X

Virus X, which is spread by leaf contact and does not need an aphid vector, exists in many strains. At one time many varieties were infected widely with mild strains of the virus which produced negligible symptoms and only very small reductions in yield. Quick serological tests for the presence of the virus can be made easily and are used on a wide scale to produce virus-free stocks.

There are several sources of resistance to virus X (Cockerham, 1970). These include a comprehensive resistance found in the USDA seedling 41956 and also in the Andigena clone CPC 1673 used in breeding for resistance to *Globodera* (*Heterodera*) *rostochiensis* (see Section 15.9.1), resistance to most virus X strains but not to the so-called virus B, and resistance to virus B but not to other strains of virus X. They are controlled respectively by three dominant genes, *Rx*, *Nx* and *Nb*. Gene *Nb* is not of much practical importance as virus B type strains of virus X are uncommon. Gene *Nx* occurs in several old British varieties, e.g. King Edward and Epicure, where it has been responsible for keeping these varieties free of virus X. Gene *Rx* is found in several recently introduced Dutch varieties.

Virus A

Virus A, an aberrant strain of potato virus Y, is spread by aphids and can cause considerable losses in yield, particularly in conjunction with virus X, infection with the two viruses causing 'crinkle'. Genes giving resistance to normal strains of virus Y also confer resistance to virus A. There is also a dominant gene *Na* in many Tuberosum varieties which produces hypersensitivity and 'field-immunity' to virus A. It is closely linked to gene *Nx* which produces a similar reaction to virus X. This close linkage of *Nx* and *Na* gives some advantage to the breeder in using gene *Nx* for resistance to the common European strains of virus X rather than using the comprehensive resistance source of USDA 41956 and CPC 1673 (see above).

Virus Y

Virus Y, which exists in many strains, is both one of the commonest viruses of potatoes in all parts of the world and also a virus causing severe reductions in yield. Although it is aphid-transmitted and can therefore be controlled by the use of good 'seed' districts, there would be an advantage in having varieties which are resistant to virus Y in that stocks of such varieties would remain virus free for several years outside good seed districts. It should also be noted that virus Y is a 'non-persistent' virus unlike leaf roll, the other major virus disease of potatoes, and infection cannot therefore be controlled by insecticide treatment of the crop.

Resistance to virus Y is of several types. A degree of tolerance is found in some varieties, e.g. in International Kidney which is widely grown in Jersey, C.I., as Jersey Royal. Resistance to infection also occurs, e.g. in Pentland Crown, the maincrop variety at present most widely grown in the U.K. (see Table 15.10b). Varieties of the future may contain dominant genes from wild species such as *S. stoloniferum* which control extreme resistance to all strains of the virus (Cockerham, 1970).

Leaf roll

There is no good source of resistance to leaf roll similar to those for resistance to viruses X, A and Y which can be used by potato breeders. This is unfortunate as leaf roll is a very widespread virus causing large reductions in yield. Being transmitted by aphids, control can be achieved by the use of good 'seed' districts. It is possible (Baerecke, 1958) to select for resistance to infection but, as this is due to a number of minor genes, the resistance disappears on crossing with the standard, susceptible varieties. It is therefore a very difficult task to breed a new variety with infection resistance to leaf roll and with the many other desirable characters necessary.

Other viruses

Among the many other viruses known to infect potatoes are virus S, mop top, tobaccoo rattle and spindle tuber. Virus S, transmitted by leaf contact, was until recently widespread in many standard varieties. It can be detected easily by serological testing. It would be possible to breed resistant varieties as there is known an Andigena clone with a dominant gene giving hypersensitivity and hence 'field-immunity' (Baerecke, 1967). No resistance to potato mop top virus is known. Tobacco rattle virus, transmitted by migratory nematodes, is important in certain areas of western Europe as the main cause of spraing. Apparently immune Tuberosum varieties are known, e.g. Bintje and Record: there are no published genetical results but the immunity may be due to a single dominant gene. No resistance to potato spindle tuber virus is known. This virus is exceptional in that it is transmitted to some extent at least through true seed and even pollen. It could therefore cause serious trouble to breeders who get their parent clones infected.

15.8.2 Bacteria

There are three important bacterial diseases of potatoes: blackleg [*Erwinia phytophthora* Appel, also designated as *Erwinia caratovora* (Jones) Bergey *et al.* var. *atroseptica* (van Hall) Dye)], ring rot [*Corynebacterium sepedonicium* (Spiedk. and Kotth.) Skapt. and Burkh.], and brown rot or bacterial wilt (*Pseudomonas solanacearum* E.K. Smith). While there may be small differences in the susceptibility of tubers of different varieties to blackleg and to ring rot, there are no known good sources of resistance to either disease and hence no breeding programmes for resistance can be formulated. Both diseases can be controlled by other means: blackleg by producing disease-free stocks from stem cuttings and ring rot by isolating disease-free tubers to produce new 'seed' stocks.

Recent work on breeding for resistance to *Pseudomonas solanacearum*, which can be a serious pathogen in warm soils, has been concentrated on using certain clones of the diploid, cultivated species, *S. phureja*, as a source of resistance. The problem is complicated by the existence of several races of *P. solanacearum*. Resistance is due to a number of dominant genes plus modifiers (Rowe *et al.*, 1972) and varies with soil temperature, varieties being more susceptible the higher the temperature.

15.8.3 Actinomycetes

Common scab, caused by *Streptomyces scabies* (Thaxt.) Waksm. and Henrici, has become of more importance recently because of an increase in the amount of potatoes sold washed in polythene bags. Although the scab pustules are usually superficial and removed with the skin, tubers with scab are not so attractive as clean tubers and may be difficult to sell. There are big differences between varieties in their susceptibilities to common scab and it is possible to breed for resistance. Crosses of susceptible by resistant may give about 20% of resistant progeny, i.e., more than one gene is involved. The development of scab depends upon having a dry period during the first part of tuber growth and, in order to have a reliable test each year, it may be necessary to have some means of covering the plots to keep the soil dry (Langton, 1972a). Similarly scab can be controlled by irrigation at the appropriate time and varies both with soil type and rainfall. Varieties with no resistance to scab can therefore be successful.

15.8.4 Fungi

Many potato diseases are caused by fungi; they even include a rust (*Puccinia pitteriana*) in Colombia and a smut (*Thecaphora solani*) in Peru. While some diseases caused by fungi occur in all or nearly all countries in which potatoes are grown, other diseases are only of regional importance, e.g. charcoal rot (*Macrophomina phaseoli*) in India.

Wart disease

Wart disease, caused by *Synchytrium endobioticum*, is found on potatoes in South America and has been taken with potatoes to many parts of the world. It is, however, not now found in the U.S.A. and has never been found in some countries, e.g. Australia and New Zealand. In the first two decades of the twentieth century it was a very important disease in western Europe. The use of so-called 'wart-immune' varieties is often quoted as being one of the triumphs of plant breeding. Certainly the use of wart-immune varieties coupled with legislative action has reduced the incidence of wart in many countries so much that it now presents a problem of little importance. The triumph is, however, probably an example of the success of integrated control; it was possible to prevent the growing of susceptible varieties in districts where wart had been found because wart-immune varieties could be grown whereas it would have been much more difficult to stop the growing of any potatoes. And the legislative control on the movement of seed potatoes must have been responsible for the small spread of the disease-producing organism.

There are ample sources of resistance to the common race of *Synchytrium endobioticum* and resistance behaves as a dominant character. In many western European countries, in spite of the fact that wart is now of negligible importance and wart-susceptible varieties, e.g. Bintje in the Netherlands and King Edward in the U.K., are widely grown, all new varieties are still required to be wart-immune. Actually no variety is absolutely immune and the so-called immune varieties are just at the extreme end of a continuous range of reactions from very highly resistant to susceptible (Noble and Glynne, 1970). In a few countries there are more than one race of the fungus, some of which can attack varieties resistant to the common race. In these countries breeding is taking place for resistance to the other races.

Blight

There must have been more work on breeding for resistance to blight (*Phytophthora infestans*), often called late blight to distinguish it from early blight caused by *Alternaria solani*, than on breeding for resistance to any other pathogen. *Phytophthora infestans*, originally found on wild potatoes in North America (Mexico), has now spread to all the major areas where potatoes are grown including South America and can cause very large reductions in yield. In some areas, however, it appears that the effects of blight attack on yields have been exaggerated; years of heavy blight attack may be years of high, not low, yield because potato yields are dependent on high rainfall (Cox and Large, 1960). Blight on foliage can also usually be controlled to a considerable extent by fungicide sprays, and tuber infection can be decreased by agronomic practices such as the destruction of blighted foliage 14 or more days before harvest.

Very much work in very many countries was devoted to breeding for resistance of foliage to blight using as the source of resistance the wild, Mexican, hexaploid species, *S. demissum*. This species contains a series of *R* genes giving hypersensitive reactions to blight infection. The first varieties bred from the *S. demissum* source, some of which are still widely grown (e.g. Kennebec of Table 15.9), contain gene R_1 giving resistance to races such as race 0, 2, 3 and 4 but susceptible to races such as 1, 1.2, 1.3 and 1.4. Clones have been bred with two or more *R* genes but there has always been found a race of *Phytophthora infestans* to which they are susceptible – for example a clone with genes R_1, R_2 and R_3, although resistant to races such as 1, 2 and 3, is susceptible to race 1.2.3. Breeding using these major *R* genes has, therefore, ceased. The relation between resistance due to dominant *R* genes and races of blight was one of the main examples used by van der Plank (1963) in discussing 'vertical' and 'horizontal' resistance.

With the cessation of breeding for foliage resistance to blight using the *R* genes from wild Mexican potatoes, breeding has shifted to 'field resistance' (horizontal resistance; non-specific resistance) which is independent of race of *Phytophthora infestans*. This type of resistance is controlled by polygenes as might be expected because it is due to a number of causes including resistance to infection, slow growth of the mycelium within the leaf, late sporulation and the production of relatively few spores. Crosses of fairly resistant × fairly resistant give about 30% of fairly resistant progeny and no very susceptible whereas crosses of normally susceptible × normally susceptible give no fairly resistant and about 20% very susceptible (Table 15.5). Breeding for 'field-resistance'

Table 15.5 Inheritance of field-resistance to Phytophthora infestans, after Black (1960)

Reaction group of parents	*% of seedlings in reaction groups*				
	1	*2*	*3*	*4*	*5*
2 × 2	0	29	56	15	0
2 × 3	0	8	70	19	3
2 × 4	0	8	26	53	13
3 × 3	0	8	18	70	4
3 × 4	0	0	13	75	12
4 × 4	0	0	2	79	19

Group 1: highly resistant; 3% of area affected
Group 2: fairly resistant; 10% of area affected
Group 3: slightly resistant; 30% of area affected
Group 4: normally susceptible; 60% of area affected
Group 5: very susceptible; 100% of area affected.

to blight has been much facilitated by using the Toluca valley, Mexico, for field tests (Niederhauser and Cobb, 1959; Lapwood, 1971).

Although differences in resistance of tubers to infection by blight exist (Cox and Large, 1960), breeding for such resistance has been on the whole neglected by breeders. One reason for this is the lack of an easy test for resistance, the only reliable tests still having to be done under field conditions and needing rather a large number of plants. It should also be noted that many of the foliage-resistant potatoes with *R* genes have very susceptible tubers.

Storage rots: dry rot and gangrene

Storage diseases of tubers present a special difficulty to breeders because they tend to handle tubers very carefully and it is only when their products become widely grown by farmers that storage rots, gaining their entrance through wounds, may become serious. Two diseases are of special importance: dry rot (mainly due to *Fusarium solani* var. *coeruleum*) and gangrene (*Phoma exigua* var. *foveata*). Varieties do vary in their susceptibilities to these two rots and, as tests are known using relatively small numbers of tubers (Boyd, 1952; Langton, 1972b), it should be possible in the future to avoid introducing a new variety very susceptible to either disease.

Other diseases

The potato suffers from many more diseases caused by fungi. These include pink rot (*Phytophthora erythroseptica*), black scurf (*Corticium solani*), skin spot (*Oospora pustulans*), verticillium wilt (*Verticillium albo-atrum*). While these are usually of relatively minor importance, a new variety which was very susceptible to one of these diseases might suffer severely and so never become widely grown. Potential new varieties, which were found to be very susceptible to powdery scab, have been withdrawn before they had completed official trials. The powdery scab organism is also of some importance as the vector of mop top virus.

15.9 Breeding objectives – resistance to pests

As with diseases caused by fungi, potato pests vary with importance in different regions. Some pests, however, are of importance in very many countries. Breeding for resistance to some pests has been successful but for most pests chemical control would still appear to be the best method of control.

15.9.1 Cyst nematodes [*Globodera* (Heterodera) species]

It is now known that there are two species of cyst nematodes found on potatoes: *Globodera rostochiensis* and *G. pallida* (Stone, 1972; and see also Chapter 12). In addition both species have different pathotypes (Howard, 1972; Kort *et al.*, 1977).

Modern work on breeding for resistance to cyst nematodes was made possible by Ellenby (1952) who screened the Commonwealth Potato Collection (CPC) for resistance. He found resistance in the wild diploid species, *S. vernei*, and also in five accessions of Andigena potatoes. One of the latter, CPC 1673, has been widely and successfully used in breeding many nematode-resistant varieties. The resistance of CPC 1673 is due to a single dominant gene H_1; this, together with an easy and reliable test for resistance, made the breeding work not too difficult. Ellenby's test for resistance was simple; tubers were planted in pots of a nematode-infested soil and after ten or so weeks the root-balls were inspected – plants with many cysts were susceptible and those with none (or very few) were resistant. This is also the criterion of resistance which has been used in all subsequent breeding work.

It was shown (e.g. Jones, 1954; Williams, 1958) that resistant potatoes bred from CPC 1673 and the other Andigena clones (CPC 1685 and CPC 1690) produce a root diffusate which causes as many larvae to hatch out of the cysts as does the root diffusate of susceptible varieties, that the larvae invade the roots of resistant and susceptible plants in similar numbers, but that in resistant plants, although mature males can develop, few or no larvae develop into mature females. This type of resistance is typical of that found generally for other species of plants with resistance to other species of nematodes (Howard and Cotten, 1977). It follows that the growing of resistant plants reduces nematode populations whereas the growing of susceptible plants increases them.

The first population of potato cyst nematode which could produce many cysts on the roots of resistant potatoes bred from CPC 1673 was found by Dunnett (1957) at Duddingston, near Edinburgh. Dunnett (1961) also found resistance to the Duddingston population in the wild species *S. multidissectum* and showed that it was due to another dominant gene, H_2. Clones with both genes H_1 and H_2 were bred but tests of these clones with populations known to produce many cysts on varieties with gene H_1 showed that these H_1H_2 clones were only resistant to a minority of such populations.

Five pathotypes of *G. rostochiensis* and three of *G. pallida* are now recognized in Europe (Kort *et al.*, 1977). Gene H_1 gives resistance to Rol (previously known as *rostochiensis* pathotype A of both Britain and Holland) and also to Ro4 (Dutch pathotype F of *rostochiensis*, not so far found in Britain). Gene *H*2 gives resistance to Pal (previously called British pathotype B of *pallida*) but not to Pa2 and Pa3. The pathotype Pa3 is the British and Dutch pathotype E of *pallida*. It is very widespread in Britain (Brown, 1970).

The problem in breeding potatoes resistant to cyst nematodes is now to obtain resistance to pathotypes other than Rol and Ro4. This is being tackled in three ways. First resort has been made to the wild diploid species, *S. vernei* which has been claimed to have polygenic resistance to all pathotypes of both species (Ross and Huijsman, 1969) – there must be some doubt about this claim as different sources of *S. vernei* differ in their resistance to the several pathotypes (Kort *et al.*, 1977). Secondly there are clones of Andigena potatoes with resistance to pathotypes of *G. pallida* (Howard *et al.*, 1970; Howard and Fuller, 1971; Fuller and Howard, 1974). Thirdly resort can be made to wild species other than *S. vernei* such as *S. oplocense*, *S. spegazzinii* (Ross and Huijsman, 1969), *S. sparsipilum*, *S. boliviense*, *S. brevicaule* and *S. megistacrolobum* (Deshmukh and Weischer, 1970). The first successes in this breeding for resistance to pathotypes other than Ro1 and Ro4 have been obtained in the Netherlands where varieties with resistance to Ro1, Ro2, Ro3 and Ro4 have been produced from *S. vernei* crosses. No varieties with resistance to pathotypes of *G. pallida* are yet available.

It should be emphasized that breeding for resistance to potato cyst nematode species is given a very high priority in many countries. There are two reasons why this should be so. The first is because potato cyst nematode species are widespread in very many areas where potatoes are grown and produce marked decreases in yield (Brown, 1969, 1970). The second is that the breeding of resistant varieties is probably the easiest way of increasing potato yields. Because resistant varieties reduce nematode populations, their use can lead to increased yields of susceptible varieties grown later.

The future control of potato nematodes may be integrated control, using both a nematicide and a resistant variety. This seems possible now that there have been developed a number of efficient and relatively cheap nematicides. Such integrated control has been practised for several years for starch potatoes in the Netherlands (Nollen and Mulder, 1969). The control should also include non-growing successive potato crops and an appropriate intervention of other crops (see also Chapter 12).

15.9.2 Other nematodes

Resistance to root-knot nematodes, *Meloidogyne* species, may be important in warmer soils. Resistance to *M. incognita* and other species has been found in several *Solanum* species (e.g. Nirula *et al.*, 1969) but no work on either the inheritance of resistance or of attempts to breed resistant varieties appears to have been published.

Other nematodes feeding on potato roots include false root-knot nematodes (*Nacobbus* spp), root-lesion nematodes (*Pratylenchus* spp) and migratory nematodes (*Trichodorus* spp). It is unlikely from comparison with work on other crop plants that there is any resistance to the migratory nematodes.

Potatoes are also attacked by a race of the stem nematode *(Ditylenchus dipsaci)*. It has been claimed that there is resistance to this nematode but, as it is rarely of any importance, no breeding work for resistance has ever been carried out.

15.9.3 Aphids

Aphids can on occasions cause direct damage to potato plants, but the main importance of resistance to aphids is usually considered to be control of the aphid-transmitted virus diseases. Resistance to both *Myzus persicae* and *Macrosiphum euphorbiae* is found in wild species, particularly in those of Mexico (Radcliffe and Lauer, 1971). Many of these wild species are, however, difficult to cross with cultivated potatoes. Recently Gibson (1971) has suggested that steps should be taken to transfer the genes which determine the possession of glandular hairs that trap aphids from the three wild species, *S. polyadenium*, *S. tarijense* and *S. berthaultii*, to cultivated potatoes. Such a task will be very difficult unless the presence of glandular hairs is due to one or two major genes.

15.9.4 Colorado beetle

Before 1950 it was considered desirable to breed for resistance to Colorado beetle *(Leptinotarsa decemlineata)*, particularly in Germany and eastern Europe. Although there are some recent publications on resistance to Colorado beetle, breeding for resistance is not now given a high priority because of successful control of the beetle by insecticides. Several species of wild potatoes have resistance to Colorado beetle, the most important of which is probably *S. chacoense*. There is a danger of resistant potatoes owing their resistance to high contents of alkaloids (Schwarze, 1963).

15.10 Breeding objectives – synthesis

15.10.1 The chance of breeding a successful variety

As every practical potato breeder knows only too well, but is apparently not always appreciated by many pure scientists, the chance that any seedling will become a successful variety is very small. The position has been put very clearly by Simmonds (1969):

"In framing the objectives of a potato breeding programme the first requirement is realism. The 'perfect potato' does not exist and there is good reason to think that it never will. The reasons for this statement are as follows. Consider, for example, the breeding of a maincrop ware variety. A population of seedlings of the appropriate maturity range is selected for a number of characters of economic importance; the characters are assumed to be genetically independent. Suppose we select at the ten % level for the following seven characters: yield, tuber shape, cracking, cooking quality, blight resistance, leaf roll resistance, and scab resistance; at the 20% level for the following seven characters: tuber size, foliage type, flesh colour, resistance to gangrene, resistance to skin spot, resistance to dry rot and resistance to blackleg; and at the 50% level for three characters: skin colour, resistance to wart and resistance to virus Y. The severest selection here is at 10% which is quite weak; and yet a clone which would satisfy all these 17 criteria would emerge only once in about 10 million seedlings. If, by contrast, all 17 characters were selected very weakly at the 50% level, then one selection would emerge in about 150 000 seedlings. The conclusion is clear; one must either select for fewer characters or select less intensively. Negative correlations between characters would make effective selection even more difficult."

15.10.2 Choice of objectives

Because the chance of breeding a successful variety is so small, it is obvious that the choice of breeding objectives is very important and must always be considered very carefully. It may be necessary to take into account several factors in deciding which are the most important problems to be tackled and which are of relatively minor importance.

As has already been pointed out (Section 15.3.1) there may be one or two deficiences in the varieties being commonly grown which are of outstanding importance. Such deficiencies will obviously vary from country to country. In those countries where potatoes are widely grown and where the agricultural extension services are well organized, it may be easy to obtain the necessary information. This, however, is not always so and it may need surveys and field experiments to determine the most serious problems which the breeder should tackle.

15.10.3 Resistance to diseases and pests

Although the use of resistant varieties is always the cheapest method of controlling a disease or pest, there may be circumstances where chemical control or other means is satisfactory and it is not necessary to give the potato breeder a difficult problem to solve. There may, for example, be no need to breed varieties resistant to virus diseases if good 'seed' districts are available. Similarly the use of stocks raised from stem cuttings may obviate the need for varieties resistant to blackleg *(Erwinia caratovara* var. *atroseptica)*, to skin spot *(Oospora pustulans)* and some other tuber diseases (Hardie, 1971; Calvert, 1973). Protein content is another possible example – it may be easier to obtain more protein per acre by breeding high yielding varieties of normal protein content and to grow legumes on the area made vacant because of the higher yields than to breed potato varieties with high protein content.

On the other hand, as has already been suggested for nematode control (Section 15.9.1), it may be that in the future there will be more use of integrated control, i.e. using both a resistant variety and some other method of control simultaneously. For example, good control of loss of yield due to blight is possible in much of England by fungicide treatment of foliage plus varieties which have a high resistance of tubers to infection (Howard *et al.*, 1970). Control of virus diseases can involve growing in a good 'seed' district where there are no, or few, aphids to spread some viruses plus either resistant varieties and/or serological testing to control the viruses spread by leaf contact.

It has often been said that potato breeding has been dominated by breeding for resistance to diseases and pests. This may also seem to be so, from much of this chapter. There are three main reasons why resistance to diseases and pests are so important in potato breeding. The first, and most important reason is that disease and pest resistance may increase yields markedly. A second is that it may make the production of healthy seed easier and so also lead to increases in yield. Thirdly, disease resistance may lead to big increases in saleable yield by making the tubers more attractive.

It is therefore obvious that disease and pest resistance are very important and that the problem for the breeder is how much importance to attach to resistance to any one disease or pest and how high the resistance needs to be. Many established and widely grown varieties do not have high resistance to common diseases, e.g. Bintje and King Edward to blight, Maris Piper to common scab. On the other hand a very high susceptibility to a disease may make a new variety unacceptable, e.g. Doon Star because of its susceptibility to dry rot (and gangrene?). It would often appear that a successful variety has good resistance to a few important diseases coupled with a not too high susceptibility to others.

15.11 A possible new breeding method – the use of dihaploids

Although there are cultivated diploid potatoes in South America (Section 2.2.1), the species, *S. tuberosum* L., which is cultivated over the largest area in S. America and is the only species cultivated in the rest of the world, is a tetraploid with a chromosome number of $2n = 48$. It is, however, now possible to produce haploid ($2n = 24$) plants of *S. tuberosum* in large numbers by a method discovered by Hougas and Peloquin (1957) – see also Chapter 8 of Howard (1970). The method is to pollinate *S. tuberosum* with certain clones of the diploid species *S. phureja*. It appears that in many ovules the pollinator contributes two 12-chromosome gametes to the endosperm leaving none available to fertilize the egg nucleus which develops parthenogenetically. Because haploids of *S. tuberosum* have a chromosome number of $2n = 24$ which is twice the basic number of 12 found in the genus, it is convenient to describe them as dihaploids. A small proportion of dihaploids are pollen fertile and many are ovule fertile; they behave as diploids, producing from crosses between two dihaploids nearly all progeny with the diploid chromosome number of $2n = 24$.

Most primary dihaploids, i.e. dihaploids produced from tetraploid. *S. tuberosum* varieties, are weak plants but, crossed with other diploid species or among themselves, they can produce vigrous offspring. The weak nature of the original dihaploids is due to inbreeding depression, the production of a dihaploid being equivalent to about three generations of inbreeding. In spite of the lack of vigour and of the relatively low seed-fertility shown by most dihaploids, it is possible that the use of dihaploids and their progeny could be very useful to the breeder.

It has been suggested that dihaploids of *S. tuberosum* would allow potato breeding to be carried out at the diploid rather than at the tetraploid level. This might be possible but it would take very many years to assemble the necessary diploid parents, particularly as the fertility of dihaploids is low and selection therefore has to be made for fertility as well as for desired characters. Of more immediate promise would appear to be the selection of dihaploid stocks with desired characters for crossing with normal tetraploid varieties. Such crosses, due to the production of a high proportion of unreduced gametes by many dihaploids, result in tetraploid, not triploid, progeny. Alternately the dihaploids can be converted into tetraploids by colchicine treatment before crossing with tetraploid varieties.

The possible advantages of using dihaploids depend upon the type of desirable characters it is wanted to transfer to tetraploids. Consider first the case of disease or pest resistance determined by major dominant genes. If only a single dominant gene is involved, then there is not a large advantage to be gained from the use of dihaploids. At the tetraploid level most crosses would be of the type simplex *(Aaaa)* × nulliplex *(aaaa)*, giving about 50% of progeny with the required dominant gene *A*. If a duplex parent were available, then 5/6 of the progeny (83%) would be expected to have gene *A*. From the cross of a dihaploid of constitution A*a* by tetraploid, the frequency of tetraploid offspring with *A* would vary according to how unreduced gametes were formed and, if due to restitution nucleus formation, on the frequency of crossing over between gene *A* and the centromere of the chromosome carrying the gene. One estimate (Ross and Langton, 1974), for the gene *Rx* giving extreme resistance to potato virus X, is 86.7% transmission of the gene to the tetraploid progeny. This increased frequency of transmission becomes important when the transmission of more than one gene is necessary. Fuller and Howard (1974), for example, suggested that in breeding for resistance to cyst nematodes (Section 15.9.1) it is necessary to have three genes, *H1*, *H2* and *H3*, for resistance to both *G. rostochiensis* and *G. pallida*. If the cross were made at the tetraploid level between a parent simplex for each of the three genes and another without any of the genes, then 12.5% of the progeny would be expected to have all three genes. The cross of a dihaploid with genes *H1*, *H2* and *H3* with a tetraploid with none of the genes would give 65% of tetraploid progeny with the three genes, assuming an 86.7% transmission for each gene. There would thus be about five times as many offspring with genes *H1*, *H2* and *H3* than obtained from crosses at the tetraploid level. Alternatively a dihaploid with all three genes could be converted by colchicine treatment into a tetraploid duplex for all three genes and this tetraploid used for breeding.

It has, however, been suggested that the use of dihaploids would be especially advantageous where a desired character is determined by polygenes. Consider for example, a desired character determined by ten unlinked polygenes, then from a cross of parent with these ten polygenes with a parent possessing none, only about 0.1% of the progeny will be expected to have the ten genes. Using a dihaploid with all ten genes could result in about 24% of progeny having the genes.

There are, however, two important conditions which have not been taken into account so far. The first is that the dihaploids with the desired characters can be obtained relatively easily. This should be possible for a small number of major genes but may not be easy for a character controlled by a number of polygenes. A tetraploid parent possessing the polygenes will only give a very small proportion of dihaploids with all the genes. Secondly it is not going to be of any advantage using dihaploids with the required genes or character if these dihaploids have not got many of the other qualities needed in a commercial potato.

Clearly the use of dihaploids to produce successful varieties will need further development. First it will be necessary to produce stocks with pollen fertility which can be used to cross with other dihaploids as they are produced from tetraploid parents with the desired genes. Secondly it will be necessary to have a high fertility in the derived dihaploids with the desired genes, and thirdly the dihaploids must also have good agronomic qualities. Such a programme will take time and need considerable resources.

15.12 Distribution and spread of new varieties

In many countries, Tables 15.6–15.9, there appears to be a reluctance to change from the old, well known varieties to more recently produced varieties with potatoes in contrast to cereals where new varieties spread quickly. At least three reasons for this can be suggested. First, once domestic consumers have become accustomed to the quality and other characteristics of a widely grown variety, there is a strong tendency for such varieties to become the standard of quality. This appears to have

Table 15.6 Popularity of varieties in the U.K. in 1976 (from Potato Marketing Board, provisional figures). Only varieties with more than 2% of the area given

Variety	*% total area*	*Date of introduction*
Pentland Crown	18.7	1959
Maris Piper	13.1	1963
King Edward VII	12.8	1902
Desiree	8.2	1962
Record	7.0	1932
Pentland Dell	6.5	1961
Maris Peer	5.1	1962
Pentland Ivory	3.9	1966
Pentland Hawk	3.6	1965
Home Guard	3.2	1942
Ulster Sceptre	2.6	1962

Majestic (introduced 1911), which was for many years the most widely grown British variety, occupied only 1.8% of the total area in 1976 (see also Table 15.10).

Table 15.7 Popularity of varieties in the German Federal Republic (DBR) in 1975. Data published in *Der Kartoffelbau* (only varieties with more than 2% of total area listed)

Variety	*% of total potato area*	*Date of introduction*
Grata	22.1	1955
Clivia	8.6	1962
Hansa	5.6	1957
Irmgard	4.1	1960
Bodenkraft	3.3	1967
Datura	2.6	1957
Cosima	2.5	1959
Hela	2.2	1964
Saskia	2.2	1946
Désirée	2.2	1962
Sieglinde	2.1	1935
Jetta	2.1	1968
Isola	2.0	1958

been so for King Edward in the U.K. and Bintje in the Netherlands. Neither variety was considered to be of good quality when first introduced but both are now the standards of good quality (Howard, 1974). Secondly, the fact that the ware crop and the seed crop are grown in different districts by different farmers makes it difficult to predict what is going to be the demand for seed of a new variety. There is a need for more and bigger trials in the ware areas so that the merits of new varieties on a farm scale

Table 15.8 Popularity of varieties in the Netherlands in 1975. % of total seed area, varieties with 2% or more (from the 51st Beschrijverde Rassenlijst voor Landbouwgewasen, Wageningen, 1976)

Variety	*% of total area*	*Date of introduction*
Bintje	32.6	1910
Sirtema	5.8	1949
Desiree	5.1	1962
Alpha	4.9	1925
Jaerla	4.8	1967
Spunta	4.4	1968
Eersteling	3.9	1900
Ostara	3.2	1962
Irene	2.2	1954
Climax	2.0	1959

Table 15.9 Popularity of varieties in Canada and U.S.A. in 1975. Percentage of total seed area (entered acreage)—only varieties with more than 2% (from certification section of PAA)

Variety	*% of total area of seed*	*Date of introduction*
Russett Burbank	32.8	Before 1890
Kennebec	19.8	1948
Katahdin	10.1	1935
Superior	6.2	1961
Sebago	5.5	1938
Norchip	5.4	1968
Red Pontiac	3.3	1949
Norgold Russett	3.1	1964
Norland	2.4	1957

can be ascertained and for the dissemination of these results to both seed and ware producers. Thirdly the rate of increase of potatoes is small; it may take $1\frac{1}{4}$ tons of seed acre^{-1} to produce only 6 tons of seed acre^{-1}. This contrasts markedly with cereals where 100 lb acre^{-1} sown could give a yield of at least 3500 lb acre^{-1}. It also explains why the most recently introduced variety in Table 15.6, where the qualification to be included is 2% of the total acreage, is Pentland Ivory introduced in 1966, i.e. 10 years before the census. Similarly in Table 15.7 the corresponding variety is Hela introduced in 1964.

It should also be noted that the positions in the U.K. and the German Federal Republic (Tables 15.6 and 15.7) differ somewhat from those in the Netherlands and the U.S.A. (Tables 15.8 and 15.9). In the latter two

Table 15.10 Changes in popularity of varieties in Great Britain from 1960–74. Data from Potato Marketing Board's reports. Only varieties with over 10 000 acres in any year included in earlies, and over 20 000 in maincrops

(a) Early varieties (first earlies plus second earlies)

	Area grown (in 1000 acres) in							
Variety	*1960*	*1962*	*1964*	*1966*	*1968*	*1970*	*1972*	*1974*
Arran Pilot	58	49	50	26	17	11	6	3
Craigs Royal	18	26	14	8	6	5	3	2
Red Craigs Royal	—	—	16	16	21	17	14	11
Home Guard	22	15	19	18	22	22	17	16
Maris Peer	—	—	—	—	13	30	28	23
Ulster Prince	7	9	11	11	11	11	10	7
Ulster Sceptre	—	—	—	—	—	—	—	11

(b) Maincrop varieties

	Area grown (in 1000 acres) in							
Variety	*1960*	*1962*	*1964*	*1966*	*1968*	*1970*	*1972*	*1974*
Desiree	—	—	—	—	5	16	24	33
Kerrs Pink	24	21	18	12	9	6	4	3
Redskin	34	34	31	29	23	14	7	3
King Edward VII	114	128	117	100	100	90	75	68
Majestic	303	249	267	232	171	109	49	19
Maris Piper	—	—	—	—	1	14	34	48
Pentland Crown	—	—	—	16	52	109	114	87
Pentland Dell	—	—	—	16	67	50	50	42
Record	7	10	25	24	30	34	28	32

countries the most widely grown variety continues to be a very old one – Bintje in the Netherlands introduced in 1910 and Russet Burbank in the U.S.A. introduced before 1890. Also it is only recently, Table 15.10, that the position has changed in the U.K. For many years prior to 1970 the only two maincrop varieties of outstanding importance were Majestic (introduced in 1911) and King Edward (introduced in 1902).

15.12.1 The importance of new varieties

New varieties are needed for several reasons. Very often they are introduced because their yields have been found in trials to be higher than those of the widely grown established varieties. Other new varieties may be of a similar yield to the established but of a better quality. Yet other new varieties have neither increased yield nor improved quality but a higher disease and pest resistance makes their culture easier than that of the standard varieties.

Table 15.11 Yields of first earlies in England and Wales, 1950–1972 (from Agricultural Statistics, M.A.F.F; published by H.M.S.O.)

Year	*Yield (tons acre^{-1})*	*5-year average, median year*	*Year*	*Yield (tons acre^{-1})*	*5-year average, median year*	*Year*	*Yield (tons acre^{-1})*	*5-year average, median year*
1950	4.8	–	1958	5.7	5.8	1966	6.5	6.6
1	4.9	–	9	5.7	5.7	7	6.2	6.8
2	5.2	5.2	1960	6.2	5.8	8	7.5	6.8
3	5.6	5.3	1	5.8	5.9	9	7.1	7.1
4	5.7	5.6	2	5.4	6.0	1970	6.7	7.3
5	5.2	5.6	3	6.2	6.1	1	7.8	7.4
6	6.3	5.6	4	6.3	6.2	2	7.4	–
1957	5.2	5.6	1965	6.7	6.4	3	7.9	

Table 15.12 Yields of second earlies plus maincrops in England and Wales, 1950–72 (from Agricultural Statistics, M.A.F.F; published by H.M.S.O.)

Year	*Yield (tons acre^{-1})*	*5-year average, median year*	*Year*	*Yield (tons acre^{-1})*	*5-year average, median year*	*Year*	*Yield (tons acre^{-1})*	*5-year average, median year*
1950	8.5	–	1958	7.1	8.3	1966	10.7	10.6
1	8.5	–	9	8.9	8.6	7	10.7	10.7
2	8.5	8.6	1960	9.1	9.0	8	10.3	10.8
3	9.3	8.3	1	9.8	9.5	9	10.4	11.1
4	8.4	8.6	2	10.0	9.7	1970	11.8	11.3
5	7.8	8.4	3	9.6	10.1	1	12.3	11.8
6	8.8	8.0	4	9.8	10.3	2	11.9	–
1957	7.7	8.1	1965	11.4	10.6	3	12.7	

Yields of both first earlies (Table 15.11) and second earlies plus maincrops (Table 15.12) in England and Wales have increased markedly during the last 20 years. This has been reported as being due at least in part to the replacement of old varieties by new. This assertion, however, is difficult to prove.

Potato yields fluctuate greatly with the weather conditions, mostly with rainfall during May, June and July (Cox and Large, 1960). Accordingly in Tables 15.11 and 15.12 reference should be concentrated on the five-year averages rather than on the yields in any particular year. For first earlies the five-year average yield for the median year 1957 was 5.6 tons acre^{-1} and for the median year 1966 it was 6.6 tons acre^{-1}. It appears to be still rising. Between 1957 and 1966 the rise was remarkably constant at about 0.1 tons per acre per year. It also appears from inspection of Table 15.10 that this rise in yield acre^{-1} was not due to the replacement of old varieties by new. Throughout the period there was little change in the areas grown of Home Guard and Ulster Prince, the two first early varieties. The other important early varieties in Table 15.10, Arran

Pilot, Craigs Royal (and its budsport Red Craigs Royal) and Maris Peer are second earlies rather than first earlies. Also Maris Peer did not become an important variety until 1968 and the acreage of Arran Pilot did not begin to decline until 1966.

The yields of maincrop potatoes in England and Wales (Table 15.12) have also shown a marked increase in the past 20 years. There are rather large fluctuations from year to year, for example from 7.1 tons $acre^{-1}$ in 1958 to 8.9 tons $acre^{-1}$ in 1959. The years 1957 and 1958 were obviously bad years as markedly higher yields were obtained in six of the seven previous years. However an increase from about 8.5 tons per acre in 1957 to 10.6 tons $acre^{-1}$ in 1966 seems a reasonable estimate of the actual improvement. Such an increase in yield again does not seem to be due to the replacement of old varieties by new, Majestic and King Edward being the dominant varieties for the whole period.

Although it is not possible to show that new varieties have made an important contribution to increased potato yields, the above considerations do not prove that they have not done so. It may be that increased yields before 1968 are due to causes other than improved varieties and that increase after about 1968 are mainly due to new varieties such as Pentland Crown and Maris Piper in the maincrop class. Perhaps the high yields of 1970–3 do reflect the increase in share of the acreage occupied by the high yielding varieties, Desiree, Maris Piper, Pentland Crown and Pentland Dell (the yield in 1974, according to provisional figures, was again higher at over 13 tons $acre^{-1}$).

15.12.2 The distribution of new varieties

The most important factor in obtaining high yields of potatoes is healthy seed, free from virus diseases and from bacteria and fungi carried on or in the tubers. Accordingly the distribution of a new variety is closely bound up with raising healthy seed under the conditions imposed by the certification schemes. As discussed earlier the multiplication rate for potatoes is very low and it is therefore necessary to make arrangements for the multiplication of certified seed stocks as early as possible. It is usually wise to do this as soon as a variety is entered into official trials, even though this may lead to relatively large amounts of expensively produced seed having to be discarded because the variety did not perform well in trials.

The recent change in Scotland and Northern Ireland to the use of the stem-cutting method of producing the highest grade of virus-tested seed from which all other grades are eventually obtained (Hardie, 1971; Calvert, 1973) has tended to delay the amount of seed available when a new variety finishes trials as the method involves going back to a few plants when stocks of several thousand tubers have been produced by the breeder. It is, however, possible to use the original breeders' stock for extended farm-scale trials which should help to ascertain what is the potential demand for the new variety.

The demand for seed of a new variety is likely to be much greater if the merits of the new variety are obvious. They are not likely to be high if the new variety has only a marginal improvement in yield or quality over existing varieties, particularly as with seed certification schemes all old varieties can now be kept in a good state of health. Breeders therefore must concentrate on producing new varieties with either a big improvement in yield or quality or with a marked increase in disease or pest resistance.

References

Baerecke, M-L. (1958). *Handb. Pflanzenzüchtung*, eds. Kappart, H. and Rudorf, W. 2nd Edn. Vol. III, pp. 97–105. Berlin and Hamburg: Paul Parey.
Baerecke, M-L. (1967). *Züchter* **37**, 281–6.
Black, W. (1953). In *The Potato in Health and Disease.* Whitehead, T., McIntosh, T.P. and Findlay, W.M. Ch. 9, pp. 95–111. Edinburgh: Oliver and Boyd.
Black, W. (1960). *Rep. Scott. Pl. Breeding Stn*, 1960, pp. 29–38.
Blomquist, A.W. and Lauer, F.I. (1962). *Am. Potato J.* **39**, 340–3.
Boyd, A.E.W. (1952). *Ann. appl. Biol.* **39**, 339–50.
Brown, E.B. (1969). *Ann. appl. Biol.* **63**, 493–502.
Brown, E.B. (1970). *Ann. appl. Biol.* **65**, 377–83.
Calvert, E.L. (1973). *The Seed Potato : J. nat. Ass. Seed Potato Merchants* **13**, 62–5.
Carson, G.P. and Howard, H.W. (1945). *J. Genet.* **46**, 358–60.
Certification Section of the Potato Association of America (1975). *Am. Potato J.* **52**, 365–72.
Cipar, M.S., Peloquin, S.J. and Hougas, R.W. (1967). *Can. J. Genet. Cytol.* **9**, 511–18.
Cockerham, G. (1970). *Heredity* **25**, 309–48.
Cox, A.E. and Large, E.C. (1960). *Potato blight epidemics throughout the world. USDA, Agric. Handb.* **174**, Washington, D.C., U.S.A.fi
Danert, S. (1956). *Kulturpflanze* **4**, 83–129.
Davidson, W.D. (1935). *J. Dep. Agric. Repub. Ire.* **33**, 1–27.
Davidson, W.D. (1939). *Jl R. agric. Soc.* **100**, 1–22.
Deshmukh, M.G. and Weischer, B. (1970). *Potato Res.* **13**, 129–38.
Diener, T.O. (1971). *Virology* **45**, 411–28.
Dodds, K.S. and Long, D.H. (1955). *J. Genet.* **53**, 136–49.
Dunnett, J.M. (1957). *Euphytica* **6**, 77–89.
Dunnett, J.M. (1961). *Rep. Scott. Pl. Breeding Stn*, 1961, pp. 39–46.
Ellenby, C. (1952). *Nature* **170**, 1016.
Engel, K.-H. (1957). *Züchter* **27**, 98–124.
Fuller, J.M. and Howard, H.W. (1974). *Ann. appl. Biol.* **77**, 121–8.
Gibson, R.W. (1971). *Ann. appl. Biol.* **68**, 245–51.
Glendinning, D.R. (1969). *Eur. Potato J.* **12**, 13–19.
Grun, P. and Aubertin, M. (1965). *Genetics* **51**, 399–409.
Grun, P., Aubertin, M. and Radlow, A. (1962). *Genetics* **47**, 1321–33.
Hardie, J.L. (1971). *The Seed Potato : J. nat. Ass. Seed Potato Merchants* **11**, 4–7.
Hawkes, J.G. (1956). *Proc. Linn. Soc. Lond.* **166**, 97–144.

Heiken, A. (1958). *Acta Agric. scand.* **8**, 319–58.
Heiken, A. (1960). *K. Vetenskapssamallets Uppsala Handl.* (7) 125.
Hougas, R.W. and Peloquin, S.J. (1957). *Nature* **180**, 1209–10.
Howard, H.W. (1958). *Am. Potato J.* **35**, 676–8.
Howard, H.W. (1963a). *Rep. Pl. Breeding Inst., Camb.* 1961–2, 5–21.
Howard, H.W. (1963b). In *The Growth of the Potato*, eds. Ivins, J.G. and Milthorpe, F.L. pp. 292–302. London: Butterworths.
Howard, H.W. (1969). *Eur. Potato J.* **12**, 278–9.
Howard, H.W. (1970). *Genetics of the Potato, Solanum tuberosum.* London: Logos Press Ltd.
Howard, H.W. (1972). *Ann. appl. Biol.* **71**, 263–6.
Howard, H.W. (1973). *Proc. Fifth Triennial Conf. Eur. Ass. Potato Res.* p. 126 (Abs.).
Howard, H.W. (1974). *Potato Res.* **17**, 491–512.
Howard, H.W. (1975). *Potato Res.* **18**, 320–1.
Howard, H.W., Cole, C.S. and Fuller, J.M. (1970). *Euphytica* **19**, 210–16.
Howard, H.W. and Cotten, J. (1977). In Plant Nematology, ed. Southey, J.F. ch. 16 Tech. Bull. (7) M. A. F. F. London:HMSO.
Howard, H.W. and Fuller, J.M. (1971). *Pl. Path.* **20**, 32–5.
Howard, H.W., Johnson, R., Russell, G.E. and Wolfe, M.S. (1970). *Ann. Rep. Pl. Breeding Inst. Camb. 1969*, 6–36.
Hudson, P.S. (1936). *The South American Potatoes and their Breeding Value.* Cambridge: Commonwealth Bureau of Plant Genetics.
Hughes, J.C. and Evans, J.L. (1967). *Eur. Potato J.* **10**, 16–36.
Jones, F.G.W. (1954). *Ann. appl. Biol.* **41**, 348–53.
Kassanis, B. and Varma, A. (1967). *Ann. appl. Biol.* **59**, 447–50.
King, J.R. (1955). *Am. Potato J.* **32**, 389–91.
Kort, J., Ross, H., Rumpenhorst, H.-J. and Stone, A.R. (1977). *Nematologica* **23**, (in press).
Krantz, F.A. (1946). *Tech. Bull. Minnesota agric. exp. Stn* (173) 24.
Langton, F.A. (1972a). *J. agric Sci., Camb.* **78**, 75–81.
Langton, F.A. (1972b). *Potato Res.* **15**, 266–9.
Lapwood, D.H. (1971). *Ann. appl. Biol.* **68**, 41–53.
Linehan, D.J. and Hughes, J.C. (1969). *J. Sci. Fd Agric.* **20**, 110–12.
Lugt, C. (1961). *Proc. First Triennial Conf. Eur. Ass. Potato Res.* 321–3.
Maris, B. (1962). *Versl. landbouwk. Onderz. RijkslandbProefstn.* (68.8) 208.
Moller, K.-H. (1956). *Züchter* **26**, 243–8.
Moller, K.-H. and Vogel, J. (1961). *Züchter* **31**, 265–7.
Niederhauser, J.S. and Cobb, W.C. (1959). *Scient. Am.* **200**, 100–12.
Nirula, K.K., Khushu, C.L. and Raj, B.T. (1969). *Am. Potato J.* **46**, 251–3.
Noble, M. and Glynne, M.D. (1970). *FAO Pl. Protection Bull.* **18**, 125–35.
Nollen, R.M. and Mulder, A. (1969). *Proc. Fifth Br. Insecticide Fungicide Conf.* 671–4.
Patterson, C.F. (1953). *Am. Potato J.* **30**, 89–91.
Peloquin, S.J. and Hougas, R.W. (1959). *Eur. Potato J.* **2**, 176–83.
Plank, J.E. van der (1963). *Disease Resistance in Plants.* New York: Academic Press.
Puskarnath (1969). *Potato in India : Varieties.* New Delhi: Indian Council of Agricultural Research.
Radcliffe, E.B. and Lauer, F.I. (1971). An appraisal of aphid-resistant tuber-bearing *Solanum* germplasm. *Tech. Bull.* (286) *agric. exp. Stn, Univ. Minnesota.*
Richardson, D.E. (1970). *J. nat. Inst. agric. Bot.* **12**, 112–18.

Robertson, I.M. (1957). *Scott. Agric.* **37**, 89–91.
Robinson, R.A. (1969). *Rev. appl. Mycol.* **48**, 593–606.
Ross, H. and Huijsman, C.A. (1969). *Züchter (Theor. appl. Genet.)* **39**, 113–22.
Ross, H. and Langton, F.A. (1974). *Nature* **247**, 378–9.
Rothacker, D. (1957). *Züchter* **27**, 232–8.
Rowe, P.R., Sequeira, L. and Gonzalez, L.C. (1972). *Phytopathology* **62**, 1093–4.
Salaman, R.N. (1926). *Potato Varieties.* Cambridge: University Press.
Salaman, R.N. (1949). *The History and Social Influence of the Potato.* Cambridge: University Press.
Schwarze, P. (1963). *Züchter* **33**, 275–81.
Simmonds, N.W. (1962). *Biol. Rev.* **37**, 422–65.
Simmonds, N.W. (1963). *Eur. Potato J.* **6**, 45–60, 69–76.
Simmonds, N.W. (1964). *Heredity* **19**, 489–504.
Simmonds, N.W. (1969). *Forty-eighth Ann. Report, 1968–9, Scott. Breeding Stn,* 18–38.
Steineck, O. (1957). *Bodenkultur* **9**, 263–74.
Stevenson, F.J. and Clark, C.F. (1937). In *Yearbook of Agriculture, U.S. Department of Agriculture,* 1937, pp. 405–44.
Stone, A.R. (1972). *Nematologica* **18**, 591–606.
Stuart, W. (1937). *The Potato: Its Culture, Uses, History and Classification.* Chicago, Philadelphia and New York: J.B. Lippincott Company.
Talburt, A. and Smith, O. (1967). *Potato Processing.* Westport Connecticut: Avi Publishing Company.
Thijn, G.A. (1954). *Euphytica* **3**, 28–34.
Toxopeus, H.J. (1952). *Euphytica* **1**, 133–9.
Ugent, D. (1968). *Econ. Bot.* **22**, 108–23.
Warren, D.S. and Woodman, J.S. (1974). *J. Sci. Fd Agric.* **25**, 129–38.
Wiersema, H.T. (1950). *Studiekring voor Plantenveredeling, 23rd Nov. 1950, Wageningen,* pp. 338–41.
Williams, T.D. (1958). *Proc. Linn. Soc. Lond.* **169**, 93–104.
Zitnak, A. and Johnston, G.R. (1970). *Am. Potato J.* **47**, 256–60.

Chapter 16

Potato production in the context of world and farm economy

G.E. DALTON

16.1 Introduction

The potato is a relatively stable part of the diet of European and North American people. As incomes have risen, the consumption of potatoes per head has remained constant or declined slowly. At the same time, history as well as household food expenditure surveys, show that the poorer section of the community obtain a greater proportion of their calories from potatoes than do richer people. Indeed, the ability of the potato to support a large number of people on a small acreage of land is a means of withstanding population pressure. In prosperous times accompanied by yield increasing technology then the acreage of potato shrinks, leaving land for use in some more profitable way. At the same time, capital intensive technology, which is both a cause of and a response to a reduction in the farm labour force, produces larger and more specialized production units. The same phenomenon can also be observed in processing units as higher incomes increase the demand for more convenience foods so that the potato has more services added to it before it reaches the housewife.

The prospect of at least an interruption in the growth of incomes in the Western world and even a decline in incomes accompanied by shortages of non-renewable forms of energy, causes us to consider the economics of the potato in a light other than adjustments to a more energy intensive agriculture against a backcloth of rising incomes. Now there is a strategic argument for considering the role of the potato 'should the worst happen' and our interest is focussed again on the relative technical efficiency of the potato in its use of scarce resources. The potato is not yet grown widely in areas of the world where hunger and malnutrition are already realities and given its bulk, even though it is cheap food, trade in potatoes will not be a means of directly alleviating the problems. Fortunately roots and tubers are an important source of food in the tropics so that lessons learned, both technical and economic, from our study of the potato may be of indirect use in improving the productivity of these alternative crops. It is perhaps a sad commentary on our state of mind in the Western world that our chief interest in manioc (cassava) has been until recently as a cheaper source of energy for feeding our animals.

It is against this background that the first part of this paper reviews the trends in world production, trade and utilisation of potatoes and other root crops. The likelihood of these trends continuing into the future is also considered in the light of knowledge about the nature of changes in demand and supply in the potato market. Adjustments to the market and new technology come about by the collective response of many individual potato producers. The nature of some of these decisions is examined in the second part of the chapter.

16.2 World trends

16.2.1 World production

According to statistics on world production (FAO, 1975a), the production of potatoes has been more or less static in recent years at around 300×10^6t. This is in marked contrast to the world production of cereals which has increased from $676\,072 \times 10^6$ t in 1961–5 to 1334×10^6 t in 1974. The production of other roots and tubers has increased slightly but the reliability of the statistics is probably not as good as those for potatoes. World production of sweet potatoes is estimated to have risen from an average of 108×10^6 t in 1961–5 to 134×10^6 t in 1974 (most of which were produced in China); over the same period world cassava production levels have increased from 76 to 105×10^6 t. The production of yams and cocoyams has remained roughly constant at around $3–4 \times 10^6$ and 18×10^6 t respectively.

Analysis of production data on a regional basis shows that potatoes are mainly a European crop (see Table 16.1). The U.S.S.R. is the world's major producer. Within Europe, using 1974 figures, the largest producer, Poland, produces over 50×10^6 t per annum followed by West Germany (14.5×10^6 t); East Germany 11.4×10^6 t); France 7.6×10^6 t) and the Netherlands and the U.K. (5.6 and 6.8×10^6 t respectively). In Asia, China is estimated to produce 38×10^6 t. Production in the U.S.A. is about 15×10^6 t while the dominant producers in South America are Argentina (1.8×10^6 t), Brazil (1.7), Columbia (1.1) and Peru (1.1).

The relative importance of the world's major starchy roots is shown in Table 16.2 on a regional basis. These data do not of course bring out the importance of a particular crop in a particular locality or changes that are occurring. Potato production, for example, is increasing in India,

Table 16.1 Regional production of potatoes

	1961–5			1970–4		
	Area × 10^6 *(ha)*	*Yield* *(kg ha^{-1})*	*Production* × 10^6 *(t)*	*Area* × 10^6 *(ha)*	*Yield* *(kg ha^{-1})*	*Production* × 10^6 *(t)*
World	23.733	11 927	283.074	22.225	13 475	299.497
Africa	0.269	7 537	2.024	0.378	8 121	3.075
North and Central America	0.742	20 192	14.990	0.729	23 579	17.178
South America	0.977	7 176	7.010	1.007	8 438	8.479
Asia excluding U.S.S.R.	4.403	8 707	38.334	5.163	9 604	49.58
Oceania	0.052	15 560	0.805	0.050	20 456	1.018
Europe excluding U.S.S.R.	8.653	15 981	138.283	6.911	18 675	128.939
U.S.S.R.	8.638	9 449	81.628	7.987	11 419	91.224

From FAO (1975a).

Table 16.2 Production of main root and tuber crops on a regional basis 1974 (000 t)

	Western Europe	*U.S.S.R.*	*North America*	*Latin America*	*Near East*	*Far East*	*China*	*Oceania*	*Africa*	*World*
Potatoes	56 937	80 700	17 869	8 826	4 054	6 677	38 027	903	1 844	293 724
Sweet Potatoes	74	–	606	3 487	570	7 944	113 100	9	5 577	134 225
Manioc	–	–	–	34 802	1 110	21 221	330	217	47 541	104 891
Cocoyam	–	–	–	–	52	90	20	257	3 397	4 352
Yam	–	–	–	277	128	30	–	187	18 643	19 137

From FAO (1975a).

especially in the states of Uttar Pradesh and West Bengal. They do show, however, the importance of the potato in the U.S.S.R. and Europe, the importance of sweet potatoes in China, of cassava in Latin America, Africa and the Far East and yams and cocoyams in Africa.

Referring to Table 16.1 it can be seen that the world production of potatoes has increased slightly over the last decade or so, mainly in the U.S.S.R. and Asia. This increased production has been achieved by higher yields on a smaller area. The world production of two of the other main root crops, sweet potatoes and cassava has increased over the same period by a much greater proportionate amount than potatoes. This increase has occurred in all the major producing regions as shown in Table 16.3. Increasing yields have been responsible for some of the rise in production but, unlike the case of potatoes, the area devoted to their production has been enlarged.

The comparison of yields in Tables 16.1 and 16.3 illustrates that for particular crops there are wide differences between regions; the highest yields of potatoes are obtained in North and Central America, Oceania and Europe, the lowest in Africa; Africa also has the lowest yields of cassava and sweet potatoes while Latin America has high relative yields of cassava. Such yields do not, however, mean very much as they are so global but they perhaps illustrate the potential contribution that technology might make to raising yields.

16.2.2 Trade

Trade can modify regional consumption patterns, but most starchy roots and tubers are so bulky and/or perishable that very little trade is carried on in proportion to production. Trade in general takes place for high value items such as for a luxury food market, in processed form as starch or meal and for seed and occasionally as ware when shortages arise. Thus, a certain amount of dried cassava products are imported from Thailand into West Germany and the Netherlands and cassava starch from Brazil is imported into the U.S.A. and Japan.

During 1974, some 3.80×10^6 t of potatoes were imported into the various countries of the world (FAO, 1975b). This amounts to only 1.27% of total world production in 1974.

Most of the trade is confined to Western Europe (2.69×10^6 t in 1974) although most countries import small amounts. In Western Europe a large proportion of trade is based on early potatoes which are shipped from Mediterranean areas to Northern European countries; although, in times of shortage, maincrop potatoes are exported, the best example of this being a substantial trade between European countries and Argentina in 1972 and Russian imports from other Comecon countries (notably Poland) in the same year. The seed potato trade is dominated by the

Table 16.3 Changes in area, yields and production in major producing regions of the world for sweet potatoes and cassava

Region/crop	*1961–5*			*1970–4*		
	Area $\times 10^6$ *(ha)*	*Yield* *(kg ha^{-1})*	*Production* $\times 10^6$ *(t)*	*Area* $\times 10^6$ *(ha)*	*Yield* *(kg ha^{-1})*	*Production* $\times 10^6$ *(t)*
Sweet potato						
World	12.977	8 320	107.964	15.905	8 304	130.980
Latin America	0.357	8 448	3.018	0.407	9 425	3.841
China	9.865	8 514	83.988	11.857	9 192	109.012
Far East	1.041	7 616	7.926	1.058	7 978	8.441
Africa	0.904	4 604	4.164	1.239	4 312	5.334
Cassava						
World	8.717	8 580	74.786	11.180	9 101	101.459
Latin America	2.027	12 704	25.754	2.649	13 578	35.958
Far East	2.145	8 072	17.312	2.293	9 285	21.299
Africa	4.404	6 924	30.491	5.970	7 108	42.431

From FAO (1975a).

Table 16.4 Dutch potato exports (000 t)

	1960–1	*1969–70*	*1970–1*	*1971–2*	*1972–3*
Ware	206	753	777	862	1107
Feed	19	4	6	3	3
Seed	223	278	290	277	328
Starch	883	1723	1555	1954	2483
Total	1331	2758	2628	3096	3921

From Wye (1975) and Agra Europe (1972a).

Netherlands. For example in 1971, the Netherlands exported 277 375 t of seed to 64 countries of which 24 were in Europe, 17 in Africa, 13 in Asia and 10 in America (Agra Europe, 1972b). Indeed the Netherlands is the most important exporter of potatoes in the world, and a breakdown of her exports is given in Table 16.4.

Trade in new potatoes is a fascinating activity. Prices are extremely volatile and depend on weather conditions, carry over stocks of maincrop potatoes, income levels in the consuming countries and import regulations. Lifting begins around February–March in North Africa, including Egypt, Tunisia, Algeria and Morocco and passes northwards through Italy and Spain as the season progresses. Particular regions in individual countries specialize in early potato production, for example Campania Puglia and Sicily in Italy, the Canary Islands, Malaga and Valencia in Spain, the Channel Islands, Cornwall, Pembroke and S.W. Scotland in the U.K. the Palatinate area of Germany where early potatoes are grown under foil, along with areas in the Rhine Main triangle, the upper Rhine plain and the lower Rhine and Hanover, and around Lake Geneva in Switzerland.

There are also strong ties between particular regions and importing countries which specialize in producing those potatoes best suited to the market (e.g. white-fleshed potatoes for the British market and yellow-fleshed oval potatoes for the German market). Morocco is the main source of French supplies, Italy supplies the bulk of West German requirements and Cyprus is the main supplier to the U.K. along with the Channel Islands, while Malta has close trade links with the Netherlands.

E.E.C. production and sources of new potatoes are shown in Table 16.5.

16.2.3 World utilization

According to the FAO Agricultural Commodity Projections (1971), roots and tubers supply some 184 cal per head of the population in the world as a whole or 7.8% of cal intake. This compares with 52.4% of cal

Table 16.5 Production and sources of early potatoes 1973 (000 t)

	Germany	*France*	*Italy*	*Netherlands*	*Belgium*	*U.K.*
Home production	833	524	305	—	114	470
Imports from:						
Italy	116	7	—	3	4	0.7
Spain	18	30.6	—	—	0.3	50
Morocco	4	78	—	0.9	0.4	—
Egypt	0.8	4.8	—	1.31	0.1	54
Cyprus	2.3	0.6	—	1.7	0.8	114

From Agra Europe (1975b).

derived from cereals. Roots and tubers only supply 2.8 g of protein per head per day of the world's population or 4.3% of the total intake, whereas cereals supply 31.1 g of protein per head per day or 47% of the total protein supply.

Starchy foods are less preferred foods, so that in countries where incomes have risen, the tendency has been for average consumption to decline. At the same time, in the case of potatoes, higher incomes have brought about a situation where more services in the form of cleaning, packaging and processing are added to the product before the point of final sale. Starchy roots are in some cases used for animal feed and also for industrial uses such as the manufacture of starch and alcohol. It must also be remembered that a good deal of waste takes place before utilization in most root crops due to disease both in the ground and during storage and also due to too large or too small tubers. Similarly in the preparation process peelings can represent a significant proportion of the crop. Finally, with potatoes and yams a high proportion of the crop is required for 'seed'.

Data from the OECD countries (OECD, 1973) demonstrate how per capita consumption levels have fallen (see Fig. 16.1) in terms of cal per head per day. The same sort of trend has occurred in the U.S.S.R., where consumption has dropped from 241 kg per head per annum in 1950 to 130 kg in 1970 (Agra Europe, 1974a). The same sort of trend has also occurred in Poland when over the period 1950–70, consumption per head has dropped from 270 to 190 kg, a fall of 69.3% (Polish Ministry of Agriculture, 1974). Whereas potato consumption has been falling, the amount of calories in the diet has increased slightly since 1953, so that the proportion of calories supplied by potatoes has declined the more so (see Table 16.6). The data in Table 16.6 also show that in most countries within the OECD, potatoes make a relatively small contribution to the total calorie intake.

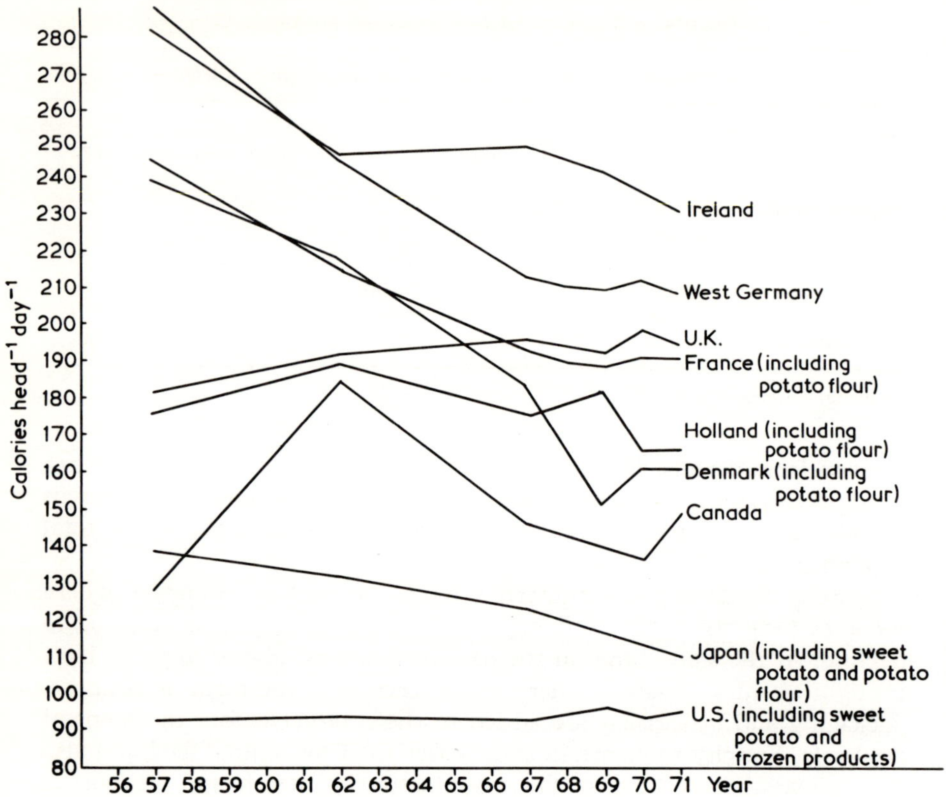

Fig. 16.1 Changes in consumption levels of potato and potato products in some OECD countries (from OECD, 1973)

This is not, however, a case for arguing that potatoes are not an important crop. Potatoes were relatively more important when income levels were lower and are still a more important part of the diets of poorer people. Thus, for example the Household Food Consumption and Expenditure surveys in the U.K. (H.M.S.O., 1973) show that among the highest income group (A1) the proportion of energy derived from carbohydrate is less than that from fat, due to increased consumption of animal products, whereas among the lowest income group (DI) carbohydrates are still the most important source of energy (see Table 16.7).

It is quite probable that the rapid developments that have taken place in the potato processing industry have maintained per capita consumption levels. In all western economies processed potatoes in various forms have become a common part of the diet. For example, in the U.K. over the period 1966–70, purchases of potato products including crisps, canned potatoes and dehydrated potatoes doubled (H.M.S.O., 1973). Most of the processed potatoes during this period in Great Britain were eaten

Table 16.6 Proportion of calories derived from the potato and other root crops in OECD countries

	1955–9	*1960–4*	*1965–9*	*1969*	*1970*	*1971*
	%	%	%	%	%	%
Austria	5.69	4.93	4.35	4.15	3.89	3.64
Belgium/Luxembourg	8.81	7.99	7.16	6.77	6.52	6.39
Canada	4.29	6.24	4.77	4.48	4.35	4.63
Denmark	6.49	6.11	5.10	4.20	4.40	4.40
Finland	6.25	6.92	6.50	6.14	5.80	5.75
France	7.86	6.79	6.04	5.94	5.76	5.78
West Germany	9.17	7.88	6.80	6.60	6.20	6.10
Greece	2.74	2.82	—	—	—	—
Ireland	8.27	7.86	7.36	7.14	6.88	6.77
Italy	3.27	3.25	2.93	2.94	2.58	2.42
Japan	6.29	5.58	4.98	4.66	4.47	4.36
Netherlands	5.55	5.80	5.32	5.18	5.01	4.96
New Zealand	3.08	3.49	3.92	4.50	4.50	—
Norway	6.65	6.63	6.36	6.34	6.34	6.36
Portugal	8.17	7.39	6.90	6.79	7.29	—
Spain	9.01	8.45	7.79	8.21	8.19	7.73
Sweden	6.30	5.53	5.49	5.88	5.78	5.73
Switzerland	4.46	4.04	3.59	3.51	3.62	—
U.K.	5.76	5.99	6.23	6.11	6.31	6.22
U.S.A.	3.11	2.97	2.88	2.88	2.83	2.84
Yugoslavia	4.29	4.21	3.89	3.84	3.82	3.90

From OECD (1973).

Table 16.7 Consumption of potatoes by income group in the U.K.

	ozs/person/week		
*Income group**	*A*	*B*	*C and D1*
1970	38.96	46.84	48.09
1971	29.93	45.85	48.06

From H.M.S.O. (1973).

*Gross income	1970	1971
Group A	> 40 (£/week for head of household)	> 45 (£/week for head of household)
Group B	23–40	27–45
Group C and D1	< 23	< 27

in the form of crisps, approximately 285 000 t (on a raw potato equivalent basis) being produced in 1965–6 out of a total processed quantity of 380 000 t and in 1973–4 almost 480 000 t out of a total usage for processing of 1 000 000 t. By 1973 the production of frozen and par-fried potatoes were using almost 400 000 t of fresh potatoes as compared with perhaps 20 000 t in 1965–6. In 1973–4, 20% of the home crop was processed as opposed to 10% in 1968–9 (Wye, 1975). Similar changes have been taking place in other European countries and in the U.S.A. although there is some specialization of product on a country basis (Commission de Communautes Europeenes, 1974).

Table 16.8 shows the extent of E.E.C. potato processing in 1971, which in turn reflects differences in taste, although the Netherlands is an important exporter of such products, as shown in Table 16.9.

The most important industrial use for potatoes is for starch production and the Netherlands is by far the most important producer of starch, supplying up to 70% of the E.E.C.'s starch needs (Agra Europe, 1971b). For example, in 1974 (Agra Europe, 1974c) 71 698 ha (45%) out of a total Dutch potato area of 158 271 ha supplied the starch market, whereas in West Germany about 3% of the total crop is made into starch which is roughly the same amount used for distilling and dry feed (Agra Europe, 1971a).

Seed requirements vary with the yields obtained, as shown in Table 16.10, but generally is something less than 10% of the total yield. The other major use, that of stock feed, is most important in West Germany, France and other continental countries, whereas in the U.K. stock feed potatoes only arise when there is a surplus of potatoes for human consumption.

16.2.4 The nature of the potato market

The potato market has been dominated in recent years by the need to match declining per capita consumption levels with increased yields per unit area. In the absence of any rapid increase in population, the solution to the problem has been to reduce the area of potatoes grown. This process has been reinforced over the same period by the fact that average income levels have risen, which has reinforced the tendency for the consumption of potatoes to fall, as potatoes are regarded as an inferior food. Various estimates by country of levels of income elasticities of demand* which have been used in OECD projection studies (see Table 16.11) demonstrate the point, the only exception to the overall trend being Ireland.

Projections as to levels of future demand carried out by OECD (OECD, 1973) and FAO (FAO, 1971) do not predict any shortage of potatoes by 1985. Indeed for Western Europe as a whole, the FAO predicts a slight falling off in the demand for starchy roots by 1985. However, these projections do assume a continued growth in the gross domestic product

*Relative change in quantity demanded/relative change in income.

Table 16.8 EEC production and processing of potatoes and potato products 1971 (000 t)

	Belgium	*Netherlands*	*West Germany*	*France*	*U.K.*	*Italy*
Home grown potatoes						
Processed	25	290	877	234	719	40
As % of total crop	2.1	13.2	9	7	14	1.8
Production						
Dehydrated products	1.7	11.5	71.5	22.1	11.4	–
Pre-cooked and deep frozen chips	2.7	73.3	126.7	5.3	92.7	–
Canned potatoes	3.9	2	9.2	1.5	30.6	–
Crisps	0.7	14	15.8	17	95.3	8.5

From Commission des Communautes Europeennes (1974).

Table 16.9 Dutch export of potato products (000 kg)

Destination	*1967*	*1968*	*1969*	*1970*
West Germany	12 475	14 557	19 637	24 281
Belgium/Luxembourg	1 813	1 994	2 092	2 449
U.K.	831	1 226	1 163	1 032
Sweden	122	530	218	348
Other Countries	1 719	953	1 380	1 125
	16 960	19 260	24 490	29 235

From Dutch Information Centre for Potatoes (1971).

Table 16.10 Utilization of potatoes: E.E.C. and Great Britain, percentages of total tonnage used*

	Year	*Stock feed* †	*Seed*	*Industrial* ‡	*Other uses*	*Human consumption*
West Germany	1967–8	53	8	3	1	34
	1968–9	51	8	3	2	36
	1969–70	45	9	4	2	39
	1970–1	44	9	4	2	39
France	1967–8	30	10	4	10	47
	1968–9	27	9	5	9	50
	1969–70	21	10	7	7	55
	1970–1	23	9	6	7	56
Italy	1967–8	18	15	1	9	58
	1968–9	18	14	§	9	59
	1969–70	18	15	§	9	58
	1970–1	18	13	§	9	59
Netherlands	1967–8	23	7	40	3	28
	1968–9	19	7	45	2	27
	1969–70	13	9	47	1	29
	1970–1	20	7	48	2	24
Belgium/Luxembourg	1967–8	23	5	§	12	60
	1968–9	16	5	§	9	70
	1969–70	10	6	§	10	74
	1970–1	20	4	§	7	68

E.E.C. total	1967–8	39	9	7	5	40
	1968–9	36	9	8	5	43
	1969–70	31	10	9	4	46
	1970–1	31	9	10	4	45
U.K.	1967–8	4	12	§	15	69
	1968–9	2	11	§	12	75
	1969–70	–	13	§	8	79
	1970–1	9	9	§	16	65

*Because of rounding, component figures do not necessarily add to 100%
†U.K. stock feed only arises in years when the Potato Marketing Board institutes a buying programme
‡Netherlands: mainly starch
§Less than 0.5 %

From NEDO (1972), E.E.C. (1973) and MAFF (1974).

Table 16.11 Estimated income elasticities of demand for selected OECD countries

U.K.	*Netherlands*	*Japan*	*Sweden*	*Germany*	
− 0.2	− 0.6	− 0.2	− 0.4	− 0.8	
France	*Australia*	*Spain*	*U.S.A.*	*Ireland*	*Denmark*
− 0.6	0.0	− 0.15	− 0.4	+ 0.4	− 0.4

From OECD (1968).

of Western European countries which at the time of writing is not as certain as it was hitherto.

At a more detailed level, the management of the potato market is a more complicated operation. Firstly it has to be remembered that potatoes are not a homogeneous product. There are considerable quality differences between varieties which make varieties specific for particular purposes, e.g. ware, earlies, processing. Secondly, quality characteristics other than variety help to determine price and various attempts have been made to grade potatoes so as to reduce marketing costs and extract the maximum possible price for the product from the market. A study by the Potato Marketing Board in the U.K. concluded that the public is willing to pay previously unsuspected prices and premiums for quality. The housewife does not like small potatoes, nor does she like carrying them long distances on a shopping expedition and she prefers good quality potatoes. Quality is defined as clean, well dressed potatoes with minimal damage, disease and blemish. This fact has resulted in the growth of washing and prepacking plants which also provide potatoes in the form required by self-service retail outlets such as supermarkets. 20% of potatoes in the U.K. were sold in prepacked form in 1970 (H.M.S.O., 1973).

Another recent trend in the potato market has been that of bulk purchases (14 lb or more). 37% of the respondents in a recent Potato Marketing Board survey (1973) bought maincrop potatoes in this way. Bulk purchases tend to be cheaper but not washed. The same survey also pointed out that 'new' potatoes tend to be regarded as a different vegetable from old potatoes.

The income elasticity of quantity purchased for processed potato products, unlike that for ware potatoes, is positive. In 1971 the U.K. household food survey estimate (H.M.S.O., 1973) for non-quick-frozen potato products was 0.21. This estimate indicates something of the revolution which has taken place in the marketing of processed potato products, illustrated by Bevan (1974) with reference to the growth in the U.K. market for potato crisps. In 1966 almost £1 000 000 was spent on media advertising on crisps in the U.K. in the wake of the adoption of large scale continuous processing facilities, the entry of a new firm into the market, the adoption of packaging techniques which increased their shelf life from 3 days to 6 months and a concentration on the household crisp market rather than the casual snack market. This led to an estimated growth in U.K. crisp sales over the period 1960–73 from 130 000 to 400 000 t (in weight of potatoes).

Short-term marketing problems (within season) also arise due to an unfortunate combination of an inelastic price demand and supply for potatoes. This tends to produce wild oscillations in prices and hence producer incomes as well as creating difficulties of using the price mechanism as a signal for the level of future supplies. However, there is no evidence to suggest that the price mechanism is failing to bring about structural change in the industry even though, for example in the E.E.C. as a whole, the average price of potatoes in 1971 was approximately half the price of potatoes in 1970 and 1972 (E.E.C., 1973) and in the 1975–6 season potato prices have quadrupled. There is also some evidence for cyclical movements in potato prices (Cohen, 1934; Ingersent, 1962) due to the fact that there is a lag in farmers' ability to respond to prices and that decisions are based on expected prices which are often too biased by more recent events. Shaw (1971) in his study of East Anglian potato production demonstrates that supplies in terms of acres are extremely sensitive to small changes in price around the equilibrium price. All these factors have resulted in various institutions being set up in many free market economies to regulate supplies and prices, the best example of which is the Potato Marketing Board in the U.K. The Potato Marketing Board has controlled supplies by means of acreage quotas and more directly by controlling the size limits of ware potatoes. It also has the power to support market prices by intervention buying and selling of surplus potatoes on the ware market for animal feeding. Similar institutions on the continent do not have such wide powers as the Potato Marketing Board but the 'Produktschap Voot Aardappelen' in the Netherlands and SNIPOT in France do undertake buying activities as well as promoting grading and quality control. Prices are also maintained by guarantee payments in the U.K. and in most countries by import levies and controls in times of surplus, especially for the new potato market. Potatoes do not come under the CAP.

In most free markets the marketing of potatoes is carried out by merchants. However the traditional farmer, merchant, wholesaler, retailer marketing chain is undergoing a rapid transformation. In the U.K. for example, Gibbons (1970) estimated that as long ago as the late 1960s eight merchandising enterprises handled over 20% of sales. Processing firms also tend to be large and this fact together with the need to observe strict control over all stages of production means that many farmers now have contracts direct with processors. The increasing demand for prepacked potatoes has also produced a move towards farmer potato groups. These are perhaps least well developed in the U.K. whereas in France, according to Butterwick and Rolfe (1971), producer groups now control most of the marketing of new and seed potatoes. Producer groups are also particularly strong in the Netherlands, being well established in the potato starch industry and also in seed potatoes (Agra Europe, 1971b, 1974b).

The situation of continual readjustment in the potato market between constant or declining demand and potentially greater supplies through increased yields contrasts sharply with the main tropical root crop, cassava. Cassava is a relatively much more important source of calories in the diet; in Africa it is estimated that cassava provides 38.6% of the total calorie requirements, in Latin America 11.7% and the Far East 6.7% (Coursey and Haynes, 1970). Moreover, estimates of the income elasticity of demand are higher than those for potatoes which together with the greater population growth rates in tropical regions means that demand is projected to increase substantially in the next decade. It is estimated by Phillips (1974) that by 1980, 500×10^6 people might well obtain half their calories from cassava. Some countries could well develop deficits in cassava production such as Columbia, Indonesia, Philippines, Vietnam, Nigeria, Sierra Leone and Zaire, while other countries could well develop exportable surpluses. It would therefore seem desirable for trade in cassava products to develop between countries in tropical regions. However, at the present time the main cassava trade is to industrialized countries where cassava is used for starch production and as animal feed. As such cassava could be regarded as a substitute for potatoes.

The main importers of cassava for starch production are the U.S.A., Canada and Japan and it would appear that, provided political constraints are removed, cassava starch faces a growing market in these countries, especially in Japan where maize starch is not such a strong competitor. Cassava starch is preferred for newsprint and cardboard production as well as for glues, for stamps and envelopes and for some food preparations. Thus Phillips (1974) projects a 1980 cassava starch demand level of between 499 000 and 1 846 000 t of fresh roots for these three countries. A much bigger potential market for cassava products exists in the European animal feed market, although this market is also subject to the uncertainty of political barriers to trade being erected. The market has been growing in recent years; in 1963 Europe imported 413 000 t, mainly from Thailand and this had grown to 1 900 000 t in 1973. The Netherlands and Germany are the most important importing countries, the trade having been stimulated by a growth in the animal feed compounding industry, artificially high cereal prices as a result of agricultural price support policies and a remarkable development of production processing and trading facilities in Thailand, the main supplier. Phillips (1974) analyses the compound feed utilization figures in the enlarged E.E.C. and on this basis expects the demand for cassava by 1980 may be from 246% to 634% greater than the 1970 demand. Such wide limits indicate that political factors make the market particularly insecure as well as price relativity movements and difficulties in forecasting the consumption of animal feed. Should this growth occur it could well be that potato products might also find their way into compounded feeds especially if surpluses occur if yields per unit area go on rising, but the substantial drying costs for potatoes has so far prevented this.

16.3 Farm economics

The potato grower is faced with a number of economic decisions. Firstly, he has to decide whether to grow potatoes at all and if so, for what type of market. Secondly, assuming he decides to grow potatoes, he has then to decide on such details as the area to be devoted to them and the appropriate method of growing them. The latter includes such decisions as the level of variable inputs to apply (seed, fertilizer, sprays), the time of operations such as lifting and whether or not to store or sell and the degree of mechanization. The solutions to all these decisions partially depend on the restrictions under which the farmer operates, such as the physical environment of the farm and the availability of labour and capital. The other major factors in these decisions are (a) the costs and rewards of pursuing a particular course of action and (b) the degree of risk and/or uncertainty attached to each alternative.

16.3.1 The choice of potatoes as a farm enterprise

The choice of potatoes as a farm enterprise depends on their anticipated profitability relative to other potential farm enterprises. It is unlikely that potatoes are complementary to any significant economic extent with other farm enterprises. Profitability needs to be defined in terms of the utilization of the most limiting resource found on the farm which need not necessarily be land. Indeed, the most binding constraints on many farms growing potatoes is the availability of labour in the autumn. The absolute profitability of potatoes depends almost entirely on the yield obtained. This is because it costs almost the same to grow a good crop as it does a poor one. Further, high yields have the effect of spreading the fixed costs so that the cost per unit of output are thereby reduced. The fixed costs of potato growing are today high relative to other crops because of the investment in machinery, buildings, handling equipment and even irrigation facilities. Potatoes are still on many farms a labour intensive crop despite advances in mechanizing various operations. The level of yield is very much dependent on the use of variable inputs and as yields have increased so have the levels of inputs. Potatoes are thus a high cost enterprise both in terms of fixed and variable costs and profits therefore depend on high yields being achieved. The inherent soil fertility has a considerable effect on potato yields which explains the movement of potatoes in recent years on to particular types of soil. Shaw (1971), for example, demonstrates that potatoes are a more stable enterprise on silt and skirt soils, where yields are higher in East Anglia in England, whereas in Holland there is a movement away from sandy soils to heavier soils (Agra Europe, 1974c). The need to avoid clods in mechanical harvesting also favours particular soil types.

It is essential to apply the concept of the opportunity cost of resources in order to understand the place of potatoes in the economy of the whole farm. For example, about 50% of the labour required by potatoes is at harvest; thus if a permanent labour force is kept solely to overcome this labour bottleneck costs are higher than if some alternative work is available at other times of the year. This means that potatoes on a mainly cereal farm are at a disadvantage since cereals also have an autumn labour peak. Thus from the point of view of labour and also machine use a careful selection of enterprises is necessary. Perhaps the best way to ascertain the costs of resources for which potatoes compete is to calculate the amounts saved were the potato enterprise dispensed with.

The acreage of potatoes grown is also a function of the area of the total farm, because of the need to maintain a given level of income within rotational restrictions and the advantages of scale in spreading the costs of indivisible items of equipment. The small farmer because of rotational restrictions finds it difficult to use the potentially high earning power of potatoes, simply because he cannot grow enough. At most, judging from Holland which has 23% of its arable acres in potatoes (van der Zaag, 1973), something like a quarter of the farm can be in potatoes. Further difficulties also arise with labour in that there is insufficient opportunity to find alternative work for regular workers, so the farmer has to rely heavily on casual labour which is increasingly difficult to find. Mechanization is not an easy option for the small farmer, especially as the size and costs of machines increase in so far as he has insufficient acres to utilize them to anything near capacity. The only alternative would be to use contractors or to co-operate with neighbours but both types of arrangement are probably more successful on the larger rather than the smaller farm (A.D.A.S., 1970). Finally, the small farmer specializing in potatoes is in a much more risky position than the larger farmer in so far as a greater proportion of his income is derived from one crop.

It is on the larger farm that there is sufficient flexibility to justify large machinery, (at least 20 ha) and expensive capital projects such as indoor storage/irrigation. Rotational opportunities are also available to grow field vegetable crops which utilize men and machinery at other times of the year. The evidence for all these forces at work is seen in all potato growing areas of the world where there is a trend towards fewer and larger potato producing units. For example, in the U.K. the average area grown per registered producer increased from 3.61 ha in 1962 to 5.11 ha in 1972 (Potato Marketing Board, annual reports).

16.3.2 The use of variable inputs

Theoretically, variable inputs such as seeds, fertilizers, sprays, should be used up to the point where the marginal revenue (price × extra output) equals marginal cost (price × extra input) assuming a response function which exhibits diminishing returns. In practice, the precise determination of the optimum level of inputs is not possible due to lack of information. For example, at the time of application, the actual response function is not known for a particular crop in so far as uncontrollable future events such as weather and disease will influence it and existing knowledge about the precise effects of particular actions is not complete. Further, the price of the product at or subsequent to harvest is difficult to predict during the growing season. Interactions between inputs and the fact that from the point of view of economic analysis the potato is a multi-product crop (ware, seed, chat) all complicate the issue, as does the fact that sufficient funds may not be available to finance the optimum application of inputs (Dillon, 1968). Nor in the case of most inputs is there a direct link between the variable under the control of the farmer and that which controls yield. In the case of seed, for example, stem numbers per acre is a more direct determinant of yield rather than seed weight (Sharpe and Dent, 1968). It is thus necessary to know the relationships between tuber weight and stem number and stem number and yield in order to be able to determine the optimum seed rate, as well as interactions with row width, variety and spacing (see Chapter 7). Quantified estimates of these relationships are generated on institutional experimental stations but as such tend not to be exactly repeatable in actual field conditions, if only because there is a tendency to get higher yields under experimental conditions (Wragg, 1970). It would be an expensive process perhaps for farmers to undertake trials on their own farms, but it may well be that on large units the cost of obtaining more precise information is less than the value. In its absence the farmer must rely either on the stability of optima under a wide range of conditions (if such optima exist) or on his experience and judgment.

In the case of fertilizer an argument put forward by Boyd (1972) is that for all practical purposes a N response function for potatoes can be represented by two straight lines the first line having a much steeper slope than the second. If such is the case then either prices of inputs/outputs are such that the farmer applies no N or he applies N at the point of inflexion. Further such points are very stable for a wide range of price ratios. In such a situation all the farmer can do to improve his efficiency is to approach this point as nearly as possible and/or to raise the overall level of the response function by seeking new varieties and techniques.

Another factor which simplifies the decision relating to the use of variable inputs is the cost of error. What are the limits of error for inputs such as herbicides, fungicides or insecticides? For example, if a fungicide is applied and a disease outbreak does not occur, the loss is merely the cost of the applied spray. However, if the fungicide is not applied and a disease such as blight occurs, the loss could be very substantial indeed. It is therefore necessary to weight the degree of uncertainty by the cost of error. Playing safe by spraying even with unreliable sprays, could in such circumstances be perfectly rational if large losses are to be avoided, especially if such losses are likely to put the farmer out of business. It could well be that similar reasoning applies to the use of other inputs, provided an overdose does not have a negative effect on the yield especially if the response function is of the shape suggested by Boyd. If so, then the use of generally accepted levels of significance used in experimental work are not appropriate for management decisions since with a treatment with a low chance of success, but with a big effect when it is successful, the expected value (outcome × probability) of it could well exceed its cost (Officer and Dillon, 1968).

16.3.3 Lifting and storing decisions

Lifting and storing decisions also illustrate the uncertainties of the potato growers' situation. The decision to lift early potatoes depends on estimates of current and future yields and prices as well as the opportunity cost of the land, none of which can be known with certainty. The current price and yield for a particular field can be estimated fairly precisely but the rate of change of future prices and yields can change quite markedly and quickly, which puts the farmer in a gambling situation. Cole (1973) provides information reproduced in Fig. 16.2 which shows the variation

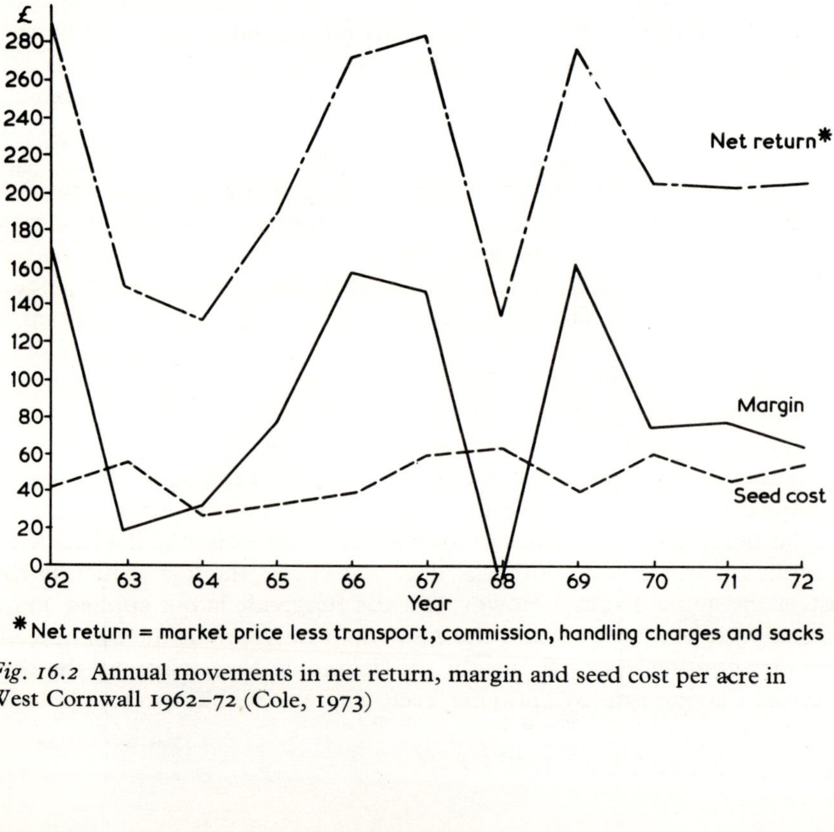

Fig. 16.2 Annual movements in net return, margin and seed cost per acre in West Cornwall 1962–72 (Cole, 1973)

Fig. 16.3 A decision tree of the choice to lift potatoes early or leave them to mature (Hardaker, 1969).

in seasonal profit over eleven years for Cornish early potato growers. The early potato grower must be able to withstand losses if he is to survive.

Lifting decisions for second earlies have been systematically described in a hypothetical example by Hardaker (1969) using a decision tree (see Fig. 16.3). In the diagram both acts and events are represented by forks, acts being represented by squares, events by circles. The outcomes from any particular act are described in terms of gross revenue (price × yield). The decision to harvest green or leave to mature depends on expectations about prices later on in the season. The way of solving the tree or the decision is to ascribe probabilities to each event fork and work out the probability of each sequence of events using the multiplicative law. The act is then chosen on the basis of the maximum expected value. The analysis as such is entirely subjective, representing the decision maker's own view of the situation in a formal way, including conditional acts and outcomes and also his degrees of belief about the likelihood of future outcomes expressed as subjective probabilities. Additional information can be incorporated into the tree using the Bayesian theory (Officer and Dillon, 1968). Should it not be possible to specify the probabilities then Games' theory would seem to be an appropriate tool in solving the tree, as for example using the minimax criterion which seeks to find the highest minimum gain no matter which particular event or sequence of events occurs (Williams, 1966).

The decision tree type of analysis also illustrates another dimension of the advantages of storage. Upton and Casey (1974) show how the existence of a potato store inherently increases the expected price of a crop simply because it increases the flexibility of the timing of sales.

The main reason for storage is to spread out supplies over the season, or in economic terms to add time utility to the potato. The first decision is of course to erect a store, which needs a considerable amount of thought and analysis nowadays in view of the high capital cost of indoor, fully insulated and ventilated buildings. It is probably true to say that very few stores can be justified purely in terms of the primary benefits of storage, namely price rises, quality improvements and labour savings since the capital, interest and handling charges as well as storage losses are often extremely large. A broader view is often necessary to understand the investment such as the fact that some kind of storage is necessary if potatoes are to be a feasible enterprise on the farm, while the particular type of store can often be explained by the availability of labour or the quality requirements of the market. Again potato stores may often have alternative uses while in inflationary times investment of this type is often a protection against the fall in the value of cash. However, as a partial aid to the decision to erect a store, some kind of calculation to work out the discounted incremental return on capital invested (including the secondary benefits) can prove helpful.

Once a store has been installed the main decision is that of the timing of sales. Contractual arrangements, cash requirements and the necessity of maintaining an even work load will determine this to some extent. However, the other main reason for spreading sales is that by so doing one avoids the worst possible situation of selling the whole stock at the lowest price but at the cost of foregoing the chance of selling all at the highest price.

16.3.4 The use of machinery

Potato production on farms of the largest producers is now so highly mechanized that it can no longer be regarded as a labour intensive enterprise. They are planted using an automatic planter, herbicides cut down or eliminate the amount of weeding to be done and complete harvesters are now available with electronic devices for sorting the potatoes from clods and stones. Perhaps the most difficult parts of the operation to mechanize economically are the handling operations for seed and for the harvested product into and out of store, as damage to the tubers can be very serious (Potato Marketing Board, 1974). All across Europe the trend is for potato growing machinery to be increasingly used, especially for planting automatically and for complete harvesting (Cole, 1973; van der Zaag, 1973; Polish Ministry of Agriculture, 1974; MAFF, 1975) in both early and maincrop production areas.

Mechanized operations can be cheaper than more labour intensive ones provided machines are used to capacity, since only then can fixed costs be spread over a sufficient output. However, capacity is a difficult entity to define since the amount of work a machine can do is subject to biological constraints, many of which are weather dependent. For example with a harvester, a farmer must have sufficient capacity to lift his potatoes in a wet autumn if he is to avoid large losses, which of course means that in average and good weather he has more than enough capacity and which in general will not bring him any benefit in such years. It may well be that insufficient capacity is available for the worst conceivable season in that the costs of the amount of machinery required exceeds the expected benefits. However, expected values are not necessarily a good guide to machinery capacity decisions in so far as large losses in a given season, particularly with a high value crop, can threaten the survival of the whole business. It is to be expected therefore that some 'insurance' machine capacity will be carried on most farms.

Some of this insurance is available by varying the method of operation, such as working machinery for longer hours or more efficiently by cutting down lost time due to breakdowns, rest periods and delays, or even by jeopardizing the quality of work. Machinery can of course improve the quality of work by more timely operations, e.g. planting and by reducing losses, e.g. harvesting and it is also a means of overcoming problems of the supply of casual labour.

Machinery use can only be fully understood as part of an integrated system. It is necessary to consider the ways in which different items of machinery fit together with each other and with the labour supply and cropping opportunities on the farm. The optimum machinery system must be defined in relation to the rest of the farm and not independent of it. At a more detailed level, the importance of adequate repair facilities and spares cannot be overemphasized. For example, if we have three machines working together in series, each with a probability of failure of 10%, then there is only a 73% chance of the whole system working at any moment in time ($0.90 \times 0.90 \times 0.90$). The likelihood of the whole system working can be increased both by increasing the reliability of individual machines and also by utilizing standby machines. Thus, in the above example if each machine could be duplicated the chances of success are increased to 97%. The same sort of reasoning applies to problems of matching the capacities of different machines; a common problem which follows the introduction of automated planters and harvesters being insufficient transport and handling facilities (A.D.A.S., 1972).

16.3.5 Irrigation

The tendency towards high yield, high cost, large scale and low unit margins in potato growing together with contracts with processors for specified tonnages means that drought and/or frost are increasingly costly. It is perhaps paradoxical that higher levels of inputs and management skill justify further investment of a risk avoiding nature, in that the farmer is less able to withstand a drop in yield, such is his commitment. As with other inputs it is therefore insufficient to evaluate investment in irrigation facilities only in terms of likely increases in yield; it is also necessary to consider the farmer's ability to withstand losses. An ideal calculation would attempt to derive a distribution of returns so that the farmer could assess the probability of achieving a given level of benefits. This procedure would seem justified in so far as the advantages of irrigation are dependent on the weather so that where future net returns are discounted to allow for the time value of money, then the sequence of the weather after the irrigation equipment has been installed, will influence the level of returns. The actual returns will be greater to irrigation if the investment is followed by a sequence of dry rather than wet seasons.

Investment in irrigation facilities is normally so expensive that it will alter not only the potato growing enterprise but also the cropping opportunities of the whole farm. Again a systems view is necessary in any evaluation if some valuable opportunity is not to be missed. Thus, irrigation may allow multiple cropping after early potatoes or the introduction of high value vegetable crops or intensification of livestock stocking rates, which in turn alters the rotational restrictions. The particular type of irrigation facilities chosen will also depend on the opportunity cost of resources, particularly those of water and labour.

16.3.6 Energy use and the potato

While potatoes have a relatively high return per unit of land they are in general inferior to cereals and crops such as sugar beet in terms of their returns per unit of labour and capital. The recent rise in the cost of energy derived from fossil fuels has also made it necessary to consider the efficiency of crops in terms of their fuel usage. Recent changes in the technology of potato growing including the substitution of labour by machines and increased applications of fertilizer and other chemicals to the crop (White, 1975) has produced a situation where the profitability and viability of potato production is very dependent on the price and viability of fossil fuels.

White (1975) has analysed the use of energy derived from fossil fuels in growing and storing a crop of potatoes up to the point of sale at the farm gate (see Table 16.12). The figures show the importance of fertilizers (especially N), fuel, grading and storage operations as users of energy. Other workers have also calculated the average efficiency of energy use for certain crops up to the farm gate and these have been summarized by Spedding and Walsingham (1975) (see Table 16.13). The calculations demonstrate that potatoes under U.K. conditions are not as efficient

Table 16.12 Energy budget for potatoes (mJ ha $^{-1}$)

Fertilizers		
N	170 kg ha $^{-1}$ × 77 mJ kg $^{-1}$	= 13 100
Phosphate	177 kg ha $^{-1}$ × 14.3 mJ kg $^{-1}$	= 2 530
Potash	244 kg ha $^{-1}$ × 8.3 mJ kg $^{-1}$	= 1 610
Herbicides	13 kg ha $^{-1}$ × 106 mJ ha $^{-1}$	= 1 380
Seed	2 500 kg ha $^{-1}$ × 1.89 mJ ha $^{-1}$ (energy cost of production)	= 4 730
Fuel used in field operations		= 5 439
Grading and storing		= 20 160
Machine depreciation and repairs		= 2 920
Total energy input		51 869
Yield of potatoes 27 500 kg ha $^{-1}$		
Energy output (metabolisable energy) = 27 500 kg ha $^{-1}$ × 2.52 mJ kg $^{-1}$		= 69 300
Energy output/energy input = E = 69 300/51 869		= 1.33

From White (1975).

Table 16.13 Energetic efficiencies for agricultural products at the farm-gate

Product	*Ratio of gross energy in product/support energy input*
Maize (U.S.A.)	2.8
Wheat (U.K.)	2.2
Cereals (U.K.)	1.4–2.1
Oats (U.K.)	2.0
Sugar beet (U.K.)	1.8
Barley (U.K.)	1.8
Potato (U.K.)	1.1
Cassava (Fiji)	71
Sweet Potatoes (Zambia)	4–9

From Spedding and Walsingham, 1975.

in gross energy terms as other major crops. However, such calculations are very dependent on the assumptions made, and do not take into account processing, distribution and cooking (downstream) energy costs. More work is required to calculate these. Some salient facts include those on wastage: 27% by weight of old potatoes is lost through peeling (MAFF, 1970) which in most households is probably thrown away. Dalton and Akwetey (1971) found in a study in South-east Ghana that cassava peelings amounted to 23% of fresh tuber weight but most of these were fed to animals. Factory processing means that waste is not transported to the housewife and there is the opportunity to find alternative uses for the waste such as animal feed although some potato wastes are an embarrassment. Waste in the production of frozen chips can amount to 55% of the fresh tuber weight (Agra Europe, 1975a). Processing itself also has a high requirement for energy which may or may not be greater than that used for food preparation in the home. For example, it is estimated that to produce a ton of dried potatoes some 88 gal (400 l) of heavy oil are required (Agra Europe, 1975a).

While more work is required on the use of energy at all stages in the production and distribution of potatoes such work will not serve as a complete guide for understanding changes since taste and food preferences are important in demand determination. Thus, it may well be that in energy terms it is inefficient to transport new potatoes from Mediterranean countries when ware from the previous season's maincrop is an alternative source of supply. Or again, prepacked graded potatoes may be more expensive in energy terms than unpacked supplies from the greengrocer. Such energy calculations do, however, help in understanding the likely consequences of changes in the price of energy and would of course be useful in planning food production and distribution if fossil fuels were for any reason limited in supply. However, most of the calculations carried out so far are expressed in terms of average efficiency which means that crops for which very few inputs are used or which have a greater yield without the use of modern industrialized inputs may have a low marginal efficiency of energy use even though the average is relatively good. In terms of managing changes in input supply, the marginal efficiency is more important. However, calculations of the marginal efficiency of energy use are not meaningful in global terms in so far as different inputs influence production in different ways. The simplest case would be that of nitrogenous fertilizer where the input is divisible and is more or less completely used up in one production cycle. The more difficult case would be machinery or storage buildings which are discrete: they are not used up in one production cycle and furthermore, can according to their quality, affect the response rate of other inputs used in production. It is only possible therefore to estimate the relative marginal efficiency of energy use for potato production in precisely defined conditions.

Cassava and sweet potato production in tropical climates have a high efficiency of energy use in so far as very few industrial inputs are used and yields of energy per unit area are high. Ratios of gross energy efficiency at the farm gate can be as high as 70 or 80 (Chandra *et al.*, 1974). Again, such information tells us very little about managing energy use on a world scale; what is required is more work on estimating the relative marginal efficiency of energy use in different forms in different situations on such crops.

16.3.7 Conclusions

Potatoes are the most important root crop in terms of world production but are confined in the main to the developed countries of the world. In these countries demand has been falling in terms of consumption per head as incomes have risen so that potatoes are not an important source of calories in the diet. Indeed, some countries would not suffer, on average, any serious reduction in nutrient levels if the potato were not grown in the future. As incomes have risen demand has grown for more and more services to be added to the potato before it reaches the final consumer. Such services ranging from grading, prepacking to complete preparation are likely to be increasingly demanded if existing trends continue and by a more thorough exploitation of all aspects of the market, may arrest the decline in per capita consumption levels. However, existing trends have been established against a background of rising income levels and there is good reason to believe that this may not continue. The interesting question then arises as to whether there will be a reversal of trends. It is probably unlikely that the processing industry will disappear, if only because the technology and plant already exist and it may be that some processed products have lower marketing and distribution costs than the more traditional ware product. Further, consumption patterns for such items as prepacked potatoes, crisps, and dehydrated products and frozen goods are also established.

Trade in potato products is for the most part confined to high value items such as new potatoes, seed and processed products such as starch, although ware may be traded between neighbouring countries and further afield in times of shortages. The bulkiness of potatoes is likely to prevent any major alteration to this state of affairs especially as the cost of dehydration is relatively more expensive given the recent changes in the pattern of energy prices.

The only flexibility of increased demand for potatoes would seem to lie in the animal feed market. This will depend more on relative costs than would be the case in the human food market. Given that imports of cassava products for animal feed are now substantial, then it would appear that there is no technical reason why potatoes should not become part of compound feeds although drying costs may well continue to prohibit this. Relative prices per unit of nutrient will also determine the feeding of fresh potatoes to animals.

Given that trends in improved yields continue against a non-expanding market, then structural change within the potato growing industry is likely to continue. There is likely to be a movement of production on to better soils. Machinery is also likely to dominate production methods more and more, which in turn will foster larger production units especially if there continues to be a further growth in the processing industry, which has an even greater need for plant and equipment. Control will be the key to success, both in production and marketing in an increasingly competitive situation as continual structural adjustment takes place in the industry. Indeed, government control of the overall market to provide price stability may become more widespread. Individual producers may also be encouraged to group together and to take out fixed price contracts with processors and large consumer outlets. The degree of control in the use of inputs for potato production may also have to be improved as farmers commit more resources per unit area of potatoes. It could well be that the adoption of new techniques by growers in the future could be determined by their ability to provide control of variations in yield as much as by increasing yields.

Criteria for adjustment have hitherto rested on returns per unit of land, capital and labour as the most limiting resources to profitable production. Concern is now being expressed concerning non-renewable energy use. More expensive energy will of course be reflected in the prices of the inputs used in agriculture and one would expect adjustment to take place according to the marginal efficiency rather than the average efficiency of energy use, provided of course the price mechanism works effectively. In terms of total energy use at the farm-gate the potato is not as efficient as cereals. However, processing costs and transport to the housewife could easily reverse this relationship. Our interest in energy use and interpretation of the calculations depends on the question asked. If energy supplies, for whatever reason, were restricted, then society would need to obtain a minimum amount of nutrients to survive from available supplies. Were more energy available than this minimum amount, valuation problems arise such as for example whether one preferred potatoes to bread or Egyptian new potatoes to pre-fried frozen chips. Only two ways of distribution in such a situation are feasible, either through the price mechanism or by some authoritarian rationing of food. In either event the relative amounts of foods available will depend on their technical efficiency of energy use.

National self-sufficiency in food is also a popular concept as the extent of world famine grows more serious. Most countries which grow potatoes are self-sufficient, potatoes having the advantage over most other crops in that they produce a large amount of dietary energy per unit of land. There would therefore seem to be distinct advantages in areas subject to land pressure of substituting more potatoes in the diet should self-sufficiency become necessary. However, self-sufficiency of itself is a rather narrow and blinkered view in that it ignores the interactions both of production and trade between different commodities and the demands these commodities make on the available supply of inputs such as fertilizers. Continued efforts to improve the yields of potatoes are important both in rich and poor areas of the world. Potato research in high income countries which increases yields and improves the presentation of a commodity which provides an insignificant amount of calories to an already well-fed people is perhaps less easy to justify. Nevertheless, the poorer sections of rich communities would benefit directly and increased yields would release resources for the production of other crops such as cereals which are an important part of the diet of all peoples and are more easily transported. Research on root crops in low income countries would appear to be an urgent and direct way of contributing to world food supplies in that little yield increasing technology has so far been applied and yet these are the areas of the world where the largest part of the world's increasing population will live. Moreover, due to the low standard of living in such countries, root crops could provide a large part of the population's energy requirements.

References

A.D.A.S. (1970). Machinery sharing in England and Wales. A review of the current situation. *Minist. Agric. Fisheries Fd.* Farm Mechanisation Studies. (19).

A.D.A.S. (1972). The utilisation and performance of potato harvesters 1971. *Joint investigation by Minist. Agric. Fisheries Fd (Agric. Development Advisory Service) and the Potato Marketing Board.* Farm Mechanisation Studies. (24).

Agra Europe (1971a). June 16th 1971. *Increasing Importance of Potato Products in West Germany.*

Agra Europe (1971b). July 28th 1971. *Dutch Potato Starch Industry.*

Agra Europe (1972a). Nov. 22nd 1972. *Progress in Dutch Potato Exports.*

Agra Europe (1972b). Dec. 6th 1972. *Changing Pattern of World Potato Trade.*

Agra Europe (1974a). April 10th 1974. *More Meat and Less Bread Consumed in the USSR.*

Agra Europe (1974b). July 1974. *ZPC: A Major Dutch Seed Potato Cooperative*

Agra Europe (1974c). Aug. 2nd 1974. *Dutch Potato Growers Plan for Export Markets.*

Agra Europe (1975a). March 21st 1975. *Potato Drying Subsidy Needs Careful Thought.*

Agra Europe (1975b). April 25th 1975. *New Potato Regime Planned.*

Bevan, A. (1974). *J. Ind. Econ.* **XXII**. (4).

Boyd, D.A. (1972). *Proc. Ninth Int. Congress Potash Inst.* 461–73.

Butterwick, M. and Rolfe, A. (1971). *Agricultural Marketing and the EEC.* London: Hutchison.

Chandra, S., De Boer, A.J. and Evenson, J.P. (1974). *Wld Crops* **26** (1) 34–7.

Cohen, R.L. (1930). Factors affecting the price of potatoes in Great Britain. *Univ. Camb. Department Agric., Farm Economics Branch.* Report (15).

Cole, H.M. (1973). Early potatoes in West Cornwall. 1962–72. An economic study. *Agricultural Enterprise Studies in England and Wales.* Economic Report (19). University of Exeter, Agricultural Economics Unit.

Commission des Communautes Europeennes (1974). Les produits derives de la pomme de terre. *Informations internes sur L'Agriculture* 133. August 1974.

Coursey, D.G. and Haynes, P.H. (1970). Root crops and their potential as food in the tropics. *Wld Crops* July/August 1970.

Dalton, G.E. and Akwetey, F. (1971). Cassava production and processing in

South East Ghana. *Department of Agric. Economics Farm Management, University of Ghana.*

Dillon, J.L. (1968). *The Analysis of Response in Crop and Livestock Production.* Oxford: Pergamon.

Dutch Information Centre for Potatoes (1971). *The Potato Processing Industry in the Netherlands.* Issued by the Ministry of Agriculture and Fisheries Foreign Information Service. (E. 123). The Hague, Holland.

EEC (1973). *Yearbook of Agricultural Statistics.*

FAO (1971). *Agricultural Commodity Projections 1970–80.* **1**, 49. Rome.

FAO (1975a). *Production Year Book 1974.* **27**, Rome.

FAO (1975b). *Trade Year Book 1974.* **27**, Rome.

Gibbons, E.T. (1970). Co-operation and the potato market. *Univ. Newcastle-upon-Tyne, Department Agric. Marketing.* Report (14).

Hardaker, J.B. (1969). Decision trees: a systematic approach to decision making under uncertainty. *Farm Management Notes.* (39). Univ. Nottingham, Department Agric. Econ.

H.M.S.O. (1973). *Household Food Consumption and Expenditure: 1970 and 1971, with a review of the five years 1966 to 1970.* London: UK. A report of the National Food Survey Committee.

Ingersent, K.A. (1962). *J. agric. Econ.* **15**, 1.

MAFF (1970). *Manual of Nutrition.* London: H.M.S.O.

MAFF (1974). *Output and Utilisation of Farm Produce in the UK. 1967/68 to 1972/73.* London: H.M.S.O.

MAFF (1975). *Agricultural Statistics. England and Wales. 1973.* London: H.M.S.O.

NEDO (1972). *UK Farming and the Common Market. Potatoes.* A report by the Economic Development Committee for Agriculture, National Economic Development Office. Millbank Tower, Millbank, London.

OECD (1968). *Agricultural Projections for 1975/85. Country Studies.* Paris.

OECD (1973). *Food Consumption Statistics. 1955–1971.* Paris.

Officer, R.R. and Dillon, J.L. (1968). *J. Aust. Inst. agric. Sci.* September, 121–9.

Phillips, T.P. (1974). *Cassava Utilisation and Potential Markets.* Ottawa, Canada: International Development Research Centre.

Polish Ministry of Agriculture (1974). *Thirty Years of Agriculture in the Polish People's Republic.* Warsaw. Agricultural Information Centre, State Publishing House for Agriculture and Forestry.

Potato Marketing Board. *Annual Reports and Statements of Accounts* London, UK.

Potato Marketing Board (1968). The marketing of quality graded potatoes. *Sutton Bridge Experimental Station Report* (4).

Potato Marketing Board (1973). *Consumer Buying Behaviour and Attitudes during 1972*–73. London, UK.

Potato Marketing Board (1974). *National Damage Survey 1973.* London, UK.

Sharpe, P.R. and Dent, J.B. (1968). *J. agric. Sci.* **70**, 123–9.

Shaw, J.R. (1971). *Optimum Farm Plans and Aggregate Supply Functions for Potatoes.* Ph.D. Thesis, University of Nottingham.

Spedding, C.R.W. and Walsingham, J.M. (1975). *Span.* **18** (1) 7–9.

Upton, M. and Casey, H. (1974). *J. agric. Econ.* **25** (2), 147–52.

van der Zaag, D.E. (1973). *Potatoes and their Cultivation in the Netherlands.* The Hague, The Netherlands: Dutch Information Centre for Potatoes. Publication (E 108).

White, D.J. (1975). *Energy in agricultural systems.* Paper presented to the Institution of Agricultural Engineers, London. May 13th 1975.

Williams, J.D. (1966). *The Compleat Strategyst.* New York: McGraw-Hill.
Wragg, S.R. (1970). *J. Agric. Econ.* **21** (1) 85–103.
Wye (1975). Survey of the potato industry in the E.E.C. *Centre for European Agricultural Studies, Report (8)* Ashford, Kent: Wye College, University of London.

Chapter 17

Application of physiological and agronomic principles to the development of the potato industry

R.K. SCOTT and S.J. WILCOCKSON

This chapter commences with an appraisal of the implications of data collected from experiments in which growth analysis has been done. It considers the extent to which yield is controlled by external or internal factors and critically appraises current thinking on maximizing yield. It then develops other lines of argument demonstrating alternative approaches to the exploitation of the principles which come from detailed studies on physiology and agronomy which relate to:

(a) efficiently using resources when output targets, i.e. when the crop is to be marketed and for what purpose, are known;

(b) exploiting and testing breeder's material to best advantage; and

(c) complementing cultural conditions and storage requirements.

17.1 The relationship between leaf growth and the build-up of yield

Since the widespread adoption of growth analysis techniques to field crops, including potatoes, many examples appear in the literature showing the time course of leaf area index and tuber weight. This implies that agronomists and crop physiologists expect a relationship between the two parameters and that growth of the whole crop, or of tubers, can be analysed on this basis, leading in some way to more effective research aimed at the control of yield. Perhaps it is opportune to ask to what extent this is so.

In conventional growth analysis the idea is to analyse growth in terms of the size and the efficiency of the photosynthetic system. The parameter L is used as a measure of the size, and E, or net assimilation rate, as a measure of the efficiency, of the photosynthetic system. At an early stage it was suggested that an advantage inherent in this analysis was that net assimilation rate would be mainly dependent on external factors and remain independent of internal components, at least during vegetative growth. This has not proved to be so and the parameter has proved extremely difficult to understand and therefore to use.

At the end of his chapter 'The physiology of growth and tuber yield', Moorby concludes 'The aim should now be a quantitative understanding of the behaviour of the crop which can be used in a predictive fashion and will allow us to make better assessments of the potentialities of unusual genotypes or the possibilities for the crop in new environments'. We believe that a parallel approach to conventional growth analysis (Monteith, 1977) offers advantages. Crop growth is analysed in terms of (a) the amount of light intercepted during the growing season and (b) the efficiency (the ratio of absorbed radiant energy stored in the chemical bonds of carbohydrate) with which intercepted light is used. The amount of energy intercepted is related to the total incident radiation and to various characteristics of the leaf canopy including *L*. Light interception is, however, not a linear but an exponential function of leaf area.

17.1.1 Data from Sutton Bonington, England

Several field experiments during the 1960s showed that, at least during the vegetative phase, crops assimilated carbon and accumulated dry matter at rates which were proportional to intercepted radiation (Monteith, 1972). Although there are many examples of leaf area growth curves for the potato crop, only the unpublished work of R.A.S. Williams at Sutton Bonington has directly measured radiation interception. In 1964 he planted 12 000, 27 000, 47 000 and 109 000 sets ha^{-1} on the flat and arranged on the square. Tube solarimeters, which measured both total (0.4–2.0 μm range) and photosynthetically active (0.4–0.7 μm) radiation (by difference) were placed above the canopy and at the base of each population except 47 000, arranged in such a way that the solarimeter spanned a cross-sectional unit of the canopy, i.e. the same proportion of the surface immediately at the centre of the hill, midway between hills, etc. Fig. 17.1 shows data for total radiation interception (0.4–2.0 μm) from the three populations up until the time when the crops lodged after a heavy storm – 54 days after emergence.

From Williams' data, with harvests at frequent intervals, it is possible to relate radiation interception to crop growth rate (Fig. 17.2). Crop growth rate and *L*, and therefore radiation intercepted, are calculated from fitted curves, the points are means of each individual treatment combination, i.e. the factorial combination of 4 plant populations and 3 fertilizer levels. Points from similar populations are joined in order of ascending fertilizer. 20 days after emergence the relationship is similar to data from sugar beet, spring barley and winter wheat growing during spring at Sutton Bonington (Biscoe and Gallagher, 1977). For all crops

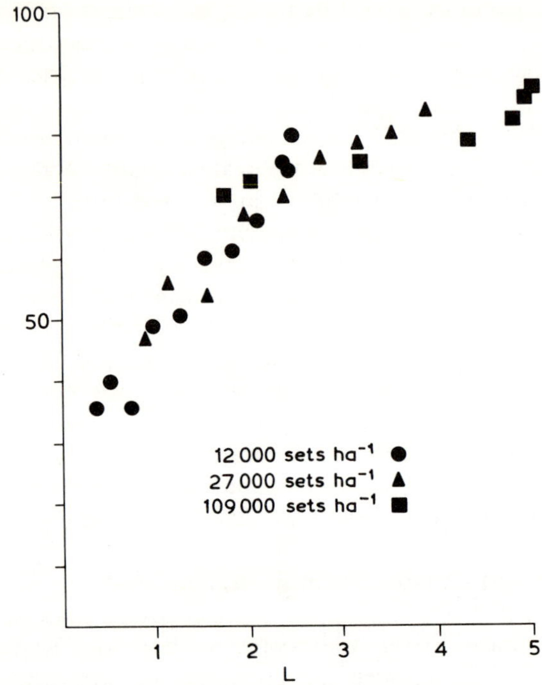

Fig. 17.1 The change in percentage total radiation interception with increasing leaf area index (L)

the rate at which leaf area expands determines growth rate. Similar results showing a linear response of growth rate to radiation interception have been obtained for maize (Williams *et al.*, 1965), soyabeans (Shibles and Webber, 1966) and wheat (Puckridge and Donald, 1967).

Unlike the crops just mentioned, potatoes grow from stem reserves and it is possible that during the early harvests growth depends, in part, on mobilization of reserves from the mother tuber. Data from Younger (1975) and Darby (1974) indicate that the maximum for which these reserves can account during the period 10–30 days after emergence would be 10% of growth, even at the earliest harvests.

In Williams' data it can be seen that 60 days after emergence the relationship between growth rate and intercepted radiation still clearly holds, although by this time many of the treatments have reached L of between 3 and 4, when most radiation is intercepted. When this condition is reached the growth rate of healthy crops, which are well fertilized and free from water stress, depends on the intensity of solar radiation. 60 days after emergence Williams' crop was well advanced in development and it is therefore not surprising that the maximum dry weights of different treatments were strongly related to the total amount of radiation they

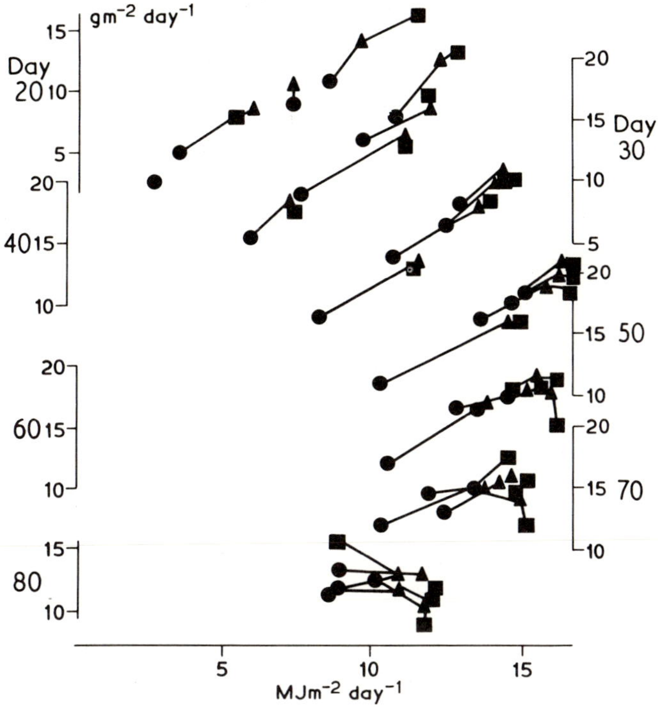

Fig. 17.2 The relationship between crop growth rate (g m^{-2} day^{-1}) and intercepted radiation (MJ m^{-2} day^{-1}) at 20 day intervals between 20 and 80 days after emergence

intercepted during the growing season (Fig. 17.3). That there is a minimal radiation quantity which is to be intercepted before positive growth occurs is consistent with the idea of the existence of a compensation point. The line represents an efficiency of between 2 and 3%. The maximum dry weights of crops recorded here are less than would truly exist for two reasons. Firstly roots are not included – Sale (1973b) includes a correction of 15% to account for this. Secondly, by the time standing dry weight is maximal at the end of linear bulking some leaves have already senesced and fallen from the plant. Data from Dyson and Watson (1971) indicate that the maximum loss might be 25%. Since the leaf dry weight itself accounts for no more than 25% of the weight of the whole crop at this time, the error involved cannot exceed 6%.

Since Williams' experiments on Majestic, work at Sutton Bonington has concentrated on Pentland Crown. Younger (1975) grew different sprouting treatments in the dullest year at Sutton Bonington since records commenced, 1972, and Darby (1974) followed growth of identical treatments in the brightest year, 1973. Fig. 17.3 shows that the maximum total standing dry weight which occurs at the end of linear bulking of

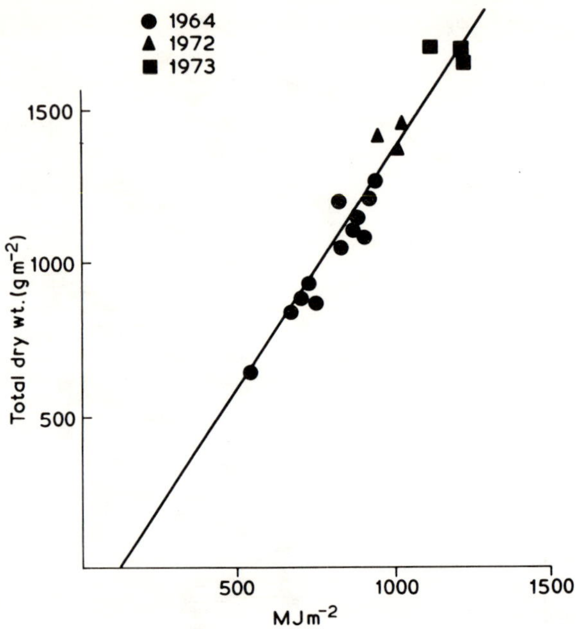

Fig. 17.3 Total dry weight and intercepted radiation data from Sutton Bonington

Pentland Crown from setts which were either cold stored until planting, given a short (3 weeks) or a long (12 weeks) sprouting period bore the same relationship to radiation interception (estimated from values of L on the basis of Williams' data) as crops of Majestic grown by Williams. It is interesting that it was in 1973 that yields of 86 ton ha^{-1} or 35 ton $acre^{-1}$ were achieved, the target in the blueprint formulated by Evans (1975), whose idea was to grow a crop in such a way as to maximize radiation interception. The year 1973 was extremely favourable for all crops, having the rare combination of predominantly bright, yet intermittently wet conditions. It should be said too that the experiments in that year were made on an extremely fertile site, which had been permanent grassland grazed by pigs for some years previously.

17.1.2 Data from the Murrumbidgee, Australia

One of the most recent and certainly the most interesting series of relevant publications comes from work done by Sale (1973a–c) in the Murrumbidgee irrigation area of New South Wales, where the daily solar input averages 25–30 MJ m^{-2} in summer compared to 15–20 MJ m^{-2} in Britain. In this region two crops a year can be produced. The variety Sebago was grown in three experiments planted on 10 August, 1970, 12 February, 1971 and 26 August, 1971 respectively. Thus two were spring

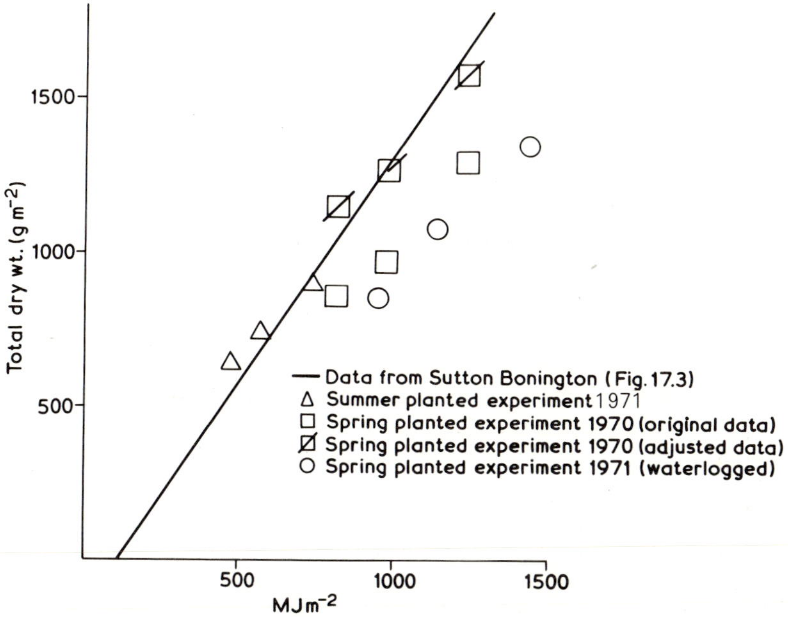

Fig. 17.4 Data from Sale's experiments compared with data from Sutton Bonington (see Fig. 17.3)

planted, and one was summer planted. Each experiment was based on a comparison of three levels of solar input, unshaded, 21% shade and 34% shade from plant emergence onwards, combined with two irrigation treatments. Data from the irrigation treatments based on restoration of the soil to field capacity with the deficit reaching 3.5 cm, which was common to all experiments, are used.

From Sale's leaf area growth curves and the histogram which he presents on radiation during the growing period, it is possible to examine the relationship of maximum dry weight of his various crops with total intercepted radiation. It can be seen that data from the summer planted experiment, i.e. the crop growing into cooler conditions, fell close to the line drawn through the Sutton Bonington data (Fig. 17.4). However, data from the spring planted experiments clearly do not. One of these, that planted on 26 August, 1971, became waterlogged when L was about 2.5 and plants were visibly affected for 2–3 weeks. This may have leached some of the available N and damaged roots, both of which could explain the small maximum leaf areas and apparently lower efficiency of the leaves in this experiment. For these reasons we have concentrated on the first spring sown experiment.

As a crop grows the proportion of dry weight represented by photosynthetically active tissue decreases. Less is known of the effects of the environment on respiration than on photosynthesis, but Sale, using canopy enclosures, demonstrated that the rate of respiration was strongly dependent on temperature. While his results show that the shape of the respiration/temperature relation was similar at all times, the magnitude of respiration occurring in one temperature appeared to depend on previous photosynthesis. On bright days respiration rates were twice those on dull days. During the 35 day period prior to topkill, when growth of the crop consisted exclusively of increase in tuber weight, daily temperatures in the spring planted experiment averaged 22.5°C while those in the summer planted experiment averaged 12.5°C. Sale (1973c) gives respiration rates for leaves plus stems and tubers plus roots for a 44 day period of bulking (67–111 days after planting). Sale's own data and that of Winkler (1971) shows that a 10°C drop in temperature over the range involved when the main period of tuber bulking is in April/May rather than November/December would halve respiration rates.

Fig. 17.4 also shows an estimate ☒ of what the standing dry weight of Sale's spring sown crop might have been had temperatures during the 35 days prior to lifting been halved. The first step was to calculate gross photosynthesis at the higher temperature, then to estimate dry weight increase with respiration rates halved. Values so obtained fell close to the line drawn through data from the U.K. It may be significant that in each of the sets of three points from the spring and the summer planted crops the point from the unshaded treatment falls below the line and the point from that most severely shaded falls above the line, suggesting that the unshaded crops use radiation less efficiently because the canopy becomes light saturated during the day.

17.1.3 Data from other crops

Thus, for different varieties grown at the same site given widely contrasting agronomic treatments, the total weights of the crops at the end of linear bulking are seen to be proportional to the amount of radiation which the foliage intercepts during growth. Moreover, data from a completely different variety grown under irrigated conditions in an area of high insolation, but shaded to differing degrees, shows a similar trend. Data for sugar beet (Fig. 17.5) also from Sutton Bonington, is shown, constituted by ten points, each representing one combination of two sowing (28 March and 5 May) and five harvest dates, stretching from September until December. Data for sugar beet are also shown for the year 1973 which was remarkable for lack of growth constraints with periods of water stress short-lived and the leaf surface maintained disease free. Maximum yields were the greatest recorded. It is very striking that so much of the data for overall productivity of sugar beet and potatoes given different treatments in different years are related to cumulative intercepted radiation by a common line. Even more remarkable, Monteith (1977) has

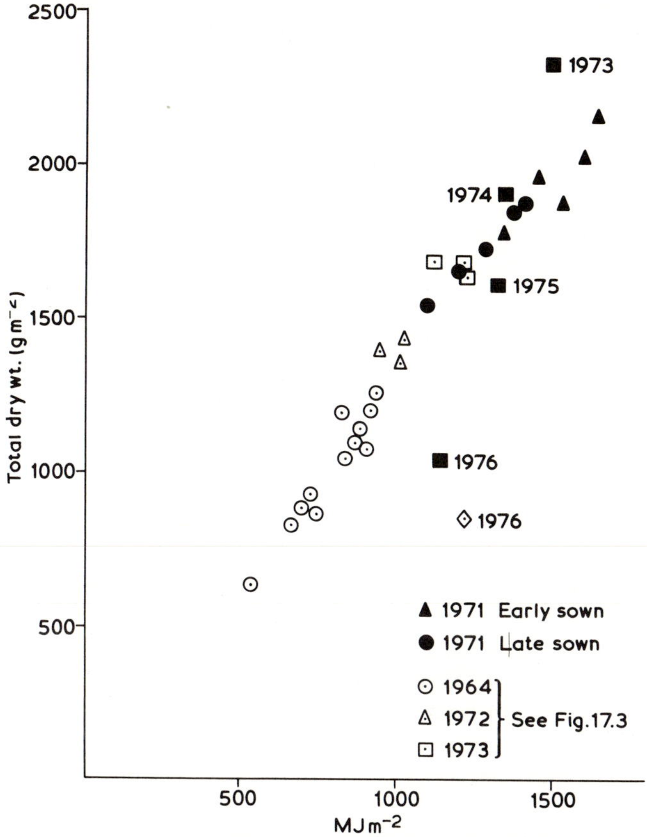

Fig. 17.5 Data from Sutton Bonington for potatoes (open) and sugar beet (closed symbols)

shown that data from Proctor barley grown at Sutton Bonington and Rothamsted, and Golden Delicious apples are surprisingly similar.

17.1.4 The implications

The message is clear: any factor which leads to increased light interception is likely to increase biological yield commensurately. Commercial yields will be increased also if harvest index is constant. For sugar beet this message can be readily translated into strategy; the longer the crop can be allowed to grow, the greater the yield. Higher yields can be expected if leaves are able to expand more rapidly in spring when growth is limited by low temperature, because even the earliest developing crops continue to increase in yield as late as the environment allows and there is no conflict because of any link between advancing early growth and hastening senescence. With potatoes, however, such a conflict does exist and maximizing radiation interception is a more subtle challenge. While there does seem scope for increasing productivity if ageing of leaves could be retarded, particularly during spells of hot, dry weather, this may only be achievable if little or no progress is made in the other direction, i.e. by hastening leaf growth in May, for this seems inevitably to be associated with premature death when the environment is otherwise favourable for growth. The wider implications of this are discussed later.

To illustrate how Fig. 17.5 provides a new basis for analysing dry matter accumulated by a crop, in terms of the light intercepted by its foliage and the efficiency with which that light is used, data for sugar beet and potatoes grown at Sutton Bonington in 1976–the year of the worst drought on record – are particularly appropriate. All the data for potatoes and sugar beet shown previously lay on a common line, indicating that the crops used light with maximum efficiency. Most failed to achieve record levels of productivity, represented by the points at the top right-hand of the line, because of dull weather or because the duration of a photosynthetically active leaf surface was curtailed by late planting, early harvest or premature senescence. However, the yields of potatoes and sugar beet grown in 1976 were clearly less than would be predicted considering the amount of radiation intercepted; water stress imposed a substantial loss of efficiency. In that year conditions were extremely bright yet the amount of intercepted radiation was substantially less for both potatoes and sugar beet than in 1971 and 1973, when total radiation receipts were less. As well as restricting efficiency the drought led to temporary defoliation and poor light interception during July and August.

17.1.5 Radiation interception and tuber growth

So far only the relationship between the growth of the leaf surface, intercepted radiation and the weight of the crop as a whole has been considered, but what can be said of the relationship with increase in tuber yield? This may be less direct on several counts. Firstly, tuber growth does not commence until some time after the growth of the whole plant. Ivins and Bremner (1965) suggested that the time when tuber growth commenced and the state of the plant at this time predetermined the subsequent course of events. Fig. 17.6 shows the course of increase of L and tuber weight of one of Williams' crops, demonstrating the features typical of other published growth curves for potatoes. Initially tuber weight increases at an increasing rate, then there is a prolonged period during which bulking rate is constant and finally a short phase when the rate of increase falls as senescence proceeds.

At first sight the apparent overall constancy of bulking rate from week to week might indicate that growth cannot depend on intercepted radiation, otherwise the influence of weeks with predominantly bright or dull days would be discernible. Moorby comments on the extent to

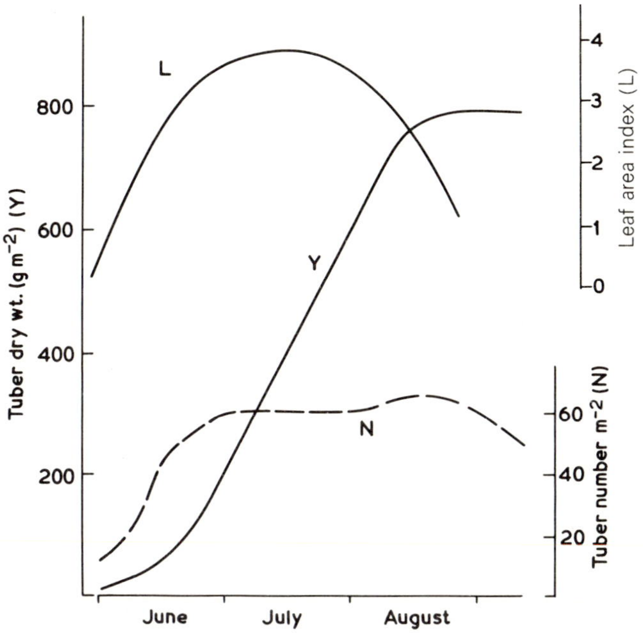

Fig. 17.6 The change with time in leaf area index, tuber number and tuber weight

which translocation from the haulm might compensate for deficiencies in the current supply of assimilates. A point which must be borne in mind is that any effects of weather on growth rate from week to week are difficult to detect because of the size of the errors involved in the estimates of tuber yields at individual harvests. Overall, numerous sets of data confirm that, for the great majority of the period during which yield is increasing, a close linear fit to the data exists. This has the advantage that having obtained an estimate of the mean bulking rate with a series of samples the experimenter, or the grower, can predict yields in the weeks ahead provided the crop remains 'green top'.

An initial phase of accelerating bulking rate might suggest that at this time growth is primarily determined by increases in cell numbers whereas a phase when yields increase at a constant rate might reflect a predominant increase in the size or weight of existing cells. That the number of tubers increases rapidly during the exponential phase and then stabilizes, indicates that an increase in numbers is involved in the initial phase. Cutter (Chapter 3) reports that during the initial stages tubers grow predominantly by increasing cell number but that for most of the growth period increase in cell size is of over-riding importance. Thus it could be that for most of the time the system on which the build-up of yield is based is a relatively simple one, in essence involving the increase in volume and weight of cells which exist at the onset of linear bulking, which coincides with the time the leaf surface approaches its maximum size. To this extent we may expect a direct relationship between the size of the leaf surface and tuber yield.

Moorby suggests that tuber growth is primarily related to current assimilates. Although at any one time 25% of assimilates moving to the tuber may be re-mobilized, over the growth period as the whole reserves previously stored elsewere in the plant do not account for more than 10% of tuber weight. Thus we might expect a relationship between bulking rate and intercepted radiation. Examination of 34 sets of data from Radley (1963), Burrage (1965), Williams (unpublished), and Dyson and Watson (1971) shows that there was not a linear relationship between maximum L and bulking rate. However, bulking rate did increase proportionately from 56 to 87 g m^{-2} $week^{-1}$ as radiation interception increased from 50 to 75%. There was, however, a considerable range in bulking rate over the many treatments which had a full canopy. No doubt this reflects both different intensities of insolation and different efficiencies of radiant energy use, depending amongst other factors on the ages and nutrient and water status of the leaves in the canopy.

Fig. 17.7 shows that Sale's shading treatment affected not only the rate, as we would expect, but also delayed the start of bulking. Plants grown in the dimmer light had larger thinner leaves, longer stems, and fewer of their stolons continued to grow to form tubers. Fig. 17.7 also shows that applying fertilizer to Williams' crops delayed the start of linear bulking. Thus the yield difference from applying 1.5 t ha^{-1} of compound fertilizer was not as great as the effect on bulking rate because of the delayed start of bulking. These inter-relationships between the growth of the haulm and the time of the onset of bulking and its overall duration, complicate the more straight-forward relationship between bulking

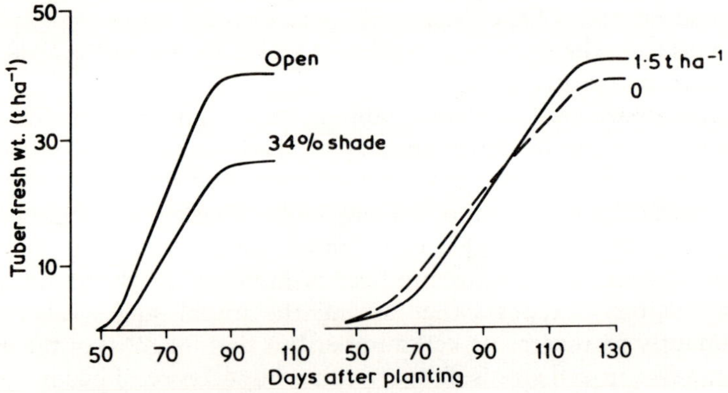

Fig. 17.7 The effect of shading (Sale) and fertilizer (Williams) treatments on the rate and duration of tuber bulking

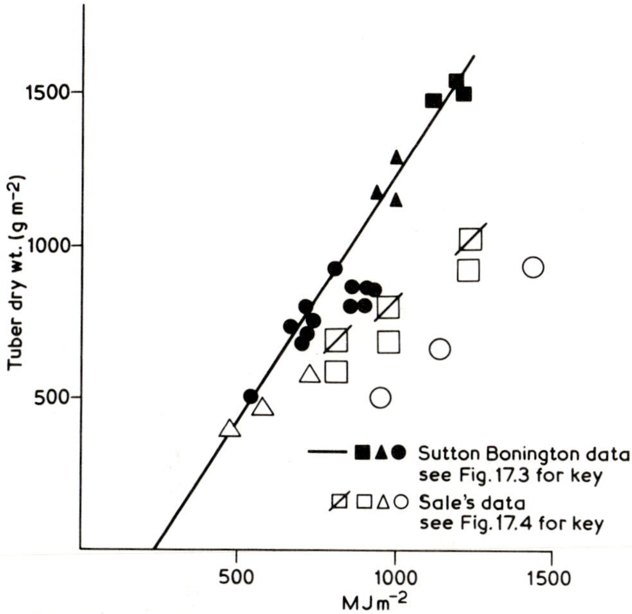

Fig. 17.8 Tuber dry weight and intercepted radiation data from Sutton Bonington and from Sale

rate and the proportion of radiation intercepted. Thus, when we look at Williams' data on Majestic we see that values of tuber dry weight from the generously fertilized crops tend to fall below the overall line relating intercepted radiation to tuber dry weight (Fig. 17.8). Sale's data for Sebago does seem to fall on a different line.

Thus it seems that what appears a relatively straight-forward picture so far as total radiation and total dry weight is concerned, becomes to some extent complicated by the considerable plasticity of the crop in partitioning assimilates between haulms and tubers and the particular bearing this has on the duration of tuber bulking. However, this does not radically alter the approach to maximizing tuber yield for there is a general, if less complete, correlation with intercepted radiation.

17.2 Approaches to maximizing radiation interception

At this stage we should consider what scope there is to promote radiation interception and maintain the efficiency of conversion to dry matter of intercepted radiant energy. The amount intercepted depends on the seasonal distribution of leaf area which, in turn depends on temperature and soil water supply. Higher yields might be achieved if leaves were able to expand more rapidly in spring when growth is limited by low temperature, but yields will not be increased unless at the other end of the season ageing of leaves can be retarded, particularly during spells of hot, dry weather.

If the annual loss of light, at best in our experience 63% in 1973 could be reduced even slightly, say to 60%, corresponding to an increase of intercepted light from 37% to 40% of annual insolation, dry matter production and possibly tuber yield should increase proportionately, i.e. by 8% of its absolute value.

What are the possible approaches to this?

17.2.1 Temperature

Early in the season the major limiting factor is temperature. Rates of leaf extension are directly proportional to temperature over the range found in spring and summer in Great Britain. Any selection by breeders for genotypes able to expand leaves more rapidly in spring would have to be associated with frost hardiness. The likelihood of frost damage is a factor which is additional to those involved with sugar beet, where Scott *et al.* (1973) have shown that, in serial sowing date experiments, maximum yields are obtained when sowing coincides with the seasonal rise in temperature. In the last decade at Sutton Bonington mid-March sowing would never have been too late, it would have been ideal in 4 years, a fortnight too early in 3 years and a month too early in the remaining 3 years. In terms of emergence, potatoes would respond in the same way but nothing will be gained if a spring frost sets the early planted crops back. Thus, planned planting operations must take account of the latest date a spring frost is probable.

As well as hastening early growth, increased temperatures also hasten senescence. Combining these responses the rate of dry matter production can be positively correlated with temperature during early growth but negatively correlated when development is more advanced. To some extent the hastened haulm decay associated with high temperatures may may be indirect; usually warm air is dry and an excessive demand of the atmosphere for water from leaves can bring about senesence.

The experiments of Wurr (personal communication) show that temperatures experienced even by the seed crop have a profound effect on growth of the succeeding ware crop, the performance of which is very much a synthesis of all that is experienced from initiation of the mother tuber onwards. He arranged to grow seed crops at elevations from close to sea level up to 365 m in the Welsh Hills. This change in altitude was associated with a temperature drop of about 4°C in the overall mean during the growing period. At seed harvest in 1975 the lowland crops had already broken dormancy and sprout growth was well advanced particularly at Wellesbourne, the warmest site. From seed harvest all tubers were kept in the same environment and treated identically to produce a ware crop at Wellesbourne, Warwick. Carry-over effects were marked and the crops from lowland seed gave greater sprout growth. However, at the final harvest they had lower yields than crops grown from upland seed. Final yields were therefore inversely related to the temperature during seed production.

One possibility – that the upland seed was capable of producing a better crop simply because it was virus free, is not the explanation – virus levels did not differ sufficiently and there are other instances of reports of similar carry-over effects associated with the temperature of the seed growing site due to seasonal weather or changing latitude, where no known virus could have been a factor.

It is interesting to speculate on whether in a year like 1976 once grown seed from the more southerly counties will be too advanced physiologically and the potential of 1977 ware crops be limited accordingly. There are many reports of unusual tuberization patterns in stored crops which clearly reflect the early dormancy break resulting from the hot days and nights of 1976. Unusual climatic conditions, as in 1976, may result in seed performing in an atypical way. For example, in that year many upland sites experienced conditions as hot as those at lowland sites and the patterns of growth and development of subsequent ware crops were similar (Allen, personal communication). In East Germany they seem to recognise how the seed crop environment can affect regrowth patterns and have devised a standard test to which seed tubers are subjected (Folsch, 1974). It might be possible to complement the innate characteristics displayed in such a test by judicious manipulation of the storage environment and choice of other cultural practices, planting date, seed rate, fertilizer rate, etc., to fit the crop best for the particular purpose for which it is intended.

17.2.2 Leaf angle, planting distances and canopy arrangement

When L is small, horizontal foliage intercepts more light than vertical foliage with the same value of L and therefore, grows faster if other factors are the same. As L increases, the advantage for prostrate leaves is gradually lost until canopies with erect leaves grow faster, theoretically at least. Monteith (1977) has modelled the performance of crops with foliage comprising leaves either predominantly prostrate or predominantly erect. He concludes that it will be difficult to achieve consistently high yields of British crops by breeding for more erect leaves. His model predicts that there is a great probability that cultivars selected for erectness will show to advantage where cloud is less and canopy photosynthesis more frequently light saturated, than in N.W.Europe.

Williams' experiment measured radiation interception over a wide range of plant density. His data are not sufficiently extensive to be confident on this point, but there is an indication that for the same value of L widely spaced crops intercepted more radiation, i.e. had more prostrate leaves than closely spaced crops. Allen in his chapter makes reference to two other factors which could affect leaf angle and light interception. He observes that wide rows produced taller plants which did not lodge and were 'probably much more effective in light interception and utilization later in the season than the lodged narrow row plots'. He also comments that increasing stem population results in taller plants with much reduced axillary branching. This affects the ability to intercept light. A further way in which canopy structure is clearly affected is by physiologically ageing seed. With physiologically old seed an almost determinate leaf habit can result. All of these effects have been observed but in no case have measurements of radiation profiles, or photosynthesis at different sections of the profile, been made to enable us to say how significant such changes in the quantity or the quality of the light penetrating the canopy could be in determining rates of growth, patterns of development and overall productivity.

Perhaps the most obvious way of promoting radiation interception early in the season is to crowd stems close together. Both Allen and Harris note that the initial advantage may be counterbalanced by increased and hastened leaf senescence, especially if water becomes limiting. As Harris explains, 'high populations increase the rate of water loss early in the season and this is not counterbalanced by a compensatory increase in rooting depth and the volume of soil from which water can be extracted'. Early irrigation may itself lead to earlier death of the haulm. Since additional N is known to delay senescence in some circumstances, it seems logical to suggest that a combination of close planting and heavy fertilizer rates might be needed; the first to advance the creation of a full leaf canopy, and the second to maintain it active. Williams' data does not show this, and as Allen comments, 'few convincing examples of this have been reported'.

17.2.3 Sprouting

One practically feasible approach with potatoes, which is not possible for other crops is to commence growth before planting by sprouting in the seed store. Sprouted crops clearly show to advantage during May and June and less radiation passes unintercepted through the canopy during the long, early summer days. At first yields are also improved and for early markets sprouting is an essential practice. Fig. 17.9 shows the changes with time in L of crops from seed sprouted in a well-lit, temperature controlled store from November, and from seed cold-stored until planting. The sprouts produced in the warmer conditions resembled those considered ideal in the days of hand planting, sturdy, dark green, about 2 cm long with leaves already expanding and root initials clearly visible. Consistently, however, leaf senescence too was advanced by sprouting and the range of treatments tested each produced a leaf surface

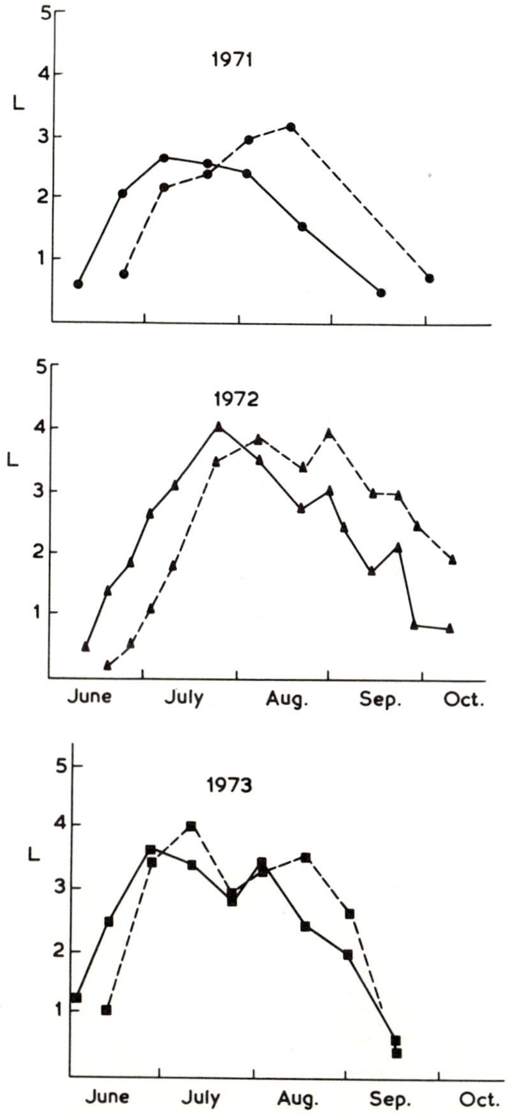

Fig. 17.9 Changes with time in leaf area index of crops from seed sprouted from November onwards (solid line) or cold stored until planting (dotted line)

which was similar in extent and duration, simply displaced in time. Each intercepted similar quantities of radiation and produced similar yields.

Even in 1973 when yields of 35 ton acre^{-1} were produced the senescence in early September almost certainly restricted yields below that which would have been possible at such a favourable site and in a season of intermittent bright and thundery spells. It is possible to estimate that had the leaf surface not decayed until October, 20% more radiation would have been intercepted and yields might have exceeded 40 ton acre^{-1}. By the standard familiar to growers of early potatoes, the technique used to produce sprouts in these experiments was not extreme. Wurr (Chapter 8) emphasizes how, by physiologically ageing seed, the crop can exhibit more determinate patterns of leaf production and have a condensed growing season. Rooting depth is also restricted. Apparently this pattern of behaviour is pre-set and largely inviolable; attempts to offset it by making water and nutrients freely available are only partially successful. So far we seem unable to devise procedures which give any guarantee of breaking the link between advancing leaf growth in spring and advanced senescence.

17.2.4 Water relations and irrigation

At Sutton Bonington one of the factors which most frequently offsets any potential benefits from sprouting is the occurrence of a June drought. We do not know whether this is primarily caused by a shortage of water supply to the roots or by an excessive water demand from the leaves when they are surrounded by very dry air. Bright conditions are often associated with a large saturation deficit. Rates of leaf expansion can be slowed or stopped after a few summer days without rain and photosynthesis rate is also slowed as stomata close. For sugar beet it is well known that leaves can wilt even when soil is moderately wet and Milford (1975) has shown that exposing the leaves to fine mist increases stomatal conductance and photosynthesis. Sparrow and Harris (unpublished data from Reading) have also shown that the application of very small quantities of water to the leaves of sugar beet around mid-day improved the efficiency of water use. This response did not occur with potatoes. As Harris comments ‘It offers interesting possibilities of increasing yields beyond those possible with other methods of applying water; however, more experimental work is needed to evaluate the magnitude of any such advantages and assess whether the extra investment in equipment would be justified’. Moorby points out that in hot, dry conditions leaf temperatures are raised dangerously higher than air temperatures, when leaves cannot maintain transpiration. Mist irrigation could provide ample supplies of water so that evaporative cooling might maintain leaf temperature close to ambient.

Regarding more general aspects of the possibilities for effective use of irrigation on potatoes, Harris concludes that the maximum benefit is only achieved where factors such as nutrient supply and plant population are not limiting yield. While this seems an eminently sensible conclusion, hard evidence to substantiate it is limited. Peeler *et al.* (1966) have reported that on the Bunter sands in Nottingham, central England, the response to irrigation increased with increasing population. In the chapter on nutrition, Harris shows that the uptake of nutrients appears to be particularly sensitive to soil water status. Perhaps the depression of growth can be attributed more to a deficiency of plant nutrients than to a lack of water. Since the greater part of the N and organic matter in soil is located in the surface layer, complete wetting of the soil profile might not be essential. However, few experiments testing the frequency of irrigation have also tested the inter-relations with nutrition, e.g. simply by supplying additional fertilizer N.

Harris argues logically that 'It would appear that the major effect of increasing plant population on the water economy of the potato crop would be to increase the rate of loss of water from the soil profile in the period before a complete cover is obtained with little compensatory increase in the volume of soil from which water could be extracted. The more rapid depletion of water in the upper nutrient rich part of the soil profile might also be expected to accentuate the need for and response to irrigation at high populations, more especially as the demand for nutrients is increased by increasing plant population'.

The fact remains that little data exists and the project of Williams was along much needed lines. He attempted to make a detailed study of radiation, water and nutrient use by four populations grown with four fertilizer levels. As population increased, there was a trend towards progressively deeper water extraction. This was almost certainly a competition effect; the greater amount of root activity per unit volume of soil at the close spacings resulted in earlier exploitation of water at shallower depths. Once this was exhausted, plants were then forced to exploit deeper soil layers in order to maintain an adequate water supply, which extended below 60 cm depth at the closest spacing.

17.2.5 The A.D.A.S. experience

The target of the 'blue-print' formulators was to maximize radiation interception. General experience with the exercise indicates that crops so grown have combined advanced leaf production with persistent growth late in the season. Neild (1977) concludes that it is this late growth, with absence of defoliation, which is the factor most responsible for increasing yield. Yields per acre can often increase by between 2-3 ton over the month of September and after the drought at Sutton Bonington in 1976 Pentland Crown increased its yield by 4 ton.

It is not immediately clear why the link between an early start and an early finish has apparently been broken. Conventional experimental work has clearly shown that there is a limit to what can be achieved by early planting, sprouting, the use of large seed and heavy seed rates. It might be claimed that the key to the answer lies in the combined use of non-limiting nutrient and water supplies associated with routine and frequent pest and disease control measures. Again conventional experiments, and this includes examples where very wide ranging treatment scales have been tested, have not consistently demonstrated that combined use of close spacing, generous fertilization and irrigation will maintain leaf persistence and give commensurate yield increases. Sometimes haulm growth has been so stimulated, apparently at the expense of tuber growth, that yields have actually been less. Perhaps Pentland Crown has innate qualities which are exceptional in this regard. Perhaps also the exercise has hit on the soils on which productivity is inherently great anyway and where crops are unusually responsive to increasing levels of input.

One of the most pressing needs is to unravel the factors of the soil and root environment which allow crops on fields in the same locality grown with similar cultural practices and experiencing similar temperature, radiation and rainfall to produce such widely differing yields when no obvious effects of pest and diseases are involved. The blue-print may well provide a safety margin; for example, where emergence is poor, or frost occurs, the compensation will be more complete than where inputs were at a conventional level.

17.3 The exploitation and testing of breeders' material

17.3.1 Exploitation

Experimental work over the last two decades, involving growth analysis techniques, has consistently demonstrated how critical date of harvest is in determining the ranking order of different varieties and of the yield advantage from a range of cultural practices applied to the same variety. Thus, decisions on choice of variety and the treatments to be given should take account of the latest date that growth can be allowed to continue. A desirable aim would be to so order events by choice of variety and management, that a succession of crops can be lifted as they mature naturally, with a consequent reduction in the susceptibility of tubers to damage and associated problems. To burn off while bulking is still underway, not only involves foregoing yield but may also result in tubers being lifted before their skins are properly set and they are fit to store; some of the consequences of this on storage potential are described in Section 17.4.2.

In this scheme, there is a need to consider how fertilizer rates, plant population and irrigation requirements need to be modified for the various combinations of variety and planned length of growing season. One piece of evidence from Sutton Bonington (Williams, unpublished data) indicates that N requirement increases by 15 kg ha^{-1} for each week of active haulm life after the onset of bulking. If confirmed, this would be the basis for promoting efficient use of fertilizer, an important objective as costs rise.

There is very clear evidence from work at Terrington Experimental Husbandry Farm (Jarvis and Shotton, 1971) that varietal response to changing planting rates differs considerably, and the order in which various varieties will yield depends on the seed rates at which they are planted. Again, rising costs make it increasingly important to change seed rates and management of that seed to meet the specific requirements of the growing period available. Allen (Chapter 7) has described how the length of growing season may affect the optimum population for a variety. Where the aim is for maximum yield irrespective of tuber size distribution, the longer the season, the lower is the population required. On the other hand, high populations are at an advantage where the season is short because of enhanced leaf growth early on, but this may be lost later because of advanced senescence caused by mutual shading or an excessive demand for water. When graded tuber yields are considered the picture changes. For early harvests of a particular size grade over a certain minimum, lower populations will result in limited inter-tuber competition and rapid attainment of low yields of the required size. For later lifts, larger and larger populations are justified, to prevent too many tubers becoming oversize, and heavier yields of the desired size grade will be achieved. It is this knowledge which is being exploited to produce crops of canning size tubers for lifting over an extended period.

17.3.2 Variety testing

The considerable range of performance that a single variety may show, depending on how the seed is managed in the store, and grown in the field has a number of implications for variety testing.

Varieties break dormancy at different times and sprouts grow at different rates, so that their physiological ages will contrast markedly by planting time; this presets the subsequent pattern of growth. Varieties with long sprouts which are physiologically old will produce favourable early yields, but at later harvests, yields may be depressed because of advanced senescence, whilst the situation is reversed with physiologically young varieties.

The present procedures of the N.I.A.B. for testing recently introduced early varieties, which have been described by Ramsbottom (1975), favour those which break dormancy early and sprout readily. Seed from the same source, which is said to be 'of the same physiological age', is grown at some or all of the centres in Cambridgeshire, Cheshire, Cornwall, Devon, Dyfed, Essex, Gwynedd and Lincolnshire, with row width and spacing etc., according to 'local practice'. Four successional lifts are made at seven day intervals, the first timed to occur before the control variety has reached a tuber yield of 5 ton acre^{-1}. Allen (personal communication) suggests that to pursue selection further along these lines is not in the best interests of growers. With early dormancy breakers, there is an increased risk of sprouting in the field before seed tubers are lifted and their long sprouts either prevent free flow or are removed during machine planting. There are possible disadvantages even if planting can be done properly. Coiled sprout and little potato disorder, which were described by Moorby in Chapter 4 may seriously depress yields, and it is also possible to so advance sprout and subsequent crop growth and development that tuber bulking ceases even before the earliest commercial lifts would be made.

In maincrop variety trials, the ranking order of yields is affected both by the seed rate at which the varieties are assessed and the time at which they are harvested. For example, Jarvis (1971) showed that ware yields of Pentland Crown and Pentland Dell increased with planting rates up to 79 000 sets ha^{-1}, but those of King Edward only responded to increasing seed rate up to 44 500 (Jarvis and Shotton, 1971), and the response varied widely from year to year. Similarly, Maris Piper, a prolific tuber producer shows limited response to heavier planting rates (Ingram, personal communication).

The decision when to harvest maincrop variety trials can to some extent determine the ranking order of yields yet the decision is fraught with problems. If the objective is to measure yield potential then the ideal would be to harvest each variety after the haulm is dead. This involves several lifting operations, possibly one per variety, and the alternative, to leave lifting until all varieties have died, is not straightforward. Firstly, some may never die off, and secondly, while those which senesce early lie in the ridge, losses may occur due to pest depredation. However, to ignore the state growth has reached and base the timing of yield assessment on some arbitrary factor such as convenience in the work programme, then to average varietal yields over seasons or sites, is potentially misleading. In the extreme it could be argued that it may hide the most useful information which the trials programme could provide. The research programme aimed to define the progress to maturity could contribute to finding a solution to these difficulties.

There would be considerable advantages from a testing scheme which demonstrates the range of performance which can be achieved by manipulating existing varieties by sprouting and contrasting cultural conditions. This would help to identify gaps which breeders should seek to fill and as Wurr (Chapter 8) puts it 'breeders might find that an assessment of the performance over a range of physiological ages would be a useful indication of the wider potential of that variety before it was released for general multiplication'.

Secondly, by carefully selecting varieties which exhibit contrasting growth features, and by subjecting these to extreme treatments, it should be possible to characterize newly released varieties against this background, so as to predict the cultural techniques required to produce their best performance. This done, it would then be possible to compare varieties, each prepared and grown in its own individually appropriate way, and perhaps avoid discarding potentially useful material because of a lack of knowledge of how best to exploit its potential.

17.4 The effects of cultural and environmental conditions on storage potential

About 85% of maincrop potatoes are stored to meet the demands of the market during winter, spring and early summer (Potato Marketing Board, 1970). Of course, the proportion which leaves the store in an acceptable state for processing and domestic consumption depends on storage management but it may also reflect treatments received from as early as seed storage and is certainly influenced by the way in which the ware crop is grown and treated at harvest.

In the past, investigations concerning growing, storing and marketing have been largely compartmentalized. Those concerned with growing have seldom extended their interest beyond the measurement of final yield in the field, whilst those concerned with storage have made only limited reference to the previous history of the bulks used in the storage experiments. The usual approach in work on storage has been to give an account of the physiological and biochemical changes in the tuber as they are affected by storage conditions. Sometimes the effects of individual cultural and seasonal variables on isolated aspects of post-harvest physiology, e.g. respiration or evaporative water loss are considered. We now require detailed information throughout the production cycle, both in the field and in the store, of a range of crops grown in contrasting ways to facilitate accurate identification of the morphological and physiological characteristics, including skin thickness and skin set, sugar status, dry matter content, cell size and number, respiration rate and the extent of foliage senescence, which determine keeping quality and storage potential.

It is the aim of a recently commenced research programme which involves the Food Research Institute, National Vegetable Research Station, University College of Wales (Aberystwyth), University of Nottingham School of Agriculture and the Potato Marketing Board's Sutton Bridge Experimental Station to provide this information. The ultimate objectives are to establish quantitative scales of tuber maturity. Eventually this may lead to improved prediction of the potential for storage of different consignments. It may also help to enable processors and growers to choose varieties and complement cultural practices to meet their storage targets or capabilities.

17.4.1 Storage losses

Loads of tubers of high storage potential eventually leave the store with minimum losses of weight and quality and retain a good appearance. In the absence of damage and disease, weight losses are the direct result of respiration, evaporative water loss and sprouting. Collectively this has been described as 'natural wastage' or 'shrinkage' (Wilson *et al.*, 1962). Deterioration proceeds in parallel with shrinkage in the form of biochemical changes. This may not contribute to weight loss but can render tubers useless for a particular purpose. For example, senescent sweetening in long-stored tubers makes them unsuitable for crisp manufacture (Burton, 1977).

It is extremely difficult to eliminate damage and disease from tubers going into store, but some attempts have been made to estimate natural wastage where damage and disease has been kept to a minimum, and storage conditions have been controlled. For example, Wilson *et al.* (1962) report that for maincrops stored in stacks for 8 months in indoor stores, with temperature controlled by forced-draught ventilation, natural wastage accounted for a 7% loss of the original weight of tubers that were put into store. Early lifted potatoes, with thin, easily damaged skins, which are suitable for canning and loaded into refrigerated commercial stores in July lose rather more – 8% after 6 months (Potato Marketing Board, 1973). More recent (unpublished) data from Sutton Bridge shows that disease free samples of Pentland Crown, carefully handlifted in early August 1976 at Sutton Bonington, and stored in small (7 kg) capacity boxes for 38 weeks at 7 °C and 95% r.h. lost as much as 17.5% of their original weight.

Storage conditions also influence the extent of natural wastage, low temperatures and high relative humidities being required to minimize it, but of course, damage and disease also contribute to weight loss. However, experiments do not seem to have been done which enable us to partition total weight losses between the effects of: (1) natural wastage; (2) damage; and (3) disease. A possible approach would be to handlift carefully samples of tubers and apply a factorial series of reproducible damage and surface sterilisation treatments before storage in controlled environments.

In this way, it should be possible to estimate natural wastage during storage, the extra weight loss which results from damage in the absence of disease and the extent to which damage permits the proliferation of disease and subsequent storage rots. However, the general impression seems to be that the effects of damage and disease on weight loss are relatively small compared with their effects on the proportion of the bulk coming out of store which is saleable.

Data from Sutton Bridge Experimental Station are given in terms of ware; these are potatoes which meet the minimum standard of grading prescribed by the Potato Marketing Board and are fit for domestic consumption. On average, only about 85% of potatoes put into store are of ware standard, with a range of 75–93%. After storage this quantity is decreased because of damage and disease. Most damage occurs during harvesting and the loading and unloading of the stores and is independent of store management and the duration of storage. Serious damage makes a potato unfit for human consumption but it may be usable for stockfeed. Any damage may allow the spread of disease and so increase indirectly the amount of stockfeed and rotten potatoes. Rotting diseases make potatoes unfit even for stockfeed and so represent a complete loss to the producer. Evidence from Sutton Bridge suggests that damage and disease may cause a loss in ware percentage of about 5% in potatoes stored in a reasonably good condition and with good store management, but disastrous losses of 10–20% can occur if there is widespread development of rotting diseases. Combining ware loss from shrinkage, damage and disease gives an expected minimum fall in the ware percentage of 6–7% of potatoes stored until the beginning of January and a total fall of between 10% and 15% for potatoes in good condition stored until March or later (Potato Marketing Board, 1970b).

17.4.2 The modifying influence of cultural conditions

In 1973, 82% of the maincrop acreage of Great Britain was deliberately defoliated. Chemical treatments accounted for 64% and mechanical methods for the other 18% (Potato Marketing Board, 1975). Presumably some of the remaining 18% were lifted 'green-top' and others when the tops had died off naturally. The timing and method of haulm destruction is perhaps one of the most obvious, and arguably the most important, cultural practice which affects tuber yield, quality and fitness for storage. In order to illustrate the extent to which this is so, some very recent data on the effects of lifting at different stages of crop growth on potential storage losses and quality deterioration are briefly described. Tubers were stored in conditions suitable for ware for domestic consumption, i.e. a holding regime of 7 °C and 95% r. h.

During tuber bulking, the normal intact periderm becomes progressively thickened although the number of layers of cells constituting it may remain relatively constant (Artschwager, 1918). Tubers lifted when the haulm is still green and active in July or August have thinner skins than those lifted in September or October, when the haulm has completely died down and ceased to function. This may affect subsequent weight loss, for there is a suggestion, but according to Burton (Chapter 14) very little evidence, that the thickness of the periderm, the barrier through which about 98% of the water is lost (Burton, 1973) may influence the rate of evaporative loss of water from tubers in store.

Perhaps more important than periderm thickness is how well the skins are set. Early lifted tubers have skins which are incompletely set; they adhere weakly to the underlying cortical tissues and are very susceptible to abrasion during harvesting and handling. Rates of evaporative water loss from scuffed tubers during the first few days of storage may be 15–100 times greater (100–600 μg cm^{-2} h^{-1} $mbar^{-1}$) than from tubers which have skins that are completely set and intact [6–10 μg cm^{-2} h^{-1} $mbar^{-1}$ (Burton, 1973)]. However, in conditions favourable for wound periderm formation (storage at 15° C and 95% r.h.) the rate of loss may subsequently decline to similar levels after 2 weeks because suberization of exposed tissues and wound healing is rapid and effective in young tubers (Lange and Rosenstock, 1964). Weight loss during the healing period may still have been considerable. For crops of Pentland Crown grown in central England, defoliated in mid-July 1976 and hand-lifted 2 weeks later, weight loss at the end of a 2 week curing period amounted to 2% of the original weight of tubers stored. For more advanced crops grown with identical husbandry conditions but in the earlier environment of Pembrokeshire the skins were thicker and well set and the loss was halved.

The morphology of the periderm and its susceptibility to damage may also affect the ingress of wound and other pathogens. Tuber infection with blight *(Phytophthora infestans)*, dry rot *(Fusarium solani* var. *caeruleum)* and skin spot *(Oospora pustulans)* (Nagdy and Boyd, 1965) is influenced by skin thickness. Spores are unable to penetrate the intact periderm and moreover, the skin and lenticels become more difficult to penetrate as the tuber ages (Nagdy and Boyd, 1965; Lacey, 1967).

Rates of tuber respiration also change during growth and this has been described by Burton in Chapter 14. Shortly after tuber initiation the rate is high but soon declines to a slowly decreasing level as the foliage dies off. Thus, tubers harvested before the virtually steady state has been reached will be respiring more rapidly than those which are lifted when the haulm has completely senesced. However, it seems that differential rates of respiration which result from lifting at different times may have little bearing on storage losses. Once in store with holding temperatures of 7°C, high rates of respiration are not maintained and CO_2 output falls to an almost trivial level (2–3 mg kg^{-1} h^{-1}) after about 2 weeks. Tuber respiration rate continues to decline steadily during storage until sprouts begin to grow, and then rises progressively (Burton *et al.*, 1955).

Tuber dry matter percentage and specific gravity also change during growth, increasing rapidly after tuber initiation, but more slowly as bulking progresses. A full account of the factors which influence dry matter percentage, including time of defoliation has been given by Gray and Hughes in Chapter 13. They point out how it may affect the rheological properties of the tuber, and hence susceptibility to damage, subsequent weight loss and rotting, when lifting occurs at different stages. Data given by the Ministry of Agriculture, Fisheries and Food (1972) from High Mowthorpe show that defoliation of maincrops in early rather than late August results in tubers with a lower dry matter content at final harvest in early October. Mechanical defoliation results in smaller yields of tubers with lower dry matter than chemical defoliation with diquat.

Where sprout suppressants are not used the interval between lifting and the start of sprout growth is usually longer for early lifted crops than late lifted ones, which occasionally have already sprouted in the field before they are harvested. Consequently, water loss from the more permeable sprouts, which may evaporate water at rates of 1 mg cm^{-2} h^{-1} $mbar^{-1}$ (Burton, 1955), 100 times faster than non-sprouted intact tubers with fully set skins, will occur sooner in late-lifted crops. At the same time respiration rate increases (Burton *et al.*, 1955) and senescent sweetening commences sooner, which although not a direct result of sprouting, proceeds in parallel with it (Burton, 1977).

17.4.3 Quantitative scales of storage potential

At present, maturity is a term which is often used to describe the suitability of tubers for storage, but seems little understood and is variously defined: when the tuber skins are set; when the total sugars are at a minimum (Burton, personal communication); or from lifts when the haulm is completely dead. Furthermore, only two extremes of the condition seem readily identifiable – immature and mature – with no working knowledge of the intermediate stages. Moorby, in Chapter 4 also recognizes the deficiencies of the term and states that 'a precise definition of an immature tuber is impossible at present, but in general terms, it is one without a fully developed periderm and in which the starch synthesizing enzymes are fully active'.

Maturity is undoubtedly a complex physiological and morphological condition with several facets – respiration and carbohydrate changes, dry matter content, moisture loss (which relates to skin thickness and skin set) dormancy and sprouting – which are inextricably linked. It seems likely, therefore, that any quantitative scale must take account of each of these characteristics. The aim of such an aid is to enable future growers to decide their storage strategies with reasonable confidence and avoid having to unload stores prematurely because qualitative assessment of storage potential was in error.

References

Artschwager, E.F. (1918). *J. agric. Res.* **14**, 221–52.

Biscoe, P.V. and Gallagher, J.N. (1977). Weather, dry matter production and yield. In '*Environmental Effects on Crop Physiology*', eds. Landsberg, J.J. and Cutting, C.V. London: Academic Press.

Burrage, W.S. (1965). The growth of early potatoes. Ph.D. Thesis, University of Nottingham.

Burton, W.G., Mann, G. and Wager, H.G. (1955). *J. Agric. Sci., Camb.* **46**, 150–63.

Burton, W.G. (1973). *Proc. Fifth Triennial Conf. Eur. Ass. Potato Res. Norwich* (1972) 63–81.

Burton, W.G. (1977). *Ann. Appl. Biol.* **85**, 433–5.

Darby, R.J. (1974). The effects of lighting, sprouting period and desprouting on the growth and yield of maincrop potatoes. B.Sc. Dissertation, University of Nottingham.

Dyson, P.W. and Watson, D.J. (1971). *Ann. appl. Biol.* **69**, 47–63.

Evans, S. (1975). *Outlook on Agric.* **8**, 184–7.

Folsch, S. (1974). *Saat-und Pflanzgut* **15** (2), 29–30.

Ivins, J.D. and Bremner, P.M. (1965). *Outlook on Agric.* **4**, 211–17.

Jarvis, R.H. (1971). *Expl. Husb.* **20**, 84–92.

Jarvis, R.H. and Shotton, F.E. (1971). *Expl. Husb.* **20**, 12–29.

Lacey, J. (1967). *Ann. appl. Biol.* **59**, 257–64.

Lange, H. and Rosenstock, G. (1964). *Phytopath. Z.* **51**, 136–52.

Milford, G.F.J. (1975). *Ann. Appl. Biol.* **20**, 247–50.

Ministry of Agriculture, Fisheries and Food (1972). Farming at High Mowthorpe —12th Annual Report (1971) 78–83.

Monteith, J.L. (1972). *J. appl. Ecol.* **9**, 747–66.

Monteith, J.L. (1977). *Proc. R. Soc., Series B.* (in press).

Nagdy, G.A. and Boyd, A.E.W. (1965). *Eur. Potato J.* **8**, 200–14.

Neild, J.R.S. (1977). *Arable Farming*, May, 18–23.

Peeler, C.H., Harvey, P.N. and Rosser, W.R. (1966). *Expl Husb.* **14**, 30–42.

Potato Marketing Board (1970a) Report on the survey of maincrop potatoes (1968). London: Potato Marketing Board.

Potato Marketing Board (1970b). The economics of potato storage. Report No. 5. Sutton Bridge Experimental Station.

Potato Marketing Board (1973). Sutton Bridge Experimental Station—Annual Review (1972) 24–26.

Potato Marketing Board (1975). Report on the Survey of Maincrop Potato Production (1973) London.

Puckridge, D.W. and Donald, C.M. (1967). *Aust. J. agric. Res.* **18**, 193–211.

Radley, R.W. (1963). Effect of planting date, variety and fertilisers on growth and yield of the potato crop. Ph.D. Thesis, University of Nottingham.

Ramsbottom, J.E. (1975). *J. nat. Inst. agric. Bot.* **13**, 312–37.

Sale, P.J.M. (1973a). *Aust. J. Agric. Res.* **24**, 733–49.

Sale, P.J.M. (1973b). *Aust. J. agric. Res.* **24**, 751–62.

Sale, P.J.M. (1973c). *Aust. J. Pl. Physiol.* **1**, 283–96.

Scott, R.K., English, S.D., Wood, D.W. and Unsworth, M.H. (1973). *J. agric. Sci., Camb.* **81**, 339–47.

Shibles, R.M. and Webber, C.R. (1966). *Crop Sci.* **6**, 55–9.

Williams, Loomis and Lepley (1965). *Crop Sci.* **5**, 211–15.

Wilson, A.R., Twiss, P.T.G. and Lessels, W.J. (1962). *Eur. Potato J.* **5**, 147–65.

Winkler, E. (1971). *Potato Res.* **14**, 1–18.

Younger, A. (1975). The significance of sprouting and irregular spacing in potato production. Ph.D. Thesis, University of Nottingham.

Index